Human Biology

Jones and Bartlett Titles in Biological Science

AIDS: The Biological Basis, Third Edition
Alcamo

AIDS: Science and Society, Fourth Edition
Fan/Conner/Villarreal

Anatomy and Physiology: Understanding the Human Body
Clark

Aquatic Entomology
McCafferty/Provonsha

Botany, Third Edition
Mauseth

The Cancer Book
Cooper

Cell Biology: Organelle Structure and Function
Sadava

Creative Evolution?!
Campbell/Schopf

Defending Evolution: A Guide to the Evolution/Creation Controversy
Alters

Early Life, Second Edition
Margulis/Dolan

Electron Microscopy, Second Edition
Bozzola/Russell

Elements of Human Cancer
Cooper

Encounters in Microbiology
Alcamo

Essential Genetics: A Genomic Perspective, Third Edition
Hartl/Jones

Essentials of Molecular Biology, Fourth Edition
Malacinski

Evolution, Third Edition
Strickberger

Experimental Techniques in Bacterial Genetics
Maloy

Exploring the Way Life Works: The Science of Biology
Hoagland/Dodson/Hauck

Alcamo's Fundamentals of Microbiology, Seventh Edition
Pommerville

Genetics: Analysis of Genes and Genomes, Sixth Edition
Hartl/Jones

Genetics of Populations, Third Edition
Hedrick

Genomic and Molecular Neuro-Oncology
Zhang/Fuller

Grant Application Writer's Handbook, Fourth Edition
Reif-Lehrer

Guide to Infectious Diseases by Body System
Pommerville

How Pathogenic Viruses Work
Sompayrac

Human Anatomy and Physiology Coloring Workbook and Study Guide, Second Edition
Anderson/Spitzer

Human Embryonic Stem Cells: An Introduction to the Science and Therapeutic Potential
Kiessling/Anderson

Human Genetics: The Molecular Revolution
McConkey

The Illustrated Glossary of Protoctista
Margulis/McKhann/Olendzenski

Introduction to the Biology of Marine Life, Eighth Edition
Sumich/Morrissey

Laboratory Research Notebooks
Jones and Bartlett Publishers

A Laboratory Textbook of Anatomy and Physiology, Eighth Edition
Donnersberger/Lesak Scott

Major Events in the History of Life
Schopf

Medical Biochemistry
Bhagavan

Microbes and Society
Alcamo

Microbial Genetics, Second Edition
Maloy/Cronan/Freifelder

Missing Links: Evolutionary Concepts and Transitions Through Time
Martin

Oncogenes, Second Edition
Cooper

100 Years Exploring Life, 1888–1988, The Marine Biological Laboratory at Woods Hole
Maienschein

The Origin and Evolution of Humans and Humanness
Rasmussen

Origin and Evolution of Intelligence
Schopf

Plant Cell Biology: Structure and Function
Gunning/Steer

Plants, Genes, and Crop Biotechnology, Second Edition
Chrispeels/Sadava

Population Biology
Hedrick

Protein Microarrays
Schena

Statistics: Concepts and Applications for Science
LeBlanc

Vertebrates: A Laboratory Text
Wessels

Fifth Edition
Human
Biology

Daniel D. Chiras
Colorado College

JONES AND BARTLETT PUBLISHERS
Sudbury, Massachusetts

BOSTON   TORONTO   LONDON   SINGAPORE

World Headquarters

Jones and Bartlett Publishers
40 Tall Pine Drive
Sudbury, MA 01776
978-443-5000
info@jbpub.com
www.jbpub.com

Jones and Bartlett Publishers Canada
2406 Nikanna Road
Mississauga, ON L5C 2W6
CANADA

Jones and Bartlett Publishers International
Barb House, Barb Mews
London W6 7PA
UK

Jones and Bartlett's books and products are available through most bookstores and online booksellers. To contact Jones and Bartlett Publishers directly, call 800-832-0034, fax 978-443-8000, or visit our website www.jbpub.com.

Substantial discounts on bulk quantities of Jones and Bartlett's publications are available to corporations, professional associations, and other qualified organizations. For details and specific discount information, contact the special sales department at Jones and Bartlett via the above contact information or send an email to specialsales@jbpub.com.

Production Credits

Chief Executive Officer: Clayton Jones
Chief Operating Officer: Don W. Jones, Jr.
President, Higher Education and Professional Publishing: Robert W. Holland, Jr.
V.P., Design and Production: Anne Spencer
V.P., Sales and Marketing: William Kane
V.P., Manufacturing and Inventory Control: Therese Bräuer
Executive Editor, Science: Stephen L. Weaver
Managing Editor, Science: Dean W. DeChambeau
Associate Editor, Science: Rebecca Seastrong
Senior Production Editor: Louis C. Bruno, Jr.
Marketing Manager: Matthew Payne
Marketing Associate: Laura M. Kavigian
Text and Cover Design: Anne Spencer
Photo Researcher: Kimberly Potvin
Illustrations: Elizabeth Morales
Composition: Graphic World, Inc.
Printing and Binding: Courier Kendallville
Cover Printing: Courier Kendallville
Cover Photo: © Stephen Simpson/Getty Images, Inc.

Library of Congress Cataloging-in-Publication Data

Chiras, Daniel D.
    Human biology / Daniel D. Chiras.—5th ed.
        p. cm.
    Includes bibliographical references and index.
    ISBN 0-7637-2899-3 (alk. paper)
1. Human biology. I. Title.

    QP34.5.c4853 2005
    612—dc22   2005043223

Printed in the United States of America
09 08 07 06 05   10 9 8 7 6 5 4 3 2 1

JONES AND BARTLETT'S COMMITMENT TO THE ENVIRONMENT
As a book publisher, Jones and Bartlett is committed to reducing its impact on the environment. Many Jones and Bartlett titles are printed using recycled post-consumer paper. We purchase our paper from manufacturers committed to sustainable, environmentally sensitive processes. We employ the Internet and office computer network technology in our effort toward sustainable solutions. New communication technology provides opportunities to reduce our use of paper and other resources through online delivery of instructors' materials and educational information—and of course, we recycle in the office.

Dedication

This book is dedicated to my family: my mother and father, for their love and support; my sons, Skyler and Forrest, for their joyous laughter and bright smiles; and Linda, for her patience, kindness, and unwavering love.

About the Author

Dr. Chiras received his Ph.D. in reproductive physiology from the University of Kansas Medical School in 1976. In September 1976, Dr. Chiras joined the Biology Department at the University of Colorado in Denver in a teaching and research position. Since then, he has taught numerous undergraduate and graduate courses, including general biology, cell biology, histology, endocrinology, and reproductive biology. Dr. Chiras also has a strong interest in environmental issues and has taught a variety of courses on the subject.

He is currently a visiting professor at Colorado College in Colorado Springs, Colorado. Most of his time, however, is spent writing books and articles.

Dr. Chiras is the author of numerous technical publications on ovarian physiology, critical thinking, sustainability, environmental education, green building, and renewable energy. He has also written numerous articles for newspapers and magazines on environmental issues. Dr. Chiras has written the environmental pollution section for World Book's annual publication *Science Year in Review* since 1993 and has written numerous articles on human biology and environmental issues for *World Book Encyclopedia, Encyclopedia Americana,* and *Grolier's Multimedia Encyclopedia.*

Dr. Chiras has published five college and high school text-books, including *Environmental Science: Creating a Sustainable Future* (Jones and Bartlett), and *Natural Resource Conservation: An Ecological Approach* (with John P. Reganold). Dr. Chiras's high school textbook, *Environmental Science: A Framework for Decision Making,* was selected as the official book of the U.S. Academic Decathlon, a nationwide competition involving thousands of American high school students in over 3000 schools.

Dr. Chiras's books for general audiences include *Beyond the Fray: Reshaping America's Environmental Response* and *Lessons from Nature: Learning to Live Sustainably on the Earth.* He has also published *Study Skills for Science Students* (West) and *Essential Study Skills* (Brooks Cole). His newest books include *The Natural House: A Complete Guide to Healthy, Energy-Efficient, Environmental Homes* (Chelsea Green); *The Solar House: Passive Heating and Cooling* (Chelsea Green); *The New Ecological Home: A Complete Guide to Green Building Options* (Chelsea Green); *EcoKids: Raising Children Who Care for the Earth* (New Society); and *The Home Energy Survival Guide* (New Society).

In addition to writing, teaching, and lecturing, Dr. Chiras is an avid bicyclist, organic gardener, river runner, and musician. He and his two sons live in Evergreen, Colorado, in a passive solar home supplied by solar and wind power and constructed from recycled materials, including 800 used automobile tires.

Brief Contents

Contents

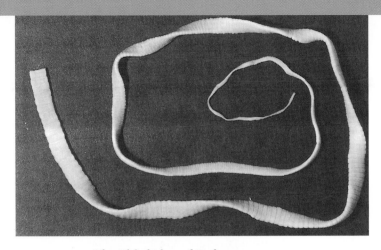

Preface

Human Biology, Fifth Edition, is written for the introductory human biology course. This book contains a wealth of information on the human body. It presents basic concepts of chemistry that will help you understand the chemistry of living things. It also explores the fascinating world of the cell and examines the structure and function of the tissues and organs that make up our bodies. The book looks at evolution, ecology, and environmental issues, too.

Like other books, this one has a central organizing theme, homeostasis. Homeostasis is a kind of internal balancing act that evolved to help regulate internal conditions vital for our survival. You will explore many fascinating processes that help control levels of chemicals in blood, body temperature, and cell function. You will see how the proper functioning of these mechanisms is essential to maintaining health. You will also see how homeostasis and personal health are dependent on a healthy environment.

In writing *Human Biology,* I had several goals in mind. First, this book was written to teach you about the human body. This knowledge, in turn, helps you understand and take care of yourselves and make informed health decisions both now and in the future. Second, this book promotes scientific literacy, that is, an understanding of how scientific information is gathered and the way it influences our lives. Furthermore, it provides background material so that you can understand new scientific discoveries and the way they influence our lives. This will aid you in understanding some of the thorniest political issues of our times. Third, this book promotes critical thinking skills, which will be valuable to you throughout your lives. Fourth, this book helps you begin to learn systems thinking—that is, thinking in terms of whole systems and understanding how the parts interact.

Organization

Human Biology is divided into six parts. Part I outlines basic biological and chemical principles vital to your understanding of the human organism. Part I contains a discussion of science and the scientific method and introduces critical thinking skills that are applied throughout the book. It closes with an overview of the cell, the basic building block of all organisms.

Part II examines the fascinating structure and function of human beings. The chapters in this chapter describe how the major organ systems operate, and introduce students to many of the most common diseases afflicting these systems. Homeostasis is emphasized in these chapters as a unifying principle of biology.

Part III is new to this edition. It includes a wealth of information on the structure and function of the immune system and disease of the immune system. This chapter also includes a chapter supplement on AIDS and a brand new chapter on infectious diseases, written with the assistance of Professor Jeffrey Pommerville of Glendale Community College. This chapter covers many important topics, including how infectious diseases spread, bioterrorism, the emergence of new infectious diseases, and the reemergence of diseases once thought to be largely under control.

Part IV discusses cell division, chromosomes, heredity, gene function, and genetic engineering with exciting real life examples. It also includes a new chapter on cancer.

Part V covers human reproduction as well as development and aging. This chapter includes a chapter supplement on common sexually transmitted diseases—their spread, their effects, and ways to prevent them.

Finally, Part VI focuses on the big picture. It looks at evolution—how we got here—and basic principles of ecology, the study of ecosystems. The final chapter includes a survey of the problems modern society has created in the natural world and offers solutions for redirecting human society onto a sustainable course

A Bold New Approach to Teaching Chemistry

Human Biology takes a bold approach to teaching chemistry to human biology students. Instead of discussing all the chemistry you need to know in one chapter early on in the book, I have introduced the basics of chemistry in chapter 2, with a brief overview of the major biological molecules.

Specific information on the important biological molecules—carbohydrates, lipids, proteins, amino acids, and nucleic acids—is provided as you need the information. For example, you learn about proteins and lipids in the chapter on plasma membranes in chapter 3, "The Life of the Cell," just before you learn about the structure of the plasma membrane. You then learn a little about carbohydrates in the discussion of energy in that chapter.

More information on carbohydrates, lipids, and proteins is presented in chapter 5, "Nutrition and Digestion." DNA and RNA are presented in detail in the discussion of molecular genetics.

This approach has at least two benefits. First, it helps prevent chemistry overload—the deluge of seemingly isolated facts about chemistry that so many students find difficult to grasp and impossible to remember. Second, it offers information on chemistry in context, so that it is easier to learn and retain. It shows the relevancy of chemistry to your understanding of important topics.

New to the Fifth Edition

Over the years, the book has expanded considerably. One of the main goals of this revision, then, has been to trim the book to a more manageable size, a size suitable for nonmajors' courses. Through careful editing, cuts, and other measures, I've reduced the length by about 20% without sacrificing content. Professors will find that although the chapters have been trimmed, they still cover essential information students need to know.

I've also updated information whenever necessary and have added or replaced many photographs and drawings in an effort to improve the pedagogical value of this book.

Special Features

Human Biology is a user-friendly book. In this edition, as in previous editions, I've tried hard to make sure the material is presented in a friendly style. Complex subjects are simplified somewhat, and analogies are used to make material more meaningful. For the most part, this book concentrates on basic information, key facts and concepts essential to students of human biology. You will find numerous examples sprinkled throughout the book, too. I have defined many terms and added pronunciations of many difficult words. The following features are included to make the student's learning experience fun, interesting, and memorable.

Scientific Discoveries That Changed the World

Science is as much a body of facts as it is a process of discovery. To highlight some of the key discoveries in human biology, I've included numerous essays called "Scientific Discoveries that Changed the World." These cover such topics as the discovery of cells, the structure of DNA, the circulation of blood, and the mechanism of gene control. This feature highlights the work of some of the world's most important scientists and illustrates how scientific discoveries have changed our view of the world. They also further a student's understanding of the scientific method and illustrate the fact that scientific advances usually require the efforts of many scientists, sometimes working in seemingly unrelated areas.

Health Notes

The Health Notes emphasize information that is vital to students both now and in the future. They present practical advice on such topics as proper diet, exercise, stress management, hair loss, heart disease, breast augmentation, cancer prevention, the perils of recreational drug use, and aging. I have updated the Health Notes and added new ones, as much of this is cutting-edge news, and all of it is information you will need throughout your life. You may want to hold on to this book and refer back to it from time to time over the coming years.

Point/Counterpoints

Many discoveries in biology have had profound impacts on our lives. Today, however, new discoveries often present profound ethical challenges and result in considerable debate. The use of genetic engineering and human embryonic stem cell research, for instance, have sparked lively debate. This book presents a number of modern-day controversies in a highly successful feature, the Point/Counterpoints. Most of the Point/Counterpoints debate social and political issues that require a good biological background; and others focus on more scientific debates.

Each Point/Counterpoint consists of two brief essays written by distinguished writers and thinkers. These essays present opposing views on important issues of our times such as the use of animals for laboratory research, genetic engineering, food irradiation, fetal cell transplantation, cancer testing, and physician-assisted suicide. There are very practical debates as well, such as the Point/Counterpoint on prioritizing medical expenditures.

Point/Counterpoints not only help inform you of current issues, they offer a chance to practice critical thinking skills. Teachers can use them to spark classroom discussions. Additional discussions, questions, and links to web pages supporting differing views on the issues are available at the Jones and Bartlett web site *Human Biology* (www.jbpub.com/humanbiology). Here's a list of the Point/Counterpoint essays:

Controversy over the Use of Animals in Laboratory Research (Chapter 1)

Controversy over Food Irradiation (Chapter 2)

Fetal Cell Transplantation (Chapter 3)

Prioritizing Medical Expenditures (Chapter 9)

Are Current Procedures for Determining Carcinogens Valid? (Chapter 18)

Controversy over Herbicide Resistance in Crops Through Genetic Engineering (Chapter 19)

Are We Facing an Epidemic of Cancer? (Chapter 20)

Physician-Assisted Euthanasia (Chapter 22)

Why Worry About Extinction? (Chapter 24)

Critical Thinking Skills

As noted earlier, chapter 1 presents a number of "rules" for improving critical thinking skills. These guidelines will help students become more discerning thinkers, a skill that could prove useful in this and many other college courses—not to mention the benefit it will have in later life.

Additional emphasis is placed on critical thinking throughout the text. Each chapter, for example, contains a Thinking Critically exercise. At the beginning of the chapter, I outline a problem or present the results of a study and ask students to apply their critical thinking skills. A brief analysis is offered at the end of the chapter.

Each exercise emphasizes one or two of the critical thinking rules presented in the first chapter. Critical thinking questions are also included after each Point/Counterpoint and in the Concept Review questions.

Health and Homeostasis Sections

Human health is dependent on homeostasis. It, in turn, is dependent on many factors, described in Health and Homeostasis sections at the end of many chapters. These sections help students gain a broader understanding of health and factors like stress and proper diet that can affect it.

In-Text Summaries

To help students learn key concepts, virtually all chapter section heads are followed by summary statements. These statements capture key concepts presented in the material that follows. These in-text summaries provide students with a way to review major concepts as they read through the material and as they prepare for exams.

End-of-Chapter Study Guides

At the end of each chapter is an extensive study guide. It includes a summary of the material in the chapter, a response to the critical thinking question posed at the beginning of the chapter, a list of key terms and concepts, a review of concepts, and a self-quiz designed to help students test their knowledge of the contents.

Summaries

Chapter summaries encapsulate the most important ideas and facts presented in each chapter. Students can use the summaries to review the most important factual information presented in the chapter after reading a chapter or when reviewing the material for tests.

Critical Thinking

This feature provides my response to the question I posed in the critical thinking exercise at the beginning of each chapter. It is meant to stimulate thought and help students become deeper, more critical thinkers.

Key Terms and Concepts

This section, new to this edition, lists the most important ideas and terms introduced in the chapter with the page on which this material occurs. Students can use it to review the material to assess their understanding of the material.

Concept Review

Each chapter contains a brief set of essay questions that enable students to assess their understanding of the material. These questions go beyond the regurgitation of facts and ask students to assimilate information they've learned.

Self-Quiz: Testing Your Knowledge

Each chapter has a short self-quiz, consisting of fill-in-the-blank questions, a feature new to this edition. The quiz can be used to see how well each student has grasped key terms.

Art Program

This book contains a remarkable collection of drawings and photographs. As noted above, we've added some new figures and refined many others in this revision. These colorful illustrations supplement the text and make the more complex concepts and processes understandable. In this edition, more of the art contains explanatory boxes describing the key processes illustrated.

Web Enhancement (www.bioscience.jbpub.com/humanbiology)

This fifth edition is linked to an extensive web site, *Human Biology,* developed exclusively for this book by Jones and Bartlett Publishers. The *Human Biology* web site offers students an unprecedented degree of integration of their text and the online world. For example, the web site introduces students to independent web sites that represent both sides of the topics debated in the text's Point/Counterpoints. *Human Biology* provides access to the latest biology news and information on the diseases introduced in the book. Each Health Notes topics is linked to sites with more information. This close linkage means the text always remains current. When there is a link to an independent site, a brief description places the site in context before the student connects to it.

To help study for exams, *Human Biology* also contains Tools for Learning, an online student review area. Students will find chapter outlines, review questions with feedback (written by the author), hundreds of flash cards, figure labeling exercises, and links to a variety of interactive tutorials.

Students can also use the web to find information to help them in class discussions and class debates. Many instructors use the Point/Counterpoints in the book to stimulate debate. Because of space limitations, however, most Point/Counterpoint authors cannot fully develop their arguments. A trip to the web can help students uncover information that will expand their knowledge.

Ancillaries

Jones and Bartlett Publishers offers an impressive variety of traditional print and interactive multimedia supplements to assist instructors and aid students in mastering human biology. Additional information and review copies of any of the following items are available through your Jones and Bartlett Sales Representative or by going to www.bioscience.jbpub.com.

For Instructors

Instructor's Tool Kit-CD-ROM Compatible with Windows and Macintosh platforms, this CD-ROM provides instructors with the following traditional ancillaries:

The Instructor's Manual is provided as a text file, contains chapter outlines, learning objectives, key terms, concept questions, and teaching tips, including ideas for using the Point/Counterpoints in class. The test bank contains over two thousand questions in a variety of formats, including goals and objectives suitable to use as handouts and classroom exercises.

The Test Bank is available as text files and as part of the Diploma™ Test Generator software included on the CD. The Test Generator software enables you to choose an appropriate variety of questions, create multiple versions of tests, even administer and grade tests on-line.

The PowerPoint™ Lecture Outline Slides presentation package provides lecture notes and images for each chapter of *Human Biology, Fifth Edition*. A PowerPoint™ viewer is provided on the CD. Instructors with the Microsoft PowerPoint™ software can customize the outlines, art, and order of presentation.

The Kaleidoscope Media Viewer provides a library of all the art, tables, and photographs in the text to which Jones and Bartlett Publishers holds the copyright or that are in the public domain. The Kaleidoscope Media Viewer uses your browser, Internet Explorer or Netscape Navigator, so you may project images from the text in the classroom, insert images into PowerPoint presentations, or print your own acetates.

Color Transparencies Qualified adopters may request a set of high-quality color transparancies. The standard set contains one hundred of the most frequently used illustrations from the text.

For Students

The Study Guide to Accompany *Human Biology,* revised by Jay Templin of Montgomery Community College, Pottstown, PA, provides a chapter on developing good study skills, chapter overviews, detailed chapter outlines, learning objectives, art labeling exercises, and practice tests.

Human Anatomy Flash Cards: Skeletal and Muscular Systems. A convenient study aid for students, this flash card set is a valuable tool designed to test and reinforce students' understanding of the skeletal and muscle systems presented in the text.

A Guide to Infectious Disease by Body System. An excellent tool for learning about microbial diseases by Jeffrey Pommerville, this book's 15 units offer a brief introduction to the human anatomical systems and the bacterial, viral, fungal, or parasitic organisms infecting each system. Anatomical illustrations are captioned with the diseases' signs and symptoms. Each unit also provides the names and brief descriptions of each disease and their causes and treatments.

Daniel G. Chiras

Evergreen, Colorado

March, 2005

Acknowledgments

A project of this magnitude is the fruit of a great many people. I wish to thank the thousands of scientists and teachers who have contributed to our understanding of human biology. A special thanks to the extraordinary teachers who have made tremendous contributions to my education, especially the late Weldon Spross, Edward Evans, the late Dr. H. T. Gier, Dr. Gilbert Greenwald, Dr. Howard Matzke, and Dr. Douglas Poorman. Their teaching techniques have made me a better teacher and textbook author, and for that I'm extremely grateful.

I am also deeply indebted to many people for their assistance during the writing of this book. A great debt of gratitude to the folks at Jones and Bartlett. Many thanks to those who assisted with the production of the book, including Dean DeChambeau, Louis Bruno, and Kathy Smith. Much appreciation for your calmness throughout, cordiality, attention to detail, and diligence. Additional thanks must be extended to Rebecca Seastrong who coordinated the supplements and to Kimberly Potvin who researched photos for the new edition. I greatly appreciate the efforts of Jones and Bartlett's many sales representatives who have helped make this book a success. It has been a pleasure and an honor to have worked with such a fine and talented group of people.

Thanks also to the many authors who contributed the Point/Counterpoints in this book. Your work will make this a more exciting journey for students as they begin to appreciate different perspectives of crucial issues.

Throughout this time, my two delightful sons, Skyler and Forrest, have offered considerable support and a counterbalance to the stresses and strains of a project of this magnitude. You're the light of my life, guys. Thanks, too, to my partner Linda Stuart for her friendship and inspiration.

Finally, a special thanks to all the reviewers on this and previous editions who offered many useful comments throughout this project. Their insights and attention to detail have been greatly appreciated. Below is a list of those who have reviewed the manuscripts.

D. Darryl Adams, Mankato State University

Donald K. Alford, Metropolitan State College

David R. Anderson, Pennsylvania State University-Fayette Campus

Felix Baerlocher, Mount Allison University

Jack Bennett, Northern Illinois University

Charles E. Booth, Eastern Connecticut State University

J. D. Brammer, North Dakota State University

Judith Byrnes-Enoch, Empire State College

Vic Chow, City College of San Francisco

Ann Christensen, Pima Community College

Peter Colverson, Mohawk Valley Community College

Francoise Cossette, University of Ottowa

John D. Cowlishaw, Oakland University

Penni Croot, SUNY Potsdam

Richard Crosby, Treasure Valley Community College

John Cummings, Waynesburg College

Jeffrey Dean, Vanderbilt University

Debby Dempsey, Northern Kentucky University

Melanie DeVores, Sam Houston State University

Stephen Freedman, Loyola University of Chicago

Sheldon R. Gordon, Oakland University

Martin Hahn, William Paterson College

John P. Harley, Eastern Kentucky University

Robert R. Hollenbeck, Metropolitan State College

Carl Johnson, Vanderbilt University

Wendel J. Johnson, University of Wisconsin, Marinette

Florence Juillerat, Indiana University-Purdue University

Ruth Logan, Santa Monica College

Charles Mays, DePauw University

David Mork, St. Cloud State University

Donald J. Nash, Colorado State University

Emily C. Oaks, State University of New York-Oswego

Lewis Peters, Northern Michigan University

Richard E. Richards, University of Colorado at Colorado Springs

Lynette Rushton, South Puget Sound Community College

Miriam Schocken, Empire State College

Richard Shippee, Vincennes University

Beverly Silver, James Madison University

Mary Vetter, University of Regina

David Weisbrot, William Paterson College

Richard Weisenberg, Temple University

Terrance O. Weitzel, Jr., Jacksonville University

Roberta Williams, University of Nevada-Las Vegas

Tommy Wynn, North Carolina State University

Study Skills

College is a demanding time. For many students, term papers, tests, reading assignments, and classes require a new level of commitment to their education. At times, the workload can become overwhelming.

Fortunately, there are many ways to lighten the load and make time spent in college more profitable. This section offers some helpful tips on ways to enhance your study skills. It teaches you how to improve your memory, how to become a better note taker, and how to get the most out of what you read. It also helps you prepare better for tests and become a better test taker.

Mastering these study skills will require some work, mostly to break old, inefficient habits. In the long run, though, the additional time you spend now learning to become a better learner will pay huge dividends. Over the long haul, improved study skills will save you lots of time and help you improve your knowledge of facts and concepts. That will no doubt lead to better grades and very likely a more fruitful life.

General Study Skills

- Study in a quiet, well-lighted space. Avoid noisy, distracting environments.
- Turn off televisions and radios.
- Work at a desk or table. Don't lie on a couch or bed.
- Establish a specific time each day to study, and stick to your schedule.
- Study when you are most alert. Many students find that they retain more if they study in the evening a few hours before bedtime.
- Take frequent breaks—one every hour or so. Exercise or move around during your study breaks to help you stay alert.
- Reward yourself after a study session with a mental pat on the back or a snack.
- Study each subject every day to avoid cramming for tests. Some courses may require more hours than others, so adjust your schedule accordingly.
- Look up new terms or words whose meaning is unclear to you in the glossaries in your textbooks or in a dictionary.

Improving Your Memory

You can improve your memory by following the PMC method. The PMC method involves three simple learning steps: (1) paying attention, (2) making information memorable, (3) correlating new information with facts you already know.

Step 1 Paying attention means taking an active role in your education—taking your mind out of neutral. Eliminate distractions when you study. Review what you already know and formulate questions about what you are going to learn before a lecture or before you read a chapter in the text. Reviewing and questioning help prime the mind.

Step 2 Making information memorable means finding ways to help you retain information in your memory. Repetition, mnemonics, and rhymes are three helpful tools.

- Repetition can help you remember things. The more you hear or read something, the more likely you are to remember it, especially if you're paying attention. Jot down important ideas and facts while you read or study to help involve all of the senses.
- Mnemonics are useful learning tools to help remember lists of things. I use the mnemonic CARRRP to remember the biological principles of sustainability: conservation, adaptability, recycling, renewable resources, restoration, and population control.
- Rhymes and sayings can also be helpful when trying to remember lists of facts.
- If you're having trouble remembering key terms, look up their roots in the dictionary. This often helps you remember their meaning.
- You can also draw pictures and diagrams of processes to help remember them.

Step 3 Correlating new information with the facts and concepts you already know helps tie facts together, making sense out of the bits and pieces you are learning.

- Instead of filling your mind with disjointed facts and figures, try to see how they relate with what you already know. When studying new concepts, spend some time tying information together to get a view of the big picture.

- After studying your notes or reading your textbook, go back and review the main points. Ask yourself how this new information affects your view of life or critical issues and how you may be able to use it.

Becoming a Better Note Taker

- Spend 5 to 10 minutes before each lecture reviewing the material you learned in the previous lecture. This is extremely important!
- Know the topic of each lecture before you enter the class and spend a few minutes reflecting on facts you already know about the subject about to be discussed.
- If possible, read the text before each lecture. If not, at least look over the main headings in the chapter, read the topic sentence of each paragraph, and study the figures. If your has a summary, read it too.
- Develop a shorthand system of your own to facilitate note taking. Symbols such as = (equals), > (greater than), < (less than), w/ (with), and w/o (without) can save lots of time so you don't miss the main points or key facts.
- Develop special abbreviations to cut down on writing time. E might stand for energy, AP might be used for air pollution, and AR could be used to signify acid rain.
- Omit vowels and abbreviate words to decrease writing time (for example: omt vwls & abbrvte wrds to dcrs wrtng tme). This will take some practice.
- Don't take down every word your professor says, but be sure your notes contain the main points, supporting information, and important terms.
- Watch for signals from your professor indicating important material that might show up on the next test (for example, "This is an extremely important point …").
- If possible, sit near the front of the class to avoid distractions.
- Review your notes soon after the lecture is over while they're still fresh in your mind. Be sure to leave room in your notes written during class so you can add material you missed. If you have time, recopy your notes after each lecture.
- Compare your notes with those of your classmates to be sure you understood everything and did not miss any important information.
- Attend all lectures.
- Use a tape recorder if you have trouble catching important points.
- If your professor talks too quickly, politely ask him or her to slow down.
- If you are unclear about a point, ask during class. Chances are other students are confused as well. If you are too shy, go up after the lecture and ask, or visit your professor during his or her office hours.

How to Get the Most Out of What You Read

- Before you read a chapter or other assigned readings, preview the material by reading the main headings or outline to see how the material is organized.
- Pause over each heading and ask a question about it.
- Next, read the first sentence of each paragraph. When you have finished, turn back to the beginning of the chapter and read it thoroughly.
- Take notes in the margin or on a separate sheet of paper. Underline or highlight key points.
- Don't skip terms that are confusing to you. Look them up in the glossary or in a dictionary. Make sure you understand each term before you move on.
- Use the study aids in your textbook, including summaries and end-of-chapter questions. Don't just look over the questions and say, "Yeah, I know that." Write out the answer to each question as if you were turning it in for a grade, and save your answers for later study. Look up answers to questions that confuse you. This book has questions that test your understanding of facts and concepts. Critical thinking questions are also included to sharpen your skills.

Preparing for Tests

- Don't fall behind on your reading assignments, and review lecture notes as often as possible.
- If you have the time, you may want to outline your notes and assigned readings. Try to prepare the outline with your book and notes closed. Determine weak areas, then go back to your text or class notes to study these areas.
- Space your study to avoid cramming. One week before your exam, go over all of your notes. Study for two nights, then take a day off that subject. Study again for a couple of days. Take another day off from that subject. Then make one final push before the exam, being sure to study not only the facts and concepts, but also how the facts are related. Unlike cramming, which puts a lot of information into your brain for a short time, spacing will help you retain information for the test and for the rest of your life.
- Be certain you can define all terms and give examples of how they are used.
- You may find it useful to write flash cards to review terms and concepts.
- After you have studied your notes and learned the material, look at the big picture—the importance of the knowledge and how the various parts fit together.
- You may want to form a study group to discuss what you are learning and to test one another.
- Attend review sessions offered by your instructor or by your teaching assistant, but study before the session and go to the session with questions.
- See your professor or class teaching assistant with questions as they arise.
- Take advantage of free or low-cost tutoring offered by your school or, if necessary, hire a private tutor to help you through difficult material. Get help quickly, though. Don't wait until you are hopelessly lost. Remember that learning is a two-way street. A tutor won't help unless you are putting in the time.
- If you are stuck on a concept, it may be that you have missed an important point in earlier material. Look back over your

notes or ask your tutor or professor what facts might be missing and causing you to be confused.

- If you have time, write and take your own tests. Include all types of questions.
- Study tests from previous years, if they are available legally.
- Determine how much of a test will come from lecture notes and how much will come from the textbook.

Taking Tests

- Eat well and get plenty of exercise and sleep before tests.
- Remain calm during the test by deep breathing.
- Arrive at the exam on time or early.
- If you have questions about the wording of a question, ask your professor.

- Skip questions you can't answer right away, and come back to them at the end of the session if you have time.
- Read each question carefully and be sure you understand its full meaning before answering it.
- For essay questions and definitions, organize your thoughts first on the back of the test before you start writing.

Now take a few moments to go back over the list. Check off those things you already do. Then, mark the new ideas you want to incorporate into your study habits. Make a separate list, if necessary, and post it by your desk or on the wall and keep track of your progress.

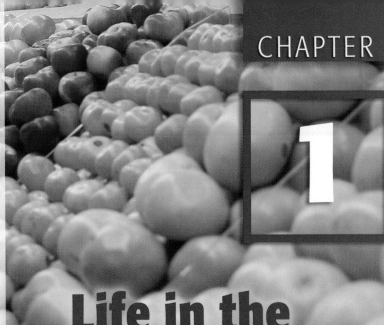

Life in the Balance: An Introduction to Human Biology

thinking critically

Your local newspaper reports the results of an experiment a student in one of the local high schools performed to test the effects of a special diet on the cholesterol content of chicken eggs. He obtained 20 chickens from two different breeders. Half of the chickens were fed his special diet; the other half were fed a diet of standard chicken feed, which he purchased at the local livestock feed store. The boy found that cholesterol levels in the eggs from the group fed his special diet were lower than the levels in eggs from the other group. The story created quite a stir in the local media—so much so that the boy's father is trying to acquire funding to market his son's new feed. Do you see any potential problems with this study?

The very first humanlike organisms roamed the grasslands of Africa about 3.5 million years ago (Figure 1-1). Scientists dubbed these creatures *Australopithecus afarensis* (aus-TRAL-owe-PITH-a-CUSS A-far-EN-suss).

Standing only three feet tall and walking upright, our earliest ancestors subsisted in large part on a diet of roots, seeds, nuts, and fruits. Studies suggest that they also supplemented their primarily vegetarian diet with **carrion** (CARE-ee-on), animals that had been killed by predators or that had died from other causes. They also may have captured and killed other animals for meat.

Weak and slow compared to other large animals, our earliest ancestors could have easily ended up as an evolutionary dead end. Fortunately, though, they possessed several anatomical features that tipped the scales heavily in their favor. Undoubtedly, one of the most important characteristics was their brain.

Today, thanks in large part to our brains, human beings inhabit a world of marvelously complex technology. These technologies make our lives easier, more convenient, and more fun. Rather than roaming in small bands, as our early ancestors did, the majority of the world's people live today in cities and towns that offer amenities our ancestors never would have dreamed possible. Rather than collecting nuts and berries from the plants around us, most people in the modern world purchase their food from grocery stores supplied by highly mechanized farms. Many farmers are now using satel-

lites and remote sensing devices, as well as computerized machinery to produce more food. Instead of being restricted to whimsical gazing into space, we travel there ourselves. Today, our scientists have begun to alter the genetic material of the cells of plants and animals to increase food production, or, in the case of animals, to produce tissues and organs that could be transplanted into human beings. Some scientists have even begun to manipulate our hereditary material in an attempt to cure diseases long thought to be incurable.

For better or for worse, humans have become a major player in **evolution**, a process of biological change that occurs in distinct groups of organisms, known as populations. Evolution results in structural, functional, and behavioral changes in organisms in populations. These changes, in turn, result in organisms better equipped to cope with their environment—that is, better able to survive and reproduce.

From most perspectives, the human experiment has been an overwhelming success. However, in the process, we are damaging the life-support system of the planet upon which we and all other species depend with dire consequences likely.

FIGURE 1-1 *Australopithecus afarensis* Current scientific evidence suggests that *Australopithecus afarensis* was the first humanlike ape. Its skeletal remains indicate that it walked upright.

Like other texts, the bulk of this book will take you on a journey through the human body. On this journey, you will learn a great deal about yourself—how you got here, how you inherited certain characteristics from your parents, and how your body functions. You will study the basics of nutrition and find out how broken bones mend. You will discover how your immune and nervous systems operate. You will learn about many common diseases and—perhaps more important—how to prevent them.

As you will soon see, the information you learn from this book will prove useful to you in many ways. It will also help you understand important political debates over issues such as genetic engineering, pollution, and vaccination. As you proceed through this book, though, be sure to take some time to marvel at the wonders of the human body—the intricate details of the cell, the fascinating structure and function of organs, and the intriguing manner in which the various parts work together.

1-1 Health and Homeostasis

In this book, you will see how human health depends on numerous internal mechanisms that have evolved over many millions of years. These internal processes help to maintain a fairly constant internal condition, a state often referred to as *homeostasis* (home-e-oh-STAY-siss).

What Is Homeostasis?

Homeostasis is a state of relative constancy.

The term **homeostasis** comes from two Greek words, *homeo*, which means "the same," and *stasis*, which means "standing." Literally translated, homeostasis means "staying the same." Thus, many people refer to homeostasis as a state of internal constancy. In reality, however, homeostasis is not a static state; rather, it is a dynamic (ever-changing) state. Let's examine body temperature as an example.

Humans are warm-blooded creatures. We generate body heat internally and maintain body temperature at a fairly constant level—about 98.6 °F. In reality, though, body temperature varies during the day, falling slightly at night when we sleep and rising during the daylight hours. It increases even more when we participate in strenuous physical activity.

Like many other internal conditions, then, body temperature fluctuates within a range. This is what is meant when we say that body temperature is in homeostasis: it is in a dynamic state, but remains more or less the same (Figure 1-2).

Homeostasis is achieved through a variety of automatic mechanisms that compensate for internal and external changes. As Chapter 4 illustrates, homeostatic mechanisms require **sensors**, structures that detect internal and external change—for example, changes in air temperature. Sensors elicit a response that offsets the change, helping to maintain a fairly constant state. On very cold days, for example, we may shiver. Shivering is a rhythmic contraction of muscles that generates body heat and is one of many homeostatic mechanisms in our bodies. Homeostatic mechanisms also maintain fairly constant levels of nutrients in the blood, which is essential for normal body function.

Homeostatic mechanisms also exist in **ecosystems**. An ecosystem is a biological system consisting of organisms and their environment. Homeostatic mechanisms help achieve balance in ecosystems.

A highly simplified example illustrates the point. In the grasslands of Kansas, rodent populations generally remain fairly constant from one year to the next. This phenomenon results, in part, from **predators** (animals that hunt and kill other organisms). Predators such as snakes, coyotes, foxes, and hawks feed on rodents and, thus, help to control rodent populations (Figure 1-3).

Although predators are a crucial element in maintaining environmental homeostasis in these grasslands and virtually all other natural systems, a host of other factors also contribute to it, such as weather and food supplies. It is the net effect of these factors that determines population sizes.

FIGURE 1-2 **Keeping Warm** The human body is remarkably able to tolerate a wide variety of conditions thanks to internal mechanisms that maintain relatively constant internal conditions. These students stay warm thanks to stepped up heat production by the body and protective clothing.

FIGURE 1-3 **Predator Control** The snake plays an important role in controlling rodent populations.

FIGURE 1-4 Old and New Concepts of Health

(a) The old concept

Poor health		Good health
Obvious disease or illness		No obvious disease or illness

(b) The new concept

Poorest health	Poor health	Good health	Best health
Obvious disease or illness			No obvious disease or illness
Many risk factors	More risk factors	A few risk factors	No risk factors
Poor fitness			Good fitness
Poor mental health			Good mental health

In this book, the term *homeostasis* is used to refer to the balance that occurs at all levels of biological organization—from cells to organisms to entire ecosystems. The abundance of homeostatic mechanisms in nature suggests their importance to life on Earth. As you shall see, maintaining "balance" is essential to the continuation of life. Without it, cells would fall into disarray, organisms would perish, and ecosystems would be destroyed!

Healthy Environments

Human health depends on maintaining healthy physical, psychological, and social environments.

Scientists have found that the health of the environment and the health of organisms, including human beings, are interdependent. Alterations in the environment—for example, adverse changes in the chemical composition of the air—can have impacts on human health. Polluted air, water, and soils take a toll on humans and other species.

The health of organisms also requires social and psychological conditions conducive to mental health. Stressful environments can lead to serious ailments in those individuals unfortunate enough to be stuck in them. Health and Homeostasis sections at the end of chapters outline some of these connections.

Although humans are the central focus of this book, it is important to note that many other species share this world with us. They, too, are affected by the condition of the environment. Scientists, for instance, are finding that many drugs that people take are excreted in their urine and end up in sewage effluent. From there they enter rivers, lakes, and streams. These chemicals are having profound effects on the growth, reproduction, and survival of aquatic species, especially fish. Scientists are also concerned about antibiotics we take that eventually end up in our waterways. They think that bacteria in surface waters such as lakes and streams that are exposed to antibiotics could evolve resistance to them. Humans who ingest these bacteria in drinking water could become deathly ill. Doctors worry that they won't have antibiotics to treat the resistant strains.

Dimensions of Health

Human health is a state of physical and mental well-being.

For many years, human health was defined as the absence of disease (Figure 1-4a). As long as a person had no obvious symptoms of a disease, that person was considered healthy. Although such a person may have had clogged arteries from a lifetime of eggs-and-bacon breakfasts, it wasn't until symptoms of heart disease—for example, chest pain—became apparent that the patient was considered unhealthy.

Today, health experts rely on a broader definition of health. It takes into account two broader categories: physical and emotional well-being.

Physical health refers to the state of the body—how well it is working. Physical health can be measured by checking temperature, blood pressure, blood sugar levels, and a number of other variables. Abnormalities in these measurements may be a signal that one's physical health is in jeopardy, even though there are no obvious symptoms of illness. Medical scientists use the term **risk factors** to refer to abnormal conditions such as high blood pressure or high blood cholesterol levels that put a person at risk for disease. The presence of one or more risk factors is a sign of less-than-perfect health. Obviously, the more risk factors there are, the worse one's physical health is (Figure 1-4b).

The new concept of health says that even though a person feels healthy and does not exhibit obvious signs of disease, such as a failing heart, the presence of risk factors indicates otherwise. As shown in Figure 1-4b, the absence of risk factors results in the best health. A few risk factors mean health is only good. More risk factors mean health is poor, even though the individual may not have exhibit any other symptoms—yet.

Scientists also use the term *risk factor* to refer to activities that make one more likely to develop diseases. Smoking, lack of exercise, and a fatty diet, for example, are risk factors for heart and artery disease.

Physical health is also measured by one's level of physical fitness. If you can't walk up a set of stairs without gasping for air, you're not considered very physically fit. You're more likely to have other problems later in life—for example, heart disease.

Emotional well-being also factors into an assessment of a person's health. Especially relevant is your ability to cope with stress. Inability to cope may lead to physical problems, such as high blood pressure and heart disease.

Mental and physical fitness are measures of our abilities to meet the demands of life. Fit people are able to cope with daily psychological stresses and are able to move about without becoming short of breath.

Maintaining good health is a lifelong job. Table 1-1 lists numerous healthy habits. By incorporating these habits into your lifestyle, you can increase your chances of living a long, healthy life.

Health and Homeostasis

Human health is dependent on maintaining homeostasis.

As pointed out earlier, physical health depends on properly functioning homeostatic mechanisms. When these controls function improperly or break down completely, illness results. Persistent stress, for example, can disrupt several of the body's homeostatic mechanisms, leading to disease. If it is prolonged, stress can increase the risk of diseases of the heart and arteries. It may also increase the risk of ulcers and weaken the immune system. Fortunately for us, stress can be alleviated by exercise, relaxation training, massage, acupuncture, and other measures discussed in Health Note 1-1.

This discussion is not meant to imply that all diseases result from homeostatic imbalance. Some are produced by genetic defects; others are caused by bacteria or viruses. Interestingly,

TABLE 1-1	Healthy Habits
Sleep seven to eight hours per day*
Eat breakfast regularly
Eat a balanced diet
Avoid snacking on junk food (sweets or fatty foods) between meals
Maintain ideal weight
Do not smoke
Avoid alcohol or use it moderately
Exercise regularly
Manage stress in your life

*Not all people need this much sleep. If you're one of them, don't try to force yourself to sleep more than you need.

though, bacteria and viruses and even genetic defects often disrupt homeostasis. An excellent example is acquired immune deficiency syndrome, or AIDS. AIDS is a sexually transmitted disease caused by a virus that attacks certain cells of the immune system. This, in turn, results in a reduction in a key protective mechanism of the body, which is vital to homeostasis.

In some diseases, temporary upsets in homeostasis may make us more susceptible to infectious agents. According to a recent study, people under stress are twice as likely to suffer from colds and the flu as those who are not. To stay healthy, we need to reduce stress levels.

1-2 Evolution and the Characteristics of Life

Homeostasis is a central theme of this book because it is so essential to life and is now threatened by modern culture. Another key concept of biology and a subtheme of this book is evolution. A few words on the subject are essential to your understanding of human biology.

All life-forms alive today exist because of evolution. In fact, every cell and every organ in the human body is a product of millions of years of evolution. Even the intricate homeostatic mechanisms evolved over long periods.

Figure 1-5 shows the five major groups or **kingdoms** of organisms that exist today. This simple diagram also illustrates evolutionary relationships. The simplest, bacterialike organisms, belonging to a group called the **monerans**, were the first to evolve. They gave rise to a more complex set of organisms, known as the protistans. The **protistans** consist of single-celled organisms such as amoebas and paramecia. During the course of evolution, the protistans gave rise to three additional major groups, or kingdoms: plants, fungi, and animals. We humans belong to the animal kingdom.

This common lineage is responsible for the striking similarities among the Earth's organisms, even in remarkably different organisms. For example, all organisms rely on the same type of genetic material. Similarities also exist on other levels besides the biochemical one. A comparison of certain anatomical features, such as the bones in a person's arm and the wings of birds, reveals a resemblance.

Common Characteristics of Living Things

Humans are similar to other organisms in many ways.

An analysis of living things turns up eight common features, referred to as the *characteristics of life*. A brief look at these characteristics not only shows our evolutionary connection with other organisms but helps us understand human beings better.

The first characteristic of life is that all organisms, including humans, are made of cells. Cells are tiny structures that are the fundamental building blocks of living organisms. Cells, in turn, consist of **molecules**, nonliving particles composed of smaller units called **atoms**. Glucose molecules, for instance, contain carbon, hydrogen, and oxygen atoms. Molecules, in turn, combine to form the parts of the cell.

The second characteristic of life is that all organisms grow and maintain their complex organization by taking in molecules and energy from their surroundings. As you will see in subsequent chapters, humans must expend considerable amounts of energy to maintain their complex internal organization. In fact, 70% to 80% of the energy adults need is used just to maintain their bodies—to transport molecules in and out of cells, to make proteins in cells, and to perform other basic body functions. The rest is used for activity—walking, running, talking, and so on.

∷∷∷ healthnote ≋≋≋≋≋≋

1-1   Maintaining Balance: Reducing Stress in Your Life

Stress: it's a normal occurrence in everyday life. It's a psychological and physical reaction we have when we are exposed to certain stimuli.

Stress can be caused by non–life-threatening situations such as a blind date or a final exam. It can also be caused by potentially life-threatening situations such as exposure to dangerous machinery in a factory.

Stressful stimuli can be real or imagined. Either way, they elicit the same response in the body: an increase in heart rate, an increase in blood flow to the muscles and a decrease in blood flow to the digestive system, a rise in blood glucose levels, and a dilation of the pupils. All of these physical changes in the body help prepare us to respond to the stress. Once the stimulus is gone, though, the body typically returns to normal.

How stress affects us, however, depends on how long we're exposed to it. If the stressful stimulus is short-lived, our bodies recover nicely. Some argue that a little stress may actually improve a person's performance.

Long-term exposure to stressful stimuli, however, can be serious. As a result, a prolonged period of stress may lead to disease. One reason this happens is that the body's immune system is often depressed by stress. The immune system protects us from bacteria and viruses that cause colds and flu and other diseases. Prolonged stress also results in changes in the blood vessels. These may accelerate the accumulation of cholesterol, which clogs the arteries and may eventually result in strokes and heart attacks.

Stress doesn't affect all people the same. Some people recognize the stress they're feeling and channel its energies into productive work.

They are better able to cope with stress. Psychologists believe that people who handle stress the best have a sense of being in control, despite the stress of their work. They typically have clear objectives and a strong sense of purpose. They view their jobs and life as a challenge, not a threat.

Unfortunately, not all people are so lucky. Many people are not in a position of control; they feel expendable and often view themselves as victims. They experience stress that can cause health problems. What can be done to deal with stress? One of the most important strategies is preventive: selecting an environment and creating a lifestyle that is as stress-free as possible. As a college student, for instance, you may want to select a realistic class load. If you must work or if you're taking hard courses, sign up for a class load that you can handle. And be sure to get plenty of sleep and spend time relaxing.

Coping with stress may also require physical and mental strategies. Let's consider the physical strategies first.

One of the easiest routes is to lessen the impact of stress through exercise. Studies show that a single workout at the gym, a bike ride, a swim, or a cross-country skiing trip reduces tension for two to five hours. A regular exercise program, however, reduces the overall stress in one's life. Individuals who are easily stressed usually find that stress levels decline after two weeks of exercise.

Exercise can be supplemented by relaxation training. As you prepare for a difficult test or get ready for a date that you are nervous about, tension often builds in your muscles. Periodically stopping to release that tension helps to reduce physical stress. Stretching or taking a walk can help. Some people find it useful to tighten their muscles forcefully, then let them relax. Massage therapy and acupuncture can also be used to reduce stress. Stress-reduction programs on tapes, videos, and CDs can teach relaxation methods, as can trained therapists.

Dealing with thoughts that exaggerate your stress can also help to eliminate it. This is a mental strategy. Start paying attention to the thoughts that provoke anxiety in your life. Are they exaggerated? If so, why? For ex-

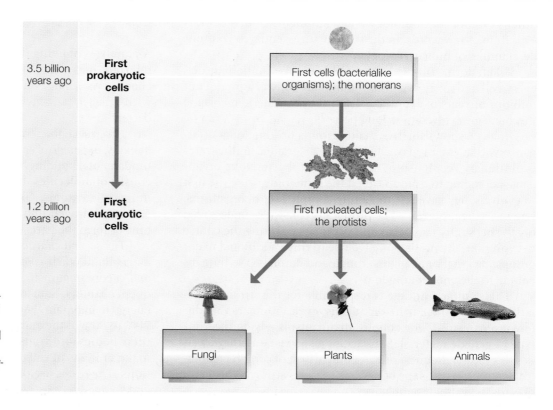

FIGURE 1-5 Evolution of the Prokaryotic and Eukaryotic Cells This diagram illustrates the evolutionary history of life. Organisms fall into one of five major groupings, or kingdoms. The first lifeforms were the monerans, which are single-celled prokaryotic organisms. They gave rise to the protistans, single-celled eukaryotic organisms. Eukaryotic protistans, in turn, gave rise to plants, fungi, and animals.

ample, are you nervous before exams? Why? Would better preparation reduce your anxiety?

Finding the source of your anxiety and taking positive action to alleviate it are helpful ways of reducing stress. However, stress reduction is not always easy. Test anxiety, for example, may be deeply rooted in feelings of insecurity and inadequacy. Many people struggle with low self-esteem their entire life. A trained psychologist can help you find the causes of your problem and assist you in learning to feel better about yourself. Psychological help is as important as medical help these days. Given the complexity and pace of our society, there is no shame in seeking counseling.

Biofeedback is another form of stress relief. A trained health care worker hooks you up to a machine that monitors heart rate, breathing, muscle tension, or some other physiological indicator of stress (Figure 1). During a biofeedback session, your trainer first will help you relax, and then perhaps discuss a stressful situation. When one of the indicators shows that you are suffering from stress, a signal is given off. Your goal is to consciously reduce the frequency of the signal. For example, if your heart started beating faster when you thought about taking an exam, the machine will make a clicking sound. By breathing deeply and relaxing, you consciously slow down your heart rate; the clicking sound slows down and then disappears. Learning to recognize the symptoms of stress and to counter them is the goal of biofeedback. Eventually, you should be able to do it without the aid of a machine.

You can also reduce tension by managing your time and your workload efficiently. Numerous books on this subject can help you learn to budget your time more effectively. See the Study Skills section at the beginning of the book for ideas on ways to be a more efficient learner.

If these techniques don't work, you may want to see a doctor. He or she can prescribe medications that relieve anxiety and muscle tension or help you sleep, or that combat depression. Herbal remedies such as Valerien root are also available. Relieving stress in our lives helps us reduce the risk of disease and helps us relax and enjoy life. It also makes us more

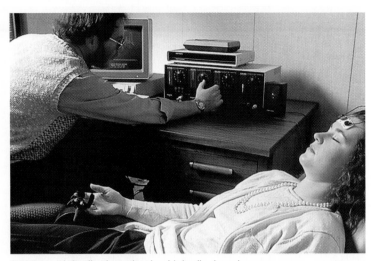

FIGURE 1 **Biofeedback** Student in a biofeedback session.

pleasant to be around. All in all, it is best to start learning early in life how to reduce or cope with stress. Lessons learned now will be useful for years to come.

www.jbpub.com/humanbiology/5e

Visit Human Biology's Internet site for links to web sites offering more information on this topic.

The third characteristic of life is that all living things exhibit a feature called metabolism. **Metabolism** refers to the chemical reactions in the cells and tissues of organisms. These reactions consist of two types—those in which food substances are built up into living tissue (anabolic reactions) and those in which food is broken down into simpler substances, often releasing energy (catabolic reactions). In human body cells, millions of reactions occur each second to maintain life.

The fourth characteristic of life is homeostasis. Simply put, all organisms rely on homeostatic mechanisms. In the constantly shifting world of many organisms, maintaining a constant internal environment is a never-ending task.

The fifth characteristic of living things is that they sense and respond to changes in their environment. The ability to perceive stimuli and respond to them is important in the day-to-day survival of all organisms.

The sixth characteristic of life is reproduction and growth. All organisms are capable of reproduction and growth. Humans produce offspring by combining sex cells, sperm and eggs from males and females, respectively.

The seventh characteristic of life is evolution. Evolution is a process that leads to structural, functional, and behavioral changes in species, known as **adaptations**. Favorable adaptations increase an organism's chances of survival and reproduction. We'll explore this topic in Chapter 23.

The eighth characteristic of life is that all organisms are part of the Earth's ecosystems. Humans, like every other plant, animal, and microorganism, are dependent on the Earth's ecosystems. They depend on them for food, fiber, water, oxygen, and a host of free services such as natural flood control. The Earth's ecosystems are not only the life support system of the planet, they are essential to the human economy.

What Makes Humans Unique?

Humans are one form of life on Earth yet have many unique characteristics.

Human beings are one of evolution's many products. Although we are like many other species, we are a unique form of life. Several features distinguish us from other species.

One of the key differences between humans and other animals is our ability to acquire and use complex languages. Another distinguishing feature is our culture. Culture has been defined in many ways. Humorist Will Cuppy remarked that culture is any-

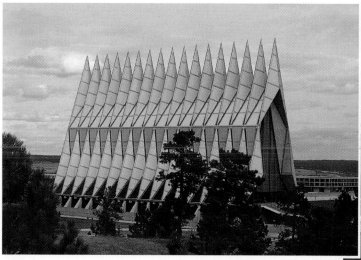

(a)

(b)

FIGURE 1-6 Human Culture: Progress and Setbacks (a) Architecture is one of our greatest achievements. But in reshaping our environment, we sometimes create enormous backlashes. (b) Earthen walls, or levees, along the Mississippi River, for example, have been used to control flooding in the past hundred years but have led to more frequent flooding in downstream areas.

thing we do that monkeys don't. On a more serious note, culture can be defined as the ideas, values, customs, skills, and arts of a given people in a given time. While other species may have some rudiments of culture—for instance, some communication skills—humans are unique in the biological world because of the complexity of our cultural achievements (Figure 1-6a).

Humans also differ from other animals in our ability to plan for the future. Although a few other animals seem to share this ability, most "planning"—like a bird's nest-building activities—is probably the result of instinct programmed by the animal's ge-

netic material. In contrast, building skyscrapers requires an extraordinary amount of forethought.

Another unique characteristic of humans is our unrivaled ability to shape the environment. Despite the benefits of our remarkable technologies, attempts to control nature sometimes backfire, creating larger problems. For instance, efforts to control upstream flooding on the Mississippi River by building levees along sections of a river have led to more frequent floods and more damage downstream (Figure 1-6b). Levees prevent water from spilling over the banks of rivers in upstream communities. However, this results in a larger slug of water delivered to downstream sections and hence increased flooding in such areas.

1-3 Understanding Science

The systematic study of the universe and its many parts today falls into the realm of science. The term *science* comes from the Latin word *scientia*, which means "to know" or "to discern." Today, **science** is defined as knowledge derived from observation, study, and experimentation. It is also a method of accumulating knowledge. In other words, it involves ways of learning facts as well as a body of knowledge.

Many people view science as an uninteresting endeavor best left in the hands of a select few (Figure 1-7). In reality, science is an exciting endeavor that often involves enormous creative energy. Because it teaches us about the workings of the world around us, it can be a source of great fascination. As the sometimes zany paleontologist Robert Bakker, a consultant to the company that made the dinosaurs for the movie *Jurassic Park*, once noted, science is "fun for the mind."

Science also has a practical side. It provides information that improves our lives in many ways. It helps us understand important phenomena such as the weather and the spread of disease. A knowledge of science makes us better voters, better able to understand many complex debates. And an understanding

FIGURE 1-7 Not Just for Scientists Science is essential to human progress and a benefit to all of us. It is a process that we all engage in.

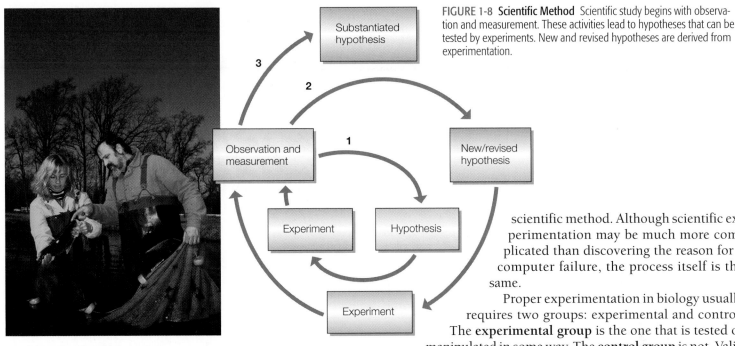

FIGURE 1-8 **Scientific Method** Scientific study begins with observation and measurement. These activities lead to hypotheses that can be tested by experiments. New and revised hypotheses are derived from experimentation.

of science, notably human biology, can help us make informed health care decisions.

The Scientific Method

> The scientific method generally starts with observations that lead to hypotheses and experiments to test them.

Scientists employ a technique called the **scientific method** to obtain information. As shown in Figure 1-8, the scientific method generally begins with observations and measurements. In some cases, these may be part of carefully conducted experiments. Others may occur more haphazardly. A scientist on vacation in the tropics, for instance, may notice a phenomenon that sparks her curiosity and leads to in-depth studies.

You may not realize that most of us use the scientific method in our day-to-day lives. To see how the scientific method works, suppose you sat down at a computer, turned on the switch, but nothing happened. You might also have noticed that the lights in the room hadn't come on. These two observations would lead to a **hypothesis** (pronounced high-POTH-eh-siss), a tentative explanation of the phenomenon. From your observations, you might hypothesize that the electricity in your house was off.

You could test your hypothesis by performing an **experiment**. An experiment is a procedure designed to test some idea. In this case, all you would need to do would be to try the light switch in the kitchen. If the lights in the kitchen worked, you would reject your original hypothesis and form a new one. Perhaps, you hypothesize, the circuit breaker to your study had been tripped. To test this idea, you would "perform" another experiment, locating the circuit breaker to see if it was turned off. If the circuit breaker was off, you would conclude that your second hypothesis was correct. To substantiate your conclusion, you would throw the switch and see if your computer worked.

This process involving observations and measurements, hypothesis, and experimentation forms the foundation of the scientific method. Although scientific experimentation may be much more complicated than discovering the reason for a computer failure, the process itself is the same.

Proper experimentation in biology usually requires two groups: experimental and control. The **experimental group** is the one that is tested or manipulated in some way. The **control group** is not. Valid conclusions come from such comparisons because, in a properly run experiment, both groups are treated identically except in one way. The difference in treatment is known as the *experimental variable*. Consider an example to illustrate this point. In order to test the effect of a new drug on laboratory mice, a good scientist would start with a group of mice of the same age, sex, weight, genetic composition, and so on (Figure 1-9). These animals would be divided into two groups, the experimental and control groups. Both groups would be treated the same throughout the experiment, receiving the same diet and being housed in the same type of cage at the same temperature. The only difference between the two should be the drug given to the experimental group. Consequently, any observed differences between the groups could be attributed to the treatment (the experimental variable).

Besides having an experimental group and a control group, good experimentation requires an adequate number of subjects to ensure that any observed differences are real. Individual variation is natural. As a rule, the smaller the number of animals in each group, the less reliable the data is—because of possible variation in response. In laboratory experiments, at least 10 test animals are required for both control and experimental groups for reliable statistical analysis. Groups larger than 10 are even better. For human health studies, much larger groups are generally used because of the wider genetic variability among people.

What Is a Theory?

> Theories are broad generalizations based on many experimental observations.

Scientific method leads to the accumulation of scientific facts (really, well-tested hypotheses). Over time, as knowledge accumulates, scientists are able to gain a broader understanding of the way the world works. These broad generalizations are known as **theories**.

FIGURE 1-9 **Experimental Design** A good experiment uses a large number of identical subjects. They are placed into two groups: control and experimental. Both groups are treated the same, except for one variable. In this case, the experimental group is given an experimental blood pressure lowering drug. The control group is given an injection but of saline, the solution the drug is dissolved in.

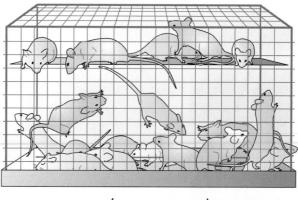

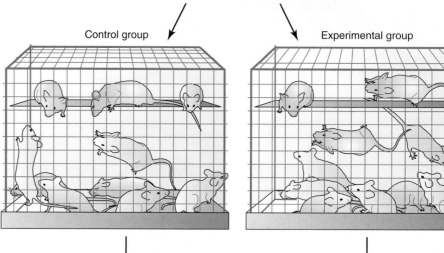

Control group   Experimental group

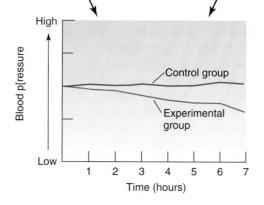

Placebo

Blood pressure drug

Theories are supported by numerous facts established by careful observation, measurement, and experimentation. Unlike hypotheses, theories cannot be tested by single experiments because they encompass many bits of information. Atomic theory, for instance, explains the structure of the atom and fits numerous observations made in different ways over many decades.

A theory commands respect in science because it has stood the test of time. This does not mean that theories always stand forever. As history bears out, numerous theories have been modified or discarded as new scientific evidence accumulated. Even widely held theories that persisted for hundreds of years have been overturned. In 140 A.D., for example, the Greek astronomer Ptolemy (TAL-eh-me) proposed a theory that placed Earth at the center of our solar system. This was called the *geocentric theory*. For nearly 1500 years, the geocentric view held sway. Many astronomers vigorously defended this position while ignoring observations that did not fit the theory. In 1580, the astronomer Nicolaus Copernicus proposed a new theory—the *heliocentric view*—which placed the sun at the center of the solar system. His work stimulated considerable controversy, but it eventually prevailed because it fit the observations.

Because theories may require modifications or rejection, scientists must be willing to analyze new evidence that throws into question their most cherished beliefs. For the most part, though, theories are talked about as if they were fact. Some people even object to calling a theory a "theory" for fear that it sounds tentative.

Another word about theories before we move on. The word *theory* is commonly misused in everyday conversation. A friend, for instance, might say, "My theory about why Jane missed the party is that she didn't want to see her ex-boyfriend." Jane's feelings aside, this is hardly worthy of the status of a theory. What your friend really meant was his "hypothesis," for his explanation was truly a tentative explanation that could be tested by experimentation—in this case, a phone call to Jane.

Inductive and Deductive Reasoning

Science relies on inductive and deductive reasoning.

Have you ever made a generalization based on your observations? For example, based on your observations of friends and family members, you may have stated that people tend to be healthier if they get a good night's sleep every night. If you have, you were engaging in a type of reasoning referred to as **inductive reasoning**.

During inductive reasoning, you create a law based on observations. Scientists do this all the time. They infer general laws based on facts and observations.

Going in the opposite direction—coming up with a specific statement or conclusion based on a general rule—is called **deductive reasoning**. Scientists engage in this kind of thinking as well, as do nonscientists. Can you think of an example in which you could arrive at a specific conclusion based on a generality? Consider the sleep example. Suppose you have an unhealthy friend who doesn't sleep well. Based on your general rule—stated above—you might deduce that his ill health might be due to his lack of sleep.

Science and Human Values

Science helps shape our lives and our values.

Science and the scientific process are essential to modern existence. We wouldn't have the microwave oven or DVD player if it weren't for science. Science can also influence political decisions regarding health care, environmental protection, and a host of other issues.

Many decisions in the public-policy arena, however, are not made solely on the basis of scientific facts. Rather, they are influenced by values—what we view as right or wrong—and economic needs. When human values are framed in the absence of scientific knowledge, however, they can lead to policies that could fail in the long run.

Although many people do not realize it, science can even influence human values. Environmental values, for instance, are influenced by information gained from the study of ecology. Ecology helps us understand the interconnectedness of living things. It helps uncover relationships that are not obvious to most people, such as the role of bacteria in recycling nutrients. It helps us understand our dependence on other species and thus helps us act more intelligently.

1-4 Critical Thinking

Critical thinking rules allow us to carefully analyze problems, issues, and information for errors of reasoning.

Another benefit of your study of science is that it can help you learn to think more critically. **Critical thinking** is not being "critical" or judgmental. Rather, it's a process that allows us to objectively analyze facts, issues, problems, and information. Ultimately, critical thinking permits people to distinguish between beliefs (what we believe to be true) and knowledge (facts well supported by research). In other words, critical thinking is a process by which we separate judgment from facts. It is our most ordered kind of thinking. It is not just thinking deeply about a subject. Critical thinking subjects facts and conclusions to careful analysis, looking for errors of reasoning. Critical thinking skills, therefore, are essential to analyzing a wide range of facts, issues, problems, and information.

Table 1-2 summarizes seven critical thinking rules that will come in handy as you read the newspaper, watch the news, listen to speeches, and study new subjects in school. Here is a brief description of each one.

The first rule of critical thinking is: gather complete information. Critical thinking requires facts. Gathering lots of information keeps you from falling into the trap of mistaking ignorance for perspective. By continually being on the lookout for new facts, you can develop an enlightened viewpoint.

The second rule of critical thinking is: understand and define all terms. Critical thinking requires a clear understanding of all terms. Understanding terms and making sure that others de-

TABLE 1-2	Critical Thinking Rules

When analyzing an issue or fact, you may find it useful to employ these rules:

1. Gather complete information.

2. Understand and define all terms.

3. Question the methods by which data and information were derived:

Were the facts derived from experiments?

Were the experiments well executed?

Did the experiment include a control group and an experimental group?

Did the experiment include a sufficient number of subjects?

Has the experiment been repeated?

4. Question the conclusions:

Are the conclusions appropriate?

Was there enough information on which to base the conclusions?

5. Uncover assumptions and biases:

Was the experimental design biased?

Are there underlying assumptions that affect the conclusions?

6. Question the source of the information:

Is the source reliable?

Is the source an expert or supposed expert?

7. Understand your own biases and values.

Scientific Discoveries that Changed the World

1-1   Debunking the Theory of Spontaneous Generation
Featuring the Work of Aristotle, Redi, and Pasteur

The Greek philosopher and scientist Aristotle (384 to 322 B.C.) proposed a theory to explain the origin of living things. It was called the *theory of spontaneous generation*. It asserted that living things arose spontaneously from innate (nonliving) matter. Mice, he believed, arose from a pile of hay and rags placed in the corner of a barn. Flies could be produced by first killing a bull, then burying it with its horns protruding from the ground. After several days, one of the horns would be sawed off and flies would emerge. People, it was said, emerged from a worm that developed from the slime in the bottom of a mud puddle.

As absurd as the idea of spontaneous generation sounds today, the view remained compelling to many scientists well into the nineteenth century, despite observations that contradicted the theory, such as the phenomenon of childbirth itself.

Debunking the theory of spontaneous generation engaged some of the best scientific minds of the day. Although many scientists were involved in debunking the theory, two scientists, Francesco Redi and Louis Pasteur, played pivotal roles.

Redi, an Italian naturalist and physician, was one of the first scientists to refute spontaneous generation through experimentation. Around 1665, Redi performed a simple but effective experiment to determine whether houseflies were spontaneously generated in rotting meat. He began by placing three small pieces of meat in three separate glass containers. The first one was covered with paper. The second was left open, and the third was covered with gauze. Left at room temperature, the meat quickly began to rot and attract flies. Soon, the meat in the open container began to seethe with maggots, larvae hatched from fly eggs. The paper-covered container showed no evidence of maggots, nor did the meat in the gauze-covered container, although maggots did appear in the gauze itself. Redi's conclusion was that maggots (which give rise to flies) do not come from the meat itself but from the eggs deposited by flies.

Redi's experiments convinced many people that flies and other organisms did not arise by spontaneous generation, but this did not put the debate to rest. Soon after Redi's now-famous experiment, a Dutch linen merchant by the name of Anton Leeuwenhoek discovered bacteria using simple microscopes he had built. This discovery revived arguments for spontaneous generation on the microscopic level. Many scientists asserted that although flies and other organisms did not arise spontaneously, microorganisms probably did. As evidence, they cited studies showing that microorganisms could arise from boiled extracts of hay or meat. According to proponents of spontaneous generation, nonliving plant and animal matter possessed a life-generating force that could give rise to microorganisms.

The debate over spontaneous generation persisted for the next 200 years. In 1861, Louis Pasteur published the results of an experiment that helped put this debate to rest. He placed sterilized broth in a sterilized swan-necked flask, a flask with a long curved neck (Figure 1). The design of the flask permitted air to enter, eliminating criticism that he had destroyed any vital forces necessary for spontaneous generation, but it blocked airborne bacteria from entering. (Airborne bacteria probably were deposited on the tube leading to the flask and were prevented from entering the broth.) In his experiments, Pasteur clearly showed that bacteria could not arise spontaneously. Only when the broth was open to the air did bacteria emerge.

This brief history points out three important lessons. First, scientific discovery is usually the result of the work of many scientists, each examining different parts of the puzzle. Second, it shows how discoveries open up new ways of thinking. Third, it illustrates the persistence of ideas that shape the way we think and the resistance people often exhibit to new ideas even in the face of contradictory evidence.

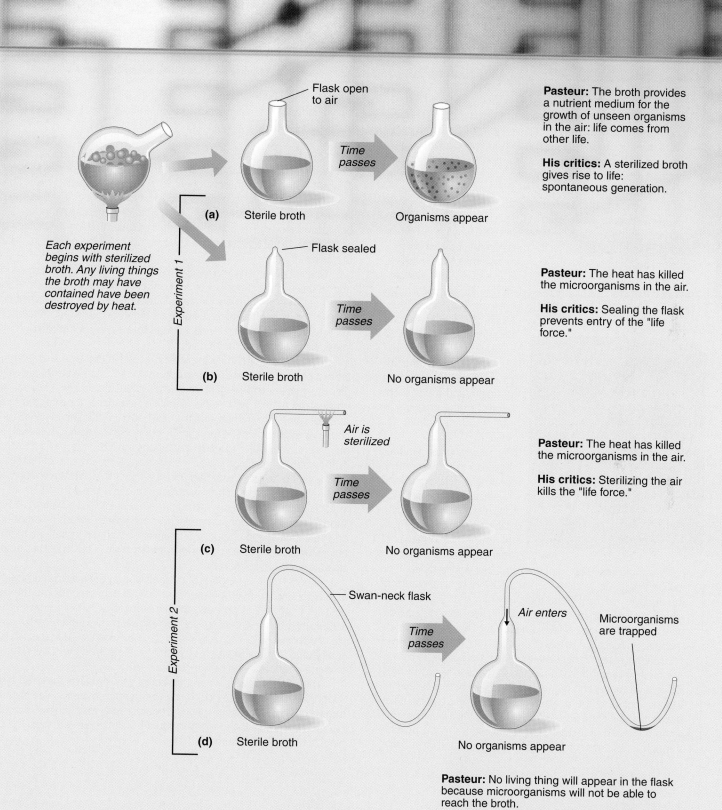

Each experiment begins with sterilized broth. Any living things the broth may have contained have been destroyed by heat.

Flask open to air

Time passes

(a) Sterile broth — Organisms appear

Pasteur: The broth provides a nutrient medium for the growth of unseen organisms in the air: life comes from other life.

His critics: A sterilized broth gives rise to life: spontaneous generation.

Flask sealed

Time passes

(b) Sterile broth — No organisms appear

Pasteur: The heat has killed the microorganisms in the air.

His critics: Sealing the flask prevents entry of the "life force."

Air is sterilized

Time passes

(c) Sterile broth — No organisms appear

Pasteur: The heat has killed the microorganisms in the air.

His critics: Sterilizing the air kills the "life force."

Swan-neck flask

Time passes

Air enters

Microorganisms are trapped

(d) Sterile broth — No organisms appear

Pasteur: No living thing will appear in the flask because microorganisms will not be able to reach the broth.

His critics: If the "life force" has free access to the flask, life will appear, given enough time.

Experiment 1

Experiment 2

FIGURE 1 Pasteur's Experiment Pasteur's simple experiment helped debunk the theory of spontaneous generation. (a) The broth was boiled to kill microorganisms. (b) Microorganisms did not appear if the flasks' necks were kept sealed. (c) No microbes grew if an open neck was sterilized. (d) These specially designed swan-necked flasks allowed air to enter, but prevented bacteria from entering the broth.

Some days later the flask is still free of any living thing. Pasteur has refuted the doctrine of spontaneous generation.

Controversy over the Use of Animals in Laboratory Research

Animal Research Is Essential to Human Health by Frankie L. Trull

Virtually every major medical advance of the last 100 years—from chemotherapy to bypass surgery, from organ transplantation to joint replacement—has depended on research with animals. Animal studies have provided the scientific knowledge that allows health care providers to improve the quality of life for humans and animals by preventing and treating diseases and disorders, and by easing pain and suffering.

Some people question animal research on the ground that data from animals cannot be extrapolated to humans. But physicians and scientists agree that the many similarities that exist provide the best insights into the complex systems of both humans and animals. Knowledge gained from animal research has contributed to a dramatically increased human life span, which has increased from 47 years in 1900 to more than 76.7 years in 1998.

Frankie L. Trull is president of the Foundation for Biomedical Research, a nonprofit educational organization dedicated to improving the quality of human and animal health by promoting public understanding and support of the ethical use of animals in scientific and medical research.

Much of this increase can be attributed to improved sanitation and better hygiene; the rest of this increased longevity is a result of health and medical advances made possible in part through animal research.

Research on animals has also led to countless treatments, techniques, and medical technologies. Animal research was indispensable in the development of immunization against many diseases, including polio, mumps, measles, diphtheria, rubella, and hepatitis. One million insulin-dependent diabetics survive today because of the discovery of insulin and the study of diabetes using dogs, rabbits, rats, and mice. Organ transplantation, considered a dubious proposition just a few decades ago, has become commonplace because of research on mice, rats, rabbits, and dogs.

Animal research has contributed immeasurably to our understanding of tumors and has led to the discoveries of most cancer treatments and therapies. Virtually all cardiovascular advances, including the heart-lung machine, the cardiac pacemaker, and the coronary bypass, could not have been possible without the use of animals. Other discoveries made possible through animal research include an understanding of DNA; X-rays; radiation therapy; hypertension; artificial hips, joints, and limbs; monoclonal antibodies; surgical dressings; ultrasound; the artificial heart; and the CAT scan.

Animal research will be essential to medical progress in the future as well. With the use of animals, researchers are gaining understanding into the cause of—and treatments for—AIDS, Alzheimer's disease, cystic fibrosis, sudden infant death syndrome, and cancer in the hopes that these problems can be eliminated. Although many nonanimal research models have been developed, no responsible scientist believes that the technology exists today or in the foreseeable future to conduct biological research without using animals.

Despite distortions and exaggerations put forth by those opposed to animal research, occurrences of poor animal care are extremely rare. Researchers care about the welfare of laboratory animals. Like everybody else, scientists don't want to see animals suffer or die. In fact, treating animals humanely is good science. Animals that are in poor health or under stress will provide inaccurate data.

Many people are under the false impression that laboratory animals are not protected by laws and regulations. In fact, many safeguards are in place to guarantee the welfare of animals used in research. A federal law, entitled the Animal Welfare Act, stipulates standards for care and treatment of laboratory animals, and the U.S. Public Health Service (PHS), the country's major source of funding for biomedical research, sets forth requirements with which research institutions must comply in order to qualify for grants for any biomedical research involving any kind of animal.

Both the Animal Welfare Act and the PHS animal welfare policy mandate review of all research by an animal care and use committee set up in each institution to ensure that laboratory animals are being used responsibly and cared for humanely. The committee, which must include one veterinarian and one person unaffiliated with the institution, has the power to reject any research proposal and stop projects if it believes proper standards are not being met.

Although animal research opponents portray the medical community as deeply divided over the merits of animal experimentation, the percentage of physicians opposed to animal research remains very small. A 1989 survey by the American Medical Association of a representative sample of all active physicians found that 99% believed animal research had contributed to medical progress, and 97% supported the continued use of animals for basic and clinical research.

The general public, when presented with the facts, has also been supportive of animal research. This support must not be allowed to erode through apathy or misconceptions. Should animal research be lost to the scientific community, the victims would be all people: our families, our neighbors, our fellow humans.

Vivisection: A Medieval Legacy by Elliot M. Katz

Vivisection, an outdated and extremely cruel form of biomedical research, is the purposeful burning, drugging, blinding, infecting, irradiating, poisoning, shocking, addicting, shooting, freezing, and traumatizing of healthy animals. In psychological studies, baby monkeys are separated from their mothers and driven insane; in smoking research, dogs have tobacco smoke forced into their lungs; in addiction studies, chimpanzees, monkeys, and dogs are addicted to cocaine, heroin, and amphetamines, then forced into convulsions and painful withdrawal symptoms; in vision research, kittens and monkeys are blinded; in spinal cord studies, kittens and cats have their spinal cords severed; in military research, cats, dogs, monkeys, goats, pigs, mice, and rats die slow, agonizing deaths after being exposed to deadly radiation, chemical, and biological agents.

Started at a time when the scientific community did not believe animals felt pain, vivisection has left a legacy of animal suffering of unimaginable proportions. Descartes, the father of vivisection, asserted that the cries of a laboratory animal had no more meaning than the metallic squeak of an overwound clock spring. Though the research community considers vivisection a "necessary evil," a growing number of scientists and health professionals see vivisection as simply evil.

As a veterinarian, I was taught that vivisection was essential to human health. My eyes were finally opened to the full horrors and futility of vivisection when years later, faculty members and campus veterinarians at the University of California informed me that animals were dying by the thousands from severe neglect and abuse; that vivisectors and campus officials were denying and concealing the abuses; and that experiments were conducted whose only benefit was to the school's finances and researchers' careers. I discovered that animal "care" committees, typically cited as assurance that animals are used responsibly, are in fact "rubber stamp" committees composed mostly of vivisectors who routinely approve each other's projects. Over the years, I have witnessed an ongoing pattern of university officials denying documented charges of misconduct, attempting instead to discredit critics of vivisection, ultimately defending even the most ludicrous and cruel experiments as necessary and humane.

I discovered that assertions by the biomedical community that vivisection is an essential and indispensable part of protecting the public's health are simply untrue. Vivisection can and should be ended. It is scientifically outdated and morally wrong. There is a plethora of modern biomedical technology that can be used to improve society's health without harming animals. The advent of sophisticated scanning technologies, including computerized tomography (CT), positron emission tomography (PET), and magnetic resonance imaging (MRI), has given scientists the ability to examine people and animals noninvasively. This technology has isolated abnormalities in the brains of patients with Alzheimer's disease, epilepsy, and autism, revolutionizing diagnosis and treatment of these diseases. Tissue and cell cultures are being increasingly used to screen cancer and AIDS drugs. Progress with AIDS has come from areas entirely unrelated to animal experimentation. Human skin cell cultures are used to test new products and drugs for toxicity and irritancy.

Why, then, is vivisection so entrenched and defended with an almost religious fervor? Dr. Murry Cohen summed it up when he stated, "Change is difficult for most people, but it is particularly painful for scientific and medical bureaucracies, which fight to maintain the status quo, especially if required change might imply admission of previously held incorrect ideas or flawed axioms." Vivisection continues today because of vested interests, habit, economics, and legal considerations, not for the real advancement of science and public health.

When presented with the facts, members of the public almost unanimously express their desire to see an end to the horrors of vivisection. Thousands of professionals like myself have reevaluated the sense, efficacy, and worth of vivisection and have formed or joined organizations working to end this outdated and cruel form of research. The ending of vivisection will lead to improved public health and restore to medicine and science much needed excellence and compassion for all beings, human and nonhuman alike.

Elliot M. Katz, DVM (here with companion, Manco) is a graduate of the Cornell University School of Veterinary Medicine. He is president and founder of In Defense of Animals, a national animal rights organization.

Sharpening Your Critical Thinking Skills

1. Summarize the main points of each author.
2. Do these authors use data or ethical, anecdotal (stories or experiences) arguments to make their cases?
3. Do you have a view on this issue? What factors weigh most heavily in making up your mind?

www.jbpub.com/humanbiology/5e

Visit Human Biology's Internet site for links to web sites offering more information on this topic.

fine them in discussions brings clarity to issues and debates. The Greek philosopher Socrates, in fact, destroyed many an argument in his time by insisting on clear, concise definitions of terms. As you analyze any information or issue, always be certain that you understand the terms, and make sure that others define their terms as well.

The third rule of critical thinking is: question the methods by which the facts are derived. In science, many debates over controversial topics hinge on the methods used to discover new information. The first question you should ask is: was the information gained from careful experimentation, or is it the result of casual, unscientific observations?

As you analyze new findings, first check to see if they were obtained from experimentation. If so, were the experiments well planned and executed? Did the experiment have a control group? Were the control and experimental groups treated identically except for the experimental variable? Did the experimenters use an adequate number of subjects? Even if all of these conditions are met, beware. In science, one experiment is rarely adequate to permit you to draw firm conclusions. Careful scrutiny, for example, may show small but significant design flaws: perhaps mice being tested in a drug study were resistant to the drug under examination or were hypersensitive to it. As a rule of thumb, wait for scientific verification of the results. A second researcher may repeat the experiment with similar results. In some cases, a new researcher may find different results.

Nowhere is caution more necessary than when you encounter announcements of scientific breakthroughs on the television news and in magazine and newspaper articles. Ever eager to showcase new scientific studies before they have been verified by others, the media sometimes does a grave disservice to the advancement of scientific knowledge; further study may show that earlier findings were invalid. Unfortunately, the media often fails to publish the results from follow-up studies that contradict previous ones. Scientific journals often favor studies that have positive results, too. Researchers who write up a study that shows no effect, may have a difficult time publishing their results. Ultimately, the public—even the scientific community—is left with a false impression of reality.

The fourth rule of critical thinking is: question the conclusions derived from facts. Surprisingly, even if an experiment is run correctly, there's no guarantee that the conclusions drawn from the results are correct. How can that be? The answer may lie in bias, ignorance, and error. Bias refers to personal beliefs that taint the interpretation of results. Ignorance is a lack of full knowledge. This, in turn, may lead a scientist to misinterpret his or her results. Finally, error does occur, in spite of our best efforts.

Two questions should be asked when one analyzes the conclusions of an experiment: (1) Do the facts support the conclusions? and (2) Are there alternative conclusions? Consider an example.

One of the earliest studies on lung cancer showed that people who consumed large quantities of table sugar (sucrose) had a higher incidence of lung cancer than those who ate moderate amounts. The researchers concluded that lung cancer was caused by sugar. Many people had trouble believing this conclusion, which forced a reexamination of the study. It, in turn, showed that the group with a higher incidence of lung cancer included a higher percentage of cigarette smokers. Subsequent studies showed that smoking, not sugar consumption, is the culprit.

Besides showing the importance of scrutinizing the conclusions of a scientific study, this example illustrates a key principle of medical research: correlation does not necessarily mean causation. In other words, two factors that appear to be related (correlated) may, in fact, not be linked at all.

The fifth rule of critical thinking is: look for assumptions and biases. This rule is related to the previous rule, questioning conclusions, but is so important that it warrants closer examination. Biases and hidden assumptions are to thinking what cyanide is to food—a poison. Unfortunately, biases and hidden assumptions run rampant in today's society.

In many contemporary debates over a wide range of issues, proponents often present information that supports their point of view. This selective inclusion of supportive data and exclusion of contradictory information is often an expression of a hidden agenda. What happens is that people make up their mind about an issue, then seek out information that supports their point of view.

The sixth rule of critical thinking is: question the source of the facts—that is, who is telling them. Closely related to the previous rule, this one calls on us to learn about the people who performed various research studies or analyses. It asks us to familiarize ourselves with the people taking various positions. Bias can influence scientific researchers. Be especially wary of people with political motives. An association with a partisan group may be a red flag, warning that bias may have influenced their conclusions. Their penchant for "putting a spin" on things often results in doctored truths.

Sometimes a study of the biographies of the people delivering the information is as instructive as an examination of their conclusions. Beware of "experts" who have a hidden agenda. Experts from industry who swear under oath about the safety of their product may be biased or even deceitful. Environmental experts may also slant the data to support their view.

Also beware of people who may not know as much as you think they should. Although we think of physicians as experts on human health, most of them received little or no training in nutrition in medical school. Many medical students still graduate without a full understanding of the role of nutrition in preventing disease and promoting good health. For questions about diet, you'd be better off consulting a registered dietitian.

The seventh rule of critical thinking is: understand your own biases, hidden assumptions, and areas of ignorance. So far, this discussion on critical thinking has concentrated on ways you can uncover mistakes in reasoning that other people make. But what about your own biases, hidden assumptions, and areas of ignorance? Do they affect your ability to think critically? Uncomfortable as it may be, it's essential to uncover—and correct—them. Only then can you become a critical thinker.

As you read this text, you will be presented with examples to help you sharpen your critical thinking skills. The Point/Counterpoint on the previous page will help you practice these rules.

SUMMARY

Health and Homeostasis

1. Humans, like all other organisms, have evolved mechanisms that ensure relative internal constancy (*homeostasis*). These homeostatic mechanisms are vital to survival and reproduction.

2. Homeostatic mechanisms exist at all levels of biological organization, from cells to organisms to ecosystems.

3. The health of all species and ecosystems is dependent on the functioning of homeostatic mechanisms. When these mechanisms break down, illnesses often result.

4. Human health has traditionally been defined as the absence of disease, but a broader definition of health is now emerging. Under this definition, good health implies a state of physical and mental well-being.

5. Physical well-being is characterized by an absence of disease or symptoms of disease, a lack of risk factors that lead to disease, and good physical fitness.

6. Mental health is also characterized by a lack of mental illness and a capacity to deal effectively with the normal stresses and strains of life.

7. Human health and the health of the many species that share this planet with us depend on a properly functioning, healthy ecosystem. Thus, alterations of the environment can have severe repercussions for all species, including humans.

Evolution and the Characteristics of Life

8. Evolution results in structural, functional, and behavioral changes in populations. These changes, in turn, result in organisms better equipped to cope with their environment—that is, better able to survive and reproduce.

9. All life-forms alive today exist because of evolution. In fact, every cell and every organ in the human body is a product of millions of years of evolution. Even the intricate homeostatic mechanisms evolved over long periods.

10. Living organisms belong to five major groups or kingdoms. Humans belong to the animal kingdom.

11. Evolution is responsible for the great diversity of life-forms. However, because the Earth's organisms evolved from early cells that arose over 3.5 billion years ago, all organisms, including humans, share many common characteristics. Items 12–18 list the common characteristics of all life-forms.

12. All organisms, including humans, are made up of cells.

13. All other organisms grow and maintain their complex organization by taking in chemicals and energy from their surroundings.

14. All living things house many chemical reactions. These reactions are collectively referred to as *metabolism.*

15. All organisms possess homeostatic mechanisms.

16. All organisms exhibit the capacity to perceive and respond to stimuli.

17. All organisms are capable of reproduction and growth.

18. All organisms are the product of evolutionary development and are subject to evolutionary change.

19. Although humans are similar to many other organisms, we also possess many unique abilities and features: culture, our ability to plan for the future, and an enormous ability to reshape the Earth through ingenuity and technology.

Science

20. Science is both a systematic method of discovery and a body of information about the world around us.

21. Scientists gather information and test ideas through the *scientific method.* The scientific method begins with observations and measurements, often made during experiments. Observations and measurements may lead to hypotheses, explanations of natural phenomena that can be tested in experiments. The results of experiments help scientists support or refute their hypotheses.

22. The body of scientific knowledge also contains theories, broad generalizations about the way the world works. Theories can change over time as new information becomes available.

Critical Thinking

23. Critical thinking is a useful tool in science and life and is defined as careful analysis that helps us distinguish knowledge from beliefs or judgments.

24. Critical thinking provides a way to analyze issues and information.

25. Table 1-2 summarizes the critical thinking rules.

 www.jbpub.com/humanbiology/5e

The site features eLearning, an online review area that provides quizzes, chapter outlines, and other tools to help you study for your class. You can also follow useful links for in-depth information, research the differing views in the Point/Counterpoints, or keep up on the latest health news.

KEY TERMS AND CONCEPTS

CONCEPT REVIEW

1. How would you define life? p. 6
2. In what ways are a rock and a living organism (for example, a bird) similar, and in what ways are they different? p. 6
3. Describe the concept of homeostasis. How does it apply to humans? How does it apply to ecosystems? Give examples from your experiences. p. 3
4. Using the definition of health and the list of healthy habits in Table 1-1, assess your own health. What areas need improvement? pp. 4–5

5. In what ways are humans different from other animals? In what ways are they similar? pp. 6–7
6. Describe the scientific method, and give some examples of how you have used it recently in your own life. p. 9
7. How do a hypothesis and a theory differ? p. 9
8. List and discuss the critical thinking skills presented in this chapter. Which skills seem to be most important for the kind of thinking you normally do? p. 11

9. A graduate student injects 10 mice with a chemical commonly found in the environment and finds that all of his animals die within a few days. Eager to publish his results, the student comes to you, his adviser. What would you suggest the student do before publishing his results? pp. 9–10
10. Given your knowledge of scientific method and critical thinking, make a list of reasons why scientists might disagree on a particular issue or research finding. p. 11

SELF-QUIZ: TESTING YOUR KNOWLEDGE

1. _____ results in structural, functional, and behavioral changes in populations of organisms that make them better able to survive and reproduce. p. 2
2. _____ is referred to as a state of relative internal constancy. p. 3
3. Abnormal conditions in the human body such as high blood pressure that may lead to ill health or serious conditions are called _____ factors. p. 4

4. Human health is dependent on physical health and _____ wellbeing. p. 4
5. Organisms are organized into five major groups called _____. p. 5
6. Chemical reactions in the body are known as _____. p. 6
7. Scientists use a process called the _____ method to learn new information and test ideas. p. 9
8. A(n) _____ is a testable assertion. p. 9

9. In many scientific experiments, two groups of subjects are used. One group, the _____ group is treated identically to the experimental group except that it is not exposed to the experimental variable. p. 9
10. _____ _____ is a process that allows us to analyze problems, assertions, conclusions, and research, distinguishing beliefs from knowledge. p. 11

critical thinking

Thinking Critically—Analysis

This analysis corresponds to the Thinking Critically scenario that was presented at the beginning of this chapter.

This experiment has several major flaws. First, the boy used chickens from two different farms, so the chickens could have been genetically dissimilar. Differences in genetics could have been responsible for the differences in cholesterol content in the eggs. Second, although differences were found in the cholesterol content of the eggs of the two groups, we don't know if they were statistically significant. Good statistical analysis is necessary to determine whether measured differences are substantial enough to be attributable to the treatment. Another problem is the small sample size. Before I donated any money to this new venture, I'd want to see it performed on a larger group of genetically similar chickens. The fourth problem is that no mention was made of the

differences between the two feeds. A careful analysis is essential to solidify one's confidence.

Clearly, this simple experiment has some flaws, but there's an even larger problem that we haven't addressed yet—notably the underlying supposition that cholesterol in eggs raises blood cholesterol levels. As it turns out, the liver produces lots of cholesterol. It's responsible for most of the cholesterol floating around in our bloodstreams. Furthermore, recent studies have shown that eggs don't contain as much cholesterol as was once thought and that eating foods rich in saturated fat (like bacon) is linked to elevated cholesterol levels in your bloodstream. (You'll learn more about this in Chapter 5 on nutrition.) Because of these findings, the American Heart Association has raised its recommendation for egg consumption from three a week to one a day for healthy people. To control cholesterol, you're better off cutting back on foods that have a high content of saturated fat like hamburgers and bacon.

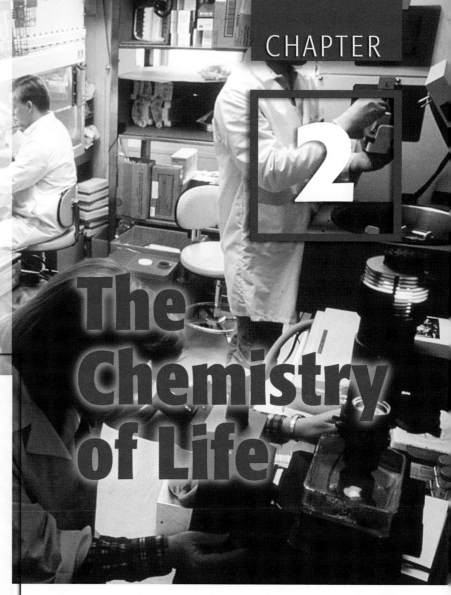

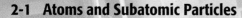

CHAPTER

2

The Chemistry of Life

thinking critically

In 1971, Joseph Chatt and his colleagues at the University of Sussex in England synthesized two compounds with the same chemical formula. However, he found that the crystals formed by one compound were sapphire-green; the crystals formed by the other were bright green.

Chatt hypothesized that he had created two compounds that have the same molecular weight and atomic composition but differ in chemical or physical properties because the atoms are linked in different ways. But, when Chatt subjected his compounds to X-ray analysis, which helps determine the structure of molecules, he found that the atoms of his compounds were in the same relative positions. The only difference appeared to be in the chemical bond between two atoms (oxygen and molybdenum), which was longer in the bright green compound than in the sapphire-green one.

Assuming that impurities are rare in single, well-formed crystals, Chatt's group concluded that the color difference in the two compounds resulted from differences in the length of a single bond, a phenomenon not previously reported.

At first, Chatt's findings were discounted, because chemists believe that molecular bonds have characteristic lengths. Most scientists concluded that his experiments were flawed. Can you find any reason(s) in the above information that might have led Chatt to an incorrect conclusion?

Chemistry is a field very few people know much about, yet it has many useful applications. In recent decades, for example, chemists have developed thousands of new drugs. These chemicals have helped humankind combat a wide range of often deadly or debilitating diseases. One of the most notable examples is antibiotics, chemicals that kill harmful bacteria. Not only have antibiotics reduced suffering, they have also greatly reduced mortality, especially among infants.

Advances in chemistry have also led to countless improvements in household products that make our lives safer. House paints are a case in point. At one time, most paints contained lead-based pigments. Because lead is toxic to the nervous system, chips of paint flaking off walls and ceilings pose a health hazard to young children, who routinely put foreign objects in their mouths as they explore their new world. Modern paints lacking lead-based pigments are a definite improvement (Figure 2-1).

The truth of the matter is that almost everything we touch or look at in our daily lives has been brought to us or improved by chemists. But chemistry is important in other ways, too. A little knowledge of chemistry, for example, can help us understand environmental problems such as ozone depletion, urban air pollution, and acid rain. On a more personal level, a knowledge of chemistry helps people who want to live long, healthy lives make informed decisions about what they eat.

But why study chemistry at the beginning of a biology course? The answer is simple. First, cells and organisms are made up of chemicals. These chemicals undergo many reactions that are the basis of life. In order to understand the structure and function of cells and organisms, you must understand some basic principles of chemistry.

This chapter introduces you to several key principles of chemistry, which form the foundation for the study of human biology. We begin with a discussion of atoms and mol-

FIGURE 2-1 Better Living Through Chemistry A new line of paint, stains, and finishes, like those being applied here from AFM in San Diego, California, help reduce exposure of workers and residents to toxic chemicals once contained in many paint products.

ecules, then look at water, acids, bases, and buffers. The major biological molecules like proteins and lipids are briefly mentioned here but will be discussed more fully in subsequent chapters.

2-1 Atoms and Subatomic Particles

The world in which we live is composed of a wide variety of physical matter such as wood, plastics, and metal. **Matter** is defined as anything that has mass and occupies space. Technically, **mass** is a measure of the amount of matter in an object. Thus, a water balloon has a greater mass than an air-filled balloon.

Atoms

Atoms are the fundamental unit of all matter.

Matter is composed of tiny particles called **atoms**. The word *atom* comes from the Greek *atomos*, which means "incapable of division." The ancient Greek philosophers who proposed the existence of atoms believed that all matter consisted of small particles that lacked any internal structure. Over time, though, evidence has shown that atoms consist of smaller particles, called **subatomic particles**.

In this chapter, we will examine three of the most important subatomic particles: electrons, protons, and neutrons (Figure 2-2). **Protons** (PRO-tauns) have the greatest mass of all and are located in the center of the atom, the **nucleus** (Figure 2-2). Each proton has a positive charge. **Neutrons** (NEW-trauns) are uncharged particles with slightly less mass than protons. Like protons, they are also found in the nucleus. Because protons are positively charged and neutrons have no charge, the nucleus of an atom is positively charged.

Surrounding the positively charged nucleus is a region called the electron cloud. It contains tiny, negatively charged parti-

cles, **electrons** (ee-LECK-trauns), that move rapidly through a large volume of space.

Most of the mass of an atom can be attributed to the protons and neutrons, which are packed into a tiny space. These particles are so heavy that a single cubic centimeter (about 1/3 teaspoon) would weigh 100 million tons. However, most of the volume of an atom is composed of the electron cloud. Consequently, the late Carl Sagan, one of America's leading astronomers, remarked that atoms are mostly space, or "electron fluff."

Atoms are electrically neutral because they contain the same number of protons and electrons. Thus, the positive charges cancel out the negative ones.

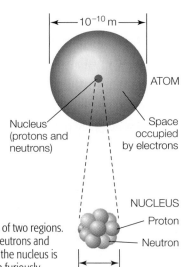

FIGURE 2-2 The Atom The atom consists of two regions. The central nucleus contains protons and neutrons and makes up 99.9% of the mass. Surrounding the nucleus is the electron cloud, where the electrons spin furiously around the nucleus. (Figure not drawn to scale.)

The Elements

Elements are made up of atoms that are composed of protons, neutrons, and electrons.

Early scientists accounted for the diversity of matter by assuming that there must exist a number of pure substances and that these substances could be combined into an infinite variety of different forms. They called these pure substances **elements**. Today, modern chemists define an element as a pure substance (for example, gold or lead) that contains only one type of atom.

At this writing, 92 naturally occurring elements are known. More than a dozen additional atoms can be made in the laboratory. Of the 92 naturally occurring elements, only about 20 are found in organisms. Four elements in this group—carbon, oxygen, hydrogen, and nitrogen (remember: COHN)—comprise 98% of the atoms of all living things.

Elements are listed on a chart called the *periodic table of elements* (Table 2-1; Appendix A). The periodic table is a handy summary of vital statistics on each element. As illustrated in Table 2-1, most elements are represented on the table by a one- or two-letter symbol. The element hydrogen, for example, is indicated by H. Oxygen is designated by O, and carbon is represented by C. Aluminum is designated by Al.

Above each symbol on the periodic table is its atomic number. The **atomic number** of an element is the number of protons found in the nuclei of the element's atoms. Because all atoms are electrically neutral, the atomic number also equals the number of electrons in the atoms.

As you can see, below the chemical symbol for each element is another number, called the *mass number*. The mass number is the average mass of the atoms of each element. The mass number is approximately equal to the number of protons and neutrons in a given atom. Because electrons have virtually no mass, they contribute very little to the mass of an atom. As shown in Table 2-1, carbon has an atomic mass of 12.

Isotopes

Isotopes are alternative forms of atoms, differing in the number of neutrons

Although elements are made of identical atoms, there are very slight differences in some of the atoms of most elements. These different forms of atoms are called **isotopes** (EYE-so-topes). Isotopes of an element differ only in the number of neutrons they contain. This does not affect the atom very much. For instance, it does not change its chemical reactivity. Hydrogen, for example, has three slightly different atomic forms. The most common hydrogen atom contains one proton and one electron, but no neutrons. The next most common form has one proton, one electron, and one neutron. The least common hydrogen atom has one proton, one electron, and two neutrons. The three forms, or isotopes, of hydrogen are chemically identical—that is, they react the same.

Although additional neutrons don't change the chemical nature of the atom, their presence makes some atoms of an element slightly heavier than others. This also makes some atoms

TABLE 2-1	A Simplified Periodic Table						
I	II	III	IV	V	VI	VII	VIII
1 H Hydrogen 1.0	← Atomic number ← Atomic symbol ← Mass number						2 He Helium 4.0
3 Li Lithium 7.0	4 Be Beryllium 9.0	5 B Boron 11.0	6 C Carbon 12.0	7 N Nitrogen 14.0	8 O Oxygen 16.0	9 F Fluorine 19.0	10 Ne Neon 20.2
11 Na Sodium 23.0	12 Mg Magnesium 24.3	13 Al Aluminum 27.0	14 Si Silicon 28.1	15 P Phosphorus 31.0	16 S Sulfur 32.1	17 Cl Chlorine 35.5	18 Ar Argon 40.0
19 K Potassium 39.1	20 Ca Calcium 40.1						

Note that each element is represented by a one- or two-letter symbol. Elements are listed according to their atomic number (the number of protons). Their atomic mass is also shown.

Scientific Discoveries that Changed the World

2-1 The Discovery of Radioactive Chemical Markers
Featuring the Work of Schoenheimer

Biochemist Rudolf Schoenheimer (1898–1941) and his colleagues offered science a chemical tool that would yield important new information and result in significant scientific and medical advances: radioactively labeled molecules—that is, common biological molecules containing one or more radioactive atoms (isotopes).

A radioactive atom incorporated in a biological molecule emits tiny bursts of radioactivity. Radioactive markers permit scientists to follow marked molecules as they progress through the body's maze of biochemical reactions. That is, they allow scientists to see where they go and how they are changed. Before the emergence of this technique, substances such as fatty acids or amino acids administered to an animal were lost almost immediately, because they were indistinguishable from molecules in the body.

Radioactive markers also permit scientists to determine where chemicals are stored and how they are removed from the body. In fact, many of the physiological processes you will read about in this book were investigated by using radioactively labeled molecules.

Before Schoenheimer's discovery, the inability to track molecules resulted in a huge gap in our knowledge about metabolism and other body functions. Futile attempts to "tag" molecules with nonradioactive molec-

a bit unstable. To achieve a more stable state, many isotopes release small bursts of energy, or tiny energetic particles from their nuclei. These emissions are known as **radiation.** Radiation carries energy and mass away from an atom's nucleus and helps atoms achieve more stable states. Radioactive isotopes are called **radionuclides** (ray-dee-oh-NEW-klides).

Radiation has provided numerous benefits to humankind. For example, scientists have discovered a wide variety of uses for radiation in medicine, including X-rays to study bones. Radioactive chemicals are also used in medical tests. One of the most important advancements is the use of radioactive isotopes as markers in experiments (Scientific Discoveries 2-1). Another extremely important application is the use of naturally occurring radioactive isotopes to date rocks and fossils. Information from this technique has helped scientists pinpoint significant evolutionary events and has given us important insights into the history of life on Earth.

Despite the many benefits of radiation, its use has not occurred without some risks. For example, radiation exposure damages molecules in body cells and causes changes in the genetic material called **mutations.** Mutations can result in birth defects and can lead to cancer in some cases. Despite these and other problems, radiation is widely used. It is now even being used to sterilize foods, thus eliminating the need for refrigeration. The Point/Counterpoint in this chapter examines the pros and cons of this procedure.

2-2 The Making of a Molecule

Atoms combine to form molecules.

The diversity of matter in the world around us is, as noted above, a result of the presence of a wide range of elements, each with its own unique properties. It also arises from the fact that atoms of various elements can combine to form compounds and molecules. The term **compound** refers to a substance made up of two or more atoms. Each compound has a unique chemical formula that indicates the ratio of atoms found in it. For instance, HCl is the chemical formula of hydrochloric acid. It contains one hydrogen atom and one chlorine atom. The chemical formula of the flammable gas methane that we cook with is CH_4. CH_4 contains one carbon atom for every four hydrogen atoms.

The term **molecule** refers to the smallest particle of a compound that still retains the properties of that compound. Methane gas given off by rotting vegetation, for instance, is a chemical compound consisting of methane molecules. Hydrogen gas is a chemical compound consisting of hydrogen molecules. A hydrogen molecule consists of two hydrogen atoms.

Chemical Bonds

Atoms bond to form more stable configurations

Atoms in molecules join together by chemical bonds. There are two types of bonds that form between atoms: ionic and covalent. In both instances, the electrons are responsible for creating the bonds that hold atoms together.

Ionic Bonds

Ionic bonds are electrostatic attractions between two oppositely charged particles.

Ionic bonds are formed between two atoms when one loses an electron and the other gains an electron. This reaction creates two charged particles, known as **ions** (EYE-ons). In this reaction, the atom that loses its electron becomes positively charged, and the one that gains an electron is negatively charged. The posi-

ular markers generally failed, because the altered molecules differed so much from their natural substances that they were treated differently by the body.

Through experimentation, Schoenheimer showed that radioactively labeled molecules were so similar to the unlabeled molecule that they reacted identically. This made them ideal markers and opened the door to many additional discoveries.

From Schoenheimer's brilliant experiments, a new picture of life has emerged. He and his colleagues showed that molecules of the body such as DNA, which had previously been considered stable, are actually in a continuous state of flux. Even apparently dormant fat deposits in the body were ever-changing. Additional studies eventually showed that amino acids were rapidly broken down and rebuilt into new ones.

Since their introduction in the 1930s, radioactive markers have also enjoyed great popularity in medicine, where they're often used to determine the functional status of various organs. Radioactive iodide ions, for instance, can be administered to a patient to assess the function of the thyroid gland, a hormone-producing gland in the neck.

Radioactively labeled molecules are helping scientists unravel the mysteries of body functions. Labeled glucose molecules, for instance, can be used to determine the relative activity of different parts of the brain, permitting scientists to map brain functions accurately.

These are but a few examples of the tremendous contribution of radioactively labeled molecules to medicine and science.

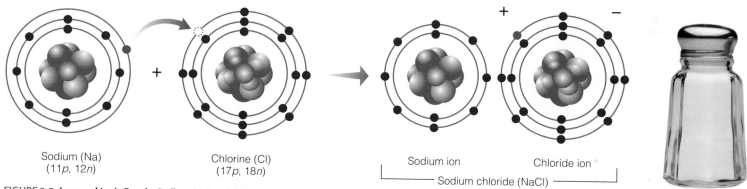

Sodium (Na)
(11p, 12n)

Chlorine (Cl)
(17p, 18n)

Sodium ion

Chloride ion

Sodium chloride (NaCl)

FIGURE 2-3 Ions and Ionic Bonds Sodium (Na) and chlorine (Cl) atoms differ in their electron affinity. Sodium has a weaker affinity and tends to give up an electron to atoms like chlorine. As a result, sodium becomes a positively charged ion and chlorine a negatively charged ion. The oppositely charged ions attract each other, forming an ionic bond.

tively charged ion is attracted to the negatively charged electron by a weak electrostatic force. This force holds them together and is referred to as an *ionic bond*. Figure 2-3 shows how an ionic bond forms between a chlorine atom and a sodium atom.

Covalent Bonds

Covalent bonds are formed by the sharing of electrons between atoms.

In contrast to ionic bonds, **covalent bonds** occur when two atoms "share" electrons. That is to say, when two atoms are bound to each other by a covalent bond, they share electrons. These electrons actually orbit around both atoms, gluing them together. This arrangement usually produces a much stronger union than an ionic bond. Figure 2-4 illustrates the formation of hydrogen gas, H_2, from two hydrogen atoms.

Methane is another compound whose molecules are held together by covalent bonds. As noted above, methane (CH_4) is a gas that consists of four hydrogen atoms and one carbon atom. The four hydrogen atoms are covalently bonded to the carbon atom. To understand how this molecule forms, we turn first to the central atom, carbon.

Carbon atoms have six electrons, two in a region or shell close to the nucleus and four in an outer shell. The four outer-shell electrons are available for bonding. To form methane, carbon reacts with four hydrogen atoms to form methane. This arrangement results in a full outer shell (for carbon) containing eight electrons.

Many other atoms react similarly, filling the outer shell during covalent bonding. When filled, the outer shell of many atoms contains eight electrons. Thus, chemists have coined the term *octet rule* (meaning a group of eight) to explain this behavior. You can use it to predict how atoms will react.

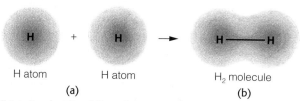

H atom

H atom

H_2 molecule

(a)

(b)

FIGURE 2-4 Covalent Bond Nonpolar covalent bonds are characterized by a sharing of electrons. (a) The shaded regions represent the hydrogen atoms' electron clouds. (b) Two atoms bound together by a covalent bond. The electrons orbit around both nuclei, holding them together to create a molecule.

Controversy over Food Irradiation

Food Irradiation: Too Many Questions by Donald B. Louria, MD

The debate about the safety of irradiating foods raises six issues:

1. **The safety issue.** Irradiated food does not become radioactive, but the radiation does cause chemical changes in food molecules. The major concerns are genetic damage in those consuming the foods and nutrient loss after irradiation.

A highly controversial, but flawed, study on small numbers of malnourished children in India suggested that consumption of freshly irradiated wheat could produce chromosomal abnormalities. However, a study of a much larger group of healthy adults in China showed no such abnormalities. The proponents of food irradiation have criticized the Indian study and pointed to the larger study they feel showed no chromosome damage. But, that does not answer the concerns. If food irradiation becomes an accepted technique, millions of malnourished people will eat irradiated foods. What is needed is a careful study of possible genetic damage involving adequate numbers of children and adults. I am prepared to believe that consumption of irradiated foods is unlikely to produce significant harm, but such a belief must be supported by proper data.

Donald B. Louria, MD teaches at the UMDNJ-New Jersey Medical School, where he is Chairman Emeritus of the Department of Preventive Medicine and Community Health.

2. **The nutrition issue.** Irradiated food loses some of its nutritional value; the extent of loss of vitamin content depends on the type of food and the dose of radiation—the higher the radiation dose, the greater the loss. Irradiation proponents suggest the effects of irradiation on vitamin content are not different from that of conventional food processing (such as heating). But, some evidence suggests that, when irradiated foods are processed (frozen, thawed, heated), there is an accelerated loss of vitamins. Furthermore, some foods that would be irradiated (fruits, vegetables) are eaten raw so the arguments about heating would not apply. Proponents also maintain that the diet in the United States contains redundant vitamins, so some loss through irradiation would not be of concern. Of course, this would not be true for millions of people in other countries, as well as those below the poverty line in the United States and those over age 60 who often have low blood vitamin concentrations.

Arguing that our diet contains adequate vitamins and that destroying the vitamin content of foods by irradiation is of no importance is not likely to be viewed well by many Americans.

3. **The necessity for this technology.** Proponents say food irradiation will reduce diarrheal illnesses. That is true, but the potential benefits have been exaggerated. There are 200 million diarrheal episodes a year in the United States; of these, only about one-third are food borne and, in most cases, the cause is unknown so food irradiation is not the answer to diarrheal disease in the United States. But, it would reduce substantially diarrheal illnesses due to several bacteria (*Salmonella, Listeria,* and *E. coli*). Proper cooking procedures are just as effective as food irradiation and that, combined with more intensive surveillance of food production and distribution, reduces the need to use irradiation.

4. **Helping to solve world hunger.** Will irradiating foods prolong their shelf lives and thus, feed the world? Shelf lives will definitely be increased; that is an important benefit for the less-developed world, but the proponents have provided no adequate data on the extent of the benefit. If the trade-off is longer shelf life, but less nutritious foods, that would be unacceptable to many people.

5. **The issue of competing technologies.** During the next decades, scientists will develop foods that have longer shelf lives and are more nutritious. Advances in biotechnology are likely to give us useful alternatives to food irradiation. Other technologic advances will greatly improve our ability to detect microbial contamination of foods before they are allowed to reach the marketplace.

6. **The environmental pollution issue.** Food irradiation is a technology that will result in numerous irradiation plants in the United States. The few irradiation plants operating in the United States have, in some cases, contaminated their workers and the environment. Imagine the potential for contamination if there were hundreds of such plants using radioactive materials.

Food irradiation does have potential benefits, but it also raises substantial concerns. Whether we should adopt the technology is obviously a matter for continuing debate. Certainly, the technology should not be adopted until the issue of potential genetic damage has been resolved and there is agreement that the public will always be informed not only that a given food has been irradiated, but also the extent, if any, of nutritional damage.

Covalent bonding can be represented by using dots and x's for the outer-shell electrons of an atom. Figure 2-5 uses this technique to show the electron sharing in a molecule of methane. In this example, carbon is represented by the letter C surrounded by four outer-shell electrons, indicated by dots. Each hydrogen atom is represented by H, and the outer-shell electron of each hydrogen atom is indicated by an x. As illustrated, each hydrogen atom shares its electron with the carbon atom, giving the carbon atom four additional electrons and thus creating a full outer shell for carbon. The shell contains eight electrons.

Another important molecule is water, H_2O. As the formula indicates, a water molecule consists of a single atom of oxygen

Food Irradiation: Safe and Sound by George G. Giddings

Irradiated foods are safe and wholesome. The ionizing radiation process applied to foods offers certain proven public health and economic benefits without significant public health risks when carried out according to well-established principles and procedures. Decades of worldwide research and testing by competent, knowledgeable, objective, and responsible scientists have led to this conclusion. This conclusion is also supported by some 30 years of experience in the radiation sterilization of medical devices and other health care products to prevent infections, plus a growing list of industrial and consumer products, including foods and their raw materials, ingredients, and packaging materials. The technology is so pollution-free that the EPA exempts it from environmental impact statements.

There has been organized political opposition to food irradiation by a network of special-interest activists serving various political agendas, notably the antinuclear/anti-irradiation one. This network includes a handful of scientists and medical professionals from other fields who act as "expert witnesses" against food irradiation to serve their hidden agendas. For example, they still point to an old Indian study on irradiated wheat and a few children as suggestive of risk even though it was dismissed in the mid-1970s. Opponents point to minor losses of certain vitamins where in fact irradiation is gentler towards vitamins and other nutrients than comparable food processes and even cooking. Their campaign is doomed to failure in the face of the unshakable facts, including a growing appreciation for public health and other proven benefits of food irradiation, and its growing worldwide regulatory approval, industrial usage, and public acceptance. The American Medical Association and World Health Organization are among the many professional bodies that have endorsed food irradiation.

Food irradiation is undoubtedly the most versatile physical process yet applied to food substances in terms of the range and variety of objectives it can accomplish. Radiation can

- inhibit the sprouting of foods, such as potatoes and onions, and delay spoilage
- rid fruits, vegetables, and grains of insect pests and prevent parasites from infecting consumers of fish and meats
- rid foods of microbial pathogens such as the salmonellae
- delay microbial spoilage of a wide variety of animal and plant products by reducing microbe levels
- sterilize packaging materials, eliminating microorganisms that would otherwise contaminate products
- sterilize a wide variety of prepared or cooked foods such as meat, poultry, and fishery products, which have already been used to feed astronauts in space and immune-compromised patients.

All of these beneficial effects and more can be, and are, being readily accomplished by the application of ionizing (gamma, electron, and X-ray) radiation according to well-established principles and procedures. Nevertheless, irradiation must compete with a number of other new technologies. As a result, it is not likely to be applied to all, or even a high percentage, of the national and world food supply. It will therefore be used in cases in which it is clearly the best all-around choice.

George G. Giddings, Ph.D., has been involved in food irradiation research and development since 1963 and has written and spoken extensively on the subject.

Sharpening Your Critical Thinking Skills

1. What is food irradiation? Why is it used?

2. List and summarize the key points of each author.

3. Using your critical thinking skills, analyze each author's position. Are you inclined to agree with either one on all issues? Why or why not?

4. In your opinion, what is the best course for the development and implementation of this technology?

5. What facts would you need to know to form a personal conclusion about this issue?

www.jbpub.com/humanbiology/5e

Visit Human Biology's Internet site for links to web sites offering more information on this topic.

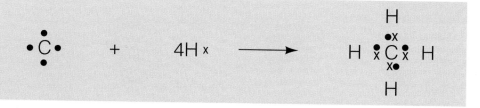

FIGURE 2-5 **Covalent Bonding Simplified** You can keep track of the electrons being shared in covalent bonds by representing the outer-shell electrons as dots or Xs. See the text for explanation.

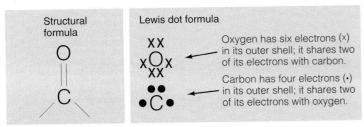

Structural formula

Lewis dot formula

Oxygen has six electrons (x) in its outer shell; it shares two of its electrons with carbon.

Carbon has four electrons (·) in its outer shell; it shares two of its electrons with oxygen.

(a) Double covalent bond

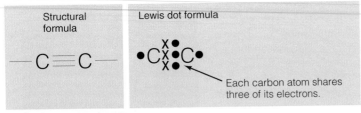

Structural formula

Lewis dot formula

Each carbon atom shares three of its electrons.

(b) Triple covalent bond

FIGURE 2-6 **Double and Triple Covalent Bonds** (a) When atoms share two pairs of electrons, a double covalent bond is formed. Chemists have devised several ways to draw these bonds. In the formula on the left, each line represents a pair of shared electrons. On the right, electrons are indicated by Xs and dots. (b) On rare occasions, atoms share three pairs of electrons, creating triple covalent bonds.

bonded to two atoms of hydrogen. Oxygen atoms contain eight electrons, two in the inner shell and six in the outer shell. By combining with two hydrogen atoms, oxygen fills its outer shell.

The bonds between carbon and hydrogen in methane and oxygen and hydrogen in water involve the sharing of one pair of electrons—one from each atom. These bonds are referred to as *single covalent bonds*. In chemical nomenclature, they are often indicated by a single line joining two atoms, as in H—H.

Some atoms share two pairs of electrons with another atom, forming *double covalent bonds*. Oxygen, for instance, can share two of its outer-shell electrons with a carbon atom, forming a double covalent bond (C=O), as shown in Figure 2-6. Other atoms can share three pairs of electrons, forming triple covalent bonds (Figure 2-6).

FIGURE 2-7 **Hydrogen Bonding** (a) Slightly unequal sharing of electrons in the water molecule creates a polar molecule. Electrons tend to spend more time around the oxygen nucleus, making it slightly negative. The hydrogen atoms are, therefore, slightly positive, as indicated by the symbols d1. (b) Because of the polarity, hydrogen atoms of one molecule are attracted to oxygen atoms of another. The attraction is called a hydrogen bond.

Polar Covalent Bonds

Polar covalent bonds occur any time there is an unequal sharing of electrons by two atoms.

The electrons shared between two atoms in a covalent bond are not always shared equally. That is to say, sometimes an electron spends more time around one atom than the other involved in a covalent bond. Why?

Not all atoms have the same affinity (attraction) for electrons. Some attract electrons more forcefully than others. For example, oxygen and hydrogen join to form a single covalent bond, but oxygen has a slightly higher affinity for electrons than hydrogen. Consequently, the electron of the hydrogen atom tends to spend more time around the oxygen atom than around the hydrogen atom. Because of this, the oxygen atom has a slightly negative charge. The hydrogen atom is visited less frequently by the electron and has a slightly positive charge (Figure 2-7a). The result is a polar covalent bond. A **polar covalent bond** is a covalent bond whose atoms bear a slight charge—either positive or negative.

The presence of polar covalent bonds in molecules makes entire molecules polar. Consider one of the most important of all polar molecules, water. As Figure 2-7a illustrates, water contains two polar covalent bonds and, therefore, has three slightly charged atoms. The slightly charged atoms are often attracted to oppositely charged atoms on neighboring molecules (Figure 2-7b).

Polar molecules are also formed when hydrogen atoms covalently bond to atoms such as nitrogen and fluorine, which, like oxygen, have a slightly higher affinity for electrons than hydrogen.

Hydrogen Bonds

Hydrogen bonds form between slightly charged atoms usually on different molecules.

Covalent bonds occur between atoms of a molecule. They are the bonds that hold the atoms together. But neighboring molecules can also be attracted to one another, thanks to the pres-

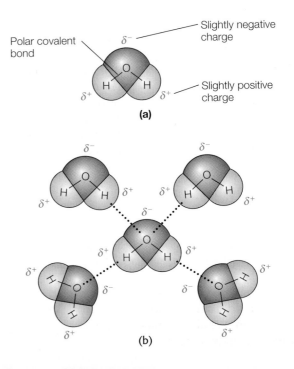

Polar covalent bond

Slightly negative charge

δ^-

Slightly positive charge

δ^+

(a)

(b)

ence of polar covalent bonds in them. Take water for an example. In a glass of water, positively charged hydrogen atoms of water molecules attract negatively charged oxygen atoms of other water molecules. The electrostatic attractions between the positively charged hydrogen atoms of one water molecule and the negatively charged oxygen atoms of another are called **hydrogen bonds**. This is an electrostatic attraction responsible for many of the unique properties of water and absolutely essential to life on Earth (discussed later). Hydrogen bonds are also found in other biologically important molecules that will be discussed in subsequent chapters.

Organic and Inorganic Compounds

Chemical compounds fall into two broad groups: organic and inorganic.

Chemists classify compounds as either organic or inorganic. **Organic compounds** contain molecules that are made primarily of carbon atoms. Proteins, for instance, are organic compounds, as are sugars.

Many organic molecules are quite large. That's because carbon atoms can be joined to one another like the beads of a necklace. These long chains of carbon atoms form the "backbone" of many organic compounds. Attached to the backbone are a number of other atoms, most often hydrogen, oxygen, and nitrogen. In organic molecules, all atoms are covalently linked.

TABLE 2-2	Comparison of Organic and Inorganic Compounds
Organic	Inorganic
Consists primarily of molecules containing carbon and hydrogen	Usually consists of positively and negatively charged ions
Atoms linked by covalent bonds	Usually consists of atoms joined by ionic bonds
Often consists of large molecules that contain many atoms	Always contains small numbers of atoms

Inorganic compounds are defined primarily by exclusion—that is, they are compounds that are not organic. Inorganic molecules are generally small molecules. They usually consist of atoms joined by ionic bonds, as opposed to organic compounds, which are always covalently bonded. Sodium chloride and magnesium chloride are examples of inorganic compounds.

As you will see in the study of biology, not all rules apply 100% of the time. In this instance, not all inorganic compounds contain ionic bonds. Water, for example, is an inorganic compound whose atoms are joined by covalent bonds. Table 2-2 summarizes the main features of each type of compound.

2-3 Water, Acids, Bases, and Buffers

Now that you've learned a few basics of chemistry, let's look at water, a biologically important molecule.

Water

Water is vital to life for many reasons.

Scientific historian and naturalist Loren Eiseley once wrote that "if there is magic on the planet, it has to be water." This remarkable substance produces billowy clouds, icicles that hang from the limbs of trees, and elegant waterfalls. Aesthetically pleasing as water is, though, it is also of great practical importance to humans and all other living organisms.

Water is a major component of all cells and organisms. In fact, nearly two-thirds of the human body is water. Thus, if you weigh 100 pounds, nearly 70 pounds of your body weight is water. Water is an important biological solvent, a fluid that dissolves other chemical substances. Human blood, for example, contains large amounts of water that dissolve and transport nutrients, hormones, and wastes throughout our bodies.

Water also participates in many chemical reactions in the body—for example, in the breakdown of protein (described later). In addition, water serves as a lubricant. Saliva, which is largely water, lubricates food in the mouth and esophagus, the tube that transports food from our mouths to our stomachs. A watery fluid in the joints enables bones to slide over one another, thus facilitating body motion.

Finally, water helps us regulate body temperature. Perspiration, for example, rids our bodies of heat, for when it evaporates, water draws off heat (Figure 2-8). People sometimes suffer heatstroke in hot, humid climates because the excess water in the atmosphere reduces evaporation. This causes heat to build up, which can cause a person to collapse.

The Dissociation of Water

Water molecules split into hydrogen and hydroxide ions.

Water molecules are fairly stable; nonetheless, some molecules break apart, forming two charged units known as *ions,* as shown in the following reaction:

$$H_2O \rightleftharpoons H^+ + OH^-$$

water   hydroxide ion   hydrogen ion

Because the hydrogen and hydroxide ions formed in this reaction can react with one another to re-form water molecules, this reaction is said to be reversible. The double arrow in the reaction formula indicates this fact.

FIGURE 2-8 Perspiration Cools the Body Perspiration is an automatic homeostatic response that helps rid the body of heat.

Although water can dissociate, the ratio of water molecules to the ions, H$^+$ and OH$^-$, in the human body is about 500 million:1. Nevertheless, even the slightest change in the hydrogen ion concentration can alter cells and organisms, shutting down biochemical pathways and sometimes killing organisms. The human body contains several homeostatic mechanisms to ensure a constant level of these ions.

Acids

Acids are substances that add hydrogen ions to solution; bases remove them.

In pure water, the concentrations of H$^+$ and OH$^-$ ions are equal. A solution containing an equal number of these ions is said to be *chemically neutral*.

Chemical neutrality can be upset by adding or removing H$^+$ or OH$^-$. For example, when hydrochloric acid (HCl) is added to water, it dissociates into hydrogen ions (H$^+$) and chloride ions (Cl$^-$). This increases the hydrogen ion concentration. Scientists call a substance that adds hydrogen ions to a solution an **acid**. Solutions with proportionately more H$^+$ than OH$^-$ are said to be acidic. **Bases** are substances that remove H$^+$ from solution. Sodium hydroxide, for example, dissociates when added to aqueous solutions, forming Na$^+$ and OH$^-$. The hydroxide ions react with hydrogen ions in the solution, forming water molecules. As sodium hydroxide is added to a solution of pure water, then, the hydrogen ion concentration declines, and the hydroxide ion concentration increases. A solution with a greater concentration of OH$^-$ than H$^+$ ions is said to be basic.

Acidity is measured on the **pH scale** (Figure 2-9). As illustrated, the pH scale ranges from 0 to 14. Neutral substances are assigned a pH of 7. Basic substances have a pH greater than 7, and acidic substances have a pH less than 7.

On the pH scale, a change in one pH unit represents a tenfold change in acidity. Thus, a solution with a pH of 3 is 10 times more acidic than one with a pH of 4 and 100 times more acidic than one with a pH of 5.

Most biochemical reactions occur at pH values between 6 and 8. Human blood, for example, has a pH of 7.4. So important is this to normal body function that a slight shift in the pH of the blood for even a short period can be fatal.

Maintaining pH: Buffers

Homeostasis is ensured in part by buffers, molecules that help maintain pH within a narrow range.

Biological systems operate within a narrow pH range, as noted above. This narrow range is maintained by buffers. **Buffers** (BUFF-firs) are chemicals found in organisms and ecosystems, particularly soil and water, that protect against shifts in pH. How do they work?

Buffers can be likened to hydrogen-ion sponges. They help maintain a constant pH by removing hydrogen ions from solution when levels increase. Buffers give back the hydrogen ions when levels fall.

One of the most common buffers is carbonic acid. Found in the blood of animals, carbonic acid is formed from water and carbon dioxide, a waste product of cellular energy production. In blood, carbonic acid dissociates into bicarbonate and hydrogen ions:

$$\underset{\substack{\text{carbonic acid}\\\text{(weak acid)}}}{H_2CO_3} \rightleftharpoons \underset{\text{hydrogen ion}}{H^+} + \underset{\substack{\text{bicarbonate}\\\text{(weak base)}}}{HCO_3^-}$$

In our blood, and in lakes and rivers where carbonic acid is also present, this chemical reaction shifts back and forth in response to changing levels of hydrogen ions. Thus, when hydrogen ions are added to water, they combine with bicarbonate ions, driving the reaction to the left. But when the hydrogen-ion concentration falls, the reaction is driven to the right. In either case, the pH of the water remains constant.

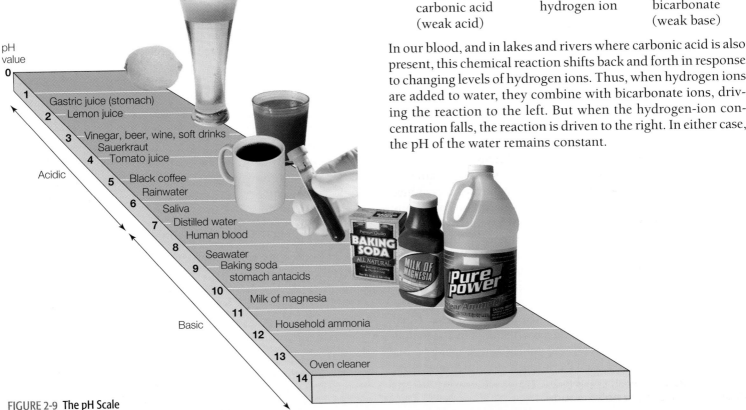

pH value

0
1 — Gastric juice (stomach)
2 — Lemon juice
3 — Vinegar, beer, wine, soft drinks
      Sauerkraut
4 — Tomato juice

Acidic

5 — Black coffee
      Rainwater
6
      Saliva
7 — Distilled water
      Human blood
8
      Seawater
9 — Baking soda
      stomach antacids
10 — Milk of magnesia
11
      Household ammonia
12
13
      Oven cleaner
14

Basic

FIGURE 2-9 **The pH Scale**

2-4 Overview of Other Biologically Important Molecules

This introduction to chemistry is intended to give you a foundation upon which to add to your chemical literacy. In your study of biology, you will encounter chemistry terms and many chemical compounds. Your studies will reveal four major groups of biological molecules: (1) carbohydrates; (2) lipids; (3) amino acids, peptides, and proteins; and (4) nucleic acids. Rather than describe each one in detail here, I will present a few important facts. These molecules are discussed in more detail in subsequent chapters.

Carbohydrates are organic molecules that range in size from very small molecules such as the blood sugar glucose, which cells use to generate energy, to very large ones such as starch. The large molecules are actually composed of many small molecules. Carbohydrates supply energy to cells. Pastas and cereals are a great source of carbohydrates.

Lipids (LIP-ids) are a rather diverse group of molecules that includes fats and steroids such as cholesterol. Some lipids are a source of energy and others are important structural components of cells. Lipids form a layer of insulation beneath the skin of many animals, including whales and humans. Lipids are found in many foods, especially meat, nuts, fried foods such as french fries, and numerous snack foods such as potato chips.

Most readers have heard the terms *amino acids*, *proteins*, and *peptides*. **Proteins** are long molecules (polymers) consisting of many smaller molecules, the **amino acids**. Found in many foods, especially meats and milk products, proteins are important structural elements of cells, as you will see in the next chapter. Many proteins are enzymes, special molecules that speed up chemical reactions in the body. **Peptides** are very small proteins—chains of amino acids.

The fourth group of biologically important molecules is the **nucleic** (new-CLAY-ick) **acids**. This structurally complex group includes DNA, the genetic material found in cells. These molecules are long chains of smaller molecules called **nucleotides** (NEW-klee-oh-tides). Another nucleic acid is RNA. Its structure and function are discussed in Chapter 17.

2-5 Health and Homeostasis

Cancer is a disease in which body cells begin to divide uncontrollably, producing growths called *tumors*. Unless destroyed, these tumors eventually kill the person. Interestingly, like many diseases, cancer sometimes results from imbalances, for example, excesses of cancer-causing agents in our environment, as the following story reveals.

For 2000 years, the people in Lin Xian, China, 250 miles south of Beijing, have been dying in record numbers from cancer of the esophagus, the muscular tube that transports food to the stomach. So prevalent is the disease that one of every four persons once succumbed to this ruthless killer, whose incidence is higher there than anywhere else in the world.

In 1959, scientists began a systematic study of the people living in the valley around Lin Xian to discover the origins of the disease and put an end to it. Scientists first found that esophageal cancer in Lin Xian was the result of a group of chemicals called nitrosamines (NYE-trose-ah-MEANS). Nitrosamines, they discovered, were being produced in the stomachs of the residents from two other chemical compounds: nitrites and amines. But where did they come from?

Research showed that residents had abnormally high levels of nitrites, which came from nitrates in the vegetables the residents ate (Figure 2-10). The amines were present in moldy bread, a delicacy in the region. But why the excess of nitrates in vegetables? And why the higher-than-normal levels of nitrites in people?

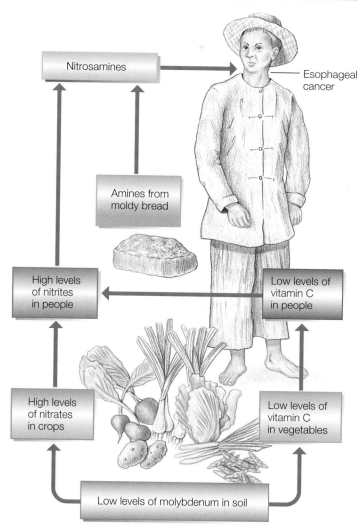

FIGURE 2-10 **Tracking a Deadly Killer** In Lin Xian, China, the link between esophageal cancer and the soil was uncovered by medical researchers.

Chemical analyses of the soil in which the region's crops were grown showed that the topsoil around Lin Xian was deficient in molybdenum (meh-LIB-deh-num), an element required by plants in minute quantities (Figure 2-10). This deficiency caused the plants to concentrate nitrates. It also reduced vitamin C production by plants. Nitrates from plants were apparently being converted to nitrites in the stomachs of the residents. Low levels of vitamin C in the plants the people ate actually promoted the conversion of nitrates to nitrites. When the residents were given vitamin C tablets, their nitrite levels dropped.

Ultimately, then, a soil deficiency, caused either by intensive agriculture, which depletes the soil, or by a natural deficiency, put the local residents who ate moldy bread rich in amines at high risk for esophageal cancer.

To reduce the risk of esophageal cancer, the villagers now coat wheat and corn seeds with molybdenum. As a result, nitrite levels in vegetables have dropped 40%, and vitamin C levels have increased 25%. Although it is still too early to tell whether restoring the soil's nutrient balance will put a halt to this deadly killer, scientists are optimistic.

SUMMARY

1. An understanding of chemistry is essential to understanding biology because all cells and organisms are composed of molecules and many life processes are nothing more than chemical reactions.

Atoms and Subatomic Particles

2. The physical world we live in is composed of matter. Matter is anything that has mass and occupies space. All matter is composed of tiny particles called *atoms*.

3. Atoms contain many *subatomic particles*, three of the most important being electrons, protons, and neutrons.

4. Negatively charged particles, *electrons*, are the smallest of these three subatomic particles; they are found in the electron cloud around the dense, central nucleus.

5. *Protons* are positively charged particles that are located in the nucleus with *neutrons*, which are uncharged.

6. Atoms are electrically neutral because the number of electrons always equals the number of protons.

7. A pure substance such as gold or lead containing only one type of atom is known as an *element*. The elements are listed on the *periodic table* by *atomic number*, which is the number of protons in the nucleus.

8. Most elements are made up of mixtures of two or more *isotopes*, atoms with slightly different atomic masses resulting from the presence of additional neutrons. Additional neutrons often make isotopes unstable. To achieve a more stable state, many isotopes release *radiation*—small bursts of energy, or tiny energetic particles—from their nuclei.

9. Radiation has proved to be a useful tool in medicine and science, but it can also damage cells of organisms, leading to birth defects and cancer.

The Making of a Molecule

10. Atoms react with one another to form organic and inorganic compounds. Their reactivity is a result of the interaction of electrons.

11. Two types of bonds are found between atoms, ionic and covalent.

12. In general, atoms tend to react with other atoms to form complete outer shells, creating more stable atomic configurations.

13. During some atomic reactions, electrons are transferred from one atom to another. Transferring an outer-shell electron from one atom to another creates two oppositely charged ions. An electrostatic attraction forms between them and is referred to as an *ionic bond*.

14. *Covalent bonds* form between atoms that share one or more pairs of electrons. This sharing of electrons holds the atoms together.

15. Unequal sharing of electrons results in the formation of a *polar covalent bond*, which results in the formation of *polar molecules*. Polar molecules are frequently attracted to one another by *hydrogen bonds*, weak electrostatic attractions between the oppositely charged atoms of neighboring molecules.

Water, Acids, Bases, and Buffers

16. *Organic molecules* are compounds made up primarily of carbon and hydrogen. The atoms of these generally large molecules are held together by covalent bonds.

17. *Inorganic molecules* are much smaller molecules whose atoms are often linked by ionic bonds.

18. Water is an important inorganic molecule and a major component of all body cells. It serves as a solvent, which transports many substances in the blood, body tissues, and cells. It also participates in many chemical reactions, serves as a lubricant, and helps regulate body temperature.

19. Although fairly stable, water molecules can dissociate, forming hydrogen and hydroxide ions. In pure water, the concentration of these ions is equal and water has a pH of 7.

20. Adding *acidic substances* increases the concentration of hydrogen ions, causing the pH to fall.

21. Substances that remove hydrogen ions from solution cause the pH to climb and are called *bases*. A solution with a pH less than 7 is acidic; a solution with a pH greater than 7 is basic.

22. Biological systems contain *buffers*, chemical compounds that offset changes in the concentration of hydrogen and hydroxide ions. Buffers are therefore an important component of chemical homeostasis in organisms and the environment.

Overview of Other Biologically Important Molecules

23. Carbohydrates, lipids, amino acids, proteins, and nucleic acids are biologically important molecules. They vary considerably in structure and function. Details of their structure and function are covered in later chapters.

Health and Homeostasis

24. Tiny imbalances of chemicals in the environment can cause upsets in agricultural crops, which, combined with other factors, can lead to cancer.

www.jbpub.com/humanbiology/5e

The site features eLearning, an online review area that provides quizzes, chapter outlines, and other tools to help you study for your class. You can also follow useful links for in-depth information, research the differing views in the Point/Counterpoints, or keep up on the latest health news.

critical thinking

Thinking Critically—Analysis

This analysis corresponds to the Thinking Critically scenario that was presented at the beginning of this chapter.

The phrase that may have raised a red flag in your mind while you were reading this exercise is "Assuming that impurities are rare in single, well-formed crystals." Beware of conclusions based on assumptions. They may prove to be wrong.

But this may not be readily apparent at first. In 1985, in fact, chemists at the Ruhr University in Germany reported the synthesis of another pair of isomers like Chatt's, but from tungsten-containing compounds. These isomers were, according to the scientists, also chemically and structurally identical except for the length of one metal-oxygen bond. Several other laboratories also reported similar findings, creating a great deal of excitement among scientists,

for it appeared that a long-held belief of the constancy of bond length was about to topple.

Theoretical chemists began proposing explanations for this exciting phenomenon. In 1988, Nobel Prize–winning chemist Roald Hoffman and his colleagues published one explanation. However, as chemists began to repeat the experiments (a process essential to good science) to learn more about the isomers, enthusiasm began to fade. In one replication of Chatt's experiment, scientists found that impurities in samples resulted in different colored crystals. After the German experiment was repeated, it was shown that those results had also resulted from contamination.

"Bond-stretch isomerism," as it was called, nearly became accepted theory based on the widely accepted, but faulty, assumption that impurities are rare in single, well-formed crystals.

KEY TERMS AND CONCEPTS

Acid, p. 28
Amino acid, p. 29
Atom, p. 20
Atomic number, p. 21
Base, p. 28
Buffer, p. 28
Carbohydrate, p. 29
Compound, p. 22
Covalent bond, p. 23
Electron, p. 20
Element, p. 21
Hydrogen bond, p. 27

Inorganic compound, p. 27
Ion, p. 22
Ionic bond, p. 22
Isotope, p. 21
Lipid, p. 29
Mass, p. 20
Matter, p. 20
Molecule, p. 22
Mutation, p. 22
Neutron, p. 20
Nucleic acid, p. 29

Nucleotide, p. 29
Nucleus, p. 20
Organic compound, p. 27
Peptide, p. 29
PH scale, p. 28
Polar covalent bond, p. 26
Protein, p. 29
Proton, p. 20
Radiation, p. 22
Radionuclide, p. 22
Subatomic particle, p. 20

CONCEPT REVIEW

1. Describe the structure of an atom, using a diagram to illustrate your answer. p. 20
2. Why is the matter on Earth so varied in its appearance? p. 21
3. Define the following terms: atomic number, mass number, and isotope. p. 21
4. How are ionic and covalent bonds different? Describe what holds the atoms together in both cases. In what ways are the bonds similar? pp. 22–23
5. Describe how polar covalent bonds form and why they result in the formation of hydrogen bonds. p. 26
6. Temperature is a measure of the speed of molecules: the higher the temperature, the higher the speed. With this information and your knowledge of water and hydrogen bonds, why does water have a higher boiling point than a nonpolar liquid like alcohol? pp. 26–27
7. Describe the biological significance of water. pp. 27–28
8. How does the body regulate H^+ levels in the blood? Describe the process. p. 28

SELF-QUIZ: TESTING YOUR KNOWLEDGE

1. Each atom has a nucleus made up of _____ and neutrons. p. 20
2. In all atoms the number of protons is equal to the number of _____. p. 20
3. The atomic number of an atom is equal to the number of _____ in the nucleus. p. 21
4. An alternative form of an atom is called an _____; it differs in the number of neutrons in the nucleus. p. 21
5. A bond formed between two atoms that share electrons is called a(n) _____ bond. p. 23
6. _____ bonds form between oppositely charged atoms of polar covalent bonds of different molecules. p. 27
7. Proteins and carbohydrates consist primarily of carbon, hydrogen, and oxygen; these two molecules are known as _____ compounds. p. 27
8. On the pH scale, a pH of __ is neutral; a solution with a pH less than this is _____ and a solution with a pH greater than this is _____. p. 28
9. Chemicals like bicarbonate that resist pH change are called _____. p. 28
10. When water molecules dissociate, they form _____ and _____ ions. pp. 27–28

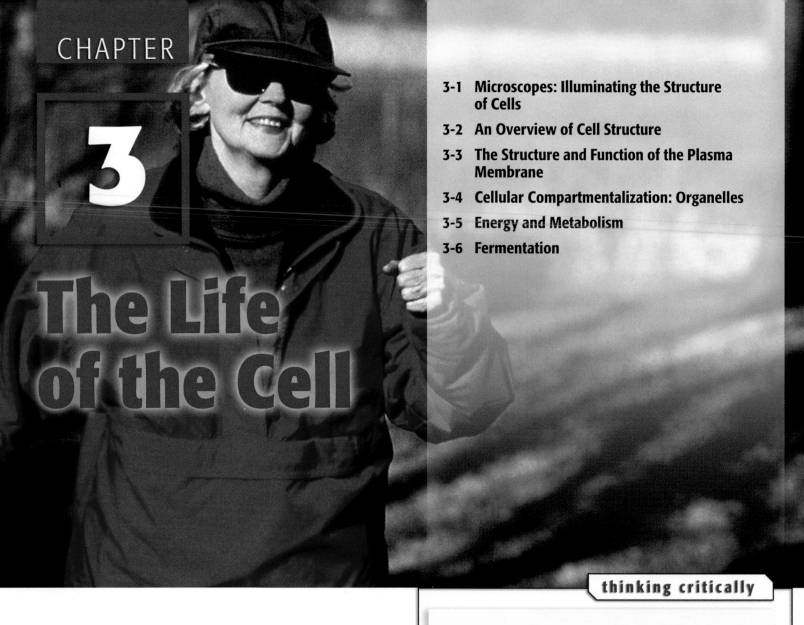

The Life of the Cell

thinking critically

Researchers at a major medical college in your area have discovered that cells live longer when cultured (grown) in petri dishes containing a certain vitamin. They suggest that this vitamin might help people live longer. A local newspaper runs with the story, and one of your friends is thinking of taking the vitamin supplement in hope of living longer. What advice would you give him before he embarks on this course of action?

In Sweden, Mexico, the United States, and a number of other countries, medical researchers are experimenting with a controversial procedure called *fetal cell transplantation.* By transplanting normal, healthy cells from human embryos into adults, medical scientists hope to be able to cure a number of chronic, debilitating diseases. Scientists believe that fetal cells can replace defective cells in the organs and tissues of adults. This procedure may someday make a significant contribution to medical science, reducing pain and suffering in tens of thousands of people worldwide—people who have little hope of a cure. Already, studies are showing that fetal cell transplantation successfully cures some patients who have a debilitating nervous system disease known as Parkinson's disease, which currently afflicts millions of people worldwide including actor Michael J. Fox of *Back to the Future* fame, the famous boxer Muhammad Ali, and Pope John Paul II.

Although the research results are promising, fetal cell transplantation poses a number of technical challenges. It also poses several significant moral and political dilemmas, discussed in the Point/Counterpoint in this chapter.

This chapter focuses on the cell and some key biological molecules. The information presented here forms a foundation on which a good understanding of human biology can be built. This material will also help you understand the details of fetal cell transplantation.

3-1 Microscopes: Illuminating the Structure of Cells

Cells are exceedingly small and, with few exceptions, cannot be seen with the naked eye. Our knowledge of them has depended on the **microscope**, a device consisting of a lens or combination of lenses that enlarges tiny objects. (For a discussion of the discovery of cells, see Scientific Discoveries 3-1.)

Microscopes fit into two categories: (1) light *microscopes*, which use ordinary visible light to illuminate the specimen un-

der study; and (2) *electron microscopes*, which use a beam of electrons to create a visual image of the specimen. Both types of microscope enlarge tiny images and allow us to see details of the cell. However, light microscopes magnify objects from 100 to 400 times their original size (Figure 3-1). Electron microscopes enlarge objects 100,000 times their original size. For a comparison, see Figures 3-1 and 3-2. One type of electron microscope, known as a *scanning electron microscope*, allows scientists to obtain three-dimensional views of cells (Figure 3-2). Photographs from both light and electron microscopes are used throughout this book.

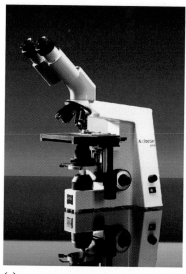

(a)

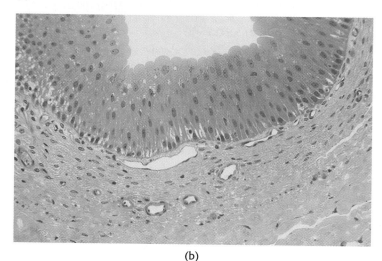

(b)

FIGURE 3-1 **The Light Microscope** (a) In the ordinary light microscope, the image is illuminated by light that is cast from a mirror or (in this case) a light bulb located below the object. Two lenses (one in the eyepiece and one just above the object) magnify the object. (b) Light micrograph of a section of the lining of the urinary bladder.

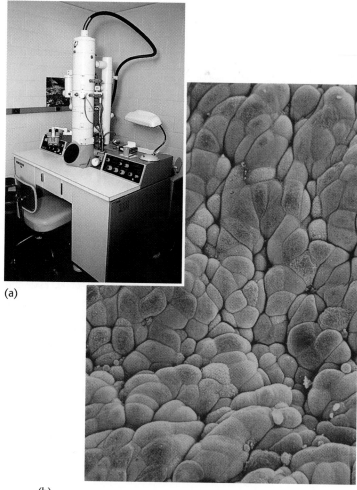

(a)

(b)

FIGURE 3-2 **The Electron Microscope** (a) Scanning electron microscope. (b) Scanning electron micrograph of the lining of the urinary bladder.

Fetal Cell Transplantation

Fetal Tissue Transplants: Auschwitz Revisited by Tom Longua

Persons who advocate medical research using tissue from aborted babies claim it has genuine medical value. But that premise is just plain false. On April 22, 1999, both the *New York Times* and *Washington Post* ran stories about a four-year trial by the National Institute of Neurological Disorder, in which a group of Parkinson patients had fetal cells implanted into their brains, while others had a "sham" operation in which no cells were implanted.

In patients over 60 (those most susceptible to Parkinson's), no difference was observed between the two groups, either on the neurologists' examinations or on the patients' own perception of their condition. Even among younger patients—for whom the treatment allegedly "worked"—there was no difference between the two groups in initiating walking, walking balance, the number of falls, tremors, "freezing" (sudden loss of all movement), and dyskinesia.

Tom Longua has been an educator for 39 years and is currently instructor of anatomy and physiology at the Denver Academy of Court Reporting. He has been active in the pro-life movement since 1974 and is a member of the board of directors of Colorado Right to Life.

But even if the claims were true, should a civilized society allow research on human beings who may be killed, albeit legally, even though they are innocent of any crime? After World War II, Nazi doctors who experimented on prisoners in concentration camps were tried at Nuremberg for "crimes against humanity." The prosecutor opened his case with this statement:

The defendants in this case are charged with. . . . Atrocities committed in the name of medical science. . . . The wrongs which we seek to condemn and punish have been so calculated, so malignant, and so devastating, that civilization cannot tolerate their being ignored, because it cannot survive their being repeated.

Yet with research involving tissue from elective abortions, we are committing the same "crimes against humanity." Even the claimed justification is the same.

- Nazi doctor August Hirt: "Wouldn't it be ridiculous to send the bodies to the crematory oven without giving them an opportunity to contribute to the progress of society?"
- Lawrence Lawn, a doctor involved in research with tissue from elective abortions: "We are simply using something which is destined for the incinerator to benefit mankind."

Proponents of such research claim that it is acceptable if it is "separated" from abortions that are freely chosen anyway. Again, this argument echoes the Nazi doctors' defense at Nuremberg: "We caused no deaths. They [the victims] were all consigned to death by legal authorities; with . . . professional correctness we tried to salvage some good from their plight."

But "separation" is a myth. As the Auschwitz research depended on the "acceptability" of killing prisoners, fetal tissue research depends on the "acceptability" of killing helpless, innocent human beings in the first place. (Certainly there is nothing wrong with using tissue from ectopic pregnancies or miscarriages; but this debate concerns babies who are purposely killed simply because they are unwanted.) Moreover, new abortion techniques have been contrived specifically for the purpose of securing usable (and salable) tissue. So much for "separation."

But besides the perverseness of the practice itself, such research will have other detrimental effects:

- It will certainly increase the number of abortions. Some medical journals claim that fetal tissue could be used to treat a vast array of diseases. In a1988 guest editorial in the *Wall Street Journal*, Dr. Emanuel Thorne estimated that the 1.6 million annual abortions in America would not be enough to keep up with the demand for such tissue. While there have been well-publicized cases of women becoming pregnant to create organ or tissue sources for their parents and other children, it is unlikely that many women would get pregnant to create tissue that might be "useful" for others (although stranger things have happened). But with money to be made, the abortion industry will certainly pressure women who are undecided about having abortions, using the argument that the tissue will serve some "benevolent" purpose. Indeed, retired Aurora abortionist James Parks bragged, "[My patients] say, 'Thank God, some good is going to come out of this.'" (*New York Times*, 11/19/89)
- In a 1991 article on this topic, this writer said, "[This research] will lead to trafficking in human 'spare parts.'" Well, eight years later, Congress launched an investigation into exactly that grisly practice, finding that it was occurring, and that it was extremely lucrative. As the Associated Press reported on October 10, 1999, Congressman Tom Tancredo (R-CO), "displayed a brochure . . . that lists prices for fetal parts, including '$50 for eyes, $150 for lungs and hearts and $999 for an eight-week brain.'"
- Advocacy of fetal tissue research will hinder other—and ethical—research to find cures for some diseases. There are many potential treatments being studied, such as autografts and use of patients' own stem cells, but some researchers have abandoned those possibilities because of the much-touted (albeit dishonestly) fetal tissue approach.

Advocates of fetal tissue research like to depict themselves as "humanitarians." But then, so did the doctors on trial at Nuremberg.

Human Fetal Tissue Should Be Used to Treat Human Disease by Curt R. Freed

Despite its legalization, abortion remains a controversial issue and will continue to stir debate in the future. In the United States, the debate centers on whether a woman has the right to control her reproduction. The future developments of this political debate are uncertain, but recent elections suggest that pro-choice candidates have been victorious when elections are based on the abortion issue.

Currently, over 1 million legal abortions are performed in the United States each year; most abortions are performed in the first trimester. For nearly all women, having an abortion is an anguishing choice filled with regret and ambivalence. Nonetheless, the difficult personal decision to terminate a pregnancy is made. After the abortion, fetal tissue is usually discarded.

As an alternative to throwing this tissue away, research has shown that fetal tissue may be useful for treating patients with disabling diseases. For over 50 years, research in animals has demonstrated that fetal tissue has a unique capacity to replace certain cellular deficiencies and so may be useful for treating some chronic diseases of humans. These diseases include Parkinson's disease, diabetes, and some immune system disorders.

Cadaver fetal tissue offers the promise of helping large numbers of Americans with crippling diseases. Parkinson's disease, for example, affects hundreds of thousands of Americans. By reducing the ability to move, the disease can end careers and turn people into invalids. The disease is caused by the death of a small number of critically important nerve cells that produce a chemical called dopamine. Experiments in animals and early experiments in humans indicate that fetal dopamine cells transplanted into the brains of these patients may restore a patient's capacity to move and may even eliminate the disease. Patients whose minds work perfectly well and whose bodies are otherwise normal may become healthy and productive citizens once again.

Using cadaver tissue to treat humans has been debated for nearly 40 years. As kidney, cornea, and other organ transplants were developed in the 1950s, many objected to recovering organs from cadavers. In the intervening decades, opinion has changed so that the practice of recovering these organs from cadavers has gone from a provocative and controversial practice to an accepted policy endorsed by most states. In fact, in most states, a check-off box on the back of driver's licenses is used to give permission for organ donation in the event of the death of the driver. Because abortions are induced, some argue that fetal tissue should be regarded differently from other cadaver tissue. Given the facts that abortion is legal and that fetal tissue is ordinarily discarded, there should be no moral dilemma in using fetal tissue for therapeutic purposes. As with the use of all human tissue for transplant, specific informed consent by the woman donating the tissue must be obtained.

Some have proposed that using fetal tissue for therapeutic purposes will increase the number of abortions. This is preposterous. It strains the imagination to think that a woman would get pregnant and have an abortion simply on the chance that the aborted fetal tissue might be used to treat a patient unknown to her. An unwanted pregnancy is an intimate and deeply personal crisis; it is inconceivable that the pregnancy would be seen primarily as a philanthropic opportunity.

Curt R. Freed, M.D., is a professor of medicine and pharmacology at the University of Colorado School of Medicine in Denver. He has written some 70 articles on medical topics as well as numerous abstracts, chapters, and reviews.

Politics and medicine have frequently mixed in the past and will continue to do so in the future. As a physician, I think it is important to try to improve the health of patients with serious diseases. Legally acquired fetal cadaver tissue that would otherwise be discarded should be used to treat humans with disabling diseases.

Sharpening Your Critical Thinking Skills

1. Summarize the positions of each author.

2. Using your critical thinking skills, analyze the view of each author. Is each stand well substantiated? Do the author's biases play a role in each argument?

3. Do you see this debate as scientific or ethical? Explain your answer.

4. Each position is based on at least one key argument. Can you pinpoint them?

5. Which viewpoint do you agree with? Why? What factors (biases) affect your decision?

www.jbpub.com/humanbiology/5e

Visit Human Biology's Internet site for links to web sites offering more information on this topic.

Scientific Discoveries that Changed the World

3-1 The Discovery of Cells
Featuring the Work of Hooke, Leeuwenhoek, Brown, Schleiden, Schwann, and Virchow

One of the fundamental principles of biology is known as the cell theory. The cell theory consists of three parts: (1) all organisms consist of one or more cells, (2) the cell is the basic unit of structure of all organisms, and (3) all cells arise from preexisting cells. Although this may seem rather elementary, it was not so obvious to early scientists, who labored with relatively crude instruments and without the benefit of many facts we now take for granted.

One of those scientific pioneers who opened our eyes to the world of cells was Robert Hooke, a seventeenth-century British mathematician, inventor, and scientist. Equipped with a relatively crude microscope, Hooke observed just about everything he could lay his hands on——which he described in his book *Micrographia*, published in 1665.

One especially useful description was that made on a thin slice of cork (Figure 1). Peering through his microscope, Hooke beheld a network of tiny, boxlike compartments that reminded him of a honeycomb. He called these compartments *cellulae*, meaning "little rooms." Today, we know them as cells.

Hooke did not really see cells but, rather, cell walls, the structures that surround the plasma membranes of plant cells. The cytoplasm and cellular organelles had disappeared.

Hooke's work was complemented a few years later by Antony van Leeuwenhoek, a Dutch shopkeeper who spent much of his free time designing simple microscopes. Like Hooke, Leeuwenhoek examined just about everything he could find and wrote extensively on his observations. Wayne Becker, a cell biologist at the University of Wisconsin writes, "His detailed reports attest to both the high quality of his lenses and his keen powers of observation." Becker continues: "They also reveal an active imagination, since at one point he reported seeing a 'homunculus' (little man) in the nucleus of a human sperm cell."

Leeuwenhoek discovered bacteria and protozoans (single-celled eukaryotic organisms). Although his microscopes were superior to any others around at that time, they were still crude by modern standards. This fact and the tendency of the scientific community back then to focus only on the observation and description of life-forms stalled progress in our understanding of cells for some time. In fact, more than a century passed before cell biology moved significantly forward.

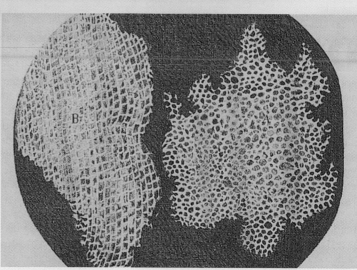

FIGURE 1 **Thin Slice of Cork** Hooke discovered tiny, boxlike compartments that reminded him of a honeycomb. He called them *cellulae*, meaning "little rooms."

Aided by improved lenses, the eighteenth-century English botanist Robert Brown noted that every plant cell he studied contained a centrally placed structure, now called a nucleus. A German colleague, Matthias Schleiden, concluded that all plant tissues consisted of cells. One year later, Theodor Schwann, a German scientist, arrived at a similar conclusion regarding animal cells. This discovery laid to rest an earlier hypothesis that plants and animals were structurally different.

Based on earlier research and his own work, Schwann proposed the first two parts of the cell theory—that all organisms consist of one or more cells and that the cell is the basic unit of structure for all organisms. Less than 20 years after the discovery of cell division, the German physiologist Rudolf Virchow added the third tenet of the cell theory, that all cells arise from preexisting cells.

Like other discoveries, the cell theory is the work of many people over many years. Although only a few people are named as having made possible this important discovery, the credit really belongs to an entire line of scientists who, through careful observation and experimentation, have changed our view of the world.

3-2 An Overview of Cell Structure

In humans, most cells perform very different tasks, and very few look alike. However, we won't be concerned about these differences. Instead, we'll focus on generalities as we examine a fictional entity, the typical animal cell, shown in Figure 3-3. Don't look for it under the microscope; it only exists on the pages of biology textbooks. You may want to refer to Table 3-1 as you study the following material and as you review this material later.

The Nuclear and Cytoplasmic Compartments

The cell consists of two main compartments.

Human cells are the product of millions of years of evolution. As illustrated in Figure 3-3, human cells consist of two basic compartments, the nuclear and the cytoplasmic. The nuclear com-

TABLE 3-1	Overview of Cell Organelles	
Organelle	Structure	Function
Nucleus	Round or oval body; surrounded by nuclear envelope	Contains the genetic information necessary for control of cell structure and function; DNA contains hereditary information.
Nucleolus	Round or oval body in the nucleus consisting of DNA and RNA	Produces ribosomal RNA
Endoplasmic reticulum	Network of membranous tubules in the cytoplasm of the cell. Smooth endoplasmic reticulum contains no ribosomes. Rough endoplasmic reticulum is studded with ribosomes.	Smooth endoplasmic reticulum (SER) is involved in the production of phospholipids and has many different functions in different cells; round endoplasmic reticulum (RER) is the site of the synthesis of lysosomal enzymes and proteins for extracellular use.
Ribosomes	Small particles found in the cytoplasm; made of RNA and protein	Aid in the production of proteins on the RER and polysomes
Polysome	Molecule of mRNA bound to ribosomes	Site of protein synthesis
Golgi complex	Series of flattened sacs usually located near the nucleus	Sorts, chemically modifies, and packages proteins produced on the RER
Secretory vesicles	Membrane-bound vesicles containing proteins produced by the RER and repackaged by the Golgi complex; contain protein hormones or enzymes	Store protein hormones or enzymes in the cytoplasm awaiting a signal for release
Food vacuole	Membrane-bound vesicle containing material engulfed by the cell	Stores ingested material and combines with lysosome
Lysosome	Round, membrane-bound structure containing digestive enzymes	Combines with food vacuoles and digests materials engulfed by cells
Mitochondria	Round, oval, or elongated structures with a double membrane. The inner membrane is thrown into folds.	Complete the breakdown of glucose, producing NADH and ATP
Cytoskeleton	Network of microtubules and microfilaments in the cell	Gives the cell internal support, helps transport molecules and some organelles inside the cell, and binds to enzymes of metabolic pathways
Cilia	Small projections of the cell membrane containing microtubules; found on a limited number of cells	Propel materials along the surface of certain cells
Flagella	Large projections of the cell membrane containing microtubules; found in humans only on sperm cells	Provide motive force for sperm cells
Centrioles	Small cylindrical bodies composed of microtubules arranged in nine sets of triplets; found in animal cells, not plants.	Help organize spindle apparatus necessary for cell division

partment, or **nucleus**, is the control center of the cell because it contains the genetic information that regulates the structure and function of the cell. As shown in Figure 3-3, the nucleus is surrounded by a double membrane, the nuclear envelope.

The cytoplasmic compartment lies between the nucleus and the plasma membrane, the outermost structure of the cell. It contains a substance known as cytoplasm. **Cytoplasm** is a semifluid material that consists of numerous molecules, including water, protein, ions, nutrients, vitamins, dissolved gases, and waste products. It forms a nutrient pool from which the cell draws chemicals it needs for metabolism. It is also a dumping ground for wastes, which are later removed from cells.

As shown in Figure 3-3, the cell contains numerous structures called **organelles**. Organelles ("little organs") carry out specific functions (Table 3-1). Many organelles are bounded by membranes and form subcompartments within the cytoplasm.

Compartmentalization is essential to the smooth functioning of complex cells because it permits cells to segregate many of their functions. This, in turn, increases the efficiency of cells and helps complex organisms perform many functions simultaneously. In a way, a cell is like a modern factory with different departments, each performing specific functions, but all working together toward one end.

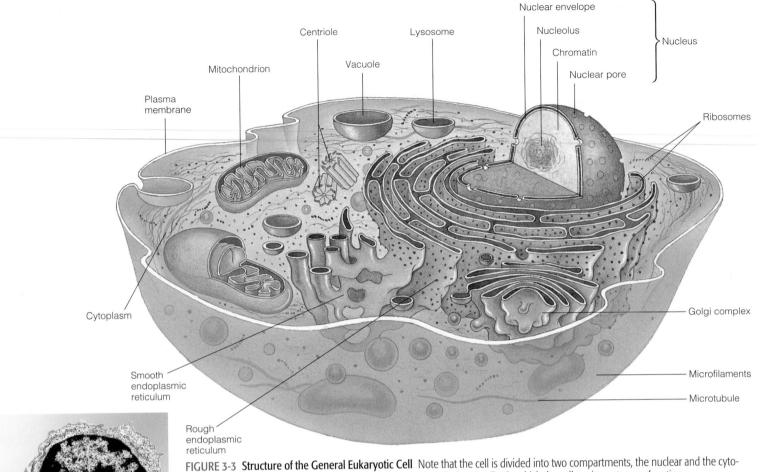

FIGURE 3-3 **Structure of the General Eukaryotic Cell** Note that the cell is divided into two compartments, the nuclear and the cytoplasmic. The cytoplasm is packed with a variety of structures called organelles in which the cell carries out many functions.

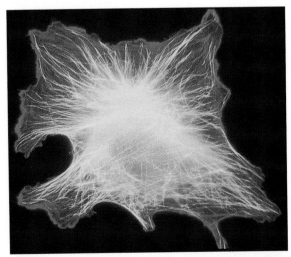

FIGURE 3-4 **The Cytoskeleton** Photomicrograph of the cytoskeleton of a human fibroblast (connective tissue cell). The microtubules are yellow, and the microfilaments are red.

FIGURE 3-5 **Metabolic Pathway** Reactions in the cell occur as parts of larger pathways where the product of one reaction becomes the reactant of another, as in this biochemical pathway. This illustrates the early steps in glycolysis, the breakdown of glucose in the cytoplasm of eukaryotic cells. Note that each reaction has its own enzyme (labeled in red). Also note that the names of these, and all other enzymes, end in -ase.

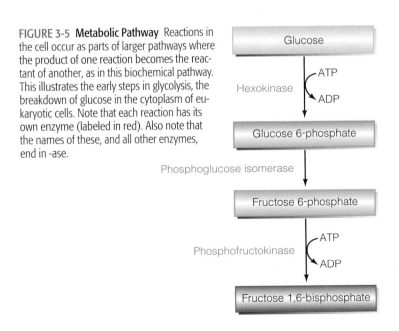

The cytoplasm of the cell also contains organelles that are not membrane-bound. These organelles perform many important functions, too (outlined shortly).

Giving shape to the cell is a network of protein tubules and filaments in the cytoplasm called the **cytoskeleton** (Figure 3-4). Besides giving support, the cytoskeleton helps organize the cell's activities, greatly increasing efficiency. The cytoskeleton facilitates this vital function by binding with enzymes. **Enzymes** are proteins that increase the rate of chemical reactions in cells. As you will see later in the chapter, many chemical reactions in cells

of the body occur in series, with one reaction giving rise to a substance that's used in the next. A series of linked reactions is called a metabolic pathway and is the cellular equivalent of an assembly line in a factory. The enzymes of the metabolic pathway are the cell's equivalent of assembly line workers.

Molecules enter one end of a metabolic pathway and are modified along their course by enzymes (Figure 3-5). Eventually, a finished product comes off the line. Because enzymes are arranged along the cytoskeleton in proper order, the product of one reaction is conveniently situated for the next reaction.

3-3 The Structure and Function of the Plasma Membrane

The **plasma membrane** is the outermost part of a cell. The plasma membrane controls what goes in and out of the cell, and thus determines the contents of cell's internal chemical environment—a function essential to cellular homeostasis and survival.

The Structure of the Plasma Membrane

The plasma membrane consists of lipids, protein, and carbohydrate.

The plasma membrane and all of the other internal membranes of the cell consist of lipids, proteins, and a small amount of carbohydrate (Figure 3-6). **Lipids** or fats are a chemically diverse group of biochemicals that form waxes, grease, and oils. All are insoluble in water. Lipids serve a variety of functions in the body. Some provide energy. Others serve as insulation and energy stor-

age. Still others play a structural role. Three different lipids are discussed in this book. The first, which is found in the membranes of cells, is the **phospholipid** (FOSS-foh-lip-id). Shown in Figure 3-7, phospholipids form lollypop-shaped molecules with a polar (charged) head and a nonpolar (uncharged) tail. Most of the lipid molecules in the plasma membrane are phospholipids.

Experimental studies suggest that the phospholipid molecules of the plasma membrane are arranged in a double layer—or bilayer. Figure 3-6 shows the most widely accepted model (theory) of the plasma membrane's structure, the **fluid mosaic model**. As shown, the polar heads of the outer layer of phospholipid molecules face the watery extracellular fluid, and the polar heads of the inner layer of molecules are directed inward toward the watery cytoplasm.

Interspersed in the lipid bilayer of human cells are large protein molecules known as *integral proteins*. **Proteins** are large mo-

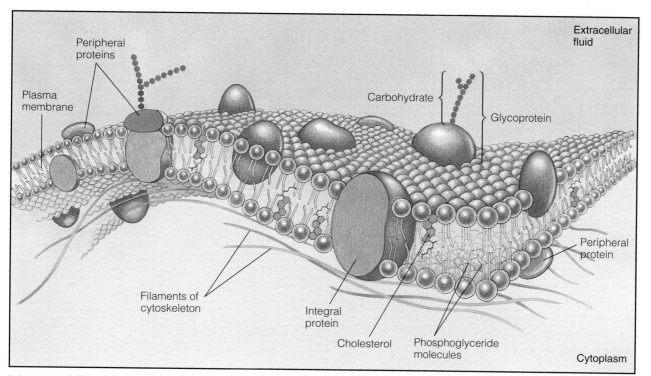

FIGURE 3-6 Fluid Mosaic Model of the Plasma Membrane of Animal Cells The fluid mosaic model is the most widely accepted theory of the plasma membrane. Phospholipids are the chief lipid component. They are arranged in a bilayer. Integral proteins float like icebergs in a sea of lipids.

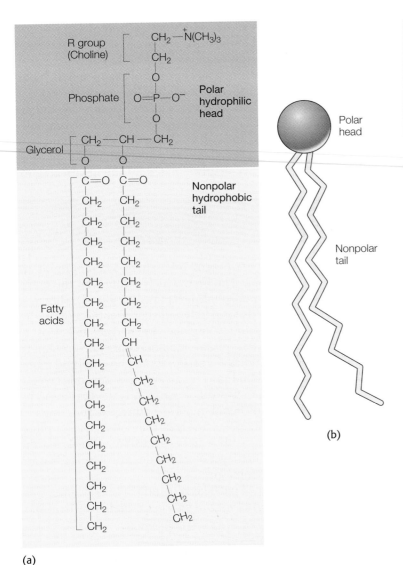

(a)

FIGURE 3-7 Phospholipids (a) Each phospholipid consists of a glycerol backbone, two fatty acids, a phosphate, and a variable group generally designated by the letter R. In this case, the R group is choline, which is polar. (b) Because of the R group, the molecule has a polar head. The nonpolar tail region is formed by the two fatty acid chains.

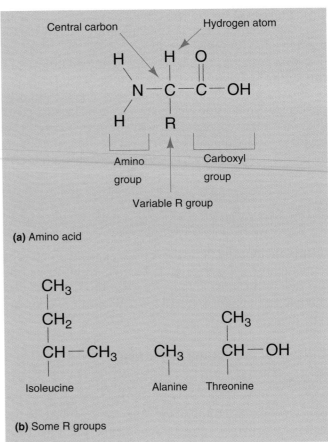

(a) Amino acid

(b) Some R groups

FIGURE 3-8 Structure of Amino Acids (a) The amino acid is a small organic molecule with four groups, one of which is variable and is designated by the letter R. (b) Some representative R groups.

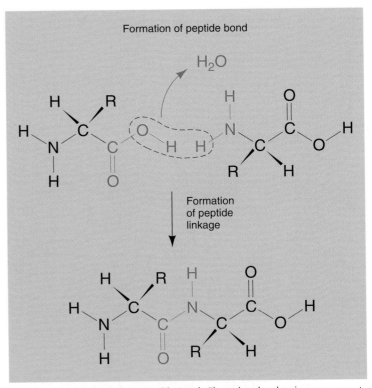

FIGURE 3-9 Formation of the Peptide Bond The carboxyl and amino groups react in such a way that a covalent bond is formed between the carbon of the carboxyl group of one amino acid and the nitrogen of the amino group of another. This bond is called a peptide bond.

lecular weight compounds made of many smaller molecules, known as amino acids. **Amino acids** contain a central carbon atom, attached to which are an amino group (NH$_2$) and a carboxylic acid group (COOH)—from which they derive their name (Figure 3-8). As shown in the figure, the central carbon atom also bonds to a hydrogen and a variable group (indicated by the letter R). About 20 amino acids are found in the proteins of humans.

Proteins serve many different functions. One important class of proteins, the enzymes, accelerates chemical reactions in the body. Without enzymes, few reactions would occur. Other proteins play a structural role. Keratin (CARE-ah-tin), for example, is a protein in human hair and nails. Collagen (coll-AH-gin), the most abundant protein in the human body, forms fibers that are found in ligaments, tendons, and bones.

Proteins are synthesized in the body one amino acid at a time, but a large protein containing hundreds of amino acids is assembled in under a minute. Figure 3-9 shows how amino acids bind. The bond between two amino acids is called a *peptide bond*. As a protein is synthesized, the chain of amino acids be-

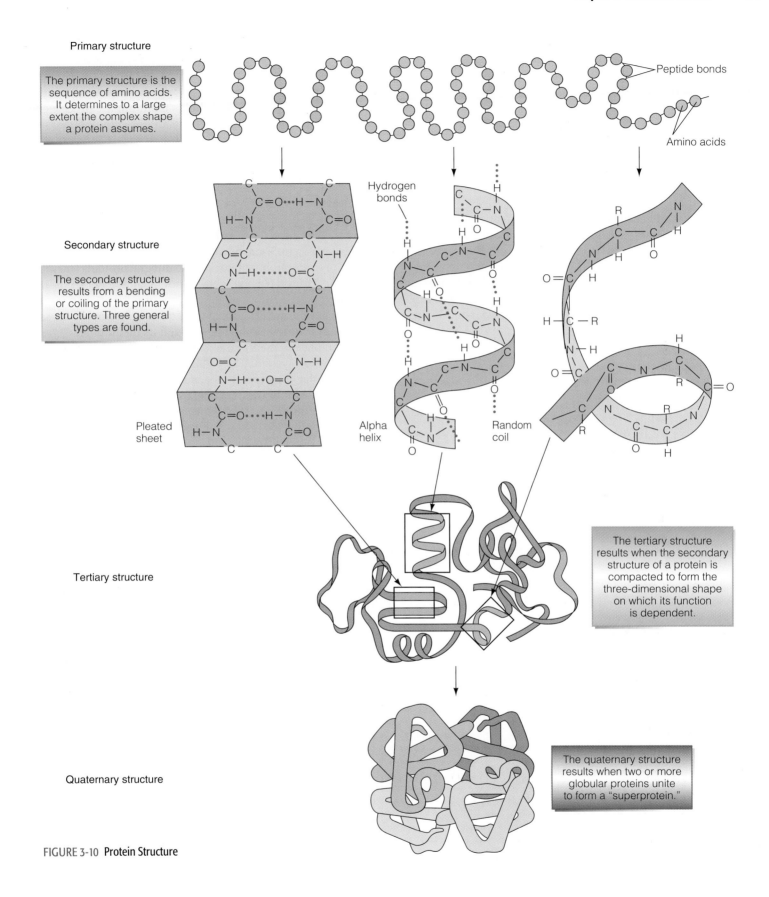

Primary structure

The primary structure is the sequence of amino acids. It determines to a large extent the complex shape a protein assumes.

Peptide bonds

Amino acids

Secondary structure

The secondary structure results from a bending or coiling of the primary structure. Three general types are found.

Hydrogen bonds

Pleated sheet

Alpha helix

Random coil

Tertiary structure

The tertiary structure results when the secondary structure of a protein is compacted to form the three-dimensional shape on which its function is dependent.

Quaternary structure

The quaternary structure results when two or more globular proteins unite to form a "superprotein."

FIGURE 3-10 **Protein Structure**

gins to twist and bend (Figure 3-10). Eventually, the amino acid chain forms an elaborate three-dimensional structure, which is often globular in nature.

The integral proteins of the plasma membrane are globular structures that float freely like giant icebergs in their sea of lipid. As in Figure 3-6, some of them completely penetrate the plasma membrane; others penetrate only partway. Several integral proteins may join to form pores that permit the movement of molecules into and out of the cell. Others may attach to the underlying cytoskeleton, anchoring the plasma membrane in place.

Another group of proteins, the **peripheral proteins**, is found in the plasma membrane, often attached to integral proteins. On the cytoplasmic side of the membrane, they may also bind to the cytoskeleton.

Plasma membranes contain small but significant amounts of carbohydrate. Most carbohydrates are oligosaccharides (short chains of simple sugars). Shown in Figure 3-6, these segments attach to integral proteins that protrude into the extracellular fluid of cells. A protein combined with a carbohydrate is a **glycoprotein**. The glycoproteins of the plasma membrane are vital to cell function.

Another type of lipid is also found in the plasma membrane. This lipid is cholesterol. Cholesterol is a steroid. Its chemical formula is shown in Figure 3-11. Cholesterol and other steroids consist of four ring structures that are joined together to form a single molecule. Cholesterol molecules are found in the plasma membranes of cells wedged in between the phospholipid molecules (Figure 3-7). Their importance is not well understood. They appear to make the membrane more elastic.

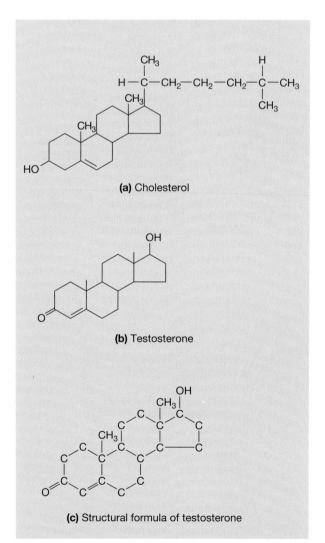

(a) Cholesterol

(b) Testosterone

(c) Structural formula of testosterone

FIGURE 3-11 **Steroids** Steroids, such as (a) cholesterol and (b) testosterone, consist of four rings joined together. Parts (a) and (b) are a shorthand way of drawing (c) the structural formula.

Maintaining Homeostasis

The plasma membrane is essential to cellular homeostasis.

The plasma membrane regulates the flow of molecules and ions into and out of the cell. This helps maintain the precise chemical concentrations inside and outside the cell—both essential for homeostasis and optimal cell function.

Because the plasma membrane regulates the molecular traffic, it is said to be selectively permeable. That is, the plasma membrane selects or controls what enters and leaves the cell. As you will learn in later chapters, the plasma membrane also plays a vital role in cellular communication. Communication is defined very broadly here to include the transmission or receipt of any kind of message. Some hormones in the tissue fluid surrounding cells, for example, attach to specific integral proteins (membrane receptors) in the plasma membranes of cells. This sends a message to the cell's interior that triggers important changes in the cell's structure and function. Consequently, hormones are said to "communicate" with the cell through its membrane.

The plasma membrane is also part of an elaborate cellular identification system. The cells in each of us have a unique protein "cellular fingerprint." Because the protein composition of the plasma membrane of the cells of each individual is unique, a person's immune system can recognize its own cells and determine when foreign cells—such as bacteria—have invaded. The immune system also identifies cancer cells. Cancer cells proliferate wildly. Their cellular fingerprint also becomes altered, making them unrecognizable or foreign. Once the immune system has identified microbial invaders or tumor cells, the body mounts an attack, often successfully eliminating them. Without this marvelous ability to recognize foreign invaders or altered cells, we would very likely perish in rapid order from infections or tumors.

TABLE 3-2	Summary of Plasma Membrane Transport
Process	Description
Simple diffusion	Flow of ions and molecules from high concentrations to low. Water-soluble ions and molecules probably pass through pores; water-insoluble molecules pass directly through the lipid layer.
Facilitated diffusion	Flow of ions and molecules from high concentrations to low concentrations with the aid of protein carrier molecules in the membrane.
Active transport	Transport of molecules from regions of low concentration to regions of high concentration with the aid of transport proteins in the cell membrane and ATP.
Endocytosis	Active incorporation of liquid and solid materials outside the cell by the plasma membrane. Materials are engulfed by the cell and become surrounded in a membrane.
Exocytosis	Release of materials packaged in secretory vesicles.
Osmosis	Diffusion of water molecules from regions of high water (low solute) concentration to regions of low water (high solute) concentrations.

Membrane Transport

Molecules move through the plasma membrane in five ways.

Cells are bathed in a liquid called **interstitial fluid** (in-ter-STICH-al), that fills the tiny spaces between them. This liquid and the cytoplasm of cells contain water and a variety of chemical substances that are dissolved or suspended in the water. Some of these substances pass through the membrane with ease; others must be transported across the membrane with the aid of special molecules. All told, five basic mechanisms exist for moving molecules and ions across plasma membranes (Table 3-2).

Diffusion

The movement of molecules from high to low concentrations Is called diffusion.

Lipid-soluble substances pass directly through the plasma membrane of cells via diffusion. Why? Because the plasma membrane is principally made of lipid, and chemical compounds readily dissolve in chemical compounds of a similar nature. Likes dissolve likes. Steroid hormones, oxygen, and carbon dioxide, for example, are all lipid-soluble and therefore pass directly and with ease through the lipid bilayer of the plasma membrane (Figure 3-12a).

Diffusion refers to any movement of molecules or ions down a concentration gradient—that is, from high to low concentration. As you will soon see, two types of diffusion exist, simple and facilitated. Movement of lipid-soluble chemicals through the membrane without assistance, as described above, is simple diffusion.

Water-soluble materials cannot pass through the lipid bilayer of the plasma membrane and must travel via other routes. Scientists believe that water and many other small water-soluble molecules and ions pass through pores in the plasma membrane formed by integral proteins (Figure 3-12b). Movement through membrane pores is another form of simple diffusion.

Carrier Proteins and Diffusion

Water-soluble molecules also diffuse through membranes with the aid of carrier molecules.

Carrier proteins are molecules that transport small water-soluble molecules and ions across the membrane in ways not yet completely understood. Carrier proteins "shuttle" molecules across membranes from regions of high concentration to regions of low concentration. This process is called **facilitated diffusion** to distinguish it from simple diffusion through pores or through the lipid bilayer (Figure 3-13a).

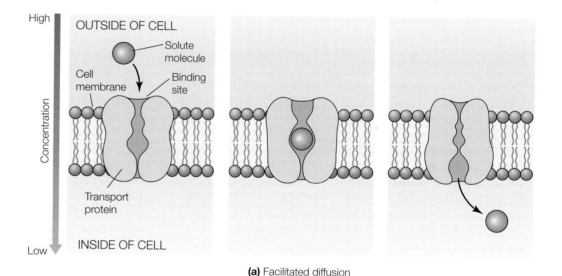

(a) Facilitated diffusion

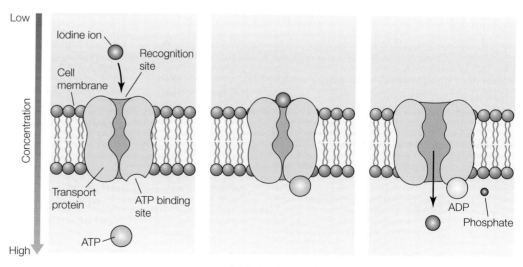

(b) Active transport

FIGURE 3-13 **Membrane Transport II** (a) Water-soluble molecules can also diffuse through membranes with the assistance of proteins in facilitated diffusion. (b) Other proteins use energy from ATP to move against concentration gradients in a process called active transport.

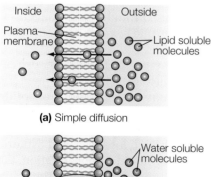

(a) Simple diffusion

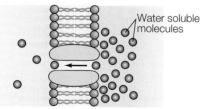

(b) Diffusion through protein pores

FIGURE 3-12 **Membrane Transport I** (a) Lipid-soluble substances pass through the membrane directly via simple diffusion. (b) Water-soluble molecules can diffuse passively through pores formed by protein molecules.

Active Transport

Molecules are also actively transported across the membrane.

Another transport mechanism in plasma membranes is known as active transport. **Active transport** is the movement of molecules across membranes with the aid of protein carrier molecules in the plasma membrane and with energy supplied by a special molecule called ATP (Figure 3-13b). **ATP** or **adenosine triphosphate** (ah-DEN-oh-seen try-FOSS-fate) consists of three smaller molecules, shown in Figure 3-14: a nitrogen-containing organic base, adenine; a five-carbon sugar, ribose; and three phosphate groups.

ATP is often likened to a form of energy currency. When the cell needs energy, say to move molecules across its membrane, ATP splits off a phosphate, forming ADP (adenosine diphosphate). The breaking of the phosphate bond yields energy that the cell can use directly. The reaction is:

$$ATP \rightarrow ADP + P_i + Energy$$

Reactions in the body that give off energy often turn the energy over to ADP. That way, new ATP can be formed.

$$ADP + P_i + Energy \rightarrow ATP$$

ATP serves as a kind of an energy shuttle, picking up energy from reactions that release it and shuttling it to reactions that require it. The energy is stored in the covalent bond between ADP and the last phosphate.

During active transport, molecules and ions are transported from regions of low concentration to regions of high concentration—movement "up," or against, the concentration gradient. Active transport occurs in cells that must concentrate chemical substances to function properly and, like the various forms of diffusion, it is essential for maintaining homeostasis. For example, cells in the thyroid gland in the neck require large quantities of the iodide ion (I^-) to manufacture the hormone thyroxine (thigh-ROX-in). But since levels of I^- in the bloodstream are fairly low, thyroid cells must actively transport iodide ions into the cytoplasm, where the iodide concentration is many times higher. Cells also use active transport to move materials out of their cytoplasm against concentration gradients.

Endocytosis

Large molecules and cells are ingested by endocytosis.

Cells also incorporate materials by endocytosis (en-doe-sigh-TOE-siss; literally, "into the cell"), as shown in Figure 3-15A. **Endocytosis** requires ATP and consists of two related activities:

phagocytosis (fag-oh-sigh-toe-siss) and pinocytosis. **Phagocytosis** ("cell eating") occurs when cells engulf larger particles, such as bacteria and viruses. In humans, phagocytosis is limited to relatively few types of cells—those involved in protecting the body against foreign invaders. **Pinocytosis** (cell drinking) occurs when cells engulf extracellular fluids and dissolved materials. Most, if not all, cells are capable of pinocytosis.

Exocytosis

Cells also regurgitate materials, releasing large molecules.

The reverse of endocytosis is **exocytosis** (meaning "out of the cell"). This process allows cells to release large molecules held in their cytoplasms (Figure 3-15b). For example, in the cells of many endocrine glands, protein hormones are packaged into tiny membrane-bound vesicles. These vesicles migrate to the plasma membranes and fuse with them. At the point of fusion, the membranes break down, and the protein hormone is released into the extracellular fluid. This process is exocytosis.

Osmosis

The diffusion of water across the plasma membrane is known as osmosis.

Like any other small molecule, water moves from one side of a plasma membrane to the other by diffusion. The diffusion of water across a selectively permeable membrane, however, is given a special name, **osmosis** (oss-MOE-siss; this word comes from a Greek word meaning "to push"). To understand osmosis, consider a simplified example in Figure 3-16.

In this example, we begin with a large bag made of a special plastic that's selectively permeable. We fill the bag with water and table sugar (sucrose) and then submerge it in a large beaker of distilled water (Figure 3-16). The bag permits the water (but not the sucrose) to move in and out. Soon after immersion in the distilled water, water molecules will begin to diffuse down the concentration gradient, and the bag will begin to swell. This swelling results from the inward movement of water molecules shown by the arrows in Figure 3-16b. In this system, water moves from a region of high water concentration (distilled water) to a region of low water concentration (inside the bag).

The concentration difference across the membrane "drives" the water across a selectively permeable membrane. Therefore, whenever two fluids with different concentrations of solute (dissolved substance) are separated by a selectively permeable membrane, water will flow from one to the other, moving down the concentration gradient; the driving force is called **osmotic pressure**. The greater the difference in water concentration, the greater the osmotic pressure, and the more quickly the water moves.

In humans and other multicellular organisms, osmosis helps regulate the concentration of fluid surrounding the cell, keeping it at the same concentration as that of the cells' cytoplasm. In such cases, the extracellular fluid is said to be *isotonic* (having the same strength).

If the fluid surrounding a cell is not isotonic, serious problems can be expected. For example, if blood cells are immersed in a solution that is more concentrated than their cytoplasm, wa-

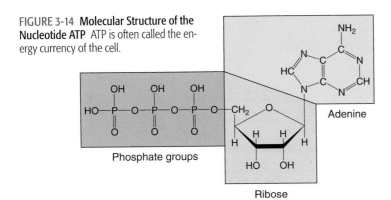

FIGURE 3-14 **Molecular Structure of the Nucleotide ATP** ATP is often called the energy currency of the cell.

NH₂

Adenine

Phosphate groups

Ribose

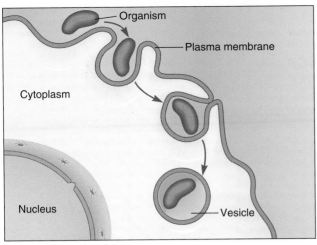

(a) PHAGOCYTOSIS

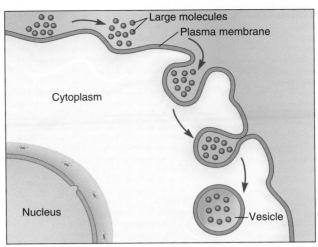

PINOCYTOSIS

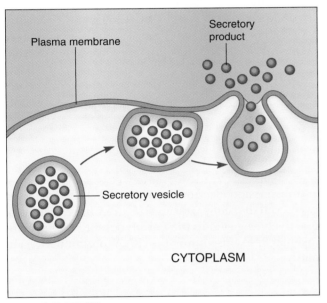

(b) EXOCYTOSIS

FIGURE 3-15 **Membrane Transport** (a) Cells can engulf large particles, cell fragments, and even entire cells via endocytosis. (b) Exocytosis, the reverse process, rids the cell of large particles.

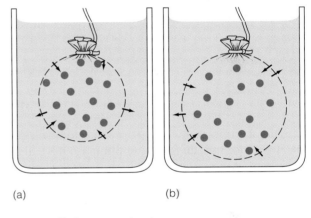

● Sucrose molecules

FIGURE 3-16 **Osmosis** Osmosis is the diffusion of water molecules from a region of higher water concentration (or low solute concentration) to one of lower water concentration (or high solute concentration) across a selectively permeable membrane. (a) To demonstrate the process, immerse a bag of sugar water in a solution of pure water. (b) Water diffuses into the bag (toward the lower concentration), causing it to swell.

ter will move out of the cells, and the cells will shrink. A solution with a solute concentration higher than the cell's cytoplasm is said to be *hypertonic* (having a greater strength). A solution with a solute concentration lower than the cell's cytoplasm is said to be *hypotonic* (having a lesser strength). If red blood cells are placed in such a solution, water will rush in, causing the cells to swell and burst.

Combined with blood pressure and other forces, osmotic pressure plays an important role in the filtering of blood in the kidney—a function essential to homeostasis (Chapter 9).

3-4 Cellular Compartmentalization: Organelles

Now that we've looked at the plasma membrane, let's turn our attention to the organelles—structures that carry out many other key functions of the cell. We begin with the nucleus.

The Nucleus

The nucleus houses the DNA and is the cell's command center.

The nucleus is one of the cell's most important organelles (Figure 3-17). It is often likened to a cellular command center. The nucleus is contained by a double membrane, the **nuclear** **envelope**, which separates the nuclear material from the cytoplasm. Minute channels in the envelope, the **nuclear pores**, allow materials to pass into and out of the nucleus (Figure 3-17b).

The bulk of the nucleus contains long, threadlike fibers of DNA and protein, called known as **chromatin** (CHROME-eh-tin). It appears as fine granules in transmission electron micrographs (Figure 3-17a). The DNA in the nucleus is the genetic material. It houses all of the information a cell needs for proper development and functioning—hence the description of the nucleus as a central command center.

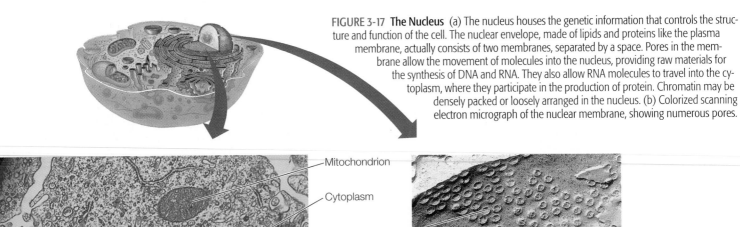

FIGURE 3-17 **The Nucleus** (a) The nucleus houses the genetic information that controls the structure and function of the cell. The nuclear envelope, made of lipids and proteins like the plasma membrane, actually consists of two membranes, separated by a space. Pores in the membrane allow the movement of molecules into the nucleus, providing raw materials for the synthesis of DNA and RNA. They also allow RNA molecules to travel into the cytoplasm, where they participate in the production of protein. Chromatin may be densely packed or loosely arranged in the nucleus. (b) Colorized scanning electron micrograph of the nuclear membrane, showing numerous pores.

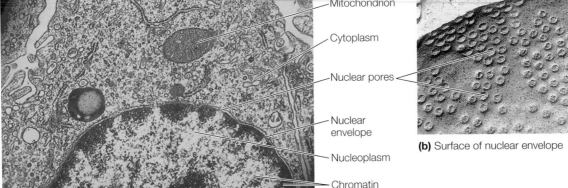

Mitochondrion

Cytoplasm

Nuclear pores

Nuclear envelope

Nucleoplasm

Chromatin

(a) Section through a typical cell

(b) Surface of nuclear envelope

A strand of chromatin with its protein is known as a **chromosome**. Just before cell division, the chromosomes coil to form short, compact bodies, highly visible through the light microscope (Figure 3-18b). This compaction greatly facilitates the cell division, making it much easier to divide the nuclear contents. A detailed discussion of DNA can be found in Chapter 17.

Proteins, water, and other small molecules and ions are also found in the nucleus, forming a semifluid material known as the **nucleoplasm** (new-KLEE-oh-plazem)—the nuclear equivalent of cytoplasm.

Also found in the nuclei of cells are structures known as **nucleoli** ("little nuclei"; singular, nucleolus). Nucleoli appear as small, clear, oval structures in light micrographs and as dense bodies in transmission electron micrographs. They are found only between cell divisions (Figure 3-3). Nucleoli are regions of the DNA actively engaged in the production of RNA, specifically **ribosomal RNA** (also known as **rRNA**; pronounced RYE-bow-zomal).

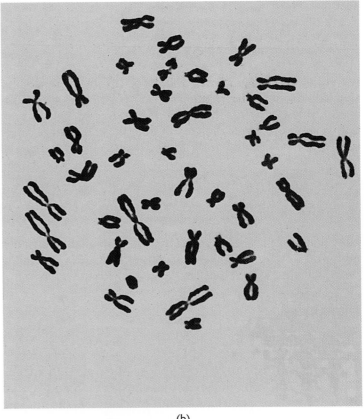

(a)

FIGURE 3-18 **Chromosomes** (a) The threadlike chromosomes, made of chromatin fibers consisting of protein and DNA, must condense before the nucleus can divide. (b) Condensed chromosomes. Can you think of any advantages to this strategy?

(b)

Ribosomal RNA gets its name from the fact that it combines with certain proteins to form ribosomes. **Ribosomes** are non-membranous organelles that appear as small, dark granules in electron micrographs and consist of two subunits. Produced in the nucleus, the subunits enter the cytoplasm through the nuclear pores. In the cytoplasm, they combine to play an important part in protein synthesis, described in Chapter 18. A detailed discussion of RNA is also presented in that chapter.

Mitochondria

Mitochondria are the site of cellular energy production.

Like a factory, a cell requires energy and a special structure to generate it. This structure is a cellular organelle known as the mitochondrion (MY-toe-CON-dree-on). The **mitochondrion**, shown in Figure 3-19, frees energy held in the chemical bonds of organic molecules, usually sugars and fats. Mitochondria (plural) do not release that energy, but rather trap it and use it to make ATP. Cells use the energy of ATP to synthesize chemical substances, transport molecules across membranes, divide, contract, and move about.

Although mitochondria vary considerably in form and number from cell to cell, they "share" several common characteristics. All mitochondria, contain outer and inner membranes. The inner membrane is thrown into folds or *cristae* (CHRIS-tee) (Figure 3-19b). This membrane therefore divides the mitochondrion into two distinct compartments, the inner compartment and the outer compartment, each with its own function. The cristae also increase the surface area on which chemical reactions take place.

During the breakdown of glucose in cells, about one-third to two-thirds of the energy released during the chemical reactions is captured by the cell to produce ATP. The rest of the energy once housed in glucose molecules is given off as heat. This "waste" product creates body heat. It, in turn, is necessary for maintaining normal enzymatic function in warm-blooded animals like you and me. Without it, many processes would cease, and we would soon die.

Protein Production

Three organelles are involved in manufacturing protein

Many cells are miniature factories synthesizing a variety of molecules vital for growth, reproduction, and day-to-day maintenance. Many of these molecules are used within the cell; others, such as hormones, are released from the cell into the bloodstream where they travel to distant cells, eliciting specific responses. This section examines three organelles involved in cellular synthesis: the endoplasmic reticulum, ribosomes, and the Golgi complex.

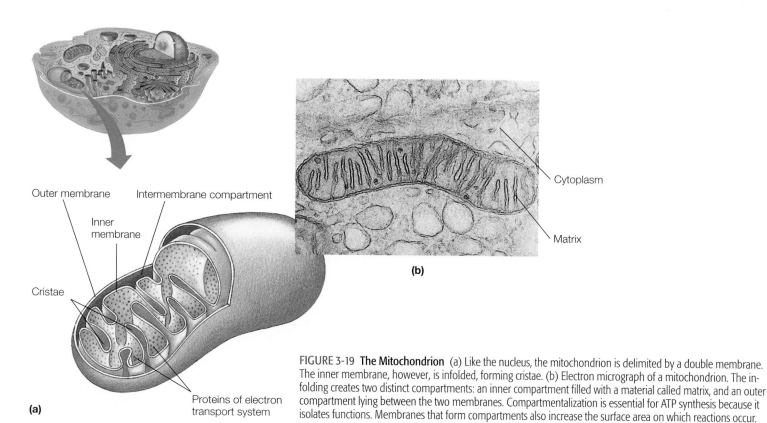

Outer membrane
Intermembrane compartment
Inner membrane
Cristae
Cytoplasm
Matrix
Proteins of electron transport system

(a)

(b)

FIGURE 3-19 **The Mitochondrion** (a) Like the nucleus, the mitochondrion is delimited by a double membrane. The inner membrane, however, is infolded, forming cristae. (b) Electron micrograph of a mitochondrion. The infolding creates two distinct compartments: an inner compartment filled with a material called matrix, and an outer compartment lying between the two membranes. Compartmentalization is essential for ATP synthesis because it isolates functions. Membranes that form compartments also increase the surface area on which reactions occur.

Endoplasmic Reticulum

Some proteins are produced on the surface of the rough endoplasmic reticulum.

Coursing throughout the cytoplasm of many cells is a branched network of membranous channels, the **endoplasmic reticulum** (END-oh-PLAZ-mick rah-TICK-u-lum; meaning intracellular network) (Figure 3-20a). These membranous channels are derived chiefly from the nuclear envelope and may be coated with ribosomes (Figure 3-20b). Ribosome-studded endoplasmic reticulum is referred to as the **rough endoplasmic reticulum (RER)**. The RER produces a variety of proteins, such as plasma membrane proteins. In the pancreas, it produces digestive enzymes. In certain endocrine glands, RER produces protein hormones.

The proteins produced by the RER are made on the outside surface of the organelle, but are quickly transferred into its interior, possibly to protect them from damage. The interior of the RER not only protects the proteins, but also contains 30 to 40 enzymes that chemically convert some proteins into glycoproteins. Proteins and glycoproteins are then transferred to yet another organelle, the Golgi complex, for final packaging.

Cells also contain endoplasmic reticulum lacking ribosomes (Figure 3-20c). Known as the **smooth endoplasmic reticulum (SER)**, this organelle specializes in the production of phospholipids, which are needed to replenish the plasma membrane. The SER performs a variety of functions in different cells. In the human liver, for instance, it detoxifies certain drugs, such as barbiturates, one type of sedative. In the stomach of humans and other mammals, the SER of certain cells produces hydrochloric acid, necessary for protein digestion. In the adrenal glands, the SER produces steroid hormones.

Ribosomes

Ribosomes are involved in all protein synthesis.

As noted earlier, ribosomes are tiny particles made of protein and ribosomal RNA (rRNA). In the cytoplasm, ribosomes attach to strands of RNA molecules known as **messenger RNA (mRNA)**. Messenger RNA molecules are produced in the nucleus of the cell and, as you shall see in Chapter 18, contain the information required to synthesize protein.

In the cytoplasm, several ribosomes attach to a single strand of mRNA, forming a structure known as a **polyribosome** or **polysome**. Together, mRNA and its attached ribosomes begin to produce protein. If it is a protein destined for inclusion in lysosomes (described shortly) or the plasma membrane or if it is intended for extracellular use (hormones and digestive enzymes), the polysome attaches to the RER, and the protein is transferred into the interior of the RER. If it is a protein for intracellular use, like those of the cytoskeleton or cytoplasmic enzymes, the polysome remains free within the cytoplasm.

The Golgi Complex

The Golgi complex modifies and packages proteins that are secreted from the cell for extracellular use.

Proteins and glycoproteins manufactured by the RER are transported internally to the ends of sacs (Figure 3-21a). As these products build up inside the RER, the ends bulge and eventually pinch off, forming tiny membrane-bound **transfer vesicles** (VESS-sick-kuls). The vesicles protect the proteins and glycoproteins as they are transported to the final packaging unit, the Golgi complex.

The **Golgi complex** (GOL-gee) consists of a series of membranous sacs (Figure 3-21b). It performs three functions. First, it

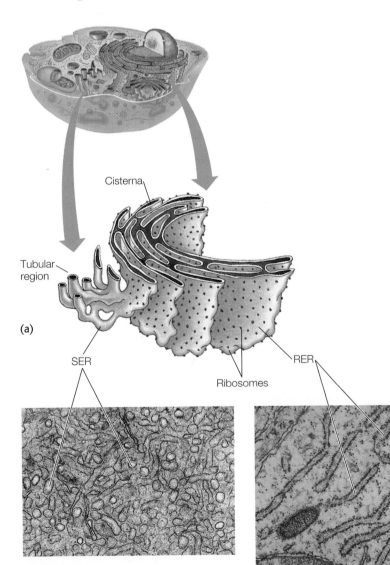

Cisterna

Tubular region

(a)

SER

Ribosomes

RER

(b)

(c)

FIGURE 3-20 The Endoplasmic Reticulum (a) Created by a network of membranes in the cytoplasm, the endoplasmic reticulum is often studded with small particles, the ribosomes, forming rough endoplasmic reticulum. This is the site of the synthesis of protein for lysosomes and extracellular use (digestive enzymes and hormones). (b) Electron micrographs of SER and (c) RER.

sorts the molecules it receives. The Golgi complex separates lysosomal enzymes from hormones and plasma membrane proteins, ensuring that products end up in the right place. Second, like the RER, the Golgi complex chemically modifies many proteins, often adding carbohydrates or other small molecules. Finally, the Golgi complex packages proteins into lysosomes or secretory vesicles (or secretory granules) (Figure 3-21a).

Secretory vesicles are membranous organelles in the cytoplasm. They serve as temporary storage sites for hormones and digestive enzymes that will eventually be excreted by the cells. When the signal for release comes, secretory vesicles migrate to the plasma membrane and fuse with it. The fused membranes soon dissolve, releasing the contents of the secretory vesicle into the extracellular space. This is an example of exocytosis.

Lysosomes

> Lysosomes are membrane-bound organelles that contain digestive enzymes.

Lysosomes (LIE-so-ZOMES; literally, digestive bodies) are membrane-bound organelles that contain numerous digestive enzymes produced by the RER and packaged by the Golgi complex. These organelles break down materials phagocytized by cells (Figure 3-22).

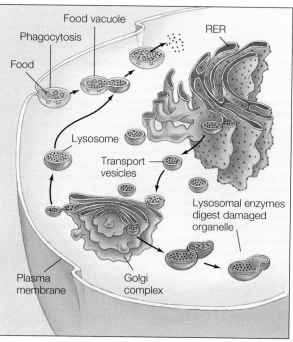

(a)

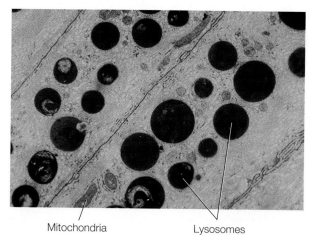

Mitochondria   Lysosomes

(b)

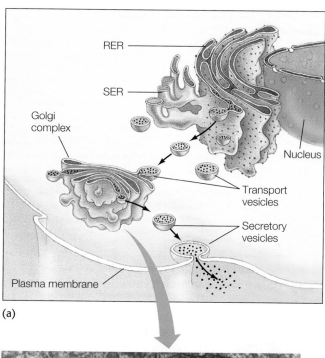

(a)

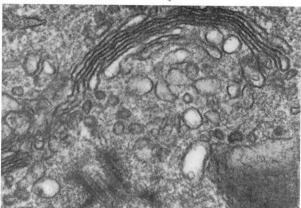

(b)

FIGURE 3-21 Protein Synthesis and Secretion (a) Protein packed in lysosomes and secretory granules (for later export) is synthesized on the RER and then transferred in tiny transport vesicles to the Golgi complex. Protein is sometimes chemically modified in the cisternae of both the RER and the Golgi complex. The Golgi complex separates protein by destination and repackages it into secretory granules, which remain in the cytoplasm until secreted by exocytosis. (b) Transmission electron micrograph of the Golgi complex.

FIGURE 3-22 Lysosomes (a) The digestive enzymes of the lysosome are produced on the RER and transported to the Golgi complex for repackaging. Lysosomes produced by the Golgi complex fuse with food vacuoles; this allows their enzymes to mix with the contents of the food vacuole. The enzymes digest the contents, which diffuse through the membrane into the cytoplasm, where they are used. The membrane surrounding the lysosome helps protect the cell from digestive enzymes. (b) Electron micrograph of lysosomes within a macrophage. The dark-staining bodies are the lysosomes, filled with digestive enzymes that are used by these cells to break down ingested material.

As illustrated in the top of Figure 3-22, material engulfed by the cell by endocytosis is enclosed in a membrane derived from the plasma membrane. The resultant structure is called a **food vacuole**. Lysosomes attach to food vacuoles and release their enzymes into them. The enzymes then break down the molecules inside the food vacuole into smaller molecules. These, in turn, diffuse into the cytoplasm, where they are often used by the cell. The undigested material left behind is expelled from the cell by exocytosis.

Besides providing the cell with nutrients, lysosomes destroy aged or defective cellular organelles, such as mitochondria. Lysosomes may also digest protein and other molecules that have become defective. This process helps cells maintain their structure and function.

Lysosomes also rid the body of injured or aged cells. In these cells, lysosomes break open, releasing their enzymes and destroying the cell. During a heart attack, for example, heart muscle cells die and are destroyed by its own lysosomes. The

lysosomes' enzymes are released into the blood as the cells disintegrate. The presence of these enzymes can be used to confirm a physician's diagnosis of a heart attack. Several other diseases can also be detected by blood enzyme levels.

Lysosomes play an important role in embryonic development. In humans, for example, the fingers of an early embryo are webbed (Figure 3-23). The cells of the webbing, however, are genetically programmed to self-destruct. At a certain point in development, the lysosomes of these cells release their enzymes, which destroy the cells and eliminate the webbing.

Most cells in the human body contain only a few lysosomes; these are used primarily to recycle outdated organelles or perhaps to put an end to the cell when its useful lifetime is over. However, certain cells contain hundreds of lysosomes. An example is the neutrophil (NEW-trowe-PHIL). Neutrophils are a type of blood cell that scavenges the blood and body tissues looking for bacteria and viruses, which they then phagocytize and digest internally with the aid of lysosomal enzymes. Thus, neutrophils play an important role in homeostasis by eliminating infectious agents.

Flagella

Flagella are organelles that permit cellular motility.

Most cells in the human body remain in the same place throughout their life as a part of a tissue or organ. One exception is the mammalian sperm cell, which must move out of the body to perform its specialized task: fertilization.

The human sperm cell moves with the aid of a **flagellum** (fla-GEL-um; plural, flagella; Latin for "whip"). As shown in Figure 3-24, a flagellum is a long, whiplike extension of the plasma membrane. It contains numerous small tubules, the **microtubules,** which are shown in Figure 3-25a. Microtubules produce the motive force. As illustrated, the nine pairs of microtubules of the flagellum are arranged around a central pair. Biologists typically refer to this as the "9 + 2" arrangement. The flagella in sperm have additional fibers outside the central 9 + 2 array, which provide extra strength.

At the base of each flagellum is an anchoring structure, the basal body (BAZ-il) (Figure 3-25b). The **basal body** gives rise to the flagellum during development and contains nine sets of microtubules (with three microtubules each), but no central pair.

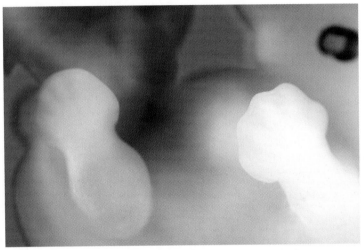

FIGURE 3-23 **The Human Embryonic Hand** The webs between the fingers present during the fetal stage are normally destroyed by lysosomes during development.

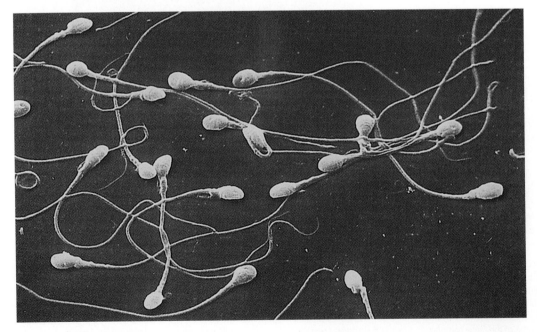

FIGURE 3-24 **The Sperm Cell** The sperm cell is a marvel of architecture, uniquely "designed" to streamline the cell for its long journey to fertilize the ovum. The nuclear material is compacted into the sperm head. The flagellum helps propel the sperm through the female reproductive tract.

FIGURE 3-25 **Flagella** (a) A sperm flagellum, showing the 9 + 2 arrangement and additional fibers thought to provide support and strength to the vigorously beating tail. (b) A basal body is found at the base of each cilium and flagellum. It consists of nine sets of triplets that give rise to the microtubule doublets of the cilium.

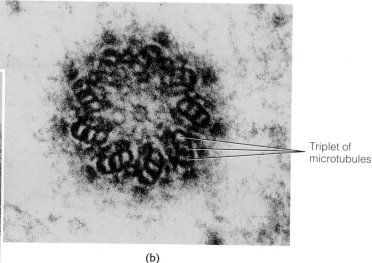

Triplet of microtubules

(b)

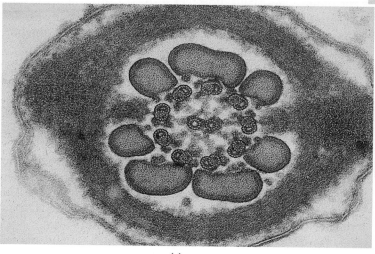

(a)

Cilia

Movement across the surface of cells is provided by cilia.

Several fixed cells in the body, such as those lining the trachea (TRAY-key-ah) or wind pipe, contain an organelle called the cilium (SILL-ee-um). **Cilia** (plural, pronounced SILL-ee-ah) are small extensions of the cell membrane (Figure 3-26). Like flagella, they have a central core of microtubules in a 9 + 2 arrangement.

Hundreds, even thousands, of cilia can be found on a single cell (Figure 3-26). In the trachea, cilia beat toward the mouth, moving mucus containing trapped dust particles away from the lungs. Here the mucus is either swallowed or expectorated (spit out). Cilia are therefore part of an automatic cleansing mechanism, vital for protecting the lungs and maintaining our health.

At the base of each cilium is a basal body, identical to those found in flagella. The basal body gives rise to the cilium during development and anchors the organelle in place.

Cilia and flagella are structurally similar, but flagella are much longer than cilia and far less numerous. Normal human sperm cells, for example, have one flagellum each, whereas each of the ciliated cells lining the respiratory system and parts of the female reproductive system (notably the Fallopian tubes) contains thousands of cilia. Flagella and cilia also beat differently. Cilia perform a stiff rowing motion with a flexible return stroke, whereas flagella beat more like whips, undulating in a continuous motion (Figure 3-27). Because cilia and flagella often beat continuously, they require a more or less constant supply of energy, provided by ATP.

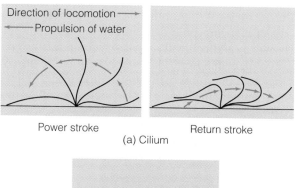

Direction of locomotion →
← Propulsion of water

Power stroke   Return stroke
(a) Cilium

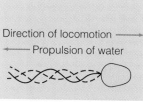

Direction of locomotion →
← Propulsion of water

Continuous propulsion
(b) Flagellum

FIGURE 3-27 **Beating of Cilia and Flagella** (a) Cilia beat with a stiff power stroke something like a rowing motion. During the return stroke, the cilium is flexible. (b) The flagellum beats like a whip, over and over, propelling the sperm cell forward.

FIGURE 3-26 **Cilia** Cilia on cells lining the trachea of the human respiratory tract.

Ameboid Movement

Some cells move by amoeboid motion.

Another type of movement is **amoeboid motion** (ah-ME-boyd). During amoeboid movement, the cell sends out many small cytoplasmic projections, or **pseudopodia** (SUE-dough-POW-dee-ah; "false feet"). Pseudopodia attach to solid surfaces "in front" of the cell (Figure 3-28). Cytoplasm flows into the pseudopodia, moving the cell forward.

In humans, most cells are stationary and don't engage in ameboid motion. There are some cells, however, that move about in the body. One of them is the neutrophil, described earlier. It is a type of white blood cell that is transported in the blood. These cells, however, also escape through the walls of thin-walled blood vessels called capillaries in body tissues. Inside the tissues, the neutrophils move about by ameboid motion, gobbling up bacteria and damaged cells.

3-5 Energy and Metabolism

Cells get the energy they need primarily from two sources: the breakdown of glucose, a type of carbohydrate, and the breakdown of triglycerides, a type of fat. Under certain circumstances, they can even capture energy from proteins (Chapter 5). In this section, we'll look briefly at how cells in humans acquire energy from the simple sugar glucose. Before we examine this process, however, a few words about glucose and other carbohydrates are in order.

As noted in Chapter 2, carbohydrates are a general class of biological molecules. **Carbohydrates** are organic compounds that range in size from simple sugars such as glucose to very large molecules such as starch. The molecular structures of two simple carbohydrates—ribose, a five-carbon sugar, and glucose, a six-carbon sugar—are shown in Figure 3-29. These are also known as **monosaccharides**.

Simple sugars such as glucose and fructose (another six-carbon sugar that is found in fruits) can unite by covalent bonds. When glucose and fructose combine, they form sucrose, commonly known as table sugar (Figure 3-30). Sucrose is a disaccharide, because it consists of two monosaccharides. As noted earlier, monosaccharides can unite to form long, branching chains (Figure 3-31). These are known as **polysaccharides** and include such important molecules as starch, glycogen, and cellulose. These molecules will be discussed in more detail elsewhere in the book.

Glucose molecules, like other organic molecules, contain energy—a fact that you could demonstrate by holding a spoonful of sugar over a candle. It will soon catch fire, releasing energy in the forms of light and heat. This energy is stored in the covalent bonds that hold the atoms of the molecule in place. Cells can tap into this energy by breaking glucose molecules apart in an or-

derly series of chemical reactions. This series begins in the cytoplasm and is completed in the mitochondrion. The complete breakdown of glucose is referred to as **cellular respiration**. This process gets its name from the fact that as in ordinary respiration, cells take up oxygen and give off carbon dioxide.

During cellular respiration, glucose, which contains six carbon atoms, is broken down into six molecules of carbon dioxide and six molecules of water. The overall reaction is

glucose $+$ oxygen $\rightarrow$ carbon dioxide $+$ water $+$ energy

$$C_6H_{12}O_6 + 6\,O_2 \rightarrow 6\,CO_2 + 6\,H_2O + \frac{38}{ATP}$$

As shown, the complete breakdown of glucose requires oxygen. Without oxygen, glucose can be only partially broken down, and only a fraction of its energy can be captured.

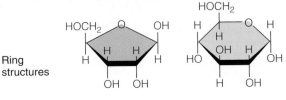

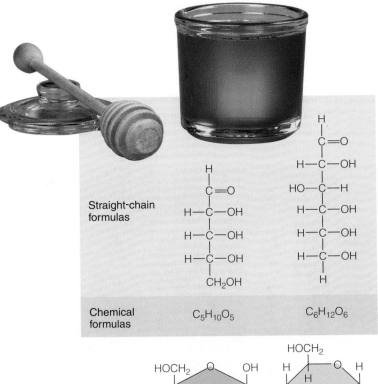

Straight-chain formulas		
Chemical formulas	$C_5H_{10}O_5$	$C_6H_{12}O_6$
Ring structures		
	(a) Ribose	(b) Glucose

FIGURE 3-29
Two Monosaccharides

FIGURE 3-28 **Amoeboid Movement** Some cells in the human body move by amoeboid movement. Studies of the amoeba (shown here) show that cells that move by amoeboid movement send out minute extensions called pseudopodia (false feet), which attach to solid surfaces. Cytoplasm then flows into the pseudopodia, moving the cell forward.

Pseudopodia

Direction of movement

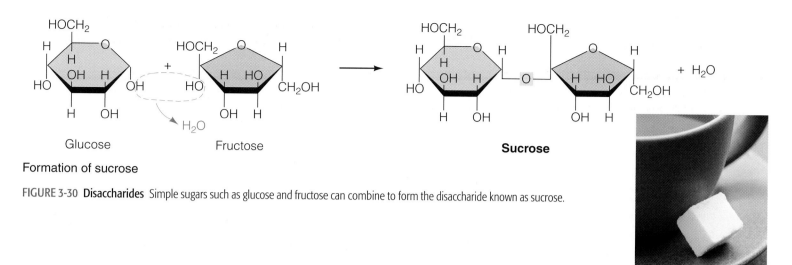

Formation of sucrose

FIGURE 3-30 Disaccharides Simple sugars such as glucose and fructose can combine to form the disaccharide known as sucrose.

FIGURE 3-31 Polysaccharides Simple sugars can also combine to form polysaccharides such as glycogen, which is a storage form of glucose in animals and is found in muscle and liver.

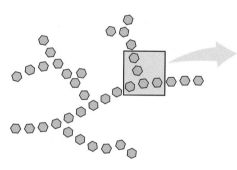

Portion of polysaccharide molecule (glycogen)

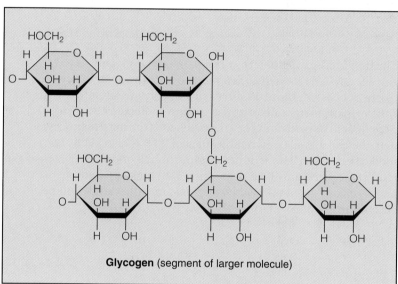

Glycogen (segment of larger molecule)

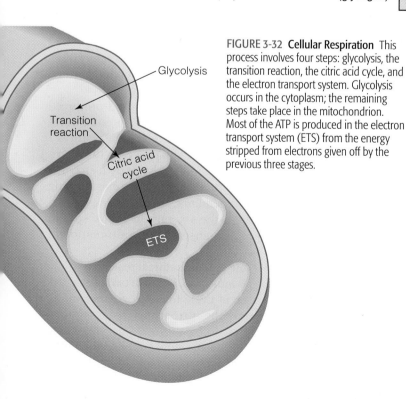

FIGURE 3-32 Cellular Respiration This process involves four steps: glycolysis, the transition reaction, the citric acid cycle, and the electron transport system. Glycolysis occurs in the cytoplasm; the remaining steps take place in the mitochondrion. Most of the ATP is produced in the electron transport system (ETS) from the energy stripped from electrons given off by the previous three stages.

The breakdown of glucose releases lots of energy. A reaction that gives off energy is known as an *exergonic reaction.* Cells can capture sizable portions of this energy and use it to synthesize ATP from inorganic phosphate (P_i) and ADP. ATP is virtually the only kind of energy cells can use. ATP synthesis is an *endergonic reaction* (energy-requiring). Exergonic and endergonic reactions are often linked so that energy from one is transferred to the other. The complete breakdown of glucose generates enough energy to produce 38 molecules of ATP from ADP and inorganic phosphate.

Cellular Respiration

The chemical breakdown of glucose occurs in four steps.

Cellular respiration is not as simple as the reaction you've just studied. In cells, the complete breakdown of glucose requires numerous chemical reactions that occur during four distinct stages: (1) glycolysis, (2) the transition reaction, (3) the citric acid cycle, and (4) the electron transport system (Figure 3-32). Table 3-3 summarizes the stages of cellular respiration and lists their products. The following paragraphs provide a more detailed summary of each process.

TABLE 3-3	Overview of Cellular Energy Production	
Reaction	Location	Description and Products
Glycolysis	Cytoplasm	Breaks glucose into two three-carbon compounds, pyruvate; nets two ATP and two NADH molecules
Transition reaction	Mitochondrion	Removes one carbon dioxide from each pyruvate, producing two acetyl CoA molecules; produces two NADH molecules
Citric acid cycle	Inner compartment of mitochondrion	Completes the breakdown of acetyl cycle CoA; produces two ATPs per glucose; produces numerous NADH and $FADH_2$ molecules
Electron transport	Inner membrane of mitochondrion	Accepts electrons from NADH and $FADH_2$, generated by previous system reactions; produces 34 ATPs

Glycolysis

Glycolysis is the breakdown of glucose into two pyruvate molecules.

The first phase of cellular respiration in cells is **glycolysis**. Glycolysis (gly-COLL-eh-siss) is a metabolic pathway in the cytoplasm of cells. As Figure 3-33 shows, during glycolysis, glucose is split in half, forming two three-carbon molecules known as pyruvic acid or pyruvate. The energy released during this process nets the cell a modest two molecules of ATP. It also yields two molecules of NADH (nicotine adenine dinucleotide). NADH is a substance (technically a coenzyme) that picks up hydrogens and electrons given off by the reactions. Its importance will be clear shortly.

The Transition Reaction

In the transition reaction, one carbon atom is cleaved off each pyruvate.

In the second stage of cellular respiration, pyruvate generated in glycolysis diffuses from the cytoplasm into the inner compartment of the mitochondrion. Here it reacts with a large molecule called coenzyme A (CoA), as shown in Figure 3-34. During this reaction, known as a **transition reaction**, one carbon atom is cut off each pyruvate molecule, forming a two-carbon compound (acetyl group). It then binds to coenzyme A (Figure 3-34). The product of this reaction is acetyl CoA (ah-SEAT-ill). The transition reaction also yields two NADH molecules.

The Citric Acid Cycle

The citric acid cycle completes the breakdown of glucose.

Two molecules of acetyl CoA produced during the transition reaction then enter the third stage of cellular respiration. This stage is a cyclic metabolic pathway located inside the inner compartment of the mitochondrion. This elaborate pathway is called the **Krebs cycle**, after the scientist who worked out many of its details, or the **citric acid cycle**, after the very first product formed in the reaction sequence.

During the cycle, each two-carbon compound produced in the transition reaction binds to a four-carbon compound known as oxaloacetate (ox-AL-oh-ass-eh-tate). The result is a six-carbon product, citric acid. Citric acid proceeds through a series of chemical reactions, each catalyzed by its own enzyme. Along the way, the molecule is modified many times. During the complex molecular rearrangements, two molecules of carbon dioxide are removed from the six-carbon chain, so that oxaloacetate is regenerated. This ensures the continuation of the cycle.

As shown in Figure 3-34, one of the chemical reactions of the citric acid cycle yields an ATP molecule. Because two molecules of pyruvate enter the citric acid cycle for each glucose molecule broken down by glycolysis, the citric acid cycle nets the cell two ATP molecules. Several other reactions release electrons, which are funneled into the fourth, and final, part of cellular respiration, the electron transport system.

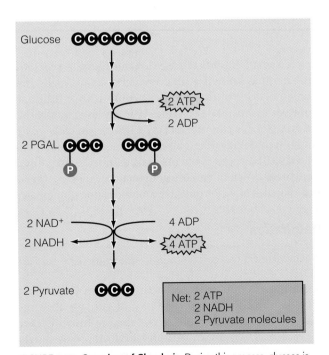

FIGURE 3-33 **Overview of Glycolysis** During this process, glucose is broken down into two molecules of PGAL, which are converted to pyruvate. The cell nets two ATP and two NADH molecules.

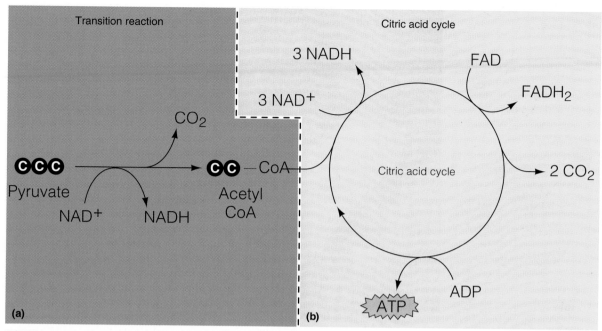

FIGURE 3-34 Overview of the Transition Reaction and the Citric Acid Cycle (a) The transition reaction cleaves off one carbon dioxide and binds the remaining two-carbon compound temporarily to a large molecule called coenzyme A. The result is a molecule called acetyl-CoA. It then enters the citric acid cycle. (b) Occurring in the matrix of the mitochondrion, the citric acid cycle liberates two carbon dioxides and produces one ATP per pyruvate molecule. Its main products, however, are NADH and FADH$_2$, bearing high-energy electrons that are transferred to the electron transport system.

The Electron Transport System

Electrons given off during the citric acid cycle and glycolysis enter the electron transport system.

The electrons liberated during certain reactions of the citric acid cycle are passed to one of the two chemical substances in the inner compartment of the mitochondrion. Those electron acceptor molecules are **nicotinamide adenine dinucleotide** or **NAD** and **flavine adenine dinucleotide**, also known as **FAD**. In so doing, NAD and FAD are converted to NADH and FADH$_2$. These molecules, in turn, transfer the electrons to a series of protein carrier molecules embedded on the inner surface of the inner membrane of the mitochondrion. The carrier proteins constitute the **electron transport system** (ETS).

Electrons entering the ETS are passed from one protein to another and are eventually given over to oxygen molecules inside the matrix. During their journey along this chain, the electrons lose energy, much like a hot potato passed along by a line of people. The energy lost along the way is used to make ATP.

All told, the electron transport system produces 34 ATP molecules per glucose molecule. Because 2 ATPs are generated by glycolysis and 2 are produced in the citric acid cycle, the total output for cellular respiration is 38 ATP molecules per glucose molecule. The "de-energized" electrons from the electron transport system eventually combine with hydrogen ions and oxygen to produce water.

Enzymes

Enzymes are essential to most chemical reactions in cells.

Before moving on, it is important to note that each of the chemical reactions of cellular respiration is regulated by an enzyme. So important are they that a missing enzyme in a metabolic pathway will shut the pathway down—in much the same way that a missing worker on an assembly line interrupts production.

Each enzyme is a large, globular protein with a small region known as the active site (Figure 3-35a). The **active site** is an indentation, or pocket, where the chemical reaction occurs. Its shape corresponds to that of the **substrate**, the molecule undergoing reaction. Because each enzyme has its own uniquely shaped active site capable of binding to one or at most a few substrates, enzymes are said to be specific. This feature allows the cells to regulate chemical reactions very precisely.

Although enzymes play an active role in the metabolism of cells, they are unchanged by the reaction. As a result, they can be used over and over again.

Enzymes can also be controlled by the cells themselves, which provides additional control over cellular metabolism. The most common regulators of enzyme activity are the end products of chemical reactions. When produced in excess, the end products often bind to specific control regions on enzymes. The control regions are called **allosteric sites** (al-oh-STAIR-ick). Allosteric sites are the molecular equivalent of switches.

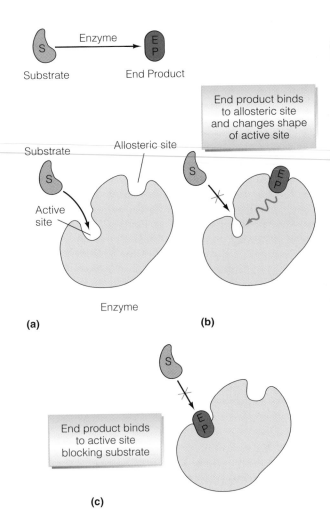

(a)

(b)

(c)

FIGURE 3-35 **Enzyme Structure and Function** Many enzymes are like factory workers on an assembly line. (a) Two-dimensional drawing of an enzyme. The active site conforms to the shape of the reacting molecule(s). (b) The allosteric site is a molecular switch that regulates enzyme activity. In some metabolic pathways, the end products bind to the allosteric site. This may turn the active site on or off, depending on the enzyme. (c) In others, the end products of a reaction bind to the active site itself, blocking it entirely.

When an end product binds to one, it turns the enzyme off, shutting down the entire metabolic pathway (Figure 3-35b). In some metabolic pathways, end products may bind directly to the active site of an enzyme. This binding also blocks substrates from entering the site, thereby inhibiting the enzyme and shutting down the metabolic pathway (Figure 3-35c).

Cellular respiration can be controlled by the buildup of chemicals such as citrate and ATP, both of which shut down a specific enzyme in the glycolytic pathway.

3-6 Fermentation

What do tired muscles, cheese, and wine have in common? The answer is that they're all products of a process called fermentation. Fermentation occurs in human cells when oxygen levels run low. It also occurs in certain microorganisms. Basically, **fermentation** is a chemical reaction in which pyruvic acid produced from glycolysis is converted to one of several other end products (Figure 3-36). These reactions are important, in part, because they produce energy. Let's look briefly at this process in body cells.

During strenuous exercise, oxygen consumption in muscle cells may exceed replenishment. When this occurs, oxygen levels inside the cells fall. (The cells are said to be anaerobic.) With oxygen no longer available to accept electrons, the electron transport system (ETS) shuts down. When the ETS falters, so does the citric acid cycle. Fortunately, glycolysis still continues. But in the absence of oxygen, the end product of glycolysis, pyruvate, is converted to lactic acid as shown in Figure 3-36a.

Thus, rather than shutting energy production down completely, muscle cells continue to generate energy via glycolysis and fermentation. Unfortunately, fermentation is exceedingly in-

efficient, netting the cell only 2 ATPs for each glucose molecule, compared with 38 during the complete breakdown.

In humans, heavy exercise, such as weight lifting and running, depletes muscle oxygen, making cells become anaerobic, which results in the buildup of lactic acid in muscle cells. This causes the muscle fatigue and soreness you feel after you've exercised heavily. Lactic acid, however, diffuses out of the muscle within hours and is carried by the blood to the liver, where it is converted first to pyruvate and then to glucose in a series of chemical reactions that is essentially the reverse of glycolysis. Some of the glucose is used to synthesize glycogen, and the rest is released into the blood, where it is redistributed to body cells and used to generate energy.

Fermentation is also the chief source of energy for many bacteria and other single-celled organisms. Many popular foods, such as cheese and yogurt, are produced by fermentation. Most cheeses, for instance, are produced from milk and selected lactic acid bacteria. Other microorganisms undergo alcoholic fermentation, which, as the name implies, produces ethanol and carbon dioxide from pyruvate (Figure 3-36b). Alcoholic fermentation is the basis of the production of beer, wine, and hard liquor.

FIGURE 3-36 **Fermentation** (a) Pyruvate, or pyruvic acid, from glycolysis is converted into lactic acid in muscle cells when oxygen levels fall. (b) Fermentation can produce a variety of products in bacteria and yeast cells. In certain microorganisms, pyruvate is converted into ethanol and carbon dioxide.

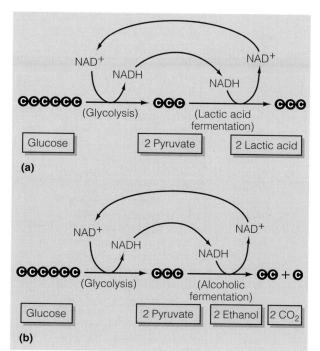

SUMMARY

1. The fundamental unit of all living organisms is the *cell*. All organisms are composed of one or more cells.

Microscopes: Illuminating the Structure of Cells

2. *Microscopes* are used to study cells. They enlarge objects and allow us to see detail. Microscopes fall into two broad categories: (1) *light microscopes*, which use ordinary visible light to illuminate the specimen; and (2) *electron microscopes*, which use beams of electrons to visualize an image.

An Overview of Cell Structure

3. Cells consist of two major compartments, the nuclear and the cytoplasmic. The nuclear compartment, or *nucleus*, contains the genetic information that regulates the structure and function of all cells. The cytoplasmic compartment contains *cytoplasm*, which, in turn, contains numerous *cellular organelles*, structures that perform specific functions.

4. The cytoplasmic compartment contains an internal supportive network known as the *cytoskeleton*. Besides providing the three-dimensional shape of cells, the cytoskeleton also binds to enzymes in metabolic pathways, increasing cellular efficiency.

The Structure and Function of the Plasma Membrane

5. The outermost boundary of the cell, the *plasma membrane*, consists of lipids, proteins, and small amounts of carbohydrate. The plasma membrane ensures the struc-

tural integrity of the cell, regulates the flow of materials into and out of the cell, and participates in cellular communication and cellular identification.

6. Because it is mostly lipid, the membrane is a natural barrier to water-soluble molecules. Lipid-soluble materials pass through with ease. Water-soluble molecules and ions pass through pores in the membrane by diffusion, the flow of a substance from an area of higher concentration to an area of lower concentration.

7. *Carrier proteins* may also transport water-soluble molecules across the membrane down the concentration gradient. This process is called *facilitated diffusion*.

8. Some molecules are actively transported across the membrane. *Active transport* mechanisms use energy to pump materials into or out of the cell and require a special class of membrane proteins, the transport molecules. Active transport permits cells to transport substances from regions of low concentration to regions of high concentration.

9. Cells engulf liquids and their dissolved materials in a process called *pinocytosis*. Cells may also engulf large molecules and cells by *phagocytosis*. Pinocytosis and phagocytosis are referred to as *endocytosis*.

10. Water is believed to move through pores in plasma membranes. The diffusion of water molecules through a selectively permeable membrane from a region of higher water concentration to a region of lower water concentration is called *osmosis*.

Cellular Compartmentalization: Organelles

11. The nucleus houses the DNA, which contains the genetic information that determines the structure and function of the cell. The nucleus is bounded by a double membrane, the *nuclear envelope*, which contains numerous pores that allow many materials to pass freely between the nucleus and the cytoplasm.

12. Energy in the cell is produced principally by the breakdown of glucose in the cytoplasm and the *mitochondria*. Energy liberated from glucose molecules is captured by ATP molecules.

13. All mitochondria have two membranes, an outer one and an inner one. The inner membrane is thrown into folds called *cristae*.

14. The *endoplasmic reticulum* is a network of membranous channels in the cytoplasm of the cell and is the main location of cellular protein synthesis. Endoplasmic reticulum may be smooth or rough. *Smooth endoplasmic reticulum* has no ribosomes. *Rough endoplasmic reticulum* contains many ribosomes on its outer surface.

15. Smooth endoplasmic reticulum produces phospholipids (needed to make more plasma membrane), and it has many additional functions in different cells. The rough endoplasmic reticulum is involved in synthesizing protein for extracellular use (for example, hormones) and in the production of enzymes that are contained in *lysosomes*.

16. Proteins produced on the surface of the RER enter the cavity, where they may be chemically modified. Proteins are transferred from

the RER to the *Golgi complex*, where they are sorted, chemically modified, and repackaged into *secretory vesicles* or lysosomes.

17. Secretory vesicles accumulate in the cytoplasm and are released when needed. Lysosomes remain in the cell, where they bind to food vacuoles or destroy malfunctioning cellular organelles.

18. Most cells in the body are fixed. Still other cells can move about in tissues via amoeboid movement. The human sperm cell utilizes another kind of motive force, the *flagellum*, a long, whiplike structure. Some cells contain numerous smaller extensions, called *cilia*, which propel fluids and materials along their surfaces. Both cilia and flagella contain nine pairs of microtubules arranged in a circle around a central pair.

Energy and Metabolism

19. Cells acquire energy from glucose, a carbohydrate. The complete breakdown of glucose is called *cellular respiration* and consists of four interdependent parts: glycolysis, the transition reaction, the citric acid cycle, and the electron transport system.

20. Glycolysis is a series of reactions occurring in the cytoplasm of cells. During glycolysis, glucose is split in two. Each molecule of glucose yields two molecules of pyruvic acid, or pyruvate. Glycolysis nets the cell two molecules of ATP and two NADH molecules containing high-energy electrons.

21. Pyruvate produced during glycolysis next diffuses into the mitochondrion and undergoes the transition reaction. This step produces acetyl CoA and two NADH molecules.

22. Acetyl CoA enters the citric acid cycle, which produces two ATPs and numerous molecules of NADH and $FADH_2$. These electron carriers accept high-energy electrons from various reactions in the citric acid cycle and move them to the electron transport system.

23. The electron transport system consists of a series of protein molecules embedded in the inner surface of the inner membrane of the mitochondrion. Electrons from NADH and $FADH_2$ are passed down the chain and are eventually given to oxygen. During this process, numerous molecules of ATP are produced.

Fermentation

24. Without oxygen, the electron transport system shuts down, as does the citric acid cycle. At this point, cells must rely on glycolysis and fermentation to generate energy. Fermentation is a chemical reaction in which pyruvate is converted into lactic acid or some other product, such as ethyl alcohol.

25. Lactic acid fermentation is also an important source of food for humans, including cheeses and yogurts. Alcoholic fermentation produces ethanol from pyruvate and occurs in certain microorganisms. Wines, beer, and hard liquor are all produced by combining plant by-products (grains or fruits) with microorganisms that are alcoholic fermenters.

critical thinking

Thinking Critically—Analysis

This analysis corresponds to the Thinking Critically scenario that was presented at the beginning of this chapter.

First, you might note that one study is never enough to test the validity of a hypothesis. Further work is needed before you can be certain that the results are valid. Some questions to ask are: Did the vitamin increase the lifespan of all cells or just certain types of cells? Were there any other factors responsible for increased lifespan? Was the test done on cells from an infant, a toddler, a youngster, a teenager, or an adult?

Second, you might also suggest that studies on animals would be needed to test this hypothesis. Tissue cultures are very artificial environments. An organism is much more complicated than some of its cells in tissue culture. Many hormones and other substances are present that might alter the effect of the vitamin.

Finally, you might want your friends to consider the potential toxic effects of the vitamins they are thinking about taking. Would vitamin supplements cause more harm than good?

KEY TERMS AND CONCEPTS

Active site, p. 55
Active transport, p. 44
Adenosine triphosphate (ATP), p. 44
Allosteric site, p. 55
Ameboid motion, p. 52
Amino acid, p. 40
Basal body, p. 50
Carbohydrate, p. 52
Carrier protein, p. 43
Citric acid cycle, p. 54
Cellular respiration, p. 52
Chromatin, p. 46
Chromosome, p. 46
Cilium, p. 51
Cytoplasm, p. 37
Cytoskeleton, p. 38
Diffusion, p. 43
Electron transport system, p. 55
Endocytosis, p. 44
Endoplasmic reticulum, p. 48
Enzyme, p. 55
Exocytosis, p. 44

Facilitated diffusion, p. 43
FAD, p. 55
Fermentation, p. 56
Flagellum, p. 50
Fluid mosaic model, p. 39
Food vacuole, p. 50
Glycolysis, p. 54
Glycoprotein, p. 42
Golgi complex, p. 48
Interstitial fluid, p. 43
Lipid, p. 39
Lysosome, p. 49
Messenger RNA, p. 48
Microscope, p. 33
Microtubule, p. 50
Mitochondrion, p. 47
Monosaccharide, p. 52
NAD, p. 55
Nuclear envelope, p. 45
Nuclear pore, p. 46
Nucleolus, p. 47
Nucleoplasm, p. 46

Nucleus, p. 36
Organelle, p. 37
Osmotic pressure, p. 44
Osmosis, p. 44
Peptide bond, p. 40
Peripheral protein, p. 42
Phagocytosis, p. 44
Phospholipid, p. 39
Pinocytosis, p. 44
Plasma membrane, p. 39
Polyribosome, p. 48
Polysaccharide, p. 52
Protein, p. 39
Pseudopodium, p. 52
Ribosomal RNA, p. 47
Ribosome, p. 47
Rough endoplasmic reticulum, p. 48
Secretory vesicle, p. 49
Smooth endoplasmic reticulum, p. 48
Substrate, p. 55
Transfer vesicle, p. 48
Transition reaction, p. 54

CONCEPT REVIEW

1. Describe the concepts of cellular and organismic homeostasis. How are they related? How does environmental homeostasis or environmental quality affect them? p. 42
2. Cellular compartmentalization allows for greater efficiency. Why? Give an example to support your argument. p. 45
3. Draw a diagram of the plasma membrane, and describe the five routes by which molecules pass through the membrane. p. 39
4. In what ways does the plasma membrane participate in cellular communication? How is this function important for the maintenance of homeostasis? p. 42
5. Define the term *osmosis*. In what way(s) is osmosis similar to diffusion, and in what way(s) is it different? p. 44
6. Describe the organelles involved in the synthesis, storage, and release of protein for extracellular use, starting with the basic instructions needed to make the protein. p. 47
7. In what ways are cilia and flagella similar, and in what ways are they different? p. 50
8. What is the main energy currency of the cell? Where is it formed? Where does it get its energy? p. 44
9. Describe the four phases of cellular energy production. p. 53

SELF-QUIZ: TESTING YOUR KNOWLEDGE

1. The cell is divided into two major compartments, the nuclear compartment and the _____ compartment. p. 36
2. The cell contains a network of protein tubules and filaments that give it shape called the _____. p. 39
3. The plasma membrane consists most of _____, a type of lipid, and protein. p. 39
4. The _____ molecules in the plasma membrane provide for cellular identity. p. 42
5. Amino acids are joined by _____ bonds to form proteins and polypeptides. p. 40
6. _____ diffusion is the movement of a substance from one side of the plasma membrane to another directly through the lipid layer or through pores in the membrane. p. 43
7. _____ transport is the movement of molecules or ions through a membrane with the aid of a carrier protein and energy in the form of _____. p. 44
8. _____ is the process in which a cell engulfs a bacterium. p. 44
9. The diffusion of water from a region of high water concentration to a region of low water concentration is known as _____. p. 44
10. Fibers made of protein and DNA that are found in the nucleus are known as _____. p. 45
11. Energy is produced primarily in an organelle known as the _____. p. 47
12. Protein for extracellular use is manufactured on the _____ endoplasmic reticulum. p. 48
13. Protein is synthesized on a template made of _____ RNA. 48
14. The _____ consists of a series of membranous sacs that modifies and packages proteins for extracellular use. p. 49
15. At the base of each cilium and flagellum is a structure known as the _____. p. 50
16. Cells that move by ameboid motion rely on small cytoplasmic projections known as _____. p. 52
17. The complete breakdown of glucose in cells is known as _____. p. 52
18. During glycolysis, glucose is broken down into two molecules of _____. p. 54
19. During the complete breakdown of glucose, the cell produces _____ molecules of ATP. p. 53
20. Label the seven parts.

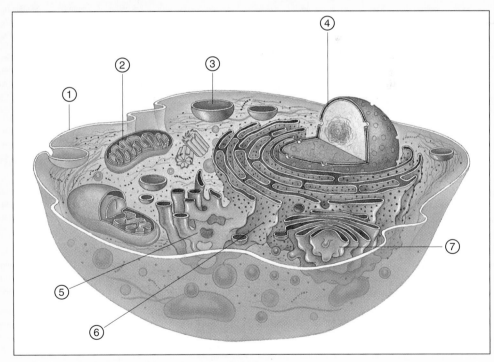

The site features eLearning, an online review area that provides quizzes, chapter outlines, and other tools to help you study for your class. You can also follow useful links for in-depth information, research the differing views in the Point/Counterpoints, or keep up on the latest health news.

Jenny Hawn has cystic fibrosis, a deadly disease that afflicts the respiratory system and pancreas. One of the main symptoms of the disease is the production of excess mucus by the cells lining the respiratory tract. Mucus traps bacteria and viruses, resulting in frequent lung infections.

thinking critically

Health care costs are skyrocketing. Some critics argue that our health care dollars are being wasted on procedures such as organ transplants for needy people, which cost taxpayers $150,000 or more each. Such expenditures, they say, are draining dollars that could be invested in preventive medicine, such as prenatal care. Analyze this view using your critical thinking skills.

Jenny's disease is genetic. Until recently, there was little hope of curing this disease, which claims most of its victims before they reach the age of 20. Today, however, there may be some hope. Geneticists are experimenting with novel ways to repair the genetic defects responsible for excessive mucus production in the cells lining the respiratory tract of patients with cystic fibrosis.

To correct this genetic defect, scientists are experimenting with certain cold viruses that infect the cells lining the respiratory tract. Normally, these viruses inject their DNA into the lining cells, causing cold symptoms. Researchers, however, are cutting out the genes of the virus that causes colds and splicing in new genes—ones that could lead to normal mucus production, hopefully creating a lasting cure for a deadly disease.

This research is just one many forms of "body work" scientists are now performing. Other researchers are working at a level slightly higher—that is, at the level of tissues and organs

(defined shortly). Some scientists, for instance, are developing genetically engineered pigs whose cells contain human genes. Their hope is that the pig-human organs could someday be transplanted into people to replace failed livers or kidneys. In fact, they speculate that an organ could be specially made with a person's own genes to prevent tissue rejection. In addition, this organ would be available in a fraction of the time it now takes to find a genetically compatible organ from a suitable donor.

Health Note 4-1 describes other remarkable medical procedures used to repair or rebuild tissues and organs. This chapter presents some basic information about body structure and function, information you will need to understand the human body and many of the exciting research projects you will hear about or read about in the news. This chapter also revisits homeostasis, elaborating on what you've already learned and setting the stage for your study of human organ systems, the subject of much of this book.

4-1 From Cells to Organ Systems

Cells unite to form tissues, and tissues combine to form organs.

The human body is made up of cells. These cells are frequently bound together by extracellular fibers and other extracellular materials, forming **tissues** (from the Latin "to weave"). Extracellular materials may be liquid (as in blood), semisolid (as in cartilage), or solid (as in bone). Tissues, in turn, combine to form **organs**, discrete structures in the body that carry out specific functions.

FIGURE 4-1 **Human Stomach** (a) The stomach, like all organs, contains all four primary tissues. (b) These are shown here in cross section.

Primary Tissues

Cells combine to form four primary tissues.

Four major tissue types are found in humans: (1) epithelial, (2) connective, (3) muscle, and (4) nervous. Table 4-1 lists them. Each tissue consists of two or three subtypes.

The tissues of the body exist in all organs, but in varying amounts. The lining of the stomach, for example, consists of a single layer of epithelial (ep-eh-THEEL-ee-ill) cells forming a protective coating (Figure 4-1). Just beneath the epithelial lining

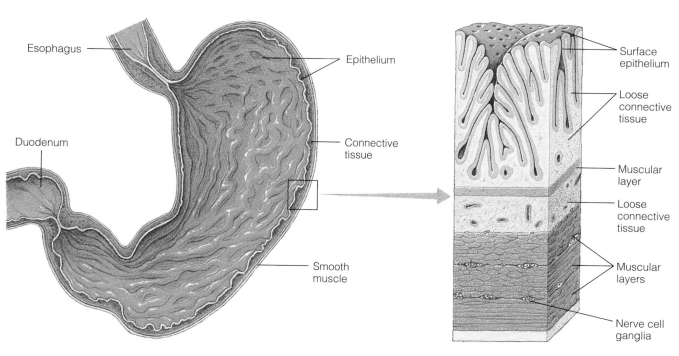

(a) Stomach

(b) Section through the wall of the stomach

·:::·healthnote

4-1   The Truth about Herbal Remedies

A male patient is treated by his physician for a painful infection of the prostate gland, one of the glands that produces a large portion of the semen. The doctor prescribes an antibiotic and advises the patient to take an herbal remedy, saw palmetto, to help prevent prostatic enlargement, a common problem in older men. A young woman suffering from restlessness and minor problems with sleep is advised by a friend to try some valerian root, another type of herbal remedy.

Herbal remedies are derived from roots, bark, seeds, fruit, flowers, leaves, and branches of plants. In China, some herbal remedies even contain animal parts. An estimated 3000 different medicinal herbs are currently used throughout the world.

Herbal remedies are sold in the United States and in other countries by herbalists and may be mixed on site. More commonly, people get their herbs in pill, capsule, or liquid form from health food stores, grocery stores, large discount stores, or through mail order or online suppliers.

Although herbs are the latest craze in the United States, the Chinese and other Asian peoples have been using them for thousands of years. In China, the history of herbal remedies goes back 5000 years. Unbeknownst to many doctors, certain herbs were used routinely in the United States and other developed nations before the advent of modern medicines. In the United States, for instance, echinacea (ek-eh-NAH-shah) was widely used to treat colds prior to the introduction of sulfa drugs (the first generation of synthetic antibiotics). In Germany, herbal remedies are considered prescription medicines and are covered by health insurance.

Herbal treatments have a long history of use. In addition, the plants they are made from have been the source of a surprisingly large proportion of the prescription and over-the-counter (OTC) drugs in use today. That is, many so-called "modern medicines" contain active ingredients originally extracted from plants. Aspirin, for instance, is derived from the inner bark of willow trees. But are herbal treatments effective?

There is no blanket answer to this question. While research shows that many claims about the effectiveness of numerous common herbal preparations appear to be unfounded, not all are. Echinacea, for instance, does appear to enhance immune system function, and it can be used as a preventative and for treatment of colds and flus. Bilberry, dried blueberries, helps maintain eyesight and improves night vision. It was even used in World War II by members of the Royal Air Force who flew night bombing missions. Valerian root is an effective, nonaddicting sleep aid and mild relaxant. Saw palmetto appears to be effective in reducing enlargement of the prostate. Ginger is effective in the treatment of nausea and morn-

---

is a layer of connective tissue. It holds tissues together. Beneath that is a thick sheet of smooth muscle cells, which forms the bulk of the stomach wall. When they contract, these muscles help to mix the contents of the stomach. Smooth muscle cells are also found in blood vessels supplying the tissues of the stomach. Nerves enter with the blood vessels and control the flow of blood.

Epithelium

> Epithelium forms the lining or external covering of organs and also forms glands.

Epithelial tissue exists in two basic forms: glandular and membranous. The membranous epithelia consist of sheets of cells tightly packed together, forming the external coverings or internal linings of organs. Membranous epithelia come in a variety of forms, each specialized to protect underlying tissues. To simplify matters, the membranous epithelia are divided into two broad categories: simple epithelia, consisting of a single layer of cells, and stratified epithelia, consisting of many layers of cells (Figure 4-2). Epithelial layers are often underlain by a basement membrane, a layer of glycoprotein that fuses with the underlying connective tissue, holding the epithelium in place.

The glandular epithelia consists of clumps of cells that form many of the glands of the body. Epithelial glands arise during embryonic development from tiny "ingrowths" of membranous epithelia, as illustrated in Figure 4-3. Some glands remain connected to the epithelium by hollow ducts and are called exocrine glands (EX-oh-crin; glands of external secretion); products of

TABLE 4-1	The Primary Tissues and Their Subtypes	

Epithelial tissue
  Membranous
  Glandular
	Connective tissue
	  Connective tissue proper
	    Loose connective tissue
	    Dense connective tissue
	  Specialized connective tissue
	    Blood
	    Bone
	    Cartilage

Muscle tissue
  Cardiac
  Skeletal
  Smooth	

Nervous tissue
  Conductive
  Supportive	

the exocrine glands flow through ducts into some other body part (Figure 4-3a). In humans, sweat glands in the skin are exocrine glands. They produce a clear, watery fluid that is released onto the surface of the skin by small ducts. This fluid evaporates from the skin and cools the body.

Some glandular epithelial cells break off completely from their embryonic source, as shown in Figure 4-3b, to form endocrine glands (EN-doh-crin; glands of internal secretion). The endocrine glands produce hormones that are released into the bloodstream, where they travel to other parts of the body (Chapter 13).

ing sickness, although it should not be used for an extended time during pregnancy. Garlic lowers serum cholesterol and may have antibacterial properties. Feverfew seems to help prevent migraine headaches. Hawthorn fights hypertension. The list goes on.

The main advantage of herbs, say some proponents, is that they have no side effects and contain no chemicals; they are also purportedly safe and effective. Unfortunately, this claim is far from true. Herbal remedies contain plant chemicals, some of which may have very serious side effects. A number of herbs, for instance, damage the liver, including comfrey—an herb that reportedly promotes bone healing. When taken for long periods, however, its active chemical ingredient is toxic to the liver.

David Kroll, a pharmacologist from the University of Colorado Health Sciences Center, notes further problems with herbal medicine. First and foremost, he argues that U.S. labeling requirements are inadequate. Herbs are sold as dietary supplements, and as such, no claims about their use are available. Individuals must often rely on advice from sales clerks, friends, or articles they read. Second, very little is known about the long-term effects of the wide variety of herbal remedies or their potential interactions with conventional medicines, although our knowledge is expanding.

Others note that without federal regulations that require strict standards for drug content in various herbal remedies, individuals can't know the exact dose they are receiving. Studies have shown that some herbal remedies have none of the active ingredient or smaller than required amounts. In Europe, however, stricter laws require much better controls on product content. Canada has also established standards for the purity and content of herbal remedies. Another bit of advice from Dr. Kroll. When given a choice between an over-the-counter remedy and an herbal one, he favors the over-the-counter treatment because they are standardized. He also notes that treatment with herbal remedies, as with OTC preparations, may delay the diagnosis of more serious conditions. The rule: See a doctor first to be certain that you have what you think you have and buy herbal products that are standardized from reputable companies.

The important lesson in all of this is buyer beware. Study an herb thoroughly before you take it, or better yet, see a qualified physician who practices both traditional and herbal medicine.

www.jbpub.com/humanbiology/5e

Visit Human Biology's Internet site for links to web sites offering more information on this topic.

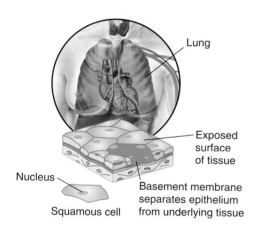

Lung

Exposed surface of tissue

Nucleus

Basement membrane separates epithelium from underlying tissue

Squamous cell

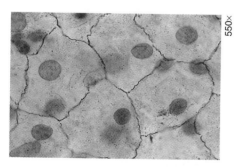

550×

Simple squamous epithelium ❶
Lining of blood vessels, air sacs of lungs, kidney tubules, and lining of body cavities

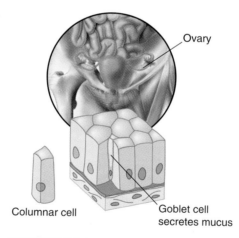

Ovary

Columnar cell

Goblet cell secretes mucus

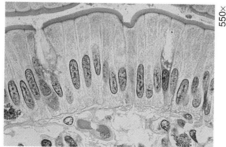

550×

Columnar epithelium ❷
Lining of the digestive tract and upper part of the respiratory tracts, auditory and uterine tubes

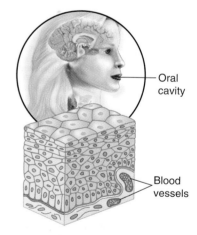

Oral cavity

Blood vessels

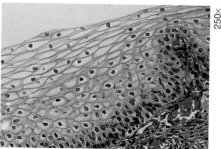

250×

Stratified squamous epithelium ❸
Skin, mouth, and throat lining; vaginal lining; anal lining; and cornea

(a)
(b)

FIGURE 4-2 **Membranous Epithelia** (a) Single-celled (simple) epithelia (1 and 2 above) and (b) stratified epithelia exist in different parts of the body.

FIGURE 4-3 **Formation of Endocrine and Exocrine Glands** (a) Exocrine glands arise from invaginations of membranous epithelia that retain their connection. (b) Endocrine glands lose this connection and thus secrete their products into the bloodstream.

The Marriage of Structure and Function

Epithelial tissues illustrate a basic biological principle: that structure closely correlates with function.

One of the basic rules of architecture is that form (the structure of a building) often follows function—in other words, architectural design reflects underlying function. This rule also applies to living things. As you study human biology, you will find a remarkable degree of correlation between structure and function in cells, tissues, and organs.

The membranous epithelia provide a good example of this phenomenon. Consider the outer layer of the human skin, the **epidermis** (ep-eh-DERM-iss). Shown in Figure 4-4a, the epider-

FIGURE 4-4 **Comparison of Two Epithelia with Different Functions** (a) A cross section of the skin showing the stratified squamous epithelium of the epidermis (above), which protects underlying skin from sunlight and desiccation. (b) The simple columnar epithelium of the lining of the small intestine, which is specialized for absorption. (c) The plasma membranes of the cells lining the intestine are thrown into folds (microvilli) that greatly increase the surface area for absorption.

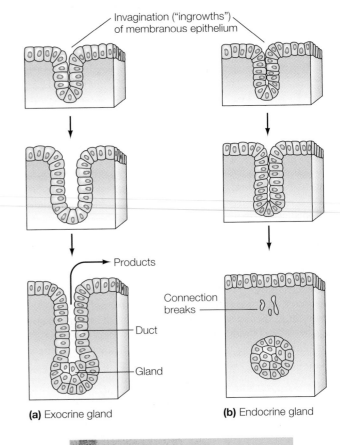

Invagination ("ingrowths") of membranous epithelium

Products

Connection breaks

Duct

Gland

(a) Exocrine gland

(b) Endocrine gland

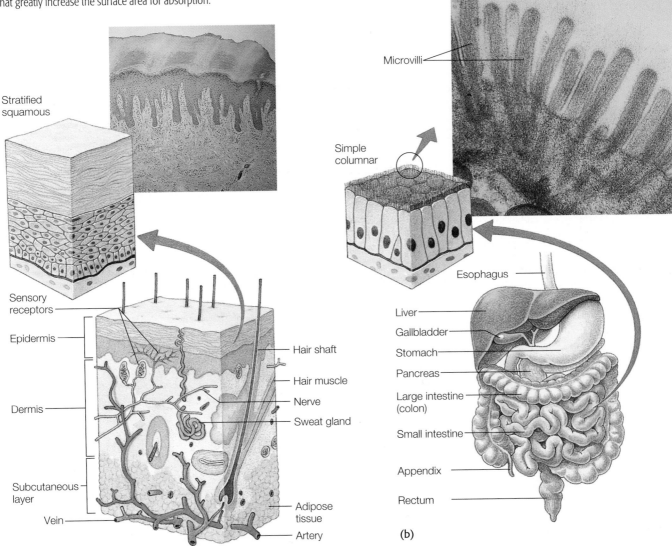

Stratified squamous

Microvilli

Simple columnar

Esophagus

Sensory receptors

Epidermis

Hair shaft

Hair muscle

Nerve

Dermis

Sweat gland

Liver

Gallbladder

Stomach

Pancreas

Large intestine (colon)

Small intestine

Subcutaneous layer

Appendix

Vein

Rectum

Adipose tissue

Artery

(a)

(b)

mis consists of numerous cell layers, known as a **stratified squamous epithelium** (SQUAW-mus). The cells of this epithelium flatten toward the surface and die. They are tightly joined by special connections. Together, the thickness of the epidermis, the adhesion of one cell to another, and the dry protective layer of dead cells reduce water loss and protect us from dehydration. They also present a formidable barrier to microorganisms. All of these structural features help maintain homeostasis.

Connective Tissue

Connective tissue binds the cells and organs of the body together.

As the name implies, **connective tissue** is the body's glue. It binds cells and other tissues together and is present in all organs in varying amounts.

The body contains several types of connective tissue, each with specific functions (Figure 4-5). Despite the differences, all connective tissues consist of two basic components: cells and varying amounts of extracellular material. Two types of connective tissue will be discussed here: connective tissue proper and specialized connective tissues—bone, cartilage, and blood.

Connective Tissue Proper

Connective tissue proper is an important structural component of organs.

Connective tissue proper consists of two types: dense connective tissue and loose connective tissue. The chief difference between them lies in the ratio of cells to extracellular fibers.

As the name implies, **dense connective tissue (DCT)** consists primarily of densely packed fibers. They are produced by

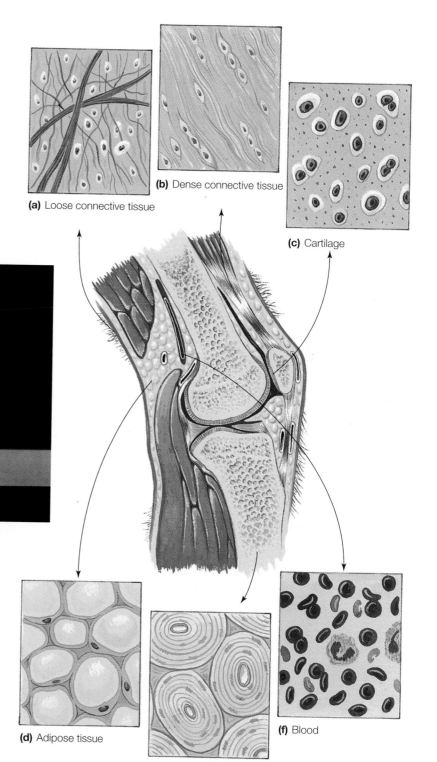

(a) Loose connective tissue

(b) Dense connective tissue

(c) Cartilage

(d) Adipose tissue

(e) Bone

(f) Blood

FIGURE 4-5 **Connective Tissue** Connective tissue consists of many diverse subtypes.

cells interspersed between the fibers. DCT is found in ligaments and tendons (Figure 4-6a). Ligaments join bones to bones at joints and provide support for joints. Tendons join muscle to bone and aid in body movement.

DCT is also found at other sites, such as the layer of the skin underlying the epidermis, known as the dermis. Although this is DCT, the fibers are less regularly arranged than those in ligaments and tendons.

As shown in Figure 4-6b, **loose connective tissue** contains cells in a loose network of collagen and elastic fibers. These fibers are made of protein. Loose connective tissue forms around blood vessels in the body and in skeletal muscles, where it binds the muscle cells together. It also lies beneath epithelial linings of the intestines and trachea, anchoring them to underlying structures.

The extracellular fibers found in dense and loose connective tissue are produced by a connective tissue cell known as the **fibroblast** (FIE-bro-blast). Fibroblasts also repair damage created by cuts or tears in body tissues. When skin is cut, for example, fibroblasts in the dermis migrate into the injured area and begin producing collagen fibers to fill the wound. This biological bandage may be overgrown entirely by the epidermis, leaving no scar if the wound is small. In larger wounds, however, the epidermal cells may be unable to cover the entire wound, leaving some of the underlying collagen exposed and producing a visible scar.

Besides binding tissues together, loose connective tissue houses several cell types that protect us against bacterial and viral infections. One of the most important of these protective cells is the **macrophage** (MACK-row-FAYGE; "big eater"). Containing numerous lysosomes for digesting foreign material, macrophages engulf (phagocytize) microorganisms that penetrate the skin and underlying loose connective tissues after an injury. This prevents bacteria from spreading to other parts of the body. Macrophages also play a role in immune protection, discussed in Chapter 14.

Some LCT contains **fat cells**. The fat cell is one of the most distinctive of all body cells. This large, conspicuous cell contains huge fat globules that occupy virtually the entire cell, pressing the cytoplasm and the nucleus to the periphery. Fat cells occur singly or in groups of varying size. Large numbers of fat cells in a given region form a modified type of LCT known as **adipose tissue** (AD-eh-poze), or, less glamorously, fat. Adipose tissue is an important storage depot for lipids, particularly triglycerides, which are used as an energy source. Fatty deposits also provide insulation for humans, many other mammals, and some birds, especially those that live in aquatic environments where heat loss can be substantial.

For humans, fatty deposits often become unsightly. To rid the body of these deposits, individuals should exercise and reduce food intake. Some individuals opt for a surgical measure called liposuction (LIE-poh-suck-shun) (Figure 4-7). In this procedure, a small incision is made in the skin through which surgeons insert a device to suck out fat deposits under the skin in various locations, such as the buttocks and abdomen. The fat cells extracted from one region can even be transferred to other regions, such as the breast, to resculpt the human body. Liposuction is a relatively safe technique but, like most surgery, not free from risk.

Specialized Connective Tissue

Specialized connective tissues perform specific functions essential to homeostasis.

The body contains three types of specialized connective tissue: cartilage, bone, and blood.

Cartilage. Cartilage consists of cells embedded in an abundant and rather impenetrable extracellular material, the matrix (Figure 4-8). Surrounding virtually all types of cartilage is a layer of dense, irregularly packed connective tissue. This layer contains the blood vessels that supply nutrients to cartilage cells through diffusion. No blood vessels penetrate the cartilage itself. Be-

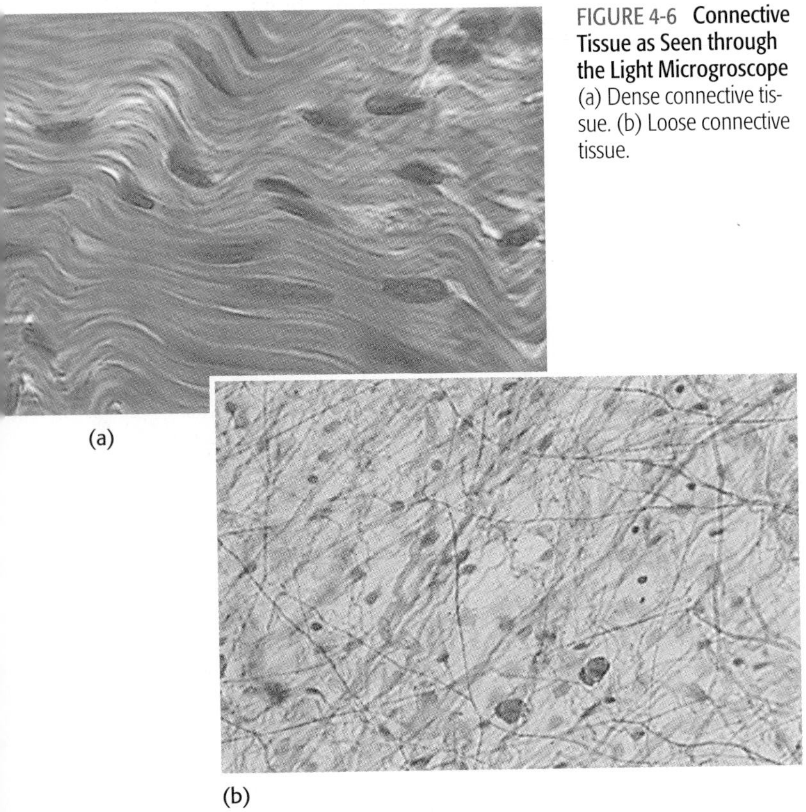

FIGURE 4-6 Connective Tissue as Seen through the Light Microgroscope (a) Dense connective tissue. (b) Loose connective tissue.

(a)

(b)

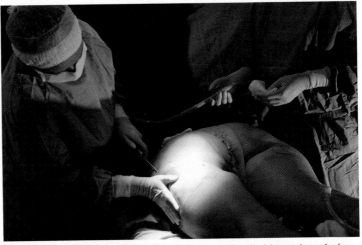

FIGURE 4-7 **Liposuction** During liposuction surgery, the physician aspirates fat from deposits lying beneath the skin, helping to reduce unsightly accumulations.

Matrix

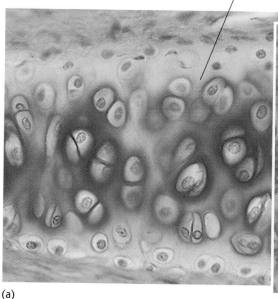

(a)

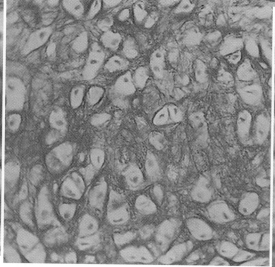

(b)

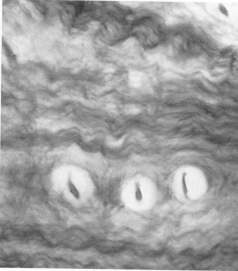

(c)

FIGURE 4-8 **Cartilage as Seen through the Light Microscope** (a) Hyaline. (b) Elastic. (c) Fibrocartilage. (d) Intervertebral disk showing location and (e) arrangement of fibrocartilage in a protective ring around the soft, spongy part of the disk that absorbs shock.

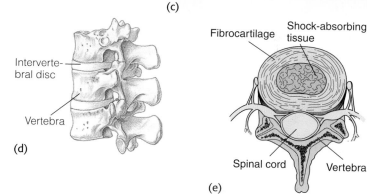

(d)

(e)

cause cartilage cells are nourished by diffusion from these blood vessels (capillaries), damaged cartilage heals very slowly. Therefore, joint injuries that involve the cartilage often take years to repair or may not heal at all.

Three types of cartilage are found in humans: hyaline, elastic, and fibrocartilage. The most prevalent type is **hyaline cartilage** (HIGH-ah-lynn) (Figure 4-8a). Hyaline cartilage contains numerous collagen fibers, which appear white to the naked eye. Found on the ends of many bones in joints, hyaline cartilage greatly reduces friction, so bones can move over one another with ease. Hyaline cartilage also makes up the bulk of the nose and is found in the larynx (voice box) and the trachea or wind pipe. The ends of the ribs that join to the sternum (breastbone) are composed of hyaline cartilage. In embryonic development, the first skeleton is made of hyaline cartilage. It is later converted to bone.

Elastic cartilage contains many wavy elastic fibers, which give it flexibility (Figure 4-8b). Elastic cartilage is therefore found in regions where support and flexibility are required—for example, in the ears.

Fibrocartilage is a rare form of cartilage. It contains strong collagen fibers and is found in areas that must withstand tension and pressure (Figure 4-8c). Fibrocartilage forms the outer layer of the shock-absorbing tissue known as the **intervertebral disks** (in-ter-VER-tah-braul) that lies between the bones of the spine, the vertebrae (Figure 4-8d). An intervertebral disk consists of a soft, cushiony central region that absorbs shock. Fibrocartilage forms a ring around the central portion of the disk, holding it in place (Figure 4-8E).

Over time, the fibrocartilage ring weakens and may tear, permitting the central part of the disk to bulge outward. This condition is referred to as a slipped or herniated disk and usually occurs in the neck or lower back, resulting in a significant amount of pain in the neck, back, or one or both legs, depending on the location of the damaged disk. Pain is generated when the disk presses against nearby spinal nerves.

This problem can be corrected by surgery, although physical therapy is often the first plan of attack. Lasers are also be-

ing used to remove herniated disks. In some cases, the entire disk is removed and the vertebrae are then fused. You can reduce your chances of "slipping" a disk in the first place by watching your weight, keeping in shape, sitting upright (not slouching) in a chair, and lifting heavy objects carefully (Figure 4-9).

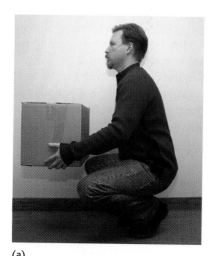

(a)

(b)

FIGURE 4-9 **Protecting Your Back** Bend your knees as the man in (a) is doing, not your back, when you're lifting. (b) With your back straight, stand up and lift with your legs.

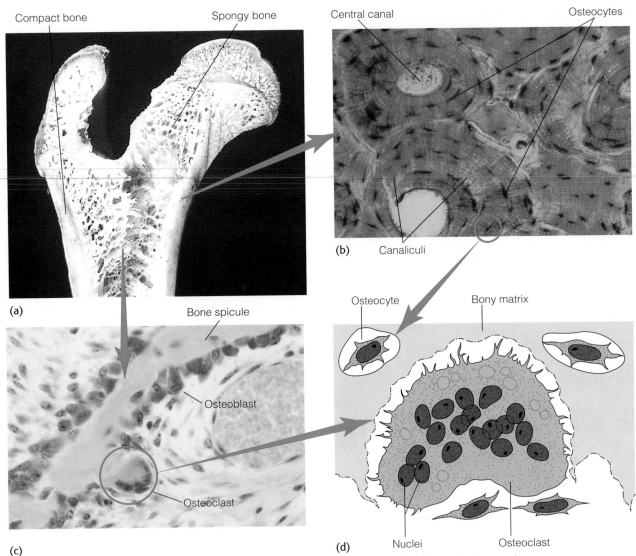

FIGURE 4-10 **Bone** (a) Compact and spongy bone, shown in a section of the humerus. (b) Light micrograph of compact bone.
(c) Photomicrograph of spongy bone, showing osteoblasts and osteoclasts. (d) Osteoclast digesting surface of bony spicule.

Bone.   Bone is a form of specialized connective tissue that provides internal support as well as protection to internal organs such as the brain. Bone also plays an important role in maintaining blood calcium levels and is, therefore, a homeostatic organ. Calcium is required for many body functions, including muscle contraction and normal nerve functioning.

Like all connective tissues, bone consists of cells embedded in an abundant extracellular matrix (Figure 4-10b). Bone matrix consists primarily of collagen fibers (which give bone its strength and resiliency), interspersed with numerous needlelike salt crystals containing calcium, phosphate, and hydroxide ions. They give bone its hardness.

Two types of bone tissue are found in the body: compact bone and spongy bone (Figure 4-10a). **Compact bone** is, as its name implies, dense and hard. As illustrated in Figure 4-10b, the cells in compact bone (osteocytes) are located in concentric rings of calcified matrix. These surround a central canal through which the blood vessels and nerves pass. Each osteocyte has numerous processes that course through tiny canals in the bony matrix, known as *canaliculi* (CAN-al-ICK-u-LIE; literally "little canals"). The

canaliculi provide a route for nutrients and wastes to flow to and from the bone cells.

Inside most bones of the body is a tissue known as *spongy bone*. **Spongy bone** consists of an irregular network of calcified collagen plates (spicules). As shown in Figure 4-10c, on the surface of the plates are numerous **osteoblasts** (OSS-tee-oh-BLASTS). Osteoblasts are a type of bone cell that produces the collagen, which later becomes calcified. Once these cells are surrounded by calcified matrix, they are referred to as osteocytes.

Blood.   Blood is another specialized form of connective tissue—although some biologists prefer to consider it a separate tissue type called *vascular tissue*. Blood consists of formed elements and a large amount of extracellular material. The formed elements of blood consist of the red and white blood cells and the platelets. They are responsible for 45% of the blood volume. The extracellular material is a fluid known as plasma. It makes up the remaining 55%.

Red blood cells transport oxygen and small amounts of carbon dioxide to and from the lungs and body tissues. **White**

FIGURE 4-11 **Three Types of Muscle as Seen through the Light Microscope** (a) Skeletal. (b) Cardiac. (c) Smooth.

blood cells rid the body of foreign organisms such as viruses and bacteria that can cause infections. **Platelets** are fragments of large cells (megakaryocytes) located in the red bone marrow, the principal site of blood cell formation. Platelets play a key role in blood clotting.

Muscle Tissue

Muscle tissue consists of specialized cells that contract when stimulated.

Muscle tissue is found in virtually every organ in the body. Muscle gets its name from the Latin word for "mouse" (mus). Early observers likened the contracting muscle of the biceps to a mouse moving under a carpet.

Muscle is an excitable tissue. When stimulated, it contracts, producing mechanical force. Muscle cells working in large numbers can create enormous forces. Muscles of the jaw, for instance, create a pressure of 200 pounds per square inch, forceful enough to snap off a finger. (Don't try this at home.) Muscle also moves body parts, propels food along the digestive tract, and expels the fetus from the uterus during birth. Heart muscle contracts and pumps blood through the 50,000 miles of blood vessels in the human body. Acting in smaller numbers, muscle cells are responsible for intricate movements, such as those required to play the piano or move the eyes.

Three types of muscle tissue are found in humans: skeletal, cardiac, and smooth. The cells in each type of muscle tissue contain two contractile protein filaments, actin and myosin. These very same fibers were first encountered in your study of the microfilamentous network lying beneath the plasma membrane of cells. This layer of fibers plays a key role in cell division discussed in Chapter 16. When stimulated, actin and myosin filaments of muscle cells slide over one another, shortening and causing contraction.

Skeletal Muscle. The majority of the body's muscle is called **skeletal muscle**—so named because it is frequently attached to the skeleton. When skeletal muscle contracts, it causes body parts (arms and legs, for instance) to move. Most skeletal muscle in the body is under voluntary, or conscious, control. Signals from the brain cause the muscle to contract. A notable exception is the skeletal muscle of the upper esophagus, which contracts automatically during swallowing.

Skeletal muscle

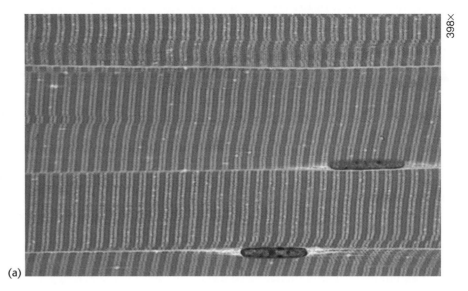

398×

(a)

Cardiac muscle

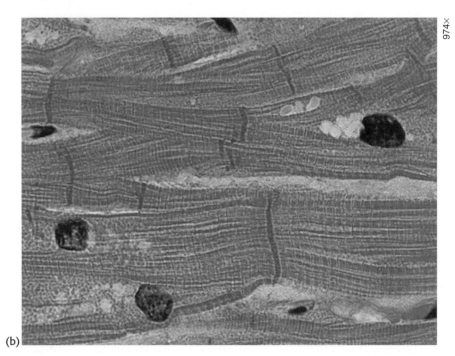

974×

(b)

Smooth muscle

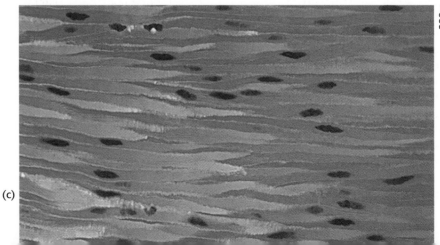

398×

(c)

Skeletal muscle cells are long cylinders formed during embryonic development by the fusion of many embryonic muscle cells. Because of this, skeletal muscle cells are usually referred to as muscle fibers. Each muscle fiber contains many nuclei (Figure 4-11a). Because of the dense array of contractile fibers in the cytoplasm of the muscle fiber, the nuclei become pressed against the plasma membrane. As a result, muscle fibers cannot divide, and damaged muscle cells cannot be replaced.

Skeletal muscle fibers appear banded, or striated, when viewed under the light microscope (Figure 4-11a). The striations result from the unique arrangement of actin and myosin filaments inside muscle cells, discussed in Chapter 12.

Cardiac Muscle.   Like skeletal muscle, **cardiac muscle** is banded (Figure 4-11b). Unlike skeletal muscle, cardiac muscle is involuntary; that is, it contracts without conscious control. Found only in the walls of the heart, each cardiac muscle cell contains a single nucleus. Cardiac muscle cells also branch and interconnect freely, and individual cells are tightly connected to one another. This adaptation helps maintain the structural integrity of the heart, an organ subject to incredible strain as it pumps blood through the body day and night. The points of connection also provide pathways for electrical impulses to travel from cell to cell, allowing the heart muscle to contract uniformly when stimulated.

Smooth Muscle.   Smooth muscle, so named because it lacks visible striations, is involuntary. Actin and myosin filaments are present in smooth muscle cells but are not organized like those found in other types (Figure 4-11c). Smooth muscle cells may occur singly or in small groups. Small rings of smooth muscle cells, for example, surround tiny blood vessels. When these cells contract, they reduce the supply of blood to tissues. Smooth muscle cells are most often arranged in sheets in the walls of organs such as the stomach, uterus, and intestines (Figure 4-1). Smooth muscle cells in the wall of the stomach churn the food, mixing the stomach contents, and force tiny spurts of liquified food into the small intestine. Smooth muscle contractions also propel the food along the intestinal tract.

Nervous Tissue

Nervous tissue contains specialized cells that conduct impulses.

Last but not least of the primary tissues is nervous tissue, which consists of two types of cells: conducting cells or neurons and nonconductive cells or neuroglia (literally, nerve glue). Together, the neurons and the neuroglial cells form the brain, spinal cord, and nerves, described in Chapter 12.

The conducting cells or **neurons** are modified to respond to specific stimuli, such as pain or temperature. Stimulation results in nerve impulses, which the neuron transmits from one region of the body to another.

The nonconductive cells of the nervous system are called **neuroglia** and were long thought to be merely a kind of nerv-

ous system connective tissue. They are by far the most prominent type of cell in the brain and spinal cord, outnumbering nerve cells by about 9 to 1. Nonconducting cells provide physical support for nerve cells. They also perform other functions. One type of neuroglial cell, for instance, engulfs bacteria, helping reduce brain infections. Another type of neuroglial cell transports nutrients from blood vessels to neurons. This cell also produces a hormone that stimulates regrowth of nerve cells. Researchers hope that one day this hormone can be used to treat diseases of the nervous system resulting from the degeneration of neurons, such as Parkinson's disease. The hormone-producing

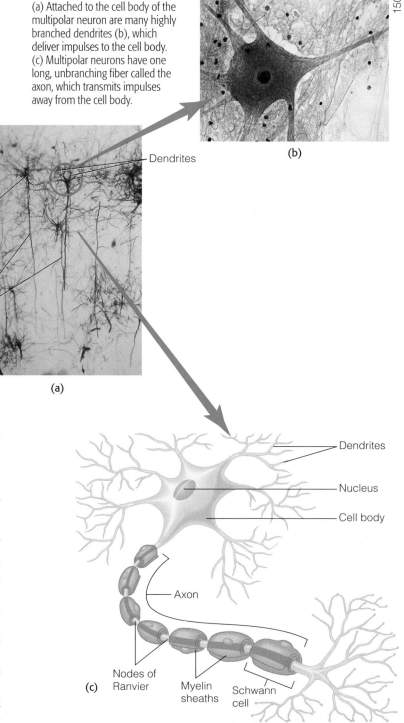

FIGURE 4-12 Multipolar Neurons (a) Attached to the cell body of the multipolar neuron are many highly branched dendrites (b), which deliver impulses to the cell body. (c) Multipolar neurons have one long, unbranching fiber called the axon, which transmits impulses away from the cell body.

150×

Dendrites

Cell body

Axons

(a)

(b)

Dendrites

Nucleus

Cell body

Axon

Nodes of Ranvier

Myelin sheaths

Schwann cell

(c)

nerve cell also guards against toxins by creating a barrier to many potentially harmful substances. Although neuroglial cells have long been considered nonconductive, new research suggests that they actually do communicate among one another and with nerve cells.

Neurons. At least three distinct types of neurons are found in the body. Despite obvious structural differences, they share several common features. We will study these similarities by looking at one of the most common nerve cells, the multipolar neuron, shown in Figure 4-12. The multipolar neuron contains a prominent cell body to which are attached several short, highly branched processes, known as dendrites (DEN-drites). (It is the presence of these processes that gives this neuron its name, multipolar.) The dendrites receive impulses from receptors or other neurons and transmit them to the cell body. Also attached to the cell body of the multipolar neuron is a large, fairly thick process, the axon (AXE-on), which transports bioelectric impulses away from the nerve cell body.

Nervous tissue plays an important role in body function. It monitors internal and external conditions through various sensors called *receptors* and causes changes that ensure our survival and well-being. Input from various sensors is fed into the brain and spinal cord via nerves. Incoming information is then compiled and appropriate responses are mounted. Some may be automatic responses, ones we're not consciously aware of. Others involve conscious responses. Both conscious and unconscious responses help maintain homeostasis.

Organs and Organ Systems

Tissues combine to form organs.

The cell contains organelles ("little organs") that carry out many of its functions in isolation from the biochemically active cytoplasm. As pointed out in Chapter 3, compartmentalization such as this is an important evolutionary adaptation. Compartmentalization also occurs at the organismic level in organs.

Organs are discrete structures that have evolved to perform specific functions, such as digestion, enzyme production, and hormone production. Most organs, however, do not function alone. Instead, they are part of a group of cooperative organs, called an **organ system**. The brain, spinal cord, and nerves are all organs that belong to an organ system known as the *nervous system*.

As you will see in upcoming chapters, components of an organ system are sometimes physically connected—as in the digestive system (Figure 4-13a). In other cases, they are dispersed throughout the body—as in the endocrine system (Figure 4-13b). In addition, some organs belong to more than one system. The pancreas, for example, produces digestive enzymes used to break down food in the small intestine. The pancreas is therefore part of the digestive system. However, some cells of the pancreas produce hormones that are released into the bloodstream where they circulate to body tissues and help control blood glucose levels. Thus, the pancreas also belongs to the endocrine system.

Many of the following chapters describe the organ systems of the human body and the functions they perform, paying special attention to their role in homeostasis. Figure 4-14 summarizes the functions of each organ system and lists the role each plays. Take a moment to read the descriptions in the blue boxes before proceeding.

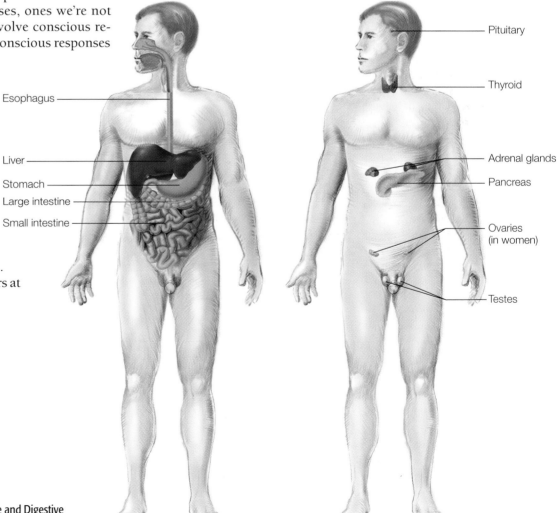

Esophagus

Liver

Stomach

Large intestine

Small intestine

Pituitary

Thyroid

Adrenal glands

Pancreas

Ovaries (in women)

Testes

FIGURE 4-13 **Comparison of the Endocrine and Digestive Systems** (a) The digestive system. (b) The endocrine system.

FIGURE 4-14 Role of the Body Systems in Maintaining Homeostasis

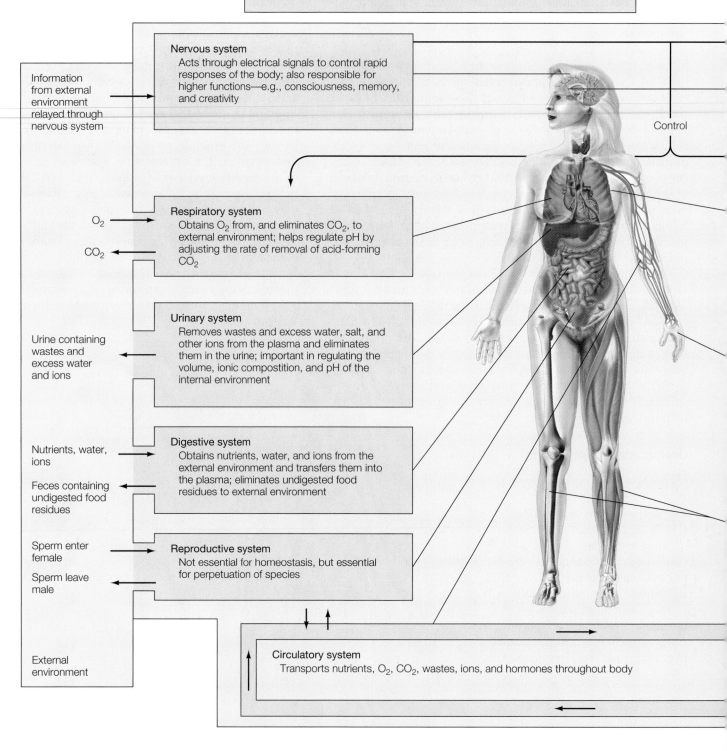

BODY SYSTEMS
Made up of cells organized by specialization to maintain homeostasis

Nervous system
Acts through electrical signals to control rapid responses of the body; also responsible for higher functions—e.g., consciousness, memory, and creativity

Information from external environment relayed through nervous system

Control

O_2

Respiratory system
Obtains O_2 from, and eliminates CO_2, to external environment; helps regulate pH by adjusting the rate of removal of acid-forming CO_2

CO_2

Urine containing wastes and excess water and ions

Urinary system
Removes wastes and excess water, salt, and other ions from the plasma and eliminates them in the urine; important in regulating the volume, ionic compostition, and pH of the internal environment

Nutrients, water, ions

Digestive system
Obtains nutrients, water, and ions from the external environment and transfers them into the plasma; eliminates undigested food residues to external environment

Feces containing undigested food residues

Sperm enter female

Reproductive system
Not essential for homeostasis, but essential for perpetuation of species

Sperm leave male

External environment

Circulatory system
Transports nutrients, O_2, CO_2, wastes, ions, and hormones throughout body

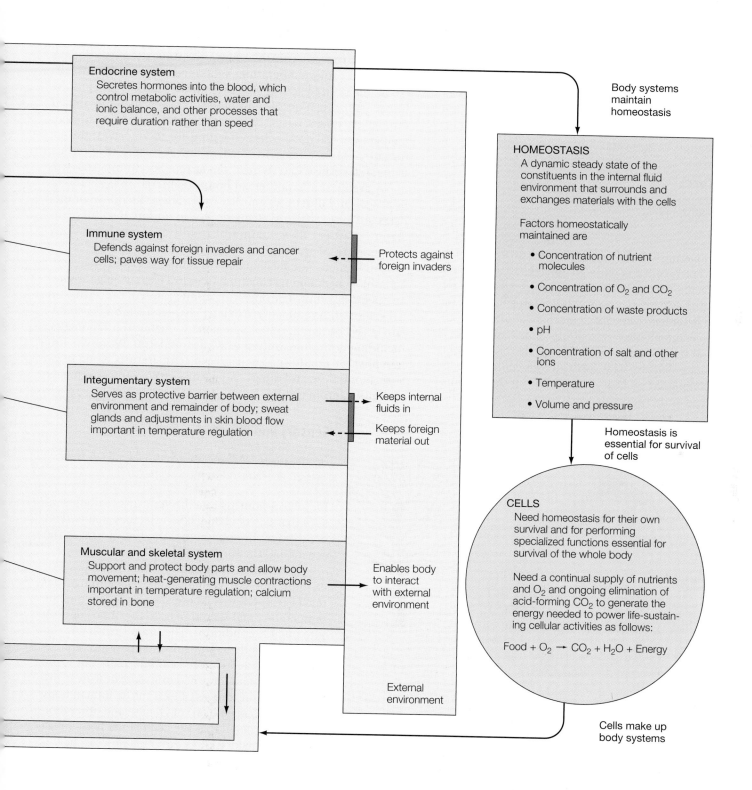

Endocrine system
Secretes hormones into the blood, which control metabolic activities, water and ionic balance, and other processes that require duration rather than speed

Immune system
Defends against foreign invaders and cancer cells; paves way for tissue repair

Protects against foreign invaders

Integumentary system
Serves as protective barrier between external environment and remainder of body; sweat glands and adjustments in skin blood flow important in temperature regulation

Keeps internal fluids in

Keeps foreign material out

Muscular and skeletal system
Support and protect body parts and allow body movement; heat-generating muscle contractions important in temperature regulation; calcium stored in bone

Enables body to interact with external environment

External environment

Body systems maintain homeostasis

HOMEOSTASIS
A dynamic steady state of the constituents in the internal fluid environment that surrounds and exchanges materials with the cells

Factors homeostatically maintained are

- Concentration of nutrient molecules
- Concentration of O_2 and CO_2
- Concentration of waste products
- pH
- Concentration of salt and other ions
- Temperature
- Volume and pressure

Homeostasis is essential for survival of cells

CELLS
Need homeostasis for their own survival and for performing specialized functions essential for survival of the whole body

Need a continual supply of nutrients and O_2 and ongoing elimination of acid-forming CO_2 to generate the energy needed to power life-sustaining cellular activities as follows:

Food + O_2 → CO_2 + H_2O + Energy

Cells make up body systems

4-2 Principles of Homeostasis

Organ systems perform many functions, virtually all of which are involved in homeostasis. Defined in Chapter 1 as a state of relative internal constancy, homeostasis occurs on a variety of levels—in cells, tissues, organs, organ systems, organisms, and even the environment. Homeostatic systems at all levels of biological organization have several common features.

Negative Feedback

> Homeostatic systems operate primarily through negative feedback mechanisms.

Feedback mechanisms control various body processes. The most common type is called **negative feedback**. To understand how negative feedback systems work, consider a familiar example, the heating system of a house. It contains many of the features present in biological feedback mechanisms (Figure 4-15a).

In the winter, the heating system of a house maintains a constant internal temperature, even though the outside temperature fluctuates sharply. Heat lost through ceilings, walls, and windows is replaced by heat generated from the combustion of natural gas or oil in the furnace (Figure 4-15a).

The furnace is controlled by a thermostat, which detects changes in room temperature. When indoor temperature falls below the desired setting, the thermostat sends a signal to the furnace, turning it on. Heat from the furnace is then distributed through the house, raising the room temperature. When the room temperature reaches the desired setting, the thermostat shuts the furnace off. Like all negative feedback mechanisms, the product of the system (heat) "feeds back" on the process, shutting it down.

A graph of a hypothetical house temperature is shown in Figure 4-15b and illustrates another important principle of homeostasis: Homeostatic systems do not maintain absolute constancy. Rather, they maintain conditions (such as body temperature) within a given range.

All homeostatic mechanisms in humans operate in a similar fashion, maintaining conditions within a narrow range around an operating point. The operating point is akin to the setting on a thermostat. As you will see in later chapters, our bodies maintain fairly constant levels of a great many chemical components, including hormones, nutrients, wastes, and ions. We also maintain physical conditions such as body temperature, blood pressure, blood flow, and others. All of this is done subconsciously and quite automatically.

Sensors and Effectors

> Homeostatic feedback mechanisms require a sensor and an effector.

Biological homeostatic mechanisms contain **sensors** that detect change and **effectors** that correct conditions. In your home, the thermostat is the sensor and the furnace is the effector.

The human body contains many sensors. For example, the skin contains nerve cell endings that detect temperature. These sensors detect outside temperature and send signals to the brain, alerting it to outside conditions. The brain then sends signals to the body to adjust for temperature, for example, to generate more heat internally if it is too cold. The options for generating more heat are many.

One of the main sources of heat in the human body is the breakdown (catabolism) of glucose and other molecules. Chapter 3 noted that the cells of the body break down glucose to make ATP. During this process, heat is given off. Each cell, then, is a tiny furnace whose heat radiates outward and is distributed throughout the body by the blood. Unlike the furnace in your home, the cellular "furnaces" cannot be turned up very quickly. They respond much more slowly than a furnace and are part of a delayed response to low temperature. As winter progresses, metabolism increases and the body produces more heat.

In order to respond to sudden changes in outside temperature, the body must rely on more rapid mechanisms. If you walked outdoors on a cold winter night dressed

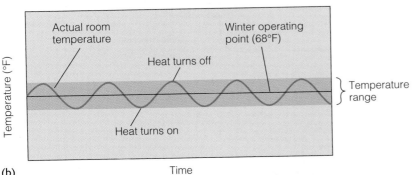

(a)

(b)

FIGURE 4-15 **Homeostasis in Our Homes** (a) Heat is maintained in a house by a furnace, which compensates for heat loss. The thermostat monitors the internal temperature and switches the furnace on and off in response to temperature changes. (b) A hypothetical temperature graph showing temperature fluctuation around the operating point.

only in a light sweater and blue jeans, for example, receptors in your skin would sense the cold and send signals to the brain. The brain, in turn, would send signals to blood vessels in the skin, causing them to constrict, reducing the flow of blood in the skin. The restriction of blood flow through the skin reduces heat loss. If it is cold enough outside, the brain may also send signals to the muscles, causing them to undergo rhythmic contractions, known as *shivering*. Shivering burns additional glucose, releasing extra heat. Many voluntary actions may also be "ordered" by the brain to reduce heat loss or generate more heat. For example, you might turn around and go back inside. In humans, conscious acts are often crucial components of homeostasis.

Upsetting Homeostasis

Homeostasis can be upset by changes in input and output.

Homeostasis can be thrown out of balance by many factors. On a hot day, for example, water escapes our bodies very rapidly through perspiration. Although perspiration helps cool us down, it can be quite damaging if water loss is severe. Severe water loss leads to dehydration, which may upset homeostasis—so much so, in fact, that death may occur.

Just as changes in the output of a substance drastically alter homeostasis, so do changes in input. For example, going without water for a prolonged period (approximately 3 days) can kill a human being. Excess input can also prove dangerous. Excess salt intake, for example, can result in high blood pressure in some people. Whatever the cause, imbalances in homeostasis can have dramatic impacts on human health.

Control of Homeostasis

Homeostatic control requires the action of nerves and hormones.

Correcting imbalances to maintain homeostasis requires a variety of body reflexes–that is, automatic responses triggered by various stimuli. Reflexes occur without conscious control. Two types of reflexes exist: nervous system and hormonal.

At any one time, the brain and spinal cord receive thousands of signals from receptors in the body. These signals alert

the nervous system to a variety of internal and external conditions. The brain and spinal cord then send out special responses, after integrating (making sense of) the various inputs they have received. The brain and spinal cord control the activity of many effectors, generally muscles and glands. You'll learn more about this in Chapter 10.

Hormones are chemical substances produced by the endocrine glands. Released into the blood, hormones travel to distant sites, where they effect some sort of change. Hormones don't just pour out of endocrine glands in a constant stream, however. They are released only when needed and then usually as part of a chemical reflex essential for maintaining homeostasis. You'll learn more about this in Chapter 13.

Nervous and endocrine feedback mechanisms generally occur over considerable distances. The body also possesses feedback mechanisms that operate over very short distances. They usually involve chemicals called *paracrines* (PEAR-ah-crins). Produced by individual cells, paracrines diffuse to neighboring cells where they elicit an effect. One example is epidermal growth factor (EGF), a paracrine produced by skin cells. It stimulates cell division when skin cells are damaged or lost, providing a degree of local control over cell growth.

One of the best-known paracrines is a group of chemical substances known as prostaglandins (PROSS-tah-GLAND-ins). Prostaglandins comprise a rather large group of molecules with diverse functions. Some stimulate blood clotting; others stimulate smooth muscle contraction.

The Link between Health and Homeostasis

Human health depends on homeostasis.

This book emphasizes a theme that is important to all of us— notably, that human health is dependent on homeostasis. Homeostasis, in turn, requires a healthy, clean environment and freedom from disease-causing organisms. The health of the environment, of course, is also dependent in part on properly functioning homeostatic systems in nature. These systems maintain conditions conducive to life.

The relationship between homeostasis and health is shown in Figure 4-16. Take a moment to study this diagram. Notice that

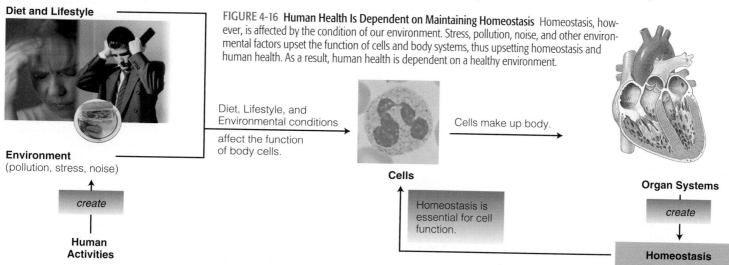

FIGURE 4-16 Human Health Is Dependent on Maintaining Homeostasis Homeostasis, however, is affected by the condition of our environment. Stress, pollution, noise, and other environmental factors upset the function of cells and body systems, thus upsetting homeostasis and human health. As a result, human health is dependent on a healthy environment.

Diet and Lifestyle

Environment
(pollution, stress, noise)

create

Human Activities

Diet, Lifestyle, and Environmental conditions affect the function of body cells.

Cells

Cells make up body.

Homeostasis is essential for cell function.

Organ Systems

create

Homeostasis

⁞⁞⁞ **healthnote**

4-2

Wide Awake at 3 A.M.: Causes and Cures of Insomnia

It's Thursday evening. The final exam in your human biology course is tomorrow. Forty percent of your grade depends on your performance. You've studied hard, but it's well past midnight and you can't get to sleep. It's happened before and you know your ability to think during the test, which is scheduled to start at 8 A.M., will be impaired. That puts more pressure on you, making it more difficult to sleep.

This frustrating problem, which affects almost everyone at some time in life, is just one of several sleep disorders that physicians call *insomnia* (in-SOM-knee-ah). Another common complaint of insomniacs is waking up in the middle of the night and not being able to fall asleep again for several hours, if at all. In other instances, individuals find that they wake up too early; still others complain of sleeping through the night but waking up feeling unrefreshed.

Insomnia is best defined as a condition of inadequate or poor quality sleep. It's a problem that affects many people. In a recent study in Canada, for example, two of every five people surveyed said they had trouble sleeping at least once a week. Insomnia affects both men and women but is more common in women and the elderly.

Insomnia may be transient (short term), intermittent (on and off), or chronic (persistent). Short-term and intermittent insomnia are generally of less concern than chronic insomnia, which can last for years. Chronic insomniacs complain of tiredness, a lack of energy, difficulty concentrating, and irritability. Insomnia can make people more susceptible to common illnesses, such as colds.

What causes insomnia? Perhaps the most common cause of insomnia is a change in one's daily routine. Starting a new semester, sleeping in a strange environment, moving into a new home, and anxiety over job interviews or tests are some possible causes. But these are generally short-lived stimuli.

Chronic insomnia is often related to more serious problems. Certain diseases that cause pain and nausea, for example, may affect sleep. Chronic anxiety, for example, caused by a project at work that may take five months to complete, with many deadlines along the way, can keep one from falling asleep.

Another very common cause of chronic insomnia is depression. Depression usually manifests itself in the wide-awake-at-3 A.M.-syndrome. A per-

the body systems (right side of diagram) maintain overall homeostasis, which is essential for proper cell function. Cells make up body systems. They also contain homeostatic mechanisms that are vital to their own function as well as to the overall economy of the organism. Notice also that environmental factors, such as pollution and disease organisms, affect the func-

tion of body systems and cells in ways that can upset the body's internal balance, thus altering human mental and physical health.

Just as human health is affected by the condition of the environment, so too is the health of many other species. For a discussion of herbal remedies that help restore health, see Health Note 4-1.

4-3 Biological Rhythms

The previous discussion may give the impression that homeostasis establishes an unwavering stability. In truth, some key bodily processes undergo rhythmic change.

Understanding Biological Rhythms

Physiological processes that fluctuate normally do so in predictable ways.

Body temperature varies during a 24-hour period by as much as 0.5°C. Blood pressure may change by as much as 20%, and the number of white blood cells, which fight infection, can vary by 50% during the day. Alertness also varies considerably. About 1:00 P.M. each day, for instance, most people go through a slump. For most of us, activity and alertness peak early in the evening, making this an excellent time to study. Daily cycles, such as these, are called **circadian rhythms** (sir-KADE-ee-an; "about a day"). Circadian rhythms are natural body rhythms linked to the 24-hour day-night cycle.

Many hormones follow daily rhythmic cycles. The male sex hormone testosterone, for example, follows a 24-hour cycle. The highest levels occur in the night, particularly during dream sleep, also known as REM sleep. Dream sleep occurs primarily in the early morning hours—the later the hour, the longer the periods of dream sleep (Figure 4-17).

Not all cycles occur over 24 hours, however. Some can be much longer. The menstrual cycle, for instance, is a recurring series of events in the reproductive functions of women that lasts, on average, 28 days. During the menstrual cycle, levels of the female sex hormone estrogen undergo dramatic but predictable shifts.

The important point is that although many chemical substances are held within a fairly narrow range by homeostatic mechanisms, others fluctuate widely in normal cycles. These cycles are part of the body's dynamic balance.

Control of Biological Rhythms

Internal biological rhythms are controlled by the brain.

Research suggests that the brain controls many internal rhythms. One region in particular, the suprachiasmatic nucleus (SUE-pra-ki-as-MAT-tick), is thought to play a major role in coordinating several key rhythms.

The suprachiasmatic nucleus (SCN) is a clump of nerve cells located at the base of the brain in a region called the *hy-*

son who is depressed may fall asleep without trouble but may wake up every evening at 2 or 3 A.M. and may stay awake for a few hours or through the entire night. Depression, which is discussed at length in **Health Note 11-1**, comes in a variety of shapes and sizes, too—and can last for many years.

What can be done to treat chronic insomnia? If you have a medical problem, treatment of that problem can be helpful. If you are feeling anxious all of the time, the stress-relieving techniques described in **Health Note 1-1** can be helpful. Depression can be treated by psychotherapy, which helps eliminate sources of the problem. Antidepressant drugs are often prescribed for insomniacs, too. It may take a few weeks, however, for depression to begin to subside. Besides these important steps, experts recommend the following do's and don't's for dealing with insomnia.

- Go to bed and get up at the same time each day.
- Maintain a comfortable temperature in your bedroom—not too hot or too cold.
- Sleep in a quiet, darkened room.
- Exercise regularly during the week to relieve stress. Exercise no later than the early afternoon.
- Engage in quiet tasks such as reading or listening to music in the hour or two before bedtime.
- Try a light bedtime snack with protein just before going to bed.

- Avoid caffeine, which is found in coffees, teas, and caffeinated sodas especially late in the day.
- Don't use alcohol to stimulate sleep.
- If you can't fall asleep within a half an hour, get out of bed, engage in some relaxing activity, and then go to bed when you feel tired.
- If you need a nap, take one early in the day, before 3 pm.

Sleeping pills can be used to treat insomnia, but they are mainly used to treat short-term insomnia that results from occasional stressors. Sleeping pills such as Ambien are usually not taken for more than two weeks because they can worsen insomnia. Ironically, several common prescription drugs used to treat insomnia-causing depression—including Prozac and Paxil—cause sleeplessness in about one-third of the patients taking them.

www.jbpub.com/humanbiology/5e

Visit Human Biology's Internet site for links to web sites offering more information on this topic.

pothalamus. The SCN is believed to regulate other control centers. As a result, the suprachiasmatic nucleus is often referred to as the "master clock." Like the automatic timer of a sprinkler system, the SCN ticks off the minutes, faithfully imposing its control on the body, turning body functions on and off.

Ultimate control of the SCN is thought to reside in a gland in the brain known as the *pineal gland* (PIE-knee-al). It secretes a hormone thought to keep the suprachiasmatic nucleus in sync with the 24-hour day-night cycle.

The study of biological rhythms has yielded some important information and insights. One practical application is a better understanding of jet lag—that drowsy, uncomfortable feeling people get from the disruption of sleeping patterns caused by long-distance airplane travel. Jet lag occurs when the body's biological clock is thrown out of sync with the day-night cycle of a traveler's new surroundings. A businesswoman who travels from Los Angeles to New York, for instance, may be wide awake at 10:00 P.M. New York time because her body is still on Los Angeles time—3 hours earlier. When the alarm goes off at 6:00 A.M., our weary traveler crawls out of bed exhausted, because as far as her body is concerned, it is 3:00 A.M. Los Angeles time.

To avoid the discomfort associated with travel, sleep researchers suggest that you abide by your internal clock, following "home time" when in a new time zone, or, barring that, try to reset your biological clock before you get there. You can do this by going to bed an hour or two earlier when you are traveling from west to east. Do just the opposite for east-to-west travel. Insomnia is another disruption of sleeping patterns. For more on this topic, see **Health Note 4-2**.

Research on body rhythms has also shown that people respond to drugs differently at different times of the day and night. By administering drugs at the body's most receptive times, physicians may be able to reduce the doses, lowering toxic side effects and fighting diseases more effectively.

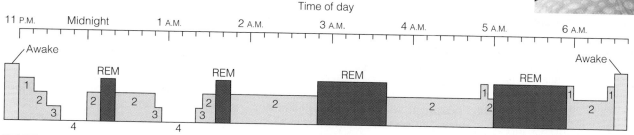

FIGURE 4-17 Stages of Sleep Numbers indicate the stages of sleep: the higher the number, the deeper the sleep. Note that around midnight sleep is deepest. As morning approaches, a person's sleep is lighter, and REM sleep, or dream sleep, occurs in longer increments.

4-4 Health and Homeostasis

Human health is dependent on maintaining homeostasis, which can be upset by factors such as disease-causing organisms. Homeostasis also requires a healthy environment, including a healthy social and psychological atmosphere with manageable levels of stress.

Unfortunately, many adults are exposed to fairly unhealthy workplaces with work schedules that can be unsettling to their internal biological rhythms. Dr. Richard Restak, a neurologist and author, notes that the "usual rhythms of wakefulness and sleep . . . seem to exert a stabilizing effect on our physical and psychological health." The greatest disrupter of our natural circadian rhythms, he says, is the variable work schedule.

Today, one out of every four working men and one out of every six working women is on a variable work schedule, shifting frequently between day and night work. In many industries, workers are at the job day and night to make optimal use of equipment and buildings. As a result, more restaurants and stores are open 24 hours a day, and more health care workers must be on duty at night to care for accident victims.

Workers on alternating shifts suffer from a higher incidence of ulcers, insomnia, irritability, depression, and tension than workers on regular shifts. Making matters worse, tired, irritable workers whose judgment is impaired by fatigue pose a threat not only to themselves but also to society. Why?

Humans are meant to sleep at night and be awake during the day. Make a person work at night when he or she normally sleeps, say sleep experts, and you can expect more accidents and lower productivity. The two most significant accidents at nuclear power plants were, many experts believe, caused in part by sleepy workers who missed important signals or made serious errors in judgment because they were working late at night, a time unsuitable for clear thinking. Many experts believe that plane crashes, auto accidents, and acts of medical malpractice can also be traced to judgment errors made by individuals who were working against their natural body rhythms.

Thanks to studies of biological rhythms, researchers are finding ways to reset the body's clock. These efforts could lessen the problems shift workers face and improve safety and productivity.

SUMMARY

From Cells to Organ Systems

1. The basic structural unit of all organisms is the cell. In humans, cells and extracellular material form tissues, and tissues, in turn, combine to form organs. Organs combine to form organ systems.
2. Four basic tissues are found in the bodies of most multicellular animals such as humans: epithelial, connective, muscle, and nervous. Organs contain all four primary tissues in varying proportions.
3. Epithelial tissues consist of two types: membranous epithelia, which form the coverings or linings of organs, and glandular epithelia, which form exocrine and endocrine glands.
4. Connective tissues bind other tissues together, provide protection, and support body structures. All connective tissues consist of cells and extracellular fibers.
5. Two types of connective tissue are found in the body: connective tissue proper (dense and loose connective tissue) and specialized connective tissue (bone, cartilage, and blood).
6. Cartilage consists of specialized cells embedded in a matrix of extracellular fibers and other extracellular material.
7. Bone is a dynamic tissue that provides internal support, protects organs such as the brain, and helps regulate blood calcium levels.

8. Bone consists of bone cells (osteocytes) and a calcified cartilage matrix. Two types of bone tissue exist: spongy and compact.
9. Blood consists of numerous blood cells and platelets and an extracellular fluid, called plasma.
10. Muscle is an excitable tissue that contracts when stimulated. Three types of muscle tissue are found in the human body: skeletal, cardiac, and smooth muscle.
11. Skeletal muscle is under voluntary control, for the most part, and forms the muscles that attach to bones. Cardiac muscle is located in the heart and is involuntary. Smooth muscle is involuntary and forms sheets of varying thickness in the walls of organs and blood vessels.
12. Nervous tissue consists of two types of cells: conducting and nonconducting (supportive). The conducting cells, called neurons, transmit impulses from one region of the body to another. The nonconducting cells are a type of nervous system connective tissue but also play many other important roles.

Principles of Homeostasis

13. The nervous and endocrine systems coordinate the functions of organ systems, helping the body maintain homeostasis.
14. Homeostatic systems do not maintain absolute constancy and can be upset by alterations in input or output. Imbalances can seriously affect an individual's health.

15. Homeostatic mechanisms are reflexes, occurring without conscious control. They operate by negative feedback through the nervous and endocrine systems.

Biological Rhythms

16. Many physiological processes undergo definite rhythmic changes. These natural rhythms may take place over a 24-hour period or over much longer or shorter periods.
17. The brain controls many biological cycles. A clump of nerve cells (the suprachiasmatic nucleus) is thought to play a major role in coordinating these cycles.

Health and Homeostasis

18. The greatest disrupter of our natural circadian rhythms is the variable work schedule, which is surprisingly common among industrialized nations.
19. Workers on alternating shifts suffer from a higher incidence of ulcers, insomnia, irritability, depression, and tension than workers on regular shifts. Making matters worse, tired, irritable workers whose judgment is impaired by fatigue pose a threat not only to themselves, but also to society.
20. Researchers are finding ways to reset the biological clock, which could lessen the problems shift workers face and thereby increase work performance, particularly of those on the graveyard shift.

Thinking Critically—Analysis

This Analysis corresponds to the Thinking Critically scenario that was presented at the beginning of this chapter.

With a little digging—a step vital to critical thinking—you will find that prenatal care consists of visits to doctors for checkups and advice on nutrition and other matters pertinent to pregnant women. Prenatal care is crucial to the development of a healthy fetus. If problems arise, they can be dealt with immediately, not when

it's too late and too costly. You will also find that the cost of one organ transplant could supply medical care for 1000 pregnant women, assistance that itself could prevent ailments that necessitate costly transplant procedures.

Given this information, should Americans spend more on prevention and less on dramatic life-saving measures, such as organ transplants? Why? After you have given your opinion on this matter, put yourself in the position of a parent whose child needs a kidney transplant to survive. How does this perspective affect your position? What lessons can be learned from this?

KEY TERMS AND CONCEPTS

Adipose tissue, p. 66
Cardiac muscle, p. 69
Cartilage, p. 66
Circadian rhythms, p. 76
Compact bone, p. 67
Connective tissue, p. 65
Connective tissue proper, p. 65
Dense connective tissue, p. 66
Effector, p. 74
Elastic cartilage, p. 67
Epidermis, p. 64
Fat cells, p. 66

Fibroblast, p. 66
Fibrocartilage, p. 67
Hyaline cartilage, p. 67
Intervertebral disk, p. 67
Loose connective tissue, p. 66
Macrophage, p. 66
Muscle tissue, p. 68
Negative feedback, p. 74
Neuroglia, p. 70
Neuron, p. 70
Organ, p. 61

Organ system, p. 71
Osteoblasts, p. 68
Platelets, p. 68
Red blood cells, p. 68
Sensor, p. 74
Skeletal muscle, p. 69
Smooth muscle, p. 70
Spongy bone, p. 68
Stratified squamous epithelium, p. 65
Tissue, p. 61
White blood cells, p. 68

CONCEPT REVIEW

1. Define the following terms: tissues, extracellular material, organs, and organ systems. pp. 61, 71
2. List the four basic tissue types and their subtypes. p. 61
3. Discuss some biological examples showing how structure reflects function. p. 64
4. Describe the two types of connective tissue and how they function. pp. 65–68

5. Why do cartilage injuries repair so slowly or not at all? Bone is repaired much more easily than cartilage. Why? p. 67
6. In what way is bone part of a homeostatic system? p. 67
7. What is an organ system? List some examples. p. 71
8. Define homeostasis, and describe the major principles of homeostasis presented in this

chapter. Use an example to illustrate your points. p. 74
9. Are biological rhythms an exception to the principle of homeostasis? p. 76
10. Describe the biological clock, and explain how it is synchronized with the 24-hour day-night cycle. How does shift work upset the biological clock? How can these problems be lessened? p. 78

SELF-QUIZ: TESTING YOUR KNOWLEDGE

1. The internal and external linings of organs are a type of tissue known as _____. p. 62
2. A simple epithelium consists of ___ layer(s) of cells. p. 61
3. Ligaments and tendons are made up of _____ connective tissue. p. 66
4. The cell responsible for making fibers in connective tissue is known as the _____. p. 66
5. Accumulations of fat cells in the body are known as _____ tissue. p. 66
6. _____ is poorly supplied with blood and therefore does not heal readily after injury. p. 67

7. Bone cells or _____ are formed when osteoblasts become surrounded by bony matrix. p. 68
8. _____ muscle is the type of muscle that is under voluntary control and is responsible for moving body parts like the arms and legs. p. 69
9. Many organs like the stomach and uterus contain thick sheets of involuntary muscle known as _____ muscle. p. 70
10. The nerve cell or _____ is a conducting cell found in the nervous system. p. 70

11. The nonconducting cells in the nervous system are known collectively as _____. p. 70
12. A collection of organs that function together is called an _____. p. 71
13. Homeostatic mechanisms in the body are typically controlled by _____ feedback. p. 74
14. All homeostatic mechanisms rely on _____ to detect changes and _____ to respond to them. p. 74
15. A biological rhythm lasting 24 hours is known as a _____ rhythm. p. 76

 www.jbpub.com/humanbiology/5e

The site features eLearning, an online review area that provides quizzes, chapter outlines, and other tools to help you study for your class. You can also follow useful links for in-depth information, research the differing views in the Point/Counterpoints, or keep up on the latest health news.

thinking critically

Kidney stones begin as tiny crystals. These crystals, made of calcium and other ions, grow larger and larger over time. In some cases, they grow so large that they eventually obstruct the flow of urine, causing considerable pain and discomfort. Because stones are made primarily of calcium, physicians often recommend that patients with stones reduce their intake of calcium by cutting back on milk and milk products, such as cheese and yogurt.

Unfortunately, a new study of 45,000 men suggests that a low-calcium diet may actually increase the risk of developing kidney stones. The researchers showed that men who ate a diet rich in calcium actually had a 34% lower risk of developing kidney stones than men who ate a low-calcium diet.

Kidney specialist Gary Curhan believes that the reason for this seemingly contradictory result is due to a chemical called *oxalate*, which is present in many foods. Oxalate binds to calcium and forms insoluble crystals that make up most kidney stones. Curhan believes that in people who ingest normal amounts of calcium oxalate binds to the mineral in the intestine. This prevents blood oxalate levels from rising and retards the formation of stones. Low-calcium diets permit an excess of oxalate to enter the body; this oxalate binds to calcium in the kidney and causes stones. How would you test this hypothesis? What lessons can be gleaned here?

It must be a law of human nature. Ask almost any couple, and they will tell you: She lies shivering under the covers on a cold winter night while he bakes. Out on a hike in winter, he stays warm in a light jacket while she bundles up in a down coat. What causes this difference between many men and women?

Part of the answer may lie in iron—not pumping iron, but dietary iron. Quite simply, many American women do not consume enough iron to offset losses that occur during menstruation, the monthly discharge of blood and tissue from the lining of the uterus. Iron deficiencies in women may reduce internal heat production.

John Beard, a researcher at Pennsylvania State University, published a study that supports this conclusion. Beard compared two groups of women, one with low levels of iron in the blood and another with normal levels. Beard found that body temperature dropped more quickly in iron-deficient women exposed to cold than in those with normal iron levels. He also found that iron-deficient women generated 13% less body heat. Beard continued his study by asking the iron-deficient women to take iron supplements for 12 weeks, after which they responded normally to cold.

These findings illustrate how important proper nutrition is to normal body function, especially homeostasis. You'll see additional examples as we explore basic human nutrition and digestion in this chapter.

5-1 An Introduction to Nutrition

If you live to be 65, you will consume over 70,000 meals in your lifetime. Because foods affect your body in many ways, what you eat will help to determine how you feel in your later years. That's how important nutrition is to your health. To perform and feel your very best, you must eat a balanced diet to acquire the energy and nutrients needed by your cells, tissues, and organs.

A balanced diet is attained by eating a variety of foods. In 1992, the U.S. Department of Agriculture released a helpful tool called a **food pyramid**, to help Americans eat better. It places foods in six major groups and prescribes allotments from each group that are necessary for proper nutrition (Figure 5-1). Take a few moments to study the pyramid and compare it to your diet.

As you can see from Figure 5-1, a healthy diet consists of breads, cereals, rice, and pasta. It also includes plenty of fruits and vegetables. Meat and milk products are required, but in lesser amounts. Fat is to be minimized. Most people in the more developed countries go wrong by eating too many fatty foods, too much meat, too much carbohydrate, and not enough fruits and vegetables.

The nutrients we need to survive and prosper physically and mentally can be divided into two broad categories: macronutrients and micronutrients. Table 5-1 lists the basic nutrients and includes some of the foods and beverages that provide them. Let's take a look at each group.

TABLE 5-1	Macronutrients and Micronutrients	
Nutrients		**Foods Containing Them**
Macronutrients	Water	All drinks and many foods
	Amino acids and proteins	Milk and milk products such as cheese; soy products such as tofu and soy milk; meat; and eggs
	Lipids	Milk and milk products, meats, eggs, nuts, oils, and seeds
	Carbohydrates	Breads, pastas, cereals, sweets
Micronutrients	Vitamins	Many vegetables, meats, and fruits
	Minerals	Many vegetables, meats, fruits, nuts, and seeds

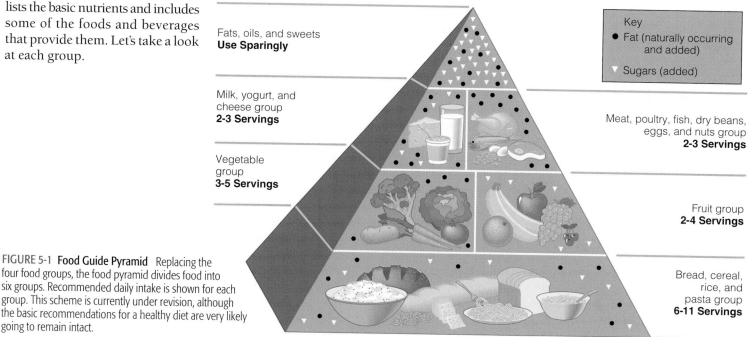

Fats, oils, and sweets
Use Sparingly

Milk, yogurt, and cheese group
2-3 Servings

Vegetable group
3-5 Servings

Key
● Fat (naturally occurring and added)
▼ Sugars (added)

Meat, poultry, fish, dry beans, eggs, and nuts group
2-3 Servings

Fruit group
2-4 Servings

Bread, cereal, rice, and pasta group
6-11 Servings

FIGURE 5-1 **Food Guide Pyramid** Replacing the four food groups, the food pyramid divides food into six groups. Recommended daily intake is shown for each group. This scheme is currently under revision, although the basic recommendations for a healthy diet are very likely going to remain intact.

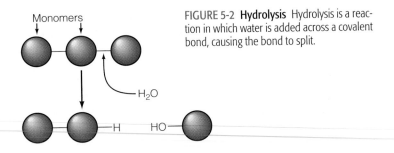

FIGURE 5-2 **Hydrolysis** Hydrolysis is a reaction in which water is added across a covalent bond, causing the bond to split.

Macronutrients

Macronutrients are needed in large quantity and include water, carbohydrates, lipids, and proteins.

Water is one of the most important of all the substances we ingest. Without it, a person can survive only about 3 days. Despite its importance, water is not included on the food pyramid. In part, that is because it is supplied in so many different ways. For example, water constitutes the bulk of the liquids we drink and is present in virtually all of the solid foods we eat (Chapter 3).

Maintaining the proper level of water in the body is important for several reasons. First, water participates in many chemical reactions in the body. For example, the breakdown of several large chemicals such as starch requires water. The breakdown of a chemical using water is called *hydrolysis* (high-DROL-ah-siss). A simplified version is shown in Figure 5-2. In these reactions, enzymes add water across covalent bonds, causing them to break apart.

Because of water's importance in the body, a decrease in its level can impair metabolism, including energy production. Athletic performance may drop significantly when the body's water level falls even slightly.

Maintaining an adequate water volume also helps to stabilize body temperature. A decline in the amount of water in your body decreases blood volume and extracellular fluid. This, in turn, causes body temperature to rise, because the heat normally produced by body cells is being absorbed by a smaller volume of water. A rise in body temperature can impair cellular function and can, if high enough, lead to death.

Maintaining proper water levels also helps individuals maintain normal concentrations of nutrients and toxic waste products in the blood and extracellular fluid. If you don't drink enough liquid, your urine will become more concentrated; the rise in the concentration of chemicals in the blood and urine increases your chance of developing kidney stones.

Carbohydrates

Carbohydrates are a major source of energy.

Carbohydrates are a group of organic compounds that includes such well-known examples as table sugar (sucrose), blood sugar (glucose), and starch. Carbohydrates consist primarily of carbon, oxygen, and hydrogen. Carbohydrates belong to three broad groups: monosaccharides, disaccharides, and polysaccharides.

Monosaccharides are the smallest carbohydrates. They are often simple sugars. These water-soluble molecules contain three to seven carbons, and many carbohydrates form ring structures. One of the most common and most important monosaccharides is **glucose** (GLUE-kose), a six-carbon sugar introduced in Chapter 3. Glucose is made by plants during photosynthesis. It is one of the main sources of energy in the human body.

Humans need a continuous supply of energy, even the most ardent couch potatoes! Interestingly, 70% to 80% of the total energy required by sedentary humans (individuals who get little exercise) is used to perform basic functions: metabolism, food digestion, absorption, and so on. The remaining energy is used to power body movements such as walking, talking, and turning on the television via the remote control. For more active people, these percentages shift considerably. A bicycle commuter, for example, uses proportionately more energy for vigorous muscular activity than an office worker who drives a car to work.

Disaccharides (dye-SACK-are-ides) are carbohydrates that consist of two monosaccharides covalently bonded to each other. The disaccharide sucrose, for instance, consists of two monosaccharides, glucose and fructose.

In animals and plants, many monosaccharides react to form much larger molecules known as *polymers*. A polymer is a general term used to describe any molecule consisting of numerous smaller molecules. Proteins, for instance, are polymers of amino acids. Starch is a polymer containing numerous glucose molecules.

Polysaccharides are carbohydrate polymers. The most common building block of polysaccharides is glucose. Plants synthesize two important polysaccharides from glucose: (1) starch and (2) cellulose (CELL-you-lose). Animals synthesize an important polysaccharide, glycogen.

Starch molecules are produced in the leaves of plants and are often stored in plant roots (for example, in potato plants) or seeds (as in wheat and rice). Starch is an important nutrient for plant-eating animals. In the digestive tract of humans, starch is broken down into glucose molecules. Glucose enters the bloodstream and is distributed to body cells, where they are broken down to release energy (Figure 5-3). Glucose can also be stored in cells of the liver and skeletal muscle for later use.

In liver and muscle cells of animals, glucose combines to form yet another polysaccharide, **glycogen** (GLYE-co-gen), which contains thousands of glucose molecules. Glycogen is often called "animal starch" because it serves a similar purpose and has a structure similar to starch found in plants.

Glycogen is an important player in the homeostatic mechanism that ensures constant blood sugar levels. When blood glucose levels fall in the periods between meals, the liver breaks down its stored supplies of glycogen. The glucose molecules derived from the breakdown of glycogen are released into the bloodstream, maintaining blood glucose concentrations needed to keep us alive and well.

The release of glucose from the liver is stimulated by a hormone called glucagon. Its release is stimulated when blood glucose levels fall. Produced by the pancreas, glucagon enters the bloodstream and circulates to the liver, where it stimulates the breakdown of glycogen. The release of glucose into the bloodstream restores blood sugar levels.

Input	Distribution in body	Output

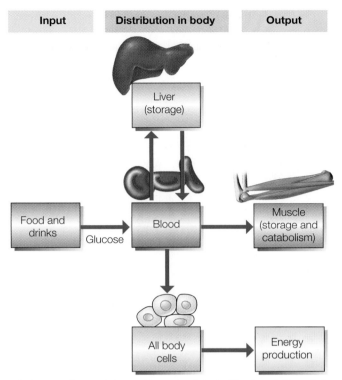

FIGURE 5-3 Glucose Balance Glucose levels in the blood result from a balance created by input and output. Input refers to glucose supplied from the digestion and absorption of food. Output results from storage in muscle and use in body cells for energy production.

Although muscle cells also store glucose in the form of glycogen, they do not participate in maintaining normal blood sugar levels. This glycogen is used only by muscle cells in times of need.

Cellulose molecules are polysaccharides found only in plants. The covalent bonds that join the glucose molecules are slightly different than those in starch. Because of this, cellulose cannot be digested by humans. We lack the enzymes needed to break the covalent bonds joining its glucose subunits. Consequently, cellulose passes through the digestive systems of humans and most other animals relatively unchanged.

Even though it is not digested, cellulose is important to normal body function. In the human diet, for example, nondigestible cellulose is a principal form of **dietary fiber**. Dietary fiber is found in fruits, vegetables, and grains. Dietary fiber exists in two basic forms: water-soluble and water-insoluble.

Water-soluble fiber consists of gummy polysaccharides in fruits, vegetables, and some grains—including apples, bananas, carrots, barley, and oats—that dissolve in water. Several studies suggest that water-soluble fiber helps lower blood cholesterol by acting as a sponge that absorbs dietary cholesterol inside the digestive tract. This prevents cholesterol from being absorbed into the bloodstream.

The second type of dietary fiber, **water-insoluble fibers,** are rigid cellulose molecules in foods such as celery, whole wheat products, and brown rice. Some foods, such as green beans and green peas, contain a mixture of both types.

Water-insoluble fiber such as that in whole wheat bread increases the water content of the feces (FEE-seas), the semisolid waste produced by the large intestine. This makes the feces softer and facilitates their transport through the large intestine. Increasing the water content also reduces constipation. Water-insoluble fiber may also decrease cancer of the colon and rectum (parts of the large intestine). You can increase your intake of water-insoluble fiber by consuming celery, whole wheat products, brown rice, green beans, and peas, among others.

Lipids

Lipids are a source of energy but also play a structural role.

Lipids are a structurally diverse group of organic molecules that are insoluble in water and soluble in nonpolar solvents such as ethyl alcohol. Lipids are the waxy, greasy, or oily compounds found in plants and animals. Fats and oils in animals are energy-rich molecules that serve as an important storage depot. Still other lipids are structural components of cells. Two biologically important lipids are discussed in this chapter: triglycerides and steroids. (Phospholipids were discussed in Chapter 3.)

Triglycerides. **Triglycerides** are known to most of us as fats and oils. Cooking oil, for example, is a triglyceride, as is the fat in a steak or the butter on a piece of bread.

Triglycerides are composed of four organic subunits: one molecule of glycerol and three fatty acid molecules (Figure 5-4). **Glycerol** is a three-carbon compound; **fatty acids** are long molecules containing many carbons and hydrogens and a COOH, or carboxyl group (car-BOX-ul), on one end.

Triglycerides contain many covalent bonds, each of which stores a small amount of energy. When cells break down triglycerides, these bonds are broken and energy is released. Energy liberated during the breakdown of triglycerides is used to drive a variety of cellular processes. Gram for gram, triglycerides yield more than twice as much energy as carbohydrates.

In humans and many other mammals, triglycerides are stored in fat cells under the skin and in other locations (Figure 5-5). In adult humans, triglycerides provide about half of the cellular energy consumed at rest, with glucose providing most of the rest. During moderate (aerobic) exercise, triglycerides provide a larger proportion of the body's energy demand, explaining why aerobic exercise like jogging and bicycling helps people lose weight.

Fatty deposits around certain organs also cushion them from damage. Horseback riders, motorcyclists, and runners, for instance, can engage in their sports for hours at a time without damaging their internal organs in part because of nature's natural cushions.

With a few exceptions, fats (from most animals) are solid at room temperature, and oils (from plants and fish) are liquid. The reason for this difference lies in their chemical structures. In fats, the carbon atoms of the fatty acids are joined by single covalent bonds (Figure 5-4). The remaining bond sites on the carbon atoms are taken up by hydrogens. Thus, the fatty acids are said to be *saturated* with hydrogens. When the carbon backbone of a fatty acid consists solely of single covalent bonds, the

Formation of a triglyceride

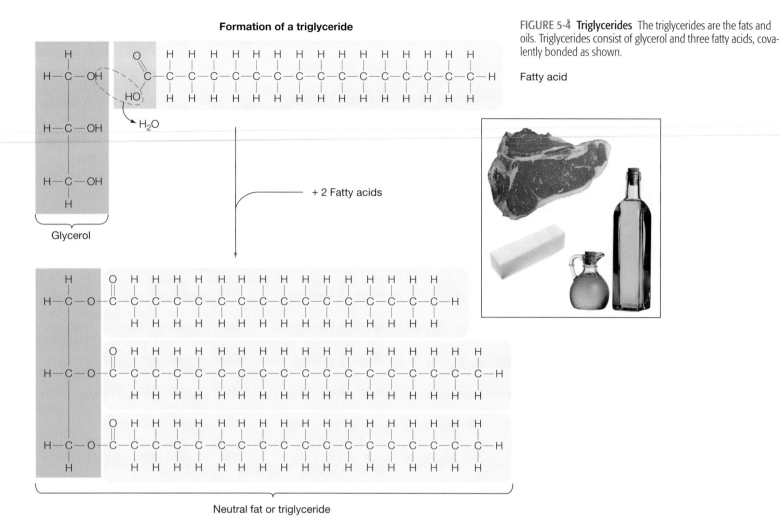

Glycerol

Fatty acid

+ 2 Fatty acids

H_2O

Neutral fat or triglyceride

FIGURE 5-4 **Triglycerides** The triglycerides are the fats and oils. Triglycerides consist of glycerol and three fatty acids, covalently bonded as shown.

structure zigzags but remains fairly linear (Figure 5-6a). This allows the triglyceride molecules of a fat to pack together, forming a solid at room temperature.

In contrast, the fatty acids in oils contain a number of double covalent bonds. These molecules are said to be *unsaturated* (Figure 5-6b). Double covalent bonds cause a bend in the fatty

acid molecules that is believed to prevent tight packing. This somewhat "looser" arrangement of triglyceride molecules produces a liquid at room temperature.

When numerous double bonds exist in a fatty acid molecule, the molecule is said to be *polyunsaturated*. Studies show that polyunsaturated fats reduce one's risk of developing ath-

Fat cells

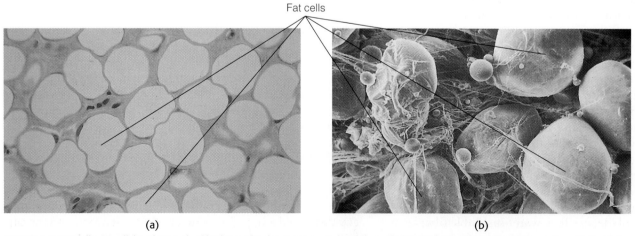

(a)

(b)

FIGURE 5-5 **Fat Cells** (a) A light micrograph of fat tissue. The clear areas are regions where the fat has dissolve during tissue preparation. Notice that the cytoplasm is reduced to a narrow region just beneath the plasma membrane. (b) A scanning electron micrograph of fat cells.

FIGURE 5-6 **The Difference a Few Double Bonds Can Make** (a) Saturated fatty acids are principally derived from animal fats. The side chains are relatively straight and allow the molecules to pack tightly together, which explains why fats are solid at room temperature. (b) Double bonds in unsaturated fatty acids in oils cause the fatty acid chains to bend and thus prohibit tight packing. Oils, derived chiefly from plants, are therefore liquid at room temperature.

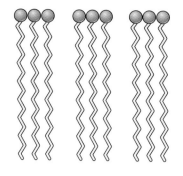
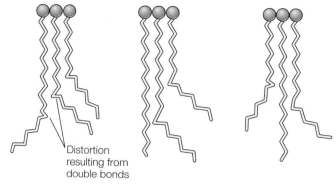

Distortion resulting from double bonds

(a) Saturated fatty acids

(b) Unsaturated fatty acids

erosclerosis (AH-ther-oh-skler-OH-siss), a disease that results from a buildup of cholesterol deposits, or atherosclerotic plaque, on the walls of arteries (Figure 5-7). Plaque restricts blood flow to the heart and brain, causing heart attacks and strokes.

For reasons not well understood, saturated fatty acids increase cholesterol production by the liver, so a diet rich in saturated fat (animal fats) tends to increase an individual's chances of developing atherosclerosis. Steak, hamburgers, cheese, chicken with its skin on, bacon, whole milk, and many common snack foods contain lots of saturated fat. In fact, fat is often added to snack foods as a flavoring.

Atherosclerosis is also more common in sedentary people and smokers. Some individuals are genetically predisposed to develop atherosclerosis. Their livers produce abnormally high levels of cholesterol.

Once thought to be a disease of adults, studies show that children, teenagers, and young adults are starting to show a surprising amount of arterial plaque. Many researchers are concerned that the high fat diets of many young children and the lack of exercise could dramatically increase heart disease in youngsters in more developed countries like the United States in the years to come.

Lowering the level of saturated fat in the diet can be accomplished by switching from whole milk to low-fat milk, reducing the consumption of red meat, trimming fat from chicken and other meats, cutting down on cheese, and cooking with unsaturated vegetable oil instead of animal fat (lard). Watch out for snack foods, too. Manufacturers list the total fat and saturated fat content of all packaged foods. Check out food labels very carefully. (For more on cholesterol, see Health Note 5-1.)

Vegetable oil can be converted to a solid (margarine) by chemically adding hydrogens to it. This process decreases the number of double bonds and tends to straighten the fatty acid chains, resulting in a solid. Many margarines still contain a large number of double bonds, that is, they still contain unsaturated fats but are still a better choice than butter. However, some margarines and food products such as crackers and baked goods made from hydrogenated polyunsaturated fats contain certain types of fatty acids called *trans fatty acids*. Research shows that trans fatty acids are just as likely or more likely to increase blood cholesterol than saturated fatty acids. Nutritionists recommend selecting products with the fewest trans fatty acids.

Wall of artery

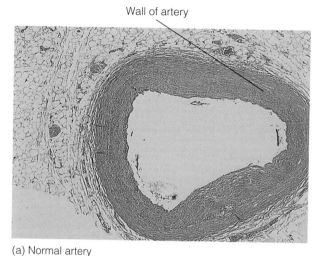

(a) Normal artery

Wall of artery   Plaque

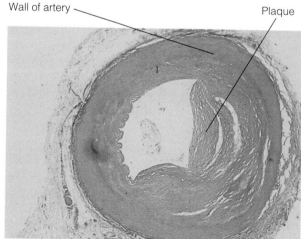

(b) Atherosclerotic artery

FIGURE 5-7 **Atherosclerosis** These cross sections of (a) a normal artery and (b) a diseased artery show how atherosclerotic plaque can obstruct blood flow.

Steroids. As a group, the **steroids** (STEER-oids) are structurally quite different from their lipid cousins, the triglycerides and phospholipids. As shown in Figure 3-11, steroids consist of three six-carbon rings and one five-carbon ring joined in one large

5-1 Lowering Your Cholesterol

Diseases of the heart and arteries are leading causes of death in the United States. Consider atherosclerosis. Atherosclerosis, the accumulation of plaque on artery walls, and the problems it creates are responsible for nearly two of every five deaths in the United States each year. Thanks to improvements in medical care and diet, the death rate from atherosclerosis has been falling steadily in recent years, but it is still a major concern. New research, in fact, shows that atherosclerotic plaque is present even in children and teenagers who eat unhealthy diets.

Researchers believe that atherosclerotic plaques begin to form after minor injuries to the lining of blood vessels. High blood pressure, they think, may damage the lining, causing cholesterol in the blood to seep through the lining. Certain cells (known as macrophages) in the region begin to absorb the cholesterol, becoming foam cells, so named because they appear foamy under the microscope. The blood vessel responds by producing cells that grow over the fatty deposit. This thickens the wall of the artery, reducing blood flow. Additional cholesterol is then deposited in the thickened wall, forming a larger and larger obstruction.

Cholesterol deposits impair the flow of blood in the heart and other organs, cutting off oxygen to tissues. Blood clots may form in the restricted sections of arteries, further reducing blood flow. When the oxygen supply to the heart is disrupted, cardiac muscle cells can die, resulting in heart attacks and death. Blood clots originating in other parts of the body may also lodge in diseased vessels, obstructing blood flow. Oxygen deprivation can weaken the heart, impairing its ability to pump blood. When the oxygen supply to the heart is restricted, the result is a type of heart attack known as a myocardial infarction. If the oxygen-deprived area is extensive, the heart may cease functioning altogether.

Atherosclerotic plaque also impairs the flow of blood to the brain. Blood clots catch in the restricted areas and block the flow of blood to vital regions of the brain. Victims may lose the ability to speak or to move limbs. If the damage is severe enough, they may die. Over time, survivors usually recover lost functions as other parts of the brain take over for the damaged regions.

Atherosclerosis and cardiovascular disease are associated with nearly 40 risk factors. Of all the risk factors, three emerge as the primary contributors to cardiovascular disease: elevated blood cholesterol, smoking, and high blood pressure.

Consider cholesterol. Cholesterol is essential to normal body function, as noted in the text. In most people, the majority of the cholesterol in the blood is produced by the liver and the concentration of cholesterol in the blood tends to stay fairly constant. If dietary input falls, the liver increases its output. If the amount of cholesterol in the diet rises, the liver reduces its production. So what's all the fuss about cholesterol in a person's diet?

structure resembling chicken wire. One of the best-known steroids is cholesterol (Figure 3-11a). Cholesterol is a component of the plasma membrane in animal cells (but not in plants). In humans, it is a raw material needed to synthesize other steroids, such as vitamin D, bile salts, and the sex hormones estrogen and testosterone. Cholesterol is also a major component of atherosclerotic plaque. In humans, cholesterol comes primarily from the liver. A lesser amount comes from the diet.

Amino Acids and Protein

> Proteins in food are broken down to amino acids, which are used to make proteins such as enzymes and hormones in the cells of the body.

Amino acids and **proteins** are important nutrients. Their structure is discussed in Chapter 3. Protein consumed in the diet is broken down in the small intestine into amino acids that are absorbed into the bloodstream. The amino acids are then used by cells to make a wide variety of body proteins including enzymes, hormones, and various structural proteins.

Proteins in the human body contain 20 different amino acids—all of which can be provided from the diet. The body, however, is capable of synthesizing 12 of the amino acids it needs from nitrogen and smaller molecules derived from carbohydrates and fats. Thus, if they are not present in the diet, they can be made. Two of the 12 are considered semiessential amino acids. They are produced in sufficient quantities in adults but not in growing children. The remaining 10 amino acids cannot be synthesized by human cells under any circumstances. They must be provided by the diet. These amino acids are called **essential amino acids**. A deficiency of even one of the essential amino acids can cause severe physiological problems. As a result, nutritionists recommend a diet containing many different protein sources so individuals receive all of the amino acids they need.

Proteins are divided into two major groups by nutritionists: complete and incomplete. **Complete proteins** contain ample amounts of all of the essential amino acids. Complete proteins are found in milk, eggs, meat, fish, poultry, cheese, and soy products, including soy milk and tofu. **Incomplete proteins** lack one or more essential amino acids and include those found in many plant products: nuts, seeds, grains, most legumes (peas and beans), and vegetables. Vegetarians who avoid all animal products, including milk and eggs, must acquire the essential amino acids they need by combining incomplete protein sources. Legumes such as beans can be combined with grains or nuts and seeds, as shown in Figure 5-8. A meal consisting of a legume, for example, baked beans, combined with wheat bread provides all of the amino acids your body needs. A bean burrito and a corn tortilla will do the same.

Even though the liver regulates cholesterol levels, it cannot work fast enough to remove dietary excesses. That is, it may simply be unable to absorb, use, and dispose of cholesterol quickly enough. Consequently, excess cholesterol circulates in the blood after a meal and is deposited in the arteries.

Cholesterol is carried in the bloodstream bound to protein. These complexes of protein and lipid fall into two groups: high-density lipoproteins (HDLs) and low-density lipoproteins (LDLs). HDLs and LDLs function very differently. LDLs, for example, transport cholesterol from the liver to blood vessels in body tissues. In contrast, HDLs are scavengers, picking up excess cholesterol and transporting it to the liver, where it is removed from the blood and excreted in the bile. Research shows that the ratio of HDL to LDL is an accurate predictor of cardiovascular disease. The higher the ratio, the lower the risk of cardiovascular disease.

High cholesterol (or hypercholesterolemia) tends to run in families. Thus, if a parent died of a heart attack or suffers from this genetic disease, his or her offspring are more likely to have high cholesterol levels.

To reduce your chances of atherosclerosis, the American Heart Association recommends that you (1) limit dietary fat to less than 30% of the total caloric intake; (2) limit dietary cholesterol to 300 milligrams per day; and (3) acquire 50% or more of one's calories from carbohydrates, especially polysaccharides (notably, starches found in potatoes, rice, vegetables and other foods). Reductions in saturated fats (animal fats) can also help lower cholesterol levels. You can cut back on saturated fat by reducing your consumption of red meat and trimming the fat off all meats before cooking. You can also increase your consumption of fruits, vegetables, and grains, letting these low-fat foods displace some of the fatty foods you might otherwise have eaten. In children under the age of 2, however, diets should not be restricted. A diet that is too restrictive may actually impair physical growth and development. What all children need is a well-balanced diet, low in fats, especially animal fat, with sufficient calories from other sources. Such a diet could encourage the eating habits necessary for good health throughout adult life.

Blood cholesterol levels can also be lowered with drugs and exercise. Research spanning several decades shows that a lower cholesterol level translates into a decline in cardiovascular disease.

www.jbpub.com/humanbiology/5e

Visit Human Biology's Internet site for links to web sites offering more information on this topic.

FIGURE 5-8 **Complementary Protein Sources** By combining protein sources, a vegetarian who consumes no animal by-products can be assured of getting all of the amino acids needed. Legumes can be combined with foods made from grains or nuts and other seeds.

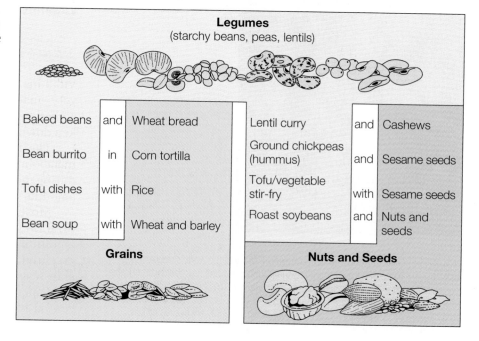

Although many people think of proteins as a source of energy, proteins are only used for this purpose when dietary intake of carbohydrates and fats is severely restricted or when protein intake far exceeds demand.

Protein deficiencies occur in millions of children throughout the world. The arms and legs of these children are often thin and wiry because their muscle cells break down protein in an effort to provide energy (Figure 5-9b). According to the United Nations, one of every six people in the world, a total of one billion people, suffer from protein malnutrition.

On the other end of the spectrum are the overfed populations of the world. The average American, for example, consumes twice the daily requirement of protein. Amino acids derived from the surplus dietary protein are broken down in the body to produce energy and fat.

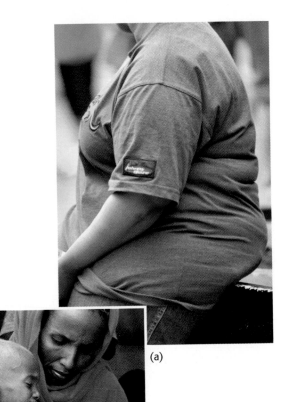

(a)

(b)

FIGURE 5-9 **Overnutrition and Undernutrition** Many of the world's people, especially those in more developed countries, suffer from overnutrition, resulting in (a) obesity while a great many others suffer from undernutrition. (b) This child suffers from severe protein deficiency (kwashiorkor; kwash-EE-or-core). His arms and legs are emaciated because muscle protein has been broken down to supply energy. His belly is slightly swollen because of a buildup of fluid in the abdomen.

Overnutrition

Overnutrition is now reaching epidemic proportions.

Residents of many countries, Americans included, ingest numerous nutrients in quantities far greater than is necessary. Excess food intake, called **overnutrition**, now rivals **undernutrition**, a lack of food. In the United States, 60% of all American adults are overweight (Figure 5-9a). The percentage of the population classified as severely overweight or obese has climbed from 15% to 23% since 1970. Even children are afflicted, with 15% now considered overweight or obese. But the United States is not alone in this regard. Fifty percent of the adults in Russia, the United Kingdom, and Germany are overweight. Even in less developed nations, overnutrition is becoming more commonplace among the wealthier classes. The effects of overeating are many: heart attacks, stroke, late-onset diabetes, and osteoarthritis. Why are so many people overweight?

Most Americans consume too much food, especially, too much carbohydrate, too much fat, and too much protein. To lower the risk of heart attack, fat intake should only be about 20% to 30%, and animal fat should be eaten in small quantities. You don't need to eliminate fat. In fact, really low fat diets can be unhealthy. Some patients develop fatty acid deficiencies. Eat fat, but consume it in moderation and limit saturated fats. To control weight you also need to consume fewer carbohydrates and less protein.

Micronutrients

Micronutrients are substances required in minute quantities and include vitamins and minerals.

With this overview of the macronutrients, we now turn to the micronutrients.

Vitamins

Vitamins are micronutrients that play a key role in many chemical reactions in the body .

Vitamins are a diverse group of organic compounds present in very small amounts in many foods. These molecules are absorbed by the lining of the digestive tract without being broken down.

The 13 known vitamins play an important role in many metabolic reactions. Because vitamins are recycled many times during metabolic reactions, they are needed only in very small amounts.

Most vitamins are not synthesized in the cells of the body or, if they can be made, they are not produced in sufficient amounts to satisfy cellular demands, making dietary intake essential. Vitamin D, for instance, is manufactured by the skin when it is exposed to sunlight. However, most Americans spend so much time indoors that dietary input is essential to good health. Table 5-2 lists the vitamins and their functions.

TABLE 5-2	Important Information on Vitamins			

Vitamin	Major Dietary Sources	Major Functions	Signs of Severe, Prolonged Deficiency	Signs of Extreme Excess
Fat-soluble				
A	Fat-containing and fortified dairy products; liver; provitamin carotene in orange and deep green fruits and vegetables	Vitamin A is a component of rhodopsin; carotenoids can serve as antioxidants; retinoic acid affects gene expression; still under intense study	Night blindness; keratinization of epithelial tissues including the cornea of the eye (xerophthalmia) causing permanent blindness; dry, scaling skin; increased susceptibility to infection	Preformed vitamin A: damage to liver, bone; headache, irritability, vomiting, hair loss, blurred vision 13-cis retinoic acid: some fetal defects; carotenoids: yellowed skin
D	Fortified and full-fat dairy products, egg yolk (diet often not as important as sunlight exposure)	Promotes absorption and use of calcium and phosphorus	Rickets (bone deformities) in children; osteomalacia (bone softening) in adults	Calcium deposition in tissues leading to cerebral, CV, and kidney damage
E	Vegetable oils and their products; nuts, seeds	Antioxidant to prevent cell membrane damage; still under intense study	Possible anemia and neurologic effects	Generally nontoxic, but at least one type of intravenous infusion led to some fatalities in premature infants; may worsen clotting defect in vitamin K deficiency
K	Green vegetables; tea	Aids in formation of certain proteins, especially those for blood clotting	Defective blood coagulation causing severe bleeding or injury	Liver damage and anemia from high doses of the synthetic form menadione
Water-soluble				
Thiamin (B-1)	Pork, legumes, peanuts, enriched or whole-grain products	Coenzyme used in energy metabolism	Nerve changes, sometimes edema, heart failure; beriberi	Generally nontoxic, but repeated injections may cause shock reaction
Riboflavin (B-2)	Dairy products, meats, eggs, enriched grain products, green leafy vegetables	Coenzyme used in energy metabolism	Skin lesions	Generally nontoxic
Niacin	Nuts, meats; provitamin tryptophan in most proteins	Coenzyme used in energy metabolism	Pellagra (multiple vitamin deficiencies including niacin)	Flushing of face, neck, hands; potential liver damage
B-6	High-protein foods in general	Coenzyme used in amino acid metabolism	Nervous, skin, and muscular disorders; anemia	Unstable gait, numb feet, poor coordination
Folic acid	Green vegetables, orange juice, nuts, legumes, grain products	Coenzyme used in DNA and RNA metabolism; single carbon utilization	Megaloblastic anemia (large, immature red blood cells); GI disturbances	Masks vitamin B-12 deficiency, interferes with drugs to control epilepsy
B-12	Animal products	Coenzyme used in DNA and RNA metabolism; single carbon utilization	Megaloblastic anemia; pernicious anemia when due to inadequate intrinsic factor; nervous system damage	Thought to be nontoxic
Pantothenic acid	Animal products and whole grains; widely distributed in foods	Coenzyme used in energy metabolism	Fatigue, numbness, and tingling of hands and feet	Generally nontoxic; occasionally causes diarrhea
Biotin	Widely distributed in foods	Coenzyme used in energy metabolism	Scaly dermatitis	Thought to be nontoxic
C (ascorbic acid)	Fruits and vegetables, especially broccoli, cabbage, cantaloupe, cauliflower, citrus fruits, green pepper, kiwi fruit, strawberries	Functions in synthesis of collagen; is an antioxidant; aids in detoxification; improves iron absorption; still under intense study	Scurvy; petechiae (minute hemorrhages around hair follicles); weakness; delayed wound healing; impaired immune response	GI upsets, confounds certain lab tests

Source: Adapted from *Nutrition for Living,* 4th ed., by J. L. Christian and L. L. Greger, Copyright © 1994 by The Benjamin/Cummings Publishing Company.

Because vitamins are needed in almost all cells of the body, a dietary deficiency in just one vitamin can cause wide-ranging effects. Vitamins also interact with other nutrients. Vitamin C, for example, increases the absorption of iron in the small intestine. Large doses of vitamin C, however, decrease copper utilization by the cells reducing the activity of key enzymes. Consequently, maintaining good health requires ingesting the proper balance of vitamins and other nutrients. Imbalances can result in upsets in homeostasis.

Vitamins fall into two broad categories: water-soluble and fat-soluble. **Water-soluble vitamins** include vitamin C and eight different forms of vitamin B. Water-soluble vitamins are transported in the blood plasma. Because they are water-soluble, they are readily eliminated by the kidneys and are not stored in the body in any appreciable amount.

Water-soluble vitamins generally work in conjunction with enzymes, promoting the cellular reactions that supply energy or synthesize cellular materials. Contrary to common myth, the vitamins themselves do not provide energy.

Many proponents of vitamin use believe that megadoses of water-soluble vitamins are harmless because these vitamins are excreted in the urine and do not accumulate in the body. However, some water-soluble vitamins, such as vitamin B-6, can be toxic when ingested in excess (Table 5-2).

The **fat-soluble vitamins** are vitamins A, D, E, and K. They perform many different functions. Vitamin A, for example, is converted to light-sensitive pigments in receptor cells of the retina, the light-sensitive layer of the eye. These pigments play an important role in vision. Another member of the vitamin A group removes harmful chemicals (oxidants) from the body.

Unlike water-soluble vitamins, the fat-soluble vitamins are stored in body fat and accumulate in the fat reserves. The accumulation of fat-soluble vitamins can have many adverse effects (Table 5-2). An excess of vitamin D, for example, can cause weight loss, nausea, irritability, kidney stones, weakness, and other symptoms. Large doses of vitamin D taken during pregnancy can cause birth defects.

Vitamin excess is encountered largely in the more affluent developed countries and usually occurs only in people taking vitamin supplements. Each year, in fact, approximately 4000 Americans are treated for vitamin supplement poisoning. To avoid problems from excess vitamins, nutritionists recommend eating a balanced diet that provides all of the vitamins the body needs, rather than taking vitamin pills. Megadoses should be avoided.

Vitamin Deficiencies.   Vitamins are important nutrients that affect many cellular processes, so deficiencies can lead to serious upsets of homeostasis and many adverse health effects. A deficiency of vitamin D, for example, can produce rickets (RICK-its), a disease that results in bone deformities. Vitamin K deficiencies can result in severe bleeding as a result of accidental injury. Vitamin C deficiency can result in delayed wound healing and reduced immunity, making people more susceptible to infectious disease.

One of the most common dietary illnesses is caused by a deficiency of vitamin A. This deficiency afflicts over 100,000 chil-

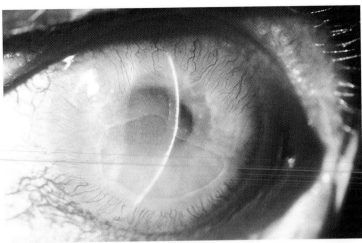

FIGURE 5-10 Vitamin A Deficiency Vitamin A deficiency causes the cornea of the eye to dry and become irritated. If the deficiency is not corrected, corneal ulcers may form and rupture, resulting in permanent blindness.

dren worldwide each year. If not corrected, vitamin A deficiency causes the eyes to dry. Ulcers may form on the eyeball and can rupture, causing blindness (Figure 5-10).

Most people afflicted by vitamin deficiencies are those who fail to get enough to eat, although even well-fed individuals may suffer from a vitamin deficiency if they are not eating a well-rounded diet. Children with insufficient vitamin intake fail to grow. How do you know if you should take vitamins?

The American Dietetic Association recommends that you eat a variety of foods to achieve good health and reduce the risk of chronic disease. Additional vitamins and minerals, they note, can be obtained from fortified foods, such as vitamin D–enriched milk, and from supplements.

If you don't eat a well-balanced diet, consider making the switch. Most people eat far too few fruits and vegetables, gorging instead on fat-rich meats (hamburgers) and desserts. While you make the shift, take a vitamin supplement.

Minerals.   On average, an adult contains about 5 pounds of minerals, naturally occurring inorganic substances vital to many life processes. Humans require about two dozen minerals, such as calcium, sodium, iron, and potassium, to carry out normal body functions.

Minerals, like vitamins, are micronutrients and are derived from the food we eat and the beverages we drink. Minerals are divided into two groups: the major minerals and the trace minerals (Table 5-3). The major minerals are present in the body in amounts larger than 5 grams; the trace minerals are found in lesser quantities. Calcium and phosphorus, for example, are major minerals, and make up three-fourths of all minerals in the human body. These two minerals form part of the dense extracellular matrix of bone. Trace minerals are required in a much smaller quantity. Zinc and copper are two trace minerals that are components of some enzymes.

The distinction between major and trace minerals is not meant to imply that one group is more important than the other.

TABLE 5-3	Important Information on Minerals			
Minerals	Major Dietary Sources	Major Functions	Signs of Severe, Prolonged Deficiency	Signs of Extreme Excess
Major Minerals				
Calcium	Milk, cheese, dark green vegetables, legumes	Bone and tooth formation; blood clotting; nerve transmission	Stunted growth; perhaps less bone mass	Depressed absorption of some other minerals; perhaps kidney damage
Phosphorus	Milk, cheese, meat, poultry, whole grains	Bone and tooth formation; acid-base balance; component of coenzymes	Weakness; demineralization of bone	Depressed absorption of some minerals
Magnesium	Whole grains, green leafy vegetables	Component of enzymes	Neurologic disturbance	Neurologic disturbances
Sodium	Salt, soy sauce, cured meats, pickles, canned soups, processed cheese	Body water balance; nerve function	Muscle cramps; reduced appetite;	High blood pressure in genetically predisposed individuals
Potassium	Meats, milk, many fruits and vegetables, whole grains	Body water balance; nerve function	Muscular weakness; paralysis	Muscular weakness; cardiac arrest
Chloride	Same as for sodium	Plays a role in acid-base balance; formation of gastric juice	Muscle cramps; reduced appetite; poor growth	High blood pressure in genetically predisposed individuals
Trace minerals				
Iron	Meats, eggs, legumes, whole grains, green leafy vegetables	Component of hemoglobin, myoglobin, and enzymes	Iron-deficiency anemia, weakness, impaired immune function	Acute: shock, death; chronic: liver damage, cardiac failure
Iodine	Marine fish and shellfish; dairy products; iodized salt; some breads	Component of thyroid hormones	Goiter (enlarged thyroid)	Iodide goiter
Fluoride	Drinking water, tea, seafood	Maintenance of tooth (and maybe bone) structure	Higher frequency of tooth decay	Acute: GI distress; chronic: mottling of teeth; skeletal deformation

Source: Adapted from *Nutrition for Living*, 4th ed., by J. L. Christian and L. L. Greger. Copyright © 1994 by The Benjamin/Cummings Publishing Company.

Table 5-3 lists some of the minerals, their function, and problems that arise when they are ingested in excess or deficient amounts.

Functional Foods

Foods contains substances that reduce cholesterol and prevent disease.

Numerous chemicals in the foods we eat can help prevent disease. These chemicals are often classified as *neutraceuticals*, a name that combines *nutrients* with *pharmaceuticals*, to indicate their healing role. Foods containing them are called functional foods. **Functional foods** are foods that may provide a health benefit beyond basic nutrition.

The number of neutraceuticals is mind boggling. Some substances are **antioxidants**. Antioxidants react with naturally occurring chemical substances in the blood that are involved in cholesterol buildup in the walls of arteries. Eating a diet rich in antioxidants may reduce heart and artery disease.

In what is good news for chocolate lovers, chocolate and cocoa powder are produced from beans that contain large quantities of natural antioxidants. They're called *flavonoids* (flay-vah-noids). Even commercially available chocolate products may contain significant amounts of these antioxidants. But you must shop carefully. Most chocolate production operations remove the flavonoids. At this writing, Mars is the only company that employs a process that preserves flavonoids. Soy milk also contains flavonoids, as do tea and red wine.

Chocolate's flavonoids not only reduce plaque buildup, they help stimulate the relaxation of smooth muscle in arteries, which is important to maintaining healthy arteries. Flavonoid-rich foods may also benefit the heart by reducing the activity of blood platelets. Platelets are tiny cell-like strutures that stimulate blood clotting. In arteries clogged by atherosclerotic plaque, blood clots can lead to heart attacks and strokes.

So does this mean you should load up on chocolate? Not really. Chocolates also contain a lot of saturated fat and calories (from sugars). Eat chocolate, but don't overindulge (as if that's possible!).

Fruits and vegetables also contain naturally occurring antioxidants. A recent study showed that a serving of salad consumed each day reduces the risk of stroke by 3 percent in women and 5 percent in men. A stroke occurs when an artery in the brain bursts or cholesterol-choked arteries develop blood clots that restrict the flow of blood to brain cells. Several foods seem to provide the most protection: broccoli, cabbage, cauliflower, green leafy vegetables, and vitamin C–rich foods such as oranges and orange juice.

Antioxidants have also been shown to reduce cancer. For example, one group of naturally occurring antioxidants, isoflaovones, which are found in soy products, have been shown to reduce prostate cancer in men. The list of beneficial chemicals is growing larger every day.

5-2 The Digestive System

The human digestive system consists of a tube opened at both ends (Figure 5-11). Food enters the mouth and is chemically and physically altered as it travels along this tube, called the gut. During its journey through the digestive system, food is absorbed through the wall of the gut, where it enters the bloodstream. Waste is excreted at the opposite end, the anus.

The Mouth

Physical breakdown of food occurs in the mouth.

The **mouth** is a complex structure in which food is broken down mechanically and, to a much lesser degree, chemically. Food taken into the mouth is sliced into smaller pieces by the sharp teeth in front; it is then ground into a pulpy mass by the flatter teeth toward the back of the mouth (Figure 5-12a). Chewing breaks food into smaller pieces and greatly increases the surface area presented to the digestive enzymes in the stomach and small intestine, which increases the efficiency of digestion.

As the food is pulverized in the mouth, it is liquefied by **saliva**, a watery secretion released by the salivary glands, three sets of exocrine glands located around the oral cavity (Figure 5-13). The release of saliva is triggered by the smell, feel, taste, and sometimes even by the thought of food.

Saliva performs at least five functions. It liquefies the food, making it easier to swallow. It kills or neutralizes some bacteria via the enzymes and antibodies it contains. It dissolves substances so they can be tasted. It begins to break down starch molecules with the aid of the enzyme amylase (AM-ah-lase). Saliva also cleanses the teeth, washing away bacteria and food particles.

The Teeth

Teeth must be kept clean to prevent plaque and cavities.

As illustrated in Figure 5-12b, each tooth consists of a hard outer layer, called *enamel*. Enamel is the hardest substance in the human body. Beneath that is a softer layer of *dentin*, which forms the bulk of the tooth. In the inside of the tooth is a space filled with nerves and blood vessels, the pulp. Each tooth is embedded in the bone of the jaw.

Because the release of saliva is greatly reduced during sleep, bacteria tend to accumulate on the surface of the teeth, where they break down microscopic food particles, producing foul-smelling chemicals that give us "dragon breath," or "morning mouth."

Controlling the bacteria that live on the teeth by regular brushing is important because these organisms secrete a sticky material called **plaque**. Plaque adheres to the surface of the teeth, trapping bacteria. They release small amounts of a weak acid that dissolves the hard outer coating of our teeth, the **enamel**. The acid forms small pits in the enamel, commonly referred to as *cavities*. Continued acid secretion may cause the cavities to deepen into the softer layer beneath the enamel. If the cavity is left untreated, an entire tooth can be lost to decay. Brushing, therefore, not only helps prevent plaque buildup, it helps reduce the incidence of cavities. Most toothpastes also contain small amounts of fluoride, which hardens the enamel and helps reduce cavities. In most cities and towns, fluoride is added to drinking water as a preventive measure. Flossing your teeth also helps reduce cavities, by removing food particles that lodge between your teeth where toothbrushes can't reach. One

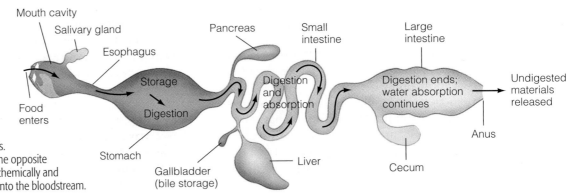

FIGURE 5-11 **The Complete Gut**
This diagram illustrates the general anatomy of the tubular digestive tract present in humans and many other animals. Food enters the mouth and is excreted at the opposite end through the anus. Along the way, it is chemically and physically broken down, then transported into the bloodstream.

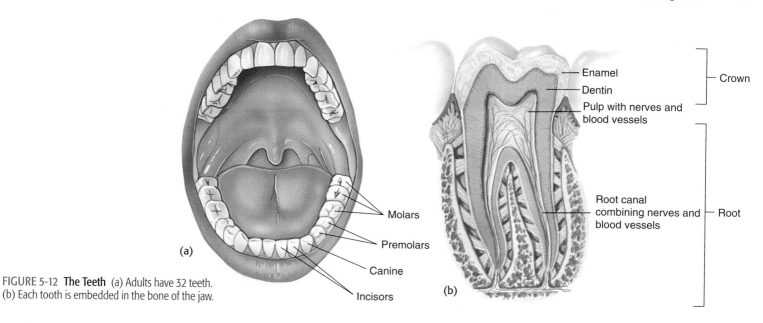

FIGURE 5-12 **The Teeth** (a) Adults have 32 teeth. (b) Each tooth is embedded in the bone of the jaw.

study showed that chewing sugarless gum increases the flow of saliva and cleanses the teeth. By chewing sugarless gum within 5 minutes after a meal, and for at least 15 minutes, the researchers showed that individuals can reduce the incidence of cavities.

Periodic teeth cleaning by a dental hygienist to remove plaque, combined with gentle brushing of your gums, is also essential to good dental health. Brushing the gums strengthens them and reduces the incidence of **gingivitis** (gin-je-VIE-tiss), a disease of the gums. In this disease, bacteria, plaque, mucus, and food particles accumulate between the swollen gums and the base of the teeth. The bacteria release an acid that gradually destroys the bone in which the teeth are embedded. Over time, the teeth may fall out.

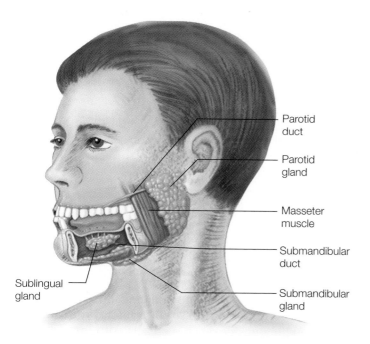

FIGURE 5-13 **Salivary Glands** Three salivary glands (parotid, submandibular, and sublingual) are located in and around the oral cavity and empty into the mouth via small ducts.

Healthy gums are pale pink or brown. They are firm and may appear speckled. If you have gingivitis, your gums will be red, soft, shiny, and swollen. They also bleed easily, even from gentle brushing.

The Tongue

The tongue plays a key role in swallowing.

After food is chewed, it must be swallowed. The tongue plays an important role in this process by pushing food to the back of the oral cavity into the **pharynx** (FAIR-inks), a funnel-shaped chamber that connects the oral cavity with the esophagus (ee-SOFF-ah-gus), a long muscular tube that leads to the stomach (Figure 5-14).

The **tongue**, which also aids in speech, contains taste receptors, or taste buds, on its upper surface. Taste buds are cocktail-onion-shaped structures located in protrusions called *papillae* (pah-pill-eye) on the upper surface of the tongue. Taste buds are described in Chapter 11.

The Epiglottis

The epiglottis prevents food from entering the trachea.

Food propelled from the pharynx into the esophagus is prevented by the **epiglottis** (ep-ah-GLOT-tis) from entering the trachea (TRAY-key-ah), or windpipe, which carries air to the lungs and lies in front of the esophagus (Figure 5-15). The epiglottis is a flap of tissue that acts like a trapdoor, closing off the trachea during swallowing.

Swallowing begins with a voluntary act—the tongue pushing food into the back of the oral cavity. Once food enters the pharynx, however, the process becomes automatic. Food in the pharynx stimulates stretch receptors in the wall of this organ. They then trigger the swallowing reflex, an involuntary contraction of the muscles in the wall of the pharynx. This forces the food into the esophagus.

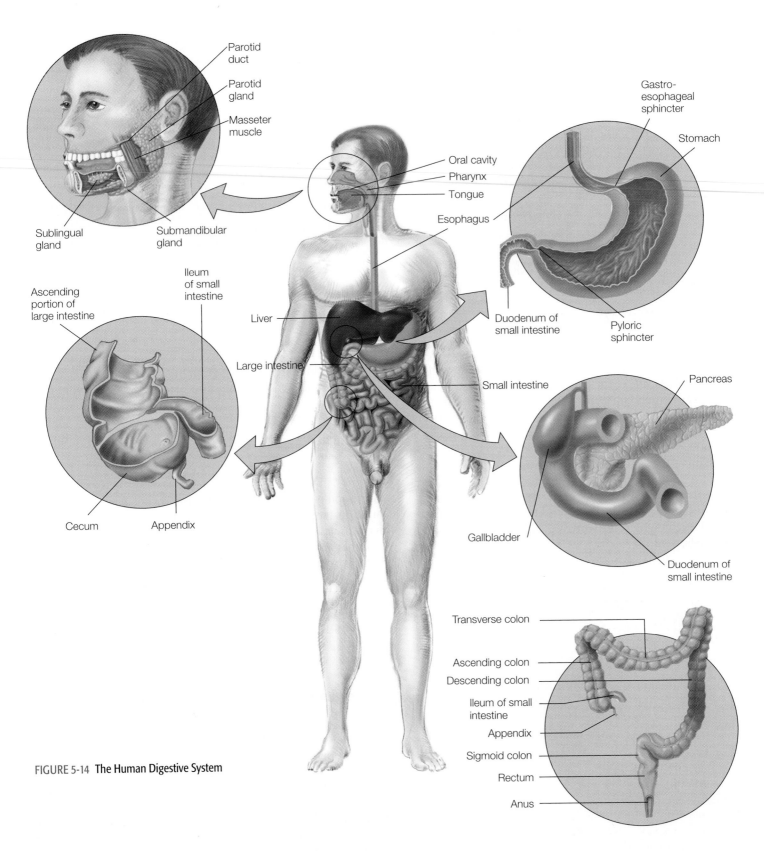

FIGURE 5-14 The Human Digestive System

The Esophagus

The esophagus transports food to the stomach via peristalsis.

Involuntary contractions of the muscular wall of the **esophagus** propel food to the stomach. As Figure 5-16 shows, the muscles of the esophagus contract above the swallowed food mass, squeez-ing it along. This involuntary muscular action is called **peri-stalsis** (pear- eh-STALL-sis). It is so powerful that you can swal-low when hanging upside down. Peristalsis also propels food (and waste) along the rest of the digestive tract.

Scientists once thought that esophageal peristalsis could proceed in the opposite direction and called this process *reverse*

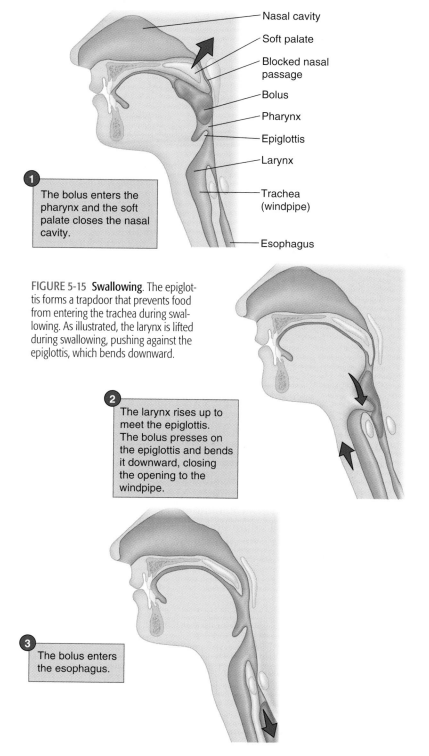

- Nasal cavity
- Soft palate
- Blocked nasal passage
- Bolus
- Pharynx
- Epiglottis
- Larynx
- Trachea (windpipe)
- Esophagus

1 The bolus enters the pharynx and the soft palate closes the nasal cavity.

FIGURE 5-15 **Swallowing**. The epiglottis forms a trapdoor that prevents food from entering the trachea during swallowing. As illustrated, the larynx is lifted during swallowing, pushing against the epiglottis, which bends downward.

2 The larynx rises up to meet the epiglottis. The bolus presses on the epiglottis and bends it downward, closing the opening to the windpipe.

3 The bolus enters the esophagus.

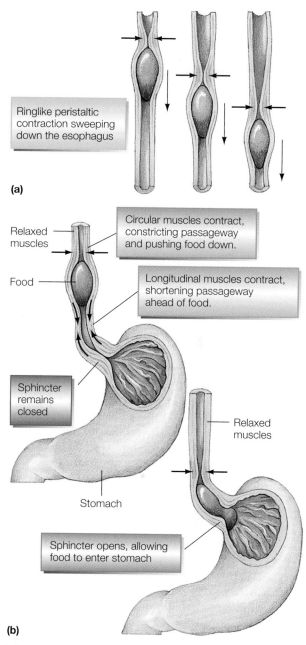

Ringlike peristaltic contraction sweeping down the esophagus

(a)

Relaxed muscles

Circular muscles contract, constricting passageway and pushing food down.

Food

Longitudinal muscles contract, shortening passageway ahead of food.

Sphincter remains closed

Relaxed muscles

Stomach

Sphincter opens, allowing food to enter stomach

(b)

FIGURE 5-16 **Peristalsis** (a) Peristaltic contractions in the esophagus propel food into the stomach. (b) When food reaches the stomach, the gastroesophageal sphincter opens, allowing food to enter.

peristalsis, or, more commonly, *vomiting*. Vomiting is a reflex action that occurs when irritants are present in the stomach. It is, therefore, believed to be a protective adaptation that allows the body to rid the stomach of bad food and harmful viruses and bacteria, with obvious survival benefits to humans. Vomiting may also be caused by (1) emotional factors, such as stress, (2) rotation or acceleration of the head, as in motion sickness, (3) stimulation of the back of the throat, and (4) elevated pressure inside the brain caused by injury.

During vomiting, food is expelled from the stomach as a result of contractions of muscles in the abdomen and diaphragm (DIE-ah-fram). (The diaphragm is a large dome-shaped muscle that separates the thoracic (chest) and abdominal cavities and plays a key role in breathing.)

The Stomach

The stomach stores food, releasing it into the small intestine in spurts.

In humans, the **stomach** (Figures 5-14 and 5-17) lies on the left side of the abdominal cavity, partly under the protection of the rib cage. Food enters the stomach via the esophagus. The opening to the stomach, however, is closed off by a sphincter, a thickened layer of smooth muscle at the juncture of the esophagus and stomach (Figure 5-17).

As food enters the lower esophagus, the sphincter opens, allowing the food to enter the stomach. The sphincter then promptly closes, preventing food and stomach acid from per-

FIGURE 5-17 **The Stomach** The stomach lies in the abdominal cavity. In its wall are three layers of smooth muscle that mix the food and force it into the small intestine, where most digestion occurs. The gastroesophageal and pyloric sphincters control the inflow and outflow of food, respectively.

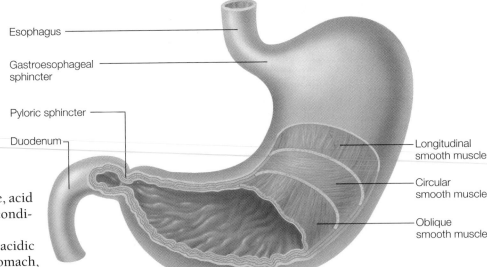

Esophagus

Gastroesophageal sphincter

Pyloric sphincter

Duodenum

Longitudinal smooth muscle

Circular smooth muscle

Oblique smooth muscle

colating upward. If the sphincter fails to close, acid rising in the esophagus causes irritation, a condition known as *heartburn*.

Inside the stomach, food is liquefied by acidic secretions of tiny glands in the wall of the stomach, the **gastric glands**. These glands produce a watery secretion called *gastric juice*. Gastric juice contains hydrochloric acid (HCl) and an enzyme precursor called *pepsinogen* (pep-SIN-oh-gen) (discussed below).

Inside the stomach, food is churned by peristalsis and mixed with gastric juice. The churning action of the stomach's muscular walls helps break down large pieces of food. Combined with the liquid from salivary glands and the gastric glands, the food becomes a rather thin, watery paste known as **chyme** (KIME). The stomach can hold 2–4 liters (2 quarts to 1 gallon) of chyme, which is gradually released into the small intestine at a rate suitable for efficient digestion and absorption.

Contrary to what many think, very little enzymatic digestion occurs in the stomach. The stomach's role is largely to prepare most food for the enzymatic digestion in the small intestine. There are some exceptions to this rule, however. Protein is one of them.

Proteins are denatured (coagulated) by HCl in the stomach. HCl also acts on pepsinogen, converting it to the active form pepsin. Pepsin is a protein-digesting enzyme that catalyzes the breakdown of proteins into large fragments. These fragments are further broken down in the small intestine, the next stop.

Besides activating pepsinogen and denaturing protein (rendering it digestible), HCl breaks down connective tissue fibers in meat and kills most bacteria, thereby protecting the body from infection.

Also contrary to popular belief, the stomach does not absorb many foodstuffs. Only a few substances, such as alcohol, can penetrate the lining of the stomach to enter the bloodstream. Alcohol consumed on an empty stomach passes quickly into the blood, often producing an immediate dizzying effect. The presence of food in the stomach, however, slows alcohol absorption.

The stomach lining is protected from destruction by a basic secretion known as *mucus*, produced by cells in the lining. Mucus coats the epithelium, and this thin coating protects the lining from acid and pepsin. Acid is further prevented from penetrating the lining of the stomach and reaching underlying tissues by tight junctions in the cells of the epithelium. These structures join the epithelial cells very tightly, creating a leakproof barrier.

For years, scientists believed that the stomach acid production could be increased by excess stress, coffee, aspirin, nicotine, and alcohol—or combinations of them. Excess acid, in

turn, was thought to erode the stomach lining, forming **ulcers**, painful, open sores in the stomach lining that bleed into the stomach. Accordingly, most ulcers that were detected early were treated by reducing stress and changing a patient's diet. Patients were advised to eat smaller amounts of food and to reduce consumption of coffee, aspirin, and alcohol to reduce acid secretion in the stomach (see Health Note 1-1). When ulcers were not detected early and when damage was severe, parts of the stomach had to be removed surgically.

Today, researchers have found that at least 70% of all ulcers are caused by a fairly common bacterium, *Heliocobacter pylori*, or *H. pylori* for short. Antibiotics can be used to rid the stomach of this agent, thus eliminating painful ulcers.

Chyme

Chyme leaves the stomach and enters the small intestine.

Chyme is ejected from the stomach into the small intestine by peristaltic muscle contractions. When the wave of contraction reaches the muscular sphincter at the juncture of the small intestine and stomach, it opens, and chyme squirts into the small intestine.

The stomach contents are emptied in 2–6 hours, depending on the size of the meal and the type of food. The larger the meal, the longer it takes the stomach to empty. Solid foods (meat) empty slower than liquid foods (milk shakes). Peristaltic contractions continue after the stomach is empty and are felt as hunger pangs.

As chyme and protein leave the stomach, gastric gland secretion declines and the stomach slowly shuts down its production of HCl until the next meal.

The Small Intestine

The small intestine serves as a site of food digestion and absorption.

The **small intestine** is a coiled tube in the abdominal cavity that is about 6 meters (20 feet) long in adults (Figure 5-14). So named because of its small diameter, the small intestine consists of three parts in the following order: the duodenum (DUE-oh-DEEN-um), jejunum (jeh-JEW-num), and ileum (ILL-ee-um).

Inside the small intestine, large molecules are broken into smaller molecules (that is, digested) with the aid of enzymes. These molecules are then transported into the bloodstream and the lymphatic system. The lymphatic system, discussed in Chapter 6, is a network of vessels that carries extracellular fluid from the tissues of the body to the circulatory system. In addition, tiny lymphatic vessels in the wall of the small intestine, known as **lacteals** (LACK-teels), absorb fats and transport them to the bloodstream.

Digestive Enzymes

Digestion in the small intestine requires enzymes from two sources.

The digestion of food molecules inside the small intestine requires enzymes produced from two different sources: the pancreas and the lining of the small intestine itself.

The **pancreas** lies beneath the stomach and is nestled in a loop formed by the first portion of the small intestine (Figure 5-19). The pancreas is a dual-purpose organ—that is, it has endocrine and exocrine functions. As an exocrine gland, it produces enzymes and sodium bicarbonate essential for the digestion of foodstuffs in the small intestine. Its endocrine function is fulfilled by special cells that produce hormones (insulin and glucagon) that help regulate blood glucose levels, thus assisting in maintaining homeostasis.

The digestive enzymes of the pancreas flow through the large pancreatic duct into the duodenum. Each day, approximately 1200–1500 milliliters (1.0–1.5 quarts) of pancreatic juice is produced and released into the small intestine. This liquid is composed of water, sodium bicarbonate, and several digestive enzymes (Table 5-4).

Sodium bicarbonate neutralizes the acidic chyme released by the stomach and thus protects the small intestine from stomach acid. It also gives the pancreatic juice a pH of about 8, creating an environment optimal for the function of the pancreatic enzymes.

The pancreatic enzymes released into the small intestine act on the large food molecules (proteins, starches, and so on), as shown in Table 5-4. As a result of pancreatic enzymatic activity, fats, proteins, and carbohydrates are broken down into smaller molecules.

Like the stomach, the pancreas secretes its enzymes in an inactive form. This protects the gland from self-destruction. **Trypsinogen** (trip-SIN-oh-gin), for example, is the inactive form of the protein-digesting enzyme **trypsin** (TRIP-sin). Produced by the pancreas, trypsinogen is activated by a substance on the epithelial lining of the small intestine. Trypsin, in turn, activates other digestive enzymes.

The final stage of digestion occurs with the aid of enzymes produced by the epithelial cells that line the small intestine. These enzymes are embedded in the membranes of the epithelial cells (Table 5-4). They cause further breakdown of the food molecules, just before the nutrients are absorbed by the cells.

TABLE 5-4	Digestive Enzymes	
Site of Production	Enzyme	Action
Salivary glands	Amylase	Digests polysaccharides in oral cavity
Stomach	Pepsin	Breaks proteins into peptides
Pancreas	Trypsin	Cleaves peptide bonds of polypeptides and proteins
	Chymotrypsin	Same as trypsin
	Carboxypeptidase	Cleaves peptide bonds on carboxyl end of polypeptides
	Amylase	Breaks starch molecules into smaller units (maltose)
	Phospholipase	Cleaves fatty acids from phosphoglycerides to form monoglycerides
	Lipase	Cleaves two fatty acids from triglycerides
	Ribonuclease	Breaks RNA into smaller nucleotide chains
	Deoxyribonuclease	Breaks DNA into smaller nucleotide chains
	Chymotrypsin	Same as trypsin
Epithelium of small intestine	Maltase	Breaks maltose into glucose subunits
	Sucrase	Breaks sucrose into glucose and fructose subunits
	Lactase	Breaks lactose into glucose and galactose subunits
	Aminopeptidase	Breaks peptides into amino acids

The Liver

The liver carries out hundreds of important functions.

The **liver** is one of the largest and most versatile organs in the body. By various estimates, the liver performs as many as 500 different functions. Situated on the right side of the abdomen under the protection of the rib cage, the liver is one of the body's storage depots for glucose (Figure 5-3). It also stores fats and many micronutrients such as iron, copper, and many vitamins. By storing glucose and lipids and releasing them as they are needed, the liver helps ensure a constant supply of energy-rich molecules for body cells. The liver also synthesizes some key blood proteins involved in clotting, and it is an efficient detoxifier of potentially harmful chemicals such as nicotine, barbiturates, and alcohol. These functions contribute significantly to homeostasis.

The liver also plays a key role in the digestion of fats through the production of a fluid called *bile*. **Bile** contains water, ions, and molecules such as cholesterol, fatty acids, and bile salts. **Bile salts** are steroids that break fat globules into smaller ones. This process is essential for lipid digestion, because lipid-digesting enzymes in the small intestine do not work well on large fat globules. Unless large fat globules can be broken into smaller ones, fat digestion is incomplete.

Produced by the cells of the liver, bile is first transported to the **gallbladder**, a sac attached to the underside of the liver (Figure 5-18). The gallbladder removes water from the bile, concentrating it. Bile is stored in the gallbladder until needed. When chyme is present in the small intestine, the gallbladder contracts and bile flows out through the duct system into the small intestine (Figure 5-18).

Bile flow to the small intestine may be blocked by gallstones, deposits of cholesterol and other materials in the gallbladder of some individuals. Gallstones may lodge in the ducts draining the organ, thus reducing—even completely blocking—the flow of bile. The lack of bile salts greatly reduces lipid digestion. Because lipid digestion is reduced, fat is not digested and absorbed.

Gallstones occur more frequently in older, overweight individuals. The incidence in the elderly is about one in five. Gallstones are usually removed surgically. This procedure requires that the entire gallbladder be removed. Bile continues to be produced, but is stored in the common bile duct, which enlarges to accommodate the bile. New drugs that dissolve gallstones in many patients may someday eliminate or reduce the need for surgery.

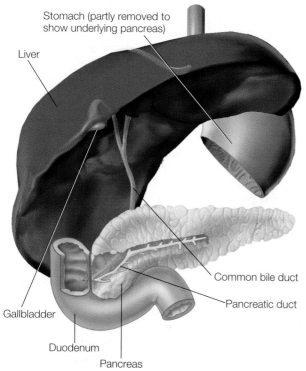

FIGURE 5-18 **The Organs of Digestion** The liver, gallbladder, and pancreas all play key roles in digestion. All empty by the common bile duct into the small intestine, in which digestion takes place.

The Intestinal Epithelium

The lining of the intestine is specially modified for absorption.

Virtually all food digestion occurs in the duodenum and jejunum. Once food molecules are digested, they are absorbed—that is, transported across the epithelial lining of the small intestine into the bloodstream or lymphatic system. In these regions, absorption is facilitated by three structural modifications of the small intestine. Figure 5-19a shows, for instance, that the lining of the small intestine is thrown into folds, which increase the overall surface area. On the surfaces of the folds are many fingerlike projections known as **villi** (VILL-eye, sing. villus). The villi also increase the surface area available for the absorption of food molecules. The intestinal surface area is further increased by **microvilli**, tiny protrusions of the plasma membranes of the epithelial cells lining the villi (Figure 5-19d). Each epithelial cell that functions in absorption contains an estimated 3000 microvilli. Two hundred million microvilli occupy a single square millimeter of intestinal lining.

Together, the circular folds, villi, and microvilli result in a dramatic increase in the surface area of the lining of the small intestine, making it 600 times greater than if it were flat. As shown in Figure 5-19e, each villus of the small intestine is endowed with a rich supply of blood and lymph capillaries. These tiny vessels absorb nutrients that have passed through the lining of the small intestine.

Numerous mechanisms are involved in absorption—they are too numerous to be discussed here. Three of the most common are diffusion, osmosis, and active transport (Chapter 3).

Although most nutrients diffuse from the epithelial cells into the blood capillaries, fatty acids and monoglycerides produced by the enzymatic digestion of triglycerides follow a different route. As shown in Figure 5-20, these molecules diffuse into the epithelial cells. Inside these cells they react with fatty acids to reform triglycerides. The triglycerides accumulate to form fat droplets. The fat droplets are then coated with a layer of lipoprotein (lipid bound to protein) that is produced by the endoplasmic reticulum. This treatment renders the fat droplets, known as *chylomicrons* (KIE-low-MIE-krons), water-soluble. The fat droplets are then released by the epithelial cells into the interstitial fluid (Figure 5-20). Because blood capillaries are relatively impermeable to chylomicrons, most of the lipid globules enter the more porous lymph capillaries in the villi. As shown in Figure 5-21, small- or medium-chain fatty acids not incorporated in triglycerides pass directly into blood capillaries in the villi.

The Large Intestine

The large intestine is the site of water resorption.

The **large intestine** is about 1.5 meters (5 feet) long and consists of four regions. Starting at the small intestine, they are the cecum, appendix, colon, and rectum (Figure 5-21). The **cecum** (SEA-come) is a pouch that forms below the juncture of the large and small intestine. A small, wormlike structure, the **appendix**, attaches to the cecum. Most of the large intestine consists of the **colon** (COE-lun). Unlike the small intestine, which

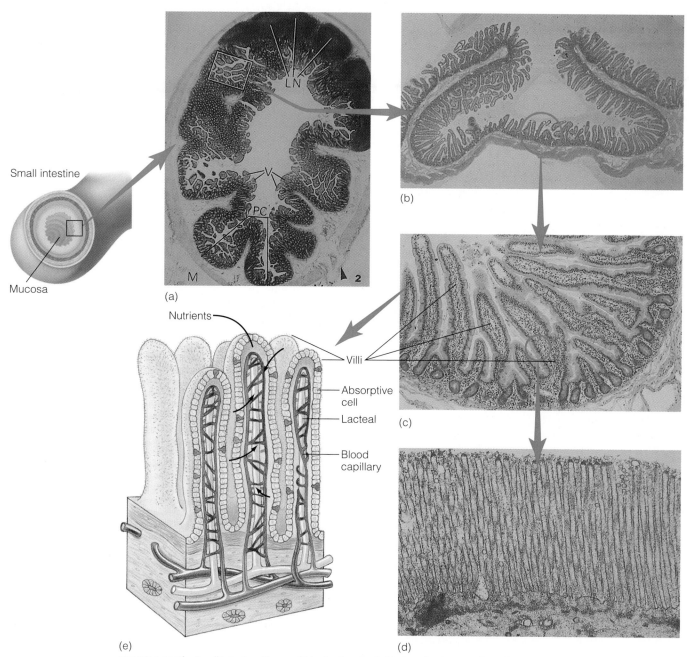

FIGURE 5-19 **The Small Intestine** The small intestine is uniquely "designed" to increase absorption. (a) A cross section showing the folds. LN means lymph nodules (aggregation of lymphocytes); V means villi; PC means plica circulares (circular fold). (b) A light micrograph of folds and villi. (c) Higher magnification of villi. (d) An electron micrograph of the surface of the absorptive cells showing the microvilli. (e) Each villus contains a loose core of connective tissue; a lacteal, or lymph, capillary; and a network of blood capillaries. Nutrients pass from the lumen of the small intestine through the epithelium and into the interior of the villi, where they are picked up by the lymph and blood capillaries.

is coiled and packed in a small volume, the colon consists of three relatively straight portions, the ascending colon, the transverse colon, and the descending colon. The colon empties into the **rectum** (WRECK-tum).

So named because of its large diameter, the large intestine receives materials from the small intestine. The material entering the large intestine consists of a mixture of water, undigested or unabsorbed food molecules, and indigestible food residues, such as cellulose. It also contains sodium and potassium ions.

The colon absorbs approximately 90% of the water and sodium and potassium ions that enter it. The undigested or unabsorbed nutrients feed a rather large population of E. coli bacteria. These bacteria synthesize several key vitamins: B-12, thiamine, riboflavin, and, most importantly, vitamin K, which is often deficient in the human diet. These vitamins are absorbed by the large intestine.

The contents of the large intestine (after water and salt have been removed) are known as the feces. The **feces** consist pri-

FIGURE 5-20 Fat Digestion and Absorption

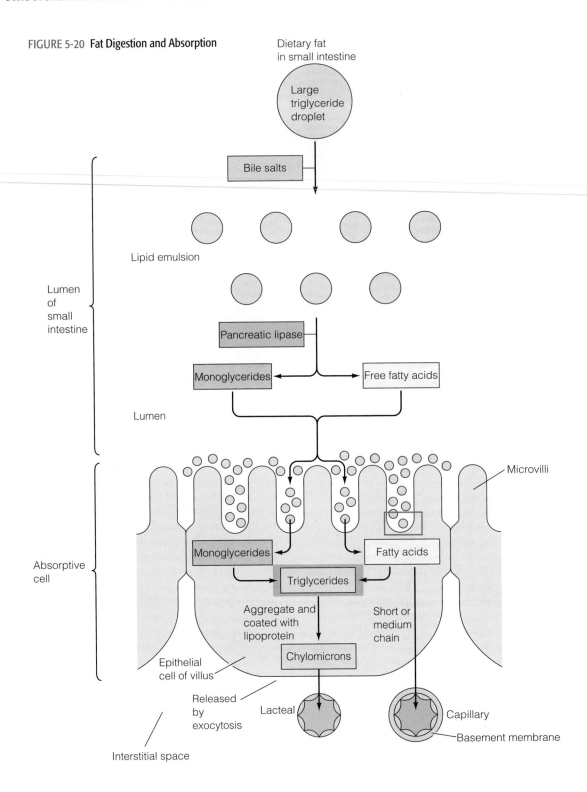

marily of undigested food, indigestible materials, and bacteria. Bacteria, in fact, account for about one-third of the dry weight of the feces.

The feces are propelled by peristaltic contractions along the colon until they reach the rectum (Figure 5-21). As fecal matter accumulates in the rectum, it distends the organ. This action, in turn, stimulates stretch receptors in the wall of the rectum. These receptors stimulate the **defecation reflex** (DEAF-eh-KA-shun). In this reflex, nerve impulses from the stretch receptors in the rectum travel via nerves to the spinal cord. In the

spinal cord, the nerve impulses stimulate nerve cells that supply the smooth muscle in the wall of the rectum, causing them to contract and expel the feces. Nerve impulses from the spinal cord also travel to the internal anal sphincter, a ring of smooth muscle, which normally keeps fecal matter from entering the anal canal. These inpulses allow the sphincter to relax.

Defecation does not occur until the external anal sphincter relaxes. This sphincter is composed of skeletal muscle and is under conscious control in all individuals except babies. If the time and place are appropriate, the external anal sphincter is re-

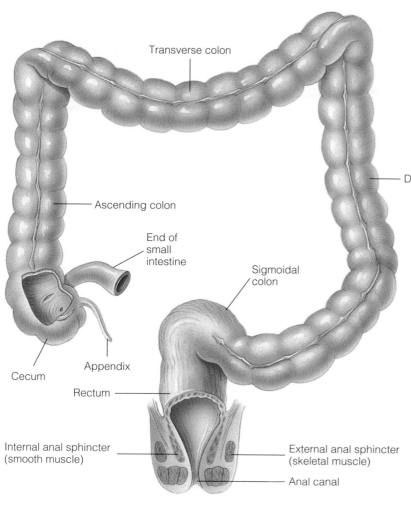

FIGURE 5-21 **The Large Intestine** This organ consists of four basic parts: the cecum, appendix, colon, and rectum.

Transverse colon

Descending colon

Ascending colon

End of small intestine

Sigmoidal colon

Cecum

Appendix

Rectum

Internal anal sphincter (smooth muscle)

External anal sphincter (skeletal muscle)

Anal canal

laxed, and defecation can occur. Defecation is usually assisted by voluntary contractions of the abdominal muscles and a forcible exhalation, both of which increase intra-abdominal pressure.

If circumstances are inappropriate, voluntary tightening of the external anal sphincter prevents defecation, despite prior activation of the reflex. When defecation is delayed, the muscle in the wall of the rectum relaxes, and the urge to defecate

subsides—at least until the next movement of feces into the rectum occurs.

If defecation is delayed too long, additional water may be removed from the feces, making them hard and dry and difficult to pass. This condition is called **constipation**.

Besides being uncomfortable, constipation may result in a dull headache, loss of appetite, and depression. Constipation is generally caused by (1) ignoring the urge to defecate; (2) decreases in colonic contractions, which occur with age, emotional stress, or low-fiber diets; (3) tumors or colonic spasms that obstruct the movement of feces; (4) certain drugs like those containing codeine; and (5) nerve injury that impairs the defecation reflex.

Constipation is not only uncomfortable, it can result in serious problems. Hardened fecal material, for instance, may become lodged in the appendix. This, in turn, may lead to inflammation of the organ, a condition called **appendicitis** (ah-PEND-eh-SITE-iss). When this occurs, the appendix becomes swollen and filled with pus and must be removed surgically to prevent the organ from bursting and spilling its contents into the abdominal cavity. Fecal matter leaking into the abdominal cavity introduces billions of bacteria and can result in a deadly infection.

5-3 Controlling Digestion

Digestion involves many different processes and is controlled largely by nerves and hormones. This section discusses some of the key events involved in the control of digestion.

Saliva and Salivation

Salivation is stimulated by a nervous reflex.

As noted earlier, digestion begins in the oral cavity with the release of saliva. The excretion of saliva may be brought on by the sight, smell, taste, and sometimes even the thought of food like a hot fudge sundae. Chewing has a similar effect. The se-

cretion of saliva is controlled by the nervous system and is largely a reflex response. For a discussion of the discovery of this phenomenon, see Scientific Discoveries 5-1.

The Gastric Glands

Gastric gland secretion is controlled by a variety of different mechanisms.

Besides activating salivary production, the stimuli listed above also cause the brain to send nerve impulses along the vagus nerve (VAG-gus) to the stomach (Figure 5-22). These nerve impulses initiate the secretion of hydrochloric acid (HCl) and pepsinogen from

Scientific Discoveries that Changed the World

5-1 ## Discovering the Nature of Digestion
Featuring the Work of van Helmont, Beaumont, and Pavlov

The foundation for the modern understanding of digestion was laid by a number of scientists. Two noteworthy contributors were a seventeenth-century Flemish physician, Jan Baptista van Helmont, and a nineteenth-century American, William Beaumont.

In van Helmont's time, most people thought that the digestion of food in the human was akin to cooking—that is, that food inside the stomach was broken down by body heat. Van Helmont, however, dismissed this theory with the simple observation that digestion occurs in cold-blooded animals such as fish. The lack of heat in such animals suggested to him that an alternative process was at work. But what was it?

The road to discovery began in a rather bizarre way. One day, a tame sparrow that often visited the scientist and would sit on his shoulder attempted to bite van Helmont's tongue. The physician detected a slight acidity in the bird's bite and hypothesized that acid must digest food.

To test this hypothesis, van Helmont tried to chemically break down meat with a vinegar solution, but he could not. Consequently, he hypothesized that the body must contain chemical substances in the stomach and small intestine that are specific for different types of food. Today, we know these chemicals as digestive enzymes.

Little insight into digestion was forthcoming in the period between van Helmont's work and the studies of the American scientist William Beaumont. Beaumont was a medical doctor. Soon after becoming a licensed physician, Beaumont joined the army, where he would make his important discoveries about digestion. He owed his great work to an accident. In 1822, a French-Canadian porter was wounded in the abdomen by a musket that had accidentally discharged at close range. The porter,

---

cells of the gastric glands. The most potent stimulus for gastric secretion, however, is the presence of protein in the stomach. Protein stimulates chemical receptors inside the stomach that activate nerves in the stomach wall. These nerves, in turn, stimulate the gastric glands to secrete HCl and pepsinogen.

Nerve impulses from the brain also stimulate the secretion of a hormone known as **gastrin**. As shown in Figure 5-22, gastrin is released into the blood. It acts on the gastric glands, stimulating additional HCl and pepsinogen secretion.

The concentration of HCl in the stomach is also regulated by a negative feedback mechanism, illustrated in Figure 5-22. If the acid content rises too high, HCl inhibits gastrin secretion, shutting down the gastric glands.

Control of Pancreatic Secretions

Pancreatic secretions are stimulated by two intestinal hormones.

While the gastric glands of the stomach are stimulated by nerve impulses and the hormone gastrin, secretion by the pancreas is controlled by two hormones. The first hormone is **secretin**. It's

release is stimulated by the presence of acidic chyme in the small intestine (Figure 5-23). Secretin is produced by the cells of the duodenum and, like all hormones, travels in the bloodstream. In the pancreas, secretin stimulates the release of sodium bicarbonate. Sodium bicarbonate, in turn, is excreted by the pancreas into the small intestine, where it neutralizes the acidic chyme and creates an environment optimal for pancreatic enzymes.

The release of pancreatic enzymes is triggered by another intestinal hormone with the tongue-twisting name **cholecystokinin** (COAL-ee-sis-toe-KIE-nin) or CCK. This hormone is produced by cells of the duodenum in the presence of chyme (Figure 5-22). CCK also stimulates the gallbladder to contract, releasing bile into the small intestine.

Interestingly, recent evidence links abnormally low secretion of CCK to **bulimia** (boo-LEEM-ee-ah), a disorder characterized by recurrent binge eating followed by intentional vomiting. Some researchers believe that CCK may trigger the feeling of being full. Approximately 4% of America's young women, and a far smaller fraction of men, suffer from bulimia. This disorder is also thought to have psychological roots.

5-4 Health and Homeostasis

There are few places where the relationship between homeostasis and human health is as evident as in nutrition. Because cells and organs involved in homeostasis require an adequate supply of nutrients, nutritional deficiencies can have noticeable impacts on body functions and health. Supporting this idea, many studies show that an unhealthy diet can increase the risk of contracting certain diseases, including cancer, heart disease, and hypertension.

Consider an example. Magnesium (a major mineral) is routinely ingested in insufficient amounts. New research suggests that such deficiencies may underlie a number of medical conditions, including diabetes, high blood pressure, problems during pregnancy, and cardiovascular disease.

Research shows that adding magnesium to the drinking water of rats with hypertension can eliminate the disease. Studies in rabbits show that magnesium reduces lipid levels in the

Alexis St. Martin, who was only 18, was brought to Beaumont for treatment. The doctor found that the bullet had ripped a hole in the man's stomach, out of which poured food he had eaten.

The patient survived, but the wound was slow to heal. For over 8 months, Beaumont tried in vain to close the hole in the stomach. During this time, it dawned on the young physician that he had stumbled across a golden opportunity to study digestion.

Beaumont's experiments consisted of feeding the patient various types of food, then studying what took place inside the man's stomach. In the next 9 years, Beaumont studied St. Martin's stomach contents with a wide assortment of foods, looking at the rate and temperature of digestion and the chemical conditions that favored different stages of the process. Through his studies, Beaumont demonstrated that digestion was a chemical phenomenon.

Beaumont did not address one important question: What caused gastric juice to be secreted? Was it the presence of food in the stomach or something else?

The answer came in 1889 from the work of Ivan Pavlov, a scientist whose experiments are legendary. In studies on dogs, Pavlov showed that the secretion of gastric juice was stimulated by the nervous system. How did he make this determination?

In one of many experiments, the Russian scientist surgically connected a fold of a dog's stomach to an opening in the animal's side; this permitted him to examine the production of gastric secretions. He next cut and tied off the esophagus of the dog, so that food the animal swallowed could not enter the stomach. Following the surgery, Pavlov gave the dog food and found that as soon as food entered the dog's mouth, gastric juice began to be secreted. These secretions continued as long as food was present, thus suggesting nervous system involvement.

Since that time, a great deal has been learned about digestion and its control. For example, as you learned in this chapter, hormones are also involved in the control of many digestive processes. But don't close the book on this subject; much more will inevitably be discovered.

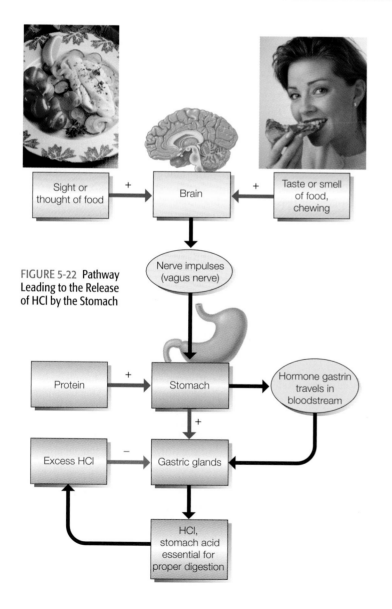

FIGURE 5-22 Pathway Leading to the Release of HCl by the Stomach

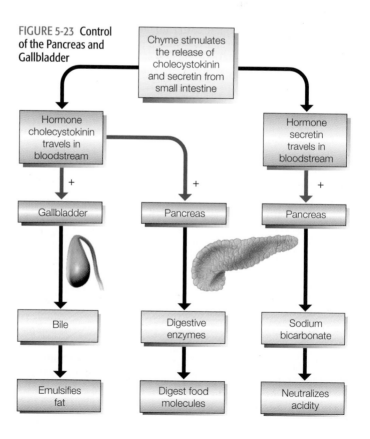

FIGURE 5-23 Control of the Pancreas and Gallbladder

blood and also reduces plaque formation in blood vessels. Rabbits on a high-cholesterol, low-magnesium diet, for example, have 80% to 90% more atherosclerotic plaque than rabbits on a high-cholesterol, high-magnesium diet.

Studies have shown that magnesium deficiencies during pregnancy result in migraine headaches, high blood pressure, miscarriages, stillbirths, and babies with low birth weight. Research suggests that magnesium deficiency causes spasms in the blood vessels of the placenta, which reduce blood flow to the fetus. This can retard fetal growth and may even kill a fetus. Magnesium supplements greatly reduce the incidence of these problems.

Researchers believe that 80% to 90% of the American public may be magnesium deficient. One reason for this deficiency is that phosphates in many carbonated soft drinks bind to magnesium in the intestine, preventing it from being absorbed into the blood. Magnesium deficiencies can be reversed by eating more green leafy vegetables, seafood, and whole grain cereals and reducing consumption of carbonated soft drinks with high phosphate levels. Mineral supplements could help as well, but they should be used with extreme caution, because excess magnesium can cause neurological disorders.

Over the years, numerous dietary recommendations have been issued to help Americans live healthier lives and reduce the risk of cancer and heart attack. Nutritionists recommend daily consumption of (1) fruits and vegetables, especially cabbage and greens, (2) high-fiber foods, such as whole-wheat bread and celery, and (3) foods high in vitamins A and C. A healthy diet also minimizes the consumption of animal fat; red meat; and salt-cured, nitrate-cured, smoked, or pickled foods, including bacon and lunch meat.

A healthy diet results largely from habit and circumstance. In the hustle and bustle of modern society, many of us ignore proper nutrition, grabbing fat-rich foods because we haven't the time to sit down to a nutritionally balanced meal. Our fast-paced world places a high premium on saving time, often ignoring the importance of eating and living right. Fast food may allow us to hurry on to our next appointment, but unless it is nutritionally sound, it probably decreases the quality of our lives in the long run.

SUMMARY

An Introduction to Nutrition

1. Humans acquire energy and nutrients from the food they eat. These nutrients fall into two categories: macronutrients, substances needed in large quantity, and micronutrients, substances required in much lower quantities.

2. The four major macronutrients are water, carbohydrates, lipids, and proteins.

3. Maintaining adequate water intake is important because water is involved in many chemical reactions in the body. It also helps to maintain body temperature and a constant level of nutrients and wastes in body fluids, all of which are vital to homeostasis. Water is contained in the liquids we drink and the foods we eat.

4. Glucose is a sugar that is broken down in the body to produce energy. In the human diet, glucose comes primarily from the polymer, starch. Glucose molecules are stored in the liver and muscle as glycogen.

5. Another important dietary carbohydrate is cellulose. Although not digested, cellulose facilitates the passage of fecal matter through the intestinal tract and reduces the incidence of colon cancer.

6. Dietary protein is chiefly a source of amino acids for building proteins, enzymes, and protein hormones. Protein is broken down into amino acids and absorbed into the bloodstream. However, the body can synthesize some amino acids. Others cannot be synthesized in the body and must be supplied in the food we eat. These are known as essential amino acids.

7. To ensure an adequate supply of all amino acids, individuals should eat complete proteins, such as those found in milk or eggs, or combine lower quality protein sources.

8. Lipids are a diverse group of organic chemicals characterized by their lack of water solubility. The biologically important lipids serve many functions. Some lipids (triglycerides) provide energy. They also form layers of heat-conserving insulation. Other lipids (phospholipids and steroids) are part of the plasma membranes of cells.

9. Triglycerides are fats and oils. Triglycerides with many double bonds in their fatty acid side chains lower one's risk of a buildup of plaque on arterial walls. Triglycerides low in saturated fatty acids tend to increase this disease; they are commonly found in animal fats.

10. Micronutrients are needed in much smaller quantities and include two groups: vitamins and minerals.

11. Vitamins are a diverse group of organic compounds that are required in relatively small quantities for normal metabolism. A deficiency or surplus of one or more vitamins may alter homeostasis, with serious effects on human health.

12. Human vitamins fit into two categories: water-soluble and fat-soluble. The water-soluble vitamins include vitamin C and the B-complex vitamins. The fat-soluble vitamins include vitamins A, D, E, and K.

13. Minerals fit into one of two groups: trace minerals, those required in very small quantity, and major minerals, those required in greater quantity. Deficiencies and excesses of both types of minerals can lead to serious health problems.

14. In recent years, scientists have discovered numerous chemicals in the foods we eat that help prevent disease. Foods containing them are called functional foods because they provide a health benefit beyond basic nutrition. One important group is known as antioxidants, chemicals that react with naturally occurring chemical oxidants, substances in the blood that are involved in cholesterol buildup in the walls of arteries as well as cancer. Eating a diet rich in antioxidants may reduce heart and artery disease as well as cancer.

The Digestive System

15. Food digestion begins in the mouth. The teeth mechanically break down the food. Saliva liquefies it, making it easier to swallow. Salivary amylase begins to digest starch molecules.

16. Food is pushed by the tongue to the pharynx, where it triggers the swallowing reflex. Peristaltic contractions propel the food down the esophagus to the stomach.

17. The stomach is an expandable organ that stores and further liquefies the food. The churning action of the stomach, created by peristaltic contractions, mixes the food, turning it into a paste known as chyme.

18. The stomach releases food into the small intestine in timed pulses, ensuring efficient digestion and absorption. Very limited chemical digestion and absorption occur in the stomach.

19. The stomach produces hydrochloric acid, which denatures protein, allowing it to be acted on by enzymes. The stomach also produces a proteolytic enzyme called pepsin, which breaks proteins into peptides.

20. The functions of the stomach are regulated by neural and hormonal mechanisms.

21. The small intestine is a long, coiled tubule in which most of the enzymatic digestion and absorption of food occur. Digestive enzymes come from the lining of the intestine and the pancreas. Pancreatic enzymes break macromolecules into smaller fragments. The intestinal enzymes break these molecules into even smaller fragments that can be absorbed by the epithelial lining of the small intestine.

22. The pancreas also releases sodium bicarbonate, which neutralizes HCl entering the small intestine with the chyme and creates an environment suitable for pancreatic enzyme function.

23. The liver plays an important role in digestion. It produces a liquid called bile that contains bile salts. Bile is stored in the gallbladder and released into the small intestine when food is present. Bile salts break fats into small globules that can be acted on by enzymes.

24. Undigested food molecules pass from the small intestine into the large intestine, which absorbs water, sodium, and potassium, as well as vitamins produced by intestinal bacteria. It also transports the waste, or feces, to the outside of the body.

Controlling Digestion

26. Digestive processes are largely controlled by the nervous and endocrine systems. The release of saliva is stimulated by the sight, smell, taste, and the thought of food. These stimuli also cause the brain to send nerve impulses to the gastric glands of the stomach, initiating the secretion of HCl and gastrin, a hormone that also stimulates HCl secretion.

Health and Homeostasis

27. Human health is dependent on good nutrition. Numerous studies suggest that a healthy, balanced diet can decrease the risk of cancer, heart disease, hypertension, and other diseases.

28. Nutritionists recommend the daily consumption of (a) fruits and vegetables; (b) high-fiber foods, such as whole-wheat bread and celery; and (c) foods high in vitamins A and C.

29. In addition, nutritionists recommend reducing the consumption of animal fat; red meat; and salt-cured, nitrate-cured, smoked, or pickled foods, including bacon and lunch meat.

critical thinking

Thinking Critically—Analysis

This Analysis corresponds to the Thinking Critically scenario that was presented at the beginning of this chapter.

To test this hypothesis, one might simply measure oxalate in the urine and blood of a large number of people eating both high- and low-calcium diets. If levels were highest in the people eating low-calcium diets, the hypothesis might very well be valid. Knowing that one study is not enough to validate a hypothesis, it might be wise to repeat it several times.

This exercise illustrates the importance of questioning basic assumptions or popularly held beliefs. It shows how dead wrong popular wisdom can be and is a startling testimony to the need for scientific testing of beliefs. One has to wonder how many other assumptions are in error.

KEY TERMS AND CONCEPTS

Amino acid, p. 86
Appendicitis, p. 101
Appendix, p. 98
Antioxidant, p. 91
Bile, p. 97
Bile salts, p. 98
Bulimia, p. 102
Cecum, p. 98
Cellulose, p. 83
Cholecystokinin, p. 102
Chyme, p. 96
Colon, p. 98
Complete proteins, p. 86
Constipation, p. 101
Defecation reflex, p. 100
Dietary fiber, p. 83
Disaccharide, p. 82
Enamel, p. 92
Epiglottis, p. 93
Esophagus, p. 94
Essential amino acid, p. 86
Fat-soluble vitamins, p. 90
Fatty acids, p. 83

Feces, p. 99
Food pyramid, p. 81
Functional food, p. 91
Gallbladder, p. 98
Gastric gland, p. 96
Gastrin, p. 102
Gingivitis, p. 93
Glucose, p. 82
Glycerol, p. 83
Glycogen, p. 82
Incomplete proteins, p. 86
Lacteals, p. 97
Large intestine, p. 98
Lipid, p. 83
Liver, p. 97
Microvilli, p. 98
Mineral, p. 90
Monosaccharide, p. 82
Mouth, p. 92
Overnutrition, p. 88
Pancreas, p. 97
Pharynx, p. 93
Peristalsis, p. 94

Plaque, p. 92
Polysaccharide, p. 82
Protein, p. 86
Rectum, p. 99
Secretin, p. 102
Saliva, p. 92
Small intestine, p. 96
Starch, p. 82
Steroid, p. 85
Stomach, p. 95
Tongue, p. 93
Triglycerides, p. 83
Trypsin, p. 97
Trypsinogen, p. 97
Ulcer, p. 96
Undernutrition, p. 88
Villus, p. 98
Vitamin, p. 88
Water, p. 82
Water-soluble fiber, p. 83
Water-insoluble fiber, p. 83
Water-soluble vitamins, p. 90

CONCEPT REVIEW

1. The body requires proper nutrient input to maintain homeostasis. Give an example, and explain how nutrition affects homeostasis. pp. 82–92
2. Describe the conditions during which protein provides cellular energy. pp. 86–88
3. If you were considering becoming a strict vegetarian, eating no animal products, even milk and eggs, how would you be assured of getting all of the amino acids your body needs? p. 86
4. Describe how the different types of dietary fiber help protect human health. p. 83
5. What are vitamins, and why are they needed in such small quantities? pp. 88–90
6. A dietary deficiency of one vitamin can cause wide-ranging effects. Why? p. 90
7. What organs physically break food down, and what organs participate in the chemical breakdown of food? pp. 92–98
8. Describe the process of swallowing. p. 93
9. Describe the function of hydrochloric acid and pepsin in the stomach. How does the stomach protect itself from these substances? p. 96
10. How do ulcers form, and how can they be treated? p. 96
11. Describe the endocrine and nervous system control of the stomach function. pp. 95–96
12. The small intestine is the chief site of digestion and absorption. Where do the enzymes needed for this process come from, and how is the release of these enzymes stimulated? What other molecules are needed for proper digestion? pp. 96–97
13. Describe the functions of the large intestine. pp. 98–101

SELF-QUIZ: TESTING YOUR KNOWLEDGE

1. Water is a type of nutrient known as a _____ because it is required in large quantities. p. 82
2. Glucose is a monosaccharide that is combined to form a nutritive polysaccharide in plants known as _____. p. 82
3. Dietary _____ may be either water soluble or water insoluble. p. 83
4. The water-insoluble type is thought to lower _____ levels in the blood. p. 83
5. _____ are a structurally diverse group including fats and oils and steroids that are all insoluble in water. p. 83
6. A _____ is made of three fatty acid molecules and a molecule of glycerol. p. 83
7. Cholesterol is produced in the _____ and ingested in some types of foods such as eggs. p. 85
8. Amino acids that cannot be synthesized by the body are called _____ amino acids. p. 86
9. A protein that contains all of the essential amino acids such as milk or tofu is known as a _____ protein. p. 86
10. _____ is a condition in which a person is severely overweight. p. 88
11. _____ are micronutrients needed in small quantity; most are not synthesized in the body and most play an important role in metabolic reactions. p. 88
12. Copper is a _____ mineral required in very small quantity. p. 90
13. A _____ food is a food that provides health benefits beyond nutrition. p. 91
14. Salivary _____ is an enzyme released in the mouth which helps break down starch molecules. p. 92
15. After food leaves the mouth, it enters the pharynx and is then passed into the _____, which transports it to the stomach. p. 93
16. Muscular contraction of the stomach and intestines is called _____. p. 94
17. Hydrochloric acid is released by the _____ glands in the stomach. p. 96
18. Bile is produced by the _____ and stored in the _____ and released into the _____ intestine. p. 98
19. The pancreas releases _____ which neutralizes the chyme and creates a pH conducive to enzyme digestion in the small intestine. p. 97
20. _____ in the large intestine produce vitamin K, which is often deficient in the human diet. p. 99

www.jbpub.com/humanbiology/5e

The site features eLearning, an online review area that provides quizzes, chapter outlines, and other tools to help you study for your class. You can also follow useful links for in-depth information, research the differing views in the Point/Counterpoints, or keep up on the latest health news.

The Circulatory System

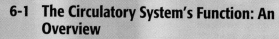

thinking critically

High blood pressure, or hypertension, is a disease that afflicts many people the world over. African Americans, however, are more prone to this affliction than most other people. In fact, they are twice as likely to contract the disease as Caucasians.

Some scientists believe that this anomaly may be the result of genetic differences in the population and that these differences arose when many blacks were captured in Africa and transported to plantations where they were held as slaves. According to this hypothesis, many African Americans are hypertensive today because of a rare genetic trait that helped a small subset of their ancestors survive the inhumane conditions of slavery. That trait is an inherited tendency to conserve salt within the body.

According to the hypothesis, the ancestors of African Americans that could conserve water were more likely to survive the grueling trip to the United States and the arduous conditions on the plantations. Conserving water meant they were less likely to die of dehydration. This trait, which proved to be an asset for slaves, may now be a deadly attribute among their modern descendants. As a scientist how could you test this hypothesis?

David McMahon, a 48-year-old New York attorney, collapsed in his office one morning and was rushed to a nearby hospital. There a team of cardiologists discovered a blood clot lodged in a narrowed section of his right coronary artery, which supplies blood to the heart. The physicians began immediate action to prevent further damage to the oxygen-starved heart muscle. Through a small incision in McMahon's groin, they inserted a tiny plastic catheter into the artery that carries blood to the leg. They then threaded the catheter up through arteries to his heart (Figure 6-1). Once they reached the heart, physicians guided the catheter into the coronary artery, where they injected an enzyme, called *streptokinase* (strep-toe-KI-nace), through the catheter. Streptokinase dissolved the blood clot, restoring blood flow to McMahon's heart muscle.

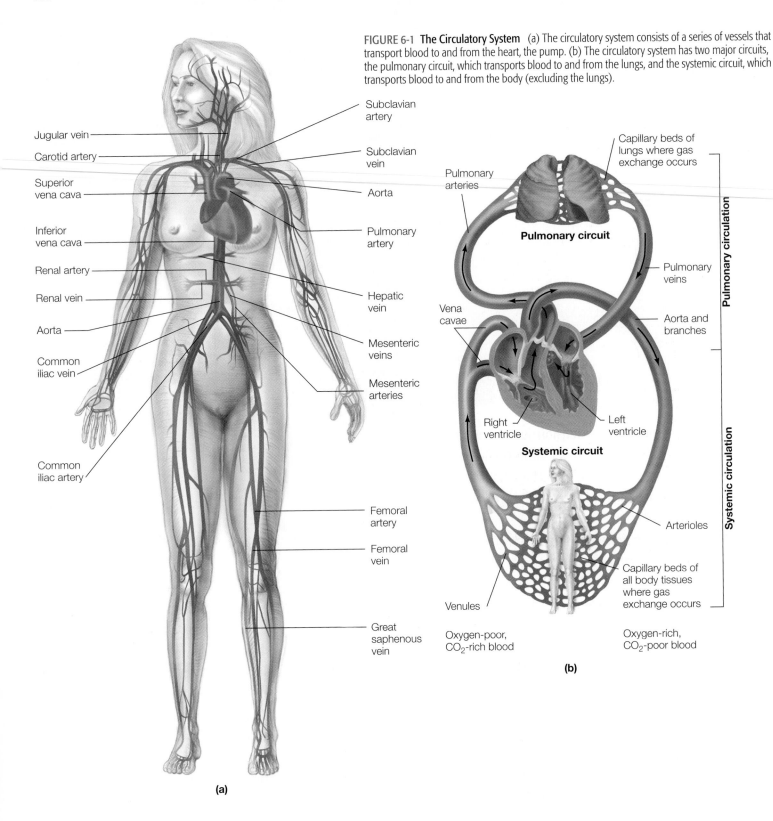

FIGURE 6-1 **The Circulatory System** (a) The circulatory system consists of a series of vessels that transport blood to and from the heart, the pump. (b) The circulatory system has two major circuits, the pulmonary circuit, which transports blood to and from the lungs, and the systemic circuit, which transports blood to and from the body (excluding the lungs).

Although David McMahon survived, many others who suffer similar attacks are not so lucky. They either arrive at the hospital too late or, if they do make it to the hospital in a timely fashion, do not receive blood-clot-dissolving agents or other treatments in time that prevent extensive damage to their heart muscle.

Heart attacks strike thousands of Americans each year. They are one of a handful of diseases of the circulatory system caused by the stressful conditions of modern life, smoking, poor eating habits, and a host of other factors. This chapter examines the circulatory system—in sickness and in health. It also discusses the lymphatic system, which functions in both circulation and immune protection.

6-1 The Circulatory System's Function: An Overview

The human circulatory system consists of two parts, a muscular pump, the heart, and a network of vessels (Figure 6-1). These vessels transport blood throughout the body, ensuring the adequate distribution of oxygen, taken in via the respiratory system, and nutrients, absorbed by the digestive system, to the cells, tissues, and organs. The circulation of blood throughout the body also ensures the distribution of body heat. Besides transporting nutrients and heat, the circulatory system picks up waste from cells in the body's tissues and transports them to the organs of excretion such as the kidneys

6-2 The Heart

The heart is a muscular organ that pumps blood throughout the body.

The **heart** is a muscular pump in the thoracic (chest) cavity. Often referred to as the workhorse of the cardiovascular system, the heart propels blood through the 50,000 miles of blood vessels in the body. Each day, the heart beats approximately 100,000 times. If you had a dollar for every heartbeat, you would be a millionaire in 10 days. The heart doesn't beat at a steady rate, however. The heart can adjust its rate to meet the changing needs of the body, for example, during exercise.

The heart, shown in Figure 6-2, is a fist-sized organ whose walls are composed of three layers, the pericardium, the myocardium, and the endocardium. The **pericardium** (pear-ah-CARD-ee-um) forms a thin, closed sac that surrounds the heart and the bases of large vessels that enter and leave the heart. The pericardial sac is filled with a clear, slippery watery fluid that reduces the friction produced by the heart's repeated contraction. The middle layer, the **myocardium** (my-oh-CARD-ee-um), is the thickest part of the wall and is composed chiefly of cardiac muscle cells (Chapter 4). The inner layer, the **endocardium** (end-oh-CARD-ee-um), is the endothelial layer, which lines the heart chambers.

The Pulmonary and Systemic Circuits

Blood flows through two distinct circuits in the cardiovascular system.

To understand how the heart's anatomy relates to its function, it is important to first know a bit about circulation. As shown in Figure 6-1b, the circulatory system consists of two distinct circuits, the **pulmonary circuit**, which carries blood to and from the lungs, and the **systemic circuit**, which transports blood to and from the rest of the body.

As you shall soon see, the pulmonary circuit has two main functions: (1) to deliver blood to the lungs so that it can be enriched with oxygen that can then be distributed via systemic circulation; and (2) to help the body get rid of carbon dioxide, a waste product of energy production in cells.

As shown in Figure 6-1b, the heart consists of four hollow chambers—two on the right side of the heart and two on the left. Blood is pumped through the pulmonary circuit by the right side of the heart—the right **atrium** and right **ventricle**. Blood is pumped through the systemic circuit by the left side of the heart—the left atrium and left ventricle.

Figure 6-2 illustrates the course that blood takes through the heart. Drawn in blue, blood low in oxygen and rich in carbon dioxide enters the right side of the heart from the **superior** and **inferior vena cavae** (VEEN-ah CAVE-ee), which are part of the systemic circulation. These veins, which carry blood that has lost most of its oxygen to the cells, tissues, and organs of the body, empty directly into the right atrium (A-tree-um), the uppermost chamber of the

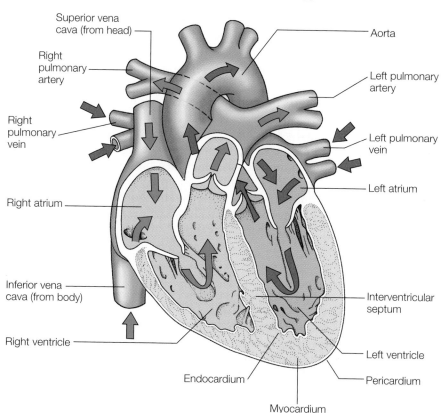

FIGURE 6-2 **Blood Flow Through the Heart** Deoxygenated (carbon-dioxide-enriched) blood (blue arrows) flows into the right atrium from the systemic circulation and is pumped into the right ventricle. The blood is then pumped from the right ventricle into the pulmonary artery, which delivers it to the lungs. In the lungs, the blood releases its carbon dioxide and absorbs oxygen. Reoxygenated blood (red arrows) is returned to the left atrium, then flows into the left ventricle, which pumps it to the rest of the body through the systemic circuit.

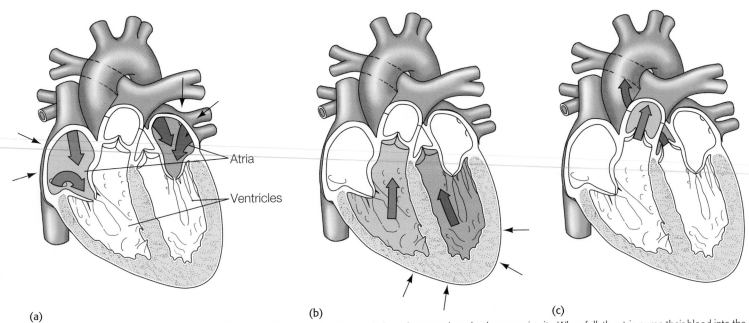

Atria

Ventricles

(a)   (b)   (c)

FIGURE 6-3 **Blood Flow Through the Heart** (a) Blood enters both atria simultaneously from the systemic and pulmonary circuits. When full, the atria pump their blood into the ventricles. (b) When the ventricles are full, they contract simultaneously, (c) delivering the blood to the pulmonary and systemic circuits.

heart. The blood is pumped from here into the right ventricle (VEN-trick-el), the lower chamber on the right side. When the right ventricle is full, the muscles in its wall contract, forcing blood into the pulmonary arteries, which lead to the lungs.

In the lungs, this blood is oxygenated, then returned to the heart via the pulmonary veins. The pulmonary veins, in turn, empty directly into the left atrium, the upper chamber on the left side of the heart. It's the first part of the systemic circuit.

Next, the oxygen-rich blood is pumped to the left ventricle. When it's full, the left ventricle's thick, muscular walls contract and propel the blood into the aorta (a-OR-tah). The **aorta** is the largest artery in the body. It carries the oxygenated blood away from the heart, delivering it to the cells and tissues of the body.

The flow of blood just described presents a slightly misleading view of the way the heart really works. As shown in Figure 6-3, both atria actually fill and contract simultaneously, delivering blood to their respective ventricles. The right and left ventricles also fill simultaneously, and when both ventricles are full, they also contract in unison, pumping the blood into the systemic and pulmonary circulations. The coordinated contraction of heart muscle is brought about by an internal timing device, or pacemaker (described later).

Heart Valves

Valves help control the flow of blood in the heart.

The human heart contains four valves that control the direction of blood flow, ensuring a steady flow from the atria to the ventricles and from the ventricles to the large vessels leading away from the heart (Figure 6-4a).

The valves between the atria and ventricles are known as **atrioventricular valves** (AH-tree-oh-ven-TRICK-u-ler). Each

valve consists of two or three flaps of tissue anchored to the inner walls of the ventricles by slender tendinous cords, the **chordae tendineae** (CORD-ee ten-DON-ee-ee), resembling the strings of a parachute (Figures 6-4a and 6-4b). The right atrioventricular valve, between the right atrium and right ventricle, is called the tricuspid valve (try-CUSS-pid), because it contains three flaps. The left atrioventricular valve is the bicuspid valve (bye-CUSS-pid); it contains two flaps. To remember the valves, imagine you are wearing a jersey with the number 32 on the front. This reminds you that the tricuspid valve is on the right side and the bicuspid valve is on the left.

Between the right and left ventricles and the arteries into which they pump blood are the semilunar valves (SEM-eye-LUNE-ir) (Figures 6-4a and 6-4c). The semilunar valves (literally, "half moon") consist of three semicircular flaps of tissue (Figures 6-4c and 6-4d).

The heart valves are one-way valves. They open when blood pressure builds on one side and close when it increases on the other. When the ventricles contract, blood forces the semilunar valves open. Blood flows out of the ventricles into the large arteries. The backflow of blood causes the valve to close, preventing blood from draining back into the ventricles. The atrioventricular valves function in similar fashion.

Heart Sounds

Heart sounds result from the closing of heart valves.

When physicians listen to a patients' hearts, they are actually listening to the sounds of the heart valves closing. The noises they hear are called the *heart sounds* and are often described as "LUB-dupp." The first heart sound (LUB) results from the closure of the atrioventricular valves. It is longer and louder than the second heart sound (dupp), produced when the semilunar valves shut.

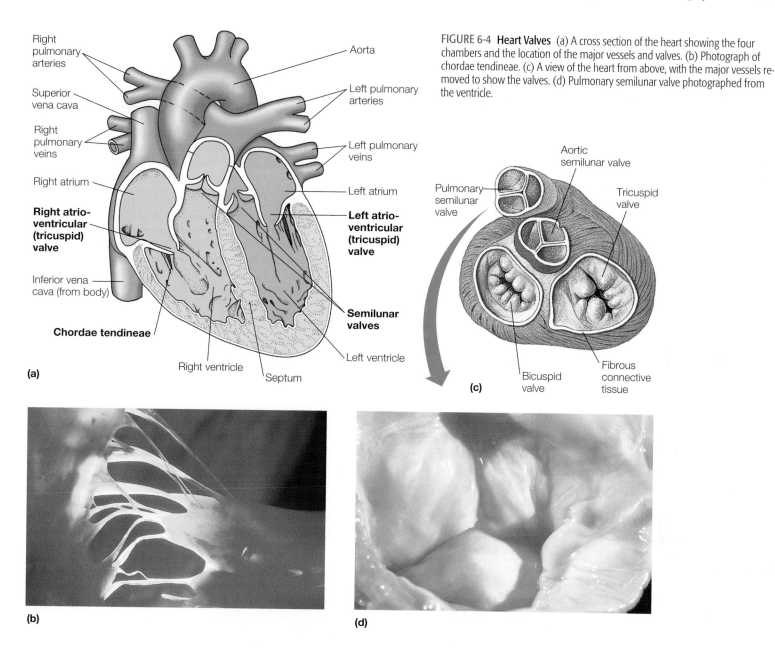

FIGURE 6-4 Heart Valves (a) A cross section of the heart showing the four chambers and the location of the major vessels and valves. (b) Photograph of chordae tendineae. (c) A view of the heart from above, with the major vessels removed to show the valves. (d) Pulmonary semilunar valve photographed from the ventricle.

Interestingly, the right and left atrioventricular valves do not close at precisely the same time. Nor do the semilunar valves. Thus, by careful placement of the stethoscope, a physician can listen to each valve individually to determine whether it is functioning properly.

In some individuals, diseases alter the function of the valves. This, in turn, may dramatically decrease the efficiency of the heart and the circulation of blood. Rheumatic (RUE-mat-tick) fever, for example, is caused by a bacterial infection and affects many parts of the body, including the heart. Although it is rare in the more affluent countries, rheumatic fever is a significant problem in the less developed countries. Rheumatic fever begins as a sore throat caused by certain types of streptococcus (STREP-toe-COCK-iss) bacteria. The sore throat—known as strep throat—is usually followed by general illness. During this infection, the immune system of the body forms antibodies to combat the bacteria. Antibodies are proteins that circulate in the bloodstream that fight off bacteria.

Unfortunately, these antibodies can damage the tissue lining of heart valves, preventing them from closing completely. This causes blood to leak back into the atria and ventricles after contraction and results in a distinct "sloshing" sound, called a *heart murmur*. This condition reduces the efficiency of the heart and causes the organ to work harder to make up for the inefficient pumping. Increased activity, in turn, causes the walls of the heart to enlarge and in severe cases, can result in heart failure. Fortunately, damaged valves can be replaced by artificial implants.

Tumors and scar tissue have an opposite effect—that is, they reduce blood flow through the heart valves. This condition is known as *valvular stenosis* (sten-OH-siss; from the Greek word *steno*, meaning narrow). Valvular stenosis prevents the ventricles from filling completely. As in valvular incompetence or heart murmur, the heart must beat faster to ensure an adequate supply of blood to the body's tissues. This acceleration also puts additional stress on the organ.

The Heart's Pacemaker

Heart rate is controlled primarily by an internal pacemaker.

The human heart functions at different rates under different conditions. At rest, it generally beats slowly. When a person is working hard, it beats faster. This variation in heart rate helps the body adjust for differences in the oxygen requirements of cells and tissues. Heart rate is controlled by a number of mechanisms.

One of the most important mechanisms is an internal pacemaker, the **sinoatrial (SA) node** (SIGN-oh-A-tree-ill noad) (Figure 6-5). Located in the wall of the right atrium, the SA node is composed of a clump of specialized cardiac muscle cells. These cells contract spontaneously and rhythmically. Each contraction produces a tiny electrical impulse, akin to those produced by nerve cells. This impulse spreads rapidly from the SA node to the cardiac muscle and then from muscle cell to muscle cell in both atria. Because cardiac muscle cells are tightly joined, and because the impulse travels quickly, the two atria contract almost simultaneously.

Left to their own devices, cardiac muscle cells would contract independently and in a disorderly way, creating an ineffective pumping action. The SA node, however, imposes a single rhythm on all of the atrial heart muscle cells.

The electrical impulse transmitted throughout the atria next passes to the ventricles. However, its passage is briefly slowed by a barrier of unexcitable tissue that separates the atria from the ventricles. The impulse is delayed approximately 1/10 second, giving the atria time to contract and fill the ventricles before they contract. After this brief delay, the impulse is channeled through a second mass of specialized muscle cells, the **atrioventricular (AV) node**, shown in Figure 6-5. From the AV node, the impulse travels along a tract of specialized cardiac muscle cells, known as the **atrioventricular bundle**.

As Figure 6-5 shows, the atrioventricular bundle divides into two branches (called the bundle branches) that travel on either side of the wall separating the ventricles. The bundle branches give off smaller branches that terminate on specialized muscle cells, **Purkinje fibers** (per-KIN-gee), named after the scientist who discovered them. These fibers terminate on the cardiac muscle cells in the walls of the ventricles, stimulating them to contract.

Controlling Heart Rate

External factors regulate the pacemaker.

The SA node of the human heart produces a steady rhythm of about 100 beats per minute when isolated from outside influences, but this is much too fast for most human activities. To bring the heart rate in line with body demand, the SA node must be slowed down. The SA node is curbed by impulses transmitted by nerves that connect the heart with a control center in the brain. At rest or during nonstrenuous activity, these impulses slow the heart to about 70 beats per minute, thus aligning heart rate with body demands. During exercise or stress, when the heart rate must increase to meet body demands, the decelerating impulses from the brain are reduced, allowing the heart rate to increase.

Other nerves also influence heart rate. These nerves carry impulses that can speed up the heart. They allow the heart to achieve rates of 180 beats or more, for example, during intense exercise when the body cells' demand for oxygen is great.

Several hormones also play a role in controlling heart rate. One of these is **epinephrine** (EP-eh-NEFF-rin), also known as **adrenalin**. The adrenal glands located on top of the kidneys secrete this hormone during stress or exercise. Epinephrine increases the heart rate, increasing the flow of blood through the body.

Measuring the Heart's Electrical Activity

The heart's electrical activity can be measured on the surface of the chest.

When the electrical impulses that cause heart muscle cells to contract arrive, they cause electrical changes in the cardiac muscle cells. The changes can be detected by electrodes, small metal

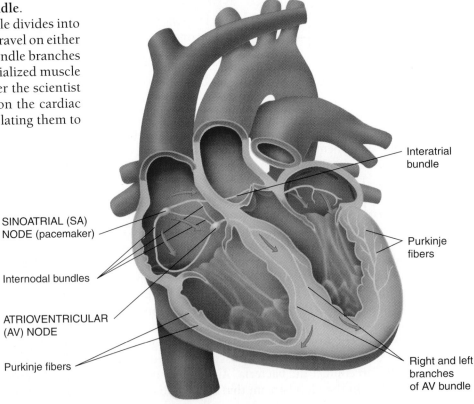

FIGURE 6-5 **Conduction of Impulses in the Heart** The sinoatrial node is the heart's pacemaker. Located in the right atrium, it sends timed impulses into the atrial heart muscles, coordinating muscle contraction. The impulse travels from cell to cell in the atria, then passes to the atrioventricular node and into the ventricles via the atrioventricular bundle and its two branches, which terminate on the Purkinje fibers.

Interatrial bundle

SINOATRIAL (SA) NODE (pacemaker)

Purkinje fibers

Internodal bundles

ATRIOVENTRICULAR (AV) NODE

Purkinje fibers

Right and left branches of AV bundle

FIGURE 6-6 **The Electrocardiogram** (a) This patient taking a treadmill test to check his heart's performance is wired to a meter that detects electrical activity produced by the heart. (b) An electrocardiogram.

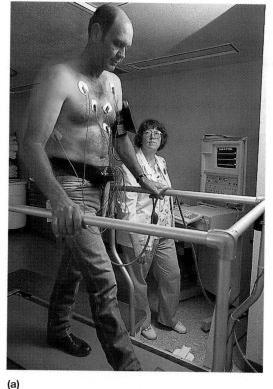

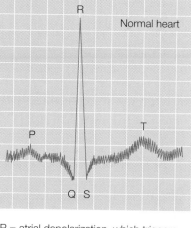

P = atrial depolarization, which triggers atrial contraction.

QRS = depolarization of AV node and conduction of electrical impulse through ventricles. Ventricular contraction begins at R.

T = repolarization of ventricles.

P to R interval = time required for impulses to travel from SA node to ventricles.

plates connected to wires and a voltage meter (Figure 6-6a). The electrodes are placed on a person's chest. The resulting reading on a volt meter is called an **electrocardiogram** (**ECG** or, more commonly, **EKG**) (Figure 6-6b).

For a normal person, the electrical tracing has three distinct waves (Figure 6-6b). The first wave, the P wave, represents the electrical changes occurring in the atria of the heart. The second wave, the QRS wave, is a record of the electrical activity taking place during ventricular contraction, and the third wave, the T wave, is a recording of electrical activity occurring as the ventricles relax. Certain diseases of the heart may disrupt one or more waves of the EKG, making the EKG a valuable diagnostic tool for heart doctors (cardiologists).

(a)

(b)

6-3 The Blood Vessels

Blood flows from the heart to the capillaries, then back to the heart via veins.

The heart pumps blood throughout the body in a series of connected blood vessels belonging to either the pulmonary or systemic circuits described earlier. (For a discussion of some of the discoveries that led to our understanding of circulation, see Scientific Discoveries 6-1.)

Arteries transport blood away from the heart and branch many times as they course through the body, forming smaller and smaller vessels that serve all of the organs of the body. The smallest of all arteries is the **arteriole** (are-TEAR-ee-ol). As shown in Figure 6-7, arterioles empty into **capillaries** (CAP-ill-air-ees), tiny, thin-walled vessels that allow nutrients to enter tissues and wastes to escape with ease.

FIGURE 6-7 **Capillary Network** A network of capillaries between the arteriole and the venule delivers blood to the cells of body tissues (not shown).

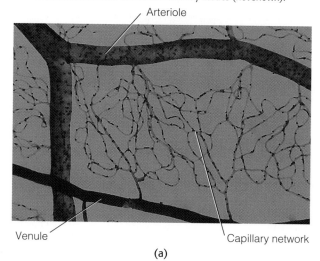

Arteriole

Venule

Capillary network

(a)

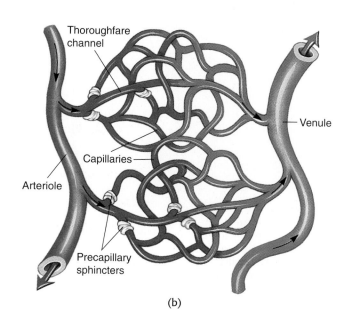

Thoroughfare channel

Venule

Capillaries

Arteriole

Precapillary sphincters

(b)

Scientific Discoveries that Changed the World

6-1 ## The Circulation of Blood in Animals
Featuring the Work of Harvey and Hales

The seventeenth-century British physician William Harvey is generally credited with the discovery of the circulation of blood in animals (Figure 1). Harvey is known as a scientist with a short temper who wore a dagger in the fashion of the day, which he reportedly brandished at the slightest provocation. He was probably not the kind of professor you might "argue" with over grades.

Temperament aside, Harvey is generally honored for his work on the role of the heart and the flow of blood in animals and is often praised as a pioneer of scientific method. His application of quantitative procedures to biology, some say, ushered in the modern age of this science.

In Harvey's medical school days, anatomists thought the intestines produced a substance called chyle. Chyle, they thought, was a fluid derived from the food people ate. It was passed from the intestines to the liver. The liver, in turn, converted the chyle to venous blood, then distributed the blood through arteries and veins.

As a medical student, Harvey was told that blood oozing through arteries and veins supplied organs and tissues with nourishment. He was

also told that the blood merely ebbed back to the heart and lungs, where impurities were removed. In other words, there was no form of circulation, just an ebb and flow similar to the tides. These ideas had been proposed by the Greek physician Galen fourteen centuries earlier and persisted nearly without challenge until Harvey's time.

As a teacher in the Royal College of Physicians in 1616, Harvey began to describe the circulation of blood, based on his experiments on and observations of animals. He described the muscular character of the heart and the origin of the heartbeat. He also demonstrated that the pulse felt in arteries resulted from the impact of blood pumped by the heart. Fur-

FIGURE 1 **William Harvey** This flamboyant scientist greatly advanced our knowledge on circulation in animals.

Blood flows out of the capillaries into the smallest of all veins, the venules. Venules, in turn, converge to form small veins. They unite with other small veins, in much the same way that small streams unite to form a river. Blood in **veins** flows toward the heart. Veins and arteries generally run side by side throughout the body.

Figure 6-8 shows a cross section of an artery and a vein. As illustrated, these two vessels are fairly similar, although veins

tend to be smaller and to have thinner walls. Both consist of three layers: (1) an external layer of connective tissue, which binds the vessel to surrounding tissues; (2) a middle layer, which is primarily made of smooth muscle; and (3) an internal layer, which is composed of a layer of flattened cells, the endothelium, and a thin, nearly indiscernible layer of connective tissue, which binds the endothelium to the middle layer. Figure 6-9 shows these layers and the scientific names used to describe

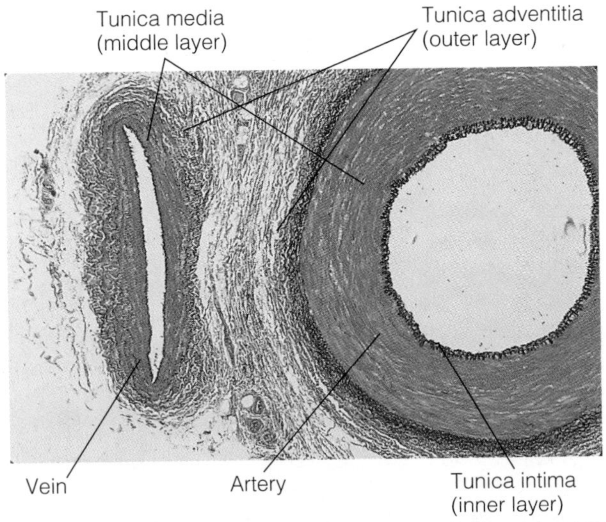

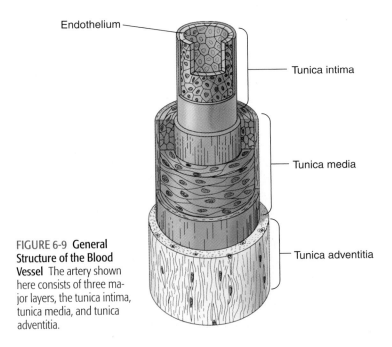

FIGURE 6-8 **Artery and Vein** A cross section through a vein shows that the muscular layer, the tunica media, is much thinner than it is in an artery. Veins typically lie alongside arteries and, in histological sections such as these, usually have irregular lumens (cavities).

FIGURE 6-9 **General Structure of the Blood Vessel** The artery shown here consists of three major layers, the tunica intima, tunica media, and tunica adventitia.

thermore, he described the pulmonary and systemic circuits and proposed that blood flowed to the tissues and organs of the body via the arteries and returned via the veins.

A brief examination of one of his experiments illustrates that even though Harvey is a key figure in the history of biological science, some of his work was based on poor assumptions and inaccurate observations. As an example, consider the work he used to rebut Galen's hypothesis that the blood was produced by the food people ate.

Harvey first approximated the amount of blood the heart ejected with each heartbeat (stroke volume), then determined the pulse rate. He called on earlier observations of a heart from a human cadaver to determine stroke volume. At that time, he had noted that the left ventricle contained more than 2 ounces of blood, and then, for reasons not entirely clear to historians of science, he hypothesized that the ventricle ejected "a fourth, a fifth, a sixth or only an eighth" of its contents. (Today, studies indicate that the heart ejects nearly all of its contents.) Based on this assumption, Harvey estimated that the stroke volume was about 3.9 grams of blood per beat. Modern estimates put it at 89 grams per beat.

Harvey also made a grave error in determining pulse. His value of 33 beats per minute is about half of the actual rate in humans. No one knows how he could have been so wrong. Armed with two erroneous measurements, Harvey derived a figure for the amount of blood that circulated through the body that was 1/36 of the lowest value accepted today.

Regardless, Harvey "proved" his point—that each half-hour the blood pumped by the heart far exceeds the total weight of blood in the body. From this he concluded that blood must be circulated. It is not, as Galen proposed, produced by the food we eat. The amount of food one eats could not produce blood in such volume.

Harvey debunked another falsehood perpetrated through the centuries—the Galenic myth that blood flowed into the extremities in both arteries and veins. Harvey first wrapped a bandage around an extremity. This obstructed the flow of blood through the veins but not the arteries. He noted that the veins swelled because, as he conjectured, blood was being pumped into them via underlying arteries and there was nowhere for the blood to go. Tightening the bandage further cut the blood flow in the arteries as well and thus prevented the veins from swelling. From these observations, Harvey correctly surmised that the arteries deliver blood to the extremities and the veins return it to the heart.

Harvey's work laid the foundation for modern cardiovascular physiology but left many questions unanswered. Many of these were addressed by the highly industrious English biologist Stephan Hales, who was born a century after Harvey. In a long series of scientifically rigorous experiments on horses, dogs, and frogs, Hales explored many aspects of the cardiovascular system. After settling many of the unanswered questions left by Harvey, Hales went on to study plant physiology and is perhaps best known for his work on the circulation of sap in plants.

them. With this overview in mind, let's examine each type of blood vessel in a little more detail.

Arteries and Arterioles

Arteries and arterioles deliver oxygen-rich blood to tissues and organs.

The largest of all arteries is the **aorta**, a massive vessel that carries oxygenated blood from the left ventricle of the heart to the rest of the body. The aorta loops over the back of the heart, then descends through the chest and abdomen, giving off large branches along its way. These branches carry blood to the head, the arms and legs, and organs of the body (Figure 6-1). The very first branches of the aorta are the coronary arteries.

The aorta and many of its chief branches contain numerous wavy elastic fibers interspersed among the smooth muscle cells of the wall. For this reason, they're called *elastic arteries*. As blood pulses out of the heart, these arteries expand to accommodate it. Like a stretched rubber band, the elastic fibers cause the arterial walls to recoil. This helps push the blood along the arterial tree, maintaining an even flow of blood through the capillaries.

The elastic arteries branch to form smaller vessels, the *muscular arteries*. Muscular arteries contain fewer elastic fibers but still expand and contract with the flow of blood. You can feel this expansion and contraction in the arteries lying near the skin's surface in your wrist and neck. It's the pulse that health care workers use to measure heart rate.

The smooth muscle of the muscular arteries responds to a number of stimuli, including nerve impulses, hormones, car-

bon dioxide, and lactic acid. These stimuli cause the blood vessels to open or close to varying degrees. This allows the body to adjust blood flow through its tissues to meet increased demands for nutrients and oxygen. Arterioles in muscles, for instance, dilate (DIE-late) when a person exercises. This increases blood flow to the muscle, providing oxygen and glucose needed for greater physical exertion.

Blood Pressure. The force that blood applies to the walls of a blood vessel is known as the **blood pressure**. Blood pressure changes in relation to a person's activity and stress levels. Blood pressure also varies throughout the cardiovascular system, being the highest in the aorta and the lowest in the capillaries and veins. Low blood pressure in the capillaries enhances the rate of exchange between the blood and the tissues.

Blood pressure is measured by using an inflatable device with the tongue-twisting name of *sphygmomanometer* (SFIG-mo-ma-NOM-a-ter), or, more commonly, *blood pressure cuff* (Figure 6-10a). The blood pressure cuff is wrapped around the upper arm. A stethoscope is positioned over the artery just below the cuff. Air is pumped into the cuff until the pressure stops the flow of blood through the artery (Figure 6-10b). The pressure in the cuff is then gradually reduced as air is released. When the blood pressure in the artery exceeds the external pressure of the cuff, the blood starts flowing through the vessel once again. This point represents the **systolic pressure** (sis-STOL-ick), the peak pressure at the moment the ventricles contract. Systolic pressure is the higher of the two numbers in a blood pressure reading (120/70, for example). The pressure at the moment the heart relaxes to let the ventricles

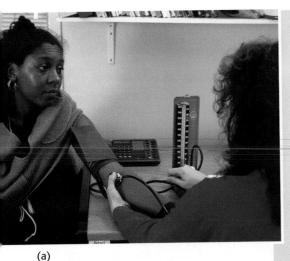

(a)

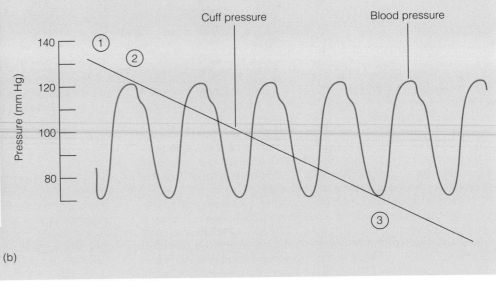

(b)

fill again is the **diastolic pressure** (DIE-ah-STOL-ick) and is the lower of the two readings. It is determined by continuing to release air from the cuff until no arterial pulsation is audible. At this point, blood is flowing continuously through the artery.

A typical reading for a young, healthy adult is about 120/70, although readings vary considerably from one person to the next. As a person ages, blood pressure tends to rise. Thus, a healthy 65-year-old might have a blood pressure reading of 140/90.

FIGURE 6-10 Blood Pressure Reading (a) A sphygmomanometer (blood pressure cuff) is used to determine blood pressure. (b) As shown, the blood pressure (indicated by the red line) rises and falls with each contraction of the heart. When the pressure in the cuff exceeds the arterial peak pressure, blood flow stops (1). No sound is heard. Cuff pressure is gradually released. When pressure in the cuff falls below the arterial pressure, blood starts flowing through the artery once again. This is the systolic pressure (2). The first sound will be heard. Cuff pressure continues to drop. When cuff pressure is equal to the lowest pressure in the artery, the artery is fully open and no sound is heard (3). This is the diastolic pressure.

Capillaries

Capillaries permit the exchange of nutrients and wastes.

As described above, the heart and arteries transport blood to the capillaries. Capillaries form branching networks, known as **capillary beds**, among the cells of body tissues. The extensive branching of capillaries brings them in close proximity to body cells. It is in these extensive capillary beds that wastes and nutrients are exchanged between the cells of the body and the blood. The walls of the capillaries consist of flattened endothelial cells. These cells permit dissolved substances to pass through them with ease.

If you could remove all of the capillaries from the body and line them up end to end, they would extend over 80,000 kilometers (50,000 miles)—enough to circle the globe at the equator two times.

As blood flows into a capillary bed, nutrients, gases, water, and hormones carried in the blood immediately begin to diffuse out of the tiny thin-walled vessels. Meanwhile, water-dissolved wastes in the tissues, such as carbon dioxide, begin to diffuse inward.

The expansion and contraction of the arterioles that "feed" the capillaries helps to control nutrient flows to body tissues. It also helps to regulate body temperature. On a cold winter day, for example, the arterioles close down, restricting blood flow through the capillaries to conserve body heat. Just the reverse happens on a warm day. The flow of blood through the skin increases, releasing body heat and often creating a pink flush. Capillaries are therefore a part of the body's system of homeostasis.

Veins and Venules

Venules drain capillaries and merge to form veins that transport blood back to the heart.

Blood leaves the capillary beds stripped of its nutrients and loaded with cellular wastes. As it drains from the capillaries, the blood enters the smallest of all veins, the **venules**. Venules converge to form small veins, which join with others to form larger and larger vessels. Eventually, all blood returning to the heart in the systemic circuit enters the superior or inferior vena cavae, the two main veins that drain into the right atrium of the heart (Figure 6-2). These vessels drain the upper and lower parts of the body, respectively.

As noted earlier, blood pressure in the veins is low, and veins have relatively thin walls with fewer smooth muscle cells than arteries (Figure 6-8). Because the veins' walls are so thin, obstructions can cause blood to pool in them. When blood pools in the obstructed veins, it forms bluish bulges called *varicose veins* (VEAR-uh-cose) (Figure 6-11). Some people inherit a tendency to develop varicose veins, but most cases can be attributed to factors that reduce the flow of blood back to the heart such as obesity, pregnancy, sedentary lifestyles, or, much less commonly, abdominal tumors.

Varicose veins can result in considerable discomfort. The restriction of blood flow, for example, may result in muscle cramps and the buildup of fluid, **edema** (uh-DEEM-ah; swelling), in the ankles and legs. Varicose veins that form in the veins in the wall of the anal canal, the internal hemorrhoidal veins (hem-eh-ROID-il), result in a condition known as *hemorrhoids* (hem-eh-ROIDS). Because these veins are supplied by numerous pain fibers, this condition can be painful.

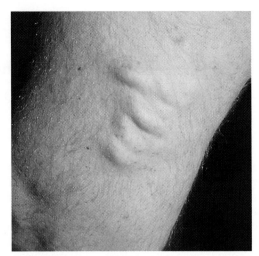

FIGURE 6-11 Varicose Veins Any restriction of venous blood flow to the heart causes veins to balloon out, creating bulges commonly known as varicose veins.

How Veins Work. Blood pressure caused by the beating heart propels blood in the arteries into the capillaries. But blood pressure, as shown in Figure 6-12, is extremely low in veins. With such low pressure, how do the veins return blood to the heart?

For blood in veins above the heart, gravity is the chief means of propulsion. For veins below the heart, return flow depends on the movement of body parts, which "squeezes" the blood upward. As you walk to class, for example, the contraction of muscles in your legs pumps the blood in the veins back to the heart.

Blood flow in veins, however, also relies on valves in the veins. Valves are flaps of tissue that span the veins and prevent the backflow of blood. As illustrated in Figure 6-13, the flaps of the veins resemble those found in the heart. Just as in the valves of the heart, blood pressure, however slight, pushes the flaps open (Figure 6-13). This allows the blood to move forward. As the blood fills the segment of the vein in front of the valve, it pushes back on the valve flaps and forces them shut.

To locate a valve, hold your arm out in front of you and make a fist. The veins should stick out or at least be apparent beneath the skin of your forearm. To locate a valve in the superficial veins on your forearm, press gently on a vein, then run your finger toward your wrist. You will note that the vein collapses behind your finger until it crosses a valve. (See Health Note 6-1 for a discussion of one novel way to restore blood flow interrupted by injury.)

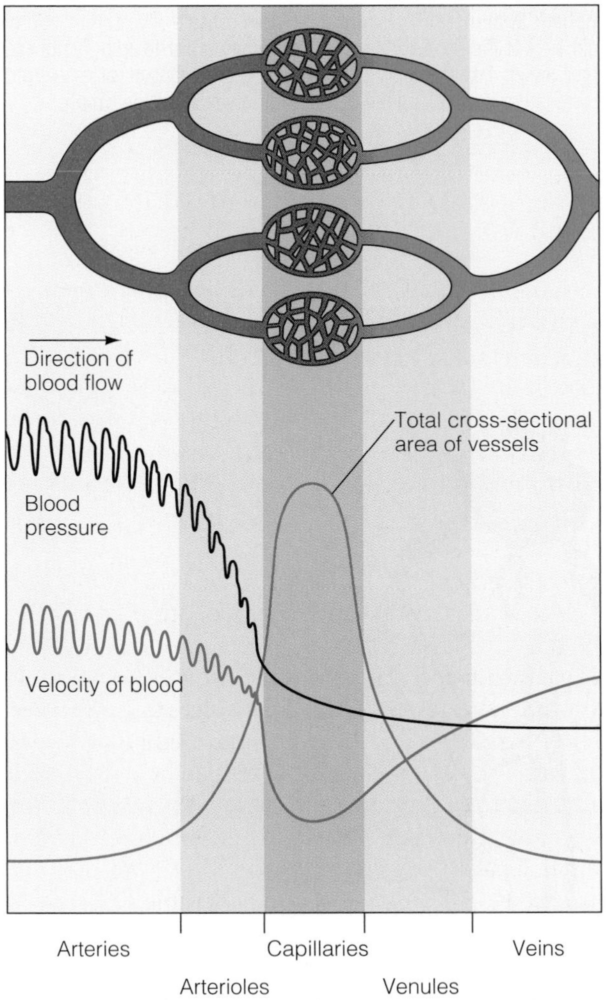

Direction of blood flow

Total cross-sectional area of vessels

Blood pressure

Velocity of blood

Arteries   Capillaries   Veins

Arterioles   Venules

FIGURE 6-12 Blood Pressure in the Circulatory System Blood pressure declines in the circulatory system as the vessels branch. Arterial pressure pulses because of the heartbeat, but pulsation is lost by the time the blood reaches the capillary networks, creating an even flow through body tissues. Blood pressure continues to decline in the venous side of the circulatory system.

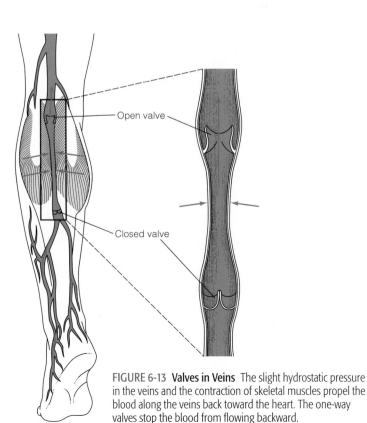

Open valve

Closed valve

FIGURE 6-13 Valves in Veins The slight hydrostatic pressure in the veins and the contraction of skeletal muscles propel the blood along the veins back toward the heart. The one-way valves stop the blood from flowing backward.

⣿ healthnote

6-1   Modern Medicine Turns to Leeches

Donna Amark is a college professor at a midwestern university. A few years ago, Donna lost her thumb in an unfortunate accident involving a neighbor's dog. Donna stopped by to visit the new neighbor. When no one answered the door, she went to the back yard. There she was confronted by the neighbor's dog, who was chained to a large maple tree. The dog seemed friendly, but as Donna approached he lunged toward her. She side-stepped the dog and grabbed the chain. Unfortunately, the chain had wrapped around her thumb. When the dog lunged at her a second tune, the sudden jerk of the chain severed her thumb, ripping it out, tendons and all.

Paramedics who rushed Donna to the hospital were unable to find the severed thumb, so she elected to have a new procedure. Doctors removed her big toe, tendons and all, and inserted it into the hand where her thumb had once been.

Her surgery went well. However, despite the surgeon's tedious work of reconnecting blood vessels, Donna's translocated thumb showed no signs of restored blood flow. Soon after, it became apparent that she was going to lose her new thumb. The surgeons in charge made what seemed at first to be a wild suggestion: use leeches to restore the blood flow.

Leeches, it turns out, have become one of modern medicine's most effective tools. They're routinely used in Great Britain and the United States to re-establish blood flow in reattached ears, fingers, and toes. A leech has a wicked looking mouth with 100 teeth. When it is attached to the skin, it bites in. But you don't feel any pain because the leech releases a tiny

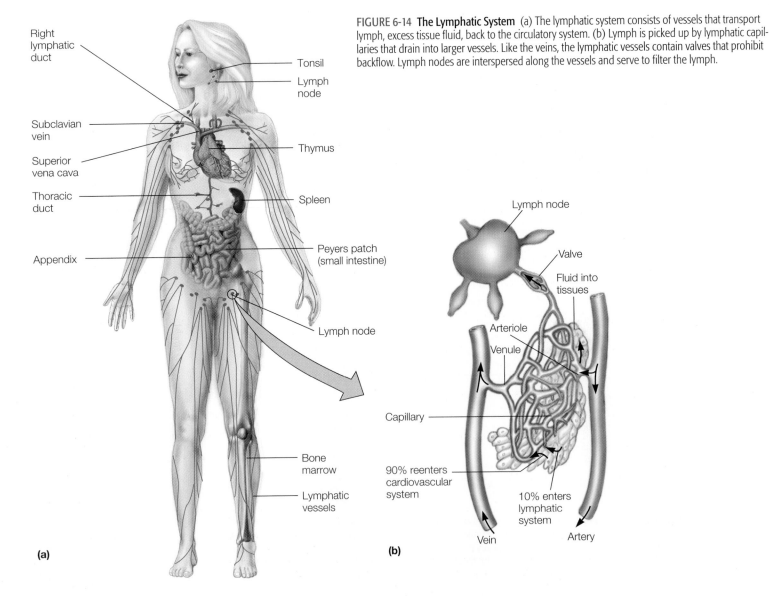

FIGURE 6-14 The Lymphatic System (a) The lymphatic system consists of vessels that transport lymph, excess tissue fluid, back to the circulatory system. (b) Lymph is picked up by lymphatic capillaries that drain into larger vessels. Like the veins, the lymphatic vessels contain valves that prohibit backflow. Lymph nodes are interspersed along the vessels and serve to filter the lymph.

(a)

Right lymphatic duct
Tonsil
Lymph node
Subclavian vein
Thymus
Superior vena cava
Thoracic duct
Spleen
Appendix
Peyers patch (small intestine)
Lymph node
Bone marrow
Lymphatic vessels

(b)

Lymph node
Valve
Fluid into tissues
Arteriole
Venule
Capillary
90% reenters cardiovascular system
10% enters lymphatic system
Vein
Artery

amount of natural anesthetic that deadens the sensory nerves in the area. The leech also releases a natural anticoagulant into the wound so blood flows uninterrupted into its mouth. When placed on a reattached finger or toe, the germ-free leech attaches and begins sucking. The patient feels nothing, but in short order oxygen-rich blood begins to flow through the finger or toe, bringing the body part back to life better than any surgeon could. The leech is left in place for some time to ensure adequate flow. After it is removed, blood continues to flow through the reattached body part. If further treatment is needed, the leech is placed on the reattached part again.

Leeches are raised in captivity in sterile environments and shipped to hospitals when needed. In Donna's case, the leeches worked miracles. She's alive and well with functional hands, thanks to the lowly leech, modern medicine's latest and greatest new tool.

www.jbpub.com/humanbiology/5e

Visit Human Biology's Internet site for links to web sites offering more information on this topic.

6-4 The Lymphatic System

The lymphatic system is a secondary system of vessels that returns excess fluid to the circulatory system.

The **lymphatic system** is an extensive network of vessels and glands (Figure 6-14). It is functionally related to two systems: the circulatory system and the immune system. This section examines the circulatory role of the lymphatic system.

You may recall from Chapter 4 that the cells of the body are bathed in a liquid called *interstitial fluid* (in-ter-STISH-il). Interstitial fluid provides a medium through which nutrients, gases, and wastes can diffuse between the capillaries and the cells.

Tissue fluid is constantly replenished by the capillaries. The flow of water out of the capillaries, however, normally exceeds the return flow by about 3 liters per day. The "excess" water is picked up by small **lymph capillaries** in tissues. Like the capillaries of the circulatory system, these vessels have thin, highly permeable walls through which water and other substances pass with ease.

Lymph drains from the capillaries into larger ducts. These vessels, in turn, merge with others, creating larger and larger ducts, that empty into the large veins at the base of the neck.

Lymph moves through the vessels of the lymphatic system in much the same way that blood is transported in veins. In the upper parts of the body, it flows by gravity. In areas below the heart, it is propelled largely by muscle contraction. Breathing pumps lymph out of the chest and walking pumps the lymph out of the extremities. Lymphatic flow is also assisted by valves similar to those in the veins.

The lymphatic system also consists of several lymphatic organs: the lymph nodes, the spleen, the thymus, and the tonsils. The lymphatic organs function primarily in immune protection and are discussed in Chapter 14. Lymph nodes, however, play a role worth considering here.

Varying in size and shape, lymph nodes are found in association with lymphatic vessels in small clusters in the armpits, groin, neck, and other locations (Figure 6-14). A lymph node consists of a network of fibers and irregular channels that slow down the flow of lymph. As lymph passes through the lymph nodes, the fibers filter out bacteria, viruses, and cellular debris. Lining the channels are numerous cells (macrophages) that engulf (phagocytize) microorganisms and other materials.

Normally, lymph is removed from tissues at a rate equal to its production. In some instances, however, lymph production exceeds the capacity of the system. A burn, for example, may cause extensive damage to blood capillaries, increasing their leakiness and overwhelming the lymphatic capillaries. This "flood" results in a buildup of fluid in tissues, that is, edema.

Lymphatic vessels may also become blocked. One of the most common causes of blockage is an infection by tiny parasitic worms transmitted to humans by mosquitoes in the tropics. The worm larvae (an immature form) enter lymph vessels and take up residence in the lymph nodes. An inflammatory reaction causes the buildup of scar tissue in the nodes that blocks the flow of lymph. After several years, the lymphatic drainage of certain parts of the body may be almost completely obstructed. A leg may swell so much, in fact, that it weighs as much as the rest of the body (Figure 6-15). This condition is known as *elephantiasis* (el-uh-FUN-TIE-ah-sis).

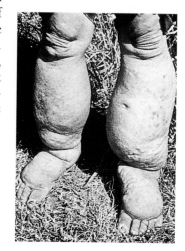

FIGURE 6-15 **Elephantiasis** A parasitic worm that invades the body stimulates the production of scar tissue, which blocks the flow of lymph through the nodes, causing tissue fluid to build up. This condition is known as elephantiasis.

6-5 Cardiovascular Diseases: Causes and Cures

The heart and blood vessels are subject to a variety of potentially life-threatening diseases.

Myocardial Infarction

Blockage of the arteries supplying heart muscle can lead to heart attacks.

Heart attacks come in two basic varieties. The most common heart attack is caused by blood clots that block one or more of the coronary arteries, usually those narrowed by atherosclerotic plaque (discussed next). Blood clots restrict the flow of blood to the heart muscle, cutting off the supply of oxygen and nutrients. This deprivation can damage and even kill the heart muscle cells. The condition is known as a **myocardial infarct**.

As noted in Health Note 5-1, the formation of atherosclerotic plaque results from a combination of factors: stress, poor diet, lack of exercise, smoking, heredity, and several others. Narrowing of a coronary artery by plaque does not usually block the flow of blood enough to cause a heart attack, however, unless it is quite severe—around 80% or 90% blockage. Nonetheless, less severe narrowing does make the vessel more susceptible to blood clots. That is, clots often form in the vessel at the site of narrowing. Also, clots that form in other parts of the body can lodge in the narrowed vessels.

The outcome of heart attacks varies. If the size of the damaged area is small and if the change in electrical activity of the heart is minor and transient, a heart attack is usually not fatal. If the damage is great or electrical activity is severely disrupted, these heart attacks can be fatal. Getting help quickly lessens the chances of severe damage to the heart. Giving a person an aspirin during a heart attack also lessens the damage, as explained shortly.

Heart attacks can occur quite suddenly, without warning, or may be preceded by several weeks of **angina** (an-GINE-ah), pain felt when the supply of oxygen to the myocardium is reduced. Angina appears in the center of the chest and can spread to a person's throat, upper jaw, back, and arms (usually just the left one). Angina is a dull, heavy, constricting pain that appears when an individual is active, then disappears when he or she ceases the activity.

Angina may also be caused by stress and exposure to carbon monoxide, a pollutant that reduces the oxygen-carrying capacity of the blood. Angina begins to show up in men at age 30 and nearly always is caused by coronary artery disease. In women, angina tends to occur at a much later age.

Fibrillation

Another type of heart attack results from a loss of electrical control.

The second major type of heart attack occurs when the SA node loses control of the heart muscle. With the SA node no longer in charge, the cardiac muscle cells beat independently, a condition known as **fibrillation**. The lack of coordination converts the heart into an ineffective, quivering mass that pumps little, if any, blood. The heart can stop beating altogether, a condition known as **cardiac arrest**.

Physicians treat fibrillation by applying a strong electrical current to the chest, a procedure known as *defibrillation*. The electrical current passes through the wall of the chest and is often sufficient to restore normal electrical activity and heartbeat. A normal heartbeat can also be restored by cardiopulmonary resuscitation (CPR), in which the heart is "massaged" externally by applying pressure to the sternum (breastbone).

Preventing Heart Attacks

Prevention is the best cure.

Proper diet, exercise, and stress management can reduce the risk of heart problems, as noted in Health Notes 1-1 and 5-1. Research also shows that a daily dose of aspirin (half a tablet a day is enough) taken over long periods can substantially reduce an individual's chances of a heart attack. Aspirin decreases the likelihood of heart attacks by reducing the formation of blood clots.

Prevention is the most effective way to control heart attacks. It could save Americans hundreds of millions of dollars each year in medical bills, lost work time, and decreased productivity and could save thousands of lives. But given human nature, the fast pace of modern life, and our inattentiveness to exercise and proper diet, heart disease will probably be around for a long time. To reduce the death rate, physicians therefore also look for ways to treat patients after they have had a heart attack.

Treating Heart Attacks

Many medicines and procedures are available to treat heart attacks.

One promising development is the use of blood-clot-dissolving agents such as streptokinase, urokinase, and TPA. When administered within a few hours of the onset of a heart attack, these agents can greatly reduce the damage to heart muscle and accelerate a patient's recovery.

In mild cases, physicians can open clogged blood vessels by inserting a small catheter with a tiny balloon attached to its tip into the affected artery. After chemical clot dissolvers are administered to a patient, the balloon is inflated, forcing the artery open and loosening the plaque from the wall. This procedure is called *balloon angioplasty* (AN-gee-oh-PLAST-ee). Scientists are also experimenting with lasers that burn away plaque in artery walls. Unfortunately, as in other techniques, within a few months cholesterol builds up again in the walls of arteries.

Yet another treatment used more and more often these days is a tiny device known as a *stent* (Figure 6-16). This tubular wire is placed in the narrowed artery, then expanded with a tiny bal-

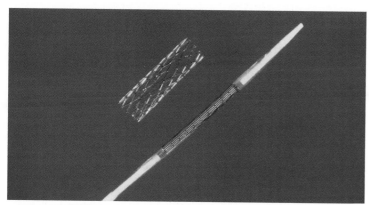

FIGURE 6-16 **Stent**

loon. After the balloon is extracted, the stent remains expanded to hold the narrowed artery open.

In severe cases where the coronary arteries are completely blocked by atherosclerotic plaque, heart surgeons often perform coronary bypass surgery to restore blood flow to the heart muscle. In this procedure, surgeons transplant small segments of veins from the leg into the heart (Figure 6-17). These venous bypasses shunt blood around the clogged coronary arteries, restoring blood flow to the heart muscle. Unfortunately, the venous grafts often fill fairly quickly with plaque. The recurrence of plaque in grafts has led researchers to turn to marginally important arteries, such as the internal mammary artery, for this procedure. Many researchers believe that arteries will prove more resistant to plaque buildup than veins.

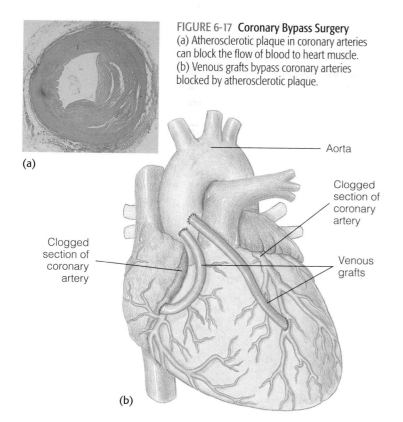

FIGURE 6-17 **Coronary Bypass Surgery**
(a) Atherosclerotic plaque in coronary arteries can block the flow of blood to heart muscle. (b) Venous grafts bypass coronary arteries blocked by atherosclerotic plaque.

(a)

Aorta

Clogged section of coronary artery

Clogged section of coronary artery

Venous grafts

(b)

Hypertension

High blood pressure develops gradually over time.

Hypertension or high blood pressure is a prolonged elevation in blood pressure. It is one of the most prevalent health problems of our time. Like other cardiovascular diseases, it has many causes, including obesity, high salt intake, heredity, and kidney disease. Nearly always a symptomless disease early on, hypertension is characterized by a gradual increase in blood pressure over time. A person may feel fine and display no physical problems whatsoever for years. Symptoms, such as headaches, rapid beating of the heart, and a general feeling of ill health, usually occur only when blood pressure is dangerously high. If untreated, blood pressure rises steadily over time. Consequently, early detection and treatment are essential to prevent serious problems, including heart attacks.

Hypertension is more common in men than in women and is twice as prevalent in African Americans as it is in Caucasians. The increased pressure in the circulatory system forces the heart to work harder. Elevated blood pressure may also damage the lining of arteries, creating a site at which atherosclerotic plaque forms. Atherosclerosis increases the risk of heart attack: A hypertensive person is six times more likely to have a heart attack than an individual with normal blood pressure. Hypertension also increases the chances of plaque buildup in the arteries supplying the brain, which can result in strokes.

Atherosclerosis

Atherosclerosis results from the buildup of cholesterol plaque in arteries.

Another common problem of modern times is **atherosclerosis**, the buildup of cholesterol plaque in the walls of arteries. Arteries clogged with cholesterol force the heart to work harder, straining it. Perhaps the most significant problem arises from blood clots that lodge in narrowed coronary arteries, reducing the flow of blood to the heart muscle.

Aneurysm

Weakening of the arterial walls can result in a rupture of the wall.

Arteries not only can clog, they also can rupture. Certain infectious diseases (such as syphilis), atherosclerotic plaque, and hypertension all can weaken the wall of arteries. This weakening causes arteries to balloon outward, a condition known as an **aneurysm** (Figure 6-18). Like a worn spot on a tire, an aneurysm can rupture when pressure builds inside or when the wall becomes too thin.

When an aneurysm breaks, blood pours out of the circulatory system. Because they hap-

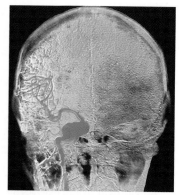

FIGURE 6-18 **Aneurysm** This X-ray shows a ballooning of one of the arteries in the brain. If untreated, an aneurysm can break, causing a stroke.

pen so quickly, most aneurysms lead to death. As in most diseases, the first line of defense against aneurysms is prevention. By reducing or eliminating the two main causes—atherosclerosis and high blood pressure—individuals can greatly lower their risk.

Physicians recommend a number of steps to reduce your chances of developing atherosclerosis and hypertension. If you smoke, stop. If you are overweight, exercise and lose weight. If you're one of those people for whom salt raises your blood pressure, cut back. If you suffer from stress at work and at home, find ways to reduce your stress levels. If you drink alcohol, consume it in moderation, or quit altogether. If you are fond of foods that are rich in fats and cholesterol, cut down on them and consume foods containing water-soluble fiber such as apples, bananas, citrus fruits, carrots, barley, and oats. These reduce cholesterol uptake by the intestine, as explained in Chapter 5.

Fish oils also protect the cardiovascular system. Fish oil contains omega-3 fatty acids that have many beneficial functions. For example, it increases high density lipoprotein (HDL), the so-called "good cholesterol." HDL binds to cholesterol in the blood and transports it to the liver for removal. This may decrease plaque buildup in arteries. Omega-3 fatty acids have also been shown to reduce blood clots.

Eating fish once a week will provide minimal benefit. Three days a week provides maximum benefit. The highest amounts of omega-3 fatty acids are found in cold water species including salmon, mackerel, lake trout, and sardines. Fish oil capsules apparently provide the same benefit.

Early detection and treatment also help combat cardiovascular disease. As noted earlier, hypertension can be detected by regular blood pressure readings. Atherosclerosis can be detected by blood tests that measure cholesterol. Once any of these diseases is detected, physicians have many options.

6-6 Health and Homeostasis

The heart and blood vessels carry out many very important functions. They distribute nutrients throughout the body. They remove wastes and distribute heat throughout the body, ensuring cellular function. These functions, in turn, help to keep the cells alive and well.

Each of these functions serves a higher purpose: to maintain the relatively constant internal conditions necessary for normal cellular function. That is, they play a key role in homeostasis. Furthermore, by keeping cells, tissues, and organs of the body alive and functioning well, the cardiovascular system ensures that homeostatic mechanisms provided by them work properly. By maintaining the functions of the kidney, for instance, the cardiovascular system ensures that it continues to perform its functions that help ensure homeostasis—for example, its removal of wastes. By maintaining the cells of the liver, the cardiovascular system ensures that it carries out its many important homeostatic roles—for example, making blood-clotting proteins.

Cardiovascular diseases upset homeostasis, however. These diseases are caused by many factors, such as stress, improper diet, and lack of exercise—the so-called "lifestyle factors." By altering the functions of the cardiovascular system, these factors threaten the broader welfare of an entire person. Keeping your cardiovascular system healthy is not just about preventing heart attacks, it's about pursuing a healthy life.

SUMMARY

The Circulatory System's Function: An Overview

1. The circulatory system is one of the body's chief homeostatic systems.

The Heart

2. The circulatory system consists of a pump, the heart, and two circuits, the pulmonary circuit, which transports blood to and from the lungs, and the systemic circuit, which delivers blood to the body and returns it to the heart.

3. The human heart consists of four chambers: two atria and two ventricles. The right atrium and right ventricle service the pulmonary circuit. The left atrium and left ventricle pump blood into the aorta and are part of the systemic circuit.

4. The right atrium receives blood from the superior and inferior vena cavae. This blood, returning from body tissues, is low in oxygen and rich in carbon dioxide.

5. From the right atrium, blood is pumped into the right ventricle, then to the lungs, where it is resupplied with oxygen and stripped of most of its carbon dioxide. Blood returns to the heart via the pulmonary veins, which empty into the left atrium.

6. Blood is pumped from the left atrium to the left ventricle, then out the aorta to the body, where it supplies cells of tissues and organs with oxygen and picks up cellular wastes.

7. Heart valves help control the direction of blood flow in the heart.

8. Cardiac muscle cells' contraction is coordinated by the heart's pacemaker, the sinoatrial (SA) node.

9. Left on its own, the SA node would produce 100 contractions per minute. Nerve impulses, however, reduce the heart rate to about 70 beats per minute when a person is inactive. During exercise or stress, the heart rate increases to meet body demands.

10. Changes in electrical activity in the heart muscle can be detected by surface electrodes. The tracing produced on the voltage meter attached to the electrodes is called an electrocardiogram (ECG). Certain diseases of the heart may disrupt one or more of the waves, making the ECG a valuable diagnostic tool.

The Blood Vessels

11. The heart pumps blood into the arteries, which distribute the blood to capillary beds. Blood is returned to the heart in the veins.

12. The aorta, which carries blood away from the heart, and many of its chief branches are elastic arteries. As blood pulses out of the heart, the elastic arteries expand to accommodate the blood, then contract, helping pump the blood and ensuring a steady flow through the capillaries. As they course through the body, the elastic arteries branch to form muscular arteries, which also expand and contract with the flow of blood.

13. The smooth muscle in the walls of muscular arteries responds to a variety of stimuli. These stimuli cause the blood vessels to open or close to varying degrees, controlling the flow of blood through body tissues.
14. Blood pressure and flow rate are highest in the aorta and drop considerably as the arteries branch. By the time the blood reaches the capillaries, its flow and pressure are greatly reduced. This dramatic decline enhances the rate of exchange between the blood and the tissues.
15. Capillaries are thin-walled vessels that form branching networks, or capillary beds, among the cells of body tissues. Cellular wastes and nutrients are exchanged between the cells of body tissues and the blood in capillary networks.
16. Blood flow through a capillary network is regulated by constriction and relaxation of the arterioles that empty into them.
17. In the veins above the heart, blood drains by gravity. Veins below the heart, however, rely on the movement of body parts to squeeze the blood upward and on valves, flaps of tissue that span the veins and prevent the backflow of blood.

The Lymphatic System

18. The lymphatic system is a network of vessels that drains excess interstitial fluid from body tissues and transports it to the blood.
19. The lymphatic system also consists of several lymphatic organs, such as the lymph nodes, the spleen, the thymus, and the tonsils, which function primarily in immune protection.
20. Along the system of lymphatic vessels are small nodular organs called lymph nodes, which filter the lymph.
21. Under normal circumstances, lymph is removed from tissues at a rate equal to its production, keeping tissues from swelling. In some cases, however, lymph production exceeds the capacity of the lymphatic capillaries, and swelling (edema) results.

Cardiovascular Diseases: Causes and Cures

22. The most common type of heart attack is a myocardial infarction, caused by a blood clot that either forms in a narrowed coronary artery or arises elsewhere and breaks loose, only to become lodged in a coronary artery. The obstruction decreases the flow of blood and oxygen to heart muscle, sometimes killing the cells.
23. Heart attacks can occur quite suddenly without warning, or they may be preceded by several weeks of angina, pain felt when blood flow to heart muscle is reduced.
24. The risk of heart attack can be reduced by proper diet, exercise, and daily doses of aspirin. Numerous treatments are available for patients who have had a heart attack.
25. Hypertension or high blood pressure is a prolonged elevation in blood pressure that typically develops over a long period without noticeable signs until it has reached a critical stage. Elevated blood pressure increases one's risk of heart attack and stroke.
26. Atherosclerosis results from the buildup of cholesterol plaque in arteries, which can become clogged by clots, leading to heart attacks, strokes, and rupture of the walls of arteries.

Health and Homeostasis

27. The cardiovascular system ensures homeostasis by distributing nutrients and ridding the body of waste. This helps to maintain cells, tissues, and organs of the body that carry out other vital homeostatic functions.

critical thinking

Thinking Critically—Analysis

This Analysis corresponds to the Thinking Critically scenario that was presented at the beginning of this chapter.

This hypothesis suggests that a selective survival process was in operation—that is, that salt-retaining (dehydration-resistant) slaves were more likely to survive the rigors of slavery than those who lacked this trait. To test this idea, one might begin by looking at hypertensive African Americans to see whether they actually have the capacity to retain excess salt and, if so, whether this capacity leads to hypertension.

Another research study that might lead to useful results would require one to locate populations of blacks consisting of individuals of African and non-African heritage. By studying differences in blood pressure and the ability to retain salt, one might be able to determine if this hypothesis is correct. One such example is the island of Barbados in the Caribbean. Its population contains some African blacks who were imported as slaves to work on the island's sugar plantations.

Interestingly, one research team found that blacks of African descent had a slightly higher incidence of high blood pressure than blacks of non-African descent. They thought that it helped support the salt retention hypothesis.

Yet another study might involve tests of blacks who reside in the regions of Africa from which slaves were drawn. Finding a low rate of hypertension in this group would suggest that a selective survival mechanism was in effect. One such study showed that, in rural parts of Nigeria, blacks did not suffer from hypertension despite the fact that they consumed large amounts of salt. The researchers concluded that if these people are descendants of the ancestors of slaves, as are African American blacks, then the hypothesis supporting selective survival seems valid.

KEY TERMS AND CONCEPTS

Adrenalin, p. 112
Aneurysm, p. 121
Angina, p. 120
Aorta, p. 110
Arteriole, p. 113
Artery, p. 113
Atheriosclerosis, p. 121
Atrioventricular bundle, p. 112
Atrioventricular node, p. 112
Atrioventricular valve, p. 110
Atrium, p. 109
Blood pressure, p. 115
Capillary, p. 113

Capillary bed, p. 116
Cardiac arrest, p. 120
Chordae tendineae, p. 110
Diastolic pressure, p. 116
Edema, p. 116
Electrocardiogram, p. 113
Endocardium, p. 109
Epinephrine, p. 112
Fibrillation, p. 120
Heart, p. 109
Hypertension, p. 121
Inferior vena cava, p. 109
Lymph capillaries, p. 119

Lymphatic system, p. 119
Myocardial infarction, p. 120
Myocardium, p. 109
Pericardium, p. 109
Pulmonary circuit, p. 109
Purkinje fibers, p. 112
Sinoatrial node, p. 112
Superior vena cava, p. 109
Systemic circuit, p. 109
Systolic pressure, p. 115
Vein, p. 114
Ventricle, p. 109
Venule, p. 116

CONCEPT REVIEW

1. List and describe several ways in which the circulatory system functions in homeostasis. pp. 109, 122
2. Describe how the pacemaker, the sinoatrial node, coordinates muscular contractions of the heart. p. 112
3. Define the pulmonary and systemic circuits. pp. 109–110
4. Describe how the valves control the direction of blood flow in the heart. What are valvular incompetence and valvular stenosis? What causes these conditions, and how do they affect the heart? pp. 110–111

5. How does atherosclerosis of the coronary arteries affect the heart? p. 121
6. Describe the general structure of the arteries and veins. How are they different? How are they similar? p. 113–114
7. How do the elastic fibers in the major arteries help ensure a continuous blood flow through the capillaries? p. 115
8. Capillaries illustrate the remarkable correlation between structure and function. Do you agree with this statement? Why or why not? p. 116

9. Describe how the movement of blood through capillary beds is controlled. Why is this homeostatic mechanism so important? Give some examples. p. 116
10. Explain how the blood is returned to the heart via the veins. p. 117
11. Explain the role of the lymphatic system in circulation. p. 119

SELF-QUIZ: TESTING YOUR KNOWLEDGE

1. The _____ circuit delivers blood from the heart to the rest of the body, excluding the lungs. p. 109
2. The muscle of the heart is known as _____. p. 109
3. The largest artery in the body is the _____. p. 110
4. The _____ node is the pacemaker of the heart. p. 112
5. A(n) _____ is a recording of the electrical activity in the heart when it contracts. p. 113
6. Blood pressure is lowest in the _____. p. 115
7. Nutrient and waste exchange in tissues occurs in _____. p. 116
8. _____ pressure is the blood pressure recorded when the ventricles contract. p. 115
9. A _____ infarction results from a blood clot in the coronary artery, which blocks blood flow to heart muscle. p. 120
10. _____ is a condition that develops slowly and results in elevated blood pressure. This condition does not become detectable until a critical state is achieved. p. 121
11. Label the drawing.

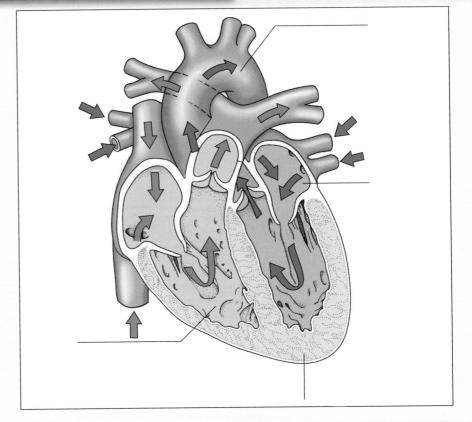

 www.jbpub.com/humanbiology/5e

The site features eLearning, an online review area that provides quizzes, chapter outlines, and other tools to help you study for your class. You can also follow useful links for in-depth information, research the differing views in the Point/Counterpoints, or keep up on the latest health news.

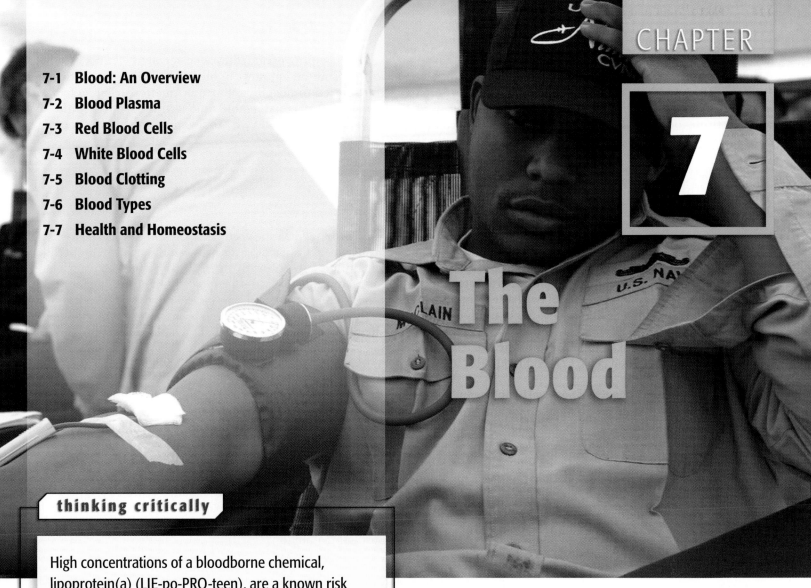

7

The Blood

thinking critically

High concentrations of a bloodborne chemical, lipoprotein(a) (LIE-po-PRO-teen), are a known risk factor of hardening of the arteries, or atherosclerosis. This molecule transports cholesterol in the blood and deposits it in arterial walls, causing a thickening of the walls. Studying muscle cells isolated from a normal human artery and grown in tissue culture, two researchers also found that lipoprotein(a) stimulates smooth muscle cells to proliferate. The researchers concluded that high concentrations of lipoprotein(a) have two effects: they deposit cholesterol and they cause muscle cells to divide—which may account for the elevated risk of cardiovascular disease they present.

Analyze this assertion based on what little you know about the experiment. What critical thinking rules helped you analyze this matter?

In 1966, Dr. Leland C. Clark of the University of Cincinnati's College of Medicine immersed a live laboratory mouse in a clear fluorocarbon solution saturated with oxygen. To the amazement of the audience, the mouse did just fine, "breathing" the oxygen-rich liquid as if it were air. Clark then extracted the mouse from the solution. After a moment, the rodent began to move about, apparently unharmed by the ordeal.

The point of this demonstration was not to show that an animal could "breathe" this fluid into its lungs and survive. Rather, it was to show that the solution held enormous amounts of oxygen, so much so that it could possibly be used as a substitute for blood. Clark and his colleagues hoped that their "artificial blood" would someday be a boon to medical science, helping paramedics keep accident victims who have lost substantial amounts of blood alive while they are being transported to hospitals. In rural America, where a trip to a hospital takes considerable time, artificial blood could save thousands of lives a year. Today, that dream is becoming a reality.

7-1 Blood: An Overview

Blood consists of plasma and formed elements, blood cells and platelets.

Human blood is a far cry from Clark's artificial substitute. The blood in our circulatory system, for example, is a water-based fluid, not a fluorocarbon, and consists of two basic components: plasma and formed elements. **Plasma** is the liquid portion of the blood and is about 90% water. It contains many dissolved substances. Three types of formed elements are suspended in the plasma: (1) white blood cells, (2) red blood cells, and (3) platelets.

Blood accounts for about 8% of our total body weight. A man weighing 70 kilograms (150 pounds), has about 5.6 kilograms (12 pounds) of blood. That's equal to 5 to 6 liters (1.3 to 1.5 gallons) of blood. On average, women have about a liter less.

Figure 7-1 shows that the plasma makes up about 55% of the blood volume and formed elements about 45%, mostly due to the presence of red blood cells.

The blood volume occupied by blood cells is known as the *hematocrit*. The hematocrit varies in individuals living at different altitudes. In people living in Denver, a mile above sea level, for example, hematocrits are typically about 5% higher than in people living at sea level. The increase in the blood cell count is the result of a homeostatic mechanism that compensates for the slightly lower level of oxygen in the atmosphere at higher altitudes. The result is an increase in red blood cells, which transport oxygen. The higher concentration of red blood cells means that more oxygen can be delivered to the cells of the body.

In this day of high-tech athletic competition, where trainers and athletes look for any competitive edge they can find, many athletes train at high altitudes. Some athletes sleep in specially designed chambers that expose them to air with slightly lower oxygen content. The rationale in both these strategies is that the higher hematocrit improves their performance at lower altitudes because their blood contains more oxygen-carrying red blood cells. In fact, it is partly for this reason that the U.S. Summer Olympic team is currently headquartered in Colorado Springs, Colorado.

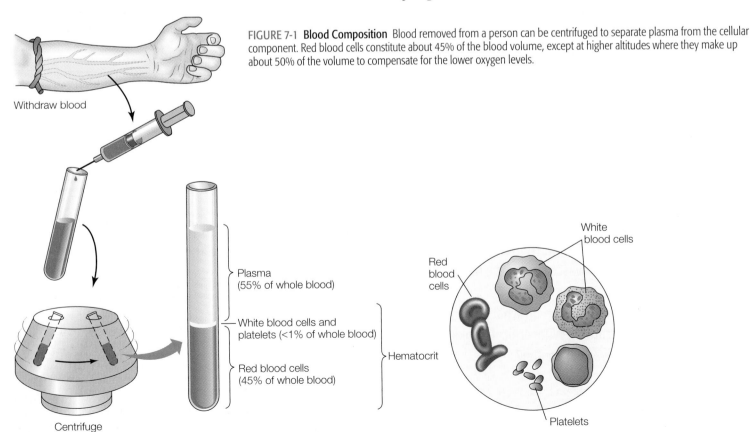

FIGURE 7-1 **Blood Composition** Blood removed from a person can be centrifuged to separate plasma from the cellular component. Red blood cells constitute about 45% of the blood volume, except at higher altitudes where they make up about 50% of the volume to compensate for the lower oxygen levels.

Withdraw blood

Centrifuge

Plasma
(55% of whole blood)

White blood cells and
platelets (<1% of whole blood)

Red blood cells
(45% of whole blood)

Hematocrit

White
blood cells

Red
blood
cells

Platelets

7-2 Blood Plasma

Plasma is a light yellow (straw-colored) fluid that plays many important functions in maintaining homeostasis.

Blood as a Transport Medium

Plasma transports many important substances.

Dissolved in the plasma are (1) gases such as nitrogen, carbon dioxide, and oxygen; (2) ions such as sodium, chloride, and calcium; (3) nutrients such as glucose and amino acids; (4) hormones; (5) proteins; (6) various wastes; and (7) lipid molecules. Lipids are either suspended in tiny globules or bound to certain plasma proteins, which transport them through the bloodstream.

Regulating pH and Osmotic Pressure

Plasma proteins help regulate blood pH and osmotic pressure.

Proteins are the most abundant of all dissolved substances in the plasma. Blood proteins contribute to osmotic pressure (described briefly in Chapter 3). Osmotic pressure helps regulate

TABLE 7-1	Summary of Plasma Proteins
Protein	Function
Albumins	Maintain osmotic pressure and transport smaller molecules, such as hormones and ions
Globulins	Alpha and beta globulins transport hormones and fat-soluble vitamins; gamma globulins (antibodies) bind to foreign substances
Fibrinogen	Converted into fibrin network that form blood clots

the flow of materials in and out of capillaries. It is therefore essential to the proper distribution of wastes and nutrients in the body, so vital to homeostasis.

Blood proteins also help to regulate the pH of the blood. They do so by binding to hydrogen ions, thus preventing the H^+ concentration in the blood from rising, also essential to homeostasis.

Transport Molecules

Some plasma proteins serve as carrier proteins.

Human blood contains three types of protein: (1) albumins, (2) globulins, and (3) fibrinogen (Table 7-1). Albumins (al-BEW-mins) and two types of globulins (GLOB-u-lins) are carrier proteins. They bind to hormones, ions, and fatty acids, and transport these molecules through the bloodstream

Carrier proteins are large, water-soluble molecules. Their binding to much smaller lipid molecules (which are not soluble in water) renders the lipids water-soluble. This, in turn, facilitates their transport through the bloodstream, which consists largely of water. Carrier proteins also protect smaller molecules from destruction by the liver. These proteins are therefore essential to maintaining chemical balance in the blood and tissue fluids.

Not all globulins are carrier proteins. For example, one group of globulins consists of antibodies, proteins that "neutralize" viruses and bacteria or target them for destruction by phagocytic cells in body tissues known as *macrophages* (Chapter 14).

Still another important blood protein is fibrinogen (fie-BRIN-oh-gin). This unique protein is converted into fibrin (FIE-brin), which forms blood clots in the walls of injured blood vessels. Clots prevent blood loss and thus also help maintain homeostasis.

7-3 Red Blood Cells

The **red blood cell** (RBC) or erythrocyte is the most abundant cell in human blood. In fact, 20 drops of blood (equal to 1 milliliter) contain approximately 5 billion RBCs. If RBCs were people, a single milliliter of blood would contain slightly more than 75% of the world population!

Structure of RBCs

Red blood cells are highly flexible cells that transport oxygen in the blood.

Human RBCs are highly flexible disks that are concave on both sides. These cells transport oxygen and, to a lesser degree, carbon dioxide in the blood (Figure 7-2; Table 7-2). The unique shape

of the human RBC, shown in Figure 7-2, increases its surface area. The greater the surface area, the more readily gases move in and out of RBCs. Like so many other structures encountered in biology, the human RBC is a remarkable example of the marriage of form and function forged during the evolution of life on Earth.

Swept along in the bloodstream, RBCs travel many times through the circulatory system each day. As they pass through the capillaries, the RBCs bend and twist. This permits them to pass through the many miles of tiny capillaries whose internal diameters are often smaller than the RBC's.

The flexibility of RBCs is an adaptation that serves us well. However, not all people are so fortunate. According to various estimates, approximately one of every 500–1000 African Americans suffers from a disease called **sickle-cell disease**. This dis-

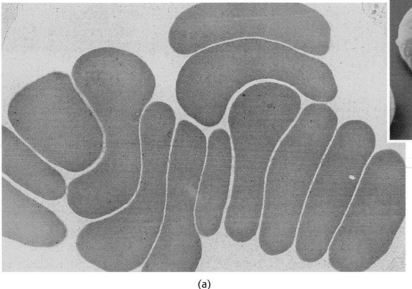

(a)

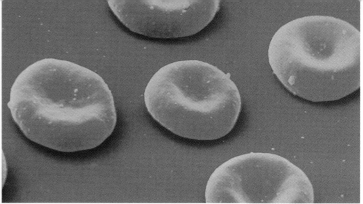

(b)

FIGURE 7-2 **Red Blood Cells** (a) Transmission electron micrograph of human RBCs showing their flexibility. (b) Scanning electron micrograph of human RBCs.

ease results in a marked decrease in the flexibility of RBCs and is caused by a genetic mutation—a slight alteration in the hereditary material (DNA) (Chapter 17). This mutation, in turn, results in the production of hemoglobin that contains one incorrect amino acid. This small defect dramatically alters the three-dimensional structure of the hemoglobin molecule, causing RBCs to transform from biconcave disks into sickle-shaped cells when they encounter low levels of oxygen in capillaries (Figure 7-3). Sickle-shaped cells are considerably less flexible than normal RBCs and are therefore unable to bend and twist. As a result, the sickle cells collect at branching points in capillary beds like logs in a logjam. This blocks blood flow, disrupting the supply of nutrients and oxygen to tissues and organs. The lack of oxygen causes considerable pain and can kill cells. Thus, blockages in the lungs, heart, and brain can be life-threatening. Blockages in the heart and brain often lead

TABLE 7-2	Summary of Blood Cells				
Name	Light Micrograph	Description	Concentration (Number of Cells/mm³)	Life Span	Function
Red blood cells (RBCs)		Biconcave disk; no nucleus	4 to 6 million	120 days	Transports oxygen and carbon dioxide
White blood cells Neutrophil		Approximately twice the size of RBCs; multi-lobed nucleus; clear-staining cytoplasm	3000 to 7000	6 hours to a few days	Phagocytizes bacteria
Eosinophil		Approximately same size as neutrophil; large pink-staining granules; bilobed nucleus	100 to 400	8–12 days	Phagocytizes antigen-antibody complex; attacks parasites
Basophil		Slightly smaller than neutrophil; contains large, purple cytoplasmic granules; bilobed nucleus	20 to 50	Few hours to a few days	Releases histamine during inflammation
Monocyte		Larger than neutrophil; cytoplasm grayish-blue; no cytoplasmic granules; U- or kidney-shaped nucleus	100 to 700	Lasts many months	Phagocytizes bacteria, dead cells, and cellular debris
Lymphocyte		Slightly smaller than neutrophil; large, relatively round nucleus that fills the cell	1500 to 3000	Can persist many years	Involved in immune protection, either attacking cells directly or producing antibodies
Platelets		Fragments of megakaryocytes; appear as small dark-staining granules	250,000	5–10 days	Play several key roles in blood clotting

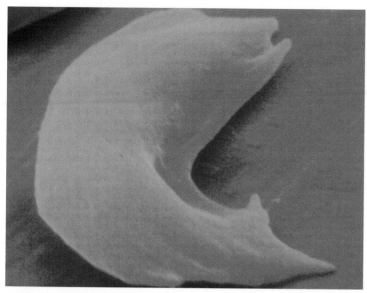

FIGURE 7-3 **Sickle-Cell Anemia** Scanning electron micrograph of a sickle cell.

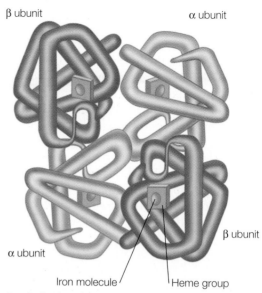

β ubunit   α ubunit

α ubunit   β ubunit

Iron molecule   Heme group

FIGURE 7-4 **Porphyrin Ring** The porphyrin ring of the hemoglobin molecule contains an iron ion that binds to oxygen and carbon monoxide.

to heart attacks and serious brain damage. Many people who have sickle-cell disease die in their late twenties and thirties; some die even earlier.

Making New RBCs

Red blood cells are replenished by stem cells in bone marrow.

On average, RBCs live about 120 days. At the end of their life span, the liver and spleen remove the aged RBCs from circulation. The iron contained in the hemoglobin, however, is recycled by these organs and used to produce new RBCs in the red bone marrow. The recycling of iron is not 100% efficient, however, so small amounts of iron must be ingested each day in the diet. Loss of blood from an injury or, in women, during menstruation, increases the body's demand for iron. Without adequate iron intake, oxygen transport may become impaired.

Human RBCs are highly specialized "cells" that lose their nuclei and organelles during cellular differentiation (Figure 7-2a). Because of this, RBCs cannot divide to replace themselves as they age. In humans, new RBCs are produced in the bone marrow. In the blood-producing bone marrow, known as *red marrow*, RBCs are produced by **stem cells**, undifferentiated cells that trace back to embryonic development. These cells give rise to 2 million RBCs per second!

In infants and children, almost all of the bone marrow is involved in the production of RBCs. As growth slows, however, the red marrow of many bones becomes inactive and gradually fills with fat cells and becomes yellow marrow, a fat storage depot. By the time an individual reaches adulthood, only a few bones, such as the hip bones, sternum (breastbone), ribs, and vertebrae still produce RBCs. In cases of severe, prolonged anemia, however, yellow marrow can be converted back into active red marrow.

The number of RBCs in the blood remains more or less constant over long periods. Maintaining a constant concentration

of RBCs is essential to homeostasis. This process is controlled by a hormone with a rather forbidding name, **erythropoietin** (eh-RITH-row-po-EAT-in) or **EPO**. As a side note, EPO is one of a dozen or more performance-enhancing chemical substances being used illegally by athletes.

Erythropoietin is produced by cells in the kidney when oxygen levels in the bloodstream decline—for example, when a person moves to high altitudes or loses a significant amount of blood in an accident. In the red bone marrow, this hormone stimulates the stem cells to divide and multiply, thus increasing RBC production. As the RBC concentration increases, oxygen supplies increase. When oxygen levels return to normal, erythropoietin levels fall, reducing the rate of RBC formation in a classical negative feedback mechanism so common among the body's homeostatic mechanisms.

Oxygen Transport

Hemoglobin is an oxygen-transporting protein found in RBCs.

Hemoglobin (HEME-oh-GLOBE-in) is a large protein molecule found exclusively in the RBCs. As shown in Figure 7-4, each of hemoglobin's four subunits contains a large, organic ring structure. In the center of the ring is iron.

When blood from the pulmonary arteries flows through the capillary beds of the lungs, oxygen diffuses into the plasma, then into the RBCs. Inside the RBC, oxygen binds to the iron in hemoglobin molecules for transport through the circulatory system. In fact, 98% of the oxygen in the blood is transported bound to iron in hemoglobin. Hemoglobin bound to oxygen is called *oxyhemoglobin*. The remaining 2% of the oxygen carried by blood is dissolved in the blood plasma.

Carbon dioxide, a waste product of cellular respiration, also binds to hemoglobin, but to a much lesser degree. Most carbon dioxide molecules are converted to bicarbonate and are transported in the plasma.

Anemia

Anemia results in a decrease in oxygen transport by the blood.

Homeostasis depends on the normal operation of the heart and blood vessels. It also requires that the blood be able to absorb a sufficient amount of oxygen as it passes through the lungs. Unfortunately, the oxygen-carrying capacity of the blood can be impaired by a number of conditions. A reduction in the oxygen-carrying capacity of the blood is known as **anemia**. This condition may result from (1) a decrease in the number of circulating RBCs, (2) a reduction in the RBC hemoglobin content, or (3) the presence of abnormal hemoglobin in RBCs.

The number of RBCs in the blood may decline because of excessive bleeding or because of the presence of a tumor in the red marrow that reduces RBC production. Several infectious diseases (such as malaria) also decrease the RBC concentration in the blood. A reduction in the amount of hemoglobin in RBCs may be caused by iron deficiency or a deficiency in vitamin B-12, protein, or copper in the diet. Abnormal hemoglobin is produced in sickle-cell anemia and other genetic disorders.

Anemia generally results in weakness and fatigue. Individuals are often pale and tend to faint or become short of breath easily. People suffering from anemia often have an increased heart rate, because the heart beats faster to offset the reduction in the oxygen-carrying capacity of the blood. As a rule, anemia is not a life-threatening condition. However, it does weaken one's resistance to other diseases and also limits a person's productivity. Therefore, it should be treated quickly.

7-4 White Blood Cells

White blood cells (WBCs), or **leukocytes**, are nucleated (they have nuclei) cells that are part of the body's protective mechanism, a homeostatic system that combats harmful microorganisms such as bacteria and viruses. White blood cells are produced in the bone marrow and circulate in the bloodstream but constitute less than 1% of the blood volume.

Function of WBCs

White blood cells are a diverse group that protects the body from infection.

Although WBCs are officially blood cells, they do most of their work combating infections outside of the bloodstream, in the body tissues. The blood and circulatory system merely serve to transport the WBCs to sites of infection. When WBCs arrive at the "scene," they escape through the walls of the capillaries by "squeezing" between the endothelial cells (Figure 7-5).

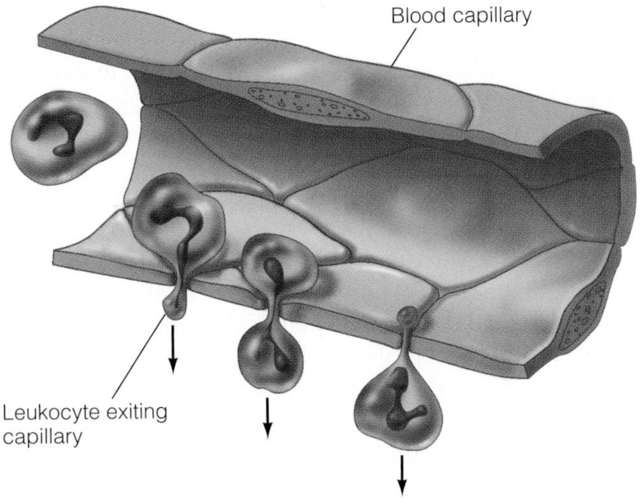

Table 7-2 lists and describes the five types of WBCs found in the blood. The three most numerous, which are discussed here, are neutrophils, monocytes, and lymphocytes.

Neutrophils

Neutrophils are phagocytic cells and are the first to arrive at an injury.

Neutrophils (NEW-trow-fills) are the most abundant of the WBCs. Approximately twice the size of RBCs, these cells are distinguished by their multi-lobed nuclei. Neutrophils circulate in the blood like a cellular police force awaiting microbial invasion. Attracted by chemicals released by infected tissue, neutrophils escape from the bloodstream, then migrate to the site of infection by amoeboid movement.

Neutrophils are usually the first WBCs to arrive on the scene. When they arrive, they immediately begin to engulf (phagocytize) microorganisms, preventing the spread of bacteria and other organisms from the site of invasion. When a neutrophil's lysosomes are used up, the cell dies and becomes part of the yellowish liquid, or pus, that exudes from wounds. Pus is a mixture of dead neutrophils, cellular debris, and bacteria, both living and dead.

Monocytes

Monocytes are a kind of clean-up crew.

Monocytes (MON-oh-sites) are also phagocytic cells. Slightly larger than neutrophils, monocytes contain distinctive U-shaped or kidney-shaped nuclei. Like neutrophils, monocytes leave the bloodstream to do their "work" and migrate through body tissues via amoeboid motion. Once at the site of an infection, they begin phagocytizing microorganisms, dead cells, and dead neutrophils. Thus, while neutrophils are the "first-line" troops, monocytes constitute something of a mop-up crew.

Monocytes also take up residence in connective tissues of the body, where they are referred to as **macrophages** (MACK-row-FAY-ges). These cells remain more or less stationary, like watchful soldiers ever ready to attack and phagocytize invaders.

FIGURE 7-5 **White Blood Cell Exit** White blood cells (leukocytes) escape from capillaries by squeezing between endothelial cells.

Blood capillary

Leukocyte exiting capillary

Lymphocytes

Lymphocytes are involved in immune reaction to microorganisms.

The second most abundant WBC is the **lymphocyte**. These cells are not phagocytic but rather mount an immune response to bacteria and viruses. Although lymphocytes are found in the blood, most of them exist outside the circulatory system in lymphoid organs (LIM-foid). This includes organs such as the spleen, thymus, lymph nodes, and lymphoid tissue. Aggregations of lymphocytes also exist beneath the lining of the intestinal and respiratory tracts. It is in these locations that lymphocytes attack microbial intruders (Figure 7-6).

Two types of lymphocytes are found in the body. The first type is the **T lymphocyte**, or **T cell**. T cells attack foreign cells such as fungi, parasites, and tumor cells directly. They are thus said to provide cellular immunity.

The second type is called the **B lymphocyte**, or **B cell**. When activated, B cells transform into another kind of cell, known as the *plasma cells*. **Plasma cells**, in turn, synthesize and release antibodies, proteins that circulate in the blood and bind to foreign substances. The binding of antibodies to "foreigners" coats them and neutralizes them. Or, it marks microorganisms and tumor cells for destruction by macrophages (Chapter 14).

Like the other formed elements of blood, the WBCs are involved in homeostasis. Their numbers can increase greatly during a microbial infection and other diseases. In fighting off a disease, they help return the body to normal function.

An increase in the number of WBCs, called *leukocytosis* (lew-co-sigh-TOE-siss), is a normal homeostatic response to intruders. It ends when the microbial invaders have been destroyed. Increases and decreases in various types of WBCs can be used to diagnose many medical disorders. For example, a dramatic increase in lymphocytes and intermittent abdominal pain, first felt in the upper abdomen or around the navel, are usually signs of appendicitis, an infection of the appendix (Chapter 5). Because variations in the WBC count accompany many diseases, a blood test is a standard procedure for patients undergoing diagnostic testing.

Leukemia

Leukemia is a cancer of the white blood cells.

Like most other cells, WBCs can become cancerous, dividing uncontrollably in the bone marrow, then entering the bloodstream. A cancer of WBCs is called **leukemia** (lew-KEEM-ee-ah; literally, "white blood"). The most serious type of leukemia is acute leukemia, so named because it kills victims quickly. Children are the primary victims of this disease.

In acute leukemia, WBCs fill the bone marrow, crowding out the cells that produce RBCs and platelets, which are involved in blood clotting (discussed shortly). This results in a decline in the production of RBCs and platelets, which leads to anemia and a reduction in platelet production, respectively. The decline in platelet production increases internal bleeding. Making matters worse, the cancerous WBCs produced in leukemia are often incapable of fighting infection. Leukemia patients typically succumb to infections and internal bleeding.

Leukemia can be treated by irradiating the bone marrow and by administering a drug called *vincristine* (VIN-chris-teen) that stops cells from dividing and thus increasing in number. In the mid-1970s, only one of every four children with leukemia survived. Today, thanks to vincristine, three of every four children with the disease survive! Vincristine was discovered in a plant known as the rosy periwinkle found in tropical rain forests (Figure 7-7). Thousands of other drugs have come from other tropical plants, underscoring the importance of protecting the rain forests.

Infectious Mononucleosis

Infectious mononucleosis is a viral disease spread through saliva.

Another common disorder of the WBCs is **infectious mononucleosis** (MON-oh-NUKE-clee-OH-siss), commonly called "mono" or "kissing disease." Mono is caused by a virus transmitted through saliva and may be spread by kissing; by sharing

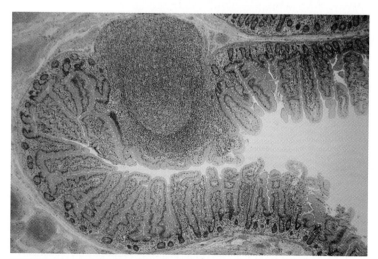

FIGURE 7-6 **Lymphoid Tissue** The loose connective tissue beneath the lining of the large intestine and other sites is often packed with lymphocytes that have proliferated in response to invading bacteria.

FIGURE 7-7 **Rosy Periwinkle** Many tropical plants contain chemicals that provide extraordinary medical benefits. One substance from the rosy periwinkle has helped physicians treat leukemia. Unfortunately, the tropical rain forests are being cut at an alarming rate, reducing our chances of finding other cures.

silverware, plates, and drinking glasses; and possibly even through drinking fountains. The virus spreads through the body. In the blood, the virus infects lymphocytes in the bloodstream, although the number of both monocytes and lymphocytes increases rapidly during an infection.

Individuals with mono complain of fatigue, aches, sore throats, and low-grade fever. Physicians recommend that victims get plenty of rest and drink lots of liquids while the immune system eliminates the virus. Within a few weeks, symptoms generally disappear, although weakness may persist for two more weeks.

7-5 Blood Clotting

The capillaries are rather delicate structures. Even minor bumps and scrapes can cause them to leak. Fortunately, leakage is quickly halted by blood clotting, one of the most intricate homeostatic systems to have evolved. A simplified version is discussed here.

Platelets

Platelets are a vital component of the blood-clotting mechanism.

Among the most important agents of blood clotting are the **platelets**, tiny formed elements produced in the bone marrow by fragmentation of a huge cell known as the **megakaryocyte** (MEG-ah-CARE-ee-oh-site) (Figure 7-8). Like RBCs, platelets lack nuclei and organelles and therefore are not true cells. Also like RBCs, platelets are unable to divide. Carried passively in the bloodstream, platelets are coated by a layer of a sticky material, which causes them to adhere to irregular surfaces such as tears in blood vessels or atherosclerotic plaque.

Clotting is a chain reaction stimulated by the release of a chemical called **thromboplastin** (THROM-bow-PLASS-tin) from injured cells lining damaged blood vessels (Figure 7-9a). Thromboplastin is a lipoprotein that acts on an inactive plasma enzyme in the blood known as **prothrombin** (produced by the

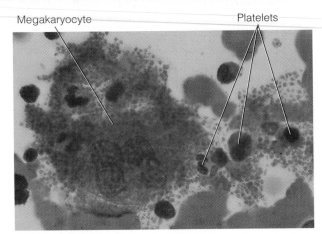

FIGURE 7-8 **Megakaryocyte** A light micrograph of a megakaryocyte, a large, multinucleated cell found in bone marrow; the megakaryocyte fragments, giving rise to platelets.

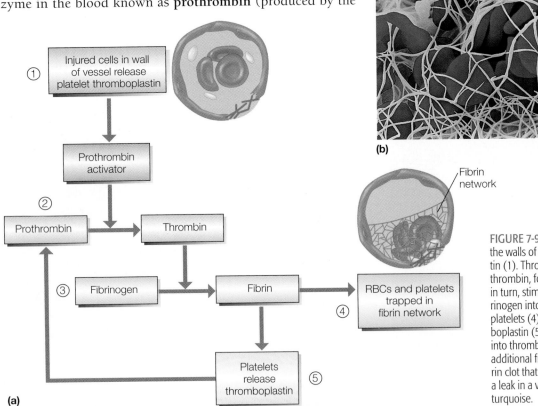

FIGURE 7-9 **Blood Clotting Simplified** (a) Injured cells in the walls of blood vessels release the chemical thromboplastin (1). Thromboplastin stimulates the conversion of prothrombin, found in the plasma, into thrombin (2). Thrombin, in turn, stimulates the conversion of the plasma protein fibrinogen into fibrin (3). The fibrin network captures RBCs and platelets (4). Platelets in the blood clot release platelet thromboplastin (5), which converts additional plasma prothrombin into thrombin. Thrombin, in turn, stimulates the production of additional fibrin. (b) A scanning electron micrograph of a fibrin clot that has already trapped platelets and RBCs, plugging a leak in a vessel. The RBCs are red, and the fibrin network is turquoise.

liver). Thromboplastin causes prothrombin to be converted into its active form, **thrombin**. Thrombin, in turn, acts on another blood protein, **fibrinogen**, also produced by the liver. When activated, fibrinogen is converted into **fibrin**, long, branching fibers that produce a weblike network in the wall of the damaged blood vessel (Figure 7-9b). The fibrin web traps RBCs and platelets, forming a plug that stops the flow of blood to the tissue. Platelets captured by the fibrin web release additional thromboplastin, known as *platelet thromboplastin*, which causes more fibrin to be laid down, thus reinforcing the fibrin network. You may want to take a look at Figure 7-9a to review these steps.

Blood clotting occurs fairly quickly. In most cases, a damaged blood vessel is sealed by a clot within 3 to 6 minutes of an injury; 30 to 60 minutes later, platelets in the clot begin to draw the clot inward, stitching the wound together. How do they perform this remarkable task?

Platelets contain contractile proteins like those in muscle cells. Contraction of the protein fibers draws the fibrin network inward, pulling the edges of the cut or damaged blood vessel together. This closes the wound like a surgical stitch.

Blood clots do not stay in place indefinitely. If they did, the circulatory system would eventually become clogged, and blood flow would come to a halt. Instead, clots are dissolved by a blood-borne enzyme known as **plasmin** (PLAZ-min).

As important as blood clots are in protecting the body, clotting can also cause problems. For example, blood clots can break loose from their site of formation, circulate in the blood, and be-come lodged in arteries, especially those narrowed by plaque. In such cases, the blood clots restrict blood flow, causing considerable damage to tissues served by the vessel. Blood clots typically lodge in the narrow vessels serving the heart muscle, brain, and other vital organs.

Hemophilia

Hemophilia is a potentially life-threatening clotting disorder.

In some people, blood clotting is severely impaired by a reduced platelet count—for example, as a result of leukemia. A reduced platelet count may also result from exposure to excess radiation, which damages red bone marrow where the megakaryocytes reside.

Reduced blood clotting can also result from liver damage, which decreases the production of blood-clotting factors. Liver damage may result from infections of the liver (hepatitis), liver cancer, or excessive alcohol consumption.

The most common cause of reduced blood clotting is a hereditary genetic defect known as **hemophilia** (he-moe-FEAL-ee-ah). In this disease, the liver fails to produce the necessary clotting factors. Problems begin early in life, and even tiny cuts or bruises can bleed uncontrollably, which can lead to death. Because of repeated bleeding into the joints, victims suffer great pain and often become disabled; they often die at a young age. Hemophiliacs can be treated by transfusions of blood-clotting factors every few days. This therapy, however, is expensive.

7-6 Blood Types

The surface of the red blood cell (RBC) membrane, like that of other cells, contains many glycoproteins that form a unique cellular fingerprint (Chapter 3). These glycoproteins are the basis of blood-typing.

The Four Blood Types

Blood types are determined by glycoproteins on RBC membranes.

Four blood types exist: A, B, AB, and O. The letters refer to one type of glycoprotein present on the plasma membrane of RBCs of an individual. As illustrated in Table 7-3, individuals with type A blood have RBCs whose plasma membranes contain the A glycoprotein. The RBCs in individuals with type B blood contain type B glycoprotein. People with AB blood have both A and B glycoproteins, and people with type O have neither one.

Physicians learned a long time ago that blood could be successfully transfused from one person to another, but only if their blood type matched. In other words, an individual with type A blood could only receive type A blood, and individuals with type B blood can only receive type B blood.

Cross-matching blood is essential to prevent life-threatening immune reactions. These reactions result from antibodies found in the blood of most people. As shown in Table 7-3, each

TABLE 7-3	Summary of Blood Types				
Blood Type	Glycoproteins (Antigens) on Plasma Membranes of RBCs	Antibodies in Blood	Safe to Transfuse	
			To	From
A	A	b*	A, AB	A, O
B	B	a	B, AB	B, O
AB	A + B	—	AB	A, B, AB, O
O	—	a + b	A, B, AB, O	O

*Lowercase *b* indicates antibody to B antigen.

blood type carries a specific type of antibody. People with type A blood, for example, naturally contain antibodies to the B glycoprotein. (This is why individuals with type A blood cannot receive type B blood.) People with type B blood contain antibodies to the A glycoprotein. For reasons not well understood, these antibodies appear in the blood during the first year of life.

Serious problems arise when incompatible blood types are mixed. For example, consider what happens if an individual with type A blood is accidentally given type B blood (Figure 7-10). The antibodies to type B blood found in the recipient bind to the transfused RBCs (which contain type B glycoproteins). This causes the type B RBCs to clump together (agglutinate) and burst. This potentially deadly response is known as the transfusion reaction (Figure 7-10).

RBC clumping restricts blood flow through capillaries, reducing oxygen and nutrient flow to cells and tissues. Massive breakdown of RBCs results in the release of large amounts of hemoglobin into the blood plasma. Hemoglobin precipitates in the kidney, blocking the tiny tubules that produce urine. This can result in acute kidney failure, which can result in death.

Because of the possibility of this potentially life-threatening reaction, successful transfusions require careful matching of the blood types of the donor and recipient. As Table 7-3 shows, RBCs from individuals with type O blood have neither A nor B glycoproteins. Therefore, type O blood can be transfused into individuals with all four blood types. Type O individuals are said to be universal donors. However, type O blood contains antibodies to both A and B glycoproteins. Therefore, individuals with type O blood can receive only type O blood.

As shown in Table 7-3, individuals with type AB blood contain RBCs with both A and B glycoproteins but no antibodies related to the ABO system. People with AB blood can therefore receive blood from all others. Because of this, they are referred to as universal recipients.

Although AB blood permits one to be a universal recipient, AB blood can be safely transfused only into individuals with the same type. If it were given to people with Type A or Type B blood, the recipients would undergo a transfusion reaction, because their blood contains antibodies that would respond to the antigens on the RBCs.

Blood Transfusion

The Rh factor must also be considered in blood transfusions.

The terms *universal donor* and *universal recipient* are somewhat misleading, because RBCs also contain other glycoproteins that can cause transfusion reactions. The most important of these is the **Rh factor**. This antigen was first identified in rhesus (REE-suss) monkeys; hence the designation *Rh*. People whose cells contain the Rh antigen, or Rh factor, are said to be Rh-positive. Those without it are Rh-negative.

Unlike the ABO system, in which blood contains antibodies without exposure to other blood types, in the Rh system, the antibodies are produced only when Rh-positive blood is transfused

into the bloodstream of a person with Rh-negative blood. The first transfusion of Rh-positive blood into an Rh-negative person generally does not result in a transfusion reaction, but a second transfusion does. To reduce the likelihood of a transfusion reaction, Rh-negative people should receive only Rh-negative blood, and Rh-positive people should receive only Rh-positive blood.

The Rh factor becomes particularly important during pregnancy. Problems can arise if an Rh-negative mother has an Rh-positive baby (Figure 7-11). Even though the maternal and fetal bloodstreams are separate, small amounts of fetal blood usually enter the maternal bloodstream at birth. Rh antibodies form in the maternal bloodstream, and the woman becomes sensitized to the Rh factor.

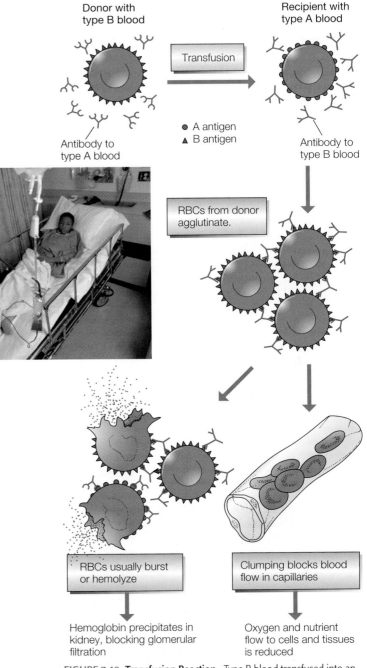

FIGURE 7-10 Transfusion Reaction Type B blood transfused into an individual with type A blood results in a transfusion reaction, characterized by agglutination and hemolysis.

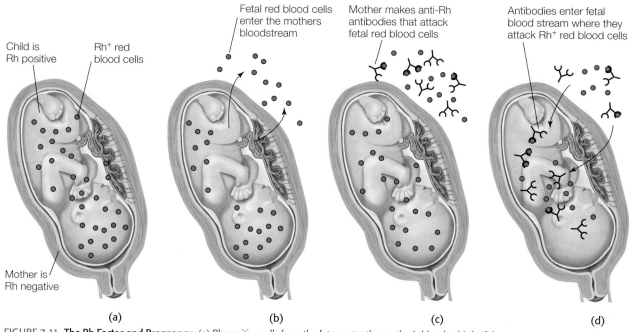

Fetal red blood cells enter the mothers bloodstream

Mother makes anti-Rh antibodies that attack fetal red blood cells

Antibodies enter fetal blood stream where they attack Rh⁺ red blood cells

Child is Rh positive

Rh⁺ red blood cells

Mother is Rh negative

(a)   (b)   (c)   (d)

FIGURE 7-11 **The Rh Factor and Pregnancy** (a) Rh-positive cells from the fetus enter the mother's blood at birth. If the mother is Rh-negative, her immune system responds, producing antibodies to the Rh-positive RBCs and destroying them. (b) Problems arise if the mother becomes pregnant again and has another Rh-positive baby. If the mother was not treated the first time, antibodies to Rh-positive RBCs cross the placenta and destroy fetal RBCs.

To prevent antibody production in Rh-negative women who give birth to Rh-positive babies, physicians routinely inject antibodies to fetal Rh-positive RBCs into the mother soon after she has given birth. (The antibody-containing serum is called RhoGAM.) These antibodies bind to Rh-positive RBCs from the fetus before a woman's immune system responds to them. This, in turn, prevents a woman from being sensitized. To be effective, however, the treatment must be given within 72 hours after the baby is born.

If the woman is not treated at the time and becomes pregnant again with an Rh-positive baby, maternal antibodies to the Rh factor will cross the placenta. In the fetal circulation, these antibodies cause fetal RBCs to agglutinate, then break down, resulting in anemia and lack of oxygen to tissues. Unless the baby receives a blood transfusion (of Rh-negative blood) before birth and several after birth, it is likely to have brain damage and may even die.

7-7 Health and Homeostasis

The blood is vitally important to homeostasis. It transports nutrients and waste to and from the cells, respectively. It protects against changes in pH. It transports excess heat to the body's surface, where it is eliminated. It plays a key role in the body's defense system, and it seals injuries in blood vessels through clots.

It is no surprise then to learn that cells depend mightily on the proper functioning of blood. Anything that upsets this function can upset cellular function throughout the body.

As you have seen in this chapter, many factors can harm the blood system, upsetting homeostasis. Genetic defects that result in a reduction in liver blood clotting agents lead to hemophilia. Leukemia, caused by exposure to cancer-causing agents, disrupts RBC and platelet production. This, in turn, decreases oxygen transport to the cells of the body and causes a marked decrease in blood clotting. Excessive alcohol consumption can damage the liver and reduce the output of blood clotting factors.

Another, even more common, disruptor of homeostasis is a common air pollutant, known as carbon monoxide. **Carbon monoxide** (CO) is a colorless, odorless gas produced by the incomplete combustion of organic fuels. It can be released by stoves, furnaces, and water heaters in our homes. It is also produced by automobiles, causing increased levels along streets and highways, in parking garages, and in tunnels. It is released by power plants and factories, polluting the air in our cities. It is even a major pollutant in tobacco smoke. According to estimates by the Environmental Protection Agency, over 40 million Americans (one of every 7 people) are exposed to levels of CO harmful to their health.

What makes CO so dangerous? Carbon monoxide, like oxygen, binds to hemoglobin. However, hemoglobin has a much greater affinity for CO than for oxygen (about 200 times greater). Consequently, CO "outcompetes"" oxygen for the binding sites on the hemoglobin molecules and thus reduces the blood's ability to carry oxygen, which is essential for energy production.

For healthy people, low levels of CO in their blood do not create much of a problem. Their bodies simply produce more RBCs or increase the heart rate to augment the flow of oxygen to tissues.

Carbon monoxide does create a problem, however, when levels are so high that the body cannot compensate. At very high levels, CO becomes deadly. Research suggests that CO levels currently deemed acceptable by federal standards can trigger chest pain (angina) in adults with coronary artery disease. Chest pains result from a lack of oxygen in the heart muscle.

In levels commonly found in and around cities, CO is also troublesome for the elderly. Carbon monoxide places additional strain on their hearts, making the already weakened organ work harder. Thus, the elderly and people suffering from cardiovascular disease are advised to stay inside on high-pollution days.

SUMMARY

Blood: An Overview

1. Blood is a watery tissue consisting of two basic components: the plasma, a fluid that contains dissolved nutrients, proteins, gases, and wastes; and the formed elements—red blood cells, white blood cells, and platelets—which are suspended in the plasma.
2. The functions of the formed elements are listed in **Table 7-2**.

Blood Plasma

3. The plasma constitutes about 55% of the volume of a person's blood, and the formed elements make up the remainder.

Red Blood Cells

4. Red blood cells (RBCs) are highly specialized cells that lack nuclei and organelles. Produced by the red bone marrow, RBCs transport oxygen in the blood.

White Blood Cells

5. White blood cells (WBCs) are nucleated cells and are part of the body's protective mechanism to combat microorganisms. WBCs are produced in the bone marrow and circulate in the bloodstream, but do most of their work outside it, in the body tissues.

6. The most abundant WBCs are the neutrophils, which are attracted by chemicals released from infected tissue. Neutrophils leave the bloodstream and migrate to the site of infection by amoeboid movement. Neutrophils are the first WBCs to arrive at the site of an infection, where they phagocytize microorganisms, preventing the spread of bacteria and other organisms.
7. The second group of cells to arrive are the monocytes, which phagocytize microorganisms, dead cells, cellular debris, and dead neutrophils.
8. Lymphocytes are the second most numerous WBCs and play a vital role in immune protection.

Blood Clotting

9. Platelets are fragments of large bone marrow cells and are involved in blood clotting. Platelets are coated by a layer of a sticky material, which causes them to adhere to irregular surfaces such as tears in blood vessels. Blood clotting is summarized in **Figure 7-9**.

Blood Types

10. Blood types are determined by the types of glycoprotein found in RBC cell membranes. Blood transfusions require careful matching of donor and recipient blood types so antibodies in a recipient's blood do not react with the A or B glycoproteins. Transfusions also require matching of the Rh factor.

Health and Homeostasis

11. Many factors influence the blood, upsetting homeostasis and leading to disease and even death.
12. Carbon monoxide is produced by the incomplete combustion of organic fuels in our homes, automobiles, factories, and power plants. Carbon monoxide binds to hemoglobin and reduces the oxygen-carrying capacity of the blood, which impairs cellular energy production, upsetting homeostasis. At high concentrations, carbon monoxide can be lethal. At lower concentrations, it is harmful to the elderly and to people with cardiovascular and lung disease.

critical thinking

Thinking Critically—Analysis

This Analysis corresponds to the Thinking Critically scenario that was presented at the beginning of this chapter.

The researchers themselves freely admit that this study only looks at the effect of lipoprotein(a) on smooth muscle cells grown in laboratory culture dishes. They have not yet determined whether the effect occurs in the body itself.

Caution is advised when examining studies of cells and tissues in culture. Although studies in tissue culture are essential to medical science, it's important to remember that a culture dish is an artificial environment. Because of this, such studies may lead to erroneous conclusions. This exercise requires one to scrutinize the experimental method.

KEY TERMS AND CONCEPTS

Anemia, p. 130
B lymphocyte, p. 131
Carbon monoxide, p. 135
Erythropoeitin, p. 129
Fibrin, p. 133
Fibrinogen, p. 133
Hemoglobin, p. 129
Hemophilia, p. 133
Infectious mononucleosis, p. 131
Leukemia, p. 131

Lymphocyte, p. 131
Macrophage, p. 130
Megakaryocyte, p. 132
Monocyte, p. 130
Neutrophil, p. 130
Plasma, p. 126
Plasma cell, p. 131
Plasmin, p. 133
Platelet, p. 132

Prothrombin, p. 132
Red blood cells, p. 127
Rh Factor, p. 134
Sickle-cell disease, p. 127
Stem cells, p. 129
T lymphocyte, p. 131
Thrombin, p. 132
Thromboplastin, p. 132
White blood cells, p. 130

CONCEPT REVIEW

1. Describe the structure and function of each of the following: red blood cells, platelets, lymphocytes, monocytes, and neutrophils. pp. 127–131

2. Define each of the following terms: leukemia, anemia, and infectious mononucleosis. pp. 130–132

3. Explain how a blood clot forms and how it helps prevent bleeding. pp. 132–133

4. Describe the many ways blood participates in homeostasis. How can these homeostatic mechanisms be upset? pp. 135–136

SELF-QUIZ: TESTING YOUR KNOWLEDGE

1. Plasma comprises _____ percent of the blood volume. p. 126

2. Plasma proteins are the most abundant dissolved substance in blood and are responsible for maintaining _____ and the osmotic concentration of the blood, among other things. p. 127

3. _____ proteins bind to smaller molecules and help transport them throughout the bloodstream. p. 127

4. The _____ blood cell is the most abundant blood cell in the body and is involved in transporting oxygen throughout the body. p. 127

5. Loss of flexibility of the RBC resulting from a genetic disorder is known as _____ disease. p. 127

6. RBCs are produced in the _____ bone marrow. The hormone _____ is responsible for controlling the production of RBCs in the body. p. 129

7. A reduction in the oxygen-carrying capacity of blood due to a decrease in RBCs or a decrease in the amount of _____ in RBCs is called _____. p. 130

8. The first cell to arrive at the site of an infection is a _____. p. 130

9. The _____ is a white blood cell involved in immune protection. p. 131

10. During blood clotting _____ becomes lodged in the _____ network formed at the wound site, stopping the flow of blood. pp. 132–133

11. People with Type A blood contain the A _____ and B antibody. They can receive blood from Type A and Type O individuals. p. 133

12. People with Type AB blood contain both A and B glycoproteins but _____ antibodies. p. 134

13. Type AB individuals are universal _____. p. 134

14. Individuals with Type O blood contain _____ glycoproteins. p. 134

15. Type O individuals can receive blood from individuals with _____ blood types. p. 134

 www.jbpub.com/humanbiology/5e

The site features eLearning, an online review area that provides quizzes, chapter outlines, and other tools to help you study for your class. You can also follow useful links for in-depth information, research the differing views in the Point/Counterpoints, or keep up on the latest health news.

The Vital Exchange: Respiration

George F. Eaton was a robust and handsome Irishman who grew up in eastern Massachusetts, married, and raised three children. To support his family, Eaton worked as a fire fighter for 30 years. Fire fighting is dangerous work, in part because it exposes men and women to smoke containing numerous potentially harmful air pollutants.

thinking critically

For years, representatives of the tobacco industry claimed that there was no proof that cigarette smoking caused cancer, despite the fact that over 40 studies on people and numerous studies on lab animals showed a strong correlation between smoking and lung cancer. Most of the studies on humans correct for other factors that might cause lung cancer such as exposure to carcinogenic chemical pollutants in the workplace.

Why did representatives from the tobacco industry claim there was no proof that smoking caused lung cancer? How would you have responded to their assertion? Can a study be devised to prove the connection? What critical thinking rules does this exercise rely on?

Upon retirement, Eaton moved to Cape Cod, but his retirement years were cut short by emphysema—a debilitating respiratory disease resulting from the breakdown of the air sacs in the lungs, where oxygen and carbon dioxide are exchanged between the air and the blood. As the walls of the alveoli degenerate, the surface area for the diffusion of oxygen and carbon dioxide gradually decreases. Emphysema is irreversible and incurable. Patients suffer from shortness of breath; eventually, even mild exertion becomes difficult, prompting one patient to describe the disease as a kind of "living hell."

The degeneration of the lungs in patients with emphysema creates a domino effect. One of the dominoes is the heart. Because oxygen absorption in the lungs declines quite substantially over time, the heart of an emphysemic patient must work harder and harder. This effort puts additional strain on this already hard-working organ and can lead to heart failure.

George Eaton died a slow death as his lungs grew increasingly more inefficient. Relatives mourning his death blamed it on the pollution to which he was exposed while working for the fire department. Fire fighting, however, was probably only part of the cause, for Eaton had smoked most of his adult life. Cigarette smoking is the leading cause of emphysema.

This chapter describes the respiratory system—its structure and function and the diseases that affect it.

8-1 Structure of the Human Respiratory System

The respiratory system supplies oxygen to the body and gets rid of carbon dioxide.

The human respiratory system functions automatically, drawing air into the lungs, then letting it out, in a cycle that repeats itself about 16 times per minute at rest—or about 23,000 times per day. If you got a dollar for each time you took a breath, you'd be a millionaire in a month and a half.

The respiratory system supplies oxygen to the body and gets rid of carbon dioxide. Oxygen, of course, is needed for cellular respiration; carbon dioxide is a waste product of this process. In its role as a provider of oxygen and a disposer of carbon dioxide waste, the respiratory system helps to maintain the constant internal environment necessary for normal cellular metabolism. Thus, like many other systems, it plays an important role in homeostasis.

Anatomy of the Respiratory System

The respiratory system consists of an air-conducting portion that delivers air to the lungs and a gas-exchange portion that allows oxygen to be transferred from inhaled air to blood.

The respiratory system consists of two basic parts: an air-conducting portion and a gas-exchange portion (Table 8-1). The air-conducting portion is an elaborate set of passageways that transports air to and from the **lungs**, two large, saclike organs in the chest (thoracic) cavity (Figure 8-1a). Like the arteries of the body, these passageways start out large. With each branching, they become progressively smaller and more numerous.

The lungs are the gas-exchange portion of the respiratory system. Each lung has millions of tiny, thin-walled air sacs or **alveoli** (singular, alveolus) (Figure 8-1b). The walls of the alveoli contain numerous capillaries that absorb oxygen from the inhaled air and release carbon dioxide (Figure 8-1b).

Air enters the respiratory system through the nose and mouth, then is drawn backward into the **pharynx** (FAIR-inks) (Figure 8-2). The pharynx is a funnel-shaped structure that opens into the nose and mouth in the front (anteriorly) and joins below with the larynx (LAIR-inks). The **larynx**, or voice box, is a rigid, hollow structure that houses the **vocal cords**, two folds of tissue that vibrate when we talk, sing, or hum (Figure 8-2). To feel it, gently put your fingers alongside your throat, then swallow. The structure that moves up and down is the larynx.

The production of sounds, called *phonation*, is critical to many members of the animal kingdom. Humans produce a wide range of sounds for communication. These sounds are produced by two elastic ligaments inside the larynx, the vocal cords, which vibrate as air is expelled from the lungs (Figure 8-3). The sounds generated by the vocal cords are modified by changing the position of the tongue and the shape of the oral cavity.

The vocal cords vary in length and thickness from one person to the next. They also vary between men and women. In men, the vocal cords are longer and thicker than vocal cords in women. This difference is due to the male sex hormone, testosterone, which is produced by the testes. Because their vocal cords are thicker than those in women, men tend to have deeper voices than most women.

TABLE 8-1	Summary of the Respiratory System
Organ	Function
Air conducting	
Nasal cavity	Filters, warms, and moistens air; also transports air to pharynx
Oral cavity	Transports air to pharynx; warms and moistens air; helps produce sounds
Pharynx	Transports air to larynx
Epiglottis	Covers the opening to the trachea during swallowing
Larynx	Produces sounds; transports air to trachea; helps filter incoming air; warms and moistens incoming air
Trachea and bronchi	Warm and moisten air; transport air to lungs; filter incoming air
Bronchioles	Control air flow in the lungs; transport air to alveoli
Gas exchange	
Alveoli	Provide area for exchange of oxygen and carbon dioxide

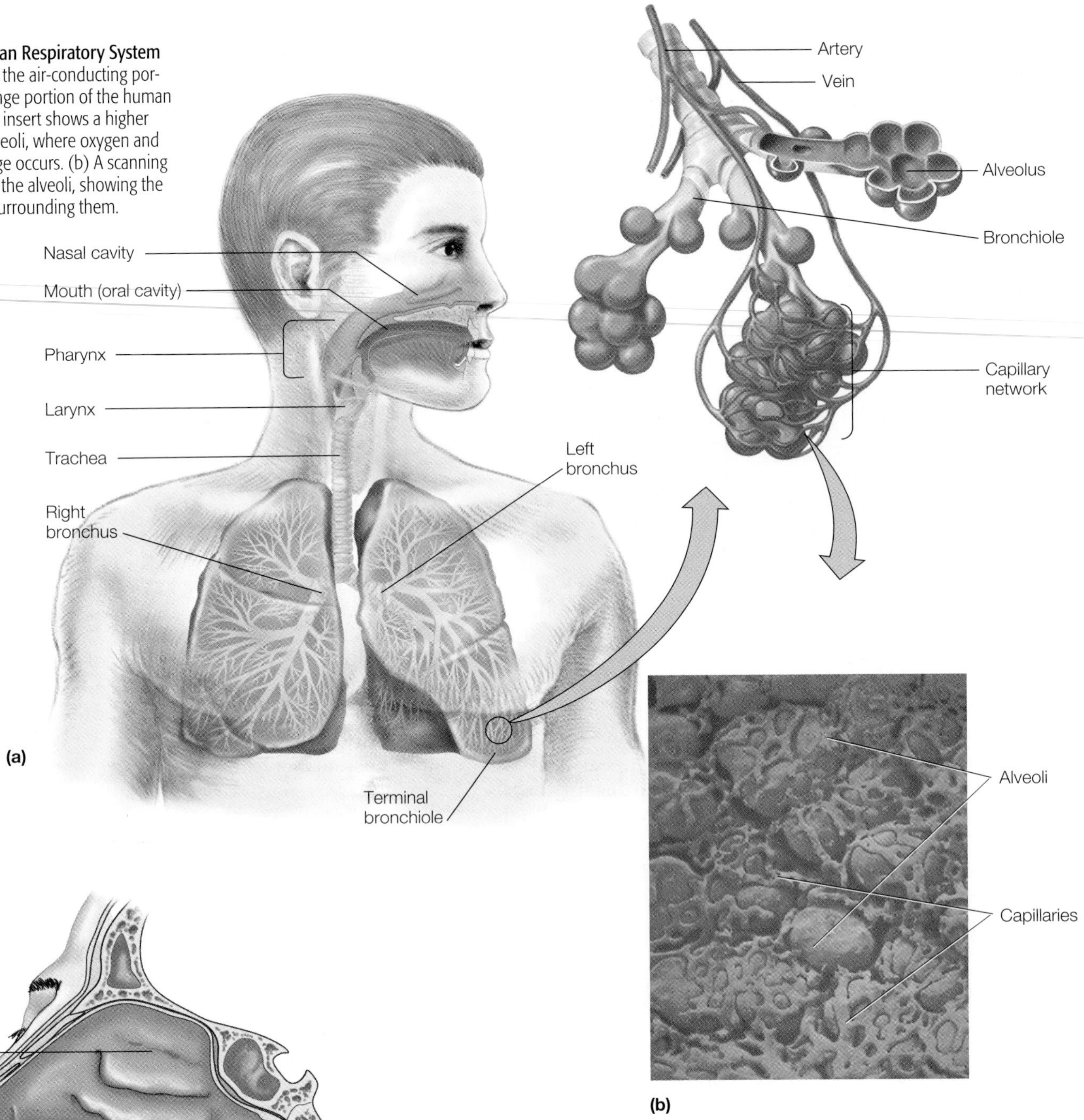

FIGURE 8-1 The Human Respiratory System
(a) This drawing shows the air-conducting portion and the gas-exchange portion of the human respiratory system. The insert shows a higher magnification of the alveoli, where oxygen and carbon dioxide exchange occurs. (b) A scanning electron micrograph of the alveoli, showing the rich capillary network surrounding them.

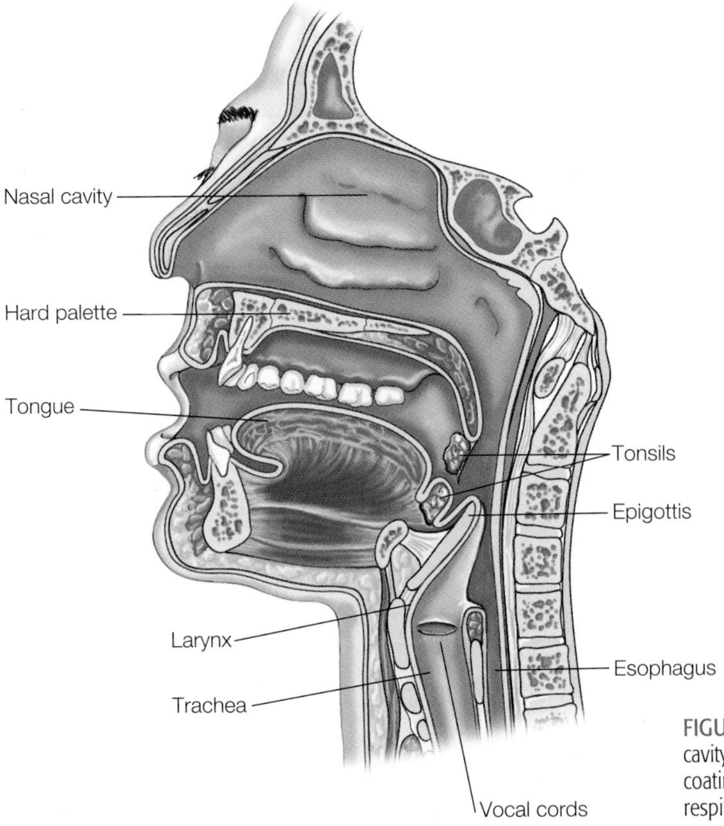

Artery

Vein

Alveolus

Bronchiole

Capillary network

Nasal cavity

Mouth (oral cavity)

Pharynx

Larynx

Trachea

Right bronchus

Left bronchus

(a)

Terminal bronchiole

Alveoli

Capillaries

(b)

Nasal cavity

Hard palette

Tongue

Tonsils

Epigottis

Larynx

Esophagus

Trachea

Vocal cords

FIGURE 8-2 Uppermost Portion of the Respiratory System   Bony protrusions into the nasal cavity (not shown here) create turbulence that causes dust particles to settle out on the mucous coating. Notice that air passing from the pharynx enters the larynx. Food is kept from entering the respiratory system by the epiglottis, which covers the laryngeal opening during swallowing.

The vocal cords are like the strings of a guitar or violin: They can be tightened or loosened, producing sounds of different pitch. The tighter the string on a guitar, the higher the note. In humans, muscles in the larynx that attach to the vocal cords make this adjustment possible. Relaxing the muscles lowers the tension on the cords, dropping the tone. Tightening the vocal cords has the opposite effect.

The larynx opens into the **trachea** (TRAY-kee-ah) or windpipe below. You can feel the trachea beneath your Adam's apple, the protrusion of the laryngeal cartilage on your neck.

As explained in Chapter 5, food is prevented from entering the larynx by the **epiglottis** (ep-eh-GLOT-tis), a flap of tissue that closes off the opening to the larynx during swallowing. Occasionally, however, food goes the wrong way, accidentally entering the larynx and trachea. This unfortunate event leads to violent coughing, a reflex that helps eject the food from the trachea. If the food cannot be dislodged by coughing, steps must be taken to remove it—and fast—or the person will suffocate. Health Note 8-1 explains what to do when a person chokes.

The trachea is a short, wide duct. Starting in the neck below the larynx, it enters the thoracic cavity, where it divides into two large branches, the right and left **bronchi** (BRON-kee). The bronchi (singular, *bronchus*) enter the lungs alongside the arteries and veins. Inside the lungs, the bronchi branch extensively, forming progressively smaller tubes that carry air to the alveoli.

The trachea and bronchi are reinforced by hyaline cartilage in the walls of these structures. This cartilage reinforcement prevents the structures from collapsing during breathing, thus ensuring a steady flow of air in and out of the lungs.

The smallest bronchi in the lungs branch to form **bronchioles** (BRON-kee-ols), small ducts that lead to the alveoli. Like the arterioles of the circulatory system, the walls of the bronchioles consist largely of smooth muscle. This permits them to open and close, and thus provides a means of controlling air flow in the lungs. During exercise or times of stress, the bronchioles open to increase the flow of air into the lungs. This homeostatic mechanism helps meet the body's need for oxygen during exercise, in much the same way that the arterioles of capillary beds dilate to let more blood into body tissues in times of need.

Filtering Incoming Air

The respiratory system filters out particles in the air we breathe.

The respiratory system is in direct contact with the external environment and is therefore quite vulnerable to infectious organisms and air pollutants present in the air all around us. The respiratory systems of animals have evolved several protective mechanisms. In humans and other animals, the air-conducting portion of the respiratory system filters airborne particles such as dust and pollen from the air we breathe.

Airborne particles or particulates come in many sizes. The large and medium-sized particles are easily removed. They drop out of the air as it travels through the nasal passageways and the upper respiratory system. Smaller particles, also known as fine particulates, however, are so tiny that they remain in the air breathed into our lungs. As a result, they can penetrate deeply into the lung. Some of these particles contain toxic metals such as mercury, which can cause lung cancer.

FIGURE 8-3 **Vocal Cords** (a) Uppermost portion of respiratory system showing location of the vocal cords. (b) Longitudinal section of the larynx showing the location of the vocal cords. Note the presence of the false vocal cord, so named because it does not function in phonation. (c) View into the larynx of a patient showing the true vocal cords from above.

(a)

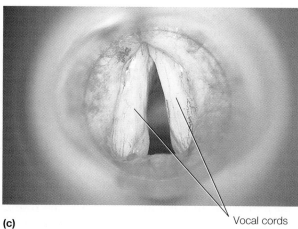

Epiglottis

Thyroid cartilage

Ventricular fold (false vocal cord)

True vocal cord

Tracheal cartilages

(b)

(c)

Vocal cords

‡‡‡‡ healthnote

8-1

First Aid for Choking That May Save Someone's Life

What do you do if you encounter a person who is choking? If the person can cough, speak, or breathe, do not interfere. Encourage the individual to continue coughing. If he or she cannot cough, speak, or breathe, **CALL OR HAVE SOMEONE CALL 911 OR YOUR LOCAL EMERGENCY NUMBER IMMEDIATELY.**

If the person is conscious:
- Stand behind him and wrap your arms around his waist.
- Make a fist with one hand (thumb outside the fist), and place the fist just above the navel in the middle of the abdomen. Be sure that your fist is well below the lower tip of the sternum.
- Wrap the other hand around the fist and perform cycles of five quick inward and upward thrusts. After every five thrusts, check the person and repeat the cycles until the person begins to cough, speak, or breathe on his own.

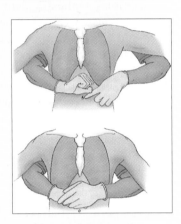

If the person is unconscious:
- Carefully position her on her back.
- Open the airway by placing one hand on the forehead and one hand on the chin and gently tilt the head backward (head-tilt/chin-lift). Place your ear next to the person's mouth, listening for air exchange; feel for air exchange with your cheek; and look to see the chest rise and fall during air exchange.
- If you do not see, hear, or feel air exchange, pinch the person's nose closed and seal your mouth around the victim's mouth and give two *slow* breaths. The chest should rise and fall gently.
- If the first breath did not go in, reposition the victim's head and attempt another breath.
- If the breaths did not go in, straddle the victim's thighs and place the heel of one hand just above the navel and place the other hand on top, interlocking your fingers.
- Give five quick inward and upward thrusts.
- Return to the victim's head; perform a finger sweep of her mouth, searching for a dislodged object; perform a head-tilt/chin-lift; and give one slow breath.
- If the breath will not go in, give five more abdominal thrusts, followed by a finger sweep, and one breath.

Particles that drop out of the inhaled air in the nose, trachea, and bronchi are trapped in a layer of **mucus** (MEW-kuss). Mucus is a thick, slimy secretion produced by certain cells, the **mucous cells**, in the epithelial lining of the nose, trachea, and bronchi (Figure 8-4).

The epithelium of the respiratory tract also contains numerous ciliated cells. **Cilia**, you may recall from Chapter 2, are small organelles on the cells lining the respiratory tract and parts of the reproductive tract of women. The cilia in the respiratory system beat upward, propelling mucus containing bacteria and dust particles toward the mouth. Operating day and night, they sweep the mucus toward the oral cavity, where it can be swallowed or expectorated (spit out). This protects the respiratory tract and lungs from bacteria and potentially harmful particulates. It is another of the body's homeostatic mechanisms.

Like all homeostatic mechanisms, the respiratory mucous trap is not invincible. Bacteria and viruses do occasionally penetrate the lining, causing respiratory infections. Interestingly, one component of cigarette smoke and urban air pollution, sulfur dioxide, temporarily paralyzes the cilia. It can even destroy

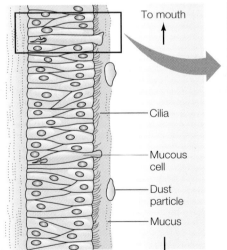

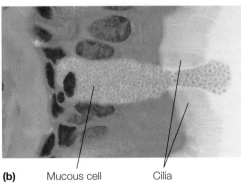

(b) Mucous cell   Cilia

FIGURE 8-4 Mucous Trap (a) Drawing of the lining of the trachea. Mucus produced by the mucous cells of the lining of much of the respiratory system traps bacteria, viruses, and other particulates in the air. The cilia transport the mucus toward the mouth. (b) Higher magnification of the lining showing a mucous cell and ciliated epithelial cells.

them. Sulfur dioxide gas in the smoke of a single cigarette, for instance, will paralyze the cilia for an hour or more, permitting bacteria and toxic particulates to be deposited on the lining of the respiratory tract and even enter the lungs. Ironically, the cilia of a smoker are paralyzed when they are needed the most!

- Continue this cycle until the victim begins to cough, speak, or breathe on his or her own.
- Perform the same technique for children as you would for an adult. Children are considered those who are over one year of age and under the age of eight. The size of the child must also be a consideration.

For infants and very small children who cannot cough, speak, or breathe:

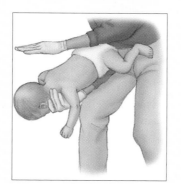

- Pick up the infant and place her face down on your forearm.
- Give the infant five back blows with the heel of your hand between the shoulder blades.
- Reposition the infant face up on the opposite forearm and give five chest thrusts using the pads of two or three fingers in the lower half of the sternum.
- Continue this cycle until the infant can breathe on his or her own or until the infant goes unconscious.

If the infant is unconscious:

- Perform a head-tilt/chin-lift and look, listen, and feel for air exchange.
- If no exchange is present, seal your mouth around the mouth and nose of the infant and give two slow breaths. The chest should rise and fall gently.

- If the first breath did not go in, reposition the infant's head and attempt another breath.
- If the breaths did not go in, position the infant for five back blows, five chest thrusts; then, check his mouth for an object and give one breath.
- Continue this cycle until the infant begins to breathe on his or her own.

In order to be skilled in these and other lifesaving skills, you should take a first aid and CPR class. Classes are offered in most communities through the National Safety Council. Call 1-800-621-7618 for information on classes in your area.

Source: Adapted by Jennifer Belcher, EMT-B from the guidelines of the American Red Cross, National Safety Council and the American Heart Association.

www.jbpub.com/humanbiology/5e

Visit Human Biology's Internet site for links to web sites offering more information on this topic.

Because smoking impairs a natural protective mechanism, it should come as no surprise that smokers suffer more frequent respiratory infections than nonsmokers. Research shows that alcohol also paralyzes the respiratory system cilia, explaining why alcoholics are also prone to respiratory infections.

Warming and Moistening Incoming Air

The conducting portion of the respiratory system moistens and warms the incoming air.

Beneath the epithelium of the respiratory tract is a rich network of capillaries. These capillaries release moisture into the incoming air, which prevents the lungs from drying out. Heat also released by the capillaries protects the lungs from cold temperatures. Except in extremely cold weather, by the time inhaled air reaches the lungs, it is nearly saturated with water and is warmed to body temperature.

On the way out of the respiratory system, much of the water that was added to the air condenses on the slightly cooler lining of the nasal cavity. This mechanism is an adaptation that conserves water and also accounts for the reason our noses tend to drip in cold weather.

The Alveoli

The alveoli are the site of gaseous exchange.

The air we breathe consists principally of nitrogen and oxygen, with small amounts of carbon dioxide and other gases (Figure 8-5). Oxygen in the atmosphere is generated by the photosynthetic activity of plants and certain single-celled organisms like algae. Oxygen is vital to humans and virtually all other living organisms. A constant supply must be delivered to body cells to sustain cellular energy production. (Remember from Chapter 2 that the complete breakdown of glucose in body cells requires oxygen!)

Oxygen is delivered to the alveoli of the lungs by the conducting portion of the respiratory system, as noted above (Figures 8-6a and 8-6b). The alveoli are the site of oxygen exchange with the bloodstream. In humans, each lung contains an estimated 150 million alveoli. The surface area they create is about the size of a tennis court, that is, about 60 to 80 square meters. This provides ample area for the exchange of oxygen.

As shown in Figure 8-7, the alveoli are lined by a single layer of flattened cells (Type I alveolar cells). Aiding in the exchange of oxygen is an extensive network of capillaries, shown in Figure 8-1b. Each capillary is made of an equally thin layer of flattened cells. Together, the cells of the alveolar wall and the

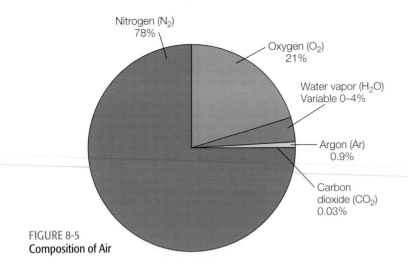

Nitrogen (N_2)
78%

Oxygen (O_2)
21%

Water vapor (H_2O)
Variable 0–4%

Argon (Ar)
0.9%

Carbon
dioxide (CO_2)
0.03%

FIGURE 8-5
Composition of Air

capillary present a fairly easy route for the passage of gases into and out of the alveoli.

The large surface area of the lungs created by the alveoli and the relatively thin barrier between the blood and the alveolar air result in a rather rapid diffusion of gases across the alveolar wall. The alveoli therefore provide another example of the marriage of form and function produced by evolution.

Another important cell in the alveoli is the **alveolar macrophage**, sometimes known as the **dust cell**. Alveolar macrophages are the cellular equivalent of a janitorial staff. They wander freely through the alveoli, phagocytizing dust, bacteria, viruses, and other particulates that manage to escape filtration by the upper portions of the respiratory system (Figure 8-7). Once they are filled with particulates, they take up residence in the connective tissue surrounding the alveoli. Over time, numerous spent cells come to reside in this region. Because there are so many particulates in tobacco smoke, a smoker's lungs are often blackened by dust cells packed with particulates. The lungs of urban residents may also be blackened by the accumulation of smoke and dust particles.

Yet another important cell in the alveoli is the Type II alveolar cell (Figure 8-7). Type II alveolar cells are large, round cells located in the alveolar epithelium. These prominent cells pro-

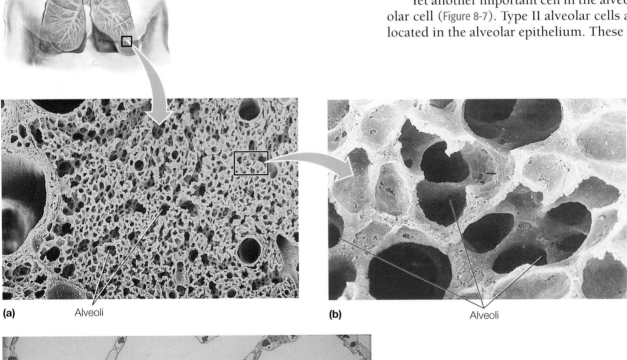

(a) Alveoli

(b) Alveoli

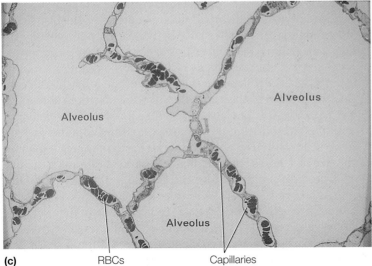

(c) RBCs Capillaries

Alveolus

Alveolus

Alveolus

FIGURE 8-6 **Alveoli** (a) A scanning electron micrograph of the lung showing many alveoli. The smallest openings are capillaries surrounding the alveoli. (b) A higher magnification scanning electron micrograph of lung tissue showing alveoli. (c) A transmission electron micrograph showing several alveoli and the close relationship of the capillaries containing RBCs (dark structures inside the capillaries). This close relationship between capillaries and alveoli ensures the rapid transport of oxygen and carbon dioxide.

FIGURE 8-7 **The Alveolar Macrophage** Drawing of the alveolus showing Type I and Type II alveolar cells and macrophages or dust cells.

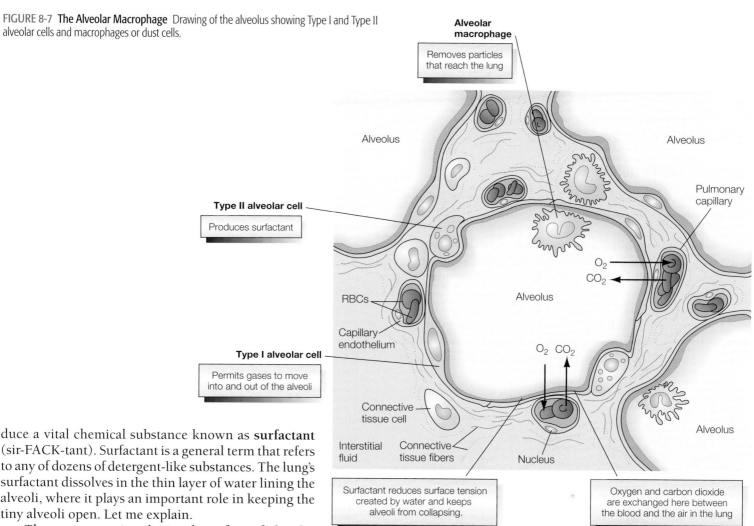

Alveolar macrophage
Removes particles that reach the lung

Alveolus

Alveolus

Pulmonary capillary

Type II alveolar cell
Produces surfactant

O_2
CO_2

Alveolus

RBCs

Capillary endothelium

O_2 CO_2

Type I alveolar cell
Permits gases to move into and out of the alveoli

Connective tissue cell

Alveolus

Interstitial fluid | Connective tissue fibers

Nucleus

Surfactant reduces surface tension created by water and keeps alveoli from collapsing.

Oxygen and carbon dioxide are exchanged here between the blood and the air in the lung

duce a vital chemical substance known as **surfactant** (sir-FACK-tant). Surfactant is a general term that refers to any of dozens of detergent-like substances. The lung's surfactant dissolves in the thin layer of water lining the alveoli, where it plays an important role in keeping the tiny alveoli open. Let me explain.

The water covering the inside surface of alveolar epithelium produces surface tension. Surface tension results from hydrogen bonds between water molecules. Discussed in Chapter 2, hydrogen bonds are intermolecular bonds. They form between the slightly charged oxygen atoms of water molecules with the slightly charged hydrogen atoms of other molecules. These hydrogen bonds draw water molecules together. At the surface of a watery fluid, the hydrogen bonds pull water molecules together more tightly than elsewhere, creating a slightly denser region referred to as *surface tension*. Surface tension on a pond permits some insects such as water striders to walk on water and is the reason a drop of water beads up on a car's windshield (Figure 8-8).

In the alveoli, surface tension tends to draw the walls of the alveoli inward. In other words, it tends to make them collapse. Surfactant, however, reduces surface tension in the alveoli, decreasing forces that might otherwise cause the alveoli to close down.

Some premature babies lack sufficient surfactant. High surface tension in the alveoli causes the larger alveoli to collapse. This, in turn, results in a dramatic, life-threatening condition known as respiratory distress syndrome or RDS. Fortunately, children born with this problem are usually treated with an artificial surfactant that keeps the alveoli open until their lungs produce enough of their own surfactant to perform this small but important function.

FIGURE 8-8 **Surface Tension** Some insects can walk on water because of surface tension, the tight packing of water molecules along the surface of a pond.

8-2 Breathing and the Control of Respiration

Air moves in and out of the lungs in much the same way that it moves in and out of the bellows that blacksmiths use to fan their fires. Breathing, however, is largely an involuntary action, controlled by the nervous system.

Inspiration and Expiration

Air moves in and out of the lungs as a result of changes in the pressure inside the chest cavity.

During breathing, air must first be drawn into the lungs. This process is known as **inspiration**, or **inhalation**. Following inspiration, air must be expelled. This is known as **expiration**, or **exhalation**.

Inhalation is an active process controlled by the brain. Nerve impulses traveling from the brain stimulate the **diaphragm**, a dome-shaped muscle that separates the abdominal and thoracic cavities (Figure 8-9a). These impulses cause the diaphragm to con-

tract. When it contracts, the diaphragm flattens and lowers. The contraction of the diaphragm draws air into the lungs much the same way that pulling the plunger of a syringe draws in air into the device.

Inhalation also involves the intercostal muscles, the short, powerful muscles that lie between the ribs. (They are the meat on barbecued ribs.) Nerve impulses traveling to these muscles cause them to contract as the diaphragm is lowered. When the intercostal muscles contract, the rib cage lifts up and out. Go ahead and try it right now. Place your hands on your chest, take a deep breath in, and feel the chest wall expand and rise. Together, the contractions of the intercostal muscles and the diaphragm increase the volume of the thoracic cavity (Figure 8-9b, bottom). This, in turn, decreases the pressure inside the lungs, known as *intrapulmonary pressure*. As a result, air naturally flows in through the mouth and nose.

In contrast to inhalation, exhalation is a passive process—that is, it does not require muscle contraction in a person at rest. Exhalation begins after the lungs have filled with air. At this point in the cycle, the diaphragm and intercostal muscles relax. The relaxed diaphragm rises and resumes its domed shape. The chest wall falls slightly inward. These changes reduce the volume of the chest cavity. This raises the internal pressure, forcing air out in much the same way that squeezing an inflated beach ball forces air out of the opening.

The lungs also contribute to passive exhalation. During inspiration, the lungs fill like balloons. When inhalation ceases, the elastic fibers inside the lungs cause the lungs to recoil, forcing air out.

Although exhalation is a passive process in an individual at rest, it can be made active by contracting the muscles of the wall of the chest and abdomen. The forceful expulsion of air is called *forced exhalation*.

Inhalation can also be consciously supplemented by a forceful contraction of the muscles of inspiration. (You can test this by taking a deep breath.) Forced inhalation increases the amount of air entering your lungs. Athletes often actively inhale and exhale just before an event to increase oxygen levels in their blood. A competitive swimmer, for example, may take several deep breaths before diving into the pool for a race. Deep breathing, while effective, can be dangerous, for reasons explained shortly.

Measuring Air Flow in the Lungs

Measurements of air flow into and out of the lungs can help physicians assess lung health.

Several measurements of lung capacity are routinely used to determine the health of a person's lungs. These measurements are taken under controlled conditions. As shown in Figure 8-10a, patients breathe into a machine that measures and records the amount of air moving in and out of the lung at various times. Figure 8-10b shows a graph of some of the common measurements.

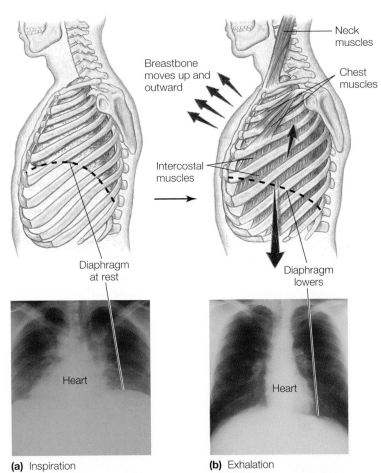

(a) Inspiration

(b) Exhalation

FIGURE 8-9 The Bellows Effect (a) The rising and falling of the chest wall through the contraction of the intercostal muscles (muscles between the ribs) is shown in the diagram, illustrating the bellows effect. Inspiration is assisted by the diaphragm, which lowers. Like pulling a plunger out on a syringe, the rising of the chest wall and the lowering of the diaphragm draw air into the lungs. (b) X-rays showing the size of the lungs in full exhalation (top) and full inspiration (bottom).

FIGURE 8-10 **Measuring Lung Capacity** (a) This machine allows health-care workers to determine tidal volume, inspiratory reserve volume, and other lung-capacity measurements to determine the health of an individual's lung. (b) This graph shows several common measurements.

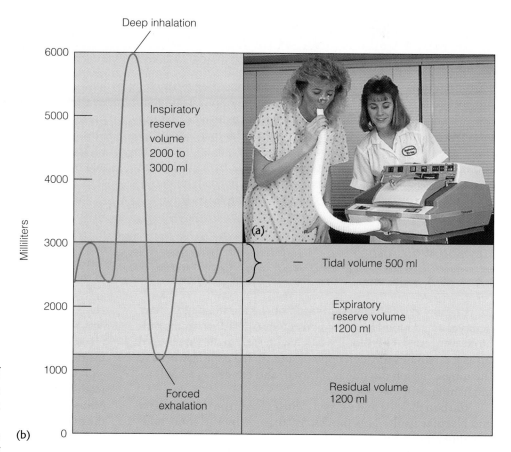

The first is the **tidal volume**, the amount of air that moves in and out during passive breathing. At rest, each breath delivers about 500 milliliters of air to the lung.

After passive exhalation, however, the lungs still contain a considerable amount of air—about 2400 milliliters. Forced exhalation will expel about half of that air. This is known as the **expiratory reserve volume**. The remaining 1200 milliliters is known as the **residual volume**.

Another important measurement is the **inspiratory reserve volume (IRV)** This is the amount of air that can be drawn into the lungs during a deep inhalation. Deep inhalation draws in four to six times more air than the tidal volume, or 2000–3000 milliliters, depending on the size of the individual. Lung diseases often result in changes in these measurements. Asthma, for example, reduces the IRV by constricting the bronchioles and allowing mucus to build up, thus blocking air flowing into the lung. Children who are exposed to tobacco smoke at home also experience a decrease in lung capacity—that is, a decrease in their ability to move air in and out of their lungs.

Control of Breathing

Breathing is controlled by the breathing center in the brain.

Breathing is controlled by a region of the brain known as the **breathing center**. It is located in the area called the brain stem (or medulla, [meh-DEW-lah]).

The breathing center contains nerve cells that generate periodic nerve impulses. These impulses stimulate inhalation by causing the periodic contraction of the intercostal muscles and the diaphragm. When the lungs fill, however, the nerve impulses cease and the muscles relax. Air is forced out of the lungs.

While the breathing center acts like a pacemaker, bringing about rhythmic contraction and relaxation of the mus-

cles of inspiration, breathing can be increased or decreased by other factors. Chemical receptors inside the brain and arteries, for example, detect levels of carbon dioxide, and send nerve impulses to the breathing center to increase or decrease breathing as needed. When levels of carbon dioxide are high, for instance, during vigorous exercise, chemoreceptors in the aorta and carotid arteries (which deliver blood to the brain) are activated. They send impulses to the breathing center that cause it to increase the depth and rate of breathing. Next time you run up a flight of stairs and start breathing hard, you'll know why.

Stretch receptors in the lung also help to control breathing. They detect when the lungs are full and send impulses back to the breathing center, causing it to cease firing. Stretch receptors probably function only during exercise, when large volumes of air are moved in and out of the lungs.

The body also contains oxygen receptors. However, these receptors are not as sensitive as the other receptors. In fact, oxygen levels must fall considerably before these receptors begin generating impulses that alter our breathing. This fact can have profound consequences for divers and swimmers. Many divers and swimmers hyperventilate before they take the plunge. Hyperventilation is repeated deep and rapid breathing that fills the blood with oxygen and makes it possible for one to hold his breath under water longer.

Unfortunately, when a diver enters the water after hyperventilating, low levels of carbon dioxide (and H^+ concentrations) suppress the urge to breathe, but oxygen levels in the blood may fall so low that the brain is deprived of oxygen. The diver may lose consciousness.

8-3 Gas Exchange in the Lungs

The respiratory system serves many functions. For example, the vocal cords, located in the larynx, produce sounds that allow us to communicate. The respiratory system also houses the olfactory membrane located in the roof of the nasal cavity. It contains sensors that detect odors (Chapter 11). In addition, the respiratory system helps maintain pH balance by its influence on carbon dioxide levels (discussed shortly). The main functions of the respiratory system, though, are to (1) replenish the blood's oxygen supply and (2) rid the blood of excess carbon dioxide.

Function of the Alveoli

Gases diffuse rapidly across the alveolar and capillary walls.

As you may recall from your study of the circulatory system, deoxygenated blood from the body enters the lungs via the pulmonary arteries. This blood is laden with carbon dioxide picked up as it circulates through body tissues. In the capillary beds of the lungs, carbon dioxide is released, and oxygen is added, replenishing supplies.

Carbon dioxide and oxygen diffuse across the thin capillary and alveolar walls. This process is "driven" by the concentration difference between the alveoli and capillaries. As illustrated in Figure 8-11, oxygen in the alveoli diffuses through the alveolar epithelium, then into the capillaries, where it enters the blood plasma (Figure 8-12). From here, oxygen molecules cross the plasma membrane of the red blood cells (RBCs). Inside the RBCs, it binds to hemoglobin molecules. About 98% of the oxygen in the blood is carried in the RBCs bound to hemoglobin; the rest is dissolved in the plasma and cytoplasm of the RBCs.

In order to understand the details of carbon dioxide diffusion in the lung, we must go back to the body cells, where CO_2 is formed. As you may recall from Chapter 2, cells produce abundant CO_2 during cellular respiration, the chemical breakdown of glucose. In body tissues, carbon dioxide diffuses out of the cells and enters the blood plasma. Some carbon dioxide (7%) remains in the plasma. The rest diffuses into the RBCs. Here, some of it (15%–25%) binds to hemoglobin. The rest reacts with water to form carbonic acid, H_2CO_3 (Figure 8-13). This reaction is catalyzed by the enzyme carbonic anhydrase, found inside RBCs.

As shown at the top of Figure 8-13, carbonic acid molecules readily dissociate to form bicarbonate ions and hydrogen ions. Many of the bicarbonate ions then diffuse out of the RBCs into the plasma, where they are carried with the blood. Hydrogen ions stay behind.

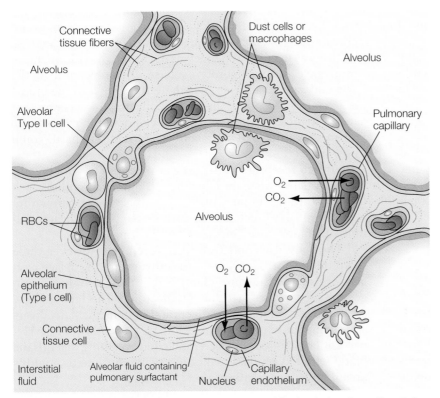

FIGURE 8-11 Close-Up of the Alveolus Oxygen diffuses out of the alveolus into the capillary. Carbon dioxide diffuses in the opposite direction, entering the alveolar air that is expelled during exhalation.

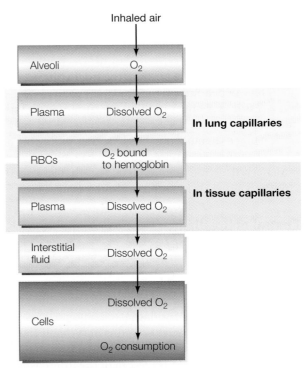

FIGURE 8-12 Oxygen Diffusion Oxygen travels from the alveoli into the blood plasma, then into the RBCs, where much of it binds to hemoglobin. When the oxygenated blood reaches the tissues, oxygen is released from the RBCs and diffuses into the plasma, then into the interstitial fluid and body cells.

FIGURE 8-13 Bicarbonate Ion Production Carbon dioxide (CO_2) diffuses out of body cells where it is produced and into the tissue fluid, then into the plasma. Although some carbon dioxide binds to hemoglobin and some is dissolved in the plasma, most is converted to carbonic acid (H_2CO_3) in the RBCs. Carbonic acid dissociates and forms hydrogen ions and bicarbonate ions. Hydrogen ions remain inside the RBCs, but most bicarbonate diffuses into the plasma where it is transported.

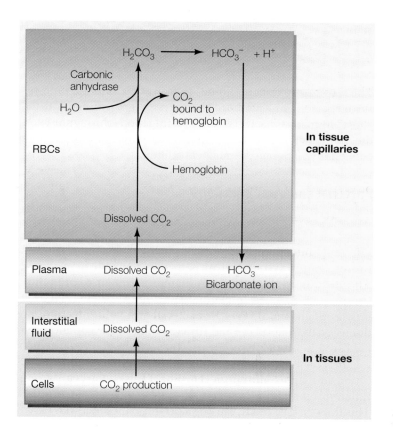

When blood rich in carbon dioxide reaches the lungs, bicarbonate ions in the plasma reenter the RBCs. They then combine with hydrogen ions to form carbonic acid (Figure 8-14). Carbonic acid, in turn, re-forms carbon dioxide. The CO_2 then diffuses out of the RBCs into the blood plasma. CO_2 then diffuses into the alveoli down a concentration gradient. Carbon dioxide is eventually expelled from the lungs during exhalation.

The uptake of oxygen and the discharge of carbon dioxide in the lungs "replenish" the blood in the alveolar capillaries. The oxygenated blood then flows back to the left atrium of the heart via the pulmonary veins. From here it empties into the left ventricle and is pumped to the body tissues via the aorta and its many branches.

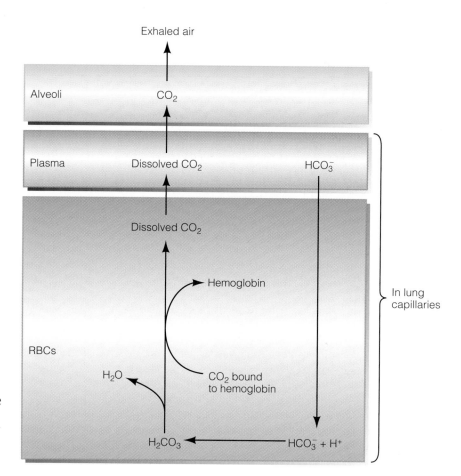

FIGURE 8-14 Carbon Dioxide Production from Bicarbonate When the carbon dioxide-laden blood reaches the lungs, bicarbonate ions diffuse back into the RBCs and combine with hydrogen ions in RBCs forming carbonic acid, which dissociates, forming carbon dioxide gas. CO_2 diffuses out of the RBCs into the plasma, then into the alveoli.

8-4 Diseases of the Respiratory System

The respiratory tract is "home" to many diseases and illnesses. With its close proximity to the external environment, it will "host" numerous bacterial and viral infections throughout your life. Some will cause considerable discomfort, and some could be fatal if you don't seek medical attention.

Bacterial and Viral Infections

Bacteria and viruses can infect many parts of the respiratory system.

Microbial infections may occur in many different locations in the respiratory system and are named by their site of residence. An infection in the bronchi is therefore known as **bronchitis** (bron-KITE-iss). An infection of the sinuses is known as **sinusitis** (sigh- nu-SITE-iss). (Bacterial and viral infections are discussed in Chapters 14 and 15; a few of the more common ones are listed in Table 8-2.)

Bacterial and viral infections of the larynx result in a condition known as **laryngitis** (lair-in-JITE-iss). Laryngitis is an inflammation of the lining of the larynx and the vocal cords. This thickens the cords, causing a person's voice to lower. Laryngitis may also be caused by tobacco smoke, alcohol, excessive talking, shouting, coughing, or singing, all of which irritate the vocal cords. In young children, inflammation results in a swelling of the larynx lining that may impede the flow of air and impair breathing, resulting in a condition called the *croup* (crewp).

Once inside the respiratory tract, bacteria, viruses, and other microorganisms can spread to other organ systems. For example, meningitis (MEN-in-JITE-iss) is a bacterial or viral infection of the meninges (meh-NIN-jees), the fibrous layers surrounding the brain and spinal cord. This potentially fatal disease usually starts out as an infection of the sinuses or the lungs. That's why you're advised to see a doctor if these infections last more than a few weeks.

The lungs are also susceptible to airborne materials, among them asbestos fibers. Asbestos is a naturally occurring fiber that has been used in thousands of products from pipe and sound insulation to car brake pads. Inside the lung, asbestos can cause enormous damage. Workers exposed to asbestos, for example, develop two types of lung cancer, one of which (mesothelioma) is highly lethal. Others have developed a debilitating disease known as *asbestosis* (ass-bes-TOE-sis). Asbestosis is a buildup of scar tissue that reduces the lung capacity. Because asbestos is believed to be dangerous, virtually all of its uses have been banned in the United States, and asbestos used for insulation and decoration is either being removed from buildings or stabilized so it won't flake off.

Asthma

Asthma is an allergic response that impairs breathing.

Another common disease of the respiratory system is **asthma**. Characterized by periodic episodes of wheezing and difficult breathing, asthma is a chronic disorder—that is, it's a disease that persists for many years.

TABLE 8-2	Common Respiratory Diseases		
Disease	Symptoms	Cause	Treatment
Emphysema	Breakdown of alveoli; shortness of breath	Smoking and air pollution	Administer oxygen to relieve symptoms; quit smoking; avoid polluted air. No known cure.
Chronic bronchitis	Coughing, shortness of breath	Smoking and air pollution	Quit smoking; move out of polluted area; if possible, move to warmer, drier climate.
Acute bronchitis	Inflammation of the bronchi; yellowy mucus coughed up; shortness of breath	Many viruses and bacteria	If bacterial, take antibiotics and decongestant tablets; use vaporizer.
Sinusitis	Inflammation of the sinuses; mucus discharge; blockage of nasal passageways; headache	Many viruses and bacteria	If bacterial, take antibiotics, cough medicine; use vaporizer.
Laryngitis	Inflammation of larynx and vocal cords; sore throat; hoarseness; mucus buildup and cough	Many viruses and bacteria	If bacterial, take antibiotics, cough medicine; avoid irritants like smoke; avoid talking.
Pneumonia	Inflammation of the lungs ranging from mild to severe; cough and fever; shortness of breath at rest; chills; sweating; chest pains; blood in mucus	Bacteria, viruses, or inhalation of irritating gases	Consult physician immediately; go to bed; take antibiotics, cough medicine; stay warm.
Asthma	Constriction of bronchioles; mucus buildup in bronchioles; periodic wheezing; difficulty breathing	Allergy to pollen, some foods, food additives; dandruff from dogs and cats; exercise	Use inhalants to open passageways; avoid irritants.

Most cases of asthma are caused by allergic reactions (abnormal immune reactions) to common stimulants such as dust, pollen, and skin cells (dander) from pets. In some individuals, it is brought on by certain foods, such as eggs, milk, chocolate, and food preservatives. Still other cases are triggered by drugs, such as antibiotics. Vigorous exercise and physiological stress are known to trigger asthma attacks, too.

In asthmatics, irritants such as pollen and dander cause a rapid increase in the production of mucus by the bronchi and bronchioles. Irritants also stimulate the constriction of the bronchioles. Mucus production and constriction of the bronchioles make breathing difficult. Asthmatics also suffer a chronic inflammation of the lining of the respiratory tract.

Although asthma is fairly common in school children, it often disappears as they grow older. But that may be changing. Studies show that the incidence of asthma in adults is on the rise. Many more children suffer from the disease, too. In fact, the number of Americans now suffering from asthma is estimated to be about 18 million, up from 6.7 million in 1980. Public health officials are concerned about the growing epidemic of asthma because periodic attacks can be quite disabling. Some even lead to death. By one estimate, 5000 Americans die each year from severe asthma attacks. Victims are generally elderly individuals who suffer from other diseases as well.

The severity of asthma attacks can be greatly lessened by proper medical treatment. One of the most common treatments is an oral spray (inhalant) containing the hormone epinephrine. This drug stimulates the bronchioles to open. Antiinflammatory drugs (steroids) can be administered to treat chronic inflammation. Screening tests can help a patient find out what substances trigger an asthmatic attack so they can be avoided.

Chronic Bronchitis and Emphysema

Chronic bronchitis is caused by an irritation of the lungs and bronchial passageways; emphysema is caused by a breakdown of the alveoli caused by chemical irritants.

Another common lung disease is **chronic bronchitis**, a persistent irritation of the bronchi. Characterized by excess mucus production, coughing, and difficulty in breathing, chronic bronchitis afflicts one out of every five American men between the ages of 40 and 60. Although the leading cause of chronic bronchitis is cigarette smoking, urban air pollution also contributes to this disease. Certain pollutants in urban air pollution irritate the lung and bronchial passages.

Another lung disease of great significance is emphysema. **Emphysema**, mentioned in the introduction, is the progressive deterioration of the alveoli caused by a breakdown of their walls. Emphysema is one of the fastest growing causes of death in the United States. Resulting principally from smoking and urban air pollution, emphysema afflicts over 1.5 million Americans, mostly men. Not surprisingly, smokers who live in polluted urban settings have the highest incidence of the disease. This condition is a progressive and incurable disease. As it worsens, lung function deteriorates, and victims eventually require supplemental oxygen to perform even routine functions, such as walking or speaking.

One of the most prevalent respiratory diseases is lung cancer caused by smoking, a topic discussed in Health Note 8-2.

8-5 Health and Homeostasis

The respiratory system plays a vital role in homeostasis. It helps regulate oxygen and carbon dioxide levels in the blood and in body tissues. It helps maintain the pH (acidity) of extracellular fluid by controlling the rate at which acid-forming carbon dioxide is removed from the blood. Furthermore, the respiratory system helps to protect us from infectious disease. The respiratory system is also vital to the proper functioning of other organ systems that play key roles in homeostasis.

Like other body systems, the respiratory system can be damaged by pathogenic organisms and chemical contaminants in the environment. Changes in respiratory function alter homeostasis and can lead to disease.

The air of the industrialized world contains a multitude of chemical pollutants that damage human health by upsetting homeostasis. Air pollution affects millions of us on a daily basis. Unfortunately, most people are unaware of the dangers of air pollution because the line between cause and effect is not always clear. Consider, for instance, the headache you experienced in traffic going home from school or work. Was it caused by tension, or could it have been caused by carbon monoxide emissions from cars, buses, and trucks? And what about the runny nose and sinus condition you experienced last winter? Were they caused by a virus or bacterium or by pollution?

One classic study of air pollution in New York City in the 1960s illustrates the relationship between air pollution and respiratory problems. Researchers found that an increase in the level of sulfur dioxide, a pollutant produced by automobiles, power plants, and factories, increased the incidence of colds, coughs, nasal irritation, and other symptoms fivefold in just a few days. Soon after the pollution levels returned to normal, the symptoms subsided. Few people knew they'd been affected by the air they were breathing. Most thought they had been stricken by some infectious agent.

No one knows the exact role of urban air pollution in human health. However, some researchers estimate that 50,000 to 60,000 Americans die each year from lung disease caused by urban air pollution.

░░░ healthnote

8-2   Smoking and Health: The Deadly Connection

Urban air pollution worries many Americans, and with good reason. However, city air that many of us breathe is benign compared with the "air" that more than 54 million Americans over the age of 18 voluntarily inhale from cigarettes. Loaded with dangerous pollutants in concentrations far greater than those of our cities, cigarette smoke takes a huge toll on citizens of the world. In the United States, for example, an estimated 419,000 people—about 1150 every day—die from the many adverse health effects of tobacco smoke, including heart attacks, lung cancer, and emphysema. Making matters worse, smoking is addictive. Nicotine, one of the many components of tobacco smoke, hooks many people on a dangerous activity.

Smoking costs society a great deal in medical bills and lost productivity. According to the Worldwatch Institute, every pack of cigarettes sold in the United States costs our society about $1.25–$3.17 in medical costs, lost wages, and reduced productivity—that's about $125–$400 billion a year!

Smoking is a principal cause of lung cancer, claiming the lives of an estimated 144,000 men and women in the United States each year. Smokers are 11–25 times more likely to develop lung cancer than nonsmokers. The more one smokes, the more risk one suffers.

Unfortunately, nonsmokers are also affected by the smoke of others. Nonsmokers inhale tobacco smoke in meetings, in restaurants, at work, and at home. Research has shown that nonsmokers—sometimes referred to as "passive smokers" because they "smoke" involuntarily—are more likely to develop lung cancer than those nonsmokers who manage to steer clear of smokers. In a study of Japanese women married to men who smoked, researchers found that the wives were as likely to develop lung cancer as people who smoked half a pack of cigarettes a day!

A report by the U.S. Environmental Protection Agency estimates that passive smoking causes 500–5000 cases of lung cancer a year in the United States. Passive smokers who are exposed to tobacco smoke for long periods also suffer from impaired lung function equal to that seen in light smokers (people who smoke under a pack a day).

Cigarette smoke in closed quarters can cause angina (chest pains) in smokers and nonsmokers afflicted by atherosclerosis of the coronary arteries. Carbon monoxide in cigarette smoke is responsible for the attacks.

Smokers are also more susceptible to colds and other respiratory infections. And smoking affects children. One study showed that children from families in which both parents smoked suffered twice as many upper respiratory infections as children from nonsmoking families. Recently, researchers from the Harvard Medical School reported finding a 7% decrease in lung capacity in children raised by mothers who smoked. The researchers believe that this change may lead to other pulmonary problems later in life.

For years, passive smokers have had little power when it came to their exposure to other people's smoke. Today, however, as a result of a growing awareness of the dangers, new regulations ban or restrict smoking in many public places and in the workplace.

The effects of smoking extend way beyond respiratory disease. Recent studies, for instance, show that smoking even decreases fertility in women. For example, women who smoke more than a pack of cigarettes a day are half as fertile as nonsmokers. Smoking may also affect the outcome of pregnancy. According to the 1985 U.S. Surgeon General's Report, women who smoke several packs a day during pregnancy are much more likely to miscarry and are also more likely to give birth to smaller children. On average, the children of these women are 200 grams (nearly 0.5 pound) lighter than children born to nonsmoking mothers. Finally, children of women who smoke heavily during pregnancy generally score lower on mental aptitude tests during early childhood than children whose mothers do not smoke.

SUMMARY

The Structure of the Human Respiratory System

1. The respiratory system consists of an air-conducting portion and a gas-exchange portion.
2. The air-conducting portion transports air from outside the body to the alveoli in the lungs, the site of gaseous exchange.
3. The alveoli are tiny, thin-walled sacs. Surrounding the alveoli are capillary beds that pick up oxygen and expel carbon dioxide.
4. The lining of the alveoli is kept moist by water. Surfactant, a chemical substance produced in the lung, reduces the surface tension inside the alveoli and prevents their collapse.

Breathing and the Control of Respiration

5. Breathing is an involuntary action controlled by the breathing center in the brain stem. Nerve cells in the breathing center send impulses to the diaphragm and intercostal muscles, which contract. This increases the volume of the thoracic cavity, which draws air into the lungs through the nose or mouth.
6. When the lungs are full, the muscles relax and the lungs recoil, expelling air.
7. The breathing center is regulated by factors such as the levels of carbon dioxide in the blood.
8. Expiration can be supplemented by enlisting the aid of abdominal and chest muscles, as can inspiration.

Gas Exchange in the Lungs

9. The respiratory system conducts air to and from the lungs, exchanges gases, and helps produce sounds. Sound is generated as air rushes past the vocal cords, causing them to vibrate. The sounds are modified by movements of the tongue and changes in the shape of the oral cavity.
10. Oxygen and carbon dioxide diffuse across the alveolar wall of the lungs, driven by concentration differences between the blood and alveolar air. Oxygen diffuses into the blood plasma, then into the RBCs, where most of it binds to hemoglobin. Carbon dioxide diffuses in the opposite direction.

Numerous studies also show that smoking is a major contributor to impotence in men. In fact, smokers are 50% more likely to be impotent than nonsmokers. Smoking also causes arterial changes that restrict blood flow to the penis. Changes begin to occur early in life. Teenage smokers could have problems in their thirties if they continue to smoke.

Tobacco smoke contains numerous hazardous substances that damage the lining of the respiratory system. Sulfur dioxide, for example, paralyzes the cilia lining the respiratory tract. Tobacco smoke is also laden with microscopic carbon particles. These carbon particles penetrate deeply into the lungs, where they accumulate in the alveolar walls, turning healthy tissue into a blackened mass that often becomes cancerous (Figure 1). Tobacco smoke may also paralyze the alveolar macrophages, making a bad situation even worse.

Toxic chemicals in cigarette smoke, many of which are known carcinogens, attach to carbon particles. Toxin-carrying particles adhere to the lungs, larynx, trachea, and bronchi. Virtually any place they stick, they can cause cancer, explaining why smokers are five times more likely than nonsmokers to develop laryngeal cancer and four times more likely to develop cancer of the oral cavity.

Nitrogen dioxide and sulfur dioxide in tobacco smoke penetrate deep into the lungs, where they dissolve in the watery layer inside the alveoli. Nitrogen dioxide is converted to nitric acid; sulfur dioxide is converted to sulfuric acid. Both acids erode the alveolar walls, leading to emphysema.

The dangers of smoking are becoming well known. As a result of widespread publicity and public pressure, smoking has dropped substantially in the United States. In 2004, for example, only 25% of American adults smoked, down from 34% in 1985 and down substantially from the 1950s and 1960s, when well over half of all men and over one-third of all women engaged in this potentially lethal habit.

Despite the downturn in smoking, more than 54 million American adults (over the age of 18) still smoke. An estimated six million teenagers and 100,000 children under the age of 13 smoke. Smoking occurs in greatest proportion in certain minority groups—particularly African American men, Native Americans, and Native Alaskans.

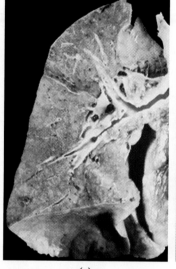
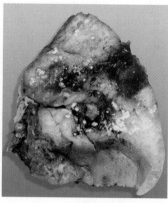

(a)   (b)

FIGURE 1 **The Normal and Cancerous Lung** (a) The normal lung appears spongy. (b) The cancerous lung from a smoker is filled with particulates and a large tumor.

www.jbpub.com/humanbiology/5e

Visit Human Biology's Internet site for links to web sites offering more information on this topic.

11. Carbon dioxide, a waste product of cellular respiration, is picked up by the blood flowing through capillaries. Some carbon dioxide is dissolved in the blood. Most of it, however, enters the RBCs in the bloodstream, where it is converted into carbonic acid, which diffuses out of the RBCs and is transported in the plasma.

12. The rate of respiration can be increased by rising blood carbon dioxide levels and by an increase in physical exercise.

Diseases of the Respiratory System

13. Bacterial and viral infections of the respiratory tract can cause considerable discomfort, and some can be fatal.

14. The lungs are also susceptible to airborne materials, among them asbestos fibers, which can cause two types of lung cancer and a debilitating disease called asbestosis.

15. Another common disease of the respiratory system is asthma, an allergy (abnormal immune reaction) to dust, pollen, and other common substances. It is characterized by periodic episodes of wheezing and difficult breathing.

16. Chronic bronchitis is a persistent irritation of the bronchi and is caused by several common air pollutants primarily in tobacco smoke, and also in urban air pollution.

17. Emphysema is a breakdown of the alveoli that gradually destroys the lung's ability to absorb oxygen. Despite the role of air pollution in causing emphysema, smoking remains the number one cause of this disease.

Health and Homeostasis

18. The respiratory system contributes mightily to homeostasis by regulating concentrations of oxygen and carbon dioxide in the blood and body tissues. It also helps control the pH of the blood and tissues. also In addition, it supports other homeostatic organs. Proper functioning of the respiratory system is therefore essential for health.

19. Respiratory function can be dramatically upset by microorganisms as well as by pollution from factories, automobiles, power plants, and even our own homes.

critical thinking

Thinking Critically—Analysis

This Analysis corresponds to the Thinking Critically scenario that was presented at the beginning of this chapter.

For years, the tobacco industry insisted that there was no link between smoking and lung cancer because, they said, there was no positive proof. Semantically, they were quite correct. Science does not prove anything. Sure, scientists have performed many studies on people that show smokers are much more likely to develop lung cancer than nonsmokers and that indicate the more an individual smokes, the higher his or her chances are of developing this disease. Technically, the scientists haven't proved anything, they've shown a correlation.

The tobacco industry was being a bit deceptive, though, for when a correlation like that between smoking and lung cancer shows up again and again or when the correlation is high, you can have a fair amount of confidence in the cause-and-effect relationship under study.

Can a study be devised to prove the connection? It's unlikely. As I mentioned earlier, science is in the business of supporting hypothesis and generating theory, not providing proof positive.

This exercise requires one to look at definitions (what constitutes proof and the difference between proving something and showing a correlation). It also shows that one usually needs to dig deeper to understand controversies. In this instance, the tobacco industry has, for years, hung its argument on semantics. This exercise also suggests the importance of looking at who's talking—and uncovering any hidden biases. The tobacco industry clearly has a vested interest in assuring people that smoking is not harmful.

KEY TERMS AND CONCEPTS

Alveoli, p. 139
Alveolar macrophage, p. 144
Asthma, p. 150
Breathing center, p. 147
Bronchi, p. 141
Bronchioles, p. 141
Bronchitis, p. 150
Chronic bronchitis, p. 151
Cilia, p. 142
Diaphragm, p. 146
Dust cell, p. 144

Emphysema, p. 151
Epiglottis, p. 141
Exhalation, p. 146
Expiration, p. 146
Expiratory reserve volume, p. 147
Inhalation, p. 146
Inspiration, p. 146
Inspiratory reserve volume, p. 147
Laryngitis, p. 150
Larynx, p. 139
Lung, p. 139

Mucous cell, p. 142
Mucus, p. 142
Pharynx, p. 139
Residual volume, p. 147
Sinusitis, p. 150
Surfactant, p. 145
Stretch receptors, p. 147
Tidal volume, p. 147
Trachea, p. 141
Vocal cord, p. 139

CONCEPT REVIEW

1. Trace the flow of air from the mouth and nose to the alveoli, and describe what happens to the air as it travels along the various passageways. pp. 141–145

2. Draw an alveolus, including all cell types found there. Be sure to show the relationship of the surrounding capillaries. Show the path that oxygen and carbon dioxide take. pp. 143–145

3. Trace the movement of oxygen from alveolar air to the blood in alveolar capillaries. Describe the forces that cause oxygen to move in this direction. Do the same for the reverse flow of carbon dioxide. pp. 143–144

4. Why would a breakdown of alveoli in emphysemic patients make it difficult for them to receive enough oxygen? p. 151

5. Describe how sounds are generated and refined. pp. 139–141

6. A baby is born prematurely and is having difficulty breathing. As the attending physician, explain to the parents what the problem is and how it could be corrected. p. 145

7. Smoking irritates the trachea and bronchi, causing mucus to build up and paralyzing cilia. How do these changes affect the lung? p. 142

8. Describe inspiration and expiration, being sure to include discussions of what triggers them and the role of muscles in effecting these actions. p. 146

9. How does the breathing center regulate itself to control the frequency of breathing? p. 147

10. Exercise increases the rate of breathing. How? p. 147

11. Turn to Figure 8–10. Explain each of the terms. p. 147

12. What is asthma? What are its symptoms? How does it affect one's inspiratory reserve volume? pp. 150–151

13. Debate this statement: Urban air pollution has very little overall impact on human health. p. 151

14. Do you agree or disagree that the single most effective way of reducing deaths in the United States would be to ban smoking? Explain. pp. 142–144, 151–152

SELF-QUIZ: TESTING YOUR KNOWLEDGE

1. The _____ are located in the larynx and are responsible for the production of sound. p. 139

2. The larynx opens below into the _____, which then splits to form the right and left bronchi. p. 141

3. The tiny air sacs in the lung that permit oxygen to be transported into the bloodstream are known as the _____. p. 143

4. Inhaled particles deposit into a layer of _____ that is then transported toward the oral cavity by cilia in the lining of the respiratory tract. p. 142

5. Particles that make it into the lung are engulfed by _____ cells, also known as alveolar macrophages. p. 144

6. Large round cells in the lining of the alveoli produce a chemical compound known as _____ that helps reduce surface tension in the lungs, keeping the alveoli open. p. 145

7. Inflammation of the vocal cords is known as _____. p. 150

8. Persistent irritation of the bronchi, which causes mucus production and coughing, is known as chronic _____. p. 151

9. The muscles of inspiration are the _____ and intercostal muscles between the ribs. p. 146

10. Breathing is controlled by the _____ center in the brain. p. 147

11. Label the diagram:

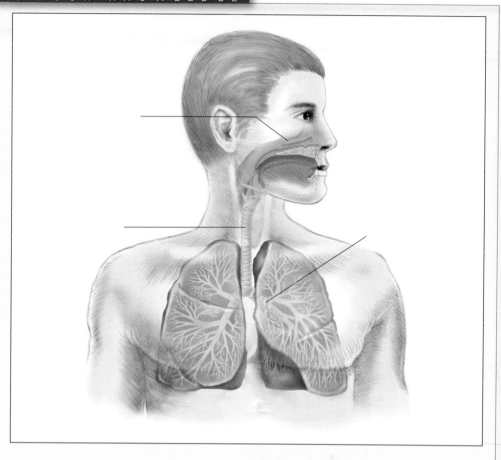

 www.jbpub.com/humanbiology/5e

The site features eLearning, an online review area that provides quizzes, chapter outlines, and other tools to help you study for your class. You can also follow useful links for in-depth information, research the differing views in the Point/Counterpoints, or keep up on the latest health news.

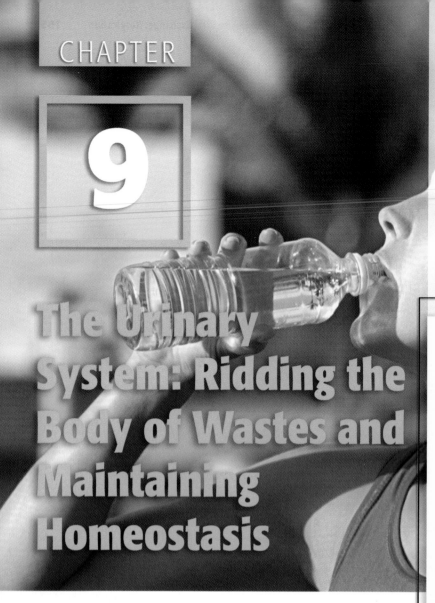

The Urinary System: Ridding the Body of Wastes and Maintaining Homeostasis

thinking critically

Each year, 24,000 new cases of kidney cancer are diagnosed in the United States. About 10,000 people a year die from this disease. Chemotherapy is virtually the only treatment available for patients whose cancers have spread to other locations, but it helps fewer than 10% of these patients.

Recently, a promising new treatment was introduced. The treatment, known as ALT, which stands for autolymphocyte therapy, is marketed by a company in Massachusetts. Costing about $22,000 per patient, ALT is designed to strengthen the immune system. Here's how it works: Blood is withdrawn from patients with kidney cancer. Lymphocytes are extracted, then treated with antibodies to the kidney tumor. This process activates the lymphocytes, causing them to produce cytokines, a chemical that boosts the activity of lymphocytes. The cytokines are divided into six batches. Once a month, for the next six months, patients donate more of their own lymphocytes. The cells are treated with cytokines, then reinjected, where they apparently go to work on the tumors.

Results of one study on the effectiveness of ALT were published in 1990 in the British medical journal *Lancet*. This study of 90 individuals showed that patients who had undergone the procedure survived 22 months after treatment, two and a half times longer than patients treated by chemotherapy. Less than a year after the publication of this clinical trial, a treatment center opened in Boston.

Assume that you are considering investing in the company that markets ALT. Using your critical thinking skills and your knowledge of scientific method, what concerns do you have, and what questions would you ask before investing in this company?

L eon Markowitz is lowered into a large pool of warm water in a special room in the hospital (Figure 9-1). Physicians position a large, cylindrical device in the water in front of one of his kidneys, the organs that filter the blood. Over the next few hours, as Markowitz listens to tapes of his favorite music, ultrasound waves, undetectable by the human ear, will smash the kidney stones that are obstructing the flow of urine and causing excruciating pain.

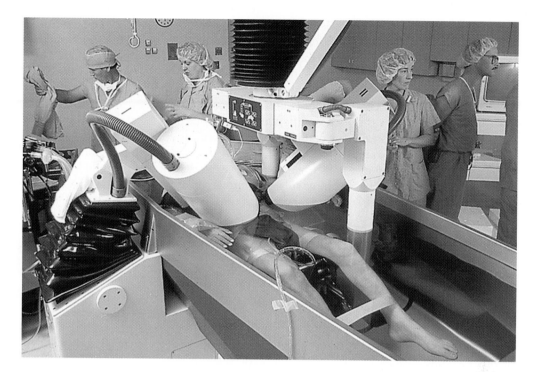

FIGURE 9-1 **Lithotripsy** Ultrasound waves, undetectable to the human ear, bombard the kidney stones in this man, smashing them into sandlike particles that can be passed relatively painlessly in the urine.

A decade earlier, surgeons would have had to cut an incision 15–20 centimeters (6–8 inches) long in Markowitz's side to access the kidney to remove the stone. Leon would have spent 7–10 days recovering in the hospital and another 8 weeks at home recuperating before returning to work. With this new technique, known as *ultrasound lithotripsy* (LITH-oh-TRIP-see) ("stone crushing"), patients usually return home within a day.

In this chapter, we will examine the kidneys and other portions of the urinary system, which function to rid our bodies of potentially harmful wastes. We will briefly examine other excretory organs, too, such as the skin and liver. Throughout the chapter, you will see how these systems contribute to homeostasis. We'll end with a discussion of common diseases, such as kidney stones, that disturb the function of this important organ system.

9-1   Organs of Excretion

Cells of the body produce a number of wastes that must be excreted to maintain homeostasis.

Ralph Waldo Emerson once said that as soon as there is life, there is risk. A biologist might say, as soon as there is life, there is waste. All living things produce waste and humans are no exception. Wastes cannot accumulate in organisms without causing harm. Thus, all organisms face a common challenge—how to get rid of internally produced wastes such as carbon dioxide.

Table 9-1 lists the major metabolic wastes and other chemicals excreted from the body. The first three entries are nitrogen-containing wastes—ammonia (ah-MOAN-ee-ah), urea (yur-EE-ah), and uric acid (YUR-ick).

Ammonia (NH_3), is a product of the breakdown of amino acids in the liver. Amino acid breakdown generally occurs when there's excess protein in the diet or a shortage of carbohydrate in our diets. This causes the body to break down protein to acquire amino acids. The amino acids are further broken down to produce energy. During the breakdown of amino acids, the amino groups (NH_2) are stripped from the molecules. The reaction is called *deamination* (dee-AM-in-A-shun). The amino groups are converted into ammonia. Ammonia is a highly toxic chemical. In the liver, most of it is converted to **urea**.

Another byproduct of metabolism in humans is uric acid. Produced by the liver, **uric acid** derives not from the breakdown

TABLE 9-1	Important Metabolic Wastes and Substances Excreted from the Body	
Chemical	Source	Organ of Excretion
Ammonia	Deamination (removal of amine group) of amino acids in liver	Kidneys
Urea	Derived from ammonia	Kidneys, skin
Uric acid	Nucleotide breakdown in liver	Kidneys
Bile pigments	Hemoglobin breakdown in liver	Liver (into small intestine)
Urochrome	Hemoglobin breakdown in liver	Kidneys
Carbon dioxide	Breakdown of glucose in cells	Lungs
Water	Food and water; breakdown of glucose	Kidneys, skin, and lungs
Inorganic ions*	Food and water	Kidneys and sweat glands

*Ions are not a metabolic waste product like the other substances shown in this table. Nonetheless, ions are excreted to maintain constant levels in the body.

of amino acids but rather from the breakdown of nucleotides. Nucleotides are the building blocks of DNA and RNA. Uric acid, like urea and ammonia, is excreted in the urine.

In adults, excess production may result in the deposition of uric acid crystals in the bloodstream and in joints, where they cause considerable pain. This condition is known as gout (gowt).

Another metabolic waste of great importance are the bile pigments. Derived from the breakdown of RBC hemoglobin in the liver, bile pigments are not excreted in the urine like the nitrogen-containing wastes just described. Rather, bile pigments are transferred from the liver to the gallbladder. The gallbladder empties into the small intestine, so bile pigments are excreted with the feces.

The liver produces another water-soluble pigment during the breakdown of hemoglobin. It is known as *urochrome* (YUR-oh-chrome). This yellow pigment is dissolved in the blood and passes to the kidneys, where it is excreted with the urine. In fact, urochrome gives urine its yellowish color.

Table 9-1 also lists inorganic ions excreted from the body. Even though they are not end-products of metabolism, inorganic ions are removed from the body by various excretory organs such as the skin and kidneys. This is essential to maintain optimal levels in the blood, tissue fluids, and the cytoplasm of cells—so vital for homeostasis.

9-2 The Urinary System

> Humans have several organ systems that rid the body of wastes; one of the most important is the urinary system.

Evolution has "provided" several avenues by which animals get rid of, or excrete, wastes. In humans, excretion of wastes occurs in the lungs, the skin, the liver, the kidneys, and even the intestines. These organs are consequently classified as excretory organs.

Of all the organs that participate in removing waste, however, the kidneys rank as one of the most important, for they rid the body of the greatest variety of dissolved wastes. In the process, the kidneys also play a key role in regulating the chemical constancy of the blood.

As Figure 9-2a shows, humans come equipped with two kidneys, They are part of the urinary system, one of the body's most important excretory systems. The kidneys don't work alone, however. They're aided by the (1) ureters (YUR-eh-ters), (2) the urinary bladder, and (3) the urethra (you-REETH-rah). The functions of these organs are described in more detail below and are listed in Table 9-2.

FIGURE 9-2 **The Urinary System** (a) Anterior view showing the relationship of the kidneys, ureters, urinary bladder, and urethra. (b) A cross section of the human kidney showing the cortex, medulla, and renal pelvis.

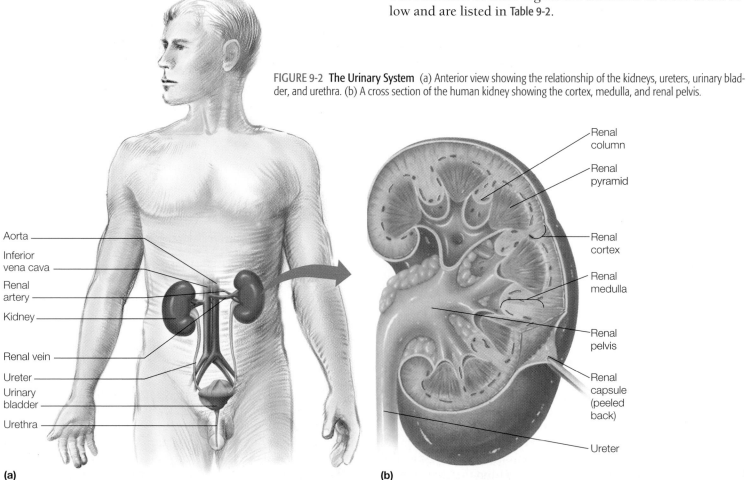

(a)   (b)

TABLE 9-2	Components of the Urinary System and Their Functions	
Component	Function
Kidneys	Eliminate wastes from the blood; help regulate body water concentration; help regulate blood pressure; help maintain a constant blood pH
Ureters	Transport urine to the urinary bladder
Urinary bladder	Stores urine; contracts to eliminate stored urine
Urethra	Transports urine to the outside of the body

The Anatomy of the Urinary System

The urinary system consists of the kidneys, ureters, bladder, and urethra.

The kidneys lie on either side of the backbone, the vertebral column. About the size of a person's fist, the kidneys are surrounded by a layer of protective fat. They are located high in the posterior (back) abdominal wall beneath the diaphragm.

(a)

(b)

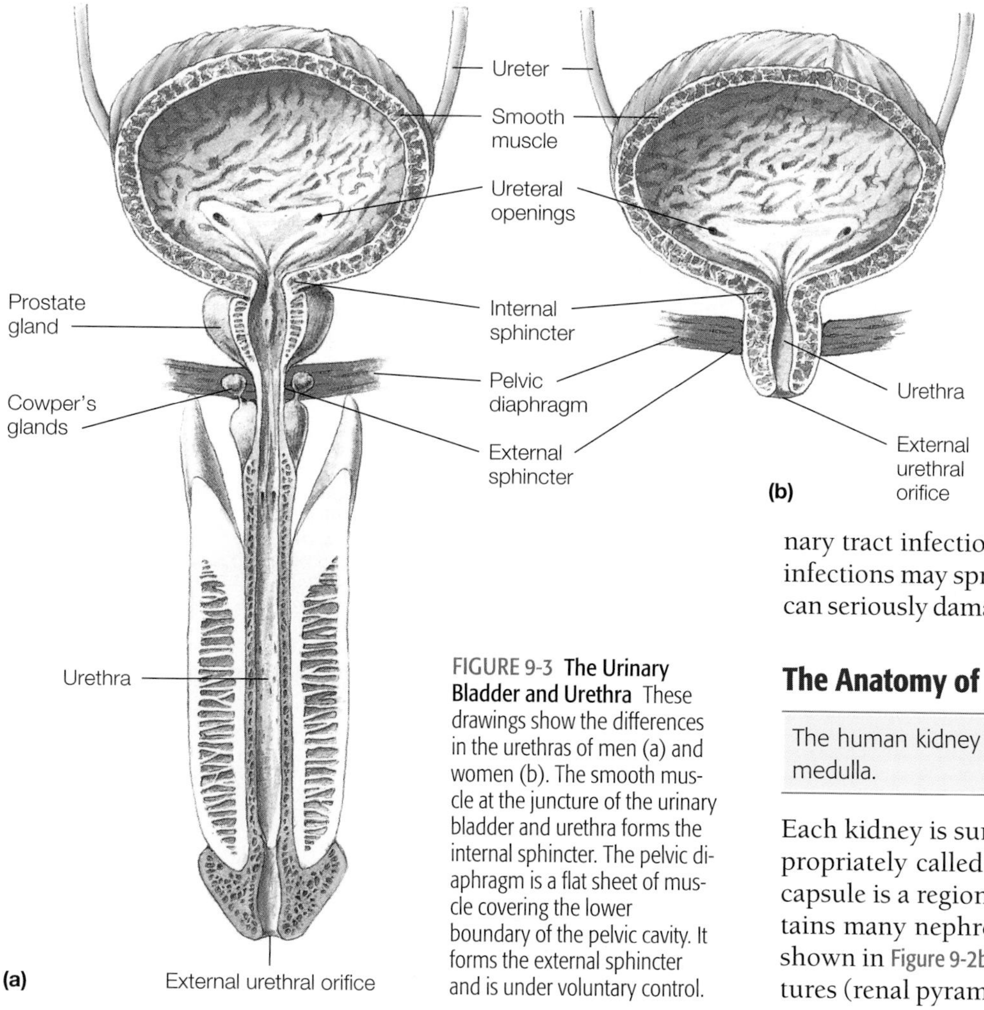

FIGURE 9-3 **The Urinary Bladder and Urethra** These drawings show the differences in the urethras of men (a) and women (b). The smooth muscle at the juncture of the urinary bladder and urethra forms the internal sphincter. The pelvic diaphragm is a flat sheet of muscle covering the lower boundary of the pelvic cavity. It forms the external sphincter and is under voluntary control.

The human kidneys are oval structures, slightly indented on one side, much like kidney beans (Figure 9-2b). (In fact, kidney beans get their name because they're shaped like this excretory organ.) As illustrated, each kidney is served by a **renal artery**, which enters at the indented region. The renal arteries are major branches of the abdominal aorta, a large blood vessel that delivers blood to the abdominal organs and lower limbs.

As you shall soon see, the kidneys filter wastes and excess chemicals from the blood entering via the renal arteries. After the blood has been filtered, it leaves the kidneys via the **renal veins**, which drain into the inferior vena cava. It transports venous blood to the heart.

Wastes extracted from the blood are eliminated in the **urine** (YUR-in), a yellowish fluid produced by the kidneys. It contains water, nitrogenous wastes (such as urea), small amounts of hormones, ions, and other substances.

Dissolved wastes are removed from the blood by numerous microscopic filtering units in the kidney, the **nephrons** (NEFF-rons). The urine produced by the nephrons is drained from the kidney by the **ureters**. They transport urine to the urinary bladder with the aid of peristaltic contractions, smooth muscle contractions. The **urinary bladder** lies in the pelvic cavity just behind the pubic bone. With walls made of smooth muscle that stretch as the organ fills, the bladder serves as a temporary receptacle for urine. When the bladder is full, its walls contract, forcing the urine out through the urethra.

The **urethra** is a narrow tube, measuring approximately 4 centimeters (1.5 inches) in women and 15–20 centimeters (6–8 inches) in men. The additional length in men largely results from the fact that the urethra travels through the penis (Figure 9-3).

The difference in the length of the urethra between men and women has important medical implications. The shorter urethra in women, for example, makes women more susceptible to bacterial infections of the bladder. Bladder infections may result in an itching or burning sensation and an increase in the frequency of urination. They may also cause blood to appear in the urine. Urinary tract infections can be treated with antibiotics. Untreated infections may spread up the ureters to the kidneys, where they can seriously damage the nephrons and impair kidney function.

The Anatomy of the Kidney

The human kidney consists of two zones, an outer cortex and inner medulla.

Each kidney is surrounded by a connective tissue capsule, appropriately called the *renal capsule*. Immediately beneath the capsule is a region known as the *renal cortex*. This region contains many nephrons. The inner zone is the *renal medulla*. As shown in Figure 9-2b, the medulla consists of cone-shaped structures (renal pyramids) and intervening tissue (renal columns).

The cone-shaped sections contain small ducts that transport urine from the nephrons into a central receiving chamber, the renal pelvis (Figure 9-2b). It joins with the ureters.

The Nephron

Nephrons consist of two parts, a glomerulus and a renal tubule.

Each kidney contains 1–2 million nephrons, visible only through a microscope. One of the marvels of evolution, each nephron consists of a tuft of capillaries, the **glomerulus** (glom-ERR-you-luss), and a long, twisted tube, the **renal tubule** (Figure 9-4b). The renal tubule consists of four segments: (1) Bowman's capsule, (2) the proximal convoluted tubule, (3) the loop of Henle, and (4) the distal convoluted tubule (Table 9-3). You may want to take a moment to study Figure 9-4b to acquaint yourself with these structures.

TABLE 9-3	Components of the Nephron and Their Function
Component	Function
Glomerulus	Mechanically filters the blood
Bowman's capsule	Mechanically filters the blood
Proximal convoluted tubule	Reabsorbs 75% of the water, salts, glucose, and amino acids
Loop of Henle	Participates in countercurrent exchange, which maintains the concentration gradient
Distal convoluted tubule	Site of tubular secretion of H^+, potassium, and certain drugs

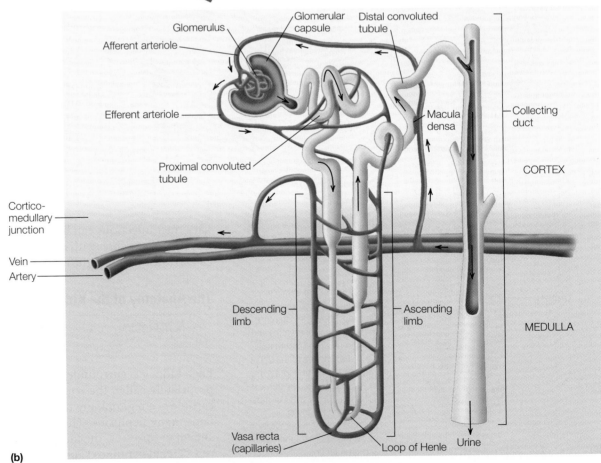

FIGURE 9-4 **The Nephron**
(a) A cross section of the kidney showing the location of the nephrons. (b) A drawing of a nephron.

As illustrated in Figure 9-4b, the glomerulus is surrounded by a saclike portion of the renal tubule, called **Bowman's capsule** (after the scientist who first described it). It is a double-walled structure; the inner wall fits closely over the glomerular capillaries and is separated from the outer wall by a small space, Bowman's space.

To understand the relationship between the glomerulus and Bowman's capsule, imagine that your fist is a glomerulus. Then imagine that you are holding a balloon in your other hand. If you push your fist (glomerulus) into the balloon, the layer immediately surrounding your fist would be the inner layer of Bowman's capsule. It is separated from the outer layer of the capsule by Bowman's space.

The outer wall of Bowman's capsule is continuous with the **proximal convoluted tubule (PCT)**, a contorted section of the renal tubule. The PCT soon straightens, then descends a bit, makes a sharp hairpin turn, and ascends. In the process, it forms a thin, U-shaped structure known as the **loop of Henle** (HEN-lee). The loops of Henle of some nephrons extend into the medulla.

The loop of Henle drains into the fourth and final portion of the renal tubule, the **distal convoluted tubule (DCT)**, another contorted segment. This structure, shown in Figure 9-4b, empties into a straight duct called a *collecting tubule*. The collecting tubules, in turn, merge to form larger ducts. They course through the cone-shaped renal pyramids and empty into the renal pelvis.

9-3 Function of the Urinary System

With the basic anatomy of the kidney and urinary system in mind, let us turn our attention to the function of the urinary system, focusing first on the nephron.

Blood Filtration in the Kidney

Blood filtration in nephrons involves three processes.

The nephrons filter enormous amounts of blood and produce from 1 to 3 liters of urine per day, depending on how much fluid a person ingests. Table 9-3 summarizes the functions of each segment of the nephron.

Glomerular Filtration

Glomerular filtration is a physical filtration process.

The first step in the purification of the blood is **glomerular filtration** (Figure 9-5). It's a rather straightforward physical filtration process not unlike that which occurs when you make coffee from ground coffee beans. Here's how it works. As blood flows through the glomerular capillaries, water and dissolved materials are forced through the walls of these tiny vessels. This liquid, called the **glomerular filtrate**, travels through the inner layer of Bowman's capsule. Red blood cells and large proteins in the blood can't escape because of the barrier formed by the inner layer of Bowman's capsule. From here, the filtrate passes into the proximal convoluted tubule, then into the loop of Henle, and finally out the distal convoluted tubule.

Tubular Reabsorption

Tubular reabsorption helps conserve valuable nutrients and ions.

Each day, approximately 180 liters (45 gallons) of filtrate is produced in the glomeruli. However, the kidneys produce only about 1 to 3 liters of urine each day. Thus, only about 1% of the filtrate actually leaves the kidneys as urine. What happens to the rest of the fluid filtered by the glomerulus?

Most of the fluid filtered by the glomeruli is reabsorbed—it passes from the renal tubule back into the bloodstream. The movement of water, ions, and molecules from the renal tubule

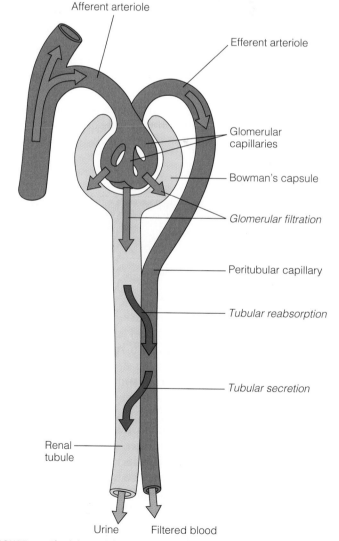

Afferent arteriole
Efferent arteriole
Glomerular capillaries
Bowman's capsule
Glomerular filtration
Peritubular capillary
Tubular reabsorption
Tubular secretion
Renal tubule
Urine   Filtered blood

FIGURE 9-5 Physiology of the Nephron The nephron carries out three processes: (1) glomerular filtration, (2) tubular reabsorption, and (3) tubular secretion. All contribute to the filtering of the blood.

TABLE 9-4	Fate of Various Substances Filtered by Kidneys	
	Filtered Substance (Average %)	
Substance	Reabsorbed	Excreted
Water	99	1
Sodium	99.5	0.5
Glucose	100	0
Waste products		
Urea	50	50
Phenol	0	100

to the bloodstream is referred to as **tubular reabsorption**. Most tubular reabsorption occurs in the proximal convoluted tubule.

During tubular reabsorption, water containing valuable nutrients and ions leaves the renal tubule and enters the networks of capillaries that surround the nephrons. As shown in Figure 9-4, these capillaries are branches of the efferent arterioles. They will eventually empty into the renal veins, returning nutrients to the body.

Tubular reabsorption is valuable because it conserves both water and important ions and nutrients that are indiscriminately filtered out in the glomerulus. Waste products like urea and uric acid, however, remain in the glomerular filtrate and will eventually be excreted in the urine. Tubular reabsorption also helps to control blood pH. Table 9-4 shows the reabsorption rates of various molecules.

Tubular Secretion

Tubular secretion is the transport of waste products from the peritubular capillaries into the renal tubule.

Waste disposal is supplemented by a third process, **tubular secretion**. This process eliminates from the blood wastes that were not filtered out by the glomerulus. It therefore removes wastes that might otherwise escape filtration. You can follow the steps in Figure 9-5.

As blood passes through the glomerular capillaries, much of it is filtered. The blood that isn't filtered passes from the efferent arterioles into the peritubular (pear-ee-TUBE-you-ler) capillaries surrounding the nephron. Some of the wastes pass through the capillary walls into the nephron. This is called *tubular secretion*. Tubular secretion not only rids the blood of wastes, it helps regulate the H^+ concentration of the blood. If the blood is too acidic, for example, H^+ ions are transported from the blood into the urine to reduce its acidity. Tubular secretion occurs in the distal convoluted tubule.

Blood draining from the capillaries surrounding the nephrons is fairly well cleansed by the time it empties into the nearby veins. These veins converge to form the renal vein, which transports the filtered blood out of the kidney and into the inferior vena cava, which carries it to the heart.

Urine produced by the nephrons flows out of the nephron into the collecting tubules. The urine leaving the nephron consists mostly of water and a variety of dissolved waste products. The collecting tubules descend through the medulla and converge to form larger ducts that empty their contents into the renal pelvis. As the collecting tubules descend through the medulla, much of the remaining water escapes by osmosis, further concentrating the urine and conserving body water.

9-4 Controlling Urination

Urination is a reflex in very young children but is controlled consciously by older children and adults.

Urine is produced continuously by the kidneys and flows down the ureters to the urinary bladder. Leakage out the bladder and into the urethra is prevented by two sphincters—muscular "valves" similar to those found in the stomach (Figure 9-3).

The first sphincter, the internal sphincter, is formed by a smooth muscle in the neck of the urinary bladder at its junction with the urethra. The second valve, the external sphincter, is a flat band of skeletal muscle that forms the floor of the pelvic cavity. When both sphincters are relaxed, urine is propelled into the urethra and out of the body. How is urination stimulated?

Urination is stimulated by the accumulation of urine inside the bladder. You need about 200–300 milliliters to start the process. When the bladder fills, the walls of the bladder stretch. Special sensors in the walls, called *stretch receptors*, detect the distension of the bladder and begin sending impulses via sensory nerves to the spinal cord. In the spinal cord, the incoming nerve impulses stimulate other nerve cells. Nerve impulses generated in these cells leave the spinal cord and travel along nerves that terminate on the muscle cells in the wall of the bladder. These impulses stimulate the smooth muscle cells to contract. Muscular contraction forces the internal sphincter to open, letting urine enter the urethra. Urination won't occur in adults, however, until the external sphincter is consciously relaxed.

In babies and very young children, urination is entirely reflex. Once the bladder expands to a certain size, it empties. Not until children grow older (2–3 years) can they begin to control urination. In older children and adults, the external sphincter is under conscious control. It will not relax until consciously permitted to do so.

Unfortunately, adults sometimes lose control over urination, resulting in a condition referred to as *urinary incontinence* (in-KAN-teh-nance). Urinary incontinence may be caused by a traumatic injury to the spinal cord that disrupts neural pathways from the brain that allow us to suppress urination. In such instances, the bladder empties as soon as it reaches a certain size, much as it does in a baby or young child.

Mild urinary incontinence is much more common. It is characterized by the escape of urine when a person sneezes or coughs. This is most common in women and usually results from damage to the external sphincter during childbirth. Childbirth stretches the skeletal muscles of the external sphincter, reduc-ing their effectiveness. To avoid this, many women undertake exercise programs to strengthen these muscles before and after childbirth. Urinary incontinence may also occur in men whose external sphincters have been injured in surgery on the prostate gland, which surrounds the neck of the urinary bladder.

9-5 Controlling Kidney Function and Maintaining Homeostasis

Concentrations of ions and other substances are controlled by regulating water levels.

The kidneys help the body control the chemical composition of the blood and maintain homeostasis. This section briefly describes the kidney's role in maintaining water balance (that is, proper levels of water).

As noted earlier, much of the water filtered out of the blood by the glomeruli is reabsorbed by the renal tubules and returned to the bloodstream. The rate of water reabsorption can be increased or decreased to alter urine production. The ability to adjust water reabsorption and urine output allows the body either to rid itself of excess water when a person is overhydrated or to conserve water when an individual is dehydrated. These functions help control the concentration of the blood, too. Two hormones play a major role in this important homeostatic process.

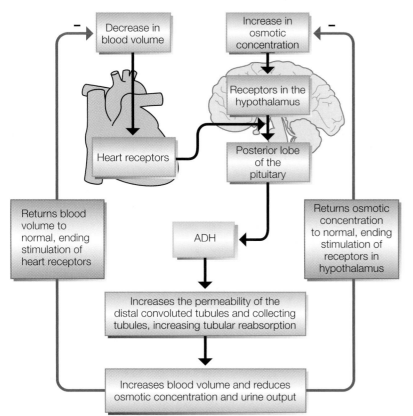

FIGURE 9-6 **ADH Secretion** ADH secretion is under the control of the hypothalamus. When the osmotic concentration of the blood rises, receptors in the hypothalamus detect the change and trigger the release of ADH from the posterior lobe of the pituitary. Detectors in the heart also respond to changes in blood volume. When it drops, they send signals to the brain, causing the release of ADH.

ADH

The hormone ADH increases water reabsorption and conserves body water.

Water reabsorption is controlled in part by **antidiuretic hormone** (ANN-tie-DIE-yur-eh-tick) or **ADH**. ADH is released by an endocrine gland at the base of the brain known as the *pituitary* (peh-TWO-eh-TARE-ee) (Chapter 13). ADH secretion is regulated by two receptors, one in the brain and one in the heart (Figure 9-6).

The first sensor is a group of nerve cells in a region of the brain called the *hypothalamus* (HIGH-poe-THAL-ah-muss). It is located just above the pituitary gland. The cells that regulate ADH monitor the osmotic concentration of the blood, as you shall soon see. The second sensor, as noted above, is located in the heart. It detects changes in blood volume, which reflect water levels.

To understand how ADH secretion is controlled, let's imagine that you're out in the hot sun. Your body's sweating and you haven't got a thing to drink. Under these conditions, the body begins to lose water. As a result, the blood volume decreases and the osmotic concentration (the concentration of dissolved substances) increases. The decrease in blood volume is detected by the heart receptors. The rise in osmotic concentration is detected by sensors in the brain. As shown in Figure 9-6, both of these stimuli trigger the release of ADH from the pituitary gland.

ADH circulates in the blood. In the kidney, this hormone stimulates tubular reabsorption of water. In other words, it stimulates the passage of water from the nephron into the peritubular capillaries. This reduces urinary output and restores the volume and osmotic concentration of the blood—if you are not too dehydrated.

Excess water intake, on the other hand, has the opposite effect. As you can imagine, drinking excess water will increase the blood volume and decrease its osmotic concentration. The sensors in the brain and heart note these changes and cause a reduction in ADH secretion. As ADH levels in the blood fall, tubular reabsorption decreases and more water is lost in the urine. As a consequence, blood volume and osmotic concentration are restored.

Aldosterone

> The hormone aldosterone stimulates water reabsorption in the kidney.

Water balance is also regulated by the hormone **aldosterone** (al-DOS-ter-own). Aldosterone is a steroid hormone produced by the adrenal glands, which sit atop the kidneys (Figure 13-1a). Aldosterone levels in the blood are controlled by three factors: (1) blood pressure, (2) blood volume, and (3) osmotic concentration.

As illustrated in Figure 9-7, aldosterone release from the kidneys is controlled by a sequence of events involving several chemical intermediaries.

To understand this process, begin at the top of Figure 9-7. As shown, when blood pressure and the volume of filtrate in the nephrons decline, certain cells in the kidney produce an enzyme called **renin** (REE-nin). In the blood, renin slices a segment off a large plasma protein called angiotensinogen (AN-gee-oh-TEN-SIN-oh-gin). The result is a small peptide molecule called angiotensin I. Angiotensin I is inactive. It must be converted into the active form, angiotensin II, by enzymes it encounters as blood flows through the lungs.

Angiotensin II stimulates aldosterone secretion by the adrenal glands. Aldosterone released from the adrenal glands circulates in the blood. In the kidney, it stimulates cells of the nephron to increase the amount of sodium they reabsorb. In other words, it increases tubular reabsorption of sodium. Water follows sodium ions out of the nephron into the peritubular capillaries. This, in turn, increases blood volume and blood pressure, shutting down this feedback loop.

Water balance is also affected by several chemicals in many people's daily diet, two of the most influential being caffeine and alcohol.

Caffeine

> Caffeine increases urine output without affecting ADH secretion.

Caffeine is a diuretic (DIE-yur-ET-ick), a chemical that increases urination. Found in coffee, (most nonherbal) teas, and many soft drinks, caffeine increases urine production in two ways. First, it increases glomerular blood pressure, which increases glomerular filtration. As a result, more filtrate is formed. Second, caffeine decreases the tubular reabsorption of sodium ions. As noted earlier, water follows sodium ions out of the renal tubule during tubular reabsorption. Thus, a decrease in sodium reabsorption results in a decline in the amount of water leaving the renal tubule and an increase in urine output.

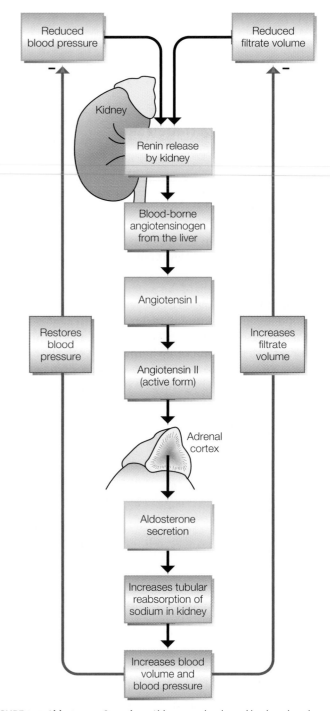

FIGURE 9-7 Aldosterone Secretion Aldosterone is released by the adrenal cortex. Its release, however, is stimulated by a chain of events that begins in the kidney.

Ethanol

Ethanol inhibits the secretion of ADH by the pituitary.

Ethyl alcohol or ethanol, for short, is also a diuretic. Ethanol is found in a variety of beverages, including beer, wine, wine coolers, and hard liquor. After being absorbed by the stomach and small intestine, ethanol enters the blood and circulates throughout the body. When it reaches the brain, it inhibits the cells that secrete ADH. When ADH levels fall, water reabsorption declines. The result is a marked increase in water loss by the kidneys, giving credence to the quip that you don't buy wine or beer, you rent it!

9-6 Diseases of the Urinary System

Like other body systems, the urinary system can malfunction, creating a homeostatic nightmare in the body. This section discusses three disorders: kidney stones, kidney failure, and diabetes insipidus.

Kidney Stones

Kidney stones can block the outflow of urine and cause severe kidney damage.

Urine contains numerous ions and dissolved wastes. Ninety percent of the dissolved waste consists of three substances: urea, sodium ions, and chloride ions. Varying amounts of other chemical substances are also present but in minute amounts.

In many ways, the urine provides a window to the chemical workings of the body. Thus, increases or decreases in the level of substances in the urine may signal underlying problems—upsets in homeostasis and possible health problems. Consequently, physicians routinely analyze the urine of their patients to test for metabolic disorders. One of the most common is diabetes mellitus (DIE-ah-BEE-tees mell-EYE-tus), or sugar diabetes. This disease results in a defect in glucose uptake by cells in the body. Because body cells cannot absorb glucose, blood levels and urine levels are elevated.

Excess chemicals and ions in the urine not only signal problems elsewhere, they can be damaging to the kidney itself. For instance, higher than normal concentrations of calcium, magnesium, and uric acid in the urine, often caused by inadequate fluid intake, may crystallize inside the kidney, most often in the renal pelvis. Small deposits enlarge by a process called *accretion* (ah-CREE-shun)—the deposition of materials on the outside of the stone, causing them to grow in much the same way that a pearl grows in an oyster (Figure 9-8). These small crystals, or **kidney stones**, eventually grow into fairly large deposits. Kidney stones are a health threat and an economic liability.

Fortunately, many kidney stones are flushed out of the kidney into the urinary bladder and out the urethra on their own. Although small stones that enter the urinary bladder are often excreted in the urine, the sharp edges of the stone can dig into the walls of the ureters and urethra, causing pain.

Problems arise when stones become lodged inside the kidney or the ureters and obstruct the flow of urine. This causes internal pressure to increase that may result in considerable damage to the nephrons if left untreated.

For years, kidney stones were removed surgically. Today, however, the relatively new medical technique of ultrasound lithotripsy is generally used. In this technique, mentioned in the opening paragraph of this chapter, physicians bombard kidneys with ultrasound waves, which shatter the stones, producing fine, sandlike grains that are passed in the urine without incident. The procedure is nearly painless and much safer than surgery. Table 9-5 lists additional urinary system disorders.

Renal Failure

Renal failure may occur either suddenly or gradually and can be treated by dialysis.

The importance of the kidneys is most obvious when they stop working, a condition known as *renal failure*. **Renal failure** may be caused by the presence of certain toxic chemicals in the blood or as a result of an immune reaction to certain antibiotics. In other

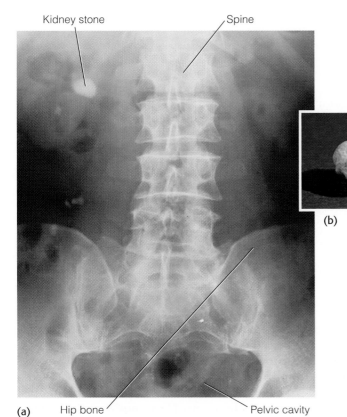

(a)   Kidney stone   Spine   Hip bone   Pelvic cavity

(b)

FIGURE 9-8 **Kidney Stones** (a) An X-ray of a kidney stone. (b) Kidney stones removed by surgery. (See pin for size.)

Prioritizing Medical Expenditures

We Need to Learn to Prioritize Medical Expenditures by Richard D. Lamm

Heath care in the United States finds itself in (to paraphrase Dickens)—the best of times and the worst of times. It is the epoch of medical miracles, yet at the same time millions go without even basic health care.

The basic dilemma of American medicine is that we have invented more health care than we can afford to deliver. Health care costs are rising at three times the rate of inflation. They are absorbing funds desperately needed elsewhere in our system to educate our kids, rebuild our infrastructure, and revitalize our industries.

Health care costs are rising for many reasons. One major reason is that no one prioritizes the myriad of procedures that health care can deliver. No one asks, "How do we buy the most health for the most people with our limited funds?" Recent policy decisions in Oregon and California, relating to the public funding of transplant operations, illustrate how two states attempted to deal with a crisis in health care.

Richard D. Lamm is the director of the Center for Public Policy and Contemporary Issues at the University of Denver, where his primary research and teaching interests have been in the area of health policy. Before assuming his present position, he served three terms as governor of Colorado, from 1975 until 1987.

Oregon received adverse publicity for its decision not to publicly fund soft tissue transplant operations. As is often the case, much of the focus has been on the handful of individuals who have been adversely affected by the policy rather than the numerous, but anonymous, people who will benefit. The Oregon policy sparked a public outcry over society's lack of compassion for individuals who desperately need transplants. Yet, when Oregon policymakers weighed the needs of a few transplant patients against the basic health care needs of the medically indigent, they decided against the former in favor of the latter. They have now set up a process that sets priorities throughout the health care system.

California took a contrary approach. Policymakers in California decided to publicly fund transplant operations. Then, one week later, they removed 270,000 low-income people from their state's medical assistance program.

Which was the wiser decision? The answer seems clear. Oregon bought more health for more of its citizens for its limited funds. Some say Oregon should have raised its taxes and funded transplants. That is very easy for nonpoliticians to say. Polls show people believe all Americans should have access to quality health care. But the polls also show that most are unwilling to accept even modest tax increases to provide the care they say they favor.

Illinois recently passed legislation giving "universal access to major organ transplants," but appropriated less than one-third of the funds needed. Politically, of course, Illinois played it far safer than Oregon but, in doing so, failed to confront one of the most pressing social issues of our time. Avoidance may be a politically expedient tactic, but sooner or later our society will be forced to allocate scarce health resources.

Clearly, medical science is inventing faster than the public is willing or able to pay. Realistically, no system of health care can avoid rationing medicine. In fact, rationing already exists. But instead of rationing medicine according to need or a patient's prospects for recovery, we ration by seniority and ability to pay. Those who pretend that we do not ration medicine forget that 43 million Americans do not have full access to health care in America. Most states have staggering numbers of medically indigent, yet they provide a small percentage with a full program of coverage. Oregon is the first state to cover 100% of the people living under the federal poverty line with basic health care.

I suggest this yardstick: Which state best asked itself, "how do we buy the most health for the most people?" Which state tried to maximize limited resources? Which state benefited the greatest number of people and which state harmed the greatest number of people?

Oregon attempted to weigh the basic health needs of the medically indigent against other programs for which the state paid. It recognized that the money the state paid for transplants and other high-cost procedures for a few people could buy more health care elsewhere, providing basic low-technology services to people not covered at that time.

Prioritizing health care does not abandon the poor; it seeks to serve the largest number with the more effective procedures.

---

cases, it is caused by severe kidney infections. Accidents in which huge amounts of blood are lost also cause renal failure due to the sudden decrease in blood flow to the kidneys.

Renal failure may occur suddenly, over a period of a few hours or a few days. This is referred to as *acute renal failure*. Kidney function may also deteriorate slowly over many years, resulting in *chronic renal impairment*. Chronic renal impairment may lead to a complete or nearly complete shutdown, known as *end-stage failure*.

Kidney failure (whether acute or end stage) is life-threatening, for when the kidneys stop working, water and toxic wastes begin to accumulate in the body. This disrupts homeostasis. If untreated, a patient will die in 2 to 3 days. Patients usually die from an increase in the concentration of potassium ions in the blood and tissue fluids. Why? Although potassium is essential for the function of heart muscle, excess potassium destroys the rhythmic contraction of the heart, causing fibrillation. As you may recall from Chapter 6, fibrillation is the

Medical Prioritization Is a Bad Idea by Arthur L. Caplan

The American health care system, the experts say, is going bust at a rapid rate. Efforts to contain our burgeoning $1 trillion-plus tab have been a total failure.

The dilemma of how to pay for health care is forcing some public officials to think the unthinkable. The state of Oregon instituted rationing policies for health care. But, before you applaud the realism, consider that Oregon rations access to health care only for the poor. Those eligible for Medicaid in Oregon are required by law to forego life-saving medical care.

Officials in Oregon note that the poor have always had less access to health care than the rich. This is true. But, our society's failure to meet the health care needs of the poor hardly justifies a public policy that asks the poor to bear the burden of rationing as a matter of law.

Who concocted this blatantly unethical scheme? Incredibly, the inspiration for the Oregon plan for pocketbook triage comes in part from those in my line of work—medical ethicists.

A California bioethics consulting firm was paid by Oregon state officials to provide moral rationales for dropping the poor out of the health care lifeboat. The consultants appeared to approach their task with gusto.

"You have to draw the line somewhere," one moralist-for-hire said in a newspaper article about rationing for the poor. "We'll provide all services to a diminishing segment of the population, and literally we'll throw the rest of the people overboard."

No hint is given of the theoretical position that would justify aiming all rationing efforts at the poor. But it is hard to think of an ethic that holds that when a nation cannot pay its doctor bills, it is the poor and only the poor who should be denied the right to see a doctor.

It is hard to understand how any ethicist could become involved in a scheme so blatantly unfair as that of rationing necessary health care only for the poor. What is worse is that the same ethicists and the officials taking their advice are not asking whether it is really necessary to institute the rationing of necessary medical care for anyone.

Before saying goodbye to the indigent, why aren't public officials in Oregon thinking about cracking down on practices that add tens of millions of dollars to state-financed health care costs each year? Before saying no to a bone marrow transplant for a three-year-old whose mother is on Medicaid, couldn't county and state legislators insist that every licensed hospital and physician be required by law to provide a fixed percentage of care for those who cannot pay?

Before creating laws that would send some of the poor to a premature demise, county and state officials ought to require private health insurers to charge subscribers an additional premium that could be used to supplement the pitifully small budgets of Medicaid and public hospitals. And would it not make some sense to insist on a luxury tax, which could be used to help meet the crucial health care needs of the poor, from the rich who avail themselves of psychotherapy, vitamins, Viagra, cosmetic surgery, diet clinics, and stress-management seminars?

It is wrong to make the poor and only the poor bear the burden of rationing. It is unethical to institute rationing of necessary health services for any group of Americans unless we have made every effort to be as efficient as we can be in spending our health care dollars.

At a time when some can indulge their wants by buying a facelift, it seems extraordinarily hard for ethicists or legislators to convincingly argue that they have no other option but to condemn the poor to die for want of money.

Arthur L. Caplan is director of the Center for Bioethics at the University of Pennsylvania.

Sharpening Your Critical Thinking Skills

1. Summarize Lamm's and Caplan's key points. List supporting information given for each position or main point.

2. Do you agree with the views of Lamm or Caplan? Explain why.

www.jbpub.com/humanbiology/5e

Visit Human Biology's Internet site for links to web sites offering more information on this topic.

---

uncoordinated contraction of heart muscle that may render the heart ineffective as a pumping organ. Patients often die.

The treatment of renal failure varies, depending on the underlying cause. If the problem is caused by an acute loss of blood, transfusions may be required. Patients whose kidneys have shut down, even temporarily, may require renal dialysis (DIE-AL-eh-siss). In this procedure, blood is drawn out of a vein and passed through a piece of tubing that transports the blood to an artificial filter that removes harmful wastes. After filtration, the blood is pumped back into the patient's bloodstream. Dialysis requires several hours and must be repeated every 2 or 3 days. Some patients have dialysis units at home and simply hook themselves up each night before they go to bed.

Another more recent method of treating renal failure is continuous ambulatory peritoneal dialysis or CAPD. In this procedure, 2 liters of dialysis fluid are injected into a person's abdomen through a permanently implanted tube or catheter (CATH-eh-ter). Waste products diffuse out of the blood vessels

TABLE 9-5 Common Urinary Disorders

Disease	Symptoms	Cause
Bladder infections	Especially prevalent in women; pain in lower abdomen; frequent urge to urinate; blood in urine; strong smell to urine.	Nearly always bacteria
Kidney stones	Large stones lodged in the kidney often create no symptoms at all; pain occurs if stones are being passed to the bladder; pains come in waves a few minutes apart.	Deposition of calcium, phosphate, magnesium, and uric acid crystals in the kidney, possibly resulting from inadequate water intake
Kidney failure	Symptoms often occur gradually; more frequent urination, lethargy, and fatigue; should the kidney fail completely, patient may develop nausea, headaches, vomiting, diarrhea, water buildup, especially in the lungs and skin, and pain in the chest and bones.	Immune reaction to some drugs, especially antibiotics; toxic chemicals; kidney infections; sudden decreases in blood flow to the kidney resulting, for example, from trauma
Pyelonephritis	Infection of the kidney's nephrons; sudden, intense pain in the lower back immediately above the waist, high temperature, and chills	Bacterial infection

into the abdominal cavity across the peritoneum (PEAR-eh-tah-KNEE-um), a thin membrane that lines the organs and wall of abdominal cavity. The fluid, containing waste products, is drained from the abdominal cavity a couple of times a day.

This form of dialysis is much simpler. Patients can take care of it themselves. In addition, it allows for more frequent filtering of the blood.

Complete kidney failure can be treated by kidney transplants. Transplants are generally most successful when they come from closely related family members. Such transplants are not rejected by the immune system, as explained in Chapter 14.

For years, the complete or nearly complete destruction of kidney function was almost always fatal. Thanks to renal dialysis and kidney transplantation, many patients today can live fairly normal, healthy lives. These procedures, especially transplants, are costly, however. The public currently picks up the tab for individuals without insurance—or they go without needed treatments and die. For a debate on prioritizing medical expenditures, see this chapter's Point/Counterpoint.

Diabetes Insipidus

Diabetes insipidus is a rare disease caused by a lack of ADH.

Hormonal imbalances can also lead to disruptions in the urinary system. Severe head injuries, for example, may halt the production of ADH, leading to a disease known as **diabetes insipidus** (DIE-ah-BE-teas in-SIP-eh-duss). This condition is not to be confused with diabetes mellitus (commonly called sugar diabetes), a disorder involving the hormone insulin.

Insufficient ADH output results in excess water loss—up to 20 liters (5 gallons) per day. Because so much water is lost, the urine is dilute and colorless. To replace this water, a person must consume large quantities of liquids.

Diabetes insipidus can be treated in a variety of ways, depending on the severity of the disorder and the cause. In patients whose urine output is only slightly elevated, dietary salt restrictions and antidiuretics (drugs that reduce urine output) work. In severe cases, patients must receive synthetic ADH, which is administered by injections or nose drops.

9-7 Health and Homeostasis

The kidneys are biological filters that help maintain the proper levels of nutrients such as blood sugar and various ions in the bloodstream. They also help to maintain the blood pH and play a major role in eliminating various cellular wastes and potentially harmful substances such as food additives, pesticides, and toxic chemicals in the food we eat and the water we drink. All of these functions are vital to maintaining homeostasis and human health.

Unfortunately, the kidneys can be seriously damaged by a many factors. An auto accident that causes a severe depletion of blood, as noted earlier, can result in kidney failure. Bacteria that enter the kidneys from the urinary bladder through the ureters can cause severe infections that impair, even destroy, kidney function. Toxic chemicals such as mercury and certain antibiotics can damage the kidneys as well.

Damage to the kidney has severe physiological repercussions, not only in the kidney but in all of the rest of the cells, tissues, and organs that depend on the kidney's ability to help maintain homeostasis. When the kidneys stop functioning, for example, the pH of the blood may be altered, reducing the effectiveness of many enzymes in the body and thus influencing many key functions. Because the kidneys control the concentrations of sodium, potassium, and calcium ions, which are essential for contraction of heart muscle and skeletal muscle and the function of the nerves, problems in the kidneys can affect vital functions performed by these tissues.

Individuals can help maintain kidney health by living a healthy lifestyle—drinking plenty of fluids; avoiding toxic chemicals, for example, by eating organic vegetables free of pesticides and herbicides; practicing safe sex; and getting plenty of rest to avoid bacterial infections.

SUMMARY

Organs of Excretion

1. All organisms face the same challenge: getting rid of wastes. They do so through organs. These organs also help regulate internal concentrations of ions and water vital for homeostasis.

2. Table 9-1 lists the major metabolic wastes and other molecules excreted from the body. These wastes are removed by the skin, lungs, liver, kidneys, and intestines.

The Urinary System

3. One of the most important organs of excretion in humans is the kidney. Kidneys remove impurities from the blood and also help regulate the water levels and ionic concentrations of the blood.

4. Blood enters the kidneys in the renal arteries whose branches deliver it to the millions of nephrons located in the kidney.

5. The nephrons produce urine, which drains from the kidneys into the ureters, slender muscular tubes that lead to the urinary bladder. Urine is stored in the urinary bladder then voided through the urethra.

6. Each nephron consists of a glomerulus, a tuft of highly porous capillaries, and a renal tubule, where urine is produced.

7. The renal tubule consists of four parts: (a) Bowman's capsule, (b) the proximal convoluted tubule, (c) the loop of Henle, and (d) the distal convoluted tubule. The distal convoluted tubules of nephrons drain into collecting tubules, which converge and empty urine into the renal pelvis.

Function of the Urinary System

8. Blood filtration is accomplished by: glomerular filtration, tubular reabsorption, and tubular secretion.

9. Glomerular filtration occurs in the glomerulus, producing a liquid called the filtrate. The filtrate is processed as it flows along the renal tubule. Water, ions, and nutrients are largely reabsorbed as they travel along the tubule. This process is called tubular reabsorption. Water and reabsorbed nutrients and ions pass into a network of capillaries, the peritubular capillaries, surrounding each nephron.

10. Not all bloodborne wastes are filtered from the blood in the glomerulus. Those that remain pass into the peritubular capillaries with the blood. These substances may be transported out of the peritubular capillaries into the renal tubule in a process known as tubular secretion. Hydrogen and potassium ions, for example, are secreted into the renal tubule.

Controlling Urination

11. Urination is a reflex in babies and very young children. In older children and adults, the urination reflex still operates, but it is overridden by a conscious control mechanism.

Controlling Kidney Function and Maintaining Homeostasis

12. The concentration of water and dissolved substances in the blood is controlled by two hormones: ADH and aldosterone.

13. ADH or antidiuretic hormone is secreted by the posterior lobe of the pituitary gland. It is released when the osmotic concentration of the blood increases or when blood volume decreases. ADH increases water reabsorption.

14. Aldosterone is produced by the adrenal cortex. Aldosterone stimulates the reabsorption of sodium ions by the nephron. Water follows the sodium out of the renal tubule, increasing blood pressure and blood volume.

Diseases of the Urinary System

15. Calcium, magnesium, and other materials can precipitate out of the urine in the renal pelvis, forming kidney stones.

16. Smaller stones may be passed along the ureters to the bladder and are often eliminated during urination. Larger stones that remain in the kidney must be removed surgically or via ultrasound lithotripsy.

17. Renal failure is a disease in which the kidneys stop working. Renal failure may occur suddenly or gradually and is a life-threatening disorder. It can be treated by dialysis.

18. Diabetes insipidus is a rare disease resulting from a lack of ADH. The disease is characterized by frequent urination and excessive fluid intake.

Health and Homeostasis

19. The kidneys help regulate the levels of nutrients and waste materials produced by the cells of the body. They also help eliminate toxins taken into the body from air, water, and food. In so doing, they help to maintain homeostasis and health.

20. Unfortunately, the kidneys can be damaged by bacteria, accidents that result in severe blood loss, and by certain toxic chemicals. These factors impairing kidney function and disrupt homeostasis with lethal consequences.

critical thinking

Thinking Critically—Analysis

This Analysis corresponds to the Thinking Critically scenario that was presented at the beginning of this chapter.

After preparing your list, compare your concerns and questions with mine to see how we match.

  Concern: A clinical trial on 90 patients is extremely small. With so few patients, the reliability of the results is in question.
  Question 1: Are more clinical tests under way?
  Question 2: Does the treatment have any adverse impacts?
  Question 3: Is the marketing company unbiased?

Question 4: Is it promoting a product that may turn out to be ineffective, opening the company to lawsuits for fraud? In other words, is it letting financial concerns outweigh the need for good scientific research and carefully controlled experiments?

  Now suppose that you have kidney cancer and have the $22,000 to pay for ALT. Would you do it? Are any of the issues above still relevant? How would you go about determining whether you should try the procedure?

  To be fair to all parties concerned, let me point out that approximately two-thirds of the insurance companies in the United States pay for ALT treatment. As a rule, most companies do not pay for experimental procedures. In other words, they must be satisfied that ALT is an effective treatment before reimbursing clients.

KEY TERMS AND CONCEPTS

CONCEPT REVIEW

1. Draw the various parts of the urinary system, and describe what each one does. pp. 158–161
2. Draw a nephron, then label its parts. pp. 160–161
3. Describe the three ways in which the kidney filters the blood. pp. 161–162
4. A drug inhibits the uptake, or reabsorption, of water by the distal convoluted tubules and collecting tubules. What effect would this drug have on urine output, urine concentration, blood pressure, blood volume, and the concentration of the blood? pp. 162–163
5. Describe how ADH controls blood pressure and the water content of the body. Describe the hormonal and physiological changes in the body that take place when excess liquid is ingested. Do the same for dehydration. pp. 163–164
6. How does urination differ between newborns and adults? Explain what is meant by this statement: In older children and adults, urination is a reflex with a conscious override. p. 162
7. Aldosterone helps regulate blood pressure and water content. In what ways is this hormone different from ADH? p. 164
8. You have just finished your residency in family medicine. A patient comes to your office complaining that he drinks water all day long and spends much of the rest of the day in the bathroom urinating. What tests would you order? What diagnosis would you suspect? p. 168

www.jbpub.com/humanbiology/5e

The site features eLearning, an online review area that provides quizzes, chapter outlines, and other tools to help you study for your class. You can also follow useful links for in-depth information, research the differing views in the Point/Counterpoints, or keep up on the latest health news.

1. _____ is a product of the breakdown of amino acids in the liver. p. 157

2 Gout is a disease caused by the buildup of _____ acid in the blood and joints. p. 158

3. Urine produced in the kidneys drains to the urinary bladder in the _____. p. 159

4. The urinary bladder is drained by the _____. p. 159

5. Each nephron consists of a tuft of capillaries known as the _____ and a long, convoluted tubule known as the renal tubule. p. 160

6. Blood is first filtered in the _____ capillaries. p. 161

7. Filtrate leaves Bowman's space and enters the _____ convoluted tubule. p. 161

8. The passage of substances from the renal tubule into the peritubular capillaries is known as tubular _____. p. 162

9. The hormone _____ from the pituitary helps conserve water by stimulating water reabsorption in the renal tubule. p. 163

10. Aldosterone produced by the _____ glands also helps control blood volume and ionic concentration of the blood. p. 164

11. Caffeine and alcohol increase urine output and are known as _____. p. 164

12. Ethanol in alcoholic drinks works by decreasing the output of the hormone _____. p. 165

13. A lack of ADH results in a disease known as diabetes _____. p. 168

14. _____ failure is a condition that results from the complete shutdown of the kidneys. p. 165

15. Renal _____ is a treatment used to filter the blood when the kidneys stop working. p. 167

16. Label the drawing:

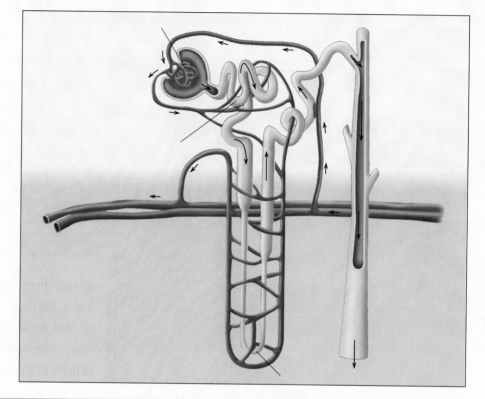

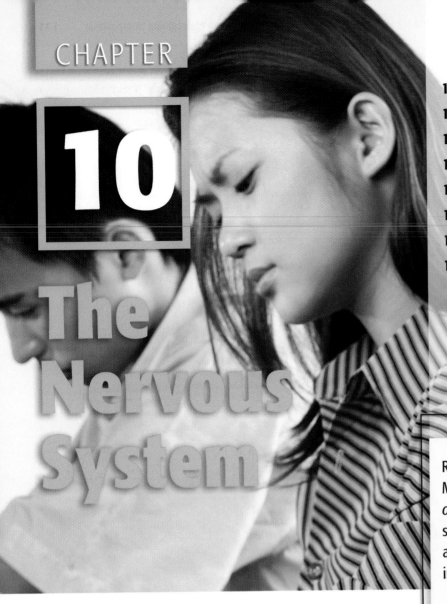

Ernest Hemingway is one of America's most celebrated writers. His novels, including *The Old Man and the Sea,* and short stories won him the Nobel Prize in literature. Despite success and widespread popularity, Hemingway was a troubled man who eventually committed suicide. What ultimately caused him to take his life no one can know, but some believe that one contributing factor was a rare and painful nervous system disorder known as *trigeminal neuralgia* (try-GEM-in-al ner-AL-gee-ah). People suffering from this disease complain of periodic, unexplained flashes of intense pain along the course of the sensory nerve that supplies the face (trigeminal nerve). A slight breeze or the pressure of a razor can set off this pain, which lasts a minute or more.

thinking critically

Researchers at the Washington University School of Medicine in St. Louis published a report in the *Journal of Neuroscience* about brain activity in people with severe depression. They compared scans of brain activity in these individuals with scans from normal individuals.

The scientists injected the subjects with radioactive oxygen, which is absorbed by metabolically active brain cells. Inside the brain, radioactive oxygen molecules emit gamma rays; these are picked up by a special device that sends signals to a computer. The computer then maps brain cell metabolic activity.

In the study under discussion, researchers compared 6 patients with depression to 18 control subjects. To compare the results, they used a computer program that combines the "brain maps" of each group to produce a composite—an average of the entire group.

The scientists found that a region of the brain known as the left prefrontal cortex was extremely active in patients suffering from depression. None of the control subjects showed a similar response. How would you interpret this data? What further studies would you perform?

Some people believe that the pain Hemingway felt may have become unbearable. Combined with other personal conflicts, it may have caused him to commit suicide.

This chapter describes the structure and function of the human nervous system. You will study the nerve that caused Hemingway so much trouble, and you will also examine ways in which the nervous system contributes to homeostasis.

10-1 An Overview of the Nervous System

> The nervous system not only controls body functions but also allows for higher functions such as thinking.

The human nervous system governs the functions of the body, exerting control over muscles, glands, and organs. It also controls heartbeat, breathing, digestion, and urination. It helps regulate blood flow as well as the concentration of chemicals in the blood. As such, the nervous system plays a major role in maintaining homeostasis.

The nervous system receives input from a large number of sources in the body. This input helps it "manage" numerous body functions. The human nervous system also provides functions not seen in other animal species. For example, the brain is the site of ideation—the formation of ideas. Our brain allows us to think about and plan for the future. It enables us to reason—that is, to judge right from wrong and logical from illogical.

The nervous system also allows us to change our environment to better serve our needs. Although actions such as draining swamps to create farmland help us, they are often detrimental to the many species that share this planet with us. In the long run, some human actions may even be detrimental to our own long-term future. Fortunately, our own hope of survival also depends on our brain—an evolutionary byproduct unrivaled in the biological world.

The Central and Peripheral Nervous Systems

> The nervous system consists of two parts, the central and peripheral nervous systems.

The human nervous system consists of three components: the brain, the spinal cord, and nerves (Figure 10-1). The brain and spinal cord constitute the **central nervous system** (CNS) and are housed in the skull and vertebral canal, respectively. Three layers of connective tissue, known as the **meninges** (men-IN-gees), surround the brain and spinal cord. The space between the middle and inner layer is filled with a liquid called cerebrospinal fluid (CSF) (sir-REE-bro-SPIE-nal), covered in more detail shortly.

Nerves consist of bundles of nerve fibers. Each fiber is part of a nerve cell, or neuron (NER-on). Nerves transport messages to and from the CNS. These nerves end on receptors. Receptors respond to a variety of internal and external stimuli. Together the nerves and sensors form the **peripheral nervous system** (**PNS**).

The brain and spinal cord receive all sensory information from the body. Right now, for instance, your brain and spinal cord are being sent a great deal of information. It comes from receptors in your body. The receptors alert you to the room tem-

perature, traffic sounds, and the touch of the page. The information is transmitted along sensory nerves. This massive inflow of information is managed by the CNS. Some information is stored in memory. Some incoming information is ignored. The brain also processes incoming information and often responds by sending nerve impulses to muscles and glands (effectors) via the nerves. The nerves therefore carry two types of information: (1) sensory impulses traveling to the CNS from sensory receptors in the body, and (2) motor impulses traveling away from the CNS to muscles and glands.

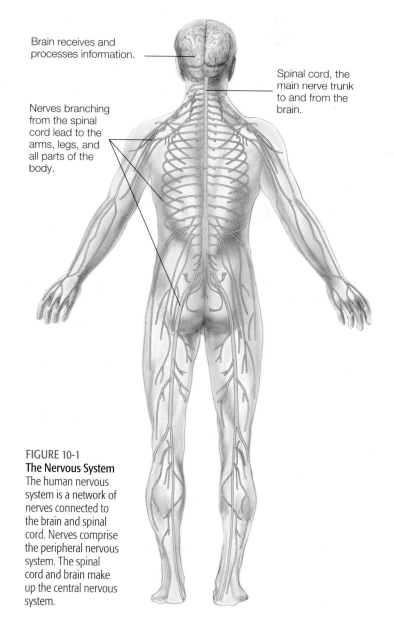

Brain receives and processes information.

Spinal cord, the main nerve trunk to and from the brain.

Nerves branching from the spinal cord lead to the arms, legs, and all parts of the body.

FIGURE 10-1
The Nervous System
The human nervous system is a network of nerves connected to the brain and spinal cord. Nerves comprise the peripheral nervous system. The spinal cord and brain make up the central nervous system.

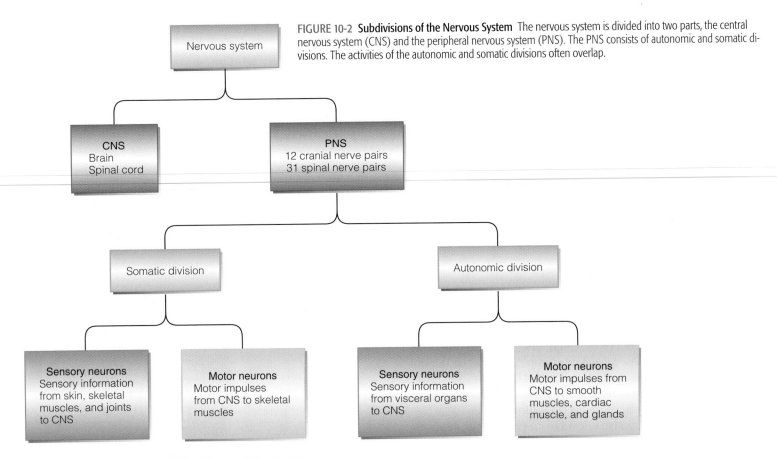

FIGURE 10-2 **Subdivisions of the Nervous System** The nervous system is divided into two parts, the central nervous system (CNS) and the peripheral nervous system (PNS). The PNS consists of autonomic and somatic divisions. The activities of the autonomic and somatic divisions often overlap.

Somatic and Autonomic Divisions of the PNS

The PNS consists of the somatic and autonomic subdivisions.

As Figure 10-2 shows, the peripheral nervous system consists of the somatic nervous system and the autonomic nervous system. The **somatic nervous system** (so-MAA-tick) is that portion of the PNS that controls voluntary functions of the body, such as the muscle contractions that cause the limbs to move.

It also controls certain involuntary reflex actions, such as the knee-jerk response.

The part of the PNS that controls the rest of the involuntary functions such as heart rate and breathing is the **autonomic nervous system** (au-toe-NOM-ick) or **ANS** (Figure 10-2). Many other functions are under the control of the autonomic nervous system, including digestion and body temperature regulation.

10-2 Structure and Function of the Neuron

The **neuron** is the fundamental structural unit of the nervous system. This highly specialized cell generates bioelectric impulses and transmits them from one part of the body to another. Such signals alert us to a variety of internal and external stimuli and permit us to respond to them.

The Structure of the Neuron

Neurons consist of a cell body and two types of processes.

Neurons come in several shapes and sizes, but they all share several characteristics. All neurons, for example, consist of a more or less spherical central portion, the cell body (Figure 10-3). It houses the nucleus, most of the cell's cytoplasm, and numerous organelles. Metabolic activities in the cell body sustain the entire neuron, providing energy and synthesizing materials necessary for proper cell function. All nerve cells contain

processes called **dendrites** (DEN-drights) that transmit bioelectric impulses toward the cell body. They also contain processes called **axons** (AXE-ons) that transmit impulses away from the cell body. When an axon reaches its destination, it often branches, giving off many small fibers. These fibers terminate in tiny swellings called **terminal boutons** (boo-TAWNS; *boutons* is the French word for buttons). Terminal boutons serve as a communication link with other neurons, muscle fibers, or glands.

The Myelin Sheath

The myelin sheath greatly increases the rate of transmission of nerve impulses.

The axons of many multipolar neurons in both the central and peripheral nervous systems are coated with a protective layer called the **myelin sheath** (MY-eh-lin) (Figure 10-4). The myelin sheath consists of many layers, and is mostly made of lipid. It appears glistening white when viewed with the naked eye. As shown in Figure 10-4, each segment of myelin is separated by a small unmyelinated segment known as a **node of Ranvier** (RON-vee-A). For reasons explained later, the myelin sheath permits nerve impulses to travel with great speed down the axons, "jumping" from node to node like a stone skipping along the surface of the water (Figure 10-4).

Although most axons are covered with myelin, some are unmyelinated. The unmyelinated axons conduct impulses much more slowly than their myelinated counterparts. As a rule, the most urgent types of information are transmitted via myelinated fibers; less urgent information is transmitted via unmyelinated fibers.

Nerve Cell Repair

Neurons lose the ability to divide.

Unlike many other cells of the body, nerve cells cannot divide. Thus, when nerve cells die as a result of injury or old age, they cannot be replaced by cell division. But not all cell death results in irreversible loss of function.

Consider what happens in the case of a stroke. A stroke is the loss of blood to a part of the brain. It can be caused by breaks in arteries of the brain. It can also be caused by blood clots in arteries of the brain that are narrowed by atherosclerosis or that are formed elsewhere but lodge in the narrowed arteries. Although blood flow to important parts of the brain may end, causing nerve cells to die, undamaged neurons elsewhere in the brain can eventually take over the function of damaged brain cells, permitting partial to complete recovery.

Nerve cells can also be damaged without permanent loss of function, but it depends on where the damage occurs. Nerves in the arm (part of the PNS), for instance, can be severed in accidents. The severed axon generally degenerates from the point of injury to the muscle or gland it supplied. The segment of the axon still attached to the cell body of the damaged neuron may elongate and replace the degenerated section. During regeneration, the axon extends along the hollow tunnel in the myelin sheath that was left by the degenerated part. Eventually, the axon reestablishes connections with the muscles or glands it once supplied, making partial or nearly complete recovery of control possible.

In the brain and spinal cord, however, severed axons cannot be repaired. Spinal cord injuries that sever nerves result in paralysis and loss of sensation in the region below the damage. Although the prognosis for spinal cord injury victims is generally poor, new procedures are being developed that could permit repair and recovery of lost function.

FIGURE 10-3 **A Neuron** The multipolar neuron resides within the central nervous system. Its multiangular cell body has several highly branched dendrites and one long axon. Collateral branches may occur along the length of the axon. When the axon terminates, it branches many times, ending on individual muscle fibers.

(a)

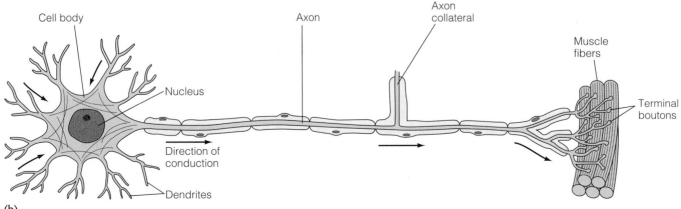

(b)

Oxygen Sensitivity of Nerve Cells

Neurons are highly susceptible to loss of oxygen.

Besides being unable to divide, nerve cells require a constant supply of oxygen and glucose for energy production. As a result, they are highly susceptible to a loss of oxygen. Thus, if the amount of oxygen flowing to the brain is drastically reduced, neurons begin to die within minutes.

To prevent brain damage from occurring in someone who has collapsed with a severe heart attack or who has drowned, rescuers must start resuscitating the victim within 4 to 5 minutes. Although victims may be revived after this crucial period, the lack of oxygen in the brain often results in permanent brain damage. Generally, the longer the deprivation, the greater the damage. The only exception is when someone has been submerged in very cold water. In such instances, resuscitation may be successful if begun within an hour. In most cases, victims recover without any detectable brain damage. Recovery is possible in such instances because cold water greatly slows brain metabolism and oxygen demand. As a result, brain cells are preserved, and brain damage is minimized or prevented.

Nerve Cell Impulses

Nerve impulses result from the flow of ions across their plasma membranes.

Nerve impulses are not like the electric current that powers computers or light bulbs, which is formed by the flow of electrons. Rather, nerve impulses are small ionic changes in the membrane of the neuron that move along the plasma membrane of a nerve cell.

To understand the **nerve impulse**, we begin by examining the plasma membrane of a neuron. We first place tiny electrodes on the outside and inside of the plasma membrane of a neuron and hook them up to a device called voltmeter, a device that measures voltage. Voltage is a measure commonly used when studying electricity. You buy batteries with different voltages that you can measure with a simple voltmeter. Voltage is rather difficult to explain. In general, it tells you the potential difference between two points. The higher the voltage, the greater the difference. In a battery, voltage is a measure of the tendency of charged particles (electrons) to flow from one pole of the battery to the other. The higher the voltage, the greater the tendency for electrons to flow through a wire connected to the poles. In the nerve cell, however, electrons do not flow from one side of the membrane to another; sodium and potassium ions do.

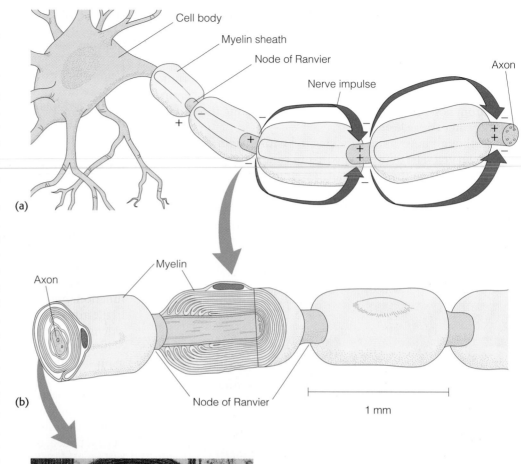

(a)

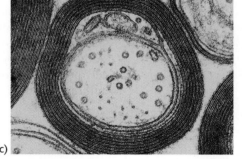

(b)

(c)

FIGURE 10-4 **The Myelin Sheath and Saltatory Conduction** (a) The myelin sheath allows impulses to "jump" from node to node, greatly accelerating the rate of transmission. (b) A drawing showing the arrangement of Schwann cell membrane in the myelin sheath. Schwann cells produce myelin during development by wrapping around axons. (c) A transmission electron micrograph of an axon in cross section showing a myelin sheath.

In neurons, the potential difference, or voltage, is a measure of the force that can drive sodium ions from one side of the membrane to the other. For now, it is important just to remember that a small voltage exists across the plasma membrane of the neuron. It is so small, in fact, that it is measured in millivolts (MILL-ee-volts). A millivolt is 1/1000 of a volt. Just to give you a point of reference, the voltage of the electricity running through the wires in your home is about 120 volts.

The potential difference in a typical neuron is about −70 millivolts. This is known as the **resting potential** because it is the voltage of a nerve cell at rest. The minus sign is added because the plasma membrane is positively charged on the outside and negatively charged on the inside.

To understand the bioelectric impulse, it is important to know that sodium ions are found in greater concentration outside the neuron. Potassium ions are found in greater concentration inside the cell. This difference is due to an active transport pump, the sodium-potassium pump, in the cell membrane of nerves. This pump consists of many membrane proteins that pump sodium ions out and potassium ions into the cell, using energy in the form of ATP.

The Action Potential

Stimulating a nerve cell increases its permeability to ions, resulting in a change in the resting potential.

The plasma membrane of a nerve cell has a built-up charge. The charge consists of sodium ions concentrated on the outside of the cell. When the neuron is stimulated, its membrane suddenly becomes more permeable to sodium ions. Neurophysiologists believe that stimulating the nerve cell causes protein pores in the plasma membrane to open. Sodium ions flow into the cell through these pores.

Electrodes implanted in a nerve cell detect the sudden inflow of positively charged sodium ions. The electrodes then register a shift in the resting potential from −70 millivolts to +30 millivolts. The change in voltage occurs at the site of stimulation and is called **depolarization**. Immediately after depolarization, the membrane returns to its previous state, which is referred to as **repolarization**.

The depolarization/repolarization of the membrane is shown graphically in Figure 10-5. This tracing is an **action potential**. It appears on a screen hooked up to the voltmeter. The action potential consists of (1) a brief upswing, depolarization, as the voltage goes from −70 millivolts to +30 millivolts, and (2) a rapid downswing, repolarization, the return to the resting potential.

The depolarization occurs so rapidly and the membrane returns to the resting state so quickly (in about 3/1000 of a second) that neurons can be stimulated in rapid succession. The quick recovery allows us to perform rapid muscle movements. Nerve cells can also transmit many impulses one after another because only a small number of sodium ions are exchanged with each nerve impulse.

While depolarization results in the rapid inflow of sodium ions, repolarization occurs as a result of two factors: (1) a sudden decrease in the membrane's permeability to sodium ions that stops the influx of sodium ions and (2) a rapid outflow of positively charged potassium ions (Figure 10-5). The efflux of potassium ions reestablishes the resting potential.

When a neuron is no longer stimulated, the sodium-potassium pump quickly reestablishes the sodium- and potassium-ion concentrations inside and outside the cell. It does this by pumping sodium ions that flowed into the cell during activation out of the axon and by pumping potassium ions that flowed out of the neuron during repolarization back in.

Nerve Cell Transmission

Nerve impulses are waves of depolarization.

Researchers have found that depolarization in one area of a neuron stimulates depolarization in adjacent regions. Depolarization in one region stimulates depolarization in neighboring regions by increasing membrane permeability. This process con-

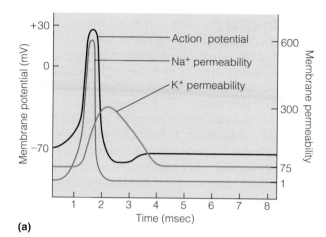

(a)

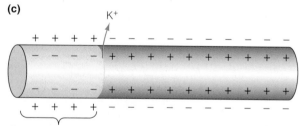

Point of stimulation

(b)

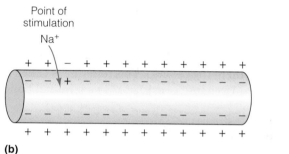

Depolarization and generation of the action potential

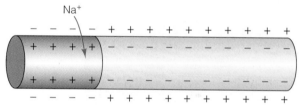

Propagation of the action potential

(c)

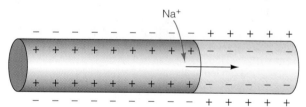

Repolarized membrane

(d)

FIGURE 10-5 **Action Potential** (a) Stimulating the neuron creates a bioelectric impulse, which is recorded as an action potential. The resting potential shifts from 270 millivolts to 130 millivolts. The membrane is said to be depolarized. This graph shows the shift in potential and the change in the permeability of sodium (Na^+) and potassium (K^+) ions, which is largely responsible for the action potential. (b) The influx of sodium ions and the depolarization that occur at the point of stimulation. (c) The impulse travels along the membrane as a wave of depolarization (gray area). (d) The efflux of potassium ions restores the resting potential, allowing the neuron to transmit additional impulses almost immediately.

tinues along the length of the axon and is responsible for the transmission of a nerve impulse along the structure.

In unmyelinated fibers in the human nervous system, nerve impulses travel like waves in water from one region to the next. In myelinated fibers, however, the depolarization "jumps" from one node of Ranvier (the section between adjacent Schwann cells) to another. Shown in Figure 10-4, impulse jumping greatly increases the rate of transmission. In fact, a nerve impulse travels along an unmyelinated fiber at a rate of about 0.5 meter per second (1.5 feet per second). In a myelinated neuron, the impulse travels 400 times faster or about 200 meters per second. That's about 450 miles per hour!

Synaptic Transmission

Nerve impulses travel from one neuron to another across synapses.

Nerve impulses travel from one neuron to another across a small space that separates them (Figure 10-6b). This juncture is called a **synapse** (SIN-apse). A synapse consists of (1) a terminal bou-

ton (or some other kind of axon terminus), (2) a gap between the adjoining neurons, the **synaptic cleft**, and (3) the membrane of the dendrite or postsynaptic cell (Figure 10-6c). The neuron that transmits the impulse is the presynaptic neuron; the one that receives the impulse is the postsynaptic neuron.

When a nerve impulse reaches a terminal bouton, depolarization of the plasma membrane of the bouton stimulates a rapid influx of calcium ions into the bouton from the extracellular fluid. Calcium ions, in turn, stimulate the release (by exocytosis) of a chemical substance stored in small vesicles in the terminal bouton. These chemicals are known as **neurotransmitters**. To date, researchers have identified approximately 30 neurotransmitters. One of the most common neurotransmitters is known as **acetylcholine** (a-SEA-tol-KO-leen). It is released by the terminal boutons of neurons that supply skeletal muscle cells.

Neurotransmitters released into the synaptic cleft bind to receptors in the plasma membrane of the postsynaptic neuron. In most instances, this stimulates a rapid increase in the permeability of the membrane of the postsynaptic cell to sodium ions. This, in turn, often creates a nerve impulse, which then travels down the new nerve cell.

Excitation and Inhibition

Neurotransmitters may excite or inhibit the postsynaptic membrane.

Although neurotransmitters stimulate the next nerve cell in the pathway, how a nerve cell responds to incoming impulses is not always so simple. In fact, in the brain a single nerve cell may have as many as 50,000 synapses. In some of these synapses, neurotransmitters stimulate the uptake of sodium ions, which slightly depolarizes the postsynaptic neuron. That is, it makes it more likely to trigger a new action potential. In other synapses, neurotransmitters make the membrane less excitable.

Whether a postsynaptic neuron fires (generates an action potential) depends on the sum of excitatory and inhibitory impulses. If the number of excitatory impulses exceeds the number of inhibitory impulses, a nerve impulse will be generated. If not, the neuron will not fire. This phenomenon provides the nervous system with a

Cell body of postsynaptic neuron

Terminal boutons of presynaptic boutons

Axon terminals

(a)

Presynaptic neuron

Direction of conduction of nerve impulse

Vesicles containing neurotransmitters

Mitochondrion

Synaptic cleft

FIGURE 10-6 The Terminal Bouton and Synaptic Transmission The arrival of the impulse stimulates the release of neurotransmitters held in membrane-bound vesicles in the axon terminals. Neurotransmitter diffuses across the synaptic cleft and binds to the postsynaptic membrane, where it elicits another action potential that travels down the dendrite to the cell body.

Postsynaptic neuron

(b)

Receptors on postsynaptic membrane bound to neurotransmitter

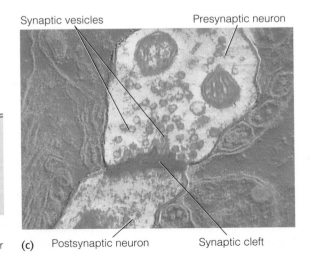

Synaptic vesicles

Presynaptic neuron

(c) Postsynaptic neuron

Synaptic cleft

way of integrating incoming information—determining a response by the kinds of input it receives.

Because only terminal boutons contain neurotransmitters and because receptors for these substances are found only in the postsynaptic membrane, transmission across a synapse occurs only in one direction.

Limiting a Neuronal Response

Neurotransmitters are quickly removed from the synapse.

Transmission across the synapse is remarkably fast, requiring only about 1 millisecond (1/1000 of a second). Synaptic transmission is also a transitory event. A short burst of neurotransmitter is released each time an impulse reaches the terminal bouton. It binds to the postsynaptic membrane, elicits a change, and is then removed from the synaptic cleft. How? Neurotransmitters can be destroyed (and hence removed) by enzymes in the synaptic cleft. Acetylcholine, for instance, is destroyed by the enzyme **acetylcholinesterase**. Neurotransmitters can also be reabsorbed by the terminal bouton for reuse or may simply diffuse away from the synapse.

Altering Synaptic Transmission

Many common chemical substances alter synaptic transmission.

The synapses are essential links in the body's system of internal communication, which is vital to homeostasis. Unfortunately, synapses are vulnerable to certain drugs and chemicals such as insecticides.

Certain insecticides, for instance, inhibit acetylcholinesterase. Because acetylcholine levels remain high in the synapses, this creates something of a nervous system overload. Unfortunately, insecticides have the same effect on people such as farm workers and pesticide applicators or others who are exposed to high levels of these toxic chemicals in the course of their work. Low doses can cause blurred vision, headaches, rapid pulse, and profuse sweating; higher doses can be fatal.

Certain **anesthetics** (an-es-THET-icks), chemicals used to deaden pain or to put people to sleep for surgery, also inhibit transmission. In so doing, they decrease transmission of pain impulses.

Many antidepressant drugs such as Prozac and Paxil decrease the uptake of the neurotransmitter serotonin in the brain. Known as *serotonin selective reuptake inhibitors*, or SSRIs, these drugs elevate serotonin levels in the brain that gradually "lifts" depression.

Caffeine and cocaine also affect synaptic transmission. Caffeine increases synaptic transmission, thus increasing overall neural activity. Cocaine, on the other hand, blocks the uptake of neurotransmitters by terminal boutons in the brain. Because the neurotransmitter remains in the synaptic cleft for a longer time, neural activity is greatly increased. This increase results in a heightened state of alertness and euphoria, commonly known as a "high," which lasts for 20 to 40 minutes. Euphoria, however, is followed by a period of depression and anxiety, which causes the user to seek another high. Excessive cocaine use can result in serious mental derangement—in particular, paranoid delusions that others are out to get the user. In this state, heavy users may become violent. For a discussion of the effects of recreational drug use see Health Note 10-1.

10-3 The Spinal Cord and Nerves

The **spinal cord** is part of the central nervous system (Figure 10-7). This long, ropelike structure—about the diameter of one's little finger—extends from the base of the brain to the lower back. It is housed in the **vertebral canal** (VER-teh-brill), a long tunnel formed by the bones of the spine, the vertebrae (ver-tah-BREE).

Spinal Nerves

The spinal cord gives off numerous nerves that supply the body.

The spinal cord gives off nerves along its course that supply the skin, muscles, bones, and organs of the body. As Figure 10-8 shows, the central portion of the spinal cord is an H-shaped zone of gray matter. Gray matter contains nerve cell bodies and is so named because it appears gray to the naked eye. Surrounding the gray matter are fiber tracts consisting of axons and a much smaller number of dendrites that travel up and down the spinal cord, carrying information to and from the brain. The fiber tracts form the white matter, whose characteristic color comes from the myelin sheaths of the axons coursing through it.

Sensory and Motor Nerves

The nerves attached to the spinal cord carry sensory and motor impulses.

Nerves that connect to the spinal cord are known as **spinal nerves**. They are part of the peripheral nervous system. These nerves carry sensory information to the spinal cord, for example, about pain, touch, pressure, and temperature. The spinal nerves also carry motor impulses from the spinal cord to muscles and various organs. Some nerves are strictly motor and some are strictly sensory, but many transmit both motor and sensory impulses.

Each spinal nerve has two roots, the dorsal (back side) and ventral (belly side). Both attach to the spinal cord (Figure 10-8). The dorsal root is the inlet for sensory information traveling to the spinal cord. On each dorsal root is a small aggregation of nerve cell bodies, the dorsal root ganglion (GANG-lee-on). The dorsal root ganglia house the cell bodies of sensory neurons. As Figure 10-8 shows, the dendrites of these cells conduct impulses from receptors in the body to the dorsal root ganglia; the axons, in turn, carry the impulse into the spinal cord along the dorsal root.

::::healthnote

10-1   Recreational Drug Use: Is It a Good Idea?

FIGURE 1   A rave

Jason Wiley was a popular young man in his college dormitory. He was friendly, generous, and fun to be with. Like others of his age, Jason sometimes used recreational drugs like Special K or ecstasy. But one day, Jason didn't return from his night of partying. When his friends found him, he was dead in the back seat of another friend's car. He had died from an overdose of Special K and methamphetamine.

Jason is just one of millions of youths the world over who uses recreational drugs. The most commonly used are (1) methamphetamine, also known as meth, crystal meth, speed, ice, and crystal; (2) phencyclidine, also called PCP, angel dust, supergrass, killer weed, embalming fluid, and rocket fuel; (3) ecstasy, (4) ketamine, also referred to as Special K or simply k; (5) cocaine; (6) LSD; (7) gamma hydroxybutyrate or GHB, liquid x, Georgia home boy, goop, gamma-oh, and grievous body harm; and (8) Ritalin, also known as Vitamin R.

Many of these drugs, such as methamphetamine, ketamine, ecstasy, GHB, cocaine, and Ritalin are stimulants. They give the user a sense of euphoria, increased well-being. These drugs increase heart rate, blood pressure, body temperature, and rate of breathing. Some are stimulants *and* hallucinogens, substances that cause the user to perceive sights, sounds, and objects that are not present. Ketamine, ecstasy, Ritalin, and GHB are examples. Some recreational drugs are addictive. Cocaine and Ritalin are examples.

All of these drugs have dangerous side effects. Methamphetamine, the recreational drug whose use is growing the fastest, is smoked, snorted, injected, or taken orally. This drug produces a state of hyperactivity, euphoria, and increased energy. High doses or chronic use of methamphetamine may cause increased nervousness, irritability, and psychosis, an altered state of mind. Chronic abusers may become paranoid and may suffer auditory and visual hallucinations, as well as erratic and violent behavior. Withdrawal from the drug can cause severe depression.

PCP is considered by many to be the most dangerous of all recreational drugs. PCP use results in frequent bad trips, psychotic reactions, and extreme violence in its users. Moderate doses cause a person to feel detached, distant, and estranged from his or her surroundings. Loss of coordination, slurred speech, and numbness are commonly observed. Auditory and visual hallucinations, severe mood disorders, and amnesia may occur. Some users of PCP report acute anxiety and feelings of impending doom. In others, severe mental disorders are observed.

Ecstasy, which is frequently used with other drugs, is often distributed at raves, night clubs, and rock concerts. This drug produces very profound positive feelings, empathy for others, and relaxation, both physical and mental. Ecstasy suppresses the need to eat, drink, or sleep, so individuals can participate in two- to three-day parties. But ignoring these basic needs can result in severe dehydration and exhaustion. Ecstasy also produces nausea, hallucinations, chills, sweating, increased body temperature, tremors, muscle cramping, and blurred vision. After-effects include anxiety, paranoia, and depression.

Overdoses of ecstasy result in high blood pressure, faintness, and panic attacks. In severe cases, it may cause loss of consciousness, seizures, and a drastic increase in body temperature. Death may occur from heart failure or extreme heat stroke. Although it less addictive than cocaine and heroin, the drug does ensnare users in a cycle of de-

pendency. New studies show that chronic use causes brain damage and may impair memory.

Cocaine, which produces a euphoric state, is addictive. It has destroyed many thousands of families and careers. Overdoses cause respiratory failure, heart failure, and cerebral hemorrhages, bleeding into the brain.

GHB, also commonly used at raves with ketamine and ecstasy, can cause seizures, severe respiratory depression, and coma. It is also called the date rape drug, because men have used it to incapacitate women, leaving them vulnerable to sexual assault.

Ketamine causes intense, but short-lived hallucinations; it is stronger and more profound than LSD, but its effects last only an hour or so. Users commonly report experiences of the mind leaving the body and floating in space, or even visions of death. Heavy use results in serious danger, including risk of suicide. Ketamine is frequently mixed with several other drugs, including heroin, cocaine, or ecstasy. Combining ketamine with these drugs produces an extremely dangerous mix.

Ritalin is one of the newest recreational drugs. Ritalin is normally prescribed to children with attention deficit disorder. Such children are fidgety and full of energy. They cannot concentrate their attention on tasks for very long and can be quite disruptive in classes. Because of this, many doctors prescribe Ritalin.

But Ritalin is also ground into a powder and snorted. Because it is a stimulant, like many other recreational drugs, those who snort the drug can stay alert and active for long periods, partying all night if the desire strikes them. Ritalin's stimulant effect also reportedly maintains alertness, so students in college or high school can study all night long to cram for tests.

Generally considered much safer than other drugs such as cocaine, Ritalin has a dark side. Overdoses of Ritalin can cause fatal heart attacks and strokes. Some health care workers warn that Ritalin tablets contain additives that were not meant to be snorted; because of this, they say the drug may be even more dangerous than cocaine.

No matter how strong the allure of recreational drugs is, it should be remembered that they're potentially dangerous, even lethal substances.

www.jbpub.com/humanbiology/5e

Visit Human Biology's Internet site for links to web sites offering more information on this topic.

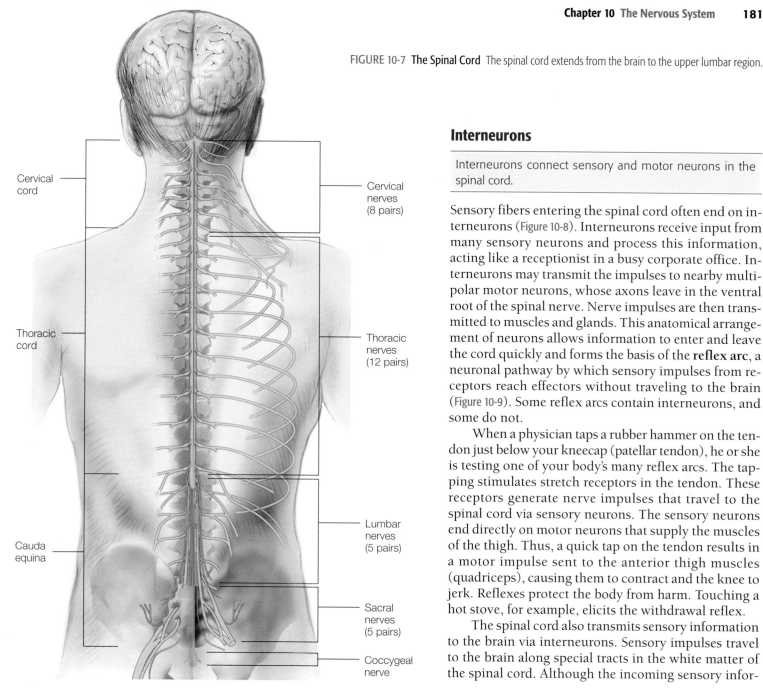

FIGURE 10-7 **The Spinal Cord** The spinal cord extends from the brain to the upper lumbar region.

Cervical cord

Cervical nerves (8 pairs)

Thoracic cord

Thoracic nerves (12 pairs)

Lumbar nerves (5 pairs)

Cauda equina

Sacral nerves (5 pairs)

Coccygeal nerve

Interneurons

> Interneurons connect sensory and motor neurons in the spinal cord.

Sensory fibers entering the spinal cord often end on interneurons (Figure 10-8). Interneurons receive input from many sensory neurons and process this information, acting like a receptionist in a busy corporate office. Interneurons may transmit the impulses to nearby multipolar motor neurons, whose axons leave in the ventral root of the spinal nerve. Nerve impulses are then transmitted to muscles and glands. This anatomical arrangement of neurons allows information to enter and leave the cord quickly and forms the basis of the **reflex arc**, a neuronal pathway by which sensory impulses from receptors reach effectors without traveling to the brain (Figure 10-9). Some reflex arcs contain interneurons, and some do not.

When a physician taps a rubber hammer on the tendon just below your kneecap (patellar tendon), he or she is testing one of your body's many reflex arcs. The tapping stimulates stretch receptors in the tendon. These receptors generate nerve impulses that travel to the spinal cord via sensory neurons. The sensory neurons end directly on motor neurons that supply the muscles of the thigh. Thus, a quick tap on the tendon results in a motor impulse sent to the anterior thigh muscles (quadriceps), causing them to contract and the knee to jerk. Reflexes protect the body from harm. Touching a hot stove, for example, elicits the withdrawal reflex.

The spinal cord also transmits sensory information to the brain via interneurons. Sensory impulses travel to the brain along special tracts in the white matter of the spinal cord. Although the incoming sensory infor-

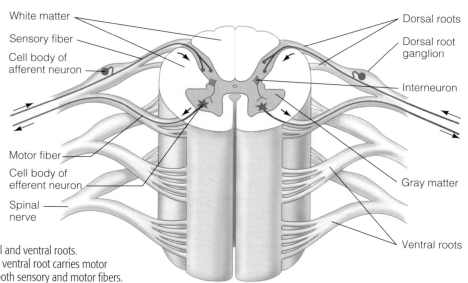

White matter

Sensory fiber

Cell body of afferent neuron

Motor fiber

Cell body of efferent neuron

Spinal nerve

Dorsal roots

Dorsal root ganglion

Interneuron

Gray matter

Ventral roots

FIGURE 10-8 **The Spinal Cord and Dorsal Root Ganglia**
Spinal nerves are attached to the spinal cord by two roots, the dorsal and ventral roots. The dorsal root carries sensory information into the spinal cord. The ventral root carries motor information out of the spinal cord. The spinal nerve often contains both sensory and motor fibers.

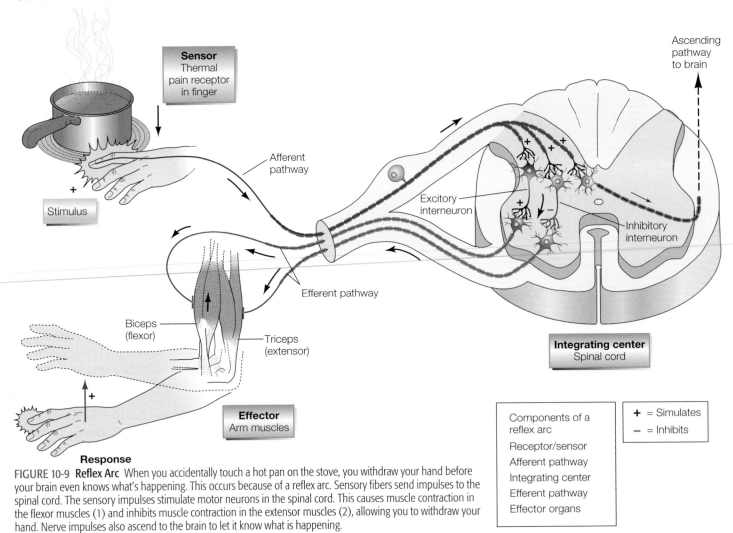

FIGURE 10-9 Reflex Arc When you accidentally touch a hot pan on the stove, you withdraw your hand before your brain even knows what's happening. This occurs because of a reflex arc. Sensory fibers send impulses to the spinal cord. The sensory impulses stimulate motor neurons in the spinal cord. This causes muscle contraction in the flexor muscles (1) and inhibits muscle contraction in the extensor muscles (2), allowing you to withdraw your hand. Nerve impulses also ascend to the brain to let it know what is happening.

mation may elicit a reflex, the brain is still informed of the problem, allowing for appropriate follow-up action.

Damage to the Spinal Cord

Injury of the spinal cord can cause permanent damage.

Automobile crashes or even bad falls can sever fiber tracts in the spinal cord. The extent of damage depends on the severity of the accident and the location of the damage. For example, if both the sensory and motor tracts of the spinal cord are severed, all of the muscles supplied by nerves below the injury be-

come paralyzed and unable to contract voluntarily. In addition, the portion of the body below the injury loses all sensation. If the spinal cord injury occurs high in the neck (above the fifth cervical vertebra), it cuts the nerve fibers traveling to the muscles that control breathing. These muscles become paralyzed, and the person dies quickly. This is how a hangman's noose kills its victim. Damage to the cord just below the fifth cervical vertebra does not affect breathing, but it paralyzes the legs and arms. This condition is known as quadriplegia (QUAD-reh-PLEE-gee-ah). If the spinal cord injury occurs below the nerves that supply the arms, the result is paraplegia (PEAR-ah-PLEE-gee-ah), paralysis of the legs.

10-4 The Brain

The **brain** is part of the central nervous system and is housed in the skull, a bony shell that protects it from injury. About the size of a cantaloupe, the human brain is an extraordinary organ, the product of many millions of years of evolutionary trial and error. Our brains have given us art and music, remarkable feats of engineering, and abstract reasoning, not to mention the ability to ponder right and wrong and to remember a vast amount of information.

Cranial Nerves

Cranial nerves are attached to the brain and supply structures in the head and upper body.

Like the spinal cord, the brain gives off numerous nerves called **cranial nerves** (Figure 10-10). The cranial nerves supply structures of the head and several key body parts such as the heart and diaphragm. The trigeminal nerves mentioned in the introduction are one of the 12 pairs of cranial nerves. Cranial nerves carry motor and sensory information.

The Cerebrum

The cerebral hemispheres function in integration, sensory reception, and motor action.

The brain is housed in the skull, a bony shell that protects it from injury. The largest and most conspicuous part of the brain is the **cerebrum** (sir-EE-brum). It constitutes about 80% of the total brain mass (Figure 10-11).

The cerebrum is divided into two halves, the right and left **cerebral hemispheres**. Each hemisphere has a thin outer layer of gray matter, the **cerebral cortex** (Figure 10-12). It contains the cell bodies of numerous multipolar neurons. Lying beneath the cerebral cortex is a thick central core of white matter containing bundles of myelinated axons that give it a white appearance. The axons of the white matter carry impulses from the cerebral cortex to the spinal cord, enabling information to flow from the brain to motor neurons that control many of the body muscles. Some axons also transmit impulses from one part of the cerebral cortex to another, permitting the integration of the activities of various parts of the cortex. As shown in Figure 10-13, the cerebral cortex is thrown into numerous folds.

The cerebral hemispheres are divided into four major portions known as *lobes* (Figure 10-11). Each of these lobes houses very specific functions.

As a general rule, the various areas of the cerebral cortex can be broadly classified into three types: motor, sensory, and association. The motor cortex stimulates muscle activity. The sensory cortex receives sensory stimuli, and the association cortex integrates information, bringing about coordinated responses.

The Primary Motor Cortex

The primary motor cortex controls voluntary movement.

The **primary motor cortex** occupies a single ridge on each hemisphere, just in front of a deep groove in the cerebral cortex, known as the *central sulcus* (Figure 10-13). This area controls voluntary motor activity—for example, the muscles in your hand that are turning the pages of your book.

The neurons in the primary motor cortex are arranged in a very specific way, so that each region of the motor cortex controls a specific body part. The neurons that control the muscles of the toes, for instance, are located in the uppermost region of the primary motor cortex. Hip muscle control occurs below that. Muscles of the hand are controlled by neurons located even lower.

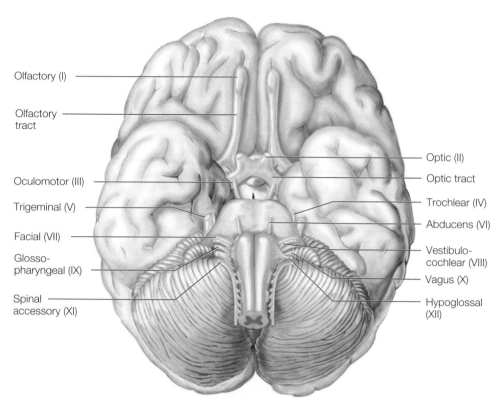

FIGURE 10-10 **Cranial Nerves** The 12 pairs of cranial nerves arise from the underside of the brain and brain stem.

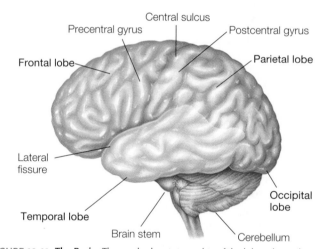

FIGURE 10-11 **The Brain** The cerebral cortex consists of the lobes shown here. The lobes, in turn, can be divided into sensory, association, and motor areas (not shown).

To bring about a voluntary movement, a conscious thought stimulates the neurons of the primary motor cortex to generate an impulse. It travels from the brain down the spinal cord to the motor neurons in the spinal cord. The axons of these neurons exit the spinal cord and terminate on the muscles.

In front of the primary motor area is the premotor cortex. The premotor cortex is also involved in controlling muscle contraction. However, the movements that it controls are fairly complex, such as the fingering required to play a musical instrument or type on a computer keyboard.

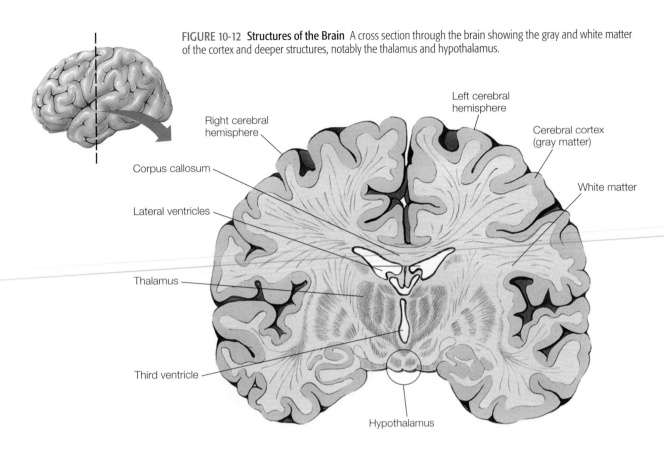

FIGURE 10-12 **Structures of the Brain** A cross section through the brain showing the gray and white matter of the cortex and deeper structures, notably the thalamus and hypothalamus.

The Primary Sensory Cortex

The primary sensory cortex receives sensory information from the body.

Just behind the central sulcus is a ridge of tissue running parallel to the primary motor area, known as the **primary sensory cortex**. It is the final destination of many sensory impulses traveling to the brain. As in the primary motor area, different parts of this ridge correspond to different parts of the body. When an impulse arrives at specific region of the primary sensory cortex it produces a sensation we interpret as pain, light touch, or pressure. Figure 10-13 also shows patches of sensory cortex for hearing (primary auditory cortex) and vision (primary visual cortex). Separate regions (not shown) exist for taste and smell.

The Association Cortex

The association cortex is the site of integration and complex intellectual activities.

The **association cortex** consists of large expanses of cerebral cortex where integration occurs. In the frontal lobe is a region of the association cortex known as the *prefrontal cortex*. It houses complex intellectual activities such as planning and ideation. This region also modifies behavior, conforming human actions with social norms.

Another important association area lies posterior (behind) to the primary sensory cortex. It interprets sensory information and stores memories of past sensations. Other association areas interpret language in written and spoken form.

Unconscious Functions

Unconscious functions are housed in the cerebellum, hypothalamus, and brain stem.

Consciousness resides in the cerebral cortex. However, many body functions occur at an unconscious level, among them heartbeat, breathing, and many homeostatic functions. One region of the brain that controls unconscious actions is the **cerebellum** (sar-ah-BELL-um), the second largest structure of the brain. As Figure 10-11 shows, it sits below the cerebrum on the brain stem.

The Cerebellum

The cerebellum controls muscle synergy and helps maintain posture.

The cerebellum plays several key roles. One of them is synergy, the coordination of skeletal muscle. The cerebellum ensures that opposing sets of muscles work together to create a smooth motion.

Problems arise if the cerebellum is damaged during childbirth—for example, if the baby's blood supply is interrupted, which sometimes happens when the umbilical cord is wrapped around the baby's neck. This reduces oxygen supply to the brain and can permanently damage the cerebellum. The result is a loss of synergistic control of the skeletal muscles. Mild damage generally causes a slight rigidness and moderately jerky motions. More severe damage causes serious impairment, with body motions becoming extremely jerky and simple tasks

FIGURE 10-13 **Functional Regions of the Cortex** (a) The cerebral cortex has three principal functions: receiving sensory input, integrating sensory information, and generating motor responses. Special sensory areas handle vision, smell, taste, and hearing. (b) A pet scan reveals the locations of increased blood flow in the brain during performance of certain tasks.

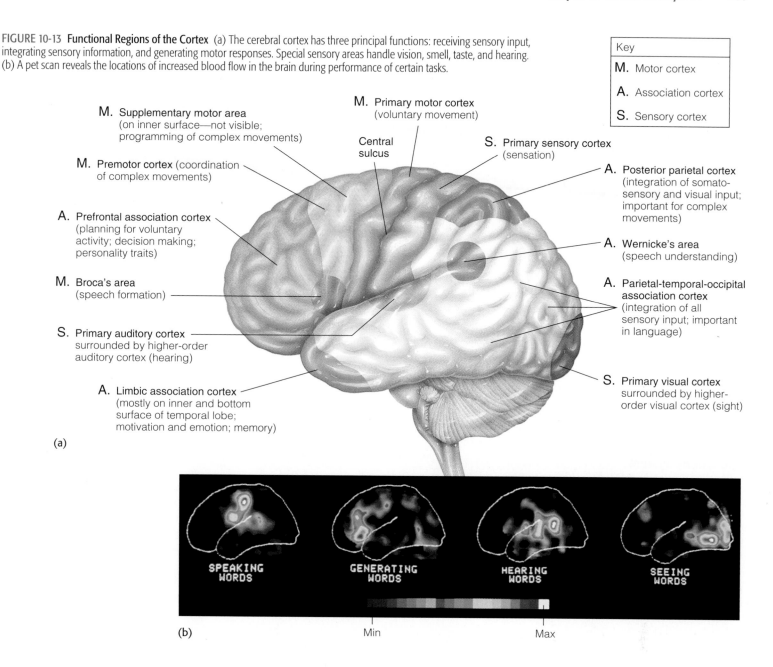

The Thalamus

The thalamus is a relay center.

Just beneath the cerebrum is a region of the brain called the **thalamus** (Figure 10-12). The thalamus is a relay center. It receives all sensory input, except for smell, then relays it to the sensory and association cortex.

requiring several attempts. This condition is known as *cerebral palsy* (PALL-zee).

Besides ensuring the smooth, coordinated contraction and relaxation of muscles, the cerebellum maintains posture. Recently, scientists have found that the cerebellum may also coordinate higher brain function, particularly thinking.

The Hypothalamus

The hypothalamus controls many autonomic functions involved in homeostasis.

Beneath the thalamus is the **hypothalamus** (high-poe-THAL-ah-muss; *hypo* = under). It consists of many groups of nerve cells. These are known as the *nuclei*, not to be mistaken for the nuclei in cells (Figure 10-14). Hypothalamic nuclei control a variety of autonomic functions, which help to maintain homeostasis. Appetite and body temperature, for example, are controlled by hypothalamic nuclei, as are water balance, blood pressure, and sexual activity.

The hypothalamus is a primitive brain center. One of its best-known functions is the control of the pituitary gland, an endocrine gland that regulates many body functions through the hormones it releases (Chapter 13).

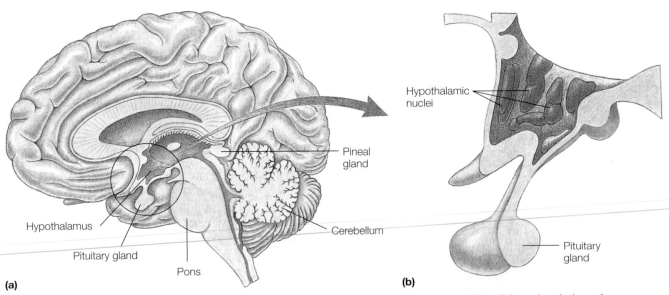

(a)

(b)

FIGURE 10-14 **The Hypothalamus** (a) The hypothalamus is clearly visible in a cross section of the brain. (b) The hypothalamus is at the base of the brain, just above the pituitary gland. It regulates many autonomic functions and plays a particularly important role in controlling the release of hormones by the pituitary.

The Limbic System

The limbic system is the site of instinctive behavior and emotion.

Instincts and several other functions reside in a complex array of structures called the **limbic system** (Figure 10-15). The limbic system operates in conjunction with the hypothalamus.

Instincts are among the most fundamental responses of organisms. They include the protective urge of a mother and the **fight-or-flight** response an animal experiences in the face of adversity.

The limbic system also plays a role in emotions—fear, anger, and so on. When researchers stimulate electrodes placed in some areas of the limbic system, they can elicit primitive rage and in other areas, calmness. Stimulation of specific regions within the limbic system of humans may elicit sensations patients describe as joy, pleasure, fear, or anxiety, depending on the site of stimulation. For a discussion of depression and its effects on emotions, see Health Note 10-2.

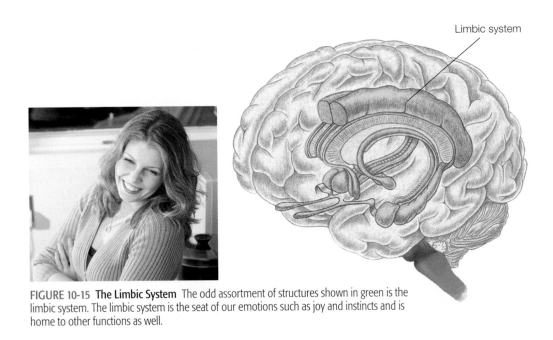

FIGURE 10-15 **The Limbic System** The odd assortment of structures shown in green is the limbic system. The limbic system is the seat of our emotions such as joy and instincts and is home to other functions as well.

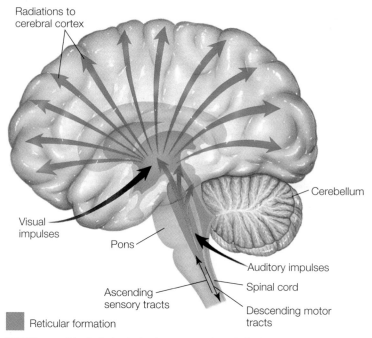

Radiations to cerebral cortex

Visual impulses

Pons

Ascending sensory tracts

Cerebellum

Auditory impulses

Spinal cord

Descending motor tracts

■ Reticular formation

FIGURE 10-16 The Reticular Activating System The reticular formation resides in the brain stem, where it receives input from incoming and outgoing neurons. Fibers projecting from the reticular formation to the cortex constitute the reticular activating system.

The Brain Stem

Many basic body functions are controlled by the brain stem.

The brain stem connects the brain to the spinal cord and gives rise to most of the cranial nerves. Like the hypothalamus, the brain stem controls basic body functions such as heart rate, blood pressure, and breathing. It also regulates swallowing, coughing, vomiting, and many digestive functions. The hypothalamus and brain stem often work in concert.

Besides containing control regions (nuclei), the brain stem is a main thoroughfare for information traveling to and from the brain. Nerve fibers passing through the brain stem, however, frequently give off branches that terminate in a special region of the brain stem known as the *reticular formation* (reh-TICK-you-ler) (Figure 10-16).

The **reticular formation** receives all incoming and outgoing information, monitoring activity like a security guard at a doorway. Nerve impulses are then sent to the cerebral cortex via special nerve fibers. These impulses activate neurons in the cortex. These nerve fibers comprise the **reticular activating system** (RAS). The reticular formation and RAS are like the alarm clock of the central nervous system. They help maintain wakefulness or alertness. The RAS not only keeps us alert during the day, it can prevent us from sleeping at night. Pain impulses, say from a bad sunburn, travel to the RAS, then on to the cortex, keeping one awake at night.

Cerebrospinal Fluid

Cerebrospinal fluid cushions the CNS.

The brain and spinal cord are housed in the skull and spinal column, respectively. These bony fortresses protect those vital and delicate organs. Additional protection is provided by a watery liquid, called **cerebrospinal fluid** (CSF) that surrounds the brain. Similar in composition to blood plasma and interstitial fluid, CSF is found between the inner and outer layers of the meninges. CSF also fills internal cavities in the brain and spinal cord (Figure 10-17a).

Continually produced, excess CSF is drained into the bloodstream. If the removal is impaired in a newborn, CSF accumulates in the brain, causing the head to enlarge and produces a condition known as *hydrocephalus* (HIGH-drow-CEFF-eh-luss; literally, "water on the brain") (Figure 10-17b). As the ventricles fill with fluid, the cortex thins and damages brain cells. If detected early, this condition can be cured.

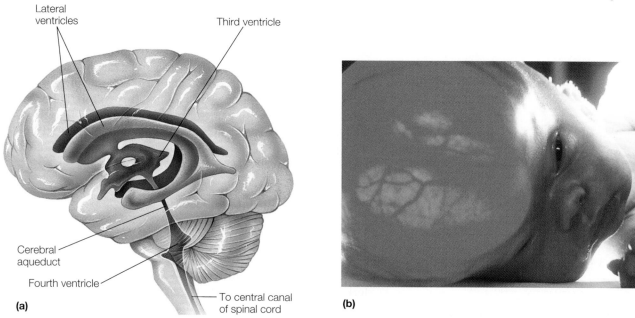

Lateral ventricles

Third ventricle

Cerebral aqueduct

Fourth ventricle

To central canal of spinal cord

(a)

(b)

FIGURE 10-17 Ventricles (a) The brain contains four internal cavities, called ventricles, which are filled with a plasma like fluid known as cerebrospinal fluid. (b) Hydrocephalus results from a blockage in the ventricles, which causes CSF to build up, thinning the cortex and causing severe brain damage.

⣿ healthnote

10-2 Feeling Blue: Understanding and Treating Depression

All people feel sad from time to time. The reasons for such sadness are many: the death of a family member or a friend; the loss of a job; the loss of a friendship; or even poor performance on an exam. All of these adversities—and a great many more—can bring on feelings commonly called depression.

To psychologists, occasional and transient feelings of sadness are a normal part of life. Psychologists reserve the term *depression* for a condition characterized by one of several symptoms, including feelings of intense sadness evoked by even minor stimuli, things that people who are not depressed seem to cope with quite well.

People suffering from depression may not feel sad but rather experience persistent numbness or emptiness. And, they often complain of feeling little or no pleasure in things that once brought them happiness. Depression can be manifested in anxiety, feelings of hopelessness, pessimism, and a general sense of worthlessness.

People suffering from depression may feel irritable and restless. They may have difficulty concentrating and making decisions. Depressed individuals may feel helpless and often suffer from sleep disorders—waking up in the middle of the night or oversleeping (see **Health Note 4-2** for a discussion of sleep disorders). Some people can't eat. Others overeat. They may complain of physical symptoms such as chronic pain and headaches that do not respond to treatment. Thoughts of suicide become prevalent.

Depression varies in intensity from mild to severe, and in duration from short term to long term. In its most extreme forms, depression can be life-threatening because severe depression can lead to suicide. Depression also comes in many forms. In this Health Note, we examine two of the many clinically recognizable forms: major depression and a less severe type.

Major depression is so severe that it often incapacitates a person. Individuals suffering from depression are unable to work, sleep, eat, or enjoy anything. Lesser depressions tend to impair work, sleep, eating, and other functions. In other words, it does not "cripple" a person; it just keeps him or her from functioning at full capacity.

FIGURE 1

FIGURE 10-18 **Electroencephalogram** Several sample tracings of brain waves during different activities and stages of sleep.

Because CSF bathes the CNS, examining small amounts of this fluid can provide physicians a means of detecting infections in the brain, spinal cord, and meninges. Samples of CSF are obtained by inserting a needle between the third and fourth lumbar vertebrae. CSF is then withdrawn.

One dangerous condition detected by this method is **meningitis** (men-in-JI-tiss), an infection of the meninges caused by certain viruses or bacteria. Meningitis usually begins with an infection in the respiratory system. In adults, meningitis is characterized by fever, headache, and nausea. Patients often vomit and complain of a stiff neck and an inability to tolerate bright light (photophobia). In babies and children, the symptoms are often less obvious, causing parents to overlook a potentially fatal condition until it is too late. In both cases, meningitis requires immediate medical attention.

Measuring Electrical Activity of the Brain

Electrical activity of the brain varies depending on activity level or level of sleep.

Electrodes applied to different parts of the scalp detect underlying electrical activity in the brain and produce a tracing known as an *electroencephalogram* (ee-LECK-trow-en-CEFF-eh-low-GRAM) or EEG. Some sample tracings of the electrical activity at different times are shown in Figure 10-18. Note that

The causes of depression are something of a mystery to medical science. Researchers know that it results from a change in brain chemistry—involving the level of certain neurotransmitters such as serotonin or acetylcholine. The same changes occur during periodic bouts of sadness. But in patients suffering from clinical depression, the changes are not reversed, even after the triggering event or stimulus is eliminated. New stresses may emerge. A person may become stressed about feeling unhappy all of the time. Lack of sleep may put emotional stress on an individual. Depression can tax relationships, creating an additional stimulus for depression. Feeling down often kills the impulse to get outside for fresh air, sunshine, or exercise—all of which can counteract depression.

When individuals become aware of their condition, most can pinpoint an event or several events responsible for their depression. Most stimuli seem to have one of three features in common: the loss of self-determination (the ability to make choices); the loss of empowerment (feeling you have power to make choices that make a difference); and the loss of self-confidence. In some instances, however, depression appears to have no link to a person's situation. The supposedly spontaneous changes in brain chemistry apparently occur without stimulus—or at least none that can be currently identified.

Although depression can be treated very effectively, many people slip into this state without even realizing it, and sink deeper and deeper in depression unaware that they are spiraling downward and are in need of help. Friends, relatives, and family physicians can help individuals realize that they are depressed.

Interestingly, in most instances, the chemical changes in the brain seem to self-correct without medical treatment, although this usually takes one to three years. For those who don't want to wait that long, psychological therapy, which addresses the root causes, can be helpful. Prescription medicines (antidepressants) can also help immensely.

Many antidepressants such as Paxil, Luvox, Prozac, and Remeron, are currently available, and they are very effective. In fact, two-thirds of those people treated by any one medication respond favorably. Those who fail to respond to one medication have a great chance of responding to another one.

Antidepressants work by elevating levels of neurotransmitters in the brain, which, for reasons unknown, seems to relieve depression. Antidepressants are not "happy" pills. They don't make one euphoric. They simply lift the feeling of depression.

Antidepressants take a while to work, usually two to three weeks, but sometimes up to two months, and have some annoying side-effects. Two of the most commonly used antidepressants, Prozac and Paxil, for instance, decrease sexual interest. Both cause nausea in about 20% of the cases and insomnia in about 15% of the patients. They often cause dry mouth, too.

There are no safeguards against depression. But eating right, getting plenty of exercise, balancing work with pleasure, and having a healthy psychological framework can all help us reduce the likelihood of this condition.

www.jbpub.com/humanbiology/5e

Visit Human Biology's Internet site for links to web sites offering more information on this topic.

the type of brain wave recorded depends on the level of cortical activity. The waves often appear irregular, but distinct patterns can sometimes be observed, especially during the different phases of sleep.

Because diseases or injuries of the cerebral cortex often give rise to altered EEG patterns, EEGs are used to diagnose brain dysfunction. Perhaps the most common disorder is epilepsy (EP-eh-LEP-see). Epilepsy is characterized by periodic seizures that occur when a large number of neurons fire spontaneously. The seizures can be accompanied by involuntary spasms of skeletal muscles and alterations in behavior. In about two-thirds of the cases, patients have no identifiable structural abnormality in the brain. In the remaining cases, the seizures can usually be traced to brain damage at birth, severe head injury, or inflammation of the brain. In some instances, epileptic seizures are caused by brain tumors.

Headaches

Headaches have many causes.

No discussion of the brain would be complete without considering headaches, the most common form of pain. Many headaches are caused by tension—sustained tightening of the muscles of the head and neck when a person is nervous, stressed, or tired. Headaches also commonly result from swelling of the membranes lining the sinuses (cavities inside the bones of the skull surrounding the nasal cavity). Such swelling usually occurs because of sinus infections or allergies. Eyestrain is another common cause of headaches, as is the dilation of cerebral blood vessels, which may be associated with high blood pressure or excessive alcohol consumption. Very serious headaches, however, may result from increased pressure inside the skull caused by a brain tumor or internal bleeding. Inflammation caused by an infection of the meninges (meningitis) or the brain itself (encephalitis), although rare, also produce intense headaches that require immediate attention.

Less threatening, but still extraordinary, are migraines. A migraine headache is a recurrent, rather severe headache. The often throbbing pain frequently occurs on one side of the head and lasts from 4 to 72 hours. In some individuals, migraines are brought about by eating certain foods such as chocolate or drinking particular beverages such as wine. Migraines may be aggravated by movement or physical activity and are typically accompanied by nausea, vomiting, and sensitivity to light and sound.

Migraines affect about one in ten people and are often hereditary. They can strike anyone at any age but are most common in young adult women. Although a cure is not known, there are treatments for stopping a migraine once it has started.

10-5   The Autonomic Nervous System

The ANS controls many body functions not under conscious control.

As noted earlier, the nervous system consists of two parts: the peripheral nervous system and the central nervous system. The PNS consists of a somatic division and an autonomic division. Each division contains motor and sensory neurons (Figure 10-2).

The somatic division consists of the nerves that receive sensory information from the skin, muscles, and joints and transmit motor impulses from the CNS back to these structures. The autonomic division is that part of the peripheral nervous system that transmits sensory information from body organs to the CNS. It also transmits motor impulses to smooth muscle, cardiac muscle, and glands—structures not under conscious control. Together, these sensory and motor nerves form the autonomic nervous system (ANS).

The ANS functions automatically, usually at a subconscious level. It helps to regulate internal organs through controls on smooth muscle, cardiac muscle, and glands.

The autonomic nervous system has two subdivisions: the sympathetic and the parasympathetic. The sympathetic division of the ANS functions in emergencies and is largely responsible for the **fight-or-flight response**. This response occurs when an individual is startled or faced with danger. A sudden scare results in sympathetic nerve impulses that bring about a whole host of responses, preparing us to fight or flee the scene. One result is a rapid increase in heart rate. The fight-or-flight response also results in pupil dilation and an increase in breathing rate. Dilation of the pupils lets more light in, and presumably helps us see better. Increased breathing delivers more oxygen to the blood. Finally, the fight-or-flight response results in a decrease in blood flow to the intestines, diverting blood from digestion, and an increase in blood flow to the skeletal muscles needed for fighting off danger or making a rapid retreat.

In contrast, the parasympathetic division of the ANS brings about internal responses associated with the relaxed state. It therefore reduces heart rate, contracts the pupils, and promotes digestion.

Most organs are supplied with both parasympathetic and sympathetic fibers. As a general rule, these fibers have opposite effects. One stimulates activity, and the other reduces activity. This provides the body with a means of fine-tuning organ function.

10-6   Learning and Memory

Learning is the acquisition of new information and skills; memory is the storage and recall of that information.

Learning is a process in which an individual acquires and retains knowledge and skills. Learning usually results from experience, instruction, or some combination of the two. It depends on one's ability to store and retrieve information in the brain, a phenomenon called **memory**.

Newly acquired information is first stored in short-term memory. **Short-term memory** holds information for periods of seconds to hours. The phone number you just looked up or the name of a new acquaintance are stashed in short-term memory. Cramming for tests places a lot of information into short-term memory. Unfortunately, soon after the test is over, the information fades into oblivion.

Long-term memory holds information for much longer periods—from days to years. It also has a much greater storage capacity than short-term memory. Transferring information from short-term to long-term memory usually requires special efforts such as repetition. In the Study Skills section at the front of this book, I described several ways to consolidate information, that is, transfer it to long-term memory, such as repetition, mnemonics, and rhymes.

The process of recalling information stored in either short- or long-term memory is called **remembering**. Short-term memory recall is generally faster than long-term memory recall unless the long-term memory you are retrieving is one that is thoroughly ingrained (for example, your place of birth). By the same token, information lost from short-term memory is generally gone forever. Information you cannot recall from long-term memory is often still there; it just takes time or a special stimulus to extract it.

Where Are Memories Stored?

Memory is stored in multiple regions of the brain.

Neurons involved in storing memories are widely distributed in the cerebral cortex, especially the temporal lobe. Other regions, including the cerebellum and the limbic system, also hold memories. The temporal lobes and a part of the limbic system known as the hippocampus (hip-poe-CAMP-us) appear essential for transferring short-term memories into long-term memory.

How Are Memories Stored?

Short-term and long-term memory may involve changes of the neurons.

Although no one knows the precise mechanism for storing memories, evidence suggests that it involves certain structural and functional changes in neurons in the brain. However, short- and long-term memories probably involve different mechanisms. Studies on slugs and snails, for instance, suggest that short-term memory may involve temporary changes in the function of synapses. Whether this mechanism holds true for humans and other animals remains to be seen.

In contrast, long-term memory appears to involve relatively permanent structural and functional changes in brain cells. Studies of lab animals exposed to sensory-rich and less-rich environments showed that the dendrites of brain cells of animals in the sensory-rich environment branched and elongated in regions of the brain thought to serve as memory storage—more than animals reared in the less-rich environment. Some studies suggest that protein synthesis may be involved in long-term memory because drugs that block protein synthesis interfere with long-term memory.

Maintaining Mental Acuity in Old Age

New evidence suggests that brain function can be retained in old age.

For years, scientists have thought that thinking and memory faded with age as brain cells died off. But there's startling new evidence that suggests the most unthinkable—that the brain can actually generate new cells, even in elderly individuals. Scientists have also found evidence to suggest that the connections among brain cells strengthen over time if individuals remain intellectually challenged. This suggests that individuals who remain mentally active can retain their brain power until the end of their lives. Although processing speed—how fast our brains store and retrieve information—does tend to decline in our later years, we can still learn and remember. We just need a little more time as we grow older. Keeping a mentally active life is all that's needed.

Scientists have also found that regular aerobic exercise such as walking helps eldering people perform certain mental functions, such as switching from one mental task to another. Combined with a lifetime of mental challenges, such as stimulating jobs or hobbies, exercise can ensure good cognitive functions. Although the research is still preliminary, many doctors tell their elderly patients to remain as active as possible.

The loss of mental function is, as you just read, not a natural consequence of aging. Losses of memory and confusion, however, do occur. Such instances are often a signal of an underlying disease, such as low thyroid hormone production, poor nutrition, or depression. Certain medications can have the same effect, for example, drugs for ulcers, high blood pressure, and Parkinson's disease. In some cases, loss of memory and confusion are caused by a serious disease known as Alzheimer's disease.

10-7 Diseases of the Brain

The brain, like other important organs, is subject to a wide assortment of diseases, many of which have dire consequences. This section reviews four: Alzheimer's disease, Parkinson's disease, multiple sclerosis, and brain tumors.

Alzheimer's Disease

Loss of memory and confusion in older individuals is often the result of Alzheimer's disease.

Alzheimer's disease is a progressive loss of mental function, which usually begins later in life. Alzheimer's afflicts an estimated 4 million Americans, mostly people over the age of 65. In fact, it is estimated that about one of ten people over the age of 65 has this disease. Alzheimer's is not life-threatening, but it does result in dramatic changes. Symptoms start out slowly. Forgetfulness is often one of the most common complaints. Irritability and lack of initiative also occur early on. As mental functions decline, patients lose the ability to remember recent events. Over time, losses may be so great that a parent may be unable to recognize his or her own children.

Scientists do not know the cause of Alzheimer's, although there is some evidence to suggest that previous severe brain injuries could be the cause in some instances. Scientists do know that a brain of a person with the disease contains deposits of a material called *plaque* and fibrous tangles. The plaque consists of a protein known as beta-amyloid. The fibrous clumps are protein from degenerated nerve cells. New research suggests that beta-amyloid binds to nerve cells in the brain and then causes cell death. This leads to the loss of memory and other symptoms. The tangles of fibers are believed to be remnants of the cells' proteins. They're formed when beta-amyloid binds to the nerve cells.

Experiments with mice suggest that antibodies to beta-amyloid could prevent or slow the development of Alzheimer's disease. When mice that had the disease were given antibodies, beta-amyloid accumulation in the brain was prevented. Existing deposits were removed. Tests on people will help scientists determine whether the antibodies work similarly. If they do, this could be great news for those who suffer from this disease or who will develop it as they age. Researchers hope that such treatments may be useful in other brain disorders such as Parkinson's disease, described shortly.

Parkinson's Disease

Parkinson's disease is a neurological disorder caused by a lack of dopamine in certain regions of the brain.

Another important nervous system disorder is Parkinson's disease. Discussed in Chapter 3, in relation to environmental pollution, **Parkinson's disease** is caused by a progressive deterioration of certain brain centers that control movement, especially semiautomatic movements such as swinging the arms when walking. It afflicts about 1 million Americans, most over the age of 60. In fact, approximately one in 100 Americans in this age group has the disease.

Parkinson's disease is characterized by tremors. In many people with the disease, the hands or the head (sometimes both) shake involuntarily. Tremors disappear or decrease, however, when one moves the afflicted part of the body intentionally—for example, when one reaches for an object. If the disease worsens, symptoms worsen and it becomes difficult to write or walk. Speaking may deteriorate too, as it becomes difficult to move one's mouth and tongue. Falls become more frequent. It may be difficult to get out of a chair. In later stages of the disease, memory and thinking may deteriorate.

Like Alzheimer's, the cause or causes of Parkinson's are unknown. Chemical pollutants may contribute to the disease. Amalgam fillings in teeth and chemicals used in wood stains may play a role in the onset of the disease. Very likely, there are many causes. Scientists know that Parkinson's is characterized by a lack of a chemical substance in the brain called *dopamine*. Because of this, patients can be treated with a drug known as *levodopa*. However, its benefits may diminish over years of use. Some patients suffer serious side effects, including severe nausea and vomiting. Several other drugs are also used.

Recent studies show that fetal cell transplants into the affected brain area may eliminate the symptoms of the disease. In one study, a year after receiving a fetal cell transplant 13 of 19 people showed elevated levels dopamine production. Surgeons can also destroy a small region of the brain that appears to be hyperactive in Parkinson's patients. This improves a patient's score on standardized movement tests, although several patients had serious side effects including facial paralysis. Electrodes can also be implanted in the brain to stop uncontrollable movements. Patients are equipped with tiny electronic stimulators, much like pacemakers, that can be switched on or off as needed.

Multiple Sclerosis

MS is caused by a destruction of the myelin sheaths of nerve cells in the CNS.

Destruction of the myelin sheath of nerve cells in the central nervous system results in a condition known as **multiple sclerosis** (skler-OH-siss). The damaged myelin results in nerve cell death that leads to numbness, slurred speech, and paralysis. Thought to be an autoimmune disease, multiple sclerosis can affect any part of the CNS. Early symptoms are generally mild weakness or a tingling or numb feeling in one part of the body. Temporary weakness may cause a person to stumble and fall. Some people report blurred vision, slurred speech, and difficulty controlling urination. In many cases, these symptoms disappear, never to return. Other individuals suffer repeated attacks. Because recovery after each attack is incomplete, patients gradu-

ally deteriorate, losing vision and becoming progressively weaker. Fortunately, many treatments are available, and only a small number of multiple sclerosis patients are crippled by the disease.

Brain Tumors

Brain tumors afflict the young and the old and may have many causes.

Two types of tumors develop in the brain tissue, benign and malignant. Benign tumors are cellular growths in the brain that do not grow uncontrollably and do not spread to other parts of the body like malignant tumors. Even as such, benign tumors can cause problems. They may, for instance, place pressure on areas of the brain, resulting in definite problems. Benign tumors have distinct borders and can be removed surgically; unlike malignant tumors, they do not recur.

Malignant tumors grow rapidly, especially late in the disease. They are likely to place pressure on neighboring structures and may invade into the tissue around them or to other parts of the body.

Brain tumors occur at any age, although the most common groups are children in the 3- to 12-year-old age bracket and adults in the 40- to 70-year-old bracket. Although researchers still do not know what causes brain tumors, there are some risk factors that increase one's likelihood for developing a cancerous tumor. Workers in certain industries, for example, oil refining, rubber manufacturing, and drug manufacturing, are at a higher risk of developing a brain tumor. Chemists and embalmers have a higher incidence of brain tumors as well. Some researchers believe that exposure to certain viruses may be responsible for brain tumors. And some brain tumors seem to run in families, so there may be a genetic cause that is passed from parents to offspring. In most cases, however, patients with brain tumors have no clear risk factors, so researchers believe that the disease is probably the result of several factors acting together.

There's a great deal of concern these days over the potential of cell phones to cause brain cancer. Although several recent studies show no link, cell phone use is a recent phenomenon. Most cancers take 5 to 30 years to develop. One thing to keep in mind is that cell phones emit radio waves. Unlike ionizing radiation, for example, from nuclear power plants, radio waves do not impart much energy in body tissues and may have little, if any, effect on cells of the body. Only time will tell.

10-8 Health and Homeostasis

A decline in blood glucose is normally prevented by homeostatic mechanisms described in Chapter 4. Serious problems can result in people whose glucose homeostasis is not working properly. Many changes may occur in the brain and behavior controlled by it. In diabetics, for instance, blood glucose levels can fall dangerously low if too much insulin is taken or if not enough glucose is ingested. Deprived of glucose, brain cells begin to falter. Individuals can become dizzy and weak. Vision may blur.

Speech may become awkward. Diabetics are sometimes mistakenly considered drunk. Chronic low blood glucose often results in severe headaches. Some diabetics become aggressive when blood sugar levels fall. In extreme cases, low blood sugar triggers convulsions and death. Other body cells are not adversely affected by a decline in blood glucose because they switch to alternative fuels, fats and proteins.

SUMMARY

An Overview of the Nervous System

1. The nervous system controls a wide range of functions and plays a key role in ensuring homeostasis. The human nervous system also performs many other functions, such as controlling muscle and higher mental functions such as planning for the future, learning, and remembering.

2. The nervous system consists of the central nervous system, made up of the brain and spinal cord, and the peripheral nervous system, which consists of the spinal and cranial nerves.

3. Receptors in the skin, skeletal muscles, joints, and organs transmit sensory input to the CNS via sensory neurons. The CNS integrates all sensory input and generates appropriate responses. Motor output leaves the CNS in motor neurons.

4. The PNS has two functional divisions. The autonomic division controls involuntary actions such as heart rate. The somatic division largely controls voluntary actions such as skeletal muscle contractions. It also provides the neural connections needed for many reflex arcs.

Structure and Function of the Neuron

5. The fundamental unit of the nervous system is the neuron. This highly specialized cell generates and transmits bioelectric impulses from one part of the body to another.

6. All neurons have more or less spherical cell bodies. Extending from the cell body are two types of processes: dendrites, which conduct impulses to the cell body, and axons, which conduct impulses away from the cell body.

7. Many axons are covered by a layer of myelin, which increases the rate of impulse transmission.

8. The terminal ends of axons branch profusely, forming numerous fibers that end in small knobs called terminal boutons.

9. During cellular differentiation, nerve cells lose their ability to divide. Because of this, neurons that die cannot be replaced by existing cells.

10. The small electrical potential across the membrane of nerve cells is known as the membrane potential or resting potential.

11. When a nerve cell is stimulated, its plasma membrane increases its permeability to sodium ions. Sodium ions rush in, causing depolarization, which spreads down the membrane.

12. Depolarization is followed by repolarization, a recovery of the resting potential stemming largely from the outflow of potassium ions. The depolarization and repolarization of the neuron's plasma membrane constitute an action potential.

13. When a nerve impulse reaches the terminal bouton, it stimulates the release of neurotransmitters contained in membrane-bound vesicles. They bind to receptors in the postsynaptic membrane.

14. Neurotransmitters may excite or inhibit the postsynaptic membrane. Whether a neuron fires depends on the sum of excitatory and inhibitory impulses it receives.

The Spinal Cord and Nerves

15. The spinal cord descends from the brain through the vertebral canal to the lower back. It carries information to and from the brain, and its neurons participate in many reflexes.

16. Two types of nerves emanate from the CNS: spinal and cranial. Spinal nerves arise from the spinal cord and may be sensory, motor, or mixed. Cranial nerves attach to the brain and supply the structures of the head and several key body parts.

The Brain

17. The brain is housed in the skull. The cerebrum with its two cerebral hemispheres is the largest part of the brain.

18. The cerebral cortex consists of many discrete functional regions, including motor, sensory, and association areas.

19. Consciousness resides in the cerebral cortex, but a great many functions occur at the unconscious level in parts of the brain beneath the cortex.

20. The cerebellum coordinates muscle movement and controls posture. The hypothalamus regulates many homeostatic functions. The limbic system houses instincts and emotions. The brain stem, like the hypothalamus, regulates basic body functions.

21. A watery fluid known as cerebrospinal fluid surrounds the brain and spinal cord. It cushions the brain and spinal cord, protecting them from traumatic injury.

22. Electrodes applied to different parts of the scalp detect electrical activity in the brain and produce a tracing known as an electroencephalogram (EEG). The type of brain wave recorded depends on one's level of cortical activity. EEGs are used to diagnose some brain dysfunctions.

23. Headaches are the most common form of pain and generally result from tension, swelling of the membranes lining the sinuses, eyestrain, or dilation of the cerebral blood vessels. Very serious headaches may result from increased intracranial pressure caused by a brain tumor, intracranial bleeding, or inflammation caused by an infection of the meninges (meningitis) or the brain itself (encephalitis). Less threatening, but still extraordinary, are migraines, a form of severe, recurrent headache.

The Autonomic Nervous System

24. The autonomic nervous system or ANS is a division of the PNS. It supplies all internal organs and has two subdivisions: the sympathetic and the parasympathetic.

25. The sympathetic division of the ANS functions in emergencies and is responsible in large part for the fight-or-flight response. The parasympathetic division of the ANS brings about a relaxed state.

Learning and Memory

26. Learning is a process in which an individual acquires knowledge and skills.

27. Newly acquired knowledge is first stored in short-term memory. Short-term memories may be transferred to long-term memory, which holds information for periods of days to years.

28. Memory is stored in multiple regions of the brain.

29. Short- and long-term memory appear to involve structural and functional changes of the neurons.

Diseases of the Brain

30. Alzheimer's disease is a progressive loss of mental function, which usually begins later in life.

31. Parkinson's disease is a neurological disorder caused by a progressive deterioration of certain brain centers that control movement. It is characterized by tremors that typically worsen as one ages. Like Alzheimer's, the cause or causes of Parkinson's are unknown.

32. Multiple sclerosis is a disease caused by degeneration of the myelin sheath of axons. Damaged to the myelin results in nerve cell death that leads to numbness, slurred speech, and paralysis.

33. Two types of tumors develop in the brain tissue, benign and malignant. Although benign tumors do not grow uncontrollably or spread to other areas, they can cause problems by placing pressure on areas of the brain.

34. Malignant tumors grow rapidly, and often place pressure on neighboring structures. They may also invade the tissue around them or may spread to other parts of the body.

Health and Homeostasis

35. Abnormal brain activity can result from a disruption of homeostasis, for example, in the mechanisms that control blood glucose levels.

critical thinking

Thinking Critically—Analysis

This Analysis corresponds to the Thinking Critically scenario that was presented at the beginning of this chapter.

The researchers hypothesized that the left prefrontal cortex of the brain may process negative thoughts. Since they found that this cortex is hyperactive in severely depressed subjects, they hypothesized that the hyperactivity results in symptoms of depression. The scientists also conjectured that chemical messengers in the brain that normally dampen the activity of this region are released in abnormally high amounts, causing the brain to malfunction.

To confirm their findings, the researchers compared another group of seven individuals with severe depression to a control group—and found similar results. Although this is encouraging, it is important to note that the sample sizes are rather small. Before one can accept these findings, it's essential that they be repeated by other researchers studying larger sample sizes.

KEY TERMS AND CONCEPTS

Acetylcholine, p. 178
Acetylcholinesterase, p. 179
Action potential, p. 177
Anesthetics, p. 179
Association cortex, p. 184
Autonomic nervous system, p. 174
Axon, p. 174
Brain, p. 182
Central nervous system, p. 173
Cerebellum, p. 184
Cerebral cortex, p. 183
Cerebral hemispheres, p. 183
Cerebrospinal fluid, p. 187
Cerebrum, p. 183
Cranial nerve, p. 182
Dendrite, p. 174
Depolarization, p. 177

Fight-or-flight response. p. 186
Hypothalamus, p. 185
Learning, p. 190
Limbic system, p. 186
Long-term memory, p. 190
Memory, p. 190
Meninges, p. 173
Meningitis, p. 188
Multiple sclerosis, p. 192
Myelin sheath, p. 174
Nerve, p. 173
Nerve impulse, p. 176
Neuron, p. 174
Neurotransmitter, p. 178
Node of Ranvier, p. 175
Parkinson's disease, p. 191
Peripheral nervous system, p. 173

Primary motor cortex, p. 183
Primary sensory cortex, p. 184
Reflex arc, p. 181
Remembering, p. 190
Repolarization, p. 177
Resting potential, p. 176
Reticular activating system, p. 187
Reticular formation, p. 187
Short-term memory, p. 190
Somatic nervous system, p. 174
Spinal cord, p. 179
Spinal nerve, p. 179
Synapse, p. 178
Synaptic cleft, p. 178
Terminal boutons, p. 174
Thalamus, p. 185
Vertebral canal, p. 179

CONCEPT REVIEW

1. The nervous system performs many functions. What functions do you think are unique to humans? p. 173
2. Describe how the resting and action potentials are generated. Describe how the plasma membrane of a neuron is repolarized. pp. 176–177
3. Draw a typical multipolar neuron, and label its parts. Show the direction in which an action potential travels. p. 175
4. Draw a typical synapse, label the parts, and explain how a nerve impulse is transmitted from one nerve cell to another. p. 178
5. Name and describe the various divisions of the nervous system. pp. 173–174
6. Draw a cross section through the spinal cord showing the spinal nerves. Label the parts, and explain a reflex arc and how nerve impulses entering a spinal nerve also travel to the brain. p. 181
7. A physician can stimulate various parts of the brain and get different responses. What effects would you expect if the electrodes were placed in the primary motor cortex? The primary sensory cortex? pp. 183–184

SELF-QUIZ: TESTING YOUR KNOWLEDGE

1. The brain and spinal cord form the _____ nervous system. p. 173

2. Three layers of connective tissue surround the brain; they're known as the _____. p. 173

3. Nerves carry _____ impulses from receptors in the body to the brain and spinal cord. p. 173

4. The spinal and cranial nerves are part of the _____ nervous system. p. 174

5. Nerve cells or neurons contain a cell body, and two processes. The one that carries impulses away from the cell body is known as the _____. p. 174

6. Axons branch at their ends, terminating in small swellings known as _____. p. 174

7. A nerve impulse is caused by a change in the permeability of the plasma membrane of nerve cells that results in an influx of _____ ions. p. 177

8. An electrical tracing of the change in the membrane potential of a nerve cell is called an _____. p. 177

9. Nerve impulses travel from one nerve cell to another thanks to the release of a chemical from the presynaptic neuron known as _____. p. 178

10. Located in the spinal canal, the _____ extends from the brain to the lower back. p. 179

11. The largest portion of the brain is known as the _____. p. 183

12. The part of the cerebral cortex that receives sensory information is known as the _____ cortex. p. 184

13. The _____ ensures muscle synergy—the coordination of muscular activity. p. 184

14. The _____ consists of many aggregations of nerve cells that control a variety of autonomic functions such as appetite and body temperature. p. 187

15. The brain is kept alert thanks to nerve impulses arising from the _____ activating system. p. 187

www.jbpub.com/humanbiology/5e

The site features eLearning, an online review area that provides quizzes, chapter outlines, and other tools to help you study for your class. You can also follow useful links for in-depth information, research the differing views in the Point/Counterpoints, or keep up on the latest health news.

Debra Cartwright noticed something strange one day while she was eating dinner: she had lost her senses of smell and taste. Doctors were puzzled at first, as Debra seemed to be in fine health. She was not suffering from a cold, sinus infection, or even any allergies that might have blocked her nasal passages and impaired her senses of smell or taste. A blood test, however, revealed the cause of her problem: low blood levels of the micronutrient zinc. Her physician prescribed a zinc supplement, and in short order her senses of smell and taste returned.

thinking critically

Imagine that you suffer from chronic pain. At times the pain in your abdomen becomes unbearable. It wakes you when you are asleep. It makes it impossible to sit still and enjoy a movie with your loved ones. It makes you irritable and short-tempered. Yet doctors can't find any cause for it.

Then, one day, your local newspaper runs a story about an experimental procedure pioneered by a researcher at a local medical school. This procedure involves the injection of chemical substances into the fluid surrounding the spinal cord. These substances, the researchers believe, bind to nerve cells that carry pain signals to your brain, blocking impulses. They've successfully used the procedure on laboratory mice and now are looking for volunteers to test the effectiveness of this procedure on humans.

Would you consider becoming a subject? If so, what questions would you want answered before you submit to the procedure?

This story illustrates the importance of a nutritionally balanced diet and the impact of a dietary deficiency on two important body functions—the senses of taste and smell. It is just one of many examples presented in this book that illustrates how body functions can be altered by disease and external factors such as stress, pollution, and diet. This chapter examines the senses, dividing the discussion into two broad categories: the general senses and the special senses.

11-1 The General and Special Senses

The human body contains receptors that detect a wide array of internal and external stimuli essential for our well-being.

Our health and human survival depend on a surveillance system that keeps track of internal and external changes. The surveillance system of the human body, like that of many other animals, consists of numerous receptors that monitor internal and external conditions. In humans, receptors are located in the skin, internal organs, bones, joints, and muscles. They detect stimuli that give rise to the **general senses**: pain, temperature, light touch, pressure, and a sense of body and limb position (Table 11-1).

The human body is also endowed with five additional senses, known as the **special senses**. They are taste, smell, vision, hearing, and balance. They are made possible by sophisticated sensory organs such as the eye. These detectors increase our ability to perceive our environment.

Receptors in humans involved in the general and special senses fit into five categories: (1) mechanoreceptors, (2) chemoreceptors, (3) thermoreceptors, (4) photoreceptors, and (4) nociceptors (pain receptors). **Mechanoreceptors** are those activated by mechanical stimulation—for example, touch or pressure. **Chemoreceptors** are activated by chemicals in the food we eat, in the air we breathe, or in our blood. **Thermoreceptors** are activated by heat and cold, and **photoreceptors** are sensitive to light. **Nociceptors** (no-see-SEP-tors) are pain receptors stimulated by pinching, tearing, or burning.

TABLE 11-1	Summary of General and Special Senses	
Sense	Stimulus	Receptor
General senses	Pain	Naked nerve endings
	Light touch	Merkel's discs; naked nerve endings around hair follicles; Meissner's corpuscles; Ruffini's corpuscles; Krause's end-bulbs
	Pressure	Pacinian corpuscles
	Temperature	Naked nerve endings
	Proprioception	Golgi tendon organs; muscle spindles; receptors similar to Meissner's corpuscles in joints
Special senses	Taste	Taste buds
	Smell	Olfactory epithelium
	Sight	Retina
	Hearing	Organ of Corti
	Balance	Crista ampularis in the semicircular canals; maculae in utricle and saccule

11-2 The General Senses

Receptors for the general body senses generally fit into two groups: naked nerve endings and encapsulated receptors.

Sit back in your chair for a moment, close your eyes, and concentrate on what you feel. You may detect the pressure of the chair on your back and heat from a reading lamp. You may feel your cat brushing against the hairs on your arm. You may detect pain from gas pressure in your intestines from the burritos you ate yesterday. Now move your arm. Even though your eyes are closed, you can feel it moving.

The various sensations you have just experienced fall into the group of general senses. Receptors for the general senses detect internal and external stimuli and relay messages to the spinal cord and brain via sensory nerves (Chapter 10). Sensory input to the central nervous system may elicit a conscious response—for example, the touch of a cat may cause you to reach forward and pet your furry friend. Some stimuli will cause unconscious responses—for example, heat may cause you to perspire. Others may simply be registered in the cerebral cortex, making you aware of the stimulus but do not elicit a response. Still other stimuli may be blocked so they are not perceived at all.

Receptors for the general body senses come in many shapes and sizes, but they generally fit into two groups based on structure: naked nerve endings and encapsulated receptors (Figure 11-1).

Naked Nerve Endings

Naked nerve endings in tissues detect pain, temperature, and light touch.

Located in the skin, bones, and internal organs and in and around joints, naked nerve endings are the terminal ends of the dendrites of sensory neurons. They are responsible for at least three sensations: pain, light touch, and temperature.

∷∷ **healthnote**

11-1 Old and New Treatments for Pain

Millions of people suffer from persistent pain. In the United States, for example, an estimated 70 to 80 million people are tormented by back pain. Another 26 million suffer from migraine headaches.

Minor aches and pains can be treated with pain killers such as acetaminophen and ibuprofen. More intense pain is treated with stronger painkillers, such as codeine. Codeine is often mixed with ibuprofen and acetaminophen. Although codeine and other strong pain killers are used today, they are addictive. One nonaddicitve option is acupuncture, used by the Chinese for thousands of years.

Acupuncture relies on thread-thin needles inserted in the skin near nerves (Figure 1). No one knows how acupuncture works. The Chinese assert that it encourages the free flow of energy through various pathways in the body whose purpose is to irrigate and nourish the tissues. They contend that blood and nervous impulses also follow these pathways. Any obstruction in the movement of the energy flow

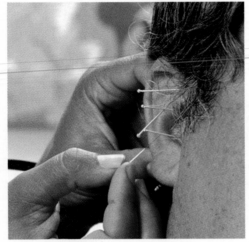

FIGURE 1 **Patient Undergoing Acupuncture** Acupuncture is generally used to alleviate pain, but it has other applications as well.

results in deficiencies of energy, blood, and nerve impulses that may lead to disease. The acupuncture needles supposedly unblock obstructions and reestablish normal flow.

Neurologists think that acupuncture blocks pain by overloading the neuronal circuitry. They note that two types of nerve fibers transmit sensory information from the body to the central nervous system: small- and large-diameter fibers. Small-diameter fibers carry pain messages. Large-diameter fibers carry many other forms of sensory information from receptors in the skin—for example, pressure and light touch. The dendrites of both small- and large-diameter sensory nerve cells often terminate in the same location in the spinal cord. From here, they send impulses to the brain, signaling pain or some other sense.

Scientists believe that acupuncture needles stimulate the large-diameter nerve fibers. This stimulation blocks nerve impulses carried by the smaller nerve fibers, blocking pain messages to the brain.

Research on the mechanism of action of acupuncture points to several other possibilities as well. Acupuncture needles, for instance, may stimulate the release of endorphins, the body's natural pain killers. Endorphins are produced in the brain.

Many U.S. physicians trained to use drugs and surgery to solve most pain remain skeptical

---

Pain. Your body perceives two basic types of pain: somatic and visceral. Somatic pain results from injuries in the skin, joints, muscles, and tendons. Somatic pain receptors respond to several types of stimuli. Some respond to cutting, crushing, and pinching. Others respond to hot or cold. Still others respond to irritating chemicals released from injured tissues.

Visceral pain results from the stimulation of naked nerve endings in body organs (the viscera). In some body organs, pain receptors are stimulated by expansion. For example, the intestinal pain you feel when you have gas results from the stretching of naked nerve fibers in the wall of the intestine. These nerve endings are mechanoreceptors. In other body organs, pain receptors are stimulated by a lack of oxygen, or anoxia (ah-NOCKS-

see-ah). For example, pain felt during a heart attack results from a lack of oxygen in the heart muscle.

Somatic pain and visceral pain are not only caused by very different stimuli, they are also perceived differently. Somatic pain is easily pinpointed. Visceral pain, however, is often vague and difficult to localize. Moreover, it is generally felt on the body surface at a site some distance from its origin. For example, pain caused by a lack of oxygen to the heart muscle appears in the chest and along the inside of the left arm (Figure 11-2).

Visceral pain that appears on the body surface away from the location of the pain is called **referred pain**. Physiologists do not know the cause of this phenomenon, but most think it is because pain fibers from internal organs enter the spinal cord at the same location that the sensory fibers from the skin enter. The brain, they hypothesize, interprets the impulses from pain fibers supplying the organs as pain from a somatic source. (Health Note 11-1 describes techniques that relieve pain.)

Light Touch. Light touch is perceived by two anatomically distinct mechanoreceptors. As shown in Figure 11-3, the first type consists of naked nerve endings (dendrites) that wrap around the base of the hair follicles. When a hair is moved—for example, by a gentle touch—these nerve fibers are stimulated.

The second light-touch mechanoreceptor, Merkel's disc, is also shown in Figure 11-3. This structure consists of small cup-shaped cells and naked nerve endings that con-

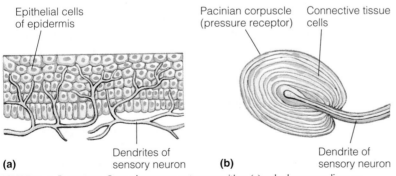

(a)

Epithelial cells of epidermis

Dendrites of sensory neuron

(b)

Pacinian corpuscle (pressure receptor)

Connective tissue cells

Dendrite of sensory neuron

FIGURE 11-1 **Receptors** General sense receptors are either (a) naked nerve endings or (b) encapsulated nerve endings.

about the usefulness of acupuncture. Over the past 20 years, however, a small but steady stream of research has confirmed the painkilling effect of this treatment. Today, most insurance companies will pay the cost of acupuncture to relieve pain caused, for example, by a car crash.

Joseph Helms, a physician with the American Academy of Acupuncture in Berkeley, California, performed acupuncture on 40 women with menstrual pain. Some women received real acupuncture treatment. Others received placebo treatments (shallow needle treatments that did not reach the acupuncture points). In the group of women receiving acupuncture, 10 out of 11 showed a marked decrease in pain. Patients reported an approximately 50% decrease in pain. In the placebo group, only 4 out of 11 reported a lessening of pain. Only 1 of 10 people given no treatment showed improvement. Acupuncture also reduced the need for painkilling drugs during treatment by over half.

Some health care practitioners are using another procedure called transcutaneous electrical nerve stimulation, or TENS. Patients are fitted with electrodes attached to the skin above the nerves that transmit pain signals to the CNS. When the pain begins, patients press a button on a battery pack that sends a tiny current to the electrode. Conducted through the skin, the current blocks the pain impulses. TENS can be used to reduce pain after surgery.

Severe pain can also be treated by surgery. Doctors may cut nerves, sever nerve tracks in the spinal cord that carry pain, or even destroy small parts of the brain to rid a patient of chronic pain. Nerves can also be destroyed by applying certain chemical substances.

Although all of these methods are effective, there are dangers in these procedures. Cutting nerve tracks in the spinal cord, a structure that's

approximately the size of your little finger, can be risky. One slip and a surgeon could destroy motor neurons. Nerves carrying pain signals also transmit sensory information. Some doctors think that patients should not risk loss of other senses or motor functions for the sake of pain relief. In addition, pain recurs in 9 of 10 patients who have undergone pain-relieving surgery, usually within a year or so, as a result of the partial regrowth of axons. Even after another operation, the pain frequently returns, often with much greater intensity.

Another promising measure is deep brain stimulation. Electrodes can be implanted in parts of the brain and stimulated to block pain impulses before they reach the sensory cortex, where pain is perceived. The electrodes are connected to a portable battery worn on the belt or implanted under the skin. When the pain begins, the patient turns on the current, blocking the pain impulses. Research shows that deep brain stimulation is a very effective blocker of even the most powerful pain stimuli.

www.jbpub.com/humanbiology/5e

Visit Human Biology's Internet site for links to web sites offering more information on this topic.

tact them. Located in the outer layer of the epidermis of the skin, these receptors are activated by gentle pressure applied to the skin.

Temperature. Certain naked nerve endings in the skin detect heat and cold within specific temperature ranges. If the temperatures are hotter or colder, pain receptors are activated.

Encapsulated Receptors

Encapsulated receptors consist of nerve endings surrounded by one or more layers of cells.

Shown in Figure 11-3, the largest encapsulated nerve ending is the Pacinian corpuscle (pah-SIN-ee-an CORE-puss-l). It consists of a naked nerve ending surrounded by numerous concentric cell layers. The Pacinian corpuscle resembles a small onion pierced by a thin wire. Located in the deeper layers of the skin, in the loose connective tissue of the body, and elsewhere, Pacinian corpuscles are stimulated by pressure such as the pressure you feel sitting in your chair.

Another common encapsulated sensory receptor is the Meissner's corpuscle (MEIZ-ners). Smaller than Pacinian corpuscles, these oval receptors contain two or three spiraling dendritic ends surrounded by a thin cellular capsule (Figure 11-3). Like the naked nerve endings wrapped around the base of hair follicles and Merkel's discs, Meissner's corpuscles are thought to respond to light touch. Located in the outermost layer of the

dermis, Meissner's corpuscles are most abundant in the sensitive parts of the body, such as the lips and the tips of the fingers.

Two additional encapsulated receptors are Krause's end-bulbs and Ruffini's corpuscles (Figure 11-3). These receptors are probably variations of the Meissner's corpuscle and are stimulated by light touch.

Proprioception (PRO-pree-oh-CEP-shun) is the sense of position. It is detected by encapsulated receptors located in the joints of the body. Resembling Meissner's corpuscles, these receptors inform us of the position of our limbs and alert us to movements of the body.

Position is also detected by sensors, the muscle spindles, located in skeletal muscles. Muscle spindles consist of several modified muscle fibers with sensory nerve endings wrapped around them. The nerve endings are stimulated when muscles are stretched. Nerve impulses generated by the spindle are then transmitted to the spinal cord, where they may ascend to the cerebral cortex, helping us remain aware of body position. Others may stimulate motor neurons in the spinal cord, eliciting a reflex contraction of the muscle being stretched. This is what happens when you fall asleep sitting upright in a chair and your head falls forward—and quickly snaps back.

Another proprioceptor is the Golgi tendon organ (GOAL-jee). These mechanoreceptors, named after the Italian scientist who discovered them, are located in tendons—the structures that connect muscles to bones. Golgi tendon organs are composed of connective tissue fibers surrounded by dendrites and encased in a capsule. When a muscle contracts, the tendon stretches

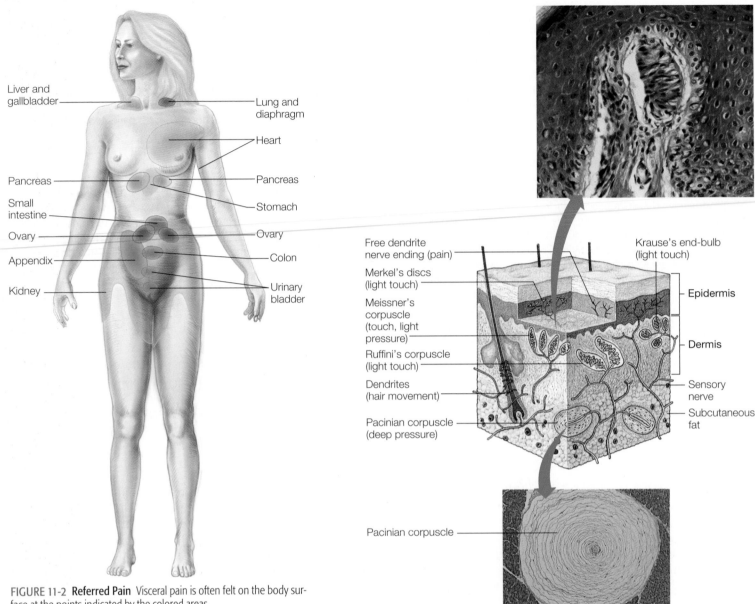

FIGURE 11-2 **Referred Pain** Visceral pain is often felt on the body surface at the points indicated by the colored areas.

FIGURE 11-3 **General Sense Receptors** The skin houses many of the receptors for general senses. Receptors fall into two categories: naked nerve endings and encapsulated receptors. The Pacinian corpuscle, often located in the dermis of the skin, detects pressure. Meissner's corpuscle, found just beneath the epidermis, detects light touch.

and stimulates the receptor, alerting the brain of movement and body position. Impulses from the Golgi tendon organ can also stimulate reflex contraction of muscles. The knee-jerk reflex described in Chapter 10 is a good example.

Adaptation

Many receptors stop generating impulses after extended exposure to stimuli.

Pain, temperature, and pressure receptors are all subject to a phenomenon called *adaptation*. This occurs when sensory receptors stop generating impulses even though the stimulus

is still present. You have probably experienced the phenomenon many times. Recall, for example, the first time you wore a ring or contact lenses. At first, the sensation may have nearly driven you mad, but after a few days, the stimulus seemed to have disappeared. What happened was that pressure receptors that were originally alerting the brain stopped generating impulses, releasing you from what could have been unrelenting discomfort.

Not all receptors adapt. Muscle stretch receptors and joint proprioceptors are two examples. Because the central nervous system must be continuously informed about muscle length and joint position to maintain posture, adaptation in these receptors could be dangerous.

11-3 Taste and Smell

The special senses are taste, smell, vision, hearing, and balance. In this section, we'll examine taste and smell.

Taste Buds

Taste buds respond to chemicals dissolved in food.

In humans and other mammals, the tongue contains receptors for taste. Known as **taste buds**, these microscopic, onion-shaped structures are located in the surface of the tongue and on small protrusions, known as *papillae* (pah-PILL-ee) (Figure 11-4a). Taste buds are also found on the roof of the oral cavity, the pharynx, and the larynx, but in smaller numbers.

Taste buds are stimulated by chemicals in the food we eat and are therefore referred to as *chemoreceptors*. Chemical substances in food and drinks dissolve in the saliva and enter the taste pores, small openings that lead to the interior of the taste bud (Figure 11-4c). Taste buds contain receptor cells, the ends of which possess large microvilli, known as *taste hairs*. Taste hairs project into the taste pore. The plasma membranes of taste hairs contain clusters of protein molecules that serve as receptors. These receptors bind to food molecules dissolved in water. This,

in turn, stimulates the receptor cells, which then stimulate the dendrites of the sensory nerves wrapped around the receptor cells. Impulses from the taste buds are then transmitted to the brain.

In the introduction to this chapter, you learned about a student who lost her sense of taste because of a dietary zinc deficiency. Zinc stimulates division of the cells in taste buds. Because cells of the taste buds are lost from normal wear and tear, a zinc deficiency reduces cellular replacement, and the taste buds eventually cease operation.

Primary Flavors

Taste buds respond preferentially to one of five primary flavors.

Humans can discriminate among thousands of taste sensations. The taste sensations are a combination of five basic flavors: sweet, sour, bitter, salty, and umami. Sweet flavors result from sugars and some amino acids. Sour flavors result from acidic substances. Salty tastes result from metal ions (like sodium in table salt). Bitter flavors result from chemical substances belonging to a group called *alkaloids*, among them caffeine, and some nonalkaloid substances such as aspirin. Umami is the

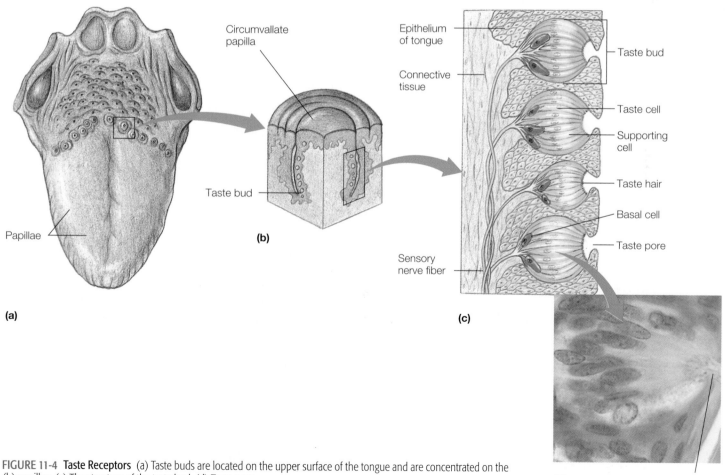

FIGURE 11-4 Taste Receptors (a) Taste buds are located on the upper surface of the tongue and are concentrated on the (b) papillae. (c) The structure of the taste bud. (d) Taste pore.

meaty taste associated with the flavorant MSG (monosodium glutamate) commonly added to Chinese food. All taste buds respond, in varying degree, to all five taste sensations, but each responds preferentially to one taste.

Taste buds are distributed unevenly on the surface of the tongue. The tip of the tongue, for instance, is most sensitive to sweet flavors as it contains a higher proportion of taste buds that respond preferentially to them. The sides of the tongue are most sensitive to sour flavors, and the back of the tongue is most sensitive to bitter flavors. Salty taste is more evenly distributed, with slightly increased sensitivity on the sides of the tongue near the front. Food contains many different flavors. What we taste, therefore, depends on the relative proportion of the five basic flavors.

The Olfactory Epithelium

The olfactory epithelium is a patch of receptor cells that detects odors.

Smell, like taste, is a chemical sense. The receptors for smell are located in the roof of each nasal cavity in a patch of cells called the **olfactory epithelium** (ole-FACK-tore-ee) (Figure 11-5). Odors are perceived by receptor cells, neurons whose dendrites extend to the surface of the olfactory membrane (Figure 11-5). In humans, there are an estimated 50 million olfactory receptor cells in the olfactory membrane. These cells terminate in six to eight long projections, called *olfactory hairs.* The olfactory hairs contain protein receptors that bind to airborne molecules. When airborne molecules are trapped on the thin, watery layer on the surface of the cell, they bind to these receptors. This activates the neurons, causing them to generate impulses that are sent to the olfactory bulb. Here, dendrites of the receptors synapse with nerve cells. Axons of the neurons in the olfactory bulb cells then travel to the brain via the olfactory nerve.

Like taste, odor discrimination is thought to depend on combinations of primary odors. Unfortunately, neuroscientists do not agree on what the primary odors are. One system of classification recognizes seven primary odors ranging from pepperminty to floral to putrid. It is thought that molecules that produce a similar odor are similarly shaped and that specific receptor sites on the olfactory hairs bind to molecules with a common shape. Various combinations of the primary odors give rise to the many odors we perceive. Some researchers hypothesize that there may be thousands of kinds of smell receptors providing odor discrimination. Research shows that there are at least 1000 different protein receptor molecules. New studies suggest that these molecules may be combined in different ways to provide the wide range of odor detection. Receptors for smell adapt within a short period—about a minute, which is a good thing, for example, if you live or work on a dairy or pig farm.

Smell and Taste

Smell influences our sense of taste, and vice versa.

Hold a piece of hot apple pie near your nose and inhale. It smells so good you can almost taste it. In fact, you are tasting it. Molecules given off by the pie enter the nose, reach the mouth through the pharynx, and dissolve in the saliva, where they stimulate taste receptors on the tongue. Just as odors stimulate taste receptors, food in our mouths also stimulates olfactory receptors. Molecules released by food enter the nasal cavities, dissolving in the water on the surface of the olfactory membrane and stimulating the receptor cells.

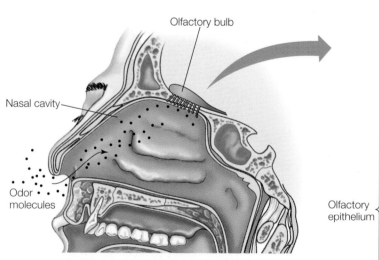

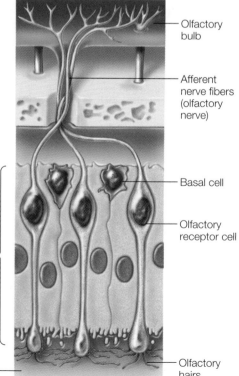

FIGURE 11-5 **Location and Structure of the Olfactory Epithelium** Olfactory receptors are located in the olfactory epithelium in the roof of the nasal cavity. Chemicals in the air dissolve in the watery fluid bathing the surface of the cells, then bind to receptors on the plasma membranes of the olfactory hairs. The olfactory receptors terminate in the olfactory bulb. From here, nerve fibers travel to the brain.

The complementary nature of taste and smell is evident when one suffers from nasal congestion. Cold sufferers often complain that they cannot taste their food. This phenomenon results from the buildup of mucus in the nasal cavities, which blocks the flow of air. Mucus may also block the olfactory hairs. Therefore, food loses its "taste" when you have a stuffy nose because your sense of smell is impaired. (You can test this by holding your nose while you eat.)

11-4 The Visual Sense: The Eye

The human eye is one of the most extraordinary products of evolution. It contains a patch of photoreceptors that permits us to perceive the remarkably diverse and colorful environment we live in.

Anatomy of the Eye

The eye consists of three distinct layers.

Human eyes are roughly spherical organs located in the eye sockets, or orbits, cavities formed by the bones of the skull. The eye is attached to the orbit by six small muscles that control eye movement.

The Sclera and Cornea. As Figure 11-6 shows, the wall of the human eye consists of three layers (Table 11-2). The outermost is a durable fibrous layer, which consists of the **sclera**, the white of the eye, and the **cornea**, the clear part in front, which lets light into the interior of the eye (Figure 11-6). Tendons of the extrinsic eye muscles attach to the sclera.

The Choroid, Ciliary Body, and Iris. The middle layer consists of three parts: the choroid, the ciliary body, and the iris. The **choroid** is the largest portion of the middle layer. It contains a large amount of melanin (MELL-ah-nin), a pigment that absorbs stray light the way the black interior of a camera does. The blood vessels of the choroid supply nutrients to the eye. Anteriorly, the choroid forms the ciliary body. The **ciliary body** contains smooth muscle fibers that control the shape of the lens, permitting us to focus incoming light. The **iris** is the colored portion of the eye visible through the cornea. Looking in a mirror, you can see a dark opening in the iris called the **pupil**. The pupil allows light to penetrate the eye. The blackness you see through the pupil is the choroid layer and the pigmented section of the retina, discussed below. Like the ciliary body, the iris contains smooth muscle cells. The smooth muscle of the iris regulates the diameter of the pupil. Opening the pupil lets more light in; narrowing it reduces the amount of light that enters. The pupils open and close reflexively in response to light intensity. This reflex is an adaptation that protects the light-sensitive inner layer, the retina. It also allows more light to

TABLE 11-2	Structures and Functions of the Eye	
Structure		Function
Wall		
Outer layer	Sclera	Provides insertion for extrinsic eye muscles
	Cornea	Allows light to enter; bends incoming light
Middle layer	Choroid	Absorbs stray light; provides nutrients to eye structures
	Ciliary body	Regulates lens, allowing it to focus images
	Iris	Regulates amount of light entering the eye
Inner layer	Retina	Responds to light, converting light to nerve impulses
Accessory structures and components	Lens	Focuses images on the retina
	Vitreous humor	Holds retina and lens in place
	Aqueous humor	Supplies nutrients to structures in contact with the anterior cavity of the eye
	Optic nerve	Transmits impulses from the retina to the brain

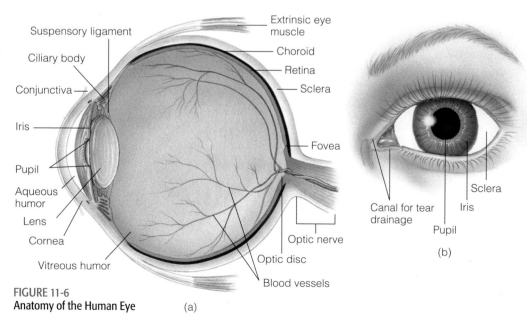

FIGURE 11-6
Anatomy of the Human Eye   (a)

enter to improve vision in dark conditions and restricts light when too much is present, reducing damage to the eye.

The Retina.   The innermost layer of the eye is the **retina**. The retina consists of an outer, pigmented layer and an inner layer consisting of photoreceptors (modified nerve cells that detect light) and associated nerve cells. The retina is weakly attached to the choroid and can become separated from it as a result of trauma to the head. A detached retina can lead to blindness if not quickly repaired by laser surgery.

FIGURE 11-7 **The Retina** (a) Cross section through the wall of the eye, showing (b) the arrangement of the cellular components of the retina. (c) The structure of the rods and cones.

The photoreceptors of the retina are highly modified nerve cells. Two types of photoreceptors are present in the retina: rods and cones (Table 11-3). The **rods**, so named because of their shape, are sensitive to low light (Figures 11-7b and 11-7c). Rods function at night and produce grayish, somewhat vague, black-and-white images. The **cones**, also named because of their shape, operate only in brighter light. They are responsible for visual acuity—sharp vision—and color vision.

As Figure 11-7b shows, the rods and cones synapse with the bipolar neurons. These, in turn, synapse with ganglion cells in the retina. The axons of the ganglion cells unite at the back of the eye to form the **optic nerve**. Because this area contains no photoreceptors, it is insensitive to light. Eye doctors call this the *blind spot*. The blood vessels that enter and leave the eye do so with the optic nerve. An ophthalmologist (eye doctor) can easily see these blood vessels and their branches by shining a light through the pupil onto the back wall of the eye.

Rods and cones are found throughout the retina, but the rods are most abundant. There are an estimated 150 million rods in each eye and only about 6 million cones. However, the cones are concentrated in a tiny region of each eye. In the center of this region is a small depression, about the size of the head of a pin, known as the **fovea centralis** (FOE-vee-ah), which contains only cones. The sharpest vision occurs at the fovea because it contains the highest concentration of cones and also be-

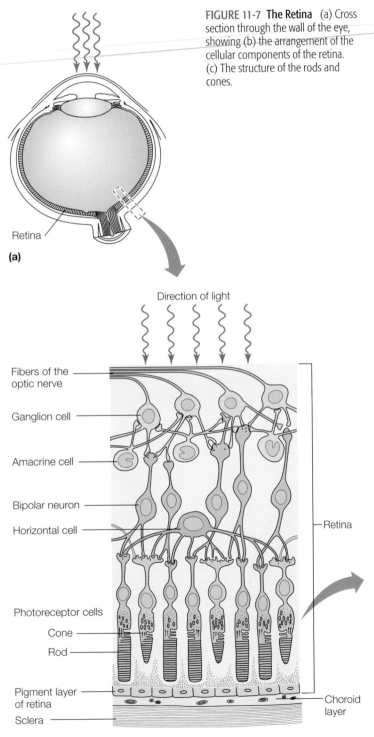

(a)

(b)

TABLE 11-3	Summary of Rods and Cones		
Photoreceptor	Day or Night	Color Vision	Location
Rods	Night vision	No	Highest concentration in the periphery of the retina
Cones	Day vision	Yes	Highest concentration in the macula and fovea

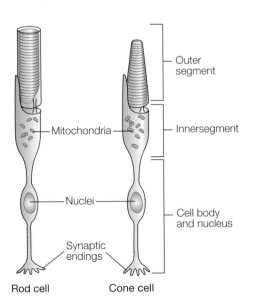

(c)

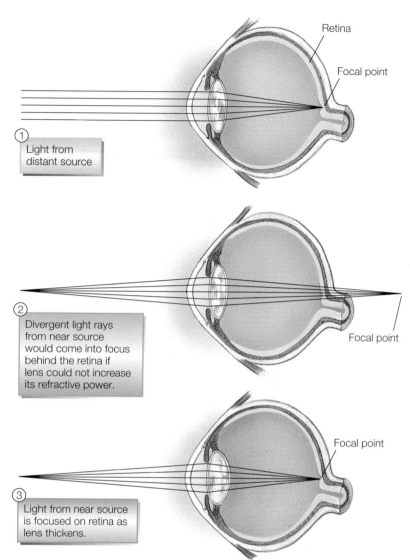

cause the bipolar neurons and ganglion cells do not cover the cones in this region as they do throughout the rest of the retina. The number of cones in the retina decreases progressively from the fovea outward, whereas the number of rods increases. Thus, the greatest concentration of rods is found in the periphery of the retina.

Images from our visual field are cast onto the retina, and impulses are transmitted to the visual cortex of the brain. Some processing of the image occurs in the retina. The rest takes place in the brain. When we focus on an object, the image is projected upside down onto the fovea. Even though the image is upside down, the brain processes the information and gives us a right-side-up image; that is, it allows us to perceive objects in their true position.

The Lens. Light is focused on the retina by the lens. The **lens** is a transparent, flexible structure that lies behind the iris. The lens is attached to the ciliary body by thin fibers. This connection allows the smooth muscle of the ciliary body to alter the shape of the lens, an action necessary for focusing the eye (discussed shortly).

As we age, the lens may develop cloudy spots, or cataracts, that obstruct vision. This condition is most commonly encountered in people who have been exposed to excessive ultraviolet radiation in sunlight. Scientists believe that the risk of developing cataracts may increase as the Earth's ozone layer, which blocks dangerous ultraviolet radiation from the sun, is eroded by chlorofluorocarbons from refrigerators, air-conditioning units, and other sources (Chapter 24). Eye surgeons correct the condition by replacing the cloudy lens with a plastic lens that helps to restore vision.

The lens separates the interior of the eye into two cavities of unequal size. Everything in front of the lens is the **anterior chamber**; everything behind it is the **posterior chamber**. The posterior cavity is filled with a clear, gelatinous material, the **vitreous humor** (VEH-tree-ous HUE-mur; "glassy liquid"). The anterior cavity contains a thin liquid, chemically similar to blood plasma, called the **aqueous humor** ("watery fluid"). The aqueous humor provides nutrients to the cornea and lens and carries away cellular wastes.

In normal, healthy individuals, aqueous humor production is balanced by absorption. If the outflow is blocked, however, the aqueous humor builds up inside the anterior chamber, creating internal pressure. This disease, called **glaucoma** (glaw-COE-mah), progresses gradually and imperceptibly. If untreated, the pressure inside the eye can damage the retina and optic nerve, causing blindness.

Focusing Light on the Retina

The cornea and lens focus light on the retina.

Incoming light is focused on the photoreceptors of the retina by the lens. However, the cornea also acts like a lens. That is, it helps focus light. Unlike the lens, though, the cornea cannot be adjusted to focus on nearby objects.

The lens is a flexible structure whose shape is controlled by the muscles in the ciliary body. These muscles are attached to

① Light from distant source

② Divergent light rays from near source would come into focus behind the retina if lens could not increase its refractive power.

③ Light from near source is focused on retina as lens thickens.

Retina · Focal point · Focal point · Focal point

FIGURE 11-8 Refraction of Light by the Cornea and Lens (1) Light rays from distant objects are parallel when they strike the eye. The refractive power of the cornea and resting lens are sufficient to bring them into focus on the retina. (2) Light rays from nearby objects are divergent. The cornea has a fixed refractive power and cannot change. The rays would focus behind the retina if the lens could not alter its refractive power. (3) To focus the image, the ciliary muscles contract. This lessens pressure on the suspensory ligaments, which allows the lens to thicken and shorten, becoming more curved and more refractive.

the lens by the suspensory ligament, tiny filamentous strands. As Figure 11-8b shows, light rays from nearby objects are divergent. To focus on nearby objects, the lens thickens and shortens, becoming more curved, in much the same way that a rubber ball flattened between your hands will return to normal when you reduce the pressure on it. This automatic adjustment in the curvature of the lens as it focuses on a nearby object is called *accommodation* (Figure 11-9).

Synchronized Eye Movement and Convergence. Six muscles located outside the eye are responsible for eye movement (Figure 11-10). As noted earlier, these muscles attach to the bony eye socket and to the sclera (the white of the eye).

The eyes generally move in unison like a pair of synchronized swimmers. Synchronized movement is an evolutionary adaptation that ensures that images are focused on the foveas

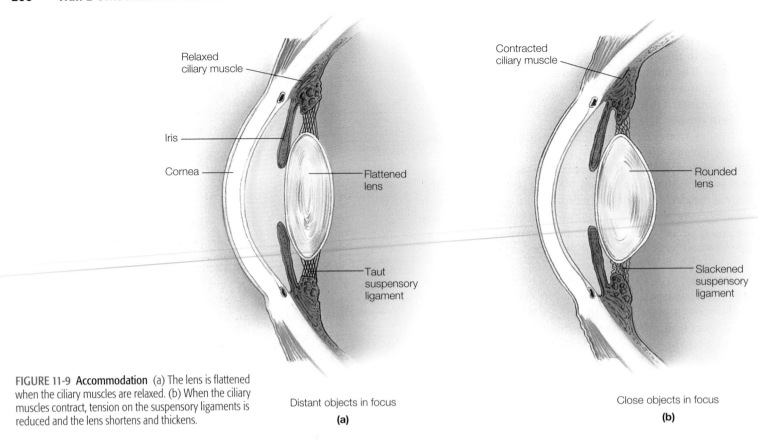

FIGURE 11-9 Accommodation (a) The lens is flattened when the ciliary muscles are relaxed. (b) When the ciliary muscles contract, tension on the suspensory ligaments is reduced and the lens shortens and thickens.

Distant objects in focus

(a)

Close objects in focus

(b)

of both eyes at the same time. When a nearby object is viewed, the eyes turn inward—that is, they converge. Convergence ensures that near images are focused on each fovea. Because convergence occurs during all near-point work—reading, writing, sewing, knitting—it tends to strain the extrinsic eye muscles, creating eyestrain and headaches.

Visual Problems

Alterations in the shape of the lens and eyeball cause the most common visual problems.

In a relaxed eye with perfect vision, objects farther than 6 meters (20 feet) away fall into perfect focus on the retina (Figure 11-11a). Many individuals, however, have imperfectly shaped eyeballs or defective lenses. These imperfections result in three visual problems: nearsightedness, farsightedness, and astigmatism.

Nearsightedness.   Nearsightedness results when the eyeball is slightly elongated (Figure 11-11b). Without corrective lenses, parallel light rays arising from distant images come into focus in front of the retina, creating a fuzzy image. In contrast, nearby images with much more divergent light rays tend to be in focus. People with nearsightedness, therefore, can see near objects without corrective lenses—hence the name. Nearsightedness may also result when the lens is too strong—that is, too concave. This lens bends the light coming from distant objects too much, causing the image to come into focus in front of the retina.

Nearsightedness is caused by many factors and tends to run in families; it generally appears around age 12, often worsening until a person reaches 20. Nearsightedness can be corrected by contact lenses or prescription glasses that cause

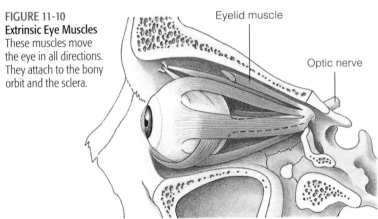

FIGURE 11-10
Extrinsic Eye Muscles
These muscles move the eye in all directions. They attach to the bony orbit and the sclera.

Eyelid muscle

Optic nerve

incoming light rays to diverge so the image comes in focus on the retina (Figure 11-11b).

Farsightedness.   Farsightedness is the opposite of nearsightedness. It results when the eyeballs are too short or the lens is too weak (too convex). In the eyes of farsighted individuals, parallel light rays from distant objects usually fall into focus on the retina, but divergent rays from nearby objects cannot be focused sufficiently. Without corrective lenses, farsighted individuals see distant objects well, but nearby objects are fuzzy. Glasses or contact lenses that bend the light inward bring near objects into sharp focus on the retina (Figure 11-11c). Farsightedness is generally present from birth and tends to run in families.

Astigmatism.   The cornea and lens are generally uniformly curved, but in some individuals they are slightly disfigured. An eye doctor explained it to me as follows. He said that the cornea

FIGURE 11-11 **Common Visual Problems**

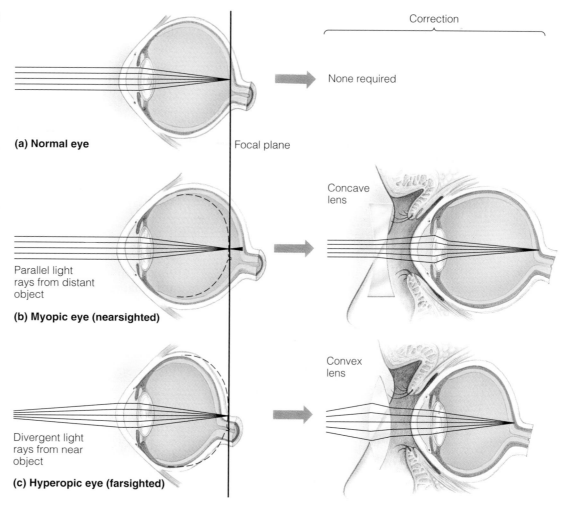

(a) **Normal eye**

Focal plane

Correction

None required

Parallel light rays from distant object

(b) **Myopic eye (nearsighted)**

Concave lens

Divergent light rays from near object

(c) **Hyperopic eye (farsighted)**

Convex lens

of a normal eye is round like a basketball. In an astigmatic eye, the cornea is shaped more like the surface of a football. This unequal curvature is called an **astigmatism** (a-STIG-mah-tiz-em). Astigmatism creates fuzzy images because light rays are bent differently by the different parts of the cornea. Astigmatism is usually present from birth and does not grow worse with age. It can also be corrected with glasses or contact lenses.

Laser Surgery. Visual defects can also be corrected by laser surgery. One of the most popular types of surgery is LASIK. In this procedure, a computer is used to map the surface of the cornea and calculate the amount of corneal tissue that needs to be removed to ensure proper vision. A thin flap of cornea is then partially shaved off using an extremely sharp blade. This creates a partial flap (Figure 11-12a). A laser is then used to remove underlying tissue and reshape the cornea (Figure 11-12b). The flap is then folded back into its original position (Figure 11-12c). It heals shortly.

Laser eye surgery is quick and accurate, with 95 percent of patients achieving 20/40 visual acuity or better. Worldwide, hundreds of thousands of patients have been successfully treated with this procedure. Problems arise in under one percent of the cases.

Presbyopia. The aging process brings with many changes: hair loss, hearing loss, and arthritis. It also results in a decline in the resiliency of the lens, making it hard to focus on nearby objects. This condition is known as **presbyopia** (PREZ-bee-OPE-ee-a). It usually begins around the age of 40 and can be corrected by glasses worn for near-point work such as reading. Presbyopia cannot be corrected by laser eye surgery, because the problem is with the lens, not the cornea.

(a)

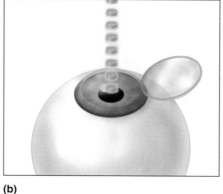

(b)

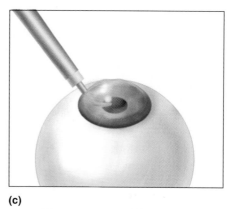

(c)

FIGURE 11-12 **LASIK Surgery** (a) Microkeratome slices off a thin layer of corneal tissue. (b) Laser burns away corneal tissue to correct eyesight. (c) Flap is restored, as is vision.

Color Blindness.   About 5 percent of the human population suffers from **color blindness**, a hereditary disorder. Color blindness is more prevalent in men than women and ranges from an inability to distinguish certain shades of color to a complete inability to perceive color. The most common form of this disorder is red-green color blindness. In individuals with red-green color blindness, the red or green cones may either be missing altogether or be present in reduced numbers. If the red cones are missing, red objects appear green. If the green cones are missing, however, green objects appear red.

Many color-blind people are either unaware of their condition or untroubled by it. They rely on a variety of visual cues such as differences in intensity to distinguish red and green objects. They also rely on position cues. In vertical traffic lights, for example, the red light is always at the top of the signal; green is on the bottom. Although the colors may appear more or less the same, the position of the light helps color-blind drivers determine whether to hit the brakes or step on the gas.

11-5   Hearing and Balance: The Cochlea and Middle Ear

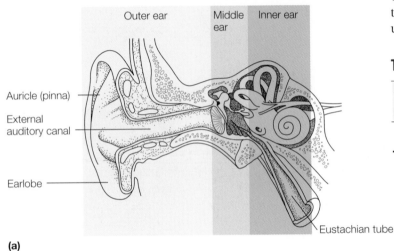

(a)

The human ear is an organ of special sense. It serves two functions: it detects sound, and it detects body position, enabling us to maintain balance (Table 11-4).

The Anatomy of the Ear

> The ear consists of three anatomically separate portions: the outer, middle, and inner ears.

The Outer Ear.   The outer ear consists of an irregularly shaped piece of cartilage covered by skin, the auricle (OR-eh-kul), and the earlobe, a flap of skin that hangs down from the auricle (Figure 11-13). The outer ear also consists of a short tube, the **external auditory canal**, which transmits airborne sound waves to the middle ear (Figure 11-13a). The external auditory canal is lined by skin containing modified sweat glands that produce earwax. Earwax traps foreign particles, such as bacteria, and contains a natural antibiotic substance that may reduce ear infections.

The Middle Ear.   The middle ear lies entirely within the skull (Figure 11-13b). The eardrum, or **tympanic membrane** (tim-PAN-ick), separates the middle ear cavity from the external auditory canal. The tympanic membrane vibrates when struck by sound waves in much the same way that a guitar string vibrates when a note is sounded by another nearby instrument.

Inside the middle ear are three minuscule bones, the **ossicles** (OSS-eh-kuls). Starting from the outside, they are the **malleus** (pronounced MAL-ee-us; hammer), **incus** (IN-cuss; anvil), and **stapes** (STAY-pees; stirrup). As illustrated in Figure 11-13b, the hammer-shaped malleus lies next to the tympanic membrane. When the membrane is struck by sound waves, it vibrates. This causes the malleus to rock back and forth. The malleus, in turn, causes the incus to vi-

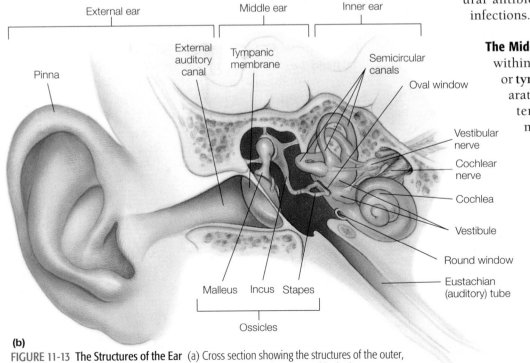

(b)

FIGURE 11-13 **The Structures of the Ear** (a) Cross section showing the structures of the outer, middle, and inner ears. (b) The receptors for balance and sound are located in the inner ear.

TABLE 11-4	Structures and Functions of the Ear	
Part	Structure	Function
Outer ear	Auricle	Funnels sound waves into external auditory canal
	Ear lobe	
	External auditory canal	Directs sound waves to the eardrum
Middle ear	Tympanic membrane, or eardrum	Vibrates when struck by sound waves
	Ossicles	Transmit sound to the cochlea in the inner ear
Inner ear	Cochlea	Converts fluid waves to nerve impulses
	Semicircular canals	Detect head movement
	Saccule and utricle	Detect head movement and linear acceleration

brate, which causes the stapes, the stirrup-shaped bone, to move in and out against the **oval window**, an opening to the inner ear covered with a membrane like the head of a drum. Thus, vibrations created in the eardrum are amplified as they are transmitted to the inner ear, where the sound receptors are located.

The Eustachian Tube

The Eustachian tube helps to equalize pressure in the middle ear.

As Figure 11-13 illustrates, the middle ear cavity opens to the pharynx via the **auditory**, or **eustachian**, **tube** (you-STAY-shun). The eustachian tube serves as a pressure valve. Normally, the eustachian tube is closed. Yawning and swallowing, however, cause it to open, allowing air to flow into or out of the middle ear cavity. This equalizes the internal and external pressure on the eardrum. You have noticed your ears "pop" when taking off in an airplane. This is the result of pressure being released by the eustachian tubes.

The Inner Ear.   The inner ear occupies a large cavity in the temporal bone of the skull and contains two sensory organs, the cochlea and the vestibular apparatus. The **cochlea** (COE-klee-ah) is shaped like a snail shell and houses the receptors for hearing.

The **vestibular apparatus** consists of two parts: the semicircular canals and the vestibule (Figure 11-13b). The **semicircular canals** are three ringlike structures set at right angles to one another. They house receptors for body position and movement. The vestibule is a bony chamber lying between the cochlea and semicircular canals. It houses receptors that respond to body position and movement.

Structure and Function of the Cochlea

Hearing requires the participation of several structures.

As noted previously, sound waves enter the external auditory canal, where they strike the eardrum. The eardrum then vibrates back and forth, causing the bones of the middle ear cavity to vibrate. This transmits sound waves to the cochlea, which houses the receptor for sound.

The cochlea is a hollow, spiral structure. A cross section through this structure reveals three fluid-filled canals (Figure 11-14a). Inside the middle cavity is the receptor for sound, the **organ of Corti**. It rests on a flexible membrane called the **basilar membrane** (BAZ-eh-ler).

The organ of Corti contains receptor cells, the hair cells (Figure 11-14b). When sound waves are transmitted into the inner ear, they create pressure waves in the fluid in the uppermost canal of the cochlea. Fluid pres-

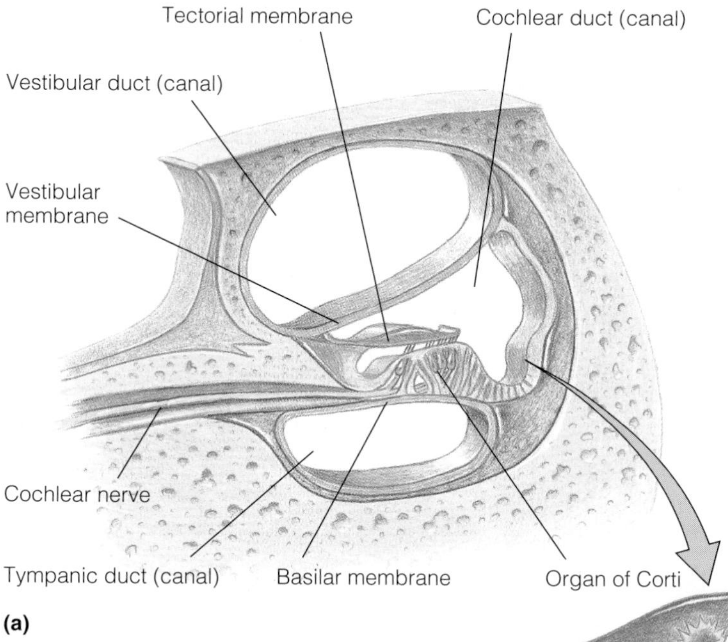

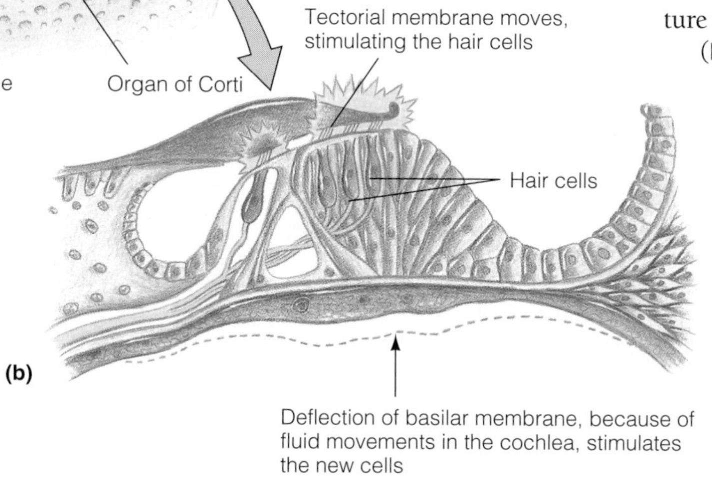

FIGURE 11-14 **Cross Section Through the Cochlea** (a) Notice the three fluid-filled canals and the central position of the organ of Corti. (b) Hair cells of the organ of Corti are embedded in the overlying tectorial membrane. When the basilar membrane vibrates, the hair cells are stimulated.

Tectorial membrane

Cochlear duct (canal)

Vestibular duct (canal)

Vestibular membrane

Cochlear nerve

Tympanic duct (canal)

Basilar membrane

Organ of Corti

(a)

Tectorial membrane moves, stimulating the hair cells

Hair cells

(b)

Deflection of basilar membrane, because of fluid movements in the cochlea, stimulates the new cells

sure waves created in the vestibular canal then travel through the vestibular membrane into the middle canal. From here, the pressure waves pass through the basilar membrane into the lowermost canal. Pressure is relieved by the outward bulging of the **round window**, an opening in the bony cochlea, which, like the oval window, is spanned by a flexible membrane.

In their course through the cochlea, the pressure waves cause the basilar membrane to vibrate. This vibration stimulates the hair cells, which contact the overlying tectorial membrane (teck-TORE-ee-al). The hair cells stimulate the dendrites wrapped around their bases and they create nerve impulses that travel to the auditory area of the brain via the **vestibulocochlear nerve** (vess-TIB-you-low-COE-klee-er), one of the 12 cranial nerves.

Distinguishing Pitch and Intensity.   A pitch pipe helps a singer find the note to begin a song. It has a range of notes from high to low. The ear can distinguish between these various pitches, or frequencies, in large part because of the structure of the basilar membrane. The basilar membrane underlying the organ of Corti is stiff and narrow at the oval window, where fluid pressure waves are first established inside the cochlea (Figure 11-15b). As the basilar membrane proceeds to the tip of the spiral, however, it becomes wider and more flexible. The change in width and stiffness results in marked differences in its ability to vibrate. The narrow, stiff end, for example, vibrates maximally when pressure waves from high-frequency sounds ("high notes") are present (Figure 11-15c). The far end of the membrane vibrates maximally with low-frequency sounds ("low notes"). In between, the membrane responds to a wide range of intermediate frequencies.

FIGURE 11-15 **The Transmission of Sound Waves Through the Cochlea** The cochlea is unwound here to simplify matters. (a) Vibrations are transmitted from the stirrup (stapes) to the oval window. Fluid pressure waves are established in the vestibular canal and pass to the tympanic canal, causing the basilar membrane to vibrate. (b) A representation of the basilar membrane, showing the points along its length where the various wavelengths of sound are perceived. Notice that the basilar membrane is narrowest at the base of the cochlea at the oval window end and widest at the apex. (c) High-frequency sounds set the basilar membrane near the base of the cochlea into motion. Hair cells send impulses to the brain, which interprets the signals as a high-pitch sound. Low-frequency sounds stimulate the basilar membrane where it is widest and most flexible.

Pressure waves caused by sound of a certain pitch stimulate one specific region of the organ of Corti. The hair cells stimulated in that region send impulses to the brain, which it interprets as a specific pitch. Each region of the organ of Corti sends impulses to a specific region of the auditory cortex in the temporal lobe of the brain.

The intensity of a sound, or its loudness, depends on how much vibration occurs in a portion of the basilar membrane. The louder the sound, the more vigorous the vibration. The more vigorous the vibration of the eardrum, the greater the deflection of the basilar membrane in the area of peak responsiveness. The greater the deflection of the basilar membrane, the more hair cells are stimulated. Loud rock music or sirens, however, cause extreme vibrations in the basilar membrane, destroying hair cells over time and causing partial deafness.

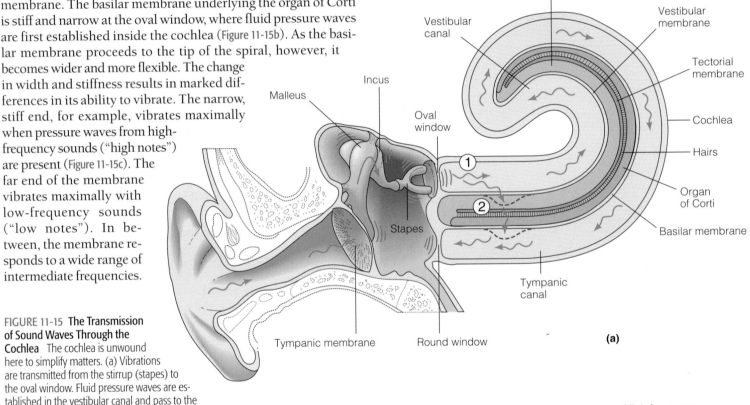

(a)

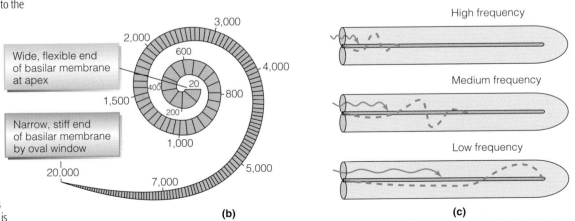

(b)   (c)

The numbers indicate the frequencies with which different regions of the basilar membrane maximally vibrate.

Hearing Loss. As people grow older, many lose their hearing gradually—so slowly, in fact, that most people are unaware until the losses are significant. In some cases, though, people lose their hearing suddenly. A loud explosion, for example, can damage the hair cells or even break the tiny ossicles.

Hearing losses may be temporary or permanent, partial or complete. Hearing loss falls into one of two categories, depending on the part of the system that is affected. The first is conduction deafness, which occurs when the conduction of sound waves to the inner ear is impaired. Conduction deafness may result from a rupture of the eardrum or damage to the ossicles. Conduction deafness most often results from infections in the middle ear. Bacterial infections may result in the buildup of scar tissue that causes the ossicles to fuse and lose their ability to transmit sound. Infections of the middle ear usually enter through the eustachian tube. Thus, an infection in the throat can easily spread to the ear, where it requires prompt treatment. Ear infections are especially common in babies and young children. If undetected, middle ear infections can slow down the development of speech. If untreated, such infections can lead to permanent deafness. Conduction deafness is treated by hearing aids that fit in the ear or just behind it.

The second type of hearing loss is neurological and is called *nerve deafness*. It results from physical damage to the hair cells, the vestibulocochlear nerve, or the auditory cortex. Explosions, extremely loud noises, and some antibiotics can damage the hair cells in the organ of Corti, creating partial to complete deafness. The auditory nerve, which conducts impulses from the organ of Corti to the cortex, may degenerate, thus ending the flow of information to the cortex. Tumors in the brain or strokes also can destroy the cells of the auditory cortex.

The Vestibular Apparatus

> The vestibular apparatus houses receptors that detect body position and movement.

The vestibular apparatus consists of the two parts, the semicircular canals and the vestibule (Figure 11-16). Both are involved in proprioception.

The Semicircular Canals. The three semicircular canals are arranged at right angles to one another. Each canal is filled with a fluid. As Figure 11-16a shows, the base of each semicircular canal expands. On the inside wall of enlarged area is a small ridge of tissue, or crista (CHRIS-tah). Each crista consists of a patch of receptor cells. Each receptor cell contains numerous microvilli and a single cilium embedded in a cap of gelatinous material, the cupula (CUE-pew-lah), shown in Figure 11-16b.

Rotation of the head causes the fluid in the semicircular canals to move. The movement of the fluid deflects the cupula, which stimulates the hair cells. They, in turn, stimulate the dendrites of sensory nerves that wrap around the base of the receptor cells. The sensory nerves send impulses to the brain, alerting it to the rotational movement of the head and body.

Because the semicircular canals are set in all three planes of space, movement in any direction can be detected. By alerting the brain to rotation and movement, the semicircular canals contribute to our sense of balance. They are therefore important to dynamic balance—helping us stay balanced when in movement.

The Utricle and Saccule. Two additional receptors play a role in balance and the detection of movement, the utricle (YOU-trehkul) and saccule (SACK-yule) (Figure 11-16). These membranous

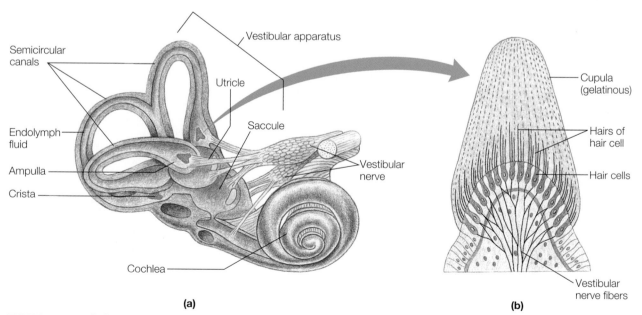

(a) **(b)**

FIGURE 11-16 **Vestibular Apparatus and Semicircular Canals** (a) This illustration shows the location of the cristae in the ampullae of the semicircular canals. The semicircular canals are filled with endolymph. (b) When the head spins, the endolymph is set into motion, deflecting the gelatinous cupula of the crista, thus stimulating the receptor cells.

compartments inside the vestibule provide input under two distinctly different conditions: at rest and during acceleration in a straight line. As a result, these receptors contribute dynamic balance and static balance (helping us stay balanced when we are not moving).

The utricle and saccule contain small receptor organs similar to the cristae of the semicircular canals. The nerve impulses generated by the receptors in the semicircular canals and the utricle and saccule are sent to a cluster of nerve cell bodies in the brain stem. Here all of the information the receptors generate on position and movement is integrated with input from both the eyes and receptors in the skin, joints, and muscles (earlier in this chapter). From this center, information flows in many directions. One major pathway leads to the cerebral cortex. This path makes us conscious of our position and movement. Another path leads to the muscles of the limbs and torso. Signals to the muscles maintain our balance and, if necessary, correct body position.

In some people, activation of the vestibular apparatus results in motion sickness, characterized by dizziness and nausea. The exact cause of motion sickness is not known.

11-6 Health and Homeostasis

Many general sense receptors play an important role in homeostasis. Mechanoreceptors that detect changes in blood pressure and chemoreceptors that respond to the ionic concentration of the blood, for example, help maintain blood pressure and proper water balance through mechanisms involving the kidney. Chemoreceptors that detect levels of carbon dioxide and hydrogen ions in the blood and cerebrospinal fluid help to regulate respiration. Chapter 13 describes additional chemoreceptors that detect levels of various hormones, nutrients, and ions in the blood and body fluids. These detectors may stimulate hormonal responses that correct potentially disruptive chemical imbalances.

The special senses help us—and other animals—acquire food and other necessities essential for normal health and homeostasis. They also stimulate behavioral responses that are essential to homeostasis.

Our environment, however, can alter homeostasis. Consider noise.

Noise damages hearing over time. It may also disturb rest and sleep, both important to maintaining homeostasis and health. Noise also raises our level of stress, which, in turn, shortens our lives. Reducing noise in our environment can therefore help us live less stressful and much healthier lives.

SUMMARY

The General and Special Senses

1. The body contains two types of receptors—those that provide general senses and those that provide special senses.

The General Senses

2. The general senses are pain, light touch, pressure, temperature, and position sense. Receptors for these senses may be naked or encapsulated nerve endings.
3. Receptors fit into five functional categories: (a) mechanoreceptors, (b) thermoreceptors, (c) photoreceptors, (d) chemoreceptors, and (e) pain receptors (nociceptors).
4. Sensory stimuli may be internal or external. Some stimuli cause reflex actions. Still others may stimulate physiological changes.
5. Many sensory receptors cease responding to prolonged stimuli, a phenomenon called sensory adaptation. Pain, temperature, pressure, and olfactory receptors all adapt.

Taste and Smell: The Chemical Senses

6. The special senses include taste, smell, vision, hearing, and balance.
7. Taste receptors called taste buds are located principally on the upper surface of the tongue. Food molecules dissolve in the saliva and bind to the membranes of the microvilli of the receptor cells inside taste buds.

8. Taste buds respond to five flavors: salty, bitter, sweet, sour, and umami.
9. The receptors for smell are located in the olfactory membrane in the roof of the nasal cavities. The receptor cells in the olfactory membrane respond to thousands of different molecules.

The Visual Sense: The Eye

10. The eye is the receptor for visual stimuli.
11. The wall of the human eye consists of three layers. The outermost layer consists of the sclera (the white of the eye) and the cornea (the clear anterior structure that lets light shine in).
12. The middle layer consists of the choroid (the pigmented and vascularized region), the ciliary body (whose muscles control the lens), and the iris (which controls the amount of light entering the eye).
13. The innermost layer is the retina, the light-sensitive layer, which contains two types of photoreceptors: rods and cones. Rods function in dim light and provide black-and-white vision, and cones operate in bright light and provide color vision. Cones are also responsible for visual acuity.
14. Light is focused on the retina by the cornea and lens.
15. The eye is divided into two cavities. The posterior cavity lies behind the lens and is filled with a gelatinous material, the vitreous humor. The anterior cavity is filled with a plasmalike fluid called the aqueous humor, which nourishes the lens and other eye structures in the vicinity. If the rate of absorption of aqueous humor decreases, however, pressure can build inside the anterior cavity, resulting in glaucoma.
16. As people age, the lens becomes less resilient and less able to focus on nearby objects.
17. Abnormalities in shape of the lens or eyeball can result in visual defects. The two most common are nearsightedness and farsightedness. Another abnormality in the cornea is astigmatism. All three conditions can be corrected by eyeglasses, contact lenses, or LASIK surgery.
18. Color blindness is a genetic disorder, more common in men, and results from a deficiency or an absence of one or more types of cones. Red-green color blindness is the most common type.

Hearing and Balance: The Cochlea and Middle Ear

19. The human ear consists of three portions: the outer, middle, and inner ears.
20. The outer ear consists of the auricle and external auditory canal, both of which direct sound to the eardrum.
21. The middle ear consists of the eardrum and three small bones that transmit vibrations to the inner ear.

22. The inner ear contains the cochlea, where the receptors for sound are located. The inner ear also houses receptors for movement and head position: the semicircular canals, utricle, and saccule.

23. Sound waves create vibrations in the eardrum and ossicles, which are transmitted to fluid in the cochlea. Pressure waves in the cochlea cause the basilar membrane to vibrate, which stimulates the hair cells of the organ of Corti. Pressure waves resulting from any given sound cause one part of the membrane to vibrate maximally. The hair cells stimulated in that region send signals to the brain, which it interprets as a specific frequency.

24. The semicircular canals are three hollow rings filled with a fluid. The receptors for head movement are located in an enlarged portion at the base of each canal. Fluid movement inside the semicircular canals deflects the gelatinous cap, stimulating the underlying receptor cells and alerting the brain to head movements.

25. Two membranous sacs in the vestibule, the utricle and saccule, contain receptors that respond to linear acceleration and tilting of the head.

Health and Homeostasis

26. Receptors that provide for the general and special senses are extremely important to homeostasis. They help us adjust internally to respond to external changes.

27. Some external stimuli such as noise can alter homeostasis. Noise, for instance, damages the structures of the internal ear. It can also disturb sleep and can lead to stress, both of which impair human health.

28. Noise damages the ears. Extremely loud noises can rupture the eardrum or break the ossicles. Less intense noises, however, generally destroy hearing gradually by damaging hair cells. In most people, hearing loss occurs so gradually as to be undetected.

29. Besides deafening us and cutting us off from the important sounds of modern life, noise disturbs communications, rest, and sleep— the last two of which are particularly important to homeostasis and health. Noise also raises our level of stress, which, in turn, shortens our lives.

critical thinking

Thinking Critically—Analysis

This Analysis corresponds to the Thinking Critically scenario that was presented at the beginning of this chapter.

If I were in this predicament, I would want to learn more about the researchers' experiments. Were they conducted properly? Did they have an adequate number of animals? Have others repeated the experiments with the same results? Did the chemical toxins affect other cells of the nervous system? What were the side effects? Did the procedure result in a lasting cure or was it just transient?

How long did they follow the rats they used to determine the long-term effectiveness and side effects of the treatment? Has this procedure or any similar procedures ever been tried on people before? What were the results? What kind of care will be provided if you suffer any side effects? Will the procedure render you insensitive to pain medications?

Clearly, there's a lot to be learned before you subject yourself to this potentially beneficial procedure. While the promise of pain relief may be overwhelming, the stakes are high.

KEY TERMS AND CONCEPTS

 www.jbpub.com/humanbiology/5e

The site features eLearning, an online review area that provides quizzes, chapter outlines, and other tools to help you study for your class. You can also follow useful links for in-depth information, research the differing views in the Point/Counterpoints, or keep up on the latest health news.

CONCEPT REVIEW

1. Define the terms general senses and special senses. p. 197
2. Using your knowledge of the senses and of other organ systems gained from previous chapters, describe the role that sensory receptors play in homeostasis. Give specific examples to illustrate your points. pp. 197–200
3. Make a list of both the encapsulated and the nonencapsulated general sense receptors. Note where each is located and what it does. pp. 197–200
4. Define the term adaptation. What advantages does it confer? Can you think of any disadvantages? p. 200
5. Describe the receptors for taste, explaining where they are located, what they look like, and how they operate. p. 201

6. Taste buds detect five basic flavors. What are they? How do you account for the thousands of different flavors that you can detect? pp. 201–202
7. Describe the olfactory membrane. How do the receptor cells operate? In what ways are taste receptors and olfactory receptors similar? In what ways are they different? p. 202
8. Explain the following statement: Taste and smell are complementary. pp. 202–203
9. Draw a cross section of the human eye and label its parts. p. 203
10. Define the following terms: retina, rods, cones, fovea centralis, and optic nerve. p. 204
11. Compare and contrast rods and cones. pp. 204–205

12. Define the following terms: nearsightedness, farsightedness, presbyopia, and astigmatism. pp. 206–207
13. What is color blindness? What is the most common type? Explain what your world would look like if you were afflicted by red-green color blindness. p. 208
14. What is a cataract, and how is it treated? p. 205
15. Describe the disease known as glaucoma. p. 205
16. Describe the anatomy of the ear and the role of the outer, middle, and inner ears in hearing. pp. 208–209
17. How do the semicircular canals operate? How do the utricle and saccule operate? pp. 211–212

SELF-QUIZ: TESTING YOUR KNOWLEDGE

1. Pain, temperature, and light touch are considered _____ senses, while vision and hearing are considered _____ senses. p. 197
2. The rods and cones of the retina are a type of receptor known as _____. p. 204
3. Pain felt on the surface of the body away from its internal source is known as _____ pain. p. 198
4. Naked nerve endings wrapped around the base of hair follicles is responsible for the sensation known as _____. p. 198
5. The sense of position is also known as _____. p. 211
6. _____ are structures that house receptors that respond to chemicals dissolved in the food we eat. p. 201
7. Smells are detected by sensors found in the _____ epithelium in the roof of the nasal cavities. p. 202
8. The white of the eye is known as the _____; it joins the clear portion of the eye in front known as the _____. p. 203
9. Light is focused on the retina by the _____ and the _____. p. 203
10. _____ are receptors that permit color vision. p. 204
11. Clouding of the lens is a condition known as _____. p. 205
12. Fluid buildup in the anterior chamber of the eye is called _____. p. 205

13. The gelatinous material in the posterior cavity is known as _____ humor. p. 205
14. An irregularly shaped cornea (football shaped) is responsible for a visual defect known as _____. p. 207
15. As one ages, the _____ becomes less elastic and unable to focus light from nearby objects resulting in a condition known as presbyopia. p. 207

16. The tiny bones in the middle ear are known as _____. They transmit sound waves from the _____ to the oval window. pp. 208–209
17. Fluid waves in the cochlea stimulate hair cells in the organ of _____. p. 209
18. The _____ canals are located in the inner ear; they are important in dynamic balance. p. 211
19. Label the diagram.

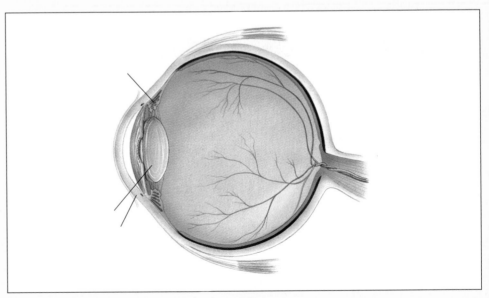

The Skeleton and Muscles

A two-year study on osteoporosis in Australian women involving 120 nonsmokers, ranging in age from 50 to 60, concluded that the best prevention against bone fractures is estrogen-replacement therapy. (Estrogen is a female sex hormone that stimulates bone formation.) During the study, the researchers examined the density of bone in the forearms of the women at 3-month intervals. Initial measurements showed very low levels of calcium, which indicates a high risk of fractures.

The women were then divided into three groups. The first group received a placebo, the second group received 1 gram of calcium per day, and the third group received daily doses of estrogen and another hormone, progesterone. All the women were encouraged to take two brisk 30-minute walks every day and engage in one low-impact aerobics class each week.

The study showed that bone loss continued in the control group (group 1, exercise only) at a rate of 2.5% per year. Bone loss also continued in group 2, which received a daily dose of calcium, but was slower than in group 1 (0.5% to 1.3% per year). In group 3, whose participants exercised and received hormone therapy, bone density increased from 0.8% to 2.7% per year.

Given the potential side effects of hormone therapy, the researchers recommended that only women with the lowest bone densities should receive hormone therapy and that others should exercise and take calcium to reduce their risk of osteoporosis.

Can you pinpoint any potential flaws in the experiment's design or execution? How could they be corrected?

Thomas Powers is a 40-year-old college professor. Like many of his friends, he likes to ski. But lately, his skiing has been curtailed by chronic knee pain from an old skiing injury. After visiting his doctor, he found that years ago, a spill on the slopes tore some cartilage in his knee. Although it never really healed, the tear caused few problems until he recently took up mountain biking. The additional strain combined with some degeneration of the knee joint itself conspired to create an intolerable situation.

This chapter discusses the human skeleton, including the joints, like those troubling Professor Powers. It also examines the skeletal muscles. Together, the bones, joints, and muscles constitute the **musculoskeletal system**. In adults, bones and muscles make up a remarkable 50%–60% of the body weight.

12-1   Structure and Function of the Human Skeleton

Besides providing internal support, bones play an important role in homeostasis.

The word **skeleton** is derived from a Greek word that means "dried-up body." Your first impression of bone might not be much different (Figure 12-1). Bone, after all, appears to be a dry, dead structure. But appearances can be deceiving. The truth be known, bone is a living, metabolically active tissue. Bone tissue contains numerous cells, known as **osteocytes** (OSS-tee-oh-SITES). These cells are embedded in an extracellular material, known as the **matrix** (MAY-tricks). The matrix consists of calcium phosphate crystals, which impart hardness and strength. It is deposited on an organic component, **collagen fibers** (COLL-ah-gin), which are made of protein. The collagen imparts flexibility.

The human skeleton consists of 206 bones, discrete structures made of bone tissue (Figure 12-2). Bones provide internal structural support, giving shape to our bodies and enabling us to maintain an upright posture. Some bones protect internal organs. The rib cage, for instance, protects the lungs and heart, and the skull protects the brain. Bones serve as the site of attachment for the tendons of many skeletal muscles whose contraction results in purposeful movements. Bones are home to cells that give rise to red blood cells, white blood cells, and platelets. Bones are a storage depot for fat, needed for cellular energy production both at work and at rest. Finally, bones are a reservoir of calcium and other minerals. Bones release and absorb calcium as needed, thus helping to maintain normal blood levels. That's important, in part, because calcium is required for muscle contraction. Disturbances in blood calcium levels can impair muscle contraction.

The Skeleton

The human skeleton consists of two parts.

The human skeleton consists of the axial and appendicular skeletons (Figure 12-2). The **axial skeleton** (AXE-ee-al) forms the long axis of the body. It consists of the skull, the vertebral column, and the rib cage. The **appendicular skeleton** (ah-pen-DICK-you-lar) consists of the bones of the arms and legs as well as the bones of the shoulders and pelvis, to which the upper and lower extremities are attached.

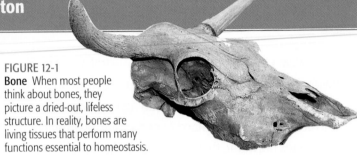

FIGURE 12-1
Bone When most people think about bones, they picture a dried-out, lifeless structure. In reality, bones are living tissues that perform many functions essential to homeostasis.

Compact and Spongy Bone

Bones have a dense outer layer and a less compact central region.

Take a moment to study the skeleton in Figure 12-2. As you examine it, you may notice that bones come in a variety of shapes and sizes. Some are long, some are short, and some are flat and irregularly shaped. Nevertheless, all bones share some characteristics. Consider one of the long bones of the body, the humerus (HUGH-mer-us), which is located in the upper arm.

Figure 12-3 illustrates the anatomy of the humerus. As shown, this bone, like other long bones of the body, consists of a long, narrow shaft, the diaphysis (die-AF-eh-siss), and two expanded ends, the epiphyses (ee-PIF-eh-seas). The ends of the epiphyses are covered with a thin layer of hyaline cartilage (Chapter 4), which reduces friction in joints. Like other bones, the humerus consists of an outer shell of dense material called **compact bone**. Inside the humerus is a mass of **spongy bone**, so named because of its spongy appearance. Spongy bone forms an interlacing network that contains many small cavities. It is therefore less dense than compact bone.

On the outer surface of the compact bone is a layer of connective tissue, the **periosteum** (PEAR-ee-oss-TEA-um; "around the bone"). The periosteum serves as the site of attachment for skeletal muscles (via tendons). It also contains cells that make new bone during remodeling or repair.

The periosteum is richly supplied with blood vessels, which enter the bone at numerous locations. Blood vessels travel through small canals in the compact bone and course through the inner spongy bone, providing nutrients and oxygen and carrying off cellular wastes such as carbon dioxide. The periosteum is also richly supplied with nerve fibers. The majority of the pain felt after a person bruises or fractures a bone results from stimulation of pain fibers in the periosteum.

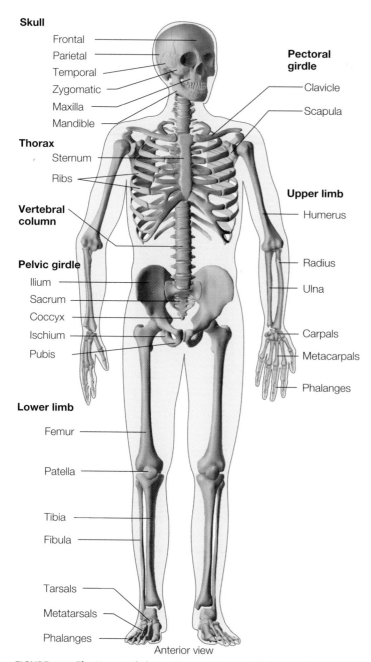

FIGURE 12-2 **The Human Skeleton** Over 200 bones of all shapes and sizes make up the human skeleton. The cartilage is shaded gray.

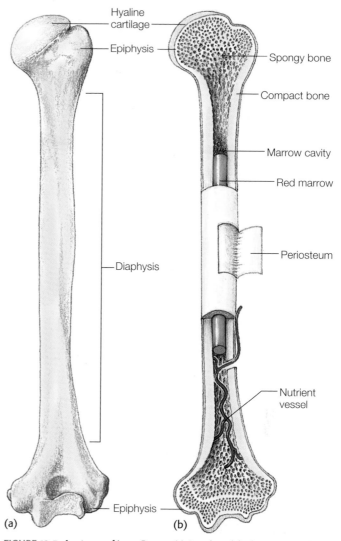

FIGURE 12-3 **Anatomy of Long Bones** (a) Drawing of the humerus. Notice the long shaft and dilated ends. (b) Longitudinal section of the humerus showing compact bone, spongy bone, and marrow.

Yellow marrow can be reactivated to produce blood cells under certain circumstances—for example, after an injury.

Meeting New Stresses

Bones are remodeled for strength when exposed to new stresses.

On the surfaces of many bony spicules of spongy bone are large, multinucleated cells called **osteoclasts** (OSS-tee-oh-KLASTS; "bone breakers") (Figure 4-10d). Osteoclasts are very large cells that digest bone with enzymes. These cells are part of a homeostatic system that ensures proper blood calcium levels. When calcium levels in the blood fall, osteoclasts are activated by the parathyroid hormone produced by the thyroid gland. Once activated, the osteoclasts digest small portions of the spongy bone. This, in turn, releases calcium into the bloodstream.

Spongy bone is "remodeled" when bones are subjected to new stresses—for example, by walking, running, or even stand-

Inside the shaft of long bones is a large cavity, the **marrow cavity** (MARE-row). The marrow cavities in most of the bones of the fetus and newborn contain red marrow, so named because of its color. **Red marrow** is tissue that serves as a blood-cell factory, producing red blood cells, white blood cells, and platelets to replace those routinely lost each day. As an individual ages, most red marrow is slowly "retired" and becomes filled with fat, becoming **yellow marrow**. Yellow marrow begins to form during adolescence and, by adulthood, is present in all but a few bones. Red blood cell formation, however, continues in the bodies of the vertebrae, the hip bones, and a few others.

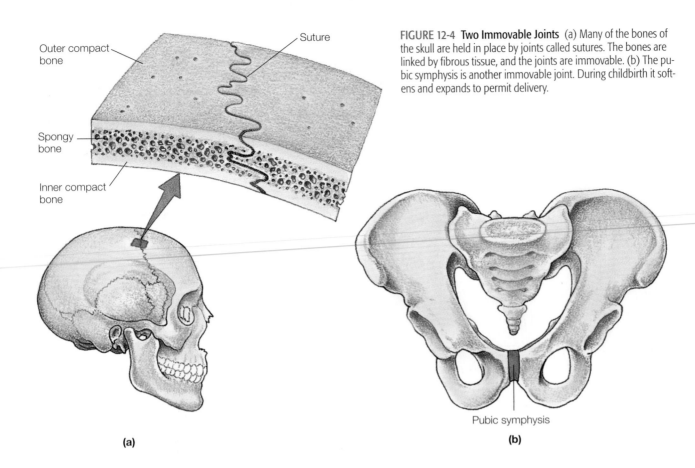

Outer compact bone

Suture

Spongy bone

Inner compact bone

FIGURE 12-4 **Two Immovable Joints** (a) Many of the bones of the skull are held in place by joints called sutures. The bones are linked by fibrous tissue, and the joints are immovable. (b) The pubic symphysis is another immovable joint. During childbirth it softens and expands to permit delivery.

(a)

(b)

Pubic symphysis

ing for longer periods. The leg bones of a desk-bound executive from Atlanta who goes on a skiing vacation in Wyoming, for instance, are remodeled to accommodate the additional demands of skiing. During this process, osteoclasts tear down some of the spongy bone, while another type of cell, the osteoblasts, rebuild new compact bone, causing it to thicken to meet the new stresses. Spongy bone is remodeled as well, helping the entire bone structure withstand greater stress. By the end of a two-week ski trip, the executive's bones have been considerably refashioned and are much stronger than when she left home. When she is back at her desk, her bones revert to their previous, weaker state. The compact bone will decrease in thickness and the spongy bone will change.

Osteoblasts (OSS-tea-oh-BLASTS) lay down new bone during the remodeling phase. When an osteoblast becomes surrounded by calcified extracellular material, it is referred to as an osteocyte.

The Joints

The joints permit varying degrees of mobility.

Mobility in humans and other animals is made possible by bone and muscles acting together. However, mobility also requires joints. **Joints** are the structures that connect the bones of the skeleton. They can be classified quite simply by the degree of movement they permit. Those that allow little movement are *slightly moveable*. Those that permit no movement are *immovable joints*, and those that allow free movement are **movable joints**. Consider some examples.

The bones of the skull are shown in Figure 12-4a. As illustrated, these bones interlock to form immovable joints, known as *sutures*. Fibrous connective tissue spans the space between the interlocking bones, holding them together. Another immovable joint is the pubic symphysis (PEW-bick SIM-fa-siss), formed by the two pubic bones (Figure 12-4b). These bones are held in place by fibrocartilage. Near the end of pregnancy, however, hormones loosen the fibrocartilage of the pubic symphysis, allowing the pelvic outlet to widen enough to permit a baby to pass through—an adaptation of enormous value to our species!

The bodies of the vertebrae are united by slightly movable joints (Figure 12-5). Each vertebra is separated from its nearest neighbors by an **intervertebral disc**. The inner portion of the disc acts as a cushion, softening the impact of walking and running. The outer, fibrous portion holds the disc in place and joins one vertebra to its nearest neighbor. The joints between the vertebrae offer some degree of movement, resulting in a fair amount of flexibility. If they did not, we would be unable to bend over to tie our shoes or to curl up on the couch for an afternoon nap.

The most common type of joint is the freely movable, or **synovial joint** (sin-OH-vee-al). The synovial joints are more complex than other types and permit varying degrees of movement. Although synovial joints differ considerably, they share several features. The first commonality is the hyaline cartilage located on the articular (joint) surfaces of the bones (Figure 12-6a). This thin cap of hyaline cartilage reduces friction, which facilitates movement. The second commonality is the joint capsule, a structure that joins one bone to another (Figure 12-6a). The outer layer of the joint capsule consists of dense connective tissue that attaches to the periosteum of adjoining bones. In many

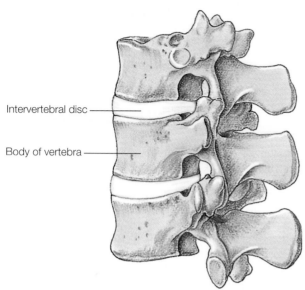

FIGURE 12-5 A Slightly Movable Joint The intervertebral discs allow for some movement, giving the vertebral column flexibility.

joints, parallel bundles of dense connective tissue fibers in the outer layer of the capsule form distinct **ligaments**. A ligament is a structure that joins bones to bones together in joints. They provide additional support to the joint.

As a rule, ligaments are fairly inflexible. However, some individuals have remarkably flexible ligaments and tendons. Tendons are connective tissue that joins muscles to bones at their points of origin and insertion. Because of the flexibility of their tendon and ligaments, some people exhibit extraordinary flexibility. For example, they may be able to extend their fingers and thumbs so much that they can touch the back of their hand. These people are said to be "double-jointed," a misnomer that suggests the flexibility is permitted by the joint when in fact it is due to the tendons and ligaments.

The inner layer of the joint capsule is the synovial membrane. It produces a fairly thick, slippery substance called *synovial fluid*. Synovial fluid provides nutrients to the articular cartilage (the hyaline cartilage on the articular surfaces of the bone) and also acts as a lubricant. Normally, the synovial membrane produces only enough fluid to create a thin film on the articular cartilage. Injuries to a joint, however, can result in a dramatic increase in synovial fluid production, causing swelling and pain in joints.

Tendons and muscles provide additional support in some joints and are commonly associated with synovial joints. In the shoulder, for example, muscles help hold the head of the humerus in place. Because muscles strengthen joints, individuals who are in poor physical shape are much more likely to suffer a dislocation on a ski trip or during exercise than someone who is in good shape. **Dislocation** is an injury in which a bone is displaced from its proper position in a joint because of a fall or some other unusual body movement. In some cases, bones slip out of place, then back in without assistance. In others, the bone can be put back in place only by a trained health care worker.

Synovial joints come in many shapes and sizes and are classified on the basis of structure. Two of the most common are the hinge joint and the ball-and-socket joint. The knee joint is a hinge joint, as are the joints in the fingers. **Hinge joints** open and close like hinges on a door. Such joints provide for movement in only one plane. The hip and shoulder joints are **ball-and-socket joints**. They allow a wider range of motion. (Compare the movements permitted by the shoulder joint to those permitted by the knee joint.)

Repairing Damaged Joints

Arthroscopy permits surgeons to repair injured joints with a minimum of trauma.

Joint injuries are common among physically active people especially athletes (Table 12-1). A hard blow to the knee of a football player, for instance, can tear the ligaments in the knee joint.

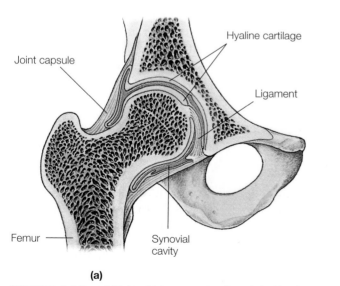

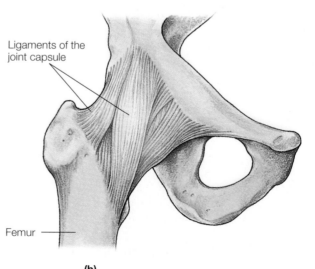

(a)   **(b)**

FIGURE 12-6 A Synovial Joint (a) A cross section through the hip joint (a ball-and-socket joint) showing the structures of the synovial joint. (b) Ligaments in the outer portion of the joint capsule help support the joint.

TABLE 12-1	Common Injuries of the Joints	
Injury	Description	Common Site
Sprain	Partially or completely torn ligament; heals slowly; must be repaired surgically if the ligament is completely torn.	Ankle, knee, lower back, and finger joints
Dislocation	Occurs when bones are forced out of a joint; often accompanied by sprains, inflammation, and joint immobilization. Bones must be returned to normal positions.	Shoulder, knee, and finger joints
Cartilage tears	Cartilage may tear when joints are twisted or when pressure is applied to them. Torn cartilage does not repair well because of poor vascularization. It is generally removed surgically; this operation sometimes makes the joint less stable.	Knee joint

Torn ligaments, tendons, and cartilage in joints heal very slowly because they have few blood vessels to bring nutrients needed for repair. For years, joint repair required major surgery that was so traumatic it put patients out of commission for several months. Today, however, new surgical techniques allow physicians to repair joints with a minimum of trauma (Figure 12-7). Through small incisions in the skin over the joint, surgeons insert a device called an **arthroscope** (ARE-throw-scope). It allows them to view the damage inside the joint and to insert instruments that remove damaged cartilage. The surgeon can thus repair damaged cartilage without making a large incision to open the joint, as was common previously. Athletes are usually back on their feet and on the playing field in a matter of weeks. New surgical techniques can rebuild torn ligaments and return joints to nearly their original state.

Arthritis

Osteoarthritis is caused by wear and tear on articular cartilage.

Virtually every time you move, you use one or more of your joints. Because of this, problems in joints, which often cause pain, are therefore often quite noticeable. One of the most common problems is called **degenerative joint disease** or **osteoarthritis**. Degenerative joint disease probably results from wear and tear on a joint. Over time, excess wear causes the articular cartilage on the ends of bones to flake and crack. As the cartilage degenerates, the bones come in contact and grind against each other during movement. This causes considerable swelling, pain, and discomfort.

Osteoarthritis occurs most often in the weight-bearing joints: the knee, hip, and spine—regions subject to the most wear over time. Osteoarthritis may also develop in the finger joints.

Osteoarthritis is extremely common. X-ray studies of people over 40 years of age show that most people have some degree of degeneration in one or more joints. Fortunately, many people do not even notice this condition, and the disease rarely becomes a serious medical problem.

Wear and tear on joints is worsened by obesity. The extra pressure wears the cartilage away more quickly. Weight control can reduce the rate of degeneration and ease the pain. Painkillers such as aspirin and other anti-inflammatory drugs are used to treat the pain and swelling. Injections of steroids may help reduce inflammation, although repeated injections often damage the joint.

Rheumatoid Arthritis

Rheumatoid arthritis is an autoimmune disease that results in an inflammation of the synovial membrane.

Another common disorder of the synovial joint is **rheumatoid arthritis**. Rheumatoid arthritis (RUE-mah-toid) is the most painful and crippling form of arthritis. It is caused by an inflammation of the synovial membrane. The inflammation often spreads to the articular cartilages. If the condition persists, rheumatoid arthritis can wear through the cartilage and cause degeneration of the underlying bone. Thickening of the synovial membrane and degeneration of the bone often disfigure the joints, reduce mobility, and cause considerable pain (Figure 12-8). Afflicted joints can be completely immobilized. In severe cases, the bones become dislocated, causing the joints to collapse.

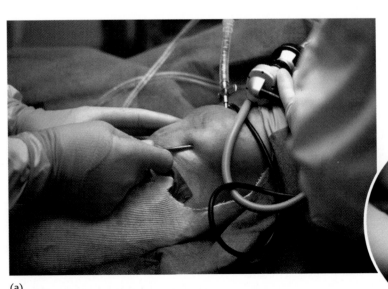

(a)

(b)

FIGURE 12-7 **Arthroscopic Surgery** (a) A physician performing arthroscopic surgery. (b) Inside view of a damaged knee joint through an arthroscope.

FIGURE 12-8 **Hand Disfigured by Rheumatoid Arthritis**
Degeneration of the finger joints causes mild to extreme disfiguration.

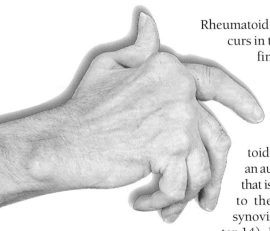

Rheumatoid arthritis generally occurs in the joints of the wrist, fingers, and feet. It can also affect the hips, knees, ankles, and neck.

Research suggests that rheumatoid arthritis results from an autoimmune reaction—that is, an immune response to the cells of one's own synovial membrane (Chapter 14). Rheumatoid arthritis occurs in people of all ages, but most commonly appears in individuals between the ages of 20 and 40. Rheumatoid arthritis is usually a permanent condition, although the degree of severity varies widely. Patients suffering from it can be treated with physical therapy, painkillers, anti-inflammatory drugs, and surgery.

Diseased joints can also be replaced by prostheses (pross-THEE-seas), artificial joints that restore mobility and reduce pain. Plastic joints are used to replace the finger joints, greatly improving the appearance of the hands by eliminating the gnarled, swollen joints. Moreover, patients regain the use of previously crippled fingers. Day-to-day chores (buttoning a shirt) that often required assistance become noticeably easier. Once impossible tasks—such as opening screw-top jars—are again feasible. Severely damaged knee and hip joints are replaced with special steel or Teflon substitutes, which, if fitted properly, may last 10–15 years (Figure 12-9).

In knee-replacement surgery, the ends of the bones are sawed off and the artificial joints are glued in place. In a hip replacement, the end of the leg bone, the femur, is sawed off and replaced with a metal substitute. The socket into which the head of the femur once fit is replaced with a teflon-coated socket.

FIGURE 12-9 **Artificial Joints** An artificial knee joint (a) and hip joint (b).

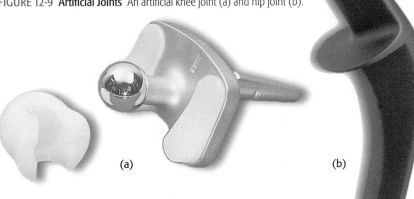

(a)   (b)

Bone Formation

Most of the bones of the human skeleton start out as hyaline cartilage.

During embryonic development, hyaline cartilage forms in the arms, legs, head, and torso where bone will eventually be. The cartilage is then converted to bone. This process is known as **endochondral ossification** (en-doh-CON-drall).

In a newborn baby, the bones of the leg (the tibia and fibula) are quite bowed. Cramped inside the mother's uterus, the baby's bones do not grow very straight. During the first 2 years of life, however, as the child begins to walk and run, the leg bones generally straighten (Figure 12-10). The bones are remodeled by osteoclasts and osteoblasts to meet the markedly different stresses placed on them by upright posture.

Maintaining Calcium Levels

Bone helps to maintain proper levels of calcium in the body.

In addition to their role in remodeling bone, osteoblasts and osteoclasts help control blood calcium levels and are therefore part of a homeostatic mechanism. They are controlled by two hormones: parathormone and calcitonin.

When blood calcium levels fall, the parathyroid glands (small glands embedded in the thyroid glands in the neck) release a hormone known as **parathormone** (PAIR-ah-THOR-moan) or **PTH** into the bloodstream. PTH travels throughout the body. When it reaches the bone, it stimulates the osteoclasts, causing them to digest bone in their vicinity. The calcium released by the activity of the osteoclasts replenishes blood calcium levels.

When calcium levels rise–for example, after a meal—the thyroid gland (a large endocrine gland in the neck) releases **calcitonin** (CAL-seh-TONE-in). This hormone inhibits osteoclasts,

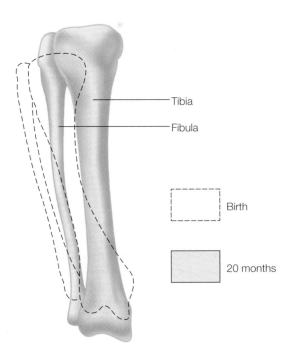

Tibia

Fibula

Birth

20 months

FIGURE 12-10 **Remodeling of a Baby's Leg Bones** Notice how dramatically the bones change to meet the changing needs of the toddler.

FIGURE 12-11
Stages in Fracture Repair

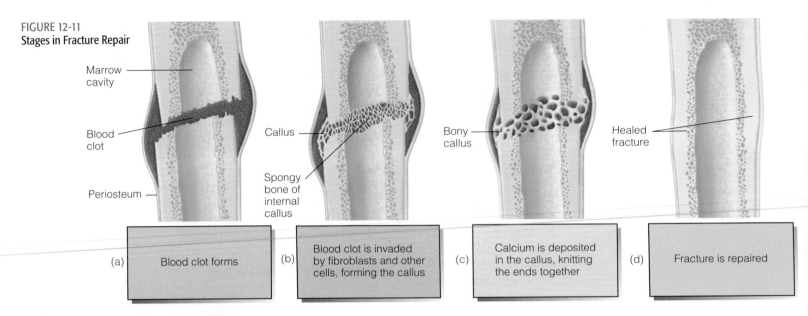

Marrow cavity

Blood clot

Periosteum

Callus

Spongy bone of internal callus

Bony callus

Healed fracture

(a) Blood clot forms

(b) Blood clot is invaded by fibroblasts and other cells, forming the callus

(c) Calcium is deposited in the callus, knitting the ends together

(d) Fracture is repaired

stopping bone destruction. It also stimulates osteoblasts, causing them to deposit new bone. The inhibition of osteoclasts and the formation of new bone help lower blood calcium levels, returning them to normal.

Repairing Bone Fractures

Bone fractures are repaired by fibroblasts and osteoblasts.

If you're like most people, chances are that sometime in your life you'll break a bone. Bone fractures can vary considerably in their severity. Some only involve hairline cracks, which mend fairly quickly. Others involve considerably more damage—a complete break, for instance—that takes much longer to repair. In some instances, the broken ends must be realigned. A plaster cast is then applied to hold the limb in position and keep the ends together to optimize healing. Severe fractures may require surgeons to insert steel pins to hold the bones together.

Bones, like other tissues, are capable of self-repair. As Figure 12-11 illustrates, after a fracture, blood from broken blood vessels in the periosteum and marrow cavity pours into the frac-

ture and forms a blood clot. Within a few days, the clot is invaded by fibroblasts, connective tissue cells from the periosteum. Fibroblasts secrete collagen fibers, thus forming a mass of cells and fibers, the **callus** (CAL-us), which bridges the broken ends internally and externally.

The callus is next invaded by osteoblasts from the periosteum. Osteoblasts slowly convert the callus to bone, thus "knitting" the broken ends together. As Figure 12-11 shows, the callus is initially much larger than the bone itself. But the excess bone is gradually removed by osteoclasts, leaving little, if any, evidence that a break has occurred.

Osteoporosis

Osteoporosis results in brittle, easy-to-break bones.

One of the most common problems adults face is **osteoporosis** (OSS-tea-oh-puh-ROE-siss; "porous bone"), a condition characterized by a progressive loss of bone calcium (Figure 12-12). In some individuals, calcium loss is so severe that bones become extremely brittle—so much so that even normal activities such as getting out of bed in the morning or doing housework cause fractures.

Osteoporosis occurs most often in postmenopausal women. Menopause (MEN-oh-pause) usually occurs between 45 and 55 years of age. It results from the loss of estrogen production by a woman's ovaries. Although estrogen is primarily a reproductive hormone, it also helps to maintain bone.

Osteoporosis also occurs in people who are immobilized for long periods. Hospital patients who are restricted to bed for 2 or 3 months, for example, show signs of osteoporosis.

Osteoporosis may also result from environmental factors. In the 1960s, women living along the Jintsu River in Japan developed a painful bone disease known as itai-itai (which literally means "ouch, ouch"). The women lived downstream from zinc and lead mines that released large amounts of the heavy metal cadmium into the river. They used river water for drinking and for irrigating rice paddies. Even though men, young women,

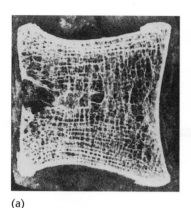
(a)

(b)

(c)

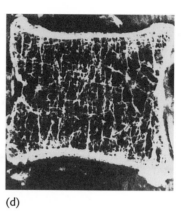
(d)

FIGURE 12-12 **Osteoporosis** The loss of estrogen or prolonged immobilization weakens bone. In these situations, bone is dissolved and becomes brittle and easily breakable. (a) A section of the body of a lumbar vertebra from a 29-year-old woman. (b) Some thinning is evident in a vertebra of a 40-year-old woman. (c) Bone loss is severe in an 84-year-old woman. (d) Bone loss is most severe in a 92-year-old woman. Osteoporosis is not inevitable and can be prevented by exercise, calcium supplements, estrogen supplements, and other measures.

and children were exposed to cadmium, 95% of the cases occurred in postmenopausal women.

Researchers hypothesized that the accelerated bone loss in the postmenopausal Japanese women was caused by cadmium. They tested this hypothesis by feeding mice diets containing various levels of cadmium chloride. The group receiving the highest dose showed significant reductions in bone calcium when their ovaries had been removed.

These findings may also explain why older women who smoke are more likely to develop osteoporosis and why they experience more bone fractures and tooth loss than nonsmokers. Cadmium is one of several harmful substances present in cigarette smoke. Smoking, however, may also decrease estrogen levels, making it doubly dangerous to a woman's health. For a discussion of ways to prevent osteoporosis, see Health Note 12-1.

12-2 The Skeletal Muscles

Skeletal muscles permit a wide range of movements.

In humans, skeletal muscles acting on bones results in movement. Skeletal muscles are under the control of the nervous system. Figure 12-13 shows some of the skeletal muscles of the body. As illustrated, most skeletal muscles cross one or more joints. When they contract, they produce movement.

As a rule, muscles generally work in groups to bring about various body movements. Groups of muscles are often arranged in such a way that one set produces one movement and another set on the opposite side of the joint causes an opposing movement. For example, the biceps (BUY-seps) is a muscle in the front of the upper arm. When it and other members of its group contract, they cause the arm to bend, a movement called *flexion* (FLECK-shun). On the back side of the upper arm is another muscle, the triceps (TRY-seps). When it contracts, it causes the arm to straighten, or extend. Such opposing muscles are called *antagonists*. When one muscle contracts to produce a movement, the antagonistic muscles relax.

Not all skeletal muscles make bones move. Some muscles simply steady joints, allowing other muscles to act. These muscles help to maintain our posture, permitting us to stand or sit upright despite the constant pull of gravity.

Another group of muscles that don't move bones are the muscles of the face. Anchored to the bones of the skull and to the skin of the face, these muscles allow us to wrinkle our skin, open and close our eyes, and move our lips.

Muscles also produce enormous amounts of heat as a byproduct of metabolism. Working muscles produce even more heat—so much, in fact, that you can cross-country ski or jog in freezing weather wearing only light clothing.

Anatomy of a Skeletal Muscle

Skeletal muscle cells are both excitable and contractile.

Skeletal muscles consist of long, unbranched **muscle fibers**. Each fiber contains many nuclei derived from individual cells that fuse during embryonic development. Viewed with the light microscope, skeletal muscle fibers appear striped or striated (STRY-a-tid) (Figure 12-14).

Like nerve cells, the muscle fiber is an excitable cell. A small potential difference (about −70 millivolts) exists across the plasma membrane of skeletal muscle fibers, as in neurons. Muscle fibers are stimulated by nerve impulses arriving at the terminal boutons of motor nerves. They stimulate the release of a neurotransmitter substance that stimulates a bioelectric impulse in the muscle fiber. This new impulse travels along the membrane of the muscle fiber in the same way a nerve impulse travels along an unmyelinated axon or dendrite.

:::: **healthnote**

12-1   Preventing Osteoporosis: A Prescription for Healthy Bones

Twenty million Americans, mostly women, suffer from osteoporosis (Figure 1). Caused by a gradual deterioration of the bone, osteoporosis results in nagging pain, discomfort, and loss of function.

Each year, nearly 60,000 Americans—again, most of them women—will die from complications resulting from this common malady. Unfortunately, most women do not realize they have the disease until it has progressed quite far.

Recent research shows that osteoporosis begins much earlier than researchers once thought—by the time a woman reaches her mid-20s. Bone demineralization occurs very rapidly. In fact, by age 30, many women have lost one-third of their bone calcium! Between the ages of 30 and 50, many women's bones continue to deteriorate, becoming extremely brittle.

Calcium loss begins so early because many weight-conscious women in their mid-20s avoid foods such as whole milk, cheese, and ice cream in an effort to control their weight. Although these foods are rich in calories, including fat, they are also a major source of calcium. Milk products are also shunned by many adults who develop an intolerance to lactose (a sugar) in milk and other dairy products. Lactose intolerance results from a sharp reduction in the production and secretion of the intestinal enzyme lactase as one ages, which makes it difficult to digest lactose. Because of these and other factors, adult women often consume only about one-half of the 1000–1500 milligrams of calcium they need every day.

Osteoporosis can be prevented and even reversed by eating calcium-rich foods, such as spinach, milk, cheese, oysters, and tofu (soybean curd). Calcium supplements can also help retard or even stop bone deterioration and restore calcium levels. Because vitamin D increases the absorption of calcium in the intestines, vitamin supplements can help.

Studies also suggest that osteoporosis can be prevented by exercise. Aerobics, jogging, walking, and tennis in conjunction with dietary changes all help prevent the disease.

Osteoporosis can be reversed by exercise even after the disease has reached the dangerous stage. Forty-five minutes of moderate exercise (walking) 3 days a week, for example, decreases the rate of calcium loss in older individuals and can stimulate the rebuilding process, replacing calcium lost in previous years.

Another effective treatment for postmenopausal women is estrogen. Low doses of estrogen promote bone formation. Because women who are given estrogen suffer an increased risk of cancer of the uterine lining, physicians often prescribe a mixed dose of estrogen and progesterone, which reduces the likelihood of this cancer.

Studies have also shown that high doses of calcium fluoride stimulate bone development. Calcium fluoride treatment increases bone mass approximately 3%–6% per year and decreases bone fractures. The average patient in one study experienced one fracture every 8 months before treatment. After treatment, that figure dropped to one fracture every 4.5 years.

Unfortunately, large doses of fluoride may erode the stomach lining, causing internal bleeding. They may also stimulate abnormal bone development and cause pain and swelling in joints. To offset these problems, researchers have developed a pill that releases the fluoride gradually.

For millions of young women, early detection and sound preventive measures, including exercise, vitamin D, dietary improvements, and fluoride treatments, can prevent this unnecessary disease.

FIGURE 1 **Bone Deterioration** An elderly woman suffering from osteoporosis. Notice the hunched back due to the collapse of vertebrae.

www.jbpub.com/humanbiology/5e

Visit Human Biology's Internet site for links to web sites offering more information on this topic.

Muscle fibers are also contractile. When stimulated, the contractile proteins inside the fibers cause the cells to shorten. Muscle fibers are also elastic and therefore capable of returning to normal length after a contraction has ended.

Muscle Fibers

Muscle fibers contain small bundles of contractile filaments.

Understanding how a muscle contracts first requires an understanding of its structure. To begin, imagine that you could tease a single skeletal muscle fiber (cell) free from a skeletal muscle (Figure 12-15a). Under the microscope, you would find that each muscle fiber is a long cylinder encased by plasma membrane and containing many nuclei. Each muscle fiber is further characterized by a series of dark and light bands (Figure 12-15b).

When the muscle fiber is viewed through an electron microscope, you will find that inside each muscle fiber are numerous threadlike filaments. They exist in tiny bundles called **myofibrils** (MY-oh-FIE-brills) (Figure 12-15b). Myofibrils contain contractile filaments (**myofilaments**) and are striated. As shown in Figure 12-15c, myofibrils contain light and dark bands arranged in a uniform pattern, which gives the myofibril and muscle fiber their striated appearance.

Figure 12-15d shows that the myofibrils contain thick and thin filaments. The thick filaments consist of the protein **myosin** (MY-oh-sin). The thin filaments are composed primarily of the protein **actin** (ACK-tin).

FIGURE 12-13
The Skeletal Muscles

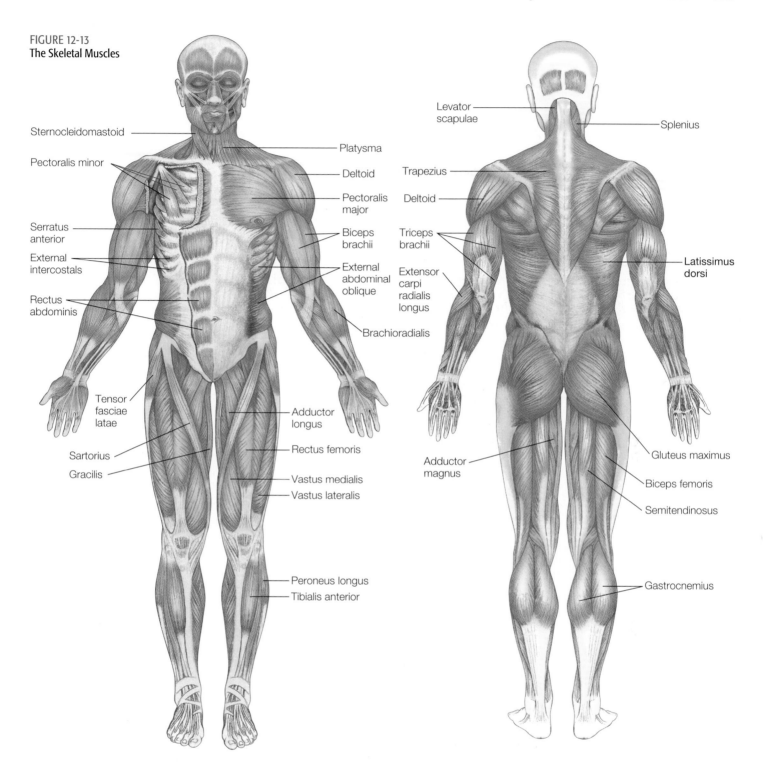

Sternocleidomastoid

Pectoralis minor

Serratus anterior

External intercostals

Rectus abdominis

Platysma

Deltoid

Pectoralis major

Biceps brachii

External abdominal oblique

Brachioradialis

Tensor fasciae latae

Sartorius

Gracilis

Adductor longus

Rectus femoris

Vastus medialis

Vastus lateralis

Peroneus longus

Tibialis anterior

Levator scapulae

Trapezius

Deltoid

Triceps brachii

Extensor carpi radialis longus

Splenius

Latissimus dorsi

Adductor magnus

Gluteus maximus

Biceps femoris

Semitendinosus

Gastrocnemius

FIGURE 12-14 **Light Micrograph of Skeletal Muscle** Notice the banding pattern on these muscle fibers.

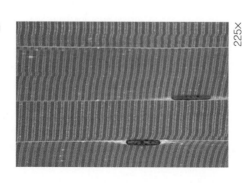

225x

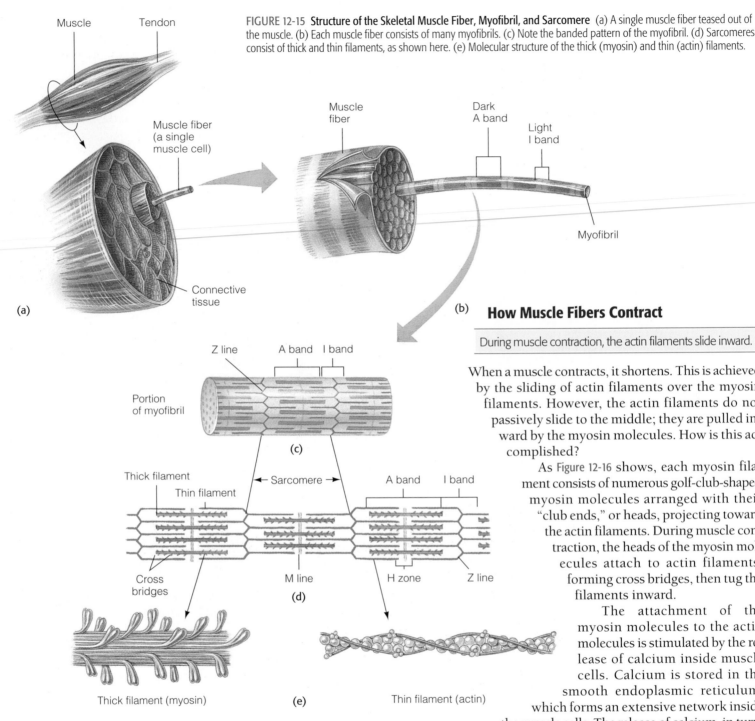

FIGURE 12-15 **Structure of the Skeletal Muscle Fiber, Myofibril, and Sarcomere** (a) A single muscle fiber teased out of the muscle. (b) Each muscle fiber consists of many myofibrils. (c) Note the banded pattern of the myofibril. (d) Sarcomeres consist of thick and thin filaments, as shown here. (e) Molecular structure of the thick (myosin) and thin (actin) filaments.

How Muscle Fibers Contract

During muscle contraction, the actin filaments slide inward.

When a muscle contracts, it shortens. This is achieved by the sliding of actin filaments over the myosin filaments. However, the actin filaments do not passively slide to the middle; they are pulled inward by the myosin molecules. How is this accomplished?

As Figure 12-16 shows, each myosin filament consists of numerous golf-club-shaped myosin molecules arranged with their "club ends," or heads, projecting toward the actin filaments. During muscle contraction, the heads of the myosin molecules attach to actin filaments, forming cross bridges, then tug the filaments inward.

The attachment of the myosin molecules to the actin molecules is stimulated by the release of calcium inside muscle cells. Calcium is stored in the smooth endoplasmic reticulum, which forms an extensive network inside the muscle cells. The release of calcium, in turn, is stimulated by impulses from the nerves. These impulses reach the muscle cell via motor nerves, which form synapses with the muscle cells.

Neurotransmitters released from the terminal boutons of the motor nerves cause impulses to be generated in muscle cells. These impulses travel along the membrane of muscle cells and enter the interior via small infoldings of the plasma membrane. These infoldings come in close contact with the smooth endoplasmic reticulum. When an impulse penetrates into the interior, it stimulates calcium release. Calcium release causes the heads of the myosin molecules to attach to the actin filaments. ATP in the muscle provides the energy needed to pull the actin filaments inward.

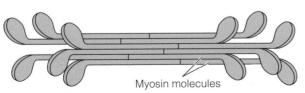

FIGURE 12-16 **Structure of Myosin Filaments** Myosin molecules join to form a myosin filament. Note the presence and orientation of the heads of the myosin molecules.

ATP and Muscle Contraction

Muscle cells use enormous quantities of ATP daily.

ATP is in relatively short supply in cells but is required in large quantity. In order to meet the cell's demands, ATP is recycled over and over again in rapid succession. In muscles, some of the ATP is regenerated by a substance called *creatine phosphate* (CREE-ah-teh-NEEN). This molecule contains a high-energy bond, indicated by the squiggly line (below). Stored in muscle in high concentrations, creatine phosphate reacts with ADP as follows:

$$creatine \sim P + ADP \rightarrow ATP + creatine$$

Creatine phosphate replenishes ATP used during muscle contraction.

ATP is also generated by glycolysis, the citric acid cycle, and the electron transport system (Chapter 3). The electron transport system is the main source of ATP. But this process requires oxygen. During vigorous muscle contraction, oxygen supplies inside muscle cells can fall precipitously. If the circulatory system cannot replace oxygen as quickly as it is being used, the citric acid cycle and electron transport system shut down.

To generate ATP, the cell must resort to fermentation—the metabolic breakdown of glucose in the absence of oxygen. Besides being inefficient, this process results in the buildup of lactic acid in muscle fibers. Exercise also depletes the muscle of glycogen, which you may recall is a long-chain molecule in muscle (and liver). It is used to store glucose. The shortage of ATP, buildup of lactic acid, and depletion of muscle glycogen that occur during vigorous exercise result in muscle fatigue. Oxygen levels are also depleted during exercise, creating an oxygen debt. The shortages of glycogen and oxygen must be remedied in order for the muscle to continue to function, just as gasoline in your car must be added to continue its functioning. Glycogen is replaced during rest by glucose from the bloodstream. Oxygen is replenished by the diffusion of oxygen from blood capillaries, but this process occurs rather quickly, often right after you exercise, which explains why you keep breathing hard for a while after you stop exercising.

Recruitment

The strength of muscle contraction is increased by stimulating many muscle fibers.

When stimulated by a nerve impulse, individual skeletal muscle fibers contract. A single contraction, followed by relaxation, is called a **twitch**. Generating the force needed to move arms and legs or eyelids, however, requires the action of many muscle fibers. The engagement of additional muscle fibers during muscle contraction is called *recruitment*.

To understand how recruitment occurs in muscle, first take a look at Figure 12-17. The drawing shows that the axons of motor neurons form many branches upon reaching the muscles they supply. A single motor neuron, in fact, may end on dozens of individual muscle fibers in a skeletal muscle. A motor neuron and the muscle fibers it supplies constitute a **motor unit**. When the neuron supplying a motor unit is activated, all of the muscle fibers in the unit are stimulated.

Motor units account for the degree of control that various muscles exert. The fewer muscle fibers in a motor unit, the finer the control. Such is the case in the muscles of the fingers, which are capable of very fine movements. In contrast, the more muscle fibers in a motor unit, the cruder the control. Thus, the neurons that supply the muscles of the leg, which produce a crude but strong propulsive force, may end on as many as 2000 muscle fibers. Each time an impulse is delivered to the motor unit, it stimulates all 2000 muscle fibers.

To increase the force of contraction in a skeletal muscle, more motor neurons are stimulated. Thus, more muscle fibers are called into action. Therefore, if you were to lift a piece of foam, only a small fraction of the motor neurons going to the arm

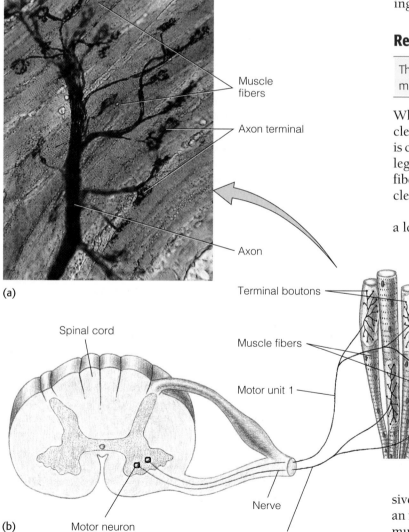

(a)

(b)

FIGURE 12-17 **The Motor Unit** (a) Light micrograph of axon branching to terminate on many muscle fibers. (b) Each axon branches at its termination, supplying a few dozen to many thousand muscle fibers. A single axon and its muscle fibers constitute a motor unit.

FIGURE 12-18 **Graph of Muscle Contraction** (a) The force generated by two separate stimuli (action potentials from motor neuron). (b) The force generated by two closely timed stimuli. (c) The force generated by many closely spaced stimuli.

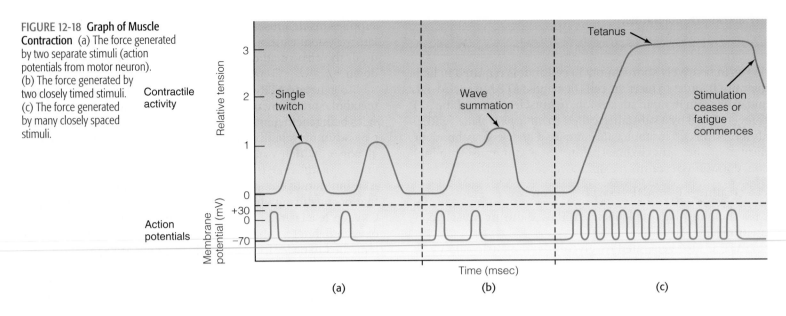

The duration of the action potentials is not drawn to scale, but is exaggerated.

muscles would be used. If you were to lift a 40-pound bag of dog food, your brain would recruit many more motor neurons and muscle fibers.

The strength of muscle contraction can also be increased by a process called **wave summation**. Let me explain what this means. Each time a muscle fiber is stimulated, it contracts to produce a twitch. Figure 12-18a shows the strength of twitch (contractions) resulting from two separate nerve impulses, the second arriving after the muscle fiber has contracted and then relaxed. Panel (b) shows what happens if nerve impulses reach a muscle fiber before it has had time to relax. As illustrated, this creates additional tension and the muscle fiber contracts more forcefully. In such instances, the second contraction "piggybacks" on the first. The waves of contraction are added up; hence the name *wave summation*.

If the nerve impulses arrive frequently at enough skeletal muscle fibers, a smooth, sustained contraction occurs in the muscle (Figure 12-18c). This is called *tetanus* (not to be confused with the serious, often fatal bacterial infection of the same name). These contractions occur in your arm muscles when you carry a bag of groceries. In such instances, the muscles contract to support the weight and remain contracted throughout the activity. Tetanic contractions eventually cause muscle fatigue and muscles stop contracting, even though the neural stimuli may continue.

Muscle Tone

Muscle tone results from the contraction of a small number of muscle fibers.

Touch one of your muscles. Even if you are not in peak physical condition, you will notice that the muscle is relatively firm. This firmness is called **muscle tone**. Muscle tone is essential for maintaining posture. Without it, you would fall into a heap on the floor. Muscle tone also generates heat in warm-blooded animals like us.

Muscle tone results from the contraction of muscle fibers during periods of inactivity. But not all fibers contract—just enough of them to keep the muscles slightly tense. Muscle tone is maintained, in part, by the muscle spindle, a receptor that monitors muscle stretching (Chapter 11). The spindle "alerts" the brain and spinal cord to the degree of stretching. When muscles relax, signals travel to the spinal cord, and then back out to motor axons, which stimulate a low level of muscle contraction to maintain muscle tone.

Slow- and Fast-Twitch Fibers

Two types of muscle fibers are found in skeletal muscle, slow- and fast-twitch.

Physiological studies have revealed the presence of two types of skeletal muscle fibers: fast-twitch and slow-twitch fibers. **Slow-twitch muscle fibers** contract relatively slowly but have incredible endurance. It is not surprising, then, that the muscles of endurance athletes (for example, long-distance runners) contain a high proportion of slow-twitch muscle fibers (Figure 12-19). These permit such athletes to perform for long periods without tiring.

Fast-twitch muscle fibers contract swiftly. The muscles of sprinters and other athletes whose performance depends on quick bursts of activity contain a high proportion of fast-twitch fibers.

Skeletal muscles generally contain a mixture of slow- and fast-twitch fibers, giving each muscle a wide range of performance abilities. However, a muscle that performs one type of function more often than another tends to have a disproportionately higher number of fibers corresponding to the type of activity it performs. The muscles of the back, for example, contain a larger number of slow-twitch fibers. These muscles operate throughout the waking hours to maintain posture. They do not need to

FIGURE 12-19 **Built to Last** An abundance of slow-twitch fibers gives the cross-country runners endurance.

contract quickly, but they must be resistant to fatigue. In contrast, the muscles of the arm are used for many quick actions—waving, playing tennis, or grasping falling objects. Fast-twitch fibers are more common in the muscles of the arm.

New research suggests that one of the reasons some people excel in certain sports may lie in the relative proportion of fast- and slow-twitch fibers in their muscles. In fact, a study of the skeletal muscles of world-class long-distance runners suggests that their physical endurance results primarily from a high proportion of slow-twitch fibers, a trait that may be genetically determined. Biochemical studies also show that the muscle cells in endurance athletes have a higher level of ATP, both at rest and during exercise, thus providing more energy for muscle contraction. Endurance athletes start out with a larger storehouse of energy and maintain a larger supply throughout exercise.

Increasing Muscle Size and Endurance

Exercise builds muscles and increases endurance.

Milo of Croton was a champion wrestler in ancient Greece who stumbled across a revolutionary way to build muscle. His method was simple. Milo borrowed a newborn calf, then proceeded to carry the animal around every day. Day after day, he faithfully followed this routine so that by the time the calf had become a bull, so had he. Milo's program worked, for he went on to win six Olympic championships in wrestling.

The Greek wrestler discovered a principle of muscle physiology familiar to weight lifters today: When muscles are made to work hard, they respond by becoming larger and stronger. The increase in size and strength results from an increase in the amount of contractile protein inside muscle cells.

Unfortunately, muscle protein is quickly made and quickly destroyed. In fact, about half of the muscle you gain in a weight-lifting program is broken down 2 weeks after you stop exercising. In order to build muscle, you must keep up with your exercise program.

High-intensity exercise such as weight lifting builds muscle, and it takes surprisingly little exercise to have an effect. Working out every other day for only a few minutes will result in noticeable changes in muscle mass. According to some sources, 18 contractions of a muscle (in three sets of six contractions each) are enough to increase muscle mass, if the contractions force your muscles to exert over 75% of their maximum capacity.

Low-intensity exercise, such as aerobics and swimming, tends to burn calories but not build muscle bulk. Aerobic exercise therefore helps increase endurance—that is, it increases one's ability to sustain muscular effort. Stamina or endurance results from numerous physiological changes. One of the most important changes occurs in the heart. The heart responds to exercise like any other muscle–it grows stronger and larger. A well-exercised heart beats more slowly but pumps more blood with each beat. The net result, then, is that the heart works more efficiently, delivering more oxygen to skeletal muscles.

Increased endurance also results from improvements in the function of the respiratory system. For example, exercise increases the strength of the muscles involved in breathing. These muscles become stronger and can operate longer without tiring. Breathing during exercise becomes more efficient.

Increased endurance may also be attributed to an increase in the amount of blood, another benefit of aerobic exercise. An increase in blood volume results in an increase in the number of RBCs, which increases the amount of oxygen available to cells. This improvement, combined with others, allows an individual to work out longer without growing tired.

When you set out on an exercise program, it is important to establish your goals first. If you are interested in increasing your endurance, you should pursue an exercise regime that works the heart and muscles at a lower intensity over longer periods such as riding a stationary bicycle. If you are after bulk, the answer lies in high-intensity exercises such as weight lifting.

Many health clubs and university gyms offer advice and exercise programs to help you achieve your goals. And many offer a variety of machines to help you build muscles or simply tone them up. Most of the exercise machines work one particular muscle group—for example, the muscles of the upper arm. In most gyms, a dozen or more machines are usually placed in a line so you can go down the line, working one set of muscles after another until you have exercised your entire body.

Exercise machines are popular because they are safer than free weights (barbells and dumbbells). These machines eliminate the chances of your dropping a weight on your toes—or someone else's! Moreover, it is almost impossible to strain your back if you make a mistake using one. In contrast, lifting free weights requires caution and training.

Exercise machines also reduce the amount of time a person needs to exercise by about half because they've been designed to require work when a joint is both flexed and extended. The biceps machine, for example, requires you to pull the weights up, then return them slowly, forcing your biceps muscles to work in both directions.

12-3 Health and Homeostasis

The skeleton and muscles provide many benefits, the most evident being mobility. These two systems also contribute to homeostasis. The skeleton, for instance, serves as a calcium reservoir, essential for controlling blood levels of calcium, which are crucial for normal muscle function. Muscle helps provide the heat necessary for the normal operation of cells. Calcium is also essential for nerve cell transmission.

Efforts to enhance the performance of the muscular system can upset homeostasis and have serious effects on other systems. Consider anabolic steroids.

Anabolic steroids are synthetic hormones that resemble the male sex hormone, testosterone. When taken in large doses, anabolic steroids stimulate muscle formation. They increase muscle size and strength by stimulating protein synthesis in muscle fibers. High doses may also reduce the inflammation that frequently results from heavy exercise, allowing athletes to work out harder and longer.

When it comes to building muscle, steroids and exercise are an unbeatable combination. Some users claim that steroids even increase aggression, which may be helpful to competitive athletes.

Despite the benefits of steroids, most physicians view them as a dangerous proposition. For example, steroids can result in mental and behavioral problems. One set of researchers interviewed 41 athletes who used steroids in doses 10 to 100 times greater than those used in medical treatment. The athletes also reported using as many as five or six steroids simultaneously in cycles lasting 4 to 12 weeks, a practice known as "stacking." In this study, researchers found that one-third of the athletes developed severe psychiatric complications.

Athletes in the study reported episodes of severe depression during and after steroid use. Some reported feelings of invincibility. One man, in fact, deliberately drove a car into a tree at 40 miles per hour while a friend videotaped him. Some subjects reported psychotic symptoms in association with steroid use, including hearing voices. Withdrawal from steroids resulted not only in depression but also in suicidal tendencies.

Other studies have shown that steroids used in excess may damage the heart and kidneys and reduce testicular size in men. In women, steroids deepen the voice and may cause enlargement of the clitoris. Steroids also cause severe acne and liver cancer. Unfortunately, there are no scientific studies on the long-term health effects of steroids.

Despite the fact that anabolic steroids are banned by the National Football League, the International Olympic Committee, and college athletic programs, athletes continue to use them. Although periodic drug tests are often made, these steroids, along with a host of other performance-enhancing drugs, often escape detection. U.S. athletes are subjected to numerous unannounced drug tests. Other countries are far more lenient.

A recent survey of 46 public and private high schools across the United States involving over 3000 teenagers suggests that steroid use is especially prevalent in high school seniors. About 1 of every 15 senior boys reported taking anabolic steroids. The study also showed that the use of anabolic steroids begins in junior high school. Two-thirds of the students surveyed said that they had used steroids by age 16. Nearly half the users said they took the drugs to boost athletic performance, and 27% said their primary motive was to improve their appearance. Researchers say that adolescents who use steroids may be putting themselves at risk of stunted growth, infertility, and psychological problems.

Steroid use in athletes is a medical problem to be sure, but it is also the result of the social environment in our highly competitive society, a social environment that may be endangering the health of our children and our athletes.

SUMMARY

Structure and Function of the Human Skeleton

1. Bones serve many functions. They provide internal support, allow for movement, and help protect internal body parts. They also produce blood cells and platelets, store fat, and help regulate blood calcium levels.
2. Most bones have an outer layer of compact bone and an inner layer of spongy bone. Inside the bone is the marrow cavity, filled with either fat cells (yellow marrow) or blood cells and blood-producing cells (red marrow) or with combinations of the two.
3. The joints unite bones. The movable joints allow for flexion, extension, and other important movements.
4. Joints are vulnerable to injury and disease. Torn ligaments and ripped cartilage are common injuries among athletes. Wear and tear on some joints can cause the articular cartilage to crack and flake off, resulting in degenerative joint disease, or osteoarthritis. Rheumatoid arthritis results from an autoimmune reaction that produces a painful inflammation and thickening of the synovial membrane, disfiguring and stiffening joints.
5. Most of the bones form from hyaline cartilage.
6. Bone is constantly remodeled after birth to accommodate changing stresses. During bone remodeling, osteoclasts destroy bone. Osteoclasts are stimulated by parathormone, produced by the parathyroid glands. Osteoclasts also participate in the homeostatic control of blood calcium levels. When activated, these cells free calcium from the bone, raising blood calcium levels.
7. Osteoblasts are bone-forming cells stimulated by calcitonin from the thyroid gland. Calcitonin secretion decreases blood calcium levels and increases the amount of calcium in bones.
8. Osteoporosis is a disease of the bone caused by progressive loss of calcium. The bones become brittle and easily broken. Osteoporosis is most common in postmenopausal women and results from the loss of the ovarian hormone estrogen. It also occurs in people who are immobilized for long periods.
9. Osteoporosis can be prevented and reversed by exercise, calcium supplements, calcium-rich foods, vitamin D, and estrogen therapy.

The Skeletal Muscles

10. Skeletal muscles are involved in body movements; they help maintain our posture and produce body heat both at rest and while we are working or exercising.

11. Each skeletal muscle consists of many muscle fibers. Muscle fibers are both excitable and contractile.

12. Inside each muscle fiber are numerous myofibrils, bundles of the contractile filaments actin and myosin.

13. Muscle contraction is stimulated by nerve impulses from motor neurons. The energy for muscle contraction comes from ATP.

14. Individual muscle fibers contract when stimulated, producing a twitch. Contractions of varying strength can be generated in whole muscles by recruitment and wave summation. Recruitment results from the involvement of many motor units. A motor unit is a motor axon and all of the muscle cells it innervates. Wave summation is a piggybacking of muscle fiber contractions that occurs when stimuli arrive before the fiber relaxes.

15. The body contains two types of skeletal muscle fibers: slow-twitch and fast-twitch fibers. Skeletal muscles generally contain a mixture of the two types, but fast-twitch fibers are found in greatest number in muscles that perform rapid movement. Slow-twitch fibers are found in muscles such as those of the back that perform slower motions or are involved in maintaining posture.

16. Muscle mass can be increased by certain forms of exercise, such as weight lifting. An increase in muscle mass results from an increase in the amount of contractile protein in muscle fibers.

17. Endurance can be increased by other forms of exercise such as aerobics.

Endurance is a function of at least three factors: the condition of the heart, the condition of the muscles of inspiration, and the blood volume. Improvement in all three factors increases the efficiency of oxygen delivery to muscles.

Health and Homeostasis

18. Many athletes are using synthetic anabolic steroids to improve performance and build muscle. Unfortunately, massive doses of steroids can increase aggression, cause psychiatric imbalance such as severe depression, and may result in damage to the heart and kidneys.

critical thinking

Thinking Critically—Analysis

This Analysis corresponds to the Thinking Critically scenario that was presented at the beginning of this chapter.

If you noted that no mention was made of dietary calcium intake by the various groups, you're correct. Unfortunately, the diets of the women in this study were not monitored to determine how much calcium was being ingested through food and beverages. Second, this study focused on the bones of the forearm, not the load-bearing bones that are at greatest risk of fractures. Walking would probably influence the bones of the legs, hips, and back. To accurately assess the effects of the exercise regime prescribed in this study, the researchers probably should have looked at bone density in all three sites, not the arms. Given these inadequacies, do you think this study's conclusions are reliable?

KEY TERMS AND CONCEPTS

CONCEPT REVIEW

1. Describe the functions of bone. In what ways does bone participate in homeostasis? p. 216

2. The synovial joints move relatively freely. Why? What structures support the joint, helping keep the bones in place? pp. 218–219

3. A young patient comes to your office with swollen joints and complains about pain and stiffness in the joints. Friends have suggested that the boy has arthritis, but his parents argue that he is too young. Only old people get arthritis, they say. How would you respond to them? pp. 220–221

4. Bone is constantly remodeled, from infancy through adulthood. Explain when bone remodeling occurs and what cells and hormones participate in the process. pp. 217–218

5. Using what you know about bone, explain why an office worker who exercises very little is more likely to break a bone on a skiing trip than a counterpart who works out every night after work. pp. 217–218

6. Describe how a bone heals after being fractured. p. 222

7. A 30-year-old friend of yours who smokes, exercises very little, and avoids milk products because she's concerned that the small amounts of fat in milk will cause her to gain weight says: "Why should I worry about osteoporosis? That's a disease of old women." Based on what you know about bone, how would you respond to her? pp. 222–223

8. Describe the major functions of skeletal muscle. p. 223

9. Describe the structure of a skeletal muscle fiber. pp. 223–225

10. Describe the molecular events involved in muscle contraction. pp. 226–227

11. Describe the two types of skeletal muscle fibers, and explain how they differ. p. 224

12. A friend comes to you complaining that he can't seem to lose weight. He works out on barbells three times a week for an hour or so each time. His diet hasn't changed much, but with the increase in exercise, he thinks he should be losing weight, rather than staying even. Why do you think he isn't losing weight? p. 229

13. A friend is thinking about using anabolic steroids to improve his performance in gymnastics. He argues that he is young and will be taking the drug only for a year or so. He points out that other young men are using steroids and they really help build muscle and endurance. What advice would you give him? p. 230

SELF-QUIZ: TESTING YOUR KNOWLEDGE

1. Bone cells, known as _____, are embedded in a hard matrix. p. 216

2. The dense bone on the outside of the arm bone is known as _____ bone. p. 216

3. The connective tissue layer surrounding bone is called the _____. p. 216

4. The spongy bone inside a bone may be filled with blood-forming tissue called the _____ bone marrow. p. 217

5. Inactive bone marrow—that is, bone marrow not involved in making blood cells—is known as _____ bone marrow. p. 217

6. The dense connective tissue structures that join bones in joints are known as _____. p. 219

7. The fluid inside joints that helps lubricate the bones is called _____ fluid. p. 219

8. _____ or degenerative joint disease results from excess wear and tear on the bones in joints. p. 220

9. Joint surgery often involves use of an instrument known as an _____, which allows doctors to peer inside the joint to fix damage without having to create large incisions. p. 220

10. The process of bone formation is known as _____ ossification. p. 221

11. A muscle cell consists of many cells fused together and is known as a muscle _____. p. 223

12. Inside the muscle cells are smaller units known as _____; they contain many smaller myofilaments. p. 224

13. Two myofilaments are involved in muscle contraction, actin, and _____. p. 224

14. Contraction results from the release of _____ ions by the smooth endoplasmic reticulum of muscle cells. p. 226

15. Energy required for muscle contraction is provided by _____. p. 227

 www.jbpub.com/humanbiology/5e

The site features eLearning, an online review area that provides quizzes, chapter outlines, and other tools to help you study for your class. You can also follow useful links for in-depth information, research the differing views in the Point/Counterpoints, or keep up on the latest health news.

The Endocrine System

thinking critically

Growth hormone, one of the hormones you'll learn about in this chapter, stimulates growth of the human body. Unfortunately, some individuals' bodies produce inadequate amounts of this hormone, resulting in mild to severe stunted growth.

With the advent of genetic engineering, though, human growth hormone is now available to physicians who sometimes use it to treat patients whose bodies produce inadequate amounts. Such treatments may stimulate normal growth. Some doctors are also treating children who are naturally small with growth hormone. Should growth hormone be given to these "normal" children to enhance their growth—for example, to make them better basketball players?

One day while on his morning jog, President George Bush (the elder) began to feel weak. His heartbeat became irregular. He felt breathless. Fearing a heart attack, his security entourage rushed him to the hospital. Here a blood test revealed elevated levels of the hormone thyroxine produced by the thyroid gland in the neck. This substance accelerates cellular energy production and affects a number of mental and physical processes. Elevated levels are indicative of a condition called hyperthyroidism (HIGH-per-THIGH-roid-izm), or Grave's disease (after its discoverer, not the outcome). This rare disease can occur at any age. It is just one of many diseases caused by a hormonal imbalance. This chapter discusses the endocrine system and presents some examples of homeostatic imbalance caused by endocrine disorders.

13-1   Principles of Endocrinology

The endocrine system produces hormones that are transported in the blood to distant sites where they influence many functions.

The human **endocrine system** consists of numerous small glands scattered throughout the body (Figure 13-1). These highly vascularized glands produce and secrete chemical substances known as **hormones** (HOAR-moans). (The word *hormone* comes from the Greek *hormon*, which means "to stimulate" or "excite.") A hormone is a chemical produced and released by cells or groups of cells forming the endocrine (ductless) glands.

All hormones are transported via the bloodstream to distant sites, where they exert some effect. The hormone insulin, for ex-

ample, is produced by the pancreas and travels in the blood to skeletal muscle and other body cells. Here the hormone stimulates glucose uptake and glycogen synthesis. The cells affected by a hormone are called its **target cells**. For a discussion of some of the early research that contributed to our understanding of hormones, see Scientific Discoveries 13-1.

Hormones function in five areas: (1) homeostasis; (2) growth and development; (3) reproduction; (4) energy production, storage, and use; and (5) behavior. At any one moment, the blood carries dozens of hormones. The cells of the body are therefore exposed to many different chemical stimuli. Do all cells respond to these stimuli?

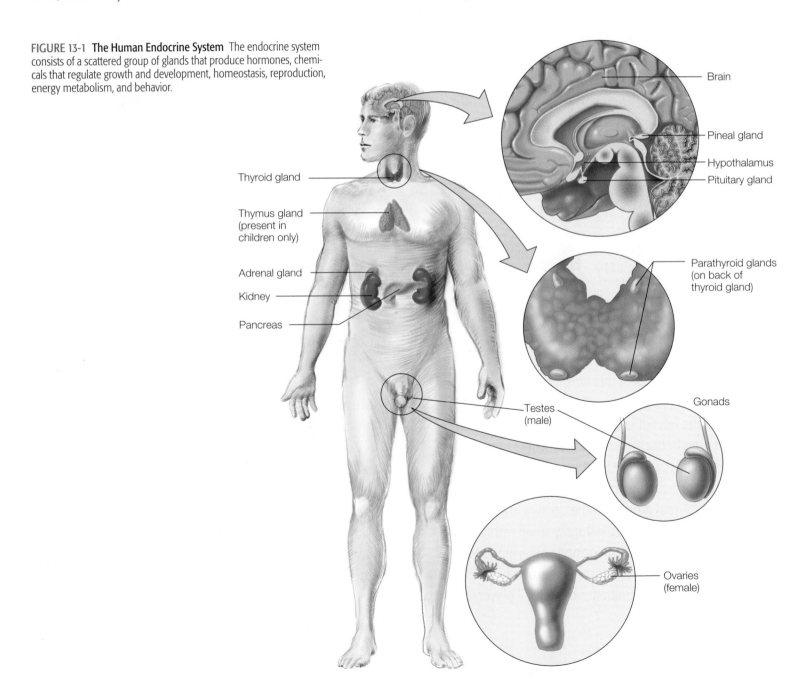

FIGURE 13-1 **The Human Endocrine System** The endocrine system consists of a scattered group of glands that produce hormones, chemicals that regulate growth and development, homeostasis, reproduction, energy metabolism, and behavior.

Target Cells

Target cells contain receptors for specific hormones.

Despite the fact that they're flooded with hormonal signals, cells of the body respond only to specific ones. This selectivity is possible because of protein receptors. In some target cells, the hormone receptors are embedded in the plasma membrane; in others, they're located in the cytoplasm. (More on this later.) Each cell contains receptors that bind to specific hormones, hormones the cell is genetically programmed to respond to.

FIGURE 13-2 Two Common Steroids

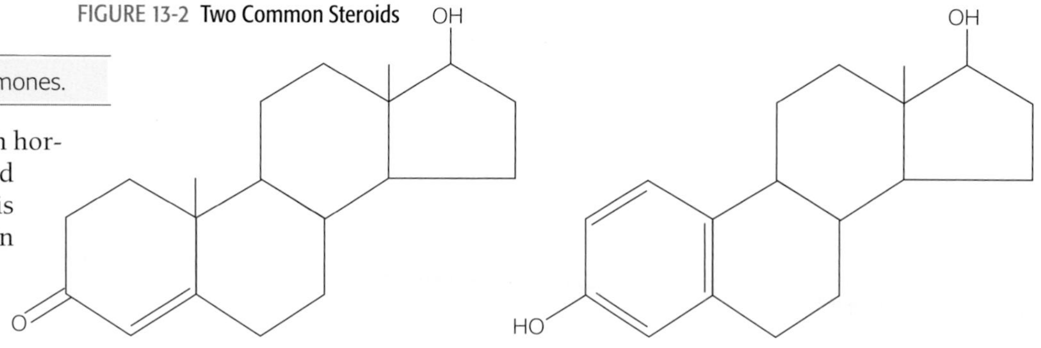

(a) Testosterone, a masculinizing hormone

(b) Estradiol, a feminizing hormone

Tropic and Nontropic Hormones

Some hormones stimulate the synthesis and release of other hormones; others activate cellular processes.

Hormones fall into two broad categories. The first are the tropic hormones (TROW-pick; "to nourish"). **Tropic hormones** stimulate other endocrine glands to produce and secrete hormones. An example is thyroid-stimulating hormone (TSH). Produced by the pituitary gland, TSH travels in the blood to the thyroid gland located in the neck on either side of the larynx. In the thyroid, TSH stimulates the release of a hormone, thyroxine (thigh-ROX-in). Thyroxine, in turn, circulates in the blood and stimulates metabolism in many types of body cells. Thyroxine is a **nontropic hormone**. Nontropic hormones stimulate cellular growth, metabolism, or other functions.

Hormones can also be classified into three groups according to their chemical composition: (1) steroids, (2) proteins and polypeptides, and (3) amines.

Steroid hormones are derivatives of cholesterol. Figure 13-2 illustrates two common steroid hormones. Only minor differences in the chemical structure of these molecules exist, but they produce profound functional differences. Protein and polypeptide hormones are the largest group of hormones. Proteins and polypeptides are polymers of amino acids. Growth hormone and insulin are two examples. Amine hormones are much smaller molecules derived from the amino acid tyrosine. Figure 13-3 shows the structure of two of the four amine hormones produced in the body.

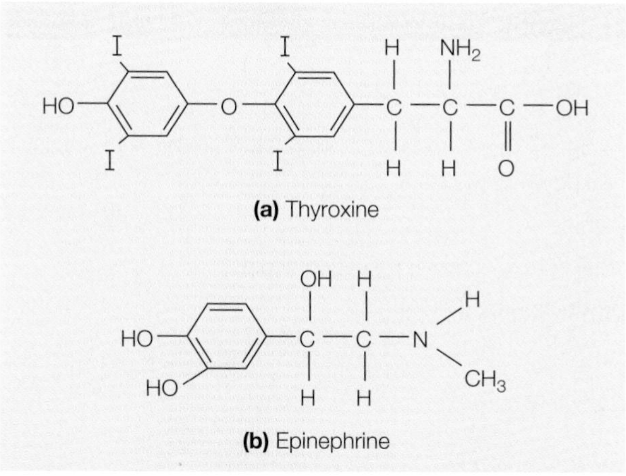

Negative Feedback Control

Hormone secretion is often controlled by negative feedback mechanisms.

Hormones help to control many homeostatic mechanisms. Not surprisingly, their production and release are generally controlled by **negative feedback loops**, described in Chapter 4. As you may recall, in negative feedback loops, the end product of biochemical process inhibits its own production. Consider the hormone glucagon (GLUE-ka-gone). Glucagon is secreted (seh-CREET-ed) by cells in the pancreas when glucose concentrations fall—for example, between meals. Glucagon, in turn, stimulates the breakdown of glycogen stored inside liver cells. Glucose released from the liver into the bloodstream restores normal levels. When normal glucose levels are achieved, glucose molecules cause the glucagon-producing cells in the pancreas to cease secretion.

Positive feedback loops are also encountered in the endocrine system, but only rarely. A **positive feedback loop** results when the hormonal product of a cell or organ stimulates the production of another hormone. This, in turn, causes more secretion and so on and so on. The noise in a restaurant is a good example of a positive feedback. As the noise level increases, people talk louder. This increases the overall noise, forcing people to talk even louder.

Positive feedback loops perform specialized functions. Ovulation (OV-you-LAY-shun), the release of the ovum from the ovary, for example, is stimulated by a positive feedback loop discussed in Chapter 21. Fortunately, each positive feedback loop has a built-in mechanism that ends the escalating cycle, preventing the response from getting completely out of hand. With this information in mind, let's turn our attention to the pituitary hormones.

(a) Thyroxine

(b) Epinephrine

FIGURE 13-3 **Representative Amine Hormones** (a) Thyroxine from the thyroid and (b) adrenalin (epinephrine) from the adrenal medulla.

Scientific Discoveries that Changed the World

13-1	**Pancreatic Function: Is It Controlled by Nerves or Hormones?** **Featuring the Work of Bayliss and Starling**

The concept of chemical control of body functions, like many other important biological ideas, is rooted in numerous experiments.

Of the many experiments crucial to our understanding of endocrinology, one by two British physiologists, W. M. Bayliss and E. H. Starling, stands out. Published in 1902, this experiment set to rest a debate about the mechanism controlling the release of pancreatic secretions.

Several experiments before the publication of this work suggested that hydrochloric acid in chyme (produced in the stomach) stimulated nerve endings in the small intestine. This, in turn, triggered a reflex that caused the release of pancreatic juices containing sodium bicarbonate, which neutralizes the acid. In these studies, researchers injected substances into the small intestine of anesthetized dogs to see what stimulated the release of pancreatic secretions. To determine the nature of the reflex, the researchers cut the nerves supplying the small intestine. Even though severing the nerves did not halt the pancreas, these researchers still argued that a nervous system reflex was involved. They hypothesized that clumps of nerve cells (ganglia) and their net-work of interconnected fibers in the wall of the intestine participated in local reflex arcs.

Bayliss and Starling repeated these experiments, but were careful to sever all nerve connections to the small intestine. They then tied off a piece of the small intestine and injected small amounts of hydrochloric acid into it. This treatment resulted in the production of pancreatic juices. Because it was previously known that acid introduced into the bloodstream had no effect on the release of pancreatic secretions, the researchers concluded that "the effect was produced by some chemical substance finding its way into the veins of the loop of jejunum in question. . . ." This substance, they wrote, was "carried in the bloodstream to the pancreatic cells."

To verify their hypothesis, the researchers tried another experiment. In this one, they scraped off the cells lining the section of the small intestine, exposed them to acid, ground them up, and then extracted the fluid, which they injected into the bloodstream. After a brief latent period, the pancreas began secreting, a clear indication that these cells were producing a chemical substance that stimulated the pancreas.

Bayliss and Starling referred to the mystery substance as "secretin," because it stimulated pancreatic secretion. This name remains in use today.

Besides settling the debate over the control of pancreatic function, Bayliss and Starling helped clarify our understanding of endocrinology, furnishing definitions for some of the most important terms in the field. One good example is the term endocrine gland, which they defined as an organ that secretes into the blood specific substances that affect some other organ or process located at a distance from the gland.

13-2 The Pituitary and Hypothalamus

The pituitary consists of two parts, the anterior and posterior lobes, and produces many hormones.

Attached to the underside of the brain by a thin stalk is the **pituitary gland** (peh-TWO-eh-TARE-ee) (Figure 13-4). About the size of a pea, the pituitary lies in a depression in the base of the skull.

The pituitary gland is divided into major parts: the **anterior pituitary** and the **posterior pituitary**. Together, they secrete a large number and variety of hormones, affecting many different body's functions. The anterior pituitary, for instance, produces seven protein and polypeptide hormones, six of which are discussed in this chapter (Table 13-1).

TABLE 13-1	Hormones Secreted by the Pituitary Gland
Hormone	**Function**
Anterior pituitary	
Growth hormone (GH)	Stimulates cell growth. Primary targets are muscle and bone, where GH stimulates amino acid uptake and protein synthesis. It also stimulates fat breakdown in the body.
Thyroid-stimulating hormone (TSH)	Stimulates release of thyroxine and triiodothyronine.
Adrenocorticotropic hormone (ACTH)	Stimulates secretion of hormones by the adrenal cortex, especially glucocorticoids.
Gonadotropins (FSH and LH)	Stimulate gamete production and hormone production by the gonads.
Prolactin	Stimulates milk production by the breast.
Melanocyte-stimulating hormone (MSH)	Function in humans is unknown.
Posterior pituitary	
Antidiuretic hormone (ADH)	Stimulates water reabsorption by nephrons of the kidney.
Oxytocin	Stimulates ejection of milk from breasts and uterine contractions during birth.

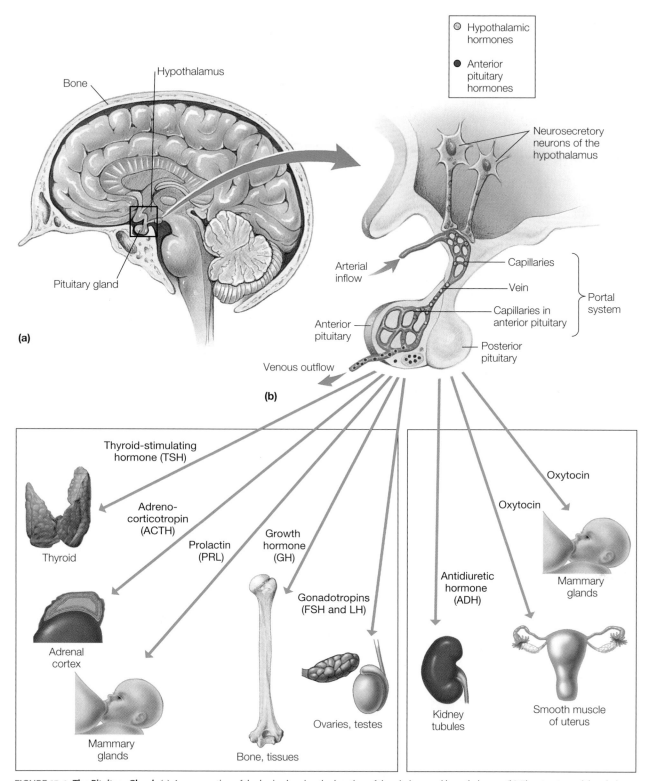

FIGURE 13-4 The Pituitary Gland (a) A cross section of the brain showing the location of the pituitary and hypothalamus. (b) The structure of the pituitary gland. (c) Releasing and inhibiting hormones travel via the portal system from the hypothalamus to the anterior pituitary, where they affect hormone secretion.

The anterior pituitary is controlled by a region of the brain known as the **hypothalamus.** Lying just above the pituitary gland, the hypothalamus contains receptors that monitor blood levels of hormones, nutrients, and ions. When activated, the receptors stimulate specialized nerve cells in the hypothalamus. These cells are a special brand of neurons called **neurosecretory neurons**—

so named because they synthesize and secrete hormones. The hormones they produce control the production and release of hormones by the anterior pituitary located nearby (Figure 13-4).

The hypothalamus produces two types of hormones—those that stimulate the release of anterior pituitary hormones, called **releasing hormones**, designated **RH**, and those that inhibit the release of hormones from the anterior pituitary. They're called **inhibiting hormones**, designated **IH**.

The releasing and inhibiting hormones are produced in the cell bodies of the neurosecretory neurons. They travel down the axons of the neurosecretory cells, which terminate in the lower part of the hypothalamus, just above the pituitary gland. The hormones are stored in the ends of the axons until needed.

When released, the hormones diffuse into nearby capillaries. As shown in Figure 13-4, these capillaries drain into a series of veins in the stalk of the pituitary. These veins drain into a capillary network located in the anterior pituitary. From here, the hormones diffuse into the anterior pituitary. They then bind to their target cells, causing them to release hormones. This unusual arrangement of blood vessels in which a capillary bed drains to a vein, which then drains into another capillary bed, is called a **portal system**.

The Anterior Pituitary

> The anterior pituitary secretes seven hormones with widely different functions.

The pituitary produces a number of hormones that profoundly influence our bodies. We will now examine the major ones produced by the anterior pituitary gland.

Growth Hormone

> Growth hormone stimulates cell growth, primarily targeting muscle and bone.

People differ in height and body build. These differences are largely attributable to **growth hormone (GH)**, a protein hormone produced by the anterior pituitary. Growth hormone stimulates growth in the body by promoting cellular enlargement (hypertrophy; HIGH-per-TROW- fee) and an increase in the number of cells through division (hyperplasia; HIGH-per-PLAY-za).

Although growth hormone affects virtually all body cells, it acts primarily on bone and muscle. In muscle, it stimulates the uptake of amino acids and protein synthesis. In bone, growth hormone stimulates cell division and protein synthesis in bone cells, resulting in an increase in both the length and width of bone.

Recent evidence shows that growth hormone does not affect bone and muscle directly. It actually operates by stimulating the production of several polypeptide hormones produced primarily by the liver. Known as *somatomedins* (so-MAT-oh-ME-dins), these hormones stimulate bone and cartilage cells.

As a rule, the more growth hormone produced during the growth phase of an individual's life cycle, the taller and heftier he or she will be. In men, body growth is also stimulated by testosterone, a sex steroid produced by the testes. It stimulates bone and muscle growth.

Growth hormone secretion undergoes a diurnal (daily) cycle. As Figure 13-5 shows, the highest blood levels are present during sleep and during strenuous exercise. It is no wonder that sleep is so important to a growing child. Growth hormone secretion declines gradually as we age.

Control of Growth Hormone Section

> The secretion of growth hormone is controlled by a releasing hormone produced by the hypothalamus.

The release of growth hormone by the pituitary is controlled by the hypothalamus in a classic negative feedback loop. When levels are low, growth hormone-RH or GH-RH is released from the hypothalamus. This causes the anterior pituitary to produce and release GH.

Growth hormone release is also stimulated directly through the nervous system. Studies show that stress and exercise both stimulate the hypothalamus via neural pathways to release GH-RH.

Deficiencies in growth hormone can result in dramatic changes in body shape and size, depending on when the deficiency occurs. Undersecretion, or hyposecretion, that occurs during the growth phase of a child is one cause of stunted growth (dwarfism) (Figure 13-6a). Oversecretion, or hypersecretion (HIGH-per-seh-KREE-shun), results in giantism (GIE-antizm) if the excess occurs during the growth phase (Figure 13-6b). If the pituitary begins producing excess growth hormone after growth is complete, the result is a relatively rare disease called **acromegaly** (AH-crow-MEG-al-ee). Facial features become coarse, and hands and feet continue to grow throughout adulthood (Figure 13-7). Growth of the vertebrae results in a hunched back. The tongue, kidneys, and liver often become quite enlarged.

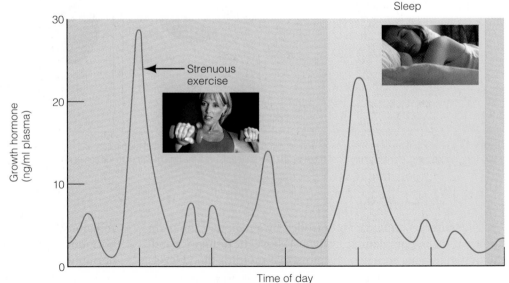

FIGURE 13-5   **Growth Hormone Secretion in an Adult**   Growth hormone stimulates muscle and bone growth and is released during exercise and sleep.

ng/ml = nanograms per mililiter

(a)

(b)

Thyroid-Stimulating Hormone

TSH stimulates the thyroid gland to produce thyroxine.

Thyroid-stimulating hormone (TSH) is a protein hormone produced by the anterior pituitary and is controlled by TSH-RH (TSH-releasing hormone) from the hypothalamus. Day-to-day levels are primarily regulated by the level of thyroxine in the blood in a negative feedback loop. Receptors in the hypothalamus monitor the level of circulating thyroxine. When circulating levels are low, these receptors signal the hypothalamus to

(a)

(b)

(c)

(d)

FIGURE 13-7 Acromegaly Hypersecretion of growth hormone in adults results in a gradual thickening of the bone, which is especially noticeable in the face, hands, and feet. There is no sign of the disorder at age 9 (a) or age 16 (b). Symptoms are evident at age 33 (c) and age 52 (d).

release TSH-RH. When the level of thyroxine increases, TSH-RH secretion declines (Figure 13-8). TSH-RH secretion, however, is also stimulated by cold and stress. When you're under a lot of stress, TSH-RH secretion will be higher than when you are calm.

TSH-RH travels to the anterior pituitary, where it stimulates the production and release of TSH. TSH then travels in the blood to the thyroid gland, located in the neck (Figure 13-1). TSH stimulates the production and release of two thyroid hormones,

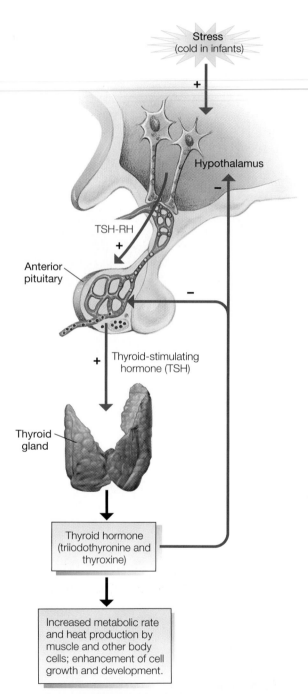

FIGURE 13-8 Negative Feedback Control of TSH Secretion Triiodothyronine (T$_3$) and thyroxine levels are detected by receptors in the hypothalamus. They also feedback on the anterior pituitary. When levels of T$_3$ and thyroxine are low, the hypothalamus releases TSH-RH, which stimulates the pituitary to release TSH. TSH, in turn, stimulates the thyroid gland to secrete T$_3$ and thyroxine. Other factors such as stress and cold influence the release of TSH via the hypothalamus. (a + denotes stimulation; a − denotes inhibition.)

thyroxine (thigh-ROX-in) or T4 and triiodothyronine (try-eye-OH-doe-THIGH-row-neen) or T3. These hormones circulate in the bloodstream and influence many body cells. One of their chief functions is to stimulate the breakdown of glucose by body cells. Because this process produces energy and heat, increased levels of these hormones raise body temperature. Secretion of T3 and T4 is therefore typically higher during cold winter months than in the summer.

Adrenocorticotropic Hormone

ACTH stimulates the release of hormones from the adrenal cortex.

Another important hormone of the anterior pituitary is **ACTH**, also known as **adrenocorticotropic hormone** (ad-REE-no-core-tick-oh-TRO-pick). Its target organ is the outer regions of the adrenal gland, the adrenal cortex. As shown in Figure 13-9, ACTH stimulates the production and release of a group of steroid hormones known as the *glucocorticoids* (GLUE-co-CORE-teh-KOIDS). **Glucocorticoids** are steroids. Their main function is to increase blood glucose levels, thus helping maintain homeostasis.

As shown in Figure 13-9, the production and release of ACTH are controlled by the hypothalamus, which releases a small peptide known as ACTH-RH. It travels to the pituitary via the portal system. ACTH secretion is controlled by several factors, two of which are discussed here. The first is the level of circulating glucocorticoids produced by the adrenal cortex. They participate in a negative feedback loop (Figure 13-9). ACTH-RH secretion is also stimulated by stress, acting through the nervous system.

Glucocorticoids increase blood glucose levels, which ensures that we have the additional energy we need to operate body systems, especially muscles, when we are under stress.

The Gonadotropins

The gonadotropins are hormones that stimulate the ovaries and testes.

Reproduction in men and women is under the control of the anterior pituitary. It produces two hormones that affect the gonads, appropriately known as *gonadotropins* (go-NAD-oh-TROW-pins). These hormones are discussed in detail in Chapter 21.

Prolactin

Prolactin stimulates milk production in the breasts of women.

Prolactin (pro-LACK-tin) is a protein hormone produced by the anterior pituitary. In women, prolactin is secreted at the end of pregnancy and acts on the mammary glands (breasts) where it stimulates milk production.

Prolactin secretion is controlled by a **neuroendocrine reflex**, a reflex involving both the nervous and endocrine systems. As Figure 13-10 shows, suckling stimulates sensory fibers in the breast, which send nerve impulses to the hypothalamus. In the hypothalamus, these impulses stimulate the release of prolactin releasing hormone (PRH). PRH, in turn, travels to the anterior pituitary in the bloodstream, where it stimulates the secretion of prolactin.

Prolactin production persists for many months, as long as suckling continues. As babies begin to eat solid food, they reduce

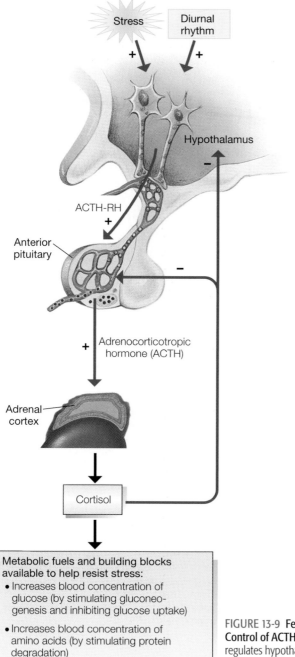

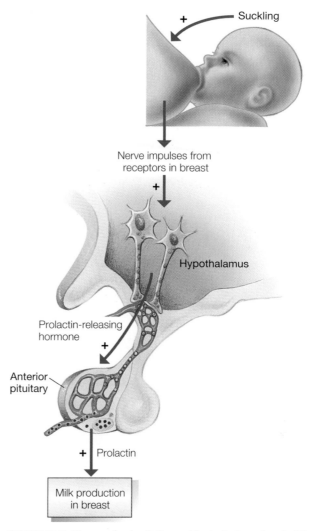

FIGURE 13-9 **Feedback Control of ACTH** Cortisol regulates hypothalamic and pituitary activity, but stress and the biological clock also influence the release of ACTH-RH.

Metabolic fuels and building blocks available to help resist stress:
• Increases blood concentration of glucose (by stimulating gluconeogenesis and inhibiting glucose uptake)
• Increases blood concentration of amino acids (by stimulating protein degradation)
• Increases blood concentration of fatty acids (by stimulating the breakdown of fat)

FIGURE 13-10 **Neuroendocrine Reflex and Prolactin Secretion** Suckling stimulates prolactin release by the anterior pituitary. Prolactin stimulates milk production by the breast. Milk release requires another hormone, oxytocin, from the posterior pituitary.

their dependence on their mother's milk. Reduced suckling shuts down prolactin secretion, causing milk production to decline.

The Posterior Pituitary

The posterior pituitary secretes two hormones.

Have you ever wondered why alcoholic beverages make a person urinate so much or what actually causes milk to flow from a woman's breast? Have you ever wondered why a woman's uterus contracts when she delivers a baby?

These phenomena can be attributed to two hormones from the posterior pituitary. Like the hypothalamus, the posterior pituitary is a neuroendocrine gland—that is, a gland made of neural tissue that produces hormones. Derived from brain tissue during embryonic development, the posterior pituitary remains connected to the brain throughout life.

The posterior pituitary secretes two hormones: **oxytocin** (OX-ee-TOE-sin) and **antidiuretic hormone** (an-tie-DIE-yur-ET-ick) (**ADH**). Both hormones consist of nine amino acids. As shown in Figure 13-11, ADH and oxytocin are produced in the cell bodies of the neurosecretory cells located in the hypothalamus. They then travel down the axons into the posterior pituitary. The posterior pituitary itself consists of the axons and terminal ends of neurosecretory cells in which these hormones are stored until they are released into the surrounding capillaries.

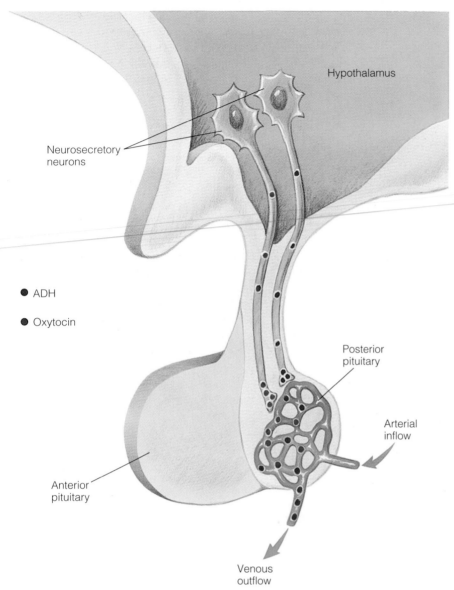

● ADH

● Oxytocin

FIGURE 13-11 **The Posterior Pituitary** Neurosecretory neurons that produce oxytocin and ADH originate in the hypothalamus and terminate in the posterior pituitary. Hormones are produced in the cell bodies of the neurons and are stored and released into the bloodstream in the posterior pituitary.

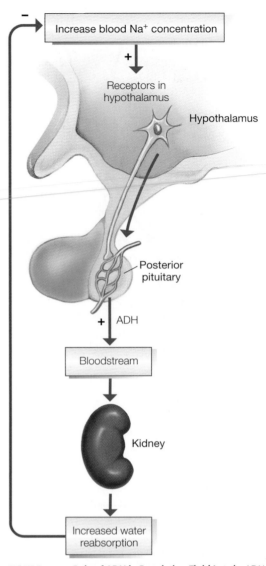

FIGURE 13-12 **Role of ADH in Regulating Fluid Levels** ADH secretion is stimulated by an increase in the blood concentration of sodium caused by dehydration. ADH increases water reabsorption in the kidney, thus eliminating the stimulus for ADH secretion.

Antidiuretic Hormone

ADH increases water absorption in the kidney.

ADH regulates water balance in humans by increasing water absorption in the nephrons of the kidneys (Figure 13-12). As a result, water reenters the bloodstream, increasing blood volume and maintaining the normal osmotic concentration of the blood.

ADH secretion is controlled by a negative feedback loop (Figure 13-12). Receptors in the hypothalamus monitor the concentration of dissolved substances (especially sodium ions) in the blood. When the concentration exceeds the normal level—for example, during dehydration—ADH is released into the bloodstream. ADH stimulates the reabsorption of water by the nephrons. When the osmotic concentration of the blood approaches homeostatic levels, ADH secretion ceases.

As Chapter 9 notes, ADH is inhibited by ethanol in alcoholic beverages. The reduction in ADH decreases water reabsorption in the kidneys and increases urine output. This, in turn, can lead to dehydration—the dry mouth and intense thirst a person may experience as part of a hangover.

Inadequate ADH secretion, for example, caused by traumatic head injuries that damage the hypothalamus, result in a condition known as *diabetes insipidus*. Patients with this disease produce several gallons of dilute urine a day and must drink enormous quantities of water to prevent dehydration.

Oxytocin

Oxytocin facilitates birth and stimulates milk let-down.

Prolactin from the anterior pituitary stimulates milk production in the breasts. But it's oxytocin from the posterior pituitary that stimulates the contraction of the smooth-muscle-like cells around the glands in the breast "pumping" the milk out. Oxytocin release is stimulated by suckling. Oxytocin then travels in the blood to the breast, where it stimulates milk let-down—the ejection of milk from the glands soon after suckling begins.

Oxytocin also plays a role in childbirth. At the end of pregnancy, in the hours prior to delivery, the walls of the uterus and cervix stretch. Impulses from the stretch receptors in the uterus travel to the cells of the hypothalamus that produce oxytocin. The impulses cause the neurosecretory cells to release oxytocin into the blood. The hormone then travels in the blood to the uterus, where it stimulates smooth muscle contraction, aiding in the expulsion of the baby.

13-3 The Thyroid Gland

The thyroid gland located in the neck produces hormones that regulate metabolism, heat production, and calcium levels in the blood.

On a 50° day in the fall, you bundle up in a sweater to feel warm. However, 50° in the spring feels warm. You may not even need a sweater. Why?

The answer lies in the thyroid gland and two of its hormones. The **thyroid gland** is a U- or H-shaped gland (it varies from one person to the next) located in the neck (Figure 13-13a). The thyroid gland produces three hormones: (1) thyroxine (tetraiodothyronine, or T_4), (2) a chemically similar compound, triiodothyronine (T_3), and (3) calcitonin. The first two are involved in controlling metabolism and heat production; the last helps regulate blood levels of calcium.

As Figure 13-13b shows, the thyroid gland consists of large, spherical structures known as *follicles* (FOLL-ick-als). Each follicle consists of a central region containing a gel-like glycoprotein called *thyroglobulin* and a surrounding layer of cuboidal follicle cells that produce thyroglobulin. **Thyroxine** and **triiodothyronine** are relatively small molecules (amines) derived from much larger thyroglobulin molecules in each follicle.

To make T_3 and T_4, the thyroid requires large quantities of iodide, I^-, which is actively transported into the cell. A shortage of iodide in the diet results in a decline in the levels of thyroxine and triiodothyronine in the blood. The hypothalamus responds by producing more TSH. TSH causes the thyroid to increase thyroglobulin production. Without sufficient iodide, however, the thyroid gland cannot produce T_3 and T_4. As a re-

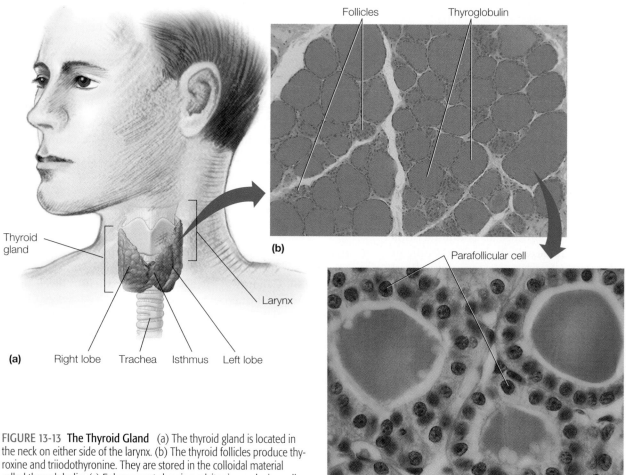

FIGURE 13-13 The Thyroid Gland (a) The thyroid gland is located in the neck on either side of the larynx. (b) The thyroid follicles produce thyroxine and triiodothyronine. They are stored in the colloidal material called thyroglobulin. (c) Enlargement showing calcitonin-producing cells known as parafollicular cells.

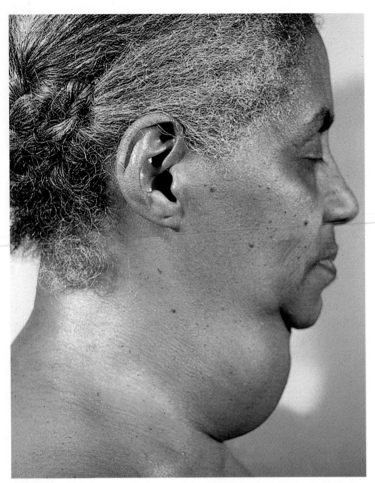

FIGURE 13-14 **Goiter** An enlargement of the thyroid gland most often results from a lack of iodine in the diet.

sult, the thyroid gland continues to produce thyroglobulin. This causes the gland to enlarge, sometimes forming softball-sized enlargements on the neck. This condition is known as *goiter* (GOY-ter) (Figure 13-14).

Goiter was once common in areas of Europe and the United States where iodine had been leached out of the soil by rain—for example, near the Great Lakes. Crops grown in iodine-poor soils failed to provide sufficient quantities. To counteract this deficiency, iodine was added to table salt. Today, few people in industrialized nations develop goiter because they eat iodized salt or ingest sufficient quantities of iodine in their food or through mineral supplements.

Actions of Thyroxine and Triiodothyronine

Thyroxine and triiodothyronine accelerate the breakdown of glucose and stimulate growth and development.

Thyroxine and triiodothyronine are virtually identical and, as a result, have nearly identical functions. Both hormones, for example, accelerate the rate of glucose breakdown in most body cells. Thyroid hormones also stimulate cellular growth and development. Bones and muscles are especially dependent on them during the growth phase. Even normal reproduction requires these hormones. A deficiency of T_3 and T_4, for example, delays sexual maturation in both sexes. In children, depressed thyroid output stunts mental as well as physical growth. If the deficiency is not detected and treated, the effects are irreversible.

In adults, reduced thyroid activity, or hypothyroidism, has less severe results. Hypothyroidism decreases the metabolic rate, making a person feel cold much of the time. People suffering from hypothyroidism also feel tired and worn out. Even simple mental tasks become difficult. Their heart rate may slow to 50 beats per minute. Hypothyroidism is treated by pills containing artificially produced thyroid hormone.

Excess thyroid activity, hyperthyroidism, in adults results in elevated metabolism, excessive sweating (due to overheating), and weight loss, despite increased food intake. The increase in thyroid hormone levels results in increased mental activity, resulting in nervousness and anxiety. People suffering from hyperthyroidism often find it difficult to sleep. Their heart rate may accelerate, as in the case of former President Bush, and they may lose their sensitivity to cold. Some people suffering from hyperthyroidism exhibit a condition called exophthalmos (EX-oh-THAL-mus), or bulging eyes brought about by a swelling of the soft tissues of the eye socket, which causes the eyes to protrude. In such individuals, the eyes may protrude so much that the eyelids cannot close completely. Exophthalmos can cause double vision or blurred vision.

Individuals with hyperthyroidism may be given antithyroid medications, drugs that block the effects of thyroid hormones. If the gland is cancerous, surgeons may remove part or all of it. The most common treatment for hyperthyroidism, however, is radioactive iodine, which is taken orally. Radioactive iodine concentrates in the thyroid gland, where it damages the cells that produce thyroglobulin, reducing the output of thyroid hormones. This procedure, while effective, may lead to other problems (notably cancer) later in life.

Calcitonin

Calcitonin decreases blood levels of calcium.

Large, round cells in the perimeter of the thyroid follicles produce **calcitonin**, or **thyrocalcitonin**, a polypeptide hormone (Figure 13-13c). Calcitonin lowers the blood calcium level, helping to maintain homeostasis. It does so in two ways (1) by inhibiting osteoclasts, bone-resorbing cells, thus reducing the release of calcium from bone, and (2) by stimulating bone-forming cells, osteoblasts, causing calcium to be deposited in bone. It also increases the excretion of calcium (and phosphate) ions by the kidneys.

Calcitonin is involved in a negative feedback loop with calcium ions in the blood. When the calcium-ion concentration increases, calcitonin secretion increases. As calcium concentrations fall, calcitonin secretion falls.

13-4 The Parathyroid Glands

The parathyroid glands produce a hormone that helps regulate serum calcium levels.

The **parathyroid glands** are four small nodules of tissue embedded in the back side of the thyroid gland. They produce a polypeptide hormone known as **parathyroid hormone**, or **parathormone** (**PTH**). PTH works with calcitonin to control calcium levels in the blood. This hormone, however, increases blood calcium levels. That is, PTH secretion is stimulated when calcium levels in the blood drop.

PTH restores blood calcium levels by increasing intestinal absorption of calcium and by stimulating bone destruction by osteoclasts. They digest bone and release calcium into the bloodstream. PTH also increases calcium reabsorption in the kidney.

Calcium levels are also influenced by vitamin D, available in some of the foods we eat, such as milk. Vitamin D is also pro-duced in the skin under the influence of sunlight. Vitamin D increases calcium absorption in the intestine. It also increases the responsiveness of bone to parathyroid hormone.

As in other glands, the parathyroid may malfunction. Excess secretion of parathyroid hormone is the most common condition. It may result from tumors in the parathyroid glands. Excess PTH from tumors or other causes results in a loss of calcium from the bones and teeth. Excess PTH also upsets several metabolic processes, resulting in indigestion and depression. Because bones contain enormous amounts of calcium, most symptoms of excess PTH (hyperparathyroidism) other than high blood calcium levels do not appear until 2–3 years after the onset of the disease. By the time the disease is discovered, kidney stones may already have formed from calcium, cholesterol, and other substances, and bones may have become more fragile and susceptible to breakage. To prevent further complications, parathyroid tumors must be removed.

13-5 The Pancreas

The pancreas produces insulin and glucagons that control blood glucose levels.

A young girl's parents complain to her doctor about their daughter's frequent urination, bladder infections, fatigue, and weakness. Tests show that she is suffering from diabetes mellitus, a disorder of the pancreas. **Diabetes mellitus** has several causes, but in young people it is generally caused by a lack of insulin production.

Insulin is a protein hormone produced by the pancreas. It is released by the pancreas within minutes after glucose levels in the blood begin to rise. Insulin has many functions, but one of the most important is that it stimulates the uptake of glucose in liver and muscle cells. Inside muscle and liver cells, glucose molecules are combined to form the polysaccharide glycogen. Glucose is said to be stored as glycogen. In healthy individuals, the amount of glycogen in the liver and muscle increases immediately after meals. Liver glycogen is broken down between meals to maintain blood glucose levels and thus energy supplies throughout the body. Muscle glycogen is broken down to supply energy to muscles.

Insulin is produced by the **pancreas** (PAN-kree-us), a dual-purpose organ located in the abdominal cavity (Figure 13-15a). The head of the pancreas lies in the curve of the duodenum, and its tail stretches to the left kidney. Most of the pancreas consists of tiny clumps of cells that produce digestive enzymes and sodium bicarbonate. Scattered throughout these enzyme-producing cells, however, are small islands of hormone-producing cells, known as the *islets of Langerhans* (EYE-lits of LANG-er-hahns) after the scientist who described them (Figure 13-15b).

Glucagon

Glucagon increases blood levels of glucose, opposing the actions of insulin.

The islets of Langerhans also produce a polypeptide hormone known as **glucagon**. This hormone has the opposite effect of insulin—that is, it increases blood levels of glucose. It does so by stimulating the breakdown of glycogen in liver cells (Figure 13-16). This process helps maintain proper glucose levels in the blood between meals. One molecule of glucagon causes 100 million molecules of glucose to be released.

Glucagon also elevates blood glucose levels by stimulating the synthesis of glucose from amino acids and glycerol molecules derived from the breakdown of one type of fat, the triglycerides. This process occurs in the liver.

Glucagon secretion, like that of insulin, is regulated by a negative feedback mechanism stimulated by glucose concentrations in the blood. When glucose levels fall, the hormone is released into the bloodstream. When glucose levels rise, glucagon secretion declines. Glucagon release is also stimulated by an increase in the concentration of amino acids in the blood. Because glucagon stimulates the conversion of amino acids to glucose, this process ensures that excess amino acids in high-protein meals are used to produce glucose.

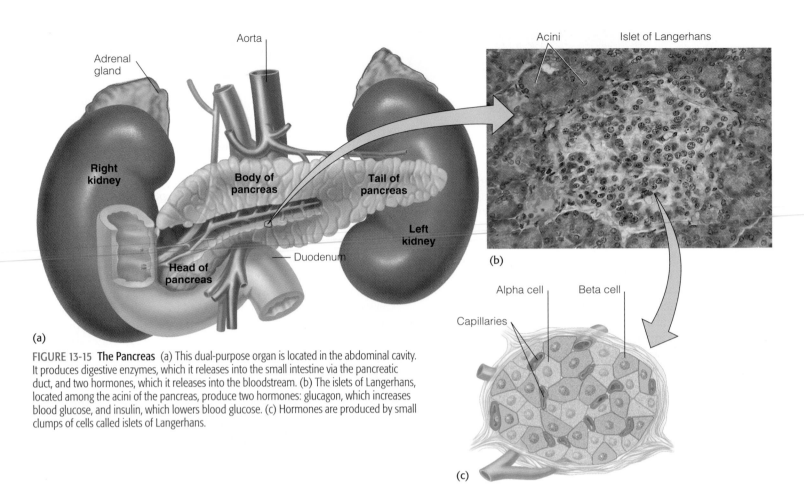

FIGURE 13-15 The Pancreas (a) This dual-purpose organ is located in the abdominal cavity. It produces digestive enzymes, which it releases into the small intestine via the pancreatic duct, and two hormones, which it releases into the bloodstream. (b) The islets of Langerhans, located among the acini of the pancreas, produce two hormones: glucagon, which increases blood glucose, and insulin, which lowers blood glucose. (c) Hormones are produced by small clumps of cells called islets of Langerhans.

Insulin

Insulin causes blood sugar levels to decline rapidly after a meal.

Figure 13-17 shows a graph of blood glucose levels in nondiabetic and diabetic people obtained during a **glucose tolerance test**. During this test, physicians give their patients an oral dose of glucose, then check blood levels at regular intervals. The blue line on the graph in Figure 13-17 represents the blood glucose levels in a normal person with a healthy pancreas. As illustrated, glucose levels increase slightly after the administration of glucose. Within 2 hours, glucose levels decrease to normal, thanks to insulin secretion.

The pink line illustrates the response in a diabetic person, who produces insufficient amounts of insulin or whose cells have become unresponsive to insulin. Glucose levels rise considerably in these people after they've been given glucose and they remain elevated for 4–5 hours. A similar response is seen in untreated diabetics after they eat. In both instances, blood glucose levels remain high for such a long period because virtually no glucose can enter skeletal muscle cells or liver cells.

Glucose levels eventually decline as this nutrient is used by brain cells, red blood cells, and others. The kidneys also excrete excess glucose. Unfortunately, this wastes a valuable nutrient and causes the kidneys to lose large quantities of water. Excess urination and constant thirst are the chief symptoms of untreated diabetes. Diabetics may have to urinate every hour, day and night.

Diabetes

Diabetes is really two diseases with similar symptoms but different causes.

Physicians recognize two types of diabetes (Table 13-2). Type I diabetes occurs mainly in young people and results from a deficiency in insulin production. Also called *juvenile diabetes* or *early-onset diabetes*, Type I diabetes is believed to be caused by damage to the insulin-producing cells of the pancreas. A leading hypothesis suggests that it is the result of an autoimmune reaction (in which the immune system attacks parts of a person's own body). In other cases, researchers speculate that Type I diabetes may be caused by a viral infection or an environmental pollutant that damages the insulin-producing cells.

Insulin production varies in patients with Type I diabetes. In some, it is only slightly reduced; in others, it is completely suppressed. In the latter, the absence of insulin starves body cells for glucose. Energy must be provided by the breakdown of fats.

Type II diabetes usually occurs in people over 40 and is also called *late-onset diabetes* (Table 13-2). In this disease, the pancreas produces normal or above-normal levels of insulin. However, because of a decline in the number of insulin receptors in target cells, cells are unresponsive to insulin.

Type II diabetes is commonly associated with obesity, a problem growing rapidly in more developed nations like the United States where people dine on high-fat, high-calorie di-

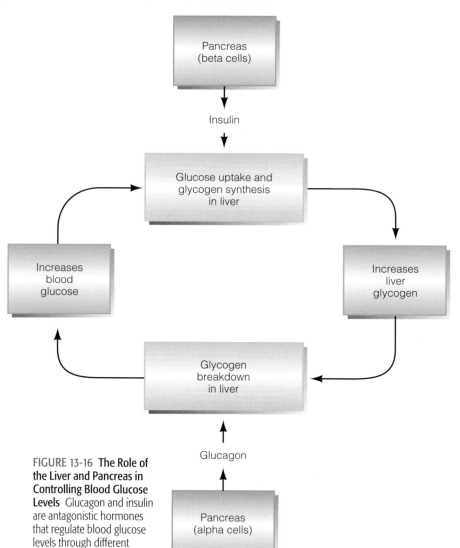

FIGURE 13-16 **The Role of the Liver and Pancreas in Controlling Blood Glucose Levels** Glucagon and insulin are antagonistic hormones that regulate blood glucose levels through different mechanisms.

TABLE 13-2	Type I and Type II Diabetes	
Characteristic	Type I Diabetes	Type II Diabetes
Level of insulin secretion	None or almost none	May be normal or exceed normal
Typical age of onset	Childhood	Adulthood
Percent of diabetics	10%–20%	80%–90%
Basic defect	Destruction of beta cells	Reduced sensitivity of insulin's target cells
Associated with obesity?	No	Usually
Speed of development of symptoms	Rapid	Slow
Development of ketosis	Common if untreated	Rare
Treatment-	Insulin injections; dietary management	Dietary control and weight reduction; occasionally oral hypoglycemic drugs

ets. In the United States, diabetes has increased dramatically in adults—and children—almost entirely from a rise in Type II diabetes and a corresponding rise in obesity. Lack of exercise and calorie-rich diets are the main culprits.

Heredity may also be an important contributor to late-onset diabetes, as it is in Type I diabetes. In fact, studies show that about one-third of the patients with Type II diabetes have a family history of the disease. Critical thinking rules in Chapter 1 suggest the need to look for other explanations. Perhaps it may be the obesity that is inherited, rather than the tendency to develop diabetes.

Although Type I and Type II diabetes have different causes, both forms of this disease exhibit similar symptoms. Excess urination and thirst are generally the first signs of trouble. Patients often feel tired, weak, and apathetic. Weight loss and blurred vision are also common. Excess glucose in the urine may result in frequent bacterial infections in the bladder.

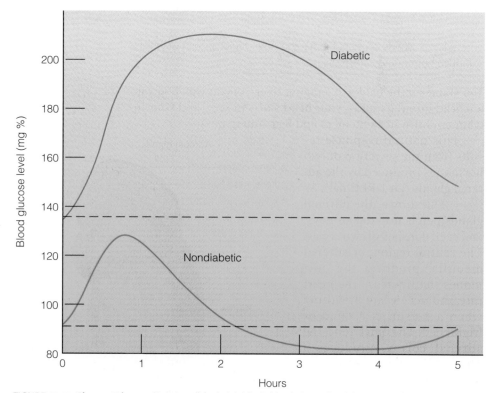

FIGURE 13-17 **Glucose Tolerance Test** In a diabetic (pink line), blood glucose levels increase rapidly and remain high after ingestion of glucose. In a normal person (blue line), blood glucose levels rise but are quickly reduced by insulin.

Treating Diabetes

Treatments for Type I and Type II diabetes are quite different.

Early-onset diabetes (Type I) is treated with insulin injections. Patients receive injections of insulin, usually two to three times per day. Patients are also required to eat meals and snacks at regular intervals to maintain constant glucose levels in the blood and to ensure that regular insulin injections always act on approximately the same amount of blood glucose.

Insulin injections are tailored to an individual's lifestyle and body demands. To mimic the body's natural release, medical researchers have developed a device called an insulin pump. This device delivers predetermined amounts of insulin. Because the pump cannot actually measure blood glucose levels, it may deliver inappropriate amounts at certain times. Medical researchers are also experimenting with ways to transplant healthy insulin-producing cells into the pancreas of diabetics.

Other researchers are working on a nasal spray to dispense insulin, and so far, clinical trials are proving to be quite promising. Researchers have also developed new ways to monitor blood glucose in diabetics. Currently, diabetics must prick their skin three or more times a day to gather a drop of blood that allows them to test blood glucose levels. New devices often advertised on TV measure blood glucose with much less pain. Another device that may prove helpful measures glucose from moisture absorbed through the skin. The device, called GlucoWatch, resembles a wrist watch. It's absorbent pad sops up moisture, while a measuring device monitors glucose continuously. Researchers are also working on ways to prevent early-onset diabetes by eliminating the autoimmune reaction that attacks the insulin-producing cells of the pancreas.

In patients with Type II diabetes, doctors typically recommend weight loss via diet and exercise as the first line of attack. Patients are also put on medications that either increase the sensitivity of their cells to insulin or stimulate insulin secretion. In severe cases, insulin may need to be administered.

Although treatments of diabetes have greatly improved, diabetics may suffer from diabetic comas, or unconsciousness, when they receive insufficient amounts of insulin or if they go without an insulin injection for a prolonged period. Without insulin, the cells of the body become starved for glucose and begin breaking down fat. Excessive fat catabolism releases toxic chemicals called *ketones* that cause the patient to lose consciousness.

Diabetics may also suffer from insulin shock as a result of an overdose of insulin. This reduces blood glucose levels. In mild cases, symptoms include tremor, fatigue, sleepiness, and the inability to concentrate. These symptoms result from a lack of glucose in the brain. In severe cases, unconsciousness and death may occur.

Over the long term, early- and late-onset diabetics also frequently suffer from loss of vision, nerve damage, and kidney failure, symptoms that appear 20–30 years after the onset of the disease, even if the diabetics are being treated. Damage to the circulatory system can cause gangrene (GANG-green), requiring amputation of limbs, especially the lower extremities. These serious complications result from the inevitable elevations in blood glucose levels that occur periodically over the years. Elevated levels of glucose can damage nerve cells and blood vessels.

13-6   The Adrenal Glands

The adrenal glands consist of inner and outer regions, both of which produce hormones.

You stand on the banks of a raging river. As you watch kayakers head into the rapids, your heart starts to race, and your intestines churn in excitement and fear. You're next.

This natural response results from the secretion of two hormones produced by the **adrenal glands** (ah-DREE-nal). As Figure 13-18 shows, the adrenal glands perch atop the kidneys, and each consists of two zones. The central region, or **adrenal medulla**, produces the hormones that increase the heart rate and accelerate breathing when a person is excited or frightened. The outer zone, the **adrenal cortex**, produces a number of steroid hormones discussed in more detail shortly.

FIGURE 13-18 **Adrenal Gland** The adrenal glands sit atop the kidney and consist of an outer zone of cells, the adrenal cortex, which produces a variety of steroid hormones, and an inner zone, the adrenal medulla. The adrenal medulla produces adrenalin and noradrenalin, the secretion of which is controlled by the autonomic nervous system.

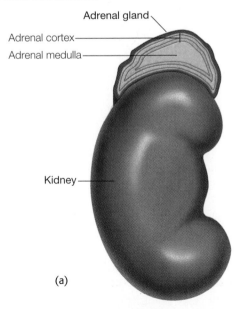

Adrenal gland
Adrenal cortex
Adrenal medulla
Kidney

(a)

(b)

Adrenal Medulla

The adrenal medulla produces stress hormones.

The adrenal medulla produces two hormones: **adrenalin** (epinephrine) and **noradrenalin** (norepinephrine). In humans, about 80% of the adrenal medulla's output is adrenalin.

Helping us meet the stresses of life, adrenalin and noradrenalin are instrumental in the **fight-or-flight response**—the physiological reactions that take place when an animal is threatened. This responce enhances our ability to fight or flee in a stressful or dangerous situation. Adrenalin and noradrenalin are secreted under all kinds of stress—for example, when a careless driver cuts in front of you in heavy traffic or as you wait outside a lecture hall, anticipating an exam. Nerve impulses traveling from the brain to the adrenal medulla trigger the release of adrenalin and noradrenalin. The response is part of the autonomic nervous system (Chapter 10).

The physiological changes these hormones cause are many. For example, these hormones elevate blood glucose levels, making more energy available to cells, particularly skeletal muscle cells. They also increase breathing rate, which provides additional oxygen to skeletal muscles and brain cells. These hormones cause heart rate to accelerate as well. This increases circulation and ensures adequate glucose and oxygen for cells that might be called into action. In addition, adrenalin and noradrenalin cause the bronchioles in the lungs to dilate, permitting greater movement of air in and out of them. Furthermore, these hormones cause blood vessels in the intestinal tract to constrict, putting digestion on temporary hold, while the blood vessels in skeletal muscles dilate, increasing flow through them. Mental alertness increases as a result of increased blood flow and hormonal stimulation. You are ready to fight or flee.

The Adrenal Cortex

The adrenal cortex produces three types of hormones with markedly different functions.

Surrounding the adrenal medulla is the adrenal cortex (Figure 13-18). The adrenal cortex produces three types of steroid hormones, each of which has a different function. The first group, the glucocorticoids, affect glucose metabolism and maintain blood glucose levels. The second group, the **mineralocorticoids** (MIN-er-al-oh-CORE-teh-KOIDS), regulate the ionic concentration of the blood and tissue fluids. The final group, the **sex steroids**, are identical to the hormones produced by the ovaries and testes. In healthy adults, the secretion of adrenal sex steroids is insignificant compared to the amounts produced by the gonads.

Glucocorticoids. The prefix *gluco* reflects the fact that these steroid hormones affect carbohydrate metabolism. (Glucose, of course, is one of many carbohydrates in living things.) The secretion of glucocorticoid is governed by ACTH from the anterior pituitary, as noted earlier. Several chemically distinct glucocorticoids are secreted, the most important being **cortisol**. Cortisol increases blood glucose by stimulating the synthesis of glucose from amino acids and glycerol. This makes more glucose available for energy production in times of stress. Cortisol

also stimulates the breakdown of proteins in muscle and bone, freeing amino acids that can then be converted to glucose in the liver.

In pharmacological doses (levels beyond those seen in the body), cortisol inhibits inflammation, the body's response to tissue damage. Such doses also depress the allergic reaction. Because they reduce inflammation, cortisol and cortisone, a synthetic glucocorticoid-like hormone, can be used to treat inflammation resulting from physical injuries or diseases such as rheumatoid arthritis. However, the benefits must be weighed carefully against the damage that can be caused by upsetting homeostasis.

Mineralocorticoids. The mineralocorticoids are involved in electrolyte or mineral salt balance. The most important mineralocorticoid is **aldosterone**. It is the most potent mineralocorticoid and constitutes 95% of the adrenal cortex's hormonal output.

Although mineralocorticoids regulate the level of several ions, their main function is to control sodium- and potassium ion concentrations. Their chief target organ is the kidney. As noted in Chapter 9, aldosterone increases the movement of sodium ions out of the nephron and into the blood, a process called *tubular reabsorption*. Aldosterone also stimulates sodium ion reabsorption in sweat glands and saliva and potassium excretion by the kidney. When sodium is shunted back into the blood in the kidney and the skin, water follows. Aldosterone therefore helps conserve body water. Aldosterone secretion is controlled by the sodium ion concentration in a negative feedback loop.

Diseases of the Adrenal Cortex. **Cushing's syndrome** generally results from pharmacological doses of cortisone. Cortisone, as noted above, is a synthetic glucocorticoid. It is used to treat rheumatoid arthritis and asthma. In a few instances, the disease may be caused by a pituitary tumor that produces excess ACTH or a tumor of the adrenal cortex that secretes excess glucocorticoid.

Patients with Cushing's disease often suffer persistent hyperglycemia—high blood sugar levels—because of the presence of high levels of glucocorticoid. Bone and muscle protein may also decline sharply, because glucocorticoids stimulate the breakdown of protein. Individuals complain of weakness and fatigue. Loss of bone protein increases the ease with which bones fracture.

Water and salt retention are also common in Cushing's patients, resulting in tissue edema (swelling). Because of this, Cushing's patients typically have a rounded "moon face." Swelling occurs because glucocorticoids, in high concentrations, have mineralocorticoid effects—that is, they stimulate the reabsorption of sodium and potassium, increasing the amount of water in the body. Because most cases of Cushing's syndrome result from steroids taken for health reasons, treatment is a simple matter: gradual reduction of the glucocorticoid dose. Tumors in the pituitary and the adrenal cortex may be treated with radiation or surgically removed.

Another disease of the adrenal cortex is **Addison's disease**. Most cases of Addison's disease are thought to be autoimmune reactions in which the cells of the adrenal cortex are recognized as foreign and are then destroyed by the body's own immune system. Addison's disease is characterized by a variety of symptoms caused by the loss of hormones from the adrenal cortex. These include loss of appetite, weight loss, fatigue, and weakness. Although insulin and glucagon are still present, the absence of cortisol upsets the body's homeostatic mechanism for controlling glucose. The body's reaction to stress is also impaired. The lack of aldosterone results in electrolyte imbalance. Because aldosterone maintains sodium levels and blood pressure, patients with Addison's disease have low blood sodium levels and low blood pressure. Addison's disease can be treated with steroid tablets that replace the missing hormones. Treatment allows patients to lead fairly normal, healthy lives.

13-7   Health and Homeostasis

Hormones orchestrate an incredible number of body functions, and help to create a dynamic balance that's vital for human health. Hormones influence homeostasis primarily by controlling the rate of various metabolic reactions and by regulating ionic balance. When this balance is altered, our health suffers (Table 13-3). The endocrine system, like other systems, is sensitive to outside factors. Stress, for example, can lead to an imbalance in adrenal hormones, resulting in high blood pressure and other complications. High blood pressure, in turn, puts strain on the heart and can result in heart attacks.

Hormonal balance can also be upset by toxic pollutants. One example is dioxin (die-OX-in), a contaminant found in some herbicides and in paper products. Scientists believe that dioxins bind to plasma membranes and cytoplasmic receptors to which hormones normally attach, upsetting body functions. Thus, some researchers are calling dioxin an "environmental hormone." One form of dioxin, TCDD, suppresses the immune system in mice at least 100 times more effectively than corticosterone, one of the body's glucocorticoids.

Studies show that a number of common herbicides (the thiocarbamates) may upset the thyroid's function and result in the formation of thyroid tumors. These herbicides are chemically similar to thyroid hormone and therefore block the secretion of TSH, resulting in goiter. At higher levels, they may cause thyroid cancer. Because hormones help regulate homeostasis, changes in their release result in upsets of this important process.

TABLE 13-3	Summary of Some Endocrine Disorders	
Disease	Cause	Symptoms
Giantism	Hypersecretion of GH starting in infancy or early life	Excessive growth of long bones
Dwarfism	Hyposecretion of GH in infancy or early life	Failure to grow
Acromegaly	Hypersecretion of GH after bone growth has stopped	Facial features become coarse; hands and feet enlarge; skin and tongue thicken
Hyperthyroidism	Overactivity of the thyroid gland	Nervousness; inability to relax; weight loss; excess body heat and sweating; palpitations of the heart
Hypothyroidism	Underactivity of the thyroid gland	Fatigue, reduced heart rate; constipation; weight gain; feel cold; dry skin
Hyperparathyroidism	Excess parathyroid hormone secretion, usually resulting from a benign tumor in the parathyroid gland	Kidney stones; indigestion; depression; loss of calcium from bones
Hypoparathyroidism	Hyposecretion of the parathyroid glands	Spasms in muscles; numbness in hands and feet; dry skin
Diabetes insipidus	Hyposecretion of ADH	Excessive drinking and urination; constipation
Diabetes mellitus	Insufficient insulin production or inability of target cells to respond to insulin	Excessive urination and thirst; poor wound healing; urinary tract infections; excess glucose in urine; fatigue and apathy
Cushing's syndrome	Hypersecretion of hormones from adrenal cortex or, more commonly, from cortisone treatment	Face and body become fatter; loss of muscle mass; weakness; fatigue; osteoporosis
Addison's disease	Gradual decrease in production of hormones from adrenal gland; most common cause is autoimmune reaction.	Loss of appetite and weight; fatigue and weakness; complete adrenal failure

SUMMARY

Principles of Endocrinology

1. The endocrine system consists of a widely dispersed set of glands that produce hormones. Hormones affect five aspects of our lives: (a) homeostasis, (b) growth and development, (c) reproduction, (d) energy production and storage, and (e) behavior.
2. Hormones act on specific cells. Specificity results from the presence of hormone receptors on target cells. Some receptors are located in the cytoplasm; others are located in the plasma membrane.
3. Three types of hormones are produced in the body: (a) steroids, (b) proteins and polypeptides, and (c) amines.
4. Hormone secretion is controlled primarily by negative feedback loops, many of which involve the hypothalamus.

The Pituitary and Hypothalamus

5. The pituitary is a pea-sized gland suspended from the hypothalamus by a thin stalk. It consists of two parts: the anterior pituitary and the posterior pituitary.
6. The anterior pituitary produces seven protein and polypeptide hormones. Their release is controlled by hormones produced by the hypothalamus, which are transported to the anterior pituitary via a portal system. The release of the hypothalamus's hormones is controlled by chemical stimuli and nerve impulses.
7. The hormones of the anterior pituitary and their functions are summarized in Table 13-1.
8. The posterior pituitary produces two hormones, antidiuretic hormone (ADH) and oxytocin, whose functions are summarized in Table 13-1.

The Thyroid Gland

9. The thyroid gland, located in the neck, produces three hormones: thyroxine (T4), triiodothyronine (T3), and calcitonin.
10. Thyroxine and triiodothyronine accelerate the rate of glucose breakdown in most cells, increasing body heat. These hormones also stimulate cellular growth and development.
11. Calcitonin lowers blood calcium levels.

The Parathyroid Glands

12. The parathyroid glands are located on the back of the thyroid gland and produce a polypeptide hormone called parathyroid hormone (PTH), or parathormone.
13. PTH increases blood calcium levels.

The Pancreas

14. The pancreas produces two hormones, insulin and glucagon, from the islets of Langerhans.
15. Insulin is the glucose-storage hormone. It stimulates the uptake of glucose by many body cells. Insulin also increases the uptake of amino acids and stimulates protein synthesis in muscle cells.
16. Glucagon raises glucose levels in the blood between meals by stimulating glycogen breakdown and the synthesis of glucose from amino acids and fats.
17. Diabetes mellitus is a disease involving insulin and blood glucose. It has two principal forms: Type I and Type II. Type I, also called early-onset diabetes, occurs early in life and may be caused by an autoimmune reaction. It can be treated by insulin injections.
18. Type II, or late-onset, diabetes, results from a reduction in the number of insulin receptors on target cells. It is caused by obesity and genetic factors and can often be treated successfully by dietary management or medication.

The Adrenal Glands

19. The adrenal glands lie atop the kidneys and consist of two separate portions: the adrenal medulla, at the center, and the adrenal cortex, a surrounding band of tissue.
20. The adrenal medulla produces two hormones under stress: adrenalin and noradrenalin. These hormones stimulate heart rate and breathing, elevate blood glucose levels, constrict blood vessels in the intestine, and dilate blood vessels in the skeletal muscles.
21. The adrenal cortex produces three classes of hormones: glucocorticoids, mineralocorticoids, and sex steroids.
22. The glucocorticoids affect carbohydrate metabolism and tend to raise blood glucose levels. The principal glucocorticoid is cortisol.
23. The chief mineralocorticoid is aldosterone. It acts on the kidneys, sweat glands, and salivary glands, causing sodium and water retention and potassium excretion.

Health and Homeostasis

24. Hormones influence an incredible number of body functions, fostering homeostasis, a dynamic balance necessary for good health. When homeostasis is altered, our health suffers.
25. The endocrine system, like others, is sensitive to upset from stress and environmental factors such as pollutants. Dioxin, a contaminant in some herbicides and paper products, may exert most of its effects through the endocrine system. A number of common herbicides (the thiocarbamates) upset the thyroid's function and may even cause thyroid tumors.

critical thinking

Thinking Critically—Analysis

This Analysis corresponds to the Thinking Critically scenario that was presented at the beginning of this chapter.

This ethical dilemma hinges on several key questions—for example, should people be tinkering with nature? It is also a medical dilemma. For example, will growth hormone injections over a long period have any adverse effects?

As it turns out, no one knows the answers to the medical questions, but some studies suggest that prolonged usage may result in diabetes or high blood pressure. A recent study showed that growth hormone therapy in normal children, while greatly increasing height and muscle mass, dramatically reduced body fat. Children appeared gangly and thin (emaciated). What effects these might have in the long term, no one knows. Some scientists believe that subtle abnormalities may develop in the cells, tissues, and organs of these individuals. Clearly, the jury is still out on the long-term effects of growth hormone therapy in children.

KEY TERMS AND CONCEPTS

Acromegaly, p. 238
Addison's disease, p. 249
Aldosterone, p. 249
Adrenal cortex and medulla, p. 248
Adrenal gland, p. 248
Anterior pituitary, p. 236
Antidiuretic hormone, p. 241
Adrenalin, p. 249
Adrenocorticotropic hormone (ACTH), p. 240
Calcitonin (or thyrocalcitonin), p. 244
Cortisol, p. 249
Cushing's syndrome, p. 249
Diabetes mellitus, p. 245
Endocrine system, p. 234
Fight-or-flight response, p. 249

Glucocorticoid, p. 240
Glucose tolerance test, p. 246
Growth hormone, p. 238
Hormone, p. 234
Hypothalamus, p. 237
Inhibiting hormone, p. 238
Insulin, p. 245
Mineralocorticoids, p. 249
Negative feedback loop, p. 235
Neuroendocrine reflex, p. 240
Neurosecretory neuron, p. 237
Nontropic hormone, p. 235
Noradrenalin, p. 249
Oxytocin, p. 241
Pancreas, p. 245

Parathyroid glands, p. 245
Parathyroid hormone or parathormone (PTH), p. 245
Pituitary gland, p. 236
Portal system, p. 238
Positive feedback loop, p. 235
Posterior pituitary, p. 236
Prolactin, p. 240
Releasing hormone, p. 237
Sex steroids, p. 249
Target cells, p. 234
Thyroid gland, p. 243
Thyroid-stimulating hormone (TSH), p. 239
Thyroxine, p. 243
Triiodothyronine, p. 243
Tropic hormone, p. 235

CONCEPT REVIEW

1. Define the following terms: endocrine system, hormone, and target cell. p. 234
2. Hormones function in five principal areas. What are they? Give some examples of each. p. 234
3. Define the terms tropic and nontropic hormones, and give several examples of each. p. 235
4. Give two examples of negative feedback loops in the endocrine system: a simple feedback mechanism and a more complex one that operates through the nervous system. p. 235
5. List the hormone(s) involved in each of the following functions: blood glucose levels, growth, milk production, milk let-down, calcium levels, and metabolic rate. pp. 238–245
6. Describe the role of the hypothalamus in controlling anterior pituitary hormone secretion. p. 237
7. Describe the ways in which the posterior pituitary differs from the anterior pituitary. pp. 238–243

8. Acromegaly and giantism are both caused by the same problem. Why are these conditions so different? p. 238
9. Describe the neuroendocrine reflex involved in prolactin secretion. pp. 240–241
10. Offer some possible explanations for the following experimental observation: Milk production occurs late in pregnancy and is thought to be stimulated by the hormone prolactin. A nonpregnant rat is injected with prolactin but does not produce milk. pp. 240–241, 242–243
11. Where is ADH produced? Where is it released? Describe how ADH secretion is controlled. What effects does this hormone have? p. 242
12. Where is oxytocin produced? Where is it released? Describe how oxytocin secretion is controlled. What effects does this hormone have? pp. 242–243
13. A patient comes into your office. She is thin and wasted and complains of excessive

sweating and nervousness. What tests would you run? p. 244
14. A patient comes into your office. He is suffering from indigestion, depression, and bone pain. An X-ray of his bone shows some signs of osteoporosis. You think that the disorder might be the result of an endocrine problem. What test would you order? p. 245
15. How are the two basic types of diabetes mellitus different? How are they similar? How are they treated, and why are these treatments chosen? pp. 246–248
16. Describe the physiological changes that occur under stress. What hormones are responsible for them? p. 249
17. Aldosterone is a mineralocorticoid. Describe its chief functions. How does it help retain body fluid? Under what conditions is aldosterone secreted? p. 249

SELF-QUIZ: TESTING YOUR KNOWLEDGE

1. The cell a hormone affects is called a _____ cell. p. 234

2. A hormone that stimulates the production and release of another hormone is known as a _____ hormone. p. 235

3. The production and release of most hormones are controlled by _____ feedback. p. 235

4. The _____ gland is attached to the underside of the brain by a thin stalk. p. 236

5. A _____ produced by the _____ travels in the bloodstream to the anterior pituitary, causing the production and releases of its hormones. p. 237

6. _____ hormone released by the anterior pituitary stimulates growth but primarily affects muscle and _____. p. 238

7. _____ is a disease caused by overproduction of growth hormone in adults and results in coarse facial features and continued growth of the hands and feet. p. 238

8. _____ and triiodothyronine are hormones produced by the thyroid gland; their production and release are controlled by _____, a hormone from the anterior pituitary. p. 240

9. ACTH stimulates the production and release of _____, a group of steroid hormones that increase blood glucose. p. 240

10. Milk production in the breasts of women is stimulated by the hormone _____, which is produced by the anterior pituitary. Its release is stimulated by _____. p. 240

11. Milk ejection from glands in the breast is controlled by the hormone _____ from the _____ pituitary. p. 242

12. _____ is a hormone of the posterior pituitary gland that stimulates water reabsorption by the kidney. p. 242

13. An insufficiency of iodine results in enlargement of the thyroid gland, a condition known as _____. p. 244

14. The pancreas produces two hormones, insulin and _____. p. 245

15. Insulin increases the uptake of glucose by liver and _____ cells, and lowers blood glucose levels. p. 246

16. Late-onset diabetes is caused primarily by _____. p. 246

17. Early-onset diabetes is treated by _____ injections. p. 248

18. The hormone _____ is released by the adrenal medulla and is responsible for physiological changes associated with the fight-or-flight response. p. 249

19. The mineralocorticoid aldosterone is produced by the adrenal _____. p. 249

20. The hormone _____ produced by the thyroid gland lowers blood calcium levels. p. 244

www.jbpub.com/humanbiology/5e

The site features eLearning, an online review area that provides quizzes, chapter outlines, and other tools to help you study for your class. You can also follow useful links for in-depth information, research the differing views in the Point/Counterpoints, or keep up on the latest health news.

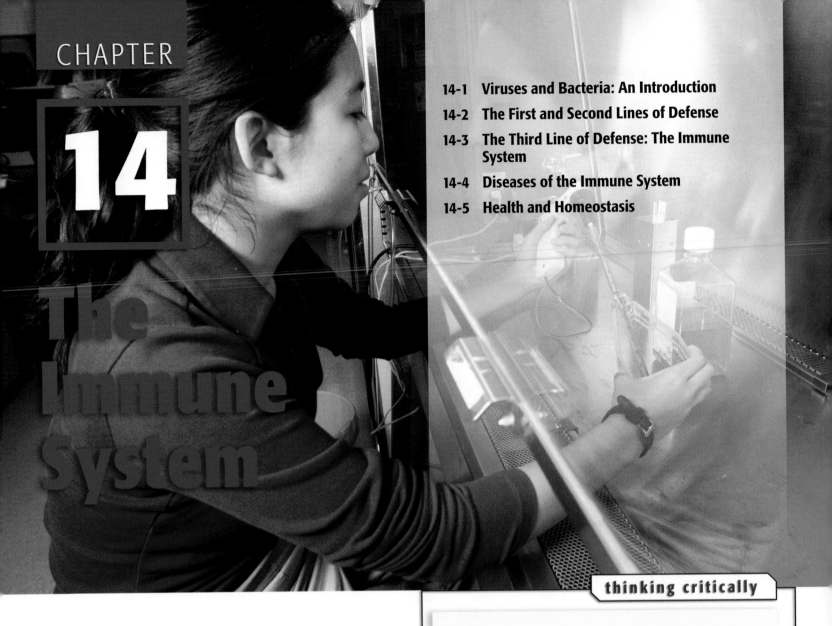

thinking critically

Floating in the air and circulating in the water are billions upon billions of microorganisms. Although the vast majority of these microbes, mostly bacteria and viruses, in our environment are innocuous, there are a few that are not, especially if you are traveling in a busy airport or walking on a crowded city street.

You're the parent of a young child and are debating whether to have your child vaccinated. Some friends tell you that it's dangerous—in fact, they say, some children have been paralyzed by vaccinations. They also tell you that vaccinations can damage the immune system and that there's no need to subject a child to this treatment because dangerous infectious diseases have been eliminated.

After reading this chapter, how would you respond? Is there merit to the claim that a vaccination damages the immune system? Have the diseases been eliminated? Do you have enough information to make an informed decision?

Why don't we succumb to these potentially harmful microbes? The answer is that humans come equipped with a system of armor (the skin and other protective layers) that protects us from potential intruders. If microorganisms break through these barriers, the immune system is ready to combat them.

In this chapter, we'll examine the body's protective measures. We will also see how the body's defense mechanisms work against cancer and tissue and organ transplants. We'll begin with a brief overview of infectious agents.

14-1   Viruses and Bacteria: An Introduction

Infectious agents come in many forms. In this section we'll look primarily at viruses and bacteria.

Viruses

> Viruses are nonliving biological agents that invade cells.

A **virus** is composed of a nucleic acid core, consisting of either DNA or RNA (Figure 14-1). Surrounding the nucleic acid core is a protein coat known as the *capsid*. Some viruses have an additional protective layer, an outer envelope that lies outside the capsid. This layer, the envelope, is structurally similar to the plasma membranes of eukaryotic cells.

Interestingly, while viruses can be deadly, they are not considered to be living organisms. That's because they do not possess many of the features we attribute to living organisms. For example, viruses cannot multiply and divide on their own like cells. They also have no metabolism.

Viruses have been likened to pirates because they invade cells and then commandeer their metabolic machinery in much the same way that pirates take over ships. Viruses have a very simple mission, however. Their sole purpose is to reproduce. Thus, after invading a cell, viruses convert their so-called host cells into miniature virus factories. That is, they use each host cell's organelles and biochemical reserves to make new protein and nucleic acid. Inside the host cell, hundreds of thousands of new viruses are formed. They soon begin to be released from the cell, spreading to other cells through the

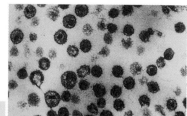

(c)

bloodstream and the lymphatic vessels. So completely spent is the host cell that it dies.

Viruses most often enter the body through the respiratory and digestive systems. However, other avenues of entry are also possible—for example, sexual contact.

Fortunately for us, the immune system kills many viruses, but usually not until we've undergone considerable suffering. Influenza viruses, for example, result in the flu, characterized by fevers, chills, headaches, and muscle pains that can last up to two weeks (Table 14-1). During this time, the immune system mounts an attack, but it usually takes 10 to 14 days before it can eliminate the virus. Colds are also caused by approximately 200 different viruses. Like the flu, colds cannot be treated with antibiotics.

Some viruses, like the two that cause fever blisters and genital herpes, evade the immune system. They take refuge in body cells, reemerging from time to time to cause problems, usually when we're under stress. The virus responsible for genital herpes (HERpees), for example, causes tiny sores that emerge periodically on the genitals, thighs, and buttocks of infected men and women.

TABLE 14-1	How Do You Know If You Have A Cold or the Flu?	
Symptoms	Cold	Flu
Onset	Gradual	Sudden
Fever	Rare	Common, may reach 101°F; may last 3 to 4 days
Headache	Rare	Common and can be severe
Cough	Hacking	Dry cough
Muscle aches and pains	Slight	Typical, often severe
Tiredness and weakness	Mild	Common, often severe
Chest discomfort	Mild to moderate	Common
Stuffy nose	Common	Sometimes
Sneezing	Usual	Sometimes
Sore throat	Common	Sometimes; may last 3 to 4 days
Caused by	Any of 200 viruses	Influenza virus

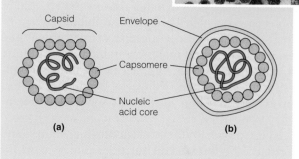

FIGURE 14-1 General Structure of a Virus (a) The virus consists of a nucleic acid core of either RNA or DNA. Surrounding the viral core is a layer of protein known as the capsid. Each protein molecule in the capsid is known as a capsomere. (b) Some viruses have an additional protective coat known as the envelope. (c) Electron micrograph of the human immunodeficiency virus (HIV).

Labels in figure: Capsid, Envelope, Capsomere, Nucleic acid core, (a), (b)

Bacteria

Many bacteria perform useful functions, while others cause serious diseases.

Unlike viruses, **bacteria** (singular, bacterium) are classified living organisms. They are capable of reproducing without taking over host cells. In fact, unlike viruses, bacteria do not enter cells. They proliferate outside of cells.

As pointed out in Chapter 3, bacteria are prokaryotes. As you may recall, the term *prokaryotes* refers to cells without nuclei. In these cells, the genetic material consists of a circular strand of DNA, not bound by any kind of membrane. Prokaryotes have cytoplasm, which is enveloped by a plasma membrane, but no cellular organelles like eukaryotic cells (Figure 14-2). Outside the plasma membrane of bacteria is a thick, rigid cell wall. Many bacteria also contain tiny circular pieces of extra-chromosomal DNA, called **plasmids**.

Like viruses, bacteria enter the body through the respiratory tract and GI tract. They may also enter through the epithelium of the urinary system. Cuts and abrasions in the skin also permit bacteria to enter the body. Inside the body, bacteria proliferate, using nutrients to make more of their kind. Some bacteria produce toxins, substances that make us ill.

Treating Bacterial and Viral Infections

Bacteria respond to antibiotics; viruses do not.

Bacteria have proven to be easily manipulated by medicines known as *antibiotics*. Antibiotics kill bacteria by inhibiting protein synthesis in bacteria. They do so without altering protein synthesis in the cells of the body.

Unfortunately, the heavy use of antibiotics (often when they're not useful, as in viral infections) has resulted in a growing problem: antibiotic resistance in bacteria. Many thousands of people die each year of infections caused by antibiotic-resistant bacteria they contract in hospitals or in other places.

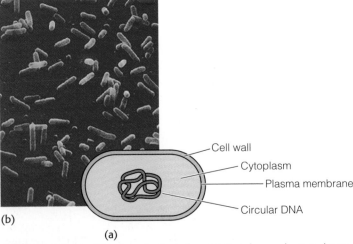

(b)

(a)

FIGURE 14-2 **General Structure of a Bacterium** (a) Bacteria come in many shapes and sizes, but all have a circular strand of DNA, cytoplasm, and a plasma membrane. Surrounding the membrane of many bacteria is a cell wall. (b) Electon micrograph of salmonella bacteria.

Nationwide, one out of every four bacterial infections today involves a resistant strain. Most of these infections occur in children under the age of five. To treat them, doctors must use stronger doses or newer antibiotics. Even then, bacteria may develop resistance.

Bacteria can become resistant to drugs in one of several ways. They can, for example, chemically alter antibiotics, rendering them ineffective. Bacteria also undergo chemical changes so that antibiotics no longer bind to them. Some resistance occurs when bacteria develop mechanisms that permit them to pump antibiotics out of their interior, lowering internal concentrations so much so that the drugs are no longer effective. Because of concern for antibiotic resistance, many doctors refuse to give antibiotics to patients suffering from viral infections, even though patients might insist they get some kind of medication. Primitive as it may sound, some doctors in Great Britain and the United States are now using maggots to clean wounds to destroy bacteria, including antibiotic-resistant strains (Health Note 14-1).

For the flu, a viral disease, doctors also prescribe medications that reduce some of the symptoms, such as acetaminophen, which reduces fever, muscle aches, and headaches. Bed rest helps patients get well, and warm fluids such as tea reduce the feeling of congestion while one's immune system battles the virus. It's wise to stay at home when sick, too, to reduce the chances of spreading the virus to others.

For colds, bed rest in a warm house is advised. Drink fluids and take over-the-counter remedies to reduce symptoms such as coughing. Mild pain relievers can rid you of muscle pain and headaches, if you have them. Decongestants can help you relieve the stuffy nose and will help you breathe better.

In recent years, pharmaceutical companies have released several anti-viral drugs. These drugs attack the influenza virus. To be effective, however, antiviral drugs must be taken early in the course of the disease, within a day or two of the onset of symptoms.

During flu season, flu shots are given to prevent people from contracting this disease. Flu shots are a vaccine that prevents the influenza virus from gaining a foothold. Vaccines are discussed in more detail later in the chapter.

Antibiotics are given to patients with bacterial pneumonia, bronchitis, sinus infections, and sore throats. Because all of these diseases can be caused by viruses, too, a doctor should be consulted. She can usually tell whether the infection is viral or bacterial. If it is bacterial, she'll prescribe antibiotics.

Individuals can help in the battle against bugs, too. When a doctor says you have a viral infection and that an antibiotic won't work, don't insist on one. But be careful, if the viral infection persists for more than two weeks, secondary bacterial infections can set in. Then, antibiotics are needed to fight them off.

Although best known for their role in causing sickness and death, most bacteria perform useful functions. Soil bacteria, for example, help recycle nutrients in rotting plants and animals, a process that is essential to the continuation of life on Earth. Without these bacteria, biological systems would quickly grind to a halt. Some of the biologically useful bacteria will be discussed in Chapter 23.

⁞⁞⁞ healthnote

14-1   Maggots: The Latest in Medical Care?

Elainor Brocknor is a diabetic who for 30 years has been injecting the hormone insulin to regulate her blood sugar. Careful as she has been, she's developed serious medical problems. The blood vessels in her feet have been damaged by periodic high concentrations of glucose. The tissues in two of her toes have begun to die and rot. She's developed gangrene and doctors fear they may have to amputate these two toes and perhaps even her foot if the infection does not stop. Making matters worse, the bacteria infecting the dying tissues in her feet don't seem to respond to antibiotics.

One day, her doctor makes a bizarre recommendation. He says he may be able to save her foot by using maggots. Maggots are the immature form of common ordinary flies. They're called larvae. They develop from eggs laid by flies. They're voracious eaters.

Elainor and her husband are mortified, at first, by the doctor's seemingly crazy suggestion. What a preposterous idea! But the doctor explains that maggots raised in a sterile environment have been used very successfully to remove dead tissue in people with gangrene. They not only feed on dead tissue, they gobble up harmful bacteria, so the infection doesn't spread.

Reluctant, Elainor agrees to try the procedure. The next day, her doctor visits her in her hospital room. He's carrying a glass jar full of writhing maggots. He removes the bandages from her feet, then prepares a new one. On the surface of the gauze, the physician sprinkles 100 tiny maggots. He then places the bandage over her wound and tapes it shut.

Elainor watches in amazement and is surprised that she feels nothing at all. A day later, the doctor returns, removes the bandage, and applies another. In a few days, he announces that the wound is clean and free of infection. They won't have to amputate.

Elainor is just one of thousands of people each year treated with maggots. In this day of high-tech medicine, such an approach may seem ridiculously primitive. Problem is, it works when high-tech solutions don't. And it is saving many people the agony of amputation.

The idea of using maggots to clean wounds is an old one. During the Civil War, doctors who treated wounded soldiers noticed that wounds in men returning from the battlefields several days away were often infested with maggots. They'd come from eggs flies laid in open wounds. Surprisingly, the maggot-filled wounds were remarkably clean.

In Africa, rural doctors noticed a similar phenomenon. In modern times, doctors have found that when they removed maggots from wounds and put a patient on antibiotics, the patients died—apparently from antibiotic-resistant bacteria that had been under control. Maggots aren't the only creepy cure. Some doctors are using leeches to restore blood flow in severed limbs or fingers that have been sewn back in place (see **Health Note 6-1**). Others are using bees and bee venom to treat patients with multiple sclerosis. Both leeches and bees are turning out to be valuable tools in the doctor's medical bag.

www.jbpub.com/humanbiology/5e

Visit Human Biology's Internet site for links to web sites offering more information on this topic.

14-2 The First and Second Lines of Defense

The First Line of Defense

> The first line of defense is a physical and chemical barrier.

The human body is like a fort with an outer barrier that wards off potential invaders. One component of this outer wall is the skin, a protective layer that repels many potentially harmful microorganisms.

As noted in Chapter 4, the skin consists of a relatively thick and impermeable layer of epidermal cells overlying the rich vascular layer, the dermis. Epidermal cells are produced by cell division in the base of the epidermis. As the basal cells proliferate, they move outward, become flattened, and die. The dead cells contain a protein called *keratin* (CARE-ah-tin). Keratin forms a fairly waterproof protective layer that not only reduces moisture loss but also protects underlying tissues from microorganisms. The cells of the epidermis are joined by special structures known as *tight junctions*, which impede water loss and microbial penetration.

Although the skin protects us from infections, the human body contains three passageways that penetrate into its interior, providing a fairly direct route for microbes to enter the body. The passages into the interior are the respiratory, digestive, and urinary tracts. All three of them are protected by epithelial linings.

The body's first line of defense also consists of chemical barriers. The skin, for example, which serves as a physical barrier, also produces slightly acidic secretions that impair bacterial growth. The stomach lining produces hydrochloric acid, which destroys many ingested bacteria. Tears and saliva contain an enzyme called *lysozyme* (LIE-so-zime) that dissolves the cell wall of bacteria, killing them. Cells in the lining of the respiratory tract produce mucus, which has antimicrobial properties.

The skin, the epithelial linings, and chemical barriers are all nonspecific. Like a castle wall, they operate indiscriminately against all invaders.

The Second Line

> The second line of defense combats infectious agents that penetrate the first line.

The first line of defense is not impenetrable. Tiny breaks in the skin or in the lining of the respiratory, digestive, or urinary tracts may permit viruses, bacteria, and other microorganisms to enter the body. Fortunately, a second line of defense exists. It involves a host of chemicals and cellular agents. Like the first line, these mechanisms are nonspecific. This section discusses four components of the second line of defense: the inflammatory response, pyrogens, interferons, and complement.

The Inflammatory Response

> The inflammatory response involves chemical and cellular responses.

Damage to body tissues triggers a series of reactions collectively referred to as an *inflammatory response*, part of the second line of defense. The **inflammatory response** is characterized by redness, swelling, and pain. (The word inflammatory refers to the heat given off by a wound.)

The inflammatory response is a kind of chemical and biological warfare waged against bacteria, viruses, and other microorganisms. It begins with the release of a variety of chemical substances by the injured tissue (Figure 14-3). Some chemicals

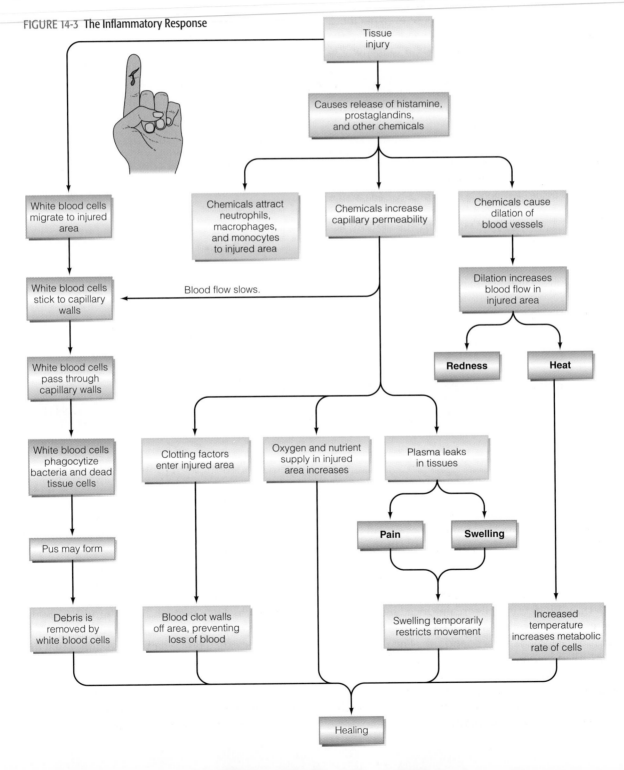

FIGURE 14-3 **The Inflammatory Response**

attract macrophages that reside in body tissues and neutrophils in the blood (Chapter 7). These cells engulf (phagocytize) bacteria that enter a wound. Soon after these cells begin to work, a yellowish fluid begins to exude from the wound. Called *pus*, it contains dead white blood cells (mostly neutrophils), microorganisms, and cellular debris that accumulate at the wound site.

Another chemical substance released by injured tissues is **histamine** (HISS-tah-mean). Histamine stimulates the arterioles in the injured tissue to dilate, causing the capillary networks to swell with blood. The increase in the flow of blood through an injured tissue is responsible for the heat and redness around a cut or abrasion. Heat, in turn, increases the metabolic rate of cells in the injured area and accelerates healing.

Still other substances released by injured tissues increase the permeability of capillaries, increasing the flow of plasma into a wounded region. Plasma carries oxygen and nutrients that aid in healing. It also carries the molecules necessary for blood clotting. The clotting mechanism walls off injured vessels and reduces blood loss.

Plasma leaking into injured tissues causes swelling, which stimulates pain receptors in the area. Pain receptors send nerve impulses to the brain. Pain also results from chemical toxins released by bacteria and from chemicals released by injured cells. One important pain-causing chemical is **prostaglandin** (PROSS-tah-GLAN-din). Aspirin and other mild painkillers work by inhibiting the synthesis and release of prostaglandins.

Inflammation even comes with its own cleanup crew in the form of late-arriving monocytes. These cells phagocytize dead cells, cell fragments, dead bacteria, and viruses.

Pyrogens, Interferons, and Complement

The second line of defense consists of three additional chemicals.

Additional protection is provided by pyrogens, interferons, and complement.

Pyrogens (PIE-rah-gins) are molecules released primarily by macrophages that have been exposed to bacteria. Pyrogens travel to the hypothalamus (a region of the brain). In the hypothalamus is a group of nerve cells that controls the body's temperature, in much the same way that a thermostat regulates the temperature of a room. Pyrogens turn the thermostat up, increasing body temperature and producing a fever. Mild fevers cause the spleen and liver to remove additional iron from the blood. Because many disease-causing bacteria require iron to reproduce, fever helps reduce their replication. Fever also increases metabolism, which facilitates healing and accelerates cellular defense mechanisms such as phagocytosis. Important as it is, fever can also be debilitating, and a severe fever (over 105°F) is potentially life-threatening because it begins to denature (coagulate) vital body proteins, especially enzymes needed for biochemical reactions in body cells.

Another chemical safeguard is a group of small proteins known as the **interferons** (in-ter-FEAR-ons). Interferons are released from cells infected by viruses. They diffuse away from the site of production through the interstitial tissue and bind to receptors on the plasma membranes of noninfected body cells (Figure 14-4). The binding of interferon to uninfected cells,

in turn, triggers the synthesis of cellular enzymes that inhibit viral replication. Thus, when viruses enter the previously uninfected cells, they cannot replicate and spread.

Interferons do not protect cells already infected by a virus; they stop the spread of viruses from one cell to another. In essence, the production and release of interferon are the dying cell's last act to protect other cells of the body. Interferons help to stop the spread of viruses while the immune response attacks and destroys the viruses outside the cells.

Another group of chemical agents that fight infection are the complement proteins. These blood proteins form the **complement system**, so named because it complements the action of antibodies.

The details of the complement system are very complex. A few points will demonstrate how this remarkable system works. Complement proteins circulate in the blood in an inactive state. When foreign cells such as bacteria invade the body, the com-

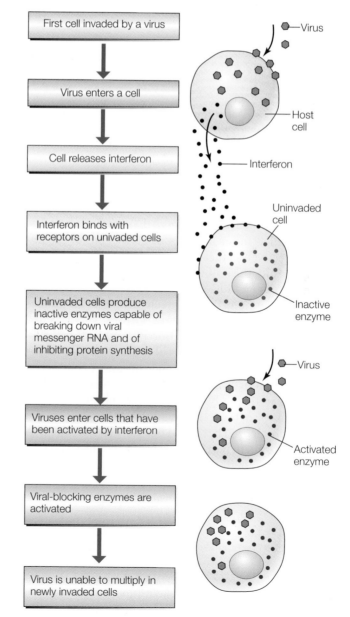

FIGURE 14-4 **How Interferon Works** Interferon protects cells from viral infection.

plement protein is activated. This triggers a chain reaction in which one complement protein activates the next.

Five proteins in the complement system join to form a large protein complex, known as the **membrane-attack complex** (Figure 14-5). The membrane-attack complex embeds in the plasma membrane of bacteria, creating an opening into which water flows. The influx of water causes bacterial cells to swell, burst, and die.

Several of the activated complement proteins also function on their own and are part of the inflammatory response. Some of them, for example, stimulate the dilation of blood vessels in an infected area, described earlier. Others increase the permeability of the blood vessels, allowing white blood cells and nutrient-rich plasma to pass more readily into an infected zone. Certain complement proteins may also act as chemical attractants, drawing macrophages, monocytes, and neutrophils to the site of infection, where they phagocytize foreign cells. Yet another complement protein (C3b) binds to microorganisms, forming a rough coat on the intruders that facilitates their phagocytosis.

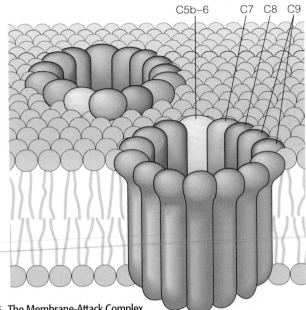

C5b–6 C7 C8 C9

FIGURE 14-5 The Membrane-Attack Complex
Five complement proteins combine and embed in a cell's membrane, causing it to leak, swell, and burst. (Source: By Dana Burns from John Ding-E Young and Zanvil A. Kohn, (How Killer Cells Kill. ©Copyright January 1988 by Scientific American, Inc. All rights reserved.)

14-3 The Third Line of Defense: The Immune System

The cells of the immune system selectively target foreign substances and foreign organisms.

The immune system is the third line of defense. Unlike the respiratory or digestive systems, the immune system is rather diffuse—spread out and indistinct. Lymphocytes, for example, circulate in the blood and lymph and also take up residence in the **lymphoid organs** such as the spleen, thymus, lymph nodes, and tonsils, as well as other body tissues. The cells of the immune system selectively target foreign substances and foreign organisms. As a result, the immune system is said to be specific.

The immune system, like the first and second lines of defense, is an important homeostatic mechanism that eliminates foreign organisms—including bacteria, viruses, single-celled fungi, and many parasites. It comes into play when foreign organisms penetrate the outer defenses of the body. The immune system also helps prevent the emergence of cancer cells.

One of the chief functions of the immune system is to identify what belongs in the body and what does not. Once a foreign substance has been detected, the immune system mounts an attack to eliminate it. Therefore, like all homeostatic systems, the immune system requires receptors to detect a change and effectors to bring about a response. In the immune system, the lymphocytes serve both functions.

Antigens

Substances that trigger an immune response are large molecular weight molecules.

The immune response is triggered by large foreign molecules, notably proteins and polysaccharides. These molecules are called **antigens** (AN-tah-gins), which is an abbreviation for antibody-generating substances. As a rule, small molecules generally do not elicit an immune reaction. In some individuals, however, small, nonantigenic molecules such as formaldehyde, penicillin, and the poison ivy toxin bind to naturally occurring proteins in the body, forming complexes. These large complexes are unique compounds that are foreign to the body that can cause an immune response.

The immune system reacts to viruses, bacteria, and single-celled fungi in the body. It also responds to parasites such as the protozoan that causes malaria. These agents elicit a response because they are enclosed by membranes, or coats. They contain large-molecular-weight proteins or polysaccharides—that is, antigens.

Cells transplanted from one person to another also elicit an immune response, because each individual's cells contain a one-of-a-kind "cellular fingerprint," resulting from the unique array of plasma membrane glycoproteins (Chapter 3). The immune system is activated by these antigens on the foreign cells. Cancer cells also present a slightly different chemical fingerprint, making them essentially foreign cells within our own bodies to which the immune system responds. Although cancer cells evoke an immune response, it is often not sufficient to stop the disease.

Antigens stimulate two types of lymphocytes: **T lymphocytes**, commonly called **T cells**, and **B lymphocytes**, also called **B cells** (Chapter 7). The "T" and "B" in their names will be explained shortly. B cells and T cells respond to different antigens and function quite differently.

As a rule, B cells recognize and react to microorganisms such as bacteria and **bacterial toxins**, chemical substances released by bacteria. B cells also respond to a few viruses. When activated, B cells produce antibodies to these antigens.

In contrast, T cells recognize and respond to our own body cells that have gone awry. This includes cancer cells as well as body cells that have been invaded by viruses. T cells also respond to transplanted tissue cells and larger disease-causing agents, such as single-celled fungi and parasites. Unlike B cells, T cells attack their targets directly.

Immunocompetence

Immature B and T cells are incapable of responding to antigens.

Lymphocytes are produced in the red bone marrow and released into the bloodstream. These immature cells circulate in the blood and lymph but are not able to function until they become immunologically competent. This process occurs in specific organs in the body. Consider the T cell.

T Cells. T cells get their name from the place where they mature, the thymus. The **thymus** (THIGH-muss) is a lymphoid organ located above the heart. After they're formed in the bone marrow, immature lymphocytes enter the bloodstream. Some of them end up in the thymus. Over the next few days, these lymphocytes mature and become functional T cells. Immunologists, scientists who study the immune system, say that the lymphocytes have developed **immunocompetence** (IM-you-know-COM-pah-tense)—that is, they've developed the capacity to respond to antigens. But not just any antigens. In ways not completely understood, T lymphocytes leave the organ transformed to respond to very specific antigens.

When the T cells leave the thymus, each of their membranes contains a unique type of membrane receptor that will bind to one—and only one—type of antigen. Over an individual's lifetime, thousands upon thousands of antigens will be encountered. Thanks to the immunocompetence developed during fetal development, each of us is equipped with an army of uniquely preprogrammed T cells that respond to the onslaught of antigens.

B Cells. B cells mature and differentiate in the bone marrow. The B in B cell, however, did not come originally from the bone marrow. It came from the bursa, an organ found in chickens. Early researchers used chickens to study lymphocytes.

After they mature, immunologically competent B cells circulate in the blood and take up residence in connective and lymphoid tissues. They therefore become part of the body's vast cellular reserve, stationed at distant outposts, awaiting the arrival of the microbial invaders.

By various estimates, several million immunologically distinct B and T cells are produced in the body early in life. Over a lifetime, only a relatively small fraction of these cells will be called into duty.

B Cells and Antibodies

B cells provide humoral immunity through the production of antibodies.

The immune response consists of two separate but related reactions: humoral immunity, provided by the B cells, and cell-mediated immunity, involving T cells (Table 14-2). Let's consider humoral immunity first.

TABLE 14-2	Comparison of Humoral and Cell-Mediated Immunity

Humoral

Principal cellular agent is the B cell.

B cell responds to bacteria, bacterial toxins, and some viruses.

When activated, B cells form memory cells and plasma cells, which produce antibodies to these antigens.

Cell-Mediated

Principal cellular agent is the T cell.

T cell responds to cancer cells, virus-infected cells, single-cell fungi, parasites, and foreign cells in an organ transplant.

When activated, T cells differentiate into memory cells, cytotoxic cells, suppressor cells, and helper cells; cytotoxic T cells attack the antigen directly.

When a bacterium first enters the body, B cells programmed to respond to the unique proteins found in the microbe's cell membrane bind to it (Figure 14-6). The B cells soon begin to divide, producing a population of immunologically similar cells known as **lymphoblasts**. As shown in Figure 14-6, some of the lymphoblasts differentiate to form another kind of cell, the **plasma cell**. Plasma cells produce antibodies. **Antibodies** are proteins that help eliminate antigens, that is, bacteria or bacterial toxins.

Released from plasma cells, antibodies circulate in the blood and lymph, where they bind to the antigens that triggered the response. Because the blood and lymph were once referred to as body "humors," this arm of the protective immune response is called **humoral immunity**. Once the invaders have been removed, no new plasma cells are formed, and existing plasma cells die.

Primary Response

The initial reaction to an antigen is slower and weaker than subsequent responses.

The first time an antigen enters the body, it elicits an immune response, but the initial reaction—or **primary response**—is relatively slow (Figure 14-7). During the primary response, antibody levels in the blood do not begin to rise until approximately the beginning of the second week after the intruder was detected, partly explaining why it takes most people about 7–10 days to combat a cold or the flu. This delay occurs because it takes time for B cells to multiply and form a sufficient number of plasma cells. Antibody levels usually peak about the end of the second week, then decline over the next three weeks.

Secondary Response

The rapidity of the secondary response results from the production of memory cells.

If the same antigen enters the body at a later date, however, the immune system acts much more quickly and more forcefully (Figure 14-7). This stronger reaction constitutes the **secondary response**. During a secondary response, antibody levels increase quickly—a few days after the antigen has entered the

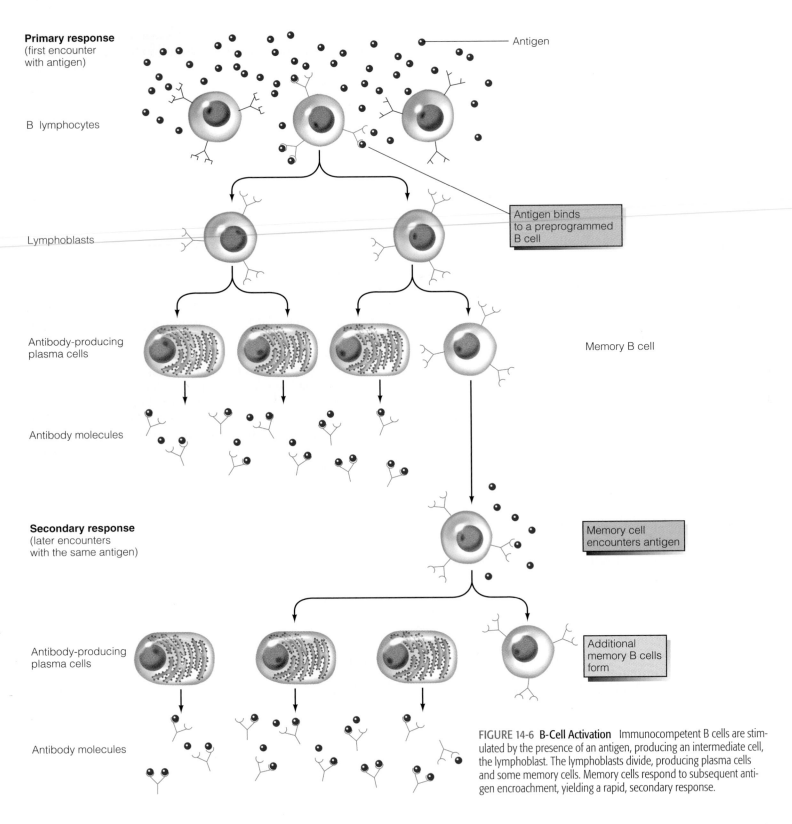

Primary response
(first encounter
with antigen)

Antigen

B lymphocytes

Antigen binds
to a preprogrammed
B cell

Lymphoblasts

Antibody-producing
plasma cells

Memory B cell

Antibody molecules

Secondary response
(later encounters
with the same antigen)

Memory cell
encounters antigen

Antibody-producing
plasma cells

Additional
memory B cells
form

Antibody molecules

FIGURE 14-6 **B-Cell Activation** Immunocompetent B cells are stimulated by the presence of an antigen, producing an intermediate cell, the lymphoblast. The lymphoblasts divide, producing plasma cells and some memory cells. Memory cells respond to subsequent antigen encroachment, yielding a rapid, secondary response.

body. The amount of antibody produced also greatly exceeds quantities generated during the primary response. Consequently, the antigen is quickly destroyed, and a recurrence of the illness is prevented.

Why is the secondary response so different? Figure 14-6 shows that during the primary response, some lymphocytes divide to produce **memory cells**. Memory cells are immunologically competent B cells. Although they do not transform into plasma cells,

they remain in the body awaiting the antigen's reentry. These cells are produced in large quantity and thus create a relatively large reserve of antigen-specific B-cells. When the antigen reappears, memory cells proliferate rapidly, producing numerous plasma cells that quickly crank out antibodies to combat the foreign invaders. Memory cells also generate additional memory cells during the secondary response. These remain in the body in case the antigen should enter the body at some later date. Im-

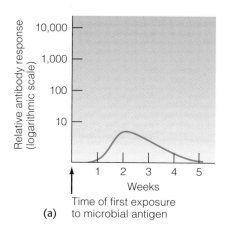

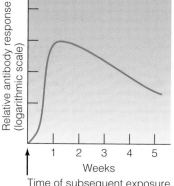

FIGURE 14-7 Primary and Secondary Responses (a) The primary (initial) immune response is slow. It takes about 10 days for antibody levels to peak. Almost no antibody is produced during the first week as plasma cells are being formed. (b) The secondary response is much more rapid. Antibody levels rise almost immediately after the antigen invades. T cells show a similar response pattern.

mune protection afforded by memory cells can last 20 years or longer, which explains why a person who has had a childhood disease such as the mumps or chicken pox is unlikely to contract it again.

How Antibodies Work

Antibodies destroy antigens in one of four ways.

Antibodies belong to a class of blood proteins called the *globulins* (GLOB-you-lynns), introduced in Chapter 7. Antibodies are specifically called **immunoglobulins** (im-YOU-know-GLOB-you-lynns).

Each antibody is a T-shaped molecule consisting of four peptide chains (Figure 14-8). The arms of the T bind to antigens.

Antibodies destroy foreign organisms and antigens via one of four mechanisms. One process is neutralization. During neutralization, antibodies bind to viruses and bacterial toxins and form a complete coating around them. This prevents viruses from binding to plasma membrane receptors of body cells and getting inside (Figure 14-9, far right), and thus they are effectively neutralized. Bacteria, bacterial toxins, and viruses neutralized by their antibody coating are eventually engulfed by macrophages and other phagocytic cells (Figure 14-9).

Other antibodies eliminate antigens by a process called *agglutination* (ah-GLUTE-tin-A-shun). Agglutination occurs when antibodies bind to numerous antigens, causing them to clump together. The antigen-antibody complexes are then phagocytized by macrophages and other phagocytic cells.

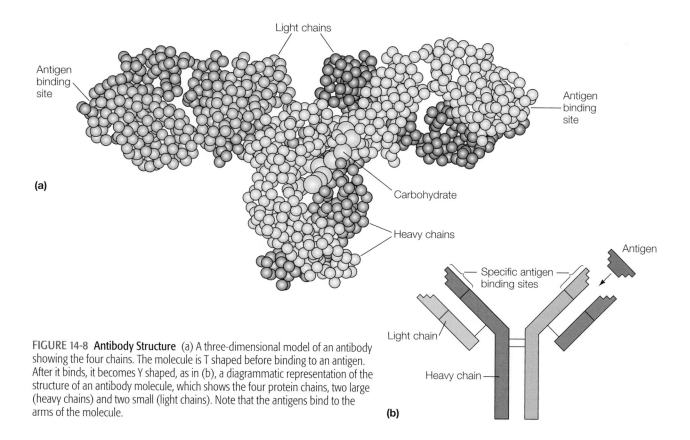

FIGURE 14-8 Antibody Structure (a) A three-dimensional model of an antibody showing the four chains. The molecule is T shaped before binding to an antigen. After it binds, it becomes Y shaped, as in (b), a diagrammatic representation of the structure of an antibody molecule, which shows the four protein chains, two large (heavy chains) and two small (light chains). Note that the antigens bind to the arms of the molecule.

FIGURE 14-9
How Antibodies Work

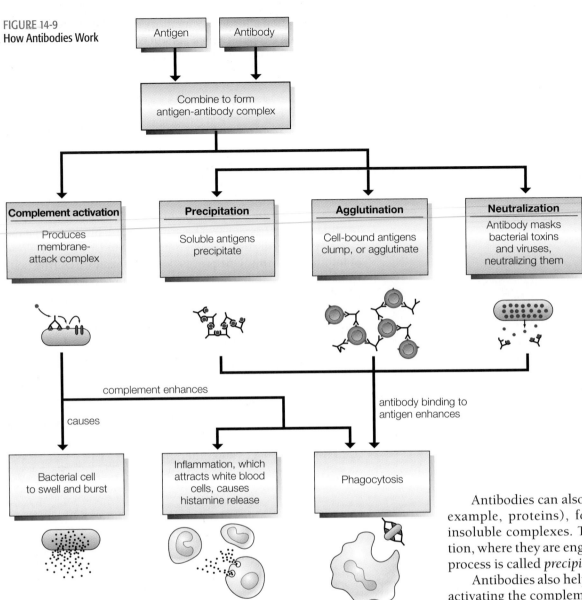

Antibodies can also bind to soluble antigens (for example, proteins), forming much larger, water-insoluble complexes. These precipitate out of solution, where they are engulfed by phagocytic cells. The process is called *precipitation*.

Antibodies also help to rid the body of bacteria by activating the complement system, described above.

T Cells

T cells attack foreign cells directly.

T cells provide a much more complex form of protection than B cells. Like B cells, they respond to the presence of antigens by undergoing rapid proliferation. T cells, however, differentiate into at least four cell types: (1) cytotoxic T cells, (2) memory T cells, (3) helper T cells, and (4) suppressor T cells (Table 14-3).

Cytotoxic T cells play a key role in cell-mediated immunity. Some cytotoxic T cells attack and kill body cells infected by viruses. Other cytotoxic T cells attack and kill bacteria, parasites, single-celled fungi, cancer cells, and foreign cells introduced during blood transfusions or tissue or organ transplants. These organisms or cells have proteins in their membranes that identify them as foreign. How do cytotoxic T cells work?

Cytotoxic T cells bind to antigenic molecules in the membranes of cells they attack. They then release a chemical substance known as **perforin-1** (purr-FOR-in). As shown in Figure 14-10,

TABLE 14-3	Summary of T Cells
Cell Type	Action
Cytotoxic T cells	Destroy body cells infected by viruses, and attack and kill bacteria, fungi, parasites, and cancer cells
Helper	Produce a growth factor that stimulates B-cell proliferation and differentiation and also stimulates antibody production by plasma cells; enhance activity of cytotoxic T cells
Suppressor	May inhibit immune reaction by T cells decreasing B- and T-cell activity and B- and T-cell division
Memory	Remain in body awaiting reintroduction T cells of antigen, at which time they proliferate and differentiate into cytotoxic T cells, helper T cells, suppressor T cells, and additional memory cells

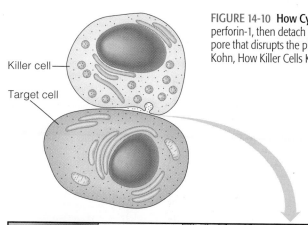

Killer cell

Target cell

FIGURE 14-10 **How Cytotoxic T Cells Work** Cytotoxic T cells, containing perforin-1 granules, bind to their target and release perforin-1, then detach in search of other invaders. Perforin-1 molecules congregate in the target plasma membrane, forming a pore that disrupts the plasma membrane, causing the cell to die. (Source: By Dana Burns from John Ding-E Young and Zanvil A. Kohn, How Killer Cells Kill. ©Copyright January 1988 by Scientific American, Inc. All rights reserved.)

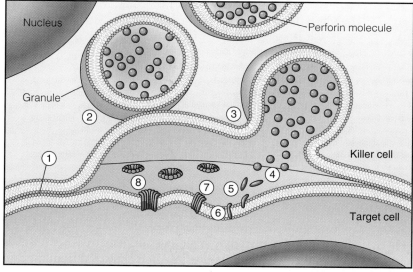

Nucleus

Perforin molecule

Granule

Killer cell

Target cell

On a side note, helper T cells are targeted by HIV, human immunodeficiency virus, which is responsible for AIDS (discussed in the chapter supplement following this chapter). Infection of helper T cells by HIV disables a person's immune system. Patients with AIDS therefore contract several infectious diseases to which they are unable to mount an effective immune response. Unless a cure can be found, all people infected with the AIDS virus will eventually die from these diseases.

Another type of T cells is the suppressor T cell. The role of **suppressor T cells** is less understood. Research suggests that they "turn off" the immune reaction as the antigen begins to disappear—that is, as the antigen is phagocytized. The activity of suppressor T cells, therefore, increases as the immune system finishes its job. Suppressor cells release chemicals that reduce T-cell division.

perforin-1 molecules are housed in membranous sacs in the cytoplasm of the cytotoxic T cells. These structures bind to the membrane of the cytotoxic T cell. The membranes at the point of contact break down and the perforin-1 molecules are released into the space between the killer cell and its prey.

Perforin molecules then embed in the plasma membrane of the target cell. There, they join to form small pores, similar to those produced by the membrane-attack complex of the complement system. These pores cause the plasma membrane to leak, destroying the target cell within a few hours. After it has delivered its lethal payload, the cytotoxic cell detaches and is free to hunt down other antigens.

Memory T cells are also produced when antigens are present. As in B cells, the memory T cells form a cellular reserve. They exist to protect the body in the event of a further incursion at some other time.

Helper T cells are also produced from T cells. Helper T cells enhance the immune response—both the humoral *and* cell-mediated immune response. Helper T cells activate T cells via the release of a chemical known as interleukin 2. It increases the activity of cytotoxic T cells.

Helper T cells are the most abundant of all the T cells (comprising 60%–70% of the circulating T cells). Some immunologists liken helper T cells to the immune system's master switch. Without them, antibody production and T-cell activity are greatly reduced.

Macrophages and Immunity

Macrophages play a very important role in humoral and cell-mediated immunity.

Macrophages play an important role in protecting the body. They even help stimulate T and B cells. They do so by phagocytizing antigens such as bacteria. They then transfer the antigens from the surface of the bacteria to their own cell membranes. Next, macrophages cluster around B cells, presenting the bacterial antigens to them. B cells programmed to respond to that antigen concentrate on the surface of the macrophage and are converted to plasma cells.

Macrophages also present antigen to helper T cells. They, in turn, produce a growth factor that stimulates the formation of plasma cells and antibody production as well.

Active and Passive Immunity

Two types of immunity are possible: active and passive.

One of the major medical advances of the 1800s was the discovery of **vaccines** (vac-SEENS) to prevent bacterial and viral infections. Vaccines contain inactivated or weakened viruses, bacteria, or bacterial toxins. When injected into the body, the "disabled" antigens in vaccines elicit an immune response. The immune system responds to the virus by producing antibodies or T cells.

⠿ healthnote

14-2 Bringing Baby Up Right: The Immunological and Nutritional Benefits of Breast Milk

A baby is born into a dangerous world in which bacteria and viruses abound. Complicating matters, the immune system of a newborn child is poorly developed. Fortunately, the newborn is protected by passive immunity—antibodies that have traveled from its mother's blood. Antibodies also travel to the infant in breast milk (Figure 1).

Several immunoglobulins are present in breast milk. One of these is called secretory IgA. It is present in very high quantities in colostrum, a thick fluid produced by the breast immediately after delivery—before the breast begins full-scale milk production. Colostrum is so important, in fact, that some hospitals give "colostrum cocktails" to newborns who are not going to be breast-fed by their mothers. Nurses remove the colostrum from the mother's breast with a breast pump and feed it to the baby in a bottle.

Colostrum, said the late Sarah McCamman, a nutritionist at the University of Kansas Medical School, coats the lining of the intestines. The IgA antibodies in colostrum prevent bacteria ingested by the infant from adhering to the epithelium and gaining entrance. Breast milk also contains lysozyme, an enzyme that breaks down the cell walls of bacteria, destroying them.

In general, breast-fed babies are healthier than bottle-fed babies. The incidence of gastroenteritis (inflammation of the intestine), otitis (ear infections), and upper respiratory infections is lower in breast-fed babies. Studies also show that children breast-fed for at least 6 months contract fewer childhood cancers than their bottle-fed counterparts. The incidence of childhood lymphoma, a cancer of the lymph glands, in bottle-fed babies is nearly double the rate in breast-fed children for reasons not yet understood.

Research also suggests that certain proteins in breast milk may stimulate the development of a newborn's immune system. In laboratory experiments, the proteins speed

FIGURE 1 Breast-Feeding Mother breast-feeding her newborn infant.

Vaccines stimulate the immune reaction because the weakened or deactivated organisms (or toxins) they contain still possess the antigenic proteins or carbohydrates that trigger B- and T-cell activation. However, because the infectious agents have been seriously weakened or deactivated, the viruses, bacteria, and bacterial toxins in vaccines do not cause disease. Some individuals may develop minor symptoms, but they're not life-threatening.

Problems may arise if the vaccines are given when a person is suffering from a cold, flu, or other sicknesses. There's some evidence, although preliminary, that exposure to vaccines during these times may be connected to chronic illnesses, such as autism. Autism is a severe developmental disability characterized by withdrawal from social relations. Autistic children have reduced motor skills and are unable to or refuse to talk.

Vaccination provides a form of protection that immunologists call **active immunity**—so named because the body actively produces memory T and B cells that protect a person against future infections. The immune response is the same as the one that occurs when a pathogenic organism enters the body. Many vaccines provide immunity or protection from microorganisms for long periods, sometimes for life. Others give only short-term protection.

Vaccinations are vital in controlling deadly diseases such as polio, typhus, and smallpox—diseases that can kill people before their immune system mounts an effective response. In fact, in the wealthier nations of the world, such as the United States, vaccines have nearly eliminated many infectious diseases such as smallpox.

The second type of immunity, called **passive immunity**, is a temporary form of protection, resulting from the injection of

immunoglobulins (antibodies to specific antigens). These antibodies are produced by injecting antigens in other animals such as sheep. The antibodies are then extracted from the animals' blood for use in humans.

Passive immunity is so named because the cells of the immune system are not activated. T cells and B cells are not called into duty. Immunoglobulins remain in the blood for a few weeks, protecting an individual from infection. Because the liver slowly removes them from the blood, a person gradually loses this protection.

Immunoglobulins are administered to prevent illnesses. Travelers to less developed nations, for instance, are often given immunoglobulins to viral hepatitis (liver infection) as a preventive measure. Immunoglobulins are also used to treat individuals who have been bitten by poisonous snakes.

Passive immunity can also be conferred naturally—for instance, from a mother to her fetus (FEE-tuss). A fetus receives a dose of antibodies via the placenta (plah-SEN-tah), the organ that transfers nutrients from the mother's bloodstream to the fetal blood. Maternal antibodies transferred via the placenta remain in the blood of a newborn infant for several months, protecting the youngster from bacteria and viruses while its immune system is developing. Mothers also transfer antibodies to their babies in breast milk. The maternal antibodies in milk attack bacteria and viruses in the intestine, protecting the infant from infection. (For more on this topic, see Health Note 14-2.)

Vaccination Fears in the United States. Vaccines have lowered the incidence of many infectious diseases in the United States and other relatively affluent industrialized nations by 99% or more. Vaccines for diphtheria, tetanus, whooping cough, polio, measles,

up the maturation of B cells and prime them for antibody production. These soluble proteins may also activate macrophages, which play a key role in the immune system.

Breast milk is also more digestible and more easily absorbed by infants than formula. Formula is a mixture of cow's milk, proteins, vegetable oils, and carbohydrates. It is only an approximation of mother's milk and is not broken down and absorbed as completely as breast milk.

Because of a growing awareness of the benefits of breast-feeding, virtually every national and international organization involved with maternal and child health supports breast-feeding, says McCamman. Breast-feeding can have a major health impact in this country. Unfortunately, the benefits are not as widely known as many people would like, even among health care professionals. Fortunately, more and more health care workers are beginning to understand the benefits of breast-feeding and are promoting this option. As a result, many middle-class American women are now choosing to breast-feed.

Unfortunately, says McCamman, there is "a huge population of low-income, poorly educated . . . women who choose not to nurse." The federal government may be playing an unwitting role in their decision. A national program aimed at improving child nutrition provides free infant formula to needy women, perhaps discouraging mothers from breast-feeding.

Another reason is economic. Low-income women often work at jobs that do not provide maternity leave. Thus, these women must return to work soon after giving birth, making breast-feeding difficult.

Still another reason for the low rate of breast-feeding among low-income women is that women need a lot of support to nurse. "People think nursing is innate, natural, and easy," says McCamman, "but this is not always the case." Getting started often requires guidance and education. Without such support, breast-feeding can be difficult and painful. Breast-feeding among all women, rich and poor, may also be discouraged by cultural attitudes and fear of embarrassment. In fact, many otherwise open-minded people find breast-feeding in public or even semipublic settings embarrassing.

Given the many benefits of breast-feeding, McCamman recommends it to all mothers who are able.

Visit *Human Biology's* Internet site for links to web sites offering more information on this topic.

mumps, and congenital rubella (German measles) have all but eliminated these deadly or crippling disease organisms.

Despite the success of vaccines, publicity concerning their rare side effects has convinced some parents not to have their children vaccinated. Another cause for the decline in vaccination stems from the success of previous immunization programs, which, as noted above, have greatly reduced the incidence of most infectious diseases. Parents who were reared in an environment free from such diseases are often unaware of the dangers of infectious disease. Having their children immunized seems unimportant. Some people, notably Christian Scientists, refuse to vaccinate children on religious grounds. Public health officials are quick to point out that pathogenic organisms that once took a huge toll on humans have not been eradicated. They fear that without widespread protection, epidemics could occur again.

Harmful side effects from conventional vaccines in the past were often caused by reactions to certain "nonessential" antigens on the injected microorganism. These antigens frequently play little or no role in immunity. By eliminating them from vaccines, researchers hope to develop safer alternatives to the vaccines in use today.

Developing a Strong Immune System

The immune system may operate optimally when exposed to antigens early on.

Because bacteria and viruses can cause illness, many people assume that the best way to keep their children healthy is to protect them from exposure early in life. Hence the outpouring of cleaning agents and other products with antibacterial agents.

Unfortunately, research is suggesting that the best way to develop a healthy immune system is to use it when you're young. That is, the best way to ensure maximum health is *not* to protect a child from exposure to non-life-threatening microbes.

Although this idea seemed preposterous when first proposed, more and more evidence is showing that children who have been sheltered from normal childhood illnesses develop more asthma and allergies—a lot more! In fact, a study released in 2000 showed that infants who go to day care centers or who have older siblings—both of which increase their risk of catching colds or the flu—are less likely to develop asthma later in childhood than those who stay at home. The reason for this is not known, but researchers believe that early exposure to common microbes, even nonpathogenic ones, may somehow start the immune system off on the right path of development.

Small families, good sanitation, and widespread use of antibiotics, say researchers, reduce childhood exposure to bacteria and viruses. This may be responsible for the dramatic increase in allergies and asthma in the United States and other developed countries. The advice given now is not to sterilize the environments we live in and not to protect children from common microbes they'd pick up at day care or school.

Tissue Transplantation

Tissue transplants often evoke cell-mediated immunity.

Although the immune system is important in protecting us from microorganisms, it presents something of a challenge to physicians and patients during tissue and organ transplants and blood transfusions (Chapter 7).

Organ and tissue transplantations are fairly common medical practices these days. Diseased hearts, kidneys, and livers, for instance, are frequently replaced when suitable donors are available. However, only three conditions exist in which a person can receive a transplant and not reject it.

One is if the tissue comes from an individual's own body. For burn victims, surgeons might use healthy skin from one part of the body to cover a badly damaged region elsewhere. Hair transplants also qualify. In hair transplants, hair from the back of the head is often moved to balding spots up front or on top.

The second instance is when a tissue is transplanted between identical twins—individuals derived from a single fertilized ovum that splits early in embryonic development to form two embryos. These individuals are genetically identical and have identical cellular antigens.

A third instance occurs when tissue rejection is inhibited by drugs that suppress the cell-mediated response mounted by the immune system. This treatment must be continued throughout the life of the patient. If not, the body will reject the transplant. Unfortunately, most immune suppressants have numerous side effects and often leave the patient vulnerable to bacterial and viral infections.

14-4 Diseases of the Immune System

The immune system, like all other body systems, can malfunction. This section looks at two disorders: allergies and autoimmune diseases. AIDS, a sexually transmitted disease that affects the immune system, is discussed in the chapter supplement.

Allergies

Allergies are an immune system overreaction to environmental substances.

An **allergy** is an overreaction to some environmental substance, such as pollen or certain foods (Figure 14-11). Those antigens that stimulate an allergic reaction are called **allergens** (AL-er-gens). Allergens cause the production of one class of immunoglobulins, the IgE antibodies, from plasma cells.

As Figure 14-11 shows, these IgE antibodies bind to mast cells, which are found in many tissues, but especially in the connective tissue surrounding blood vessels. Mast cells have large cytoplasmic vesicles containing the chemical histamine.

When allergens are present, they bind to the IgE antibodies that are attached to the mast cells. This binding triggers the release of histamine from the vesicles via exocytosis (Figure 14-11). Histamine, in turn, causes nearby arterioles to dilate. Histamine released in the lungs causes the bronchioles to constrict, reducing airflow and making breathing difficult. This condition is called **asthma** (AS-ma).

Allergic reactions usually occur in specific body tissues, where they create local symptoms that, while irritating, are not life-threatening. For example, an allergic response may occur in the eyes, causing redness and itching. Or, it may occur in the nasal passageway, causing stuffiness.

The allergic response can also occur in the bloodstream, where it can be fatal if not treated quickly. For example, the presence of penicillin or bee venom in the bloodstreams of certain people causes a massive release of histamine and other chemicals. This, in turn, causes extensive dilation of blood vessels in the skin and other tissues. The blood pressure then falls quickly, shutting down the circulatory system. Histamine released by mast cells also causes severe constriction of the bronchioles (the ducts in the lungs that open onto the alveoli), making breathing difficult. The decline in blood pressure and constriction of the bronchioles result in **anaphylactic shock** (ANN-ah-fah-LACK-tic). Death may follow if measures are not taken to reverse this catastrophic event. One such measure is an injection of the hormone epinephrine (commonly known as adrenalin), which rapidly reverses the constriction of the bronchioles.

Allergies are treated in three ways. First, patients are advised to avoid allergens—for instance, to avoid milk and milk products or stay clear of dogs and cats. Second, patients may also be given antihistamines (an-tea-HISS-tah-means), drugs that counteract the effects of histamine. Third, patients may also be given allergy shots, injections of increasing quantities of the offending allergen. In many cases, this treatment makes an individual less and less sensitive to the allergen. Desensitization results from the production of another class of antibodies, the IgG antibodies, which bind to allergens. This, in turn, prevents the allergen from binding to the plasma cells, thus reducing or eliminating the release of histamine and other chemical substances responsible for the allergic reaction.

Autoimmune Diseases

Autoimmune diseases result from an immune attack on the body's own cells.

Occasionally, the immune system mounts an attack on the body's own cells, resulting in an **autoimmune disease**. Autoimmune diseases result from many causes. For example, in some instances, normal body proteins can be modified by environmental pollutants, viruses, or genetic mutations so that they are no longer recognizable by the body as self.

In other cases, normal body proteins usually isolated from the immune system enter the bloodstream and evoke an immune response. For example, a protein called *thyroglobulin* is produced by the thyroid gland in the neck. Thyroglobulin is

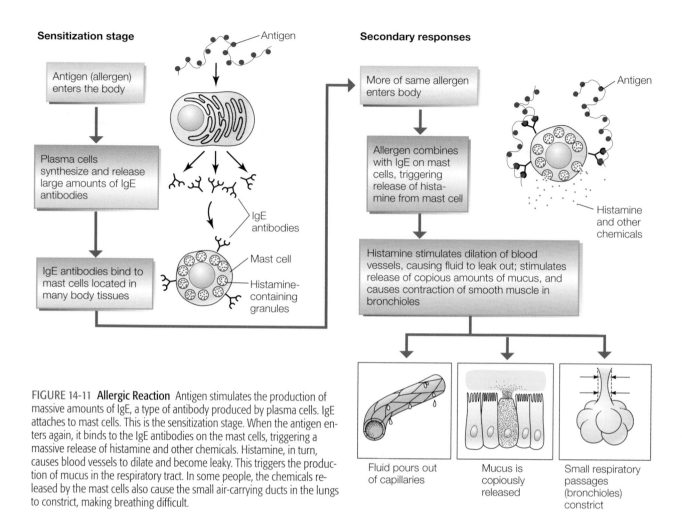

Sensitization stage

Antigen

Antigen (allergen) enters the body

Plasma cells synthesize and release large amounts of IgE antibodies

IgE antibodies bind to mast cells located in many body tissues

IgE antibodies

Mast cell

Histamine-containing granules

Secondary responses

More of same allergen enters body

Allergen combines with IgE on mast cells, triggering release of histamine from mast cell

Histamine stimulates dilation of blood vessels, causing fluid to leak out; stimulates release of copious amounts of mucus, and causes contraction of smooth muscle in bronchioles

Antigen

Histamine and other chemicals

Fluid pours out of capillaries

Mucus is copiously released

Small respiratory passages (bronchioles) constrict

FIGURE 14-11 **Allergic Reaction** Antigen stimulates the production of massive amounts of IgE, a type of antibody produced by plasma cells. IgE attaches to mast cells. This is the sensitization stage. When the antigen enters again, it binds to the IgE antibodies on the mast cells, triggering a massive release of histamine and other chemicals. Histamine, in turn, causes blood vessels to dilate and become leaky. This triggers the production of mucus in the respiratory tract. In some people, the chemicals released by the mast cells also cause the small air-carrying ducts in the lungs to constrict, making breathing difficult.

stored inside the gland and not exposed to cells of the immune system. If the gland is injured, however, thyroglobulin may enter the bloodstream. Lymphocytes encountering this protein may then mount an immune response to it.

Yet another cause of autoimmune reaction is exposure to antigens that are nearly identical to body proteins. The bacterium that causes strep throat, for example, contains an antigen structurally similar to one of the proteins found in the plasma membranes of the cells lining the heart valves of some individuals. The body mounts an attack on the bacterium, but antibodies may also bind to the lining of the heart valve, causing a local inflammation and scar tissue to develop. This can damage the valve, resulting in valvular incompetence, discussed in Chapter 6.

14-5 Health and Homeostasis

Human beings would not survive past early infancy without some means of protection against the potentially harmful viruses and microorganisms that abound in our world. But what does protection have to do with homeostasis?

The answer is, plenty. Without the protective mechanisms you've studied in this chapter, homeostasis could not be maintained. Bacteria and viruses would take over, sapping us of our strength and damaging vital organs that play a key role in homeostasis. The immune system and other the protective mechanisms

of our bodies, therefore, play a vital role in maintaining internal constancy.

Despite its prowess, the immune system is not impenetrable. Stress can suppress the immune system, making us more prone to infectious diseases. And some chemical pollutants in the environment—for example, certain pesticides—can affect our immune systems, too, making us more susceptible to infectious agents.

SUMMARY

Viruses and Bacteria: An Introduction

1. Two of the most important common infectious agents are viruses and bacteria. Viruses consist of a nucleic acid core, made up of either DNA or RNA, and an outer protein coat. Viruses must invade cells to reproduce.

2. Viruses most often enter the body through the respiratory and digestive systems and spread from cell to cell in the bloodstream and lymphatic system. However, other avenues of entry are also possible—for example, sexual contact.

3. Bacteria are single-celled microorganisms that consist of a circular strand of DNA and cytoplasm which is enclosed by a plasma membrane.

4. Unlike viruses, bacteria can be killed with antibiotics. Unfortunately, overuse of antibiotics in medicine has created a growing problem: antibiotic resistance in bacteria. Many thousands of people die each year of infections caused by antibiotic-resistant bacteria they contract in hospitals or in other places.

5. Although they are best known for their role in causing sickness and death, most bacteria perform useful functions.

The First and Second Lines of Defense

6. The first line of defense against infectious agents is the skin and the epithelia of the respiratory, digestive, and urinary systems. Some epithelia also produce protective chemical substances that kill microorganisms.

7. The second line of defense consists of cells and chemicals that the body produces to combat infectious agents that penetrate the epithelia. One of the chief agents in the second line of defense is the macrophage. Located in connective tissue beneath epithelia, macrophages phagocytize infectious agents, preventing their spread. Neutrophils and monocytes also invade infected areas from the bloodstream and destroy bacteria and viruses.

8. Another combatant in the second line of defense consists of the chemicals released by damaged tissue, which stimulate arterioles in the infected tissue to dilate. The increase in blood flow raises the temperature of the wound. Heat stimulates macrophage metabolism, accelerating the rate of the destruction of infectious agents. Heat also speeds up the healing process.

9. Still other chemicals increase the permeability of the capillaries, causing plasma to flow into the wound and increasing the supply of nutrients for macrophages and other protective cells.

10. Another part of the secondary line of defense are the pyrogens, chemicals released primarily by macrophages exposed to bacteria, which raise body temperature and lower iron availability, thus decreasing bacterial replication.

11. Interferons, a group of proteins released by cells infected by viruses, are also part of the second line of defense. Interferons travel to other virus-infected cells, where they inhibit viral replication.

12. The blood also contains the complement proteins, which circulate in an inactive state, becoming activated only when the body is invaded by bacteria. Some of the complement proteins stimulate the inflammatory response. Others embed in the plasma membrane of bacteria, creating holes and killing these pathogens.

The Third Line of Defense: The Immune System

13. The immune system consists of billions of lymphocytes that circulate in the blood and lymph and take up residence in the lymphoid organs and lymphoid tissues. Lymphocytes recognize antigens—foreign cells and foreign molecules, mostly proteins and large-molecular-weight polysaccharides.

14. T and B cells are two types of lymphocytes produced in red bone marrow. Immunocompetent B cells encounter antigens (often presented to them by macrophages) to which they are programmed to respond. They then begin to divide, forming plasma cells and memory cells. Plasma cells produce antibodies. The memory cells enable the body to respond more quickly to future invasions by the same antigen.

15. Antibodies are small protein molecules that bind to specific antigens, destroying them either by precipitation, agglutination, or neutralization, or by activation of the complement system.

16. When T and B cells first encounter an antigen, they react slowly. The initial response is called the primary response. Because the body responds slowly at first, there is often a period of illness before the pathogen is removed by the immune system.

17. Numerous memory cells produced during the first assault ensure that a reappearance of the antigen will elicit a much faster and more powerful reaction, the secondary response. The resistance created by a response to an antigen is called immunity.

18. When activated by an antigen, T cells multiply and differentiate, forming memory T cells, cytotoxic T cells, helper T cells, and suppressor T cells, whose functions are summarized in Table 14-3.

19. A solution containing a dead or weakened virus, bacterium, or bacterial toxin that is injected into people to create active immunity is called a vaccine.

20. Passive immunity can be achieved by injecting antibodies into a patient or by the transfer of antibodies from a mother to her baby through the bloodstream or breast milk. Passive immunity is short-lived, lasting at most only a few months, compared to active immunity, which lasts for years.

21. Tissue transplantation requires careful matching of donor and recipient. In such instances, only cells from the same individual or an identical twin will be accepted. All others are rejected by the T cells, unless the system is suppressed with drugs.

Diseases of the Immune System

22. The most common malfunctions of the immune system are allergies, extreme reactions to some antigens. Allergies are caused by IgE antibodies, produced by plasma cells. IgE antibodies bind to mast cells, which causes them to release histamine and other chemical substances that induce the symptoms of an allergy—production of mucus, sneezing, and itching.

23. Autoimmune diseases, another immune system disorder, result from an immune attack on the body's own cells. Autoimmune diseases may occur when normal proteins are modified by chemicals or genetic mutations so that they are no longer recognizable as self. Other possible causes are the sudden presence of proteins that are normally isolated from the immune system and exposure to antigens that are nearly identical to body proteins.

Health and Homeostasis

24. The immune system and protective mechanisms that constitute the first and second lines of defense protect the body from harmful viruses and microbes. In a sense, they help maintain a constant internal state either directly, by warding off infectious agents, or secondarily, by protecting other body cells that are essential to homeostasis. Chemicals can damage the immune system, upsetting homeostasis.

critical thinking

Thinking Critically–Analysis

This Analysis corresponds to the Thinking Critically scenario that was presented at the beginning of this chapter.

This chapter offers a brief overview of the topic of vaccination. The first conclusion that you might draw is that the text does not provide enough information to analyze the claims made by your friends about vaccination. You'd be advised to study an immunology book as well as the scientific literature. Once you've read some more, you will be better able to discern whether vaccination is worth the relatively small risk.

Despite the brief coverage, this chapter does point out that infectious diseases have not been eliminated. Vaccination has only kept them under control by reducing the potential population of hosts. Thus, if large numbers of people are unvaccinated against infectious disease, outbreaks could occur. The results could be quite dramatic because we're living on a crowded planet, and infectious organisms spread quickly in crowded environments.

KEY TERMS AND CONCEPTS

Active immunity, p. 266
Allergens, p. 268
Allergy, p. 268
Anaphylactic shock, p. 268
Antibodies, p. 261
Antigens, p. 260
Asthma, p. 268
Autoimmune disease, p. 268
Bacterium, p. 256
Bacterial toxin, p. 260
B lymphocyte, p. 260
Complement system, p. 259
Cytotoxic T cell, p. 264

Helper T cell, p. 265
Histamine, p. 259
Humoral immunity, p. 261
Immunocompetence, p. 261
Immunoglobulin, p. 263
Inflammatory response, p. 258
Interferon, p. 259
Lymphoblast, p. 261
Lymphocyte, p. 260
Lymphoid organs, p. 260
Membrane-attack complex, p. 260
Memory cell, p. 262
Passive immunity, p. 266

Perforin-1, p. 264
Plasma cell, p. 261
Plasmid, p. 256
Primary response, p. 261
Prostaglandin, p. 259
Pyrogen, p. 259
Secondary response, p. 261
Suppressor T cell, p. 265
Thymus, p. 261
T lymphocyte, p. 260
Vaccine, p. 265
Virus, p. 255

CONCEPT REVIEW

1. Describe the structure of a typical virus. p. 255
2. Based on your previous studies (Chapter 3), in what ways are bacteria similar to human cells, and in what ways are they different? p. 256
3. The human body consists of three lines of defense. Describe what they are and how they operate. pp. 257–268
4. Describe the inflammatory response, and explain how it protects the body. pp. 258–259
5. Define each of the following terms, and explain how they help protect the body: pyrogen, interferon, and complement proteins. p. 259
6. The first and second lines of defense differ substantially from the third line of defense. Describe the major differences. pp. 257–268
7. How does the immune system detect foreign substances? pp. 260–261
8. Describe how the B cell operates. Be sure to include the following terms in your discussion: bone marrow, immunocompetence, plasma membrane receptors, primary response, plasma cell, antigen, antibody, and secondary response. pp. 260–261
9. Describe the four mechanisms by which antibodies "destroy" antigens. p. 263
10. Describe the events that occur after a T cell encounters its antigen. pp. 264–265
11. What is the difference between active and passive immunity? pp. 265–266
12. A child is stung by a bee, swells up, and collapses, having great difficulty breathing. What has happened? What can be done to save the child's life? p. 268
13. What is an autoimmune disease? Explain the reasons it forms. pp. 268–269

SELF-QUIZ: TESTING YOUR KNOWLEDGE

1. Viruses contain a nucleic acid core consisting of either _____ or _____. p. 255
2. Bacteria are living cells but lack organelles and true _____. p. 256
3. _____ respond to antibiotics, but viruses do not. p. 256
4. The _____ and epithelial linings of the respiratory, urinary, and digestive tracts are part of the first line of defense against potentially harmful microbes. p. 257
5. The _____ response is a series of changes characterized by redness, swelling, and pain that protect us from potentially harmful microbes after a cut or abrasion in the skin. p. 258

6. _____ is a pain-causing chemical released by injured cells. p. 259
7. _____ is a chemical released by virus-infected cells. It prevents viruses from successfully invading neighboring cells. p. 259
8. The membrane-attack complex is formed by proteins of the _____ system. This structure creates pores in the membranes of bacteria, causing them to leak, swell, and burst. p. 260
9. _____ lymphocytes are responsible for cell-mediated immunity, while _____ lymphocytes are responsible for humoral immunity. p. 260

10. Lymphocytes cannot respond to antigens until they become _____. p. 261
11. B lymphocytes are converted to _____ cells, which produce antibodies. p. 261
12. The _____ response is more powerful and swifter than the first response to an antigen. p. 261
13. Antibodies belong to a group of plasma proteins known as _____. p. 263
14. Helper T cells _____ the immune response. p. 264
15. Vaccines provide _____ immunity, while injections of antibodies provide _____ immunity. p. 265

www.jbpub.com/humanbiology/5e

The site features eLearning, an online review area that provides quizzes, chapter outlines, and other tools to help you study for your class. You can also follow useful links for in-depth information, research the differing views in the Point/Counterpoints, or keep up on the latest health news.

Acquired Immunodeficiency Syndrome

In 1985, Damion Knight, a bright, young cabinetmaker, began to lose weight and experience bouts of unexplained fever. His lymph nodes became swollen. He complained of feeling weak and drowsy. A doctor found that Damion had **acquired immunodeficiency syndrome**, commonly known as **AIDS**. Like millions of other people worldwide, Knight died several years after his diagnosis.

The Human Immunodeficiency Virus

HIV is a sexually transmitted virus that causes AIDS, a lethal immune system disease.

AIDS is a deadly sexually transmitted disease caused by a virus (Figure 1a). This virus, known as the **human immunodeficiency virus** or **HIV** for short, attacks and weakens the immune system.

In specific, HIV is an RNA virus that attacks helper T cells, described in Chapter 14. It severely impairs the immune system. AIDS patients grow progressively weaker and generally fall victim to other infectious agents. For example, many die from an otherwise rare form of pneumonia.

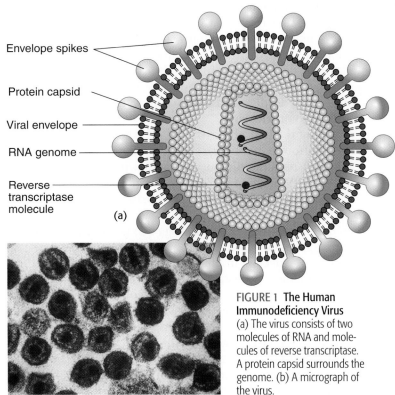

Envelope spikes

Protein capsid

Viral envelope

RNA genome

Reverse transcriptase molecule

(a)

(b)

FIGURE 1 **The Human Immunodeficiency Virus**
(a) The virus consists of two molecules of RNA and molecules of reverse transcriptase. A protein capsid surrounds the genome. (b) A micrograph of the virus.

Where Did HIV Come From?

HIV very likely originated in chimpanzees that suffer from a similar virus in west-central Africa.

AIDS rose to infamy in the early 1980s. The first documented case was that of a young Missouri boy who died at age 15 in 1969. However, studies of tissue samples of a British sailor revealed that he probably died of AIDS ten years before the death of the Missouri boy. In 1998, scientists announced the presence of HIV in a blood sample of a Bantu man from Africa's Democratic Republic of the Congo who died in 1959.

New research suggests that AIDS came to us via chimpanzees in west-central Africa. Chimpanzees contain a strain of simian immunodeficiency virus (SIV) that is very similar to HIV-1, one of the most common strains in humans. The virus was apparently transferred by exposure to blood of chimps killed by humans living the area. New studies also show that AIDS may have begun spreading among people as long ago as 1930.

The Rising Incidence of AIDS

The number of cases of AIDS in the United States increased dramatically since the early 1980s.

HIV infections and AIDS are on the rise, prompting one researcher to liken AIDS to the bubonic plague that spread through Europe in the fourteenth and fifteenth centuries. In the 1300s, the plague killed one quarter of the adult population and untold children. By comparison, as many as 50% of the residents in some African villages today are testing HIV-positive. More than one-third of the population of Botswana and Swaziland have AIDS.

HIV infection is a global epidemic. As of December 2003, the United Nations estimated that 38 million people were infected worldwide. Women account for nearly 50% of all people living with HIV. Over 20 million people worldwide have died from the disease since 1981, according to the UN. Over 14 million children have been orphaned as a result of the death of their parents to AIDS.

According to the United Nations, worldwide approximately 5 million people were infected with the AIDS virus in 2003 alone. The hardest hit area is sub-Saharan Africa. Three of every 5 new infections take place here.

Other areas of the world are also witnessing a rise in AIDS. Myanmar, Viet Nam, Cambodia, and India are experiencing a rapid increase in AIDS cases. Eastern Europe has witnessed a dramatic increase as well.

(a)

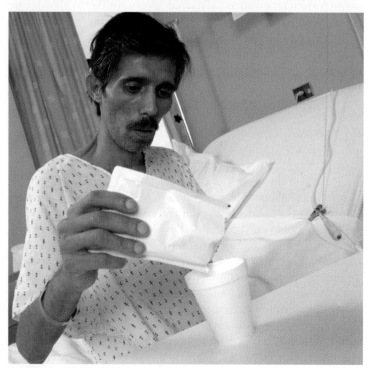

(b)

FIGURE 2 **Phases of AIDS** (a) A patient in the acute phase showing no symptoms of AIDS. (b) An AIDS sufferer in the full-blown phase.

In the United States, approximately 890,000 cases of AIDS have been reported, including more than 9,300 children (under age 13), since 1981. It is believed that as many as one million people are infected with HIV. To date, over 500,000 have died, including over 5,300 children. Although mortality from HIV infection has dropped dramatically in the United States and in Europe thanks to new drugs that treat the virus and the accompanying infections, better medical care, prevention, and other factors, an estimated 40,000 Americans will contract HIV this year alone.

The hardest hit group is men; two-thirds of all new HIV infections occur in African American and Hispanic men. AIDS is now a leading killer of African American males. According to the U.S. Centers for Disease Control and Prevention, AIDS affects six times more African Americans and three times more Hispanics than Caucasians.

Stages of AIDS

AIDS is a progressive disease that exhibits three distinct phases.

AIDS progresses through three distinct phases: acute phase, chronic phase, and full-blown AIDS (Figure 2). Interestingly, however, some individuals develop AIDS within 3 to 5 years of being infected. In others, the disease may take 10 to 12 years to manifest itself. Researchers have now isolated a group of genes in human chromosomes that apparently account for this difference. No matter what the time until onset, all AIDS victims follow a similar progression.

Figure 3 shows several key physiological parameters and symptoms during each phase. The top panel, for instance, plots the number of helper T cells, also called T4 cells. Symptoms are listed in the middle panel. The bottom panel shows the concentration of HIV and the HIV antibody levels.

As illustrated in Figure 2a, most patients show no symptoms during the first few months after infection. A few, perhaps 1%–2%, may exhibit symptoms similar to those of infectious mononucleosis—that is, fever, chills, aches, and swollen lymph nodes. However, these symptoms vanish shortly thereafter, and individuals go on about their business, unaware that they have contracted a deadly disease—and unaware that they're transmitting it to others.

During the first phase, the T4-cell count is high (Figure 3). As the bottom panel in Figure 3 shows, the number of viruses (blue line) rises rapidly during the first phase of AIDS. Antibodies to HIV begin to increase as a result.

During the second phase of this disease, T4-cell count begins to fall in response to the rise in HIV. Antibody levels rise initially, then fall as the immune system begins to falter. During the second phase, patients begin to show outward signs of the disease caused by the decline in the number of T4 cells. Severe fatigue and unexplained, persistent fever are two common symptoms. Some patients complain of a persistent cough and loss of memory, difficulty thinking, and depression. When recurring infections set in, the third and final phase of AIDS is about to begin.

The final phase, known as full-blown AIDS, is characterized by persistent infections, extreme loss of weight, and weakness. Most patients succumb to one of a handful of noninfectious organisms—microbes not ordinarily capable of producing

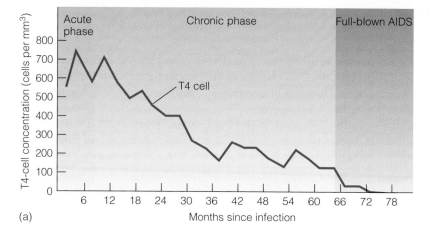

(a)

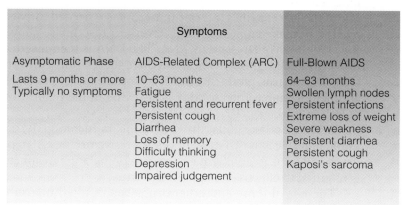

(b)

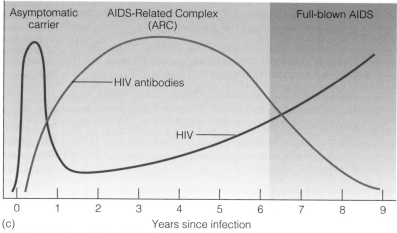

(c)

FIGURE 3 Tracking a Killer (a) T4 cell concentration, (b) symptoms, and (c) HIV and antibody levels. (Source: Data for parts (a) and (c) from R. R. Redfield, and D. S. Burke, "HIV Infection: The Clinical Picture" in *Scientific American,* October 1988.)

serious infections. When a patient is in a state of extreme immune compromise, though, these organisms take hold and become life-threatening.

Cancer, Brain Deterioration, and HIV

HIV also causes cancer and produces a substance that may cause deterioration of brain function.

HIV affects more than a person's immune system. AIDS patients, for example, often contract a rare form of skin cancer called *Kaposi's sarcoma* (kah-PO-sees sar-KOME-ah) (Figure 4). Kaposi's sarcoma can also attack internal organs.

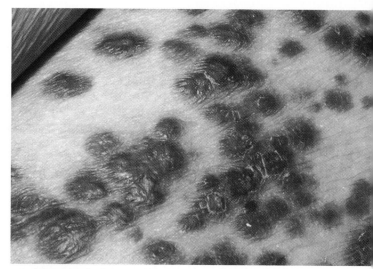

FIGURE 4 Kaposi's Sarcoma This cancer is a serious opportunistic disease of the HIV-infected person.

New research suggests that Kaposi's sarcoma is caused by a strain of the herpes virus known as herpes virus 8. It is believed to be transmitted by deep kissing, kissing that involves a lot of contact with saliva. When the immune system is impaired, this virus proliferates, and eventually leads to Kaposi's sarcoma.

AIDS patients also experience a number of neurological disorders, beginning in the second phase. These include memory loss and progressive mental deterioration. Why? Researchers have found that inside helper T cells, HIV produces several proteins that are incorporated into the viral capsid. One of those proteins is known as gp120. This protein, researchers believe, travels in the blood to the brains of some patients, where it kills neurons, thus producing neurological defects.

Loss of mental function may also be caused by a single-celled parasite that is normally found in cats but sets up residence in people whose immune systems are compromised by HIV. This parasite causes a brain infection (encephalitis) that leads to a loss of brain cells, seizures, and weakness.

HIV Transmission

HIV is transmitted in many ways, but not by casual contact.

Research has shown that HIV is passed from one person to the next primarily by three routes: (1) sexual contact, (2) blood transfusions, and (3) contaminated needles shared by intravenous drug abusers. Homosexual men, hemophiliacs, and drug addicts are the primary victims in the United States. But HIV is also transmitted among the heterosexual population and can even be transmitted from an infected mother to her baby through the placenta. An HIV-infected mother who is nursing her infant can transmit the virus in her milk, too. A recent study showed that a man infected with AIDS is many times more likely to transmit the disease to a female partner than vice versa.

There's no evidence that HIV can be transmitted in saliva or by sharing utensils, plates, or cups. In fact, laboratory studies suggest that even though HIV can be found in the saliva of infected individuals, saliva has natural components that reduce HIV's ability to infect cells. At this writing, researchers do not know whether deep kissing or oral sex increases the risk of infection.

Individuals with the genetic disorder hemophilia were once at risk for AIDS. Hemophiliacs are given clotting factors from pooled human plasma. Before 1984, blood donors were not screened for HIV. Consequently, many of the blood clotting preparations were contaminated with this deadly virus. As a result, a majority of the estimated 15,000 hemophiliacs in the United States who received clotting factors between 1975 and 1984 have HIV antibodies in their blood. Many of these people have died.

To prevent the spread of AIDS through blood transfusions, blood is now routinely tested for HIV. Tissues and organs for transplantation are also tested. Improvements in screening have dramatically reduced the risk of HIV infection. However, despite improvements in screening, blood transfusion is not a fail-safe proposition. Individuals who will need blood for an operation are therefore encouraged to donate some of their own blood for their own transfusion ahead of time.

Preventing the Spread of AIDS

AIDS can be prevented by practicing sexual abstinence, using condoms, and other measures.

HIV, although lethal, does not spread as readily as the flu virus or cold viruses; individuals can protect themselves by practicing sexual abstinence before marriage, by engaging in safe sex (using condoms, for example), and by avoiding multiple sexual partners. To prevent the spread of HIV among intravenous drug users, some countries and some U.S. cities distribute clean hypodermic needles to addicts.

Pregnant women infected with HIV can be treated with the drug AZT (described shortly) during pregnancy. If the baby is delivered by cesarean section (through an incision made in the lower abdomen), its chances of contracting the disease can be reduced dramatically. Studies are underway on yet another drug treatment that is given to the mother during labor and then the baby within three days of birth. Preliminary results show that it is even more effective than AZT.

Promising Development

The battle against AIDS has been facilitated by new screening tests and by drugs that slow down the development of the disease.

Health care workers determine the presence of HIV via an immunologic test, which detects antibodies to HIV in the blood. Scientists have also developed a new, more sensitive test that detects antigens—proteins in the HIV coat—that has greatly improved the screening of blood and tissue. There are even tests individuals can use at home, which require a pin prick to draw blood or a mouth swab to sop up antibodies to the HIV.

Although no cure has been discovered, a drug called AZT (zidovudine) may prolong the lives of people who have tested HIV-positive. AZT inhibits viral replication and is effective in people at all stages of AIDS. As noted above, it is even used to prevent the transmission of HIV from mothers to fetuses and newborns (through breast milk). Unfortunately, AZT is costly and may be carcinogenic.

To date, around three dozen drugs are available to combat HIV and to treat infections, Kaposi's sarcoma, and other complications including weight loss. One of the newest treatments that has proved very useful in prolonging the lifespan of people infected with HIV is a three-drug regime. All three drugs are enzyme inhibitors.

The treatment consists of AZT, another similar drug, and a more recent drug known as a protease inhibitor. AZT blocks one of the key enzymes (reverse transcriptase) needed early on by HIV to make copies of itself. Reverse transcriptase allows the HIV to make DNA on its RNA. The DNA is then inserted into a host cell's chromosomes, and there successfully converts the cell to a virus-producing factory. Protease inhibitors block another enzyme (protease) that is also vital in the replication of HIV in infected cells.

AZT and other similar drugs slow the proliferation of HIV in the body. Because HIV can develop a resistance to AZT-type drugs and protease inhibitors, however, combination treatment is required to effectively suppress the virus.

The use of these and other drugs has resulted in a substantial decrease in deaths from AIDS in the United States and other more developed nations where there is money to pay for the costly treatment. Even in such nations, drug costs can run as high as $15,000 a year. In less developed countries, few individuals can afford the treatment.

Although the three-drug treatment is greatly prolonging lives, it does present problems, other than its high cost. Researchers recently reported that long-term use of AIDS drugs, in particular, the protease inhibitors, can cause severe bone loss. It also causes other metabolic disorders, some very severe, when used for long periods. Some patients don't respond to the treatment, cannot tolerate the toxic effects of the drugs, or have difficulty complying with the complex dosing schedules.

Researchers have discovered two new chemicals, however, that could be added to the list of AIDS drugs. These two compounds sabotage yet another of the enzymes HIV needs to successfully reproduce. The enzyme is called *integrase* and it inserts DNA that is made on HIV's RNA into the host cell's chromosomes. Under direction of this integrated viral DNA, the cell becomes an HIV-producing factory. Complete takeover of the host cell requires about 24 hours.

Studies in the laboratory show that the new compounds stop the splicing of virally produced DNA into the host cell's DNA. They also prevent the spread of HIV in cell cultures and even stop strains of HIV that are resistant to other enzyme inhibitors.

AIDS will undoubtedly remain a significant public health threat in the world for many years. To bring this disease under control, more intensive efforts are needed to educate all people, especially individuals in high-risk groups, on ways to prevent the disease. Successes are occurring. In Thailand, Senegal, Tanzania, and Uganda, infection rates have slowed or declined thanks to sex education, the distribution of free condoms, needle exchange programs, and support from religious and civic leaders. Many scientists believe that a vaccine is the only way to ultimately bring this disease under control.

HIV Vaccine

Although some researchers are optimistic about finding a vaccine for HIV, not all share their view.

HIV is notorious for its ability to mutate, a feature that is making the task of developing a vaccine extremely difficult, if not impossible. Even within the body, HIV mutates fairly freely. In one study, for example, researchers analyzed viruses isolated from two infected patients. Over a 16-month period, they found 9 to 17 different varieties, all thought to have been formed from the original virus.

Making matters worse, HIV may be able to hide in the body. Research suggests that HIV may take up residence in bone marrow stem cells—that is, cells in bone marrow that give rise to lymphocytes. If this is true, the virus can then be transmitted to new white blood cells by cell division. Thus, once the virus is in the body, it may be there forever. Eliminating the virus from the body may be virtually impossible.

One of the ramifications of these findings is that if HIV can go into hiding, AIDS-infected blood donors may escape detection, even with the new genetic tests. AIDS-infected blood cells could unknowingly be passed to thousands of patients over the coming years.

Despite these discouraging findings, researchers remain determined to find both a cure for AIDS and a way to prevent it. At present, over two dozen vaccines are under study. Unfortunately, success has been elusive. In 2000, researchers studying a new vaccine for AIDS were dealt a major blow. The scientists had administered the vaccine to adult and juvenile rhesus monkeys. The monkeys were given a vaccine containing HIV that had been deactivated. The researchers had stripped the genes from the virus that were required for it to multiply inside infected cells. Much to their surprise, they found that virtually all of the monkeys died or showed symptoms of AIDS. Somehow, the virus managed to replicate. A second study by a different group of researchers showed identical results. Although the mechanism responsible for the virus's ability to regain virulence is unknown, hopes for developing a vaccine appear even dimmer.

On another front, researchers have developed a genetically engineered weapon that could kill cells infected with HIV, possibly eliminating the disease after it has developed. As noted earlier, cells infected with the AIDS virus produce a protein

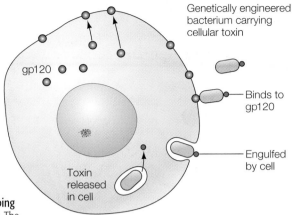

FIGURE 5 Duping the AIDS Virus The AIDS virus has a protein, gp120, in its capsid. When it infects cells, it produces more gp120 to make new capsids. However, some gp120 ends up in the infected cell's plasma membrane, thereby marking it. By genetically engineering a bacterium that can locate the infected cells through the gp120 marker, medical researchers may be able to hunt down and kill infected cells, stopping the spread of the virus.

known as gp120, which is part of the capsid (Figure 5). This protein also ends up in the plasma membrane of infected cells.

Researchers have genetically engineered a bacterium that binds to the gp120 protein. The bacterium carries with it a toxin that kills the HIV-infected cells. Preliminary studies indicate that noninfected cells are unharmed by this treatment. Although initial studies were disappointing, researchers hope that the technique can be improved or modified, making it possible to kill enough infected cells in AIDS patients to halt the disease. One question that must be answered before this procedure can be tried in people is whether AIDS-infected cells killed by this technique will degenerate and release active AIDS viruses that then spread to other body cells.

Researchers are also looking into the DNA of chimpanzees for clues for a cure. As mentioned at the beginning of this section, many wild chimps carry a virus called SIV that is very similar to HIV. Studies suggest that the virus has been in the chimp population for over 10,000 years. Interestingly, though, chimps are not affected by it. Because chimps contain 98% of the genes we have, studying the genetic differences of the two species could lead to a better understanding of why humans are so devastatingly affected by this virus. This, too, could help researchers develop a cure.

SUMMARY

1. AIDS is a disease of the immune system caused by HIV, a virus that attacks helper T cells (T4 cells), severely impairing a person's immune system.
2. AIDS progresses through three stages: acute, chronic, and full-blown AIDS.
3. During the first phase, no symptoms appear, although an individual is highly infectious—able to transmit the disease to others.
4. During the second phase, patients grow progressively weaker as their immune system falters. Lymph nodes swell and patients report persistent or recurrent fevers and a persistent cough. Mental deterioration may also occur.

5. During the last phase, patients suffer from severe weight loss and weakness. Many develop cancer and bacterial infections because of their diminished immune response.
6. AIDS is spread through body fluids during sexual contact and blood transfusions or through needles shared by drug users.
7. Stopping the virus has proved difficult, in large part because symptoms of AIDS do not appear until several months to several years after the initial HIV infection.
8. Numerous researchers are developing vaccines that they hope will protect people and eventually eradicate the virus, but so far efforts to

develop a vaccine have not proven successful. One reason is that HIV is highly mutable.
9. Although progress is being made in more developed nations, many poorer nations such as those in Africa, are witnessing an epidemic of AIDS. Lack of education and money to buy expensive treatments available in the West could result in the death of many millions of people in these nations. Even though AIDS is a serious problem in many poor nations, some have dramatically reduced the incidence of new infections through a variety of measures.

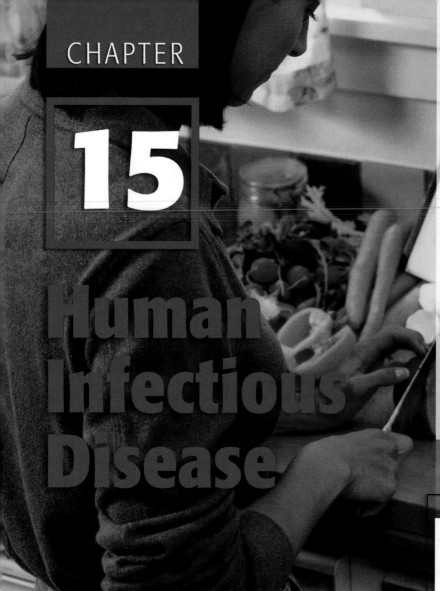

Special thanks to Jeffrey Pommerville, Professor of Biology and Microbiology at Glendale Community College, for providing the primary material in this chapter.

thinking critically

You work out regularly at a local gym and jog three times a week with a friend. One day you and your friend see a story on the local TV news about an outbreak in your city of a bacterial disease called *listeriosis*. Discovered in some packages of a brand of luncheon meat sold in your local supermarket, the bacterium, if ingested, may cause serious complications including stillbirth in pregnant women late in pregnancy. It may also cause meningitis (infection of the connective tissue layers surrounding the brain) in the elderly. After hearing the story, your friend announces that he has a package of this luncheon meat in his refrigerator and, in fact, ate some of it two days ago.

Since he is a healthy 20-year-old and is not sick, he dismisses the story. He says the package must be fine, because he hasn't gotten sick. He announces that he is going to have some of the meat in a sandwich for lunch and invites you to join him for lunch. What advice would you give your friend?

Microorganisms (commonly called *microbes*) are found in all environments on Earth. They are found in the air, on food and plants, on and in animals, in the soil and water, and on just about every surface. Microbes range in size from microscopic single-celled organisms to parasitic worms that grow to several feet in length. Although microorganisms are everywhere, few of them cause disease in humans. Those that do usually are dealt with by our immune systems, discussed in the previous chapter. Even though the vast majority of microbes are harmless, some of those that cause disease can be quite deadly. Disease-causing microorganisms make us ill because they start causing symptoms before our immune systems can respond. Making matters worse, some microorganisms undergo mutations, creating new forms that are able to breach our immune system's defenses, causing disease, even death. Disease may occur in normal healthy individuals, but those with weakened immune systems are particularly vulnerable.

In this chapter, we will learn about the agents that cause infectious disease in humans. You will also learn about the stages of human diseases, discover how microorganisms cause disease, and study how microbes are transmitted from one person to the next. In addition, you will learn about newly emerging infectious diseases and bioterrorism. Before we examine these topics, however, let's review what you learned about infectious agents—the viruses and bacteria—described in the last chapter and expand your understanding of these organisms.

15-1   The Infectious Agents of Human Disease

When most of us hear the words bacteria and virus, we usually think of infections and the diseases they cause such as the flu or the common cold. In reality, most of these microorganisms are harmless to humans. Only a few actually cause disease. Microbes that cause disease are called **pathogens**. Microbes are considered **infectious** if they invade and grow in the tissues of the body. They are considered to be **contagious** if they can be readily transmitted from one organism to another—for example, from one person to another.

FIGURE 15-1 **Beak Doctors** Some physicians during the time of plague believed that wearing a beak-like mask would protect them from the disease. Such physicians were called "Beak Doctors."

Disease-Causing Agents

Many infectious diseases remain dangerous today.

History is full of accounts of infectious diseases that periodically flared up, creating huge epidemics that killed millions of people. Infectious diseases, such as the plague, cholera, and smallpox, were once known as "slate-wipers" because they killed millions of people in Europe and other countries. Knowledge of impending epidemics often generated shock and terror in populations.

Until the 1850s, the fear of such diseases was complicated by ignorance because no one knew the cause of these diseases. To combat these diseases, people often dressed in strange costumes that were thought to ward off disease (Figure 15-1). Many strange potions and procedures, such as bloodletting, were used to cure people who had become ill. Unfortunately, few of these "cures" were effective, and some methods such as bloodletting often made a bad situation even worse.

Before the turn of the twentieth century, the work of scientists such as Louis Pasteur, Robert Koch, and others had shown quite clearly that certain microbes caused many of the devastating diseases of the time. In fact, the research of Pasteur and Koch culminated in the formulation of the **germ theory of disease** (as described in Scientific Discoveries That Changed the World 15-1). Thanks to their pioneering work, we now know the cause of most infectious diseases.

Once researchers and public officials understood the cause of infectious diseases, they set out to find ways to control or eliminate them. Over the years, better sanitation, improvements in water treatment, and the development of antibiotics and vaccines all helped control, even eliminate, infectious disease. Despite these advances, however, infectious disease remains a major threat to human health worldwide. Moreover, some diseases still bring anxiety and alarm to the public. Just think about the fear that AIDS, severe acute respiratory syndrome (SARS), and West Nile virus have caused in the United States and other countries. Figure 15-2 shows the human impact of some of the most prolific deadly diseases that exist worldwide today.

Viruses and Bacteria

Viruses and bacteria are among the most common infectious agents.

Many of the most common infections are caused by viruses and bacteria. As the term suggests, most microorganisms are tiny and cannot be seen without the aid of a microscope. To put their

Scientific Discoveries that Changed the World

15-1   The Germ Theory of Infectious Disease
Featuring the Work of Louis Pasteur and Robert Koch

Throughout history, infectious diseases such as plague, tuberculosis, typhoid fever, and diphtheria ravaged peoples and populations around the world. Neither royalty nor common folk were spared from the devastation. Fear was rampant during epidemics because no one knew how to cure people afflicted with the disease. Ways to eradicate such diseases were equally elusive.

Although cures for infectious diseases were not available in the nineteenth century, scientists were beginning to suspect infectious agents ("germs") as the cause. The link between these diseases and infectious agents, however, was an idea that many in the 1800s found hard to believe because it did not fit with the perceptions at the time. Most scientists believed that infectious diseases, such as the plague and malaria, were spread by an altered chemical quality of the atmosphere or a poisoning of the air (the word malaria comes from *mala aria*, meaning "bad air"). These qualities of the air might arise from decaying or diseased bodies.

In the mid 1850s, evidence for germs (microorganisms) as the cause of infectious disease gained stronger footing thanks to the work of the French scientist Louis Pasteur. Pasteur was a chemist by training. In 1854, at the age of 32, he was appointed Professor of Chemistry at the University of Lille in northern France. In 1857, he was asked to unravel the mystery of why local French wines were turning sour. Pasteur observed that although both good and soured wines contained tiny yeast cells, only the soured wines contained populations of barely visible rod-like organisms, known then and now as bacteria. His studies showed that yeast cells and bacteria are tiny, living factories in which important chemical changes take place. Pasteur's work also drew attention to microorganisms as agents of change because bacteria appeared to make the wine "sick." He surmised that, if microorganisms could sour wine, perhaps others could make people ill. In 1857, Pasteur published a short paper on wine souring by bacteria. In the paper, he implied that germs were related to human illness.

Other studies, discussed in **Scientific Discoveries 1-1**, furthered Pasteur's understanding of microbes. He came to view microorganisms as entities that were everywhere. Some of these ubiquitous organisms, he concluded, might be agents of human disease. So sure was he of this fact that he formulated the **germ theory of infectious disease.** The germ theory holds that microorganisms are responsible for infectious diseases. Between 1865 and 1868, Pasteur showed that a mysterious disease of silkworms that was sweeping through France was caused by a protozoan that infected both the silkworms and the mulberry leaves fed to them. Pasteur showed that the disease could be controlled by separating the healthy silkworms from the diseased silkworms and their food. In quelling the spread of the silkworm disease, he strengthened the germ theory of infectious disease.

Pasteur made yet another important contribution to science and medicine in the 1870s. During this period, huge numbers of sheep and cattle in his home country were dying of anthrax, a deadly blood disease. Pasteur's research team showed that the disease was passed from animal to

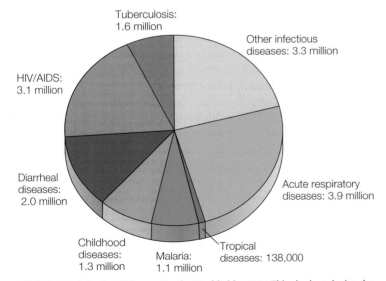

FIGURE 15-2 Infectious Disease Deaths Worldwide—2004 This pie chart depicts the leading causes of infectious disease and the number of worldwide deaths they caused in 2004, as reported by the World Health Organization. Tropical diseases include schistosomiasis and filariasis.

Tuberculosis: 1.6 million
Other infectious diseases: 3.3 million
HIV/AIDS: 3.1 million
Acute respiratory diseases: 3.9 million
Diarrheal diseases: 2.0 million
Childhood diseases: 1.3 million
Malaria: 1.1 million
Tropical diseases: 138,000

size into perspective, consider this analogy: If a virus were the size of a baseball, a typical bacterium would be the size of the pitcher's mound. On this scale, one of your body's cells that might be infected by the virus or bacterium would be the size of the entire ballpark.

Viruses are the simplest of all traditional infectious agents (Figure 15-3). As you may recall from Chapter 14, viruses contain genetic material—either DNA or RNA. The genetic information, containing genes, is surrounded by a protein coat, called the **capsid.** In some viruses, the capsid is covered by a membrane-like structure known as the **viral envelope** (see Figure 14-1).

As pointed out in Chapter 14, viruses are not cells, and they are not even considered living organisms because they lack the enzymes and cellular structures required for metabolism and growth. Because of this, viruses in the air or soil cannot replicate (make copies of themselves). To reproduce, they must invade living cells. Within these **host** cells, they find the chemicals and energy required to multiply.

All viruses go through similar life cycles. Consider, for example, the virus that causes chickenpox, one type of herpesvirus,

animal by infectious organisms, and he was confident that they could find a way of stopping the spread of anthrax by producing a vaccine.

Unfortunately, the work proved more difficult than he and his coworkers thought. Anthrax is highly infectious, and healthy animals have a strong chance of catching the disease simply by grazing over the burial site of anthrax victims. Still, he used the bacterium to produce an untested vaccine. Things came to a head in 1881 when a French veterinarian challenged Pasteur to test out his vaccine. Rather than appear unsure of his work, Pasteur accepted the challenge. His vaccine worked—his vaccinated sheep survived the injection of live anthrax spores. Again, Pasteur saved another French industry.

In 1885, Pasteur reached the zenith of his career when he successfully immunized a young boy against the dreaded disease rabies. Although he never saw the causative agent of rabies, Pasteur grew it in the brains of animals and injected the boy with bits of the brain tissue. Once again, success presented itself. Many monetary rewards followed, including a generous gift from the Russian government after Pasteur immunized 20 peasants against rabies. The funds helped establish the Pasteur Institute in Paris, one of the world's foremost scientific institutions. Pasteur presided over the Institute until his death in 1895.

Pasteur's work stimulated others to investigate the nature of microorganisms and their association with disease. One of the most significant investigators was Robert Koch, a country doctor from East Prussia (now part of Germany). Koch's work provided the crucial evidence for complete acceptance of the germ theory.

In 1876, Koch observed the anthrax-causing bacteria with a microscope and identified the bacterium as that found in animals that died from anthrax. Koch watched for hours as the rod-shaped bacteria multiplied, formed tangled threads, and finally reverted to highly resistant spores. He then took several spores on a sliver of wood and injected them into healthy mice. The symptoms of anthrax appeared within hours. Koch autopsied the animals, found their blood swarming with the same rod-shaped bacteria, and then reisolated the bacteria. The cycle was now complete. The rod-shaped bacteria definitely caused anthrax. Koch established a set of criteria (called Koch's postulates) for proving that a specific microorganism caused a specific disease.

Koch's work on anthrax verified the germ theory of infectious disease. In 1881, he outlined his methods at an international medical congress, and several days later, Koch received a personal letter of congratulations from Pasteur.

Koch also reached the height of his influence in the 1880s. In 1883, he interrupted his work on tuberculosis to lead groups studying cholera in Egypt and India. In both countries, Koch isolated the infectious bacterium by following his previous methods. In 1891, he became Director of Berlin's Institute for Infectious Diseases. At various times, he studied malaria, plague, and sleeping sickness, but his work with tuberculosis ultimately gained him the 1905 Nobel Prize in Physiology or Medicine. He died of a stroke in 1910 at the age of 66.

By the end of the nineteenth century, the pioneering discoveries of Pasteur and Koch created almost universal acceptance of the germ theory of infectious disease. With the passing of Pasteur and Koch, a new generation of international microbiologists stepped in to expand our understanding of microorganisms and infectious diseases.

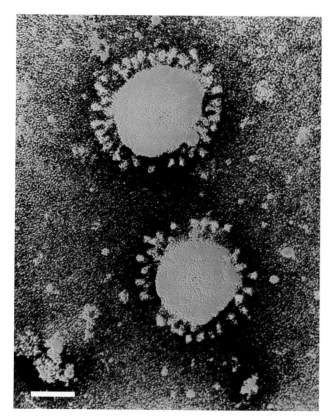

FIGURE 15-3 Coronaviruses False color transmission electron micrograph of two human coronaviruses. In humans, coronavirus typically causes colds. The spikes can be seen clearly extending from the viral envelope. Viruses similar to these are responsible for severe acute respiratory syndrome (SARS). (Bar = 60 nm.)

which is shown in Figure 15-4. To gain entry into a cell, the virus first attaches to protein receptors in the plasma membrane of target cells; in this case, it attaches to cells of the skin. The virus is then brought into the cell, often by phagocytosis, and its nucleic acid (DNA for the herpesvirus) is released into the cytoplasm. The virus' nucleic acid then takes control of the host cell's metabolic machinery, causing it to produce numerous copies of the virus' proteins. Once the components of new viruses are made, they are assembled into new viruses. In this process, an infected cell may produce hundreds of new viruses. After the viruses are assembled, they are released, usually when the cell bursts open. The host cell is typically destroyed in the process. Table 15-1 includes several examples of common diseases caused by a variety of different viruses.

TABLE 15-1	Human Diseases and Their Infectious Agent					
	Infectious Agent					
Disease	Virus	Bacterium	Fungus	Protozoan	Helminth	Other Agent
AIDS	✓					
Anthrax		✓				
Athlete's foot			✓			
Botulism		✓				
Chickenpox	✓					
Cholera		✓				
Common cold	✓					
Creutzfeldt-Jacob disease (variant)						✓
Cryptosporidiosis				✓		
Diarrhea	✓	✓		✓		
Diphtheria		✓				
Dengue fever	✓					
Ebola hemorrhagic fever	✓					
Filariasis					✓	
Genital herpes	✓					
Gonorrhea		✓				
Hantavirus pulmonary syndrome	✓					
Hepatitis	✓					
Histoplasmosis			✓			
Hookworm					✓	
Infectious mononucleosis	✓					
Influenza	✓					
Leprosy		✓				
Listeriosis		✓				
Lyme disease		✓				

TABLE 15-1	Human Diseases and Their Infectious Agent—*continued*					
			Infectious Agent			
Disease	Virus	Bacterium	Fungus	Protozoan	Helminth	Other Agent
Malaria				✓		
Measles	✓					
Monkeypox	✓					
Mumps	✓					
Oral thrush			✓			
Pinworm					✓	
Plague		✓				
Pneumonia	✓	✓	✓			
Rabies	✓					
Ringworm			✓			
Rubella	✓					
Salmonellosis		✓				
SARS	✓					
Schistosomiasis					✓	
Smallpox	✓					
Staph infections		✓				
Strep throat		✓				
Tetanus		✓				
Toxoplasmosis				✓		
Trichinosis					✓	
Tuberculosis		✓				
Typhoid fever		✓				
Valley fever			✓			
West Nile fever	✓					
Whooping cough (pertussis)		✓				

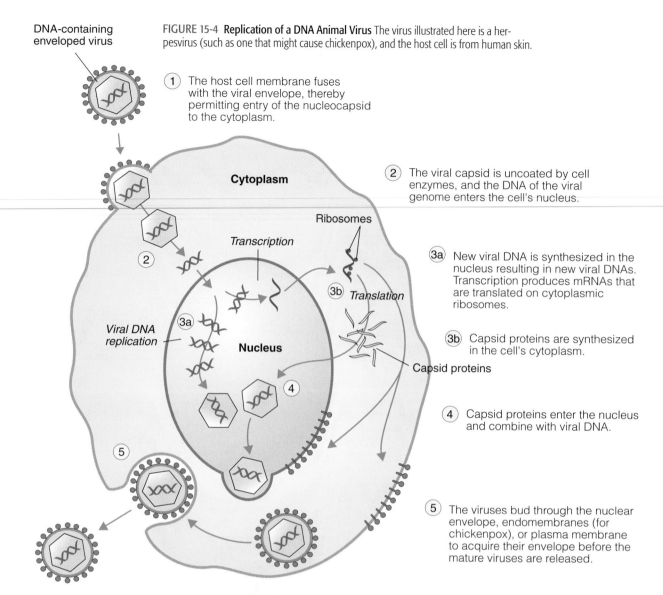

DNA-containing enveloped virus

FIGURE 15-4 Replication of a DNA Animal Virus The virus illustrated here is a herpesvirus (such as one that might cause chickenpox), and the host cell is from human skin.

(1) The host cell membrane fuses with the viral envelope, thereby permitting entry of the nucleocapsid to the cytoplasm.

Cytoplasm

(2) The viral capsid is uncoated by cell enzymes, and the DNA of the viral genome enters the cell's nucleus.

Ribosomes

Transcription

(3a) New viral DNA is synthesized in the nucleus resulting in new viral DNAs. Transcription produces mRNAs that are translated on cytoplasmic ribosomes.

(3b) Translation

Viral DNA replication

Nucleus

(3b) Capsid proteins are synthesized in the cell's cytoplasm.

Capsid proteins

(4) Capsid proteins enter the nucleus and combine with viral DNA.

(5) The viruses bud through the nuclear envelope, endomembranes (for chickenpox), or plasma membrane to acquire their envelope before the mature viruses are released.

Bacteria, first described in Chapter 14, are single-celled organisms that usually are visible with a light microscope. Even so, they are so small that it would take a thousand of them laid end to end to span the diameter of a pencil eraser. Bacteria may be shaped like short rods, spheres, or spirals, depending on the species (Figure 15-5). Because bacteria are prokaryotes, they lack most of the internal membranes found in human cells. Nonetheless, most bacteria can reproduce and grow on their own—that is, they can reproduce without invading and taking over the metabolic functions of host cells as viruses do. During reproduction, bacteria replicate their DNA, then split into two identical cells, each with one DNA copy (Figure 15-6a). This process often occurs very rapidly, producing large populations of living cells in a very short time (Figure 15-6b).

As noted earlier, very few bacteria are harmful to humans. In fact, less than 1% of the bacteria known to science cause disease. Some bacteria that live in our

bodies are beneficial. For example, a number of bacterial species live on our skin surface and protect us from potentially infectious pathogens. Likewise, many species of harmless bacteria live in our intestines and help us digest food, provide us with nutrients, and protect us against pathogens.

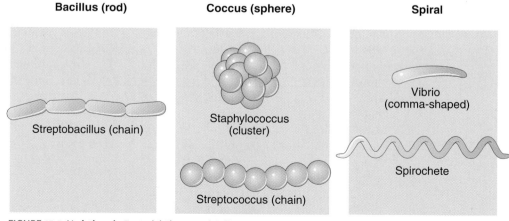

Bacillus (rod)

Streptobacillus (chain)

Coccus (sphere)

Staphylococcus (cluster)

Streptococcus (chain)

Spiral

Vibrio (comma-shaped)

Spirochete

FIGURE 15-5 Variations in Bacterial Shape and Cell Arrangements.

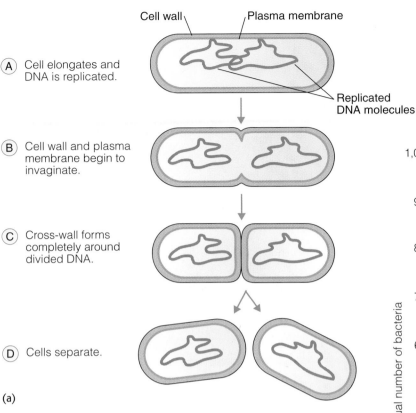

(A) Cell elongates and DNA is replicated.

Cell wall · Plasma membrane · Replicated DNA molecules

(B) Cell wall and plasma membrane begin to invaginate.

(C) Cross-wall forms completely around divided DNA.

(D) Cells separate.

(a)

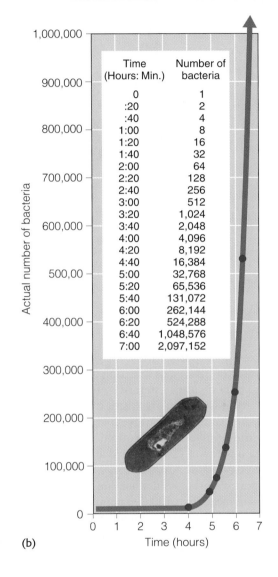

FIGURE 15-6 **Bacterial Cell Division** (a) As a result of DNA replication and cell division, two cells are formed, each genetically identical to the parent cell. (b) The number of bacteria progresses from 1 cell to 2 million cells in a mere 7 hours. The J-shaped growth curve gets steeper and steeper as the hours pass. Only a depletion of food, buildup of waste, or some other limitation will halt the progress of the curve. The photomicrograph is of a *Listeria* cell dividing.

Time (Hours: Min.)	Number of bacteria
0	1
:20	2
:40	4
1:00	8
1:20	16
1:40	32
2:00	64
2:20	128
2:40	256
3:00	512
3:20	1,024
3:40	2,048
4:00	4,096
4:20	8,192
4:40	16,384
5:00	32,768
5:20	65,536
5:40	131,072
6:00	262,144
6:20	524,288
6:40	1,048,576
7:00	2,097,152

(b)

Pathogenic bacteria, on the other hand, can infect the body, causing illness, even death. Illness results from some combination of their rapid growth, their production of poisonous chemicals known as toxins, and our immune system's response to the infection. *Listeria*, the bacterium that causes human listeriosis, mentioned in the critical thinking exercise at the beginning of the chapter, is such an example (Figure 15-6). **Table 15-1** lists several other common diseases caused by bacteria.

Eukaryotic Pathogens and Parasites

Fungi, protozoa, and helminthes also cause disease.

Not all infectious diseases are caused by bacteria and viruses. Several eukaryotic microbes such as fungi, certain protozoa, and helminthes can cause infections and some very serious diseases in humans (Table 15-1).

Fungi consist of the yeasts, molds, and a group you are probably most familiar with—mushrooms. Yeasts are single-celled organisms. Molds are multicellular and often are visible on spoiled foods, such as cheese or bread, or on overripe fruit. Mushrooms also are multicellular. Although they are not infectious organisms, some species of mushroom can produce powerful toxins that can be deadly if consumed.

Fungi live in the air, water, and soil, and on plants. Some can live in our bodies, usually without causing illness. Some species of fungi are beneficial. For example, the antibiotic penicillin is produced by a fungus. This drug kills many species of infectious bacteria. Several species of fungi also are important in making foods such as bread, cheese, and yogurt. Yeasts are used in the production of beer and wine.

A few fungi cause human illness and disease. For example, athlete's foot is caused by one species of fungi that grows in the skin. A yeast-like fungus known as *Candida* (Figure 15-7) can infect the mouth, causing oral thrush in infants, in people taking antibiotics, and in people with weakened immune systems. *Candida* is also responsible for most types of infection-induced diaper rash and is the organism responsible for "yeast infections" in the vaginas of adult women. Fungal pathogens also can produce respiratory diseases, such as valley fever in the American Southwest and histoplasmosis in the Ohio and Mississippi River valleys.

Protozoa also are single-celled eukaryotic organisms. Many protozoans inhabit the intestinal tracts of humans. Although most protozoans are harmless, some protozoans are pathogenic.

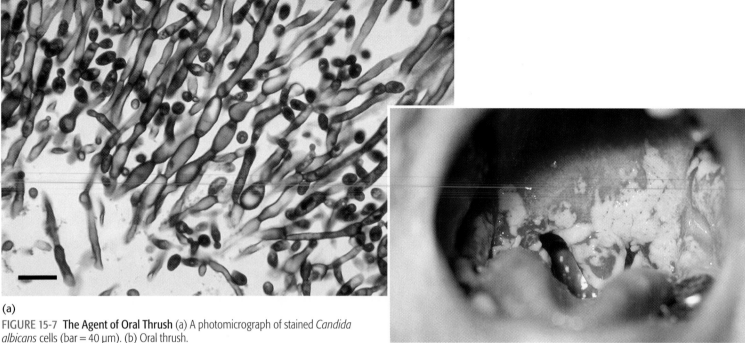

(a)

(b)

FIGURE 15-7 **The Agent of Oral Thrush** (a) A photomicrograph of stained *Candida albicans* cells (bar = 40 µm). (b) Oral thrush.

Pathogenic protozoans spend part of their life outside of humans, living in soil, water, or insects, or infecting other animals. If they infect the human body, they often live off body fluids, surviving as parasites. A **parasite** is an organisms that live within the bodies of others, often causing some level of damage.

Protozoa invade the human body through contaminated food or water. One such parasite, *Cryptosporidium parvum*, invaded the Milwaukee city water supply in 1993 and was responsible for the largest waterborne disease outbreak in U.S. history, causing illness in more than 400,000 people. Other protozoa are transmitted through sexual contact. Still others are carried by **vectors**, other organisms such as mosquitos and ticks that transmit the protozoa from one person to the next. Malaria is an example. Malaria is perhaps the most prevalent and deadly of all protozoan diseases in the world, killing more than 1 million people (mostly children in tropical countries) each year. The malaria-causing protozoan, called *Plasmodium*, is transmitted from person to person by mosquitoes (Figure 15-8). Mosquitoes suck blood from one individual infected with the protozoan. When that same mosquito attacks another person, it may transfer some of the protozoa to the uninfected person.

Helminths are parasitic worms. (The term *helminth* comes from the Greek for "worm.") Most helminths are visible to the naked eye. Nonetheless, they are designated as "microorganisms" because they can cause infectious disease.

The most common helminths are flatworms and roundworms. When adult helminths or their microscopic eggs enter the body, they live parasitically off the body's nutrients within the intestinal tract, lungs, liver, skin, or brain, depending on the species. **Flatworms** range in length from 6 inches to more than 25 feet! One of the most prevalent flatworms, *Schistosoma*, is responsible for a disease known as *schistosomiasis*, which affects more than 250 million people worldwide. Victims suffer from fever, muscle pain, diarrhea, coughing, vomiting, and a burning sensation during urination. **Tapeworms** are one type of flatworms made up of hundreds of segments, each of which is capable of breaking off and developing into a new tapeworm (Figure 15-9a).

Another type of helminth is the **roundworm.** Roundworms damage their hosts by forming large masses of worms in blood vessels, lymphatic vessels, or the intestines. The roundworm responsible for the disease filariasis, for example, blocks lymphatic vessels after many years of infection. The lymphatic vessels swell and become distorted with fluid. The condition is called *elephantiasis* (Figure 15-9b).

The most common roundworm disease in the U.S. is pinworm disease. Surprisingly, an estimated 30 percent of children and 16 percent of adults serve as hosts. Among the other roundworm diseases are trichinosis in pork and hookworm disease, which infects hundreds of thousands worldwide, causing a dry cough, mild fever due to the presence of worms in the lung, and abdominal pain due to the presence of worms in the intestine.

Besides these "traditional" microbial pathogens, other infectious agents exist. The most newsworthy nontraditional infectious agent is the prion. Prions are responsible for several diseases, including mad cow disease, and are described in Health Note 15-1.

FIGURE 15-8 **The Malaria Cycle**
Malaria continues to be among the most widespread infectious diseases in the world.

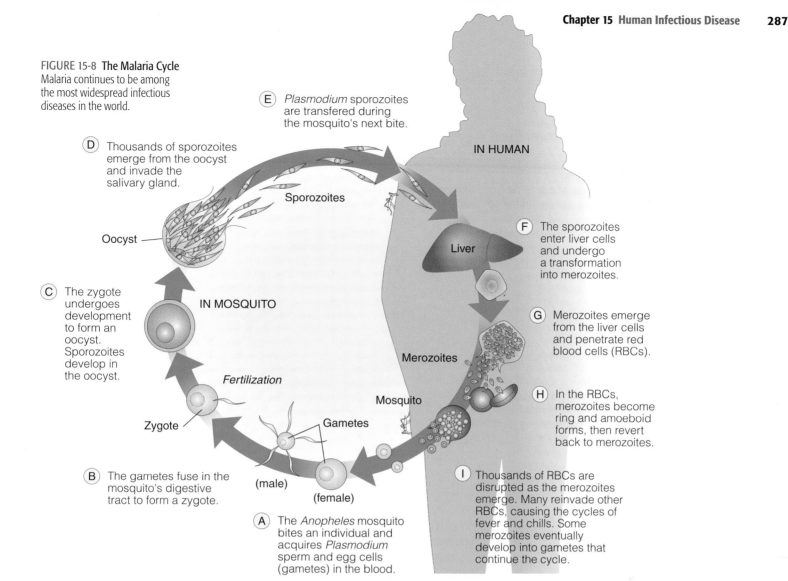

(E) *Plasmodium* sporozoites are transfered during the mosquito's next bite.

IN HUMAN

(D) Thousands of sporozoites emerge from the oocyst and invade the salivary gland.

Sporozoites

Oocyst

(C) The zygote undergoes development to form an oocyst. Sporozoites develop in the oocyst.

IN MOSQUITO

Fertilization

Zygote

Gametes

(B) The gametes fuse in the mosquito's digestive tract to form a zygote.

(male)

(female)

(A) The *Anopheles* mosquito bites an individual and acquires *Plasmodium* sperm and egg cells (gametes) in the blood.

Liver

(F) The sporozoites enter liver cells and undergo a transformation into merozoites.

(G) Merozoites emerge from the liver cells and penetrate red blood cells (RBCs).

Merozoites

Mosquito

(H) In the RBCs, merozoites become ring and amoeboid forms, then revert back to merozoites.

(I) Thousands of RBCs are disrupted as the merozoites emerge. Many reinvade other RBCs, causing the cycles of fever and chills. Some merozoites eventually develop into gametes that continue the cycle.

FIGURE 15-9 **Helminths** (a) An unmagnified view of the tapeworm *Taenia saginata.* This flatworm is long and ribbon-like, and consists of hundreds of visible segments. *T. saginata* infects the human intestinal tract. (b) Elephantiasis of the leg caused by the roundworm *Wuchereria bancrofti.* The parasite breeds in the tissues of the lymphatic vessels and damages them. As fluid accumulates, the legs swell and become distorted.

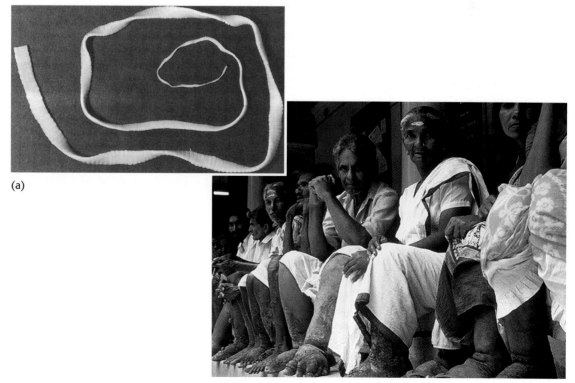

(a)

(b)

15-1   Prions: Virus-Like Infectious Agents

When viruses were discovered, scientists believed they had found the smallest infectious particles. In recent years, however, that perception has changed. Scientists have discovered an apparently new class of disease agents—virus-like agents called *prions* (PREE-on). What are prions, how were they discovered, and what diseases do they cause?

To understand prions, we begin with a story of a mysterious illness that cropped up in 1986 in Great Britain. During that year, farmers found that some of their cattle were losing weight and losing control over their muscles. Many became aggressive. All of them eventually died. This disease was dubbed *mad cow disease*. Scientists thought that the disease was caused by a virus.

In the early 1990s, several young people died of a human brain disorder that resembled mad cow disease. Symptoms included dementia, weakened muscles, and loss of balance. After many studies, health officials hypothesized that the human disease was caused by eating beef that had been processed from cattle that had been afflicted with mad cow disease. They postulated that the disease agent was transmitted from cattle to humans.

Since then, the disease has produced an economic disaster to the beef export industry in Britain and caused a political crisis throughout the European Union and beyond. In January 2001, over 1,000 cattle in Texas were suspected of being the first North American cattle to suffer from mad cow disease. Luckily, tests conducted by the U.S. Food and Drug Administration (FDA) indicated that there was no mad cow disease present and no cause for alarm. In May 2003, however, the first mad cow was discovered in Alberta, Canada. In December, 2003, one case of mad cow disease was reported in the State of Washington.

Mad cow disease, or bovine spongiform encephalopathy (BSE), is a fatal brain disorder (encephalopathy). Infected animals develop sponge-like holes in their brains (spongiform) where brain cells have been destroyed in large numbers (Figure 1). Similar neurological degenerative diseases have also been discovered in other animals and humans. These include scrapie in sheep and goats, wasting disease in elk and deer, and Creutzfeldt-Jakob disease in humans. All are examples of a group of rare diseases called *transmissible spongiform encephalopathies* (TSEs) because, like BSE, they can be transmitted to other animals of the same species and possibly to other animal species, including humans.

At first, many scientists believed the disease agent responsible for these diseases was a new type of virus. However, to date, no virus or nucleic acid characteristic of a virus has been discovered in any of the ani-

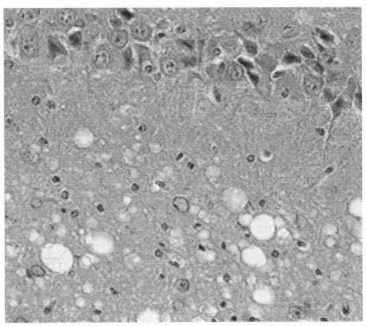

FIGURE 1 A photomicrograph showing the vacuolar degeneration of gray matter characteristic of human and animal prion diseases.

15-2   The Course of a Human Disease

Up to this point, we have been using the terms *infection* and *disease*. What exactly is meant by them?

To a microbiologist and a doctor, the term **infection** is used to refer to a state of being infected by a microorganism or the process of being infected. A **disease** is a change from the healthy state of the body. An infectious disease results from an infection that has not been initially repelled by the immune system or halted by medical intervention, for example, by antibacterial or antiviral drugs. An infectious disease occurs when body cells are damaged or destroyed because of an infectious agent or by a product they produce.

Pathogens and Disease

Pathogens differ in their ability to cause disease.

The ability of a pathogenic organism to cause disease depends on its pathogenicity. **Pathogenicity** refers to the ability of a microorganism or virus to gain entry into the host's tissues and then bring about a physiological or anatomical change that results in poor health. Pathogens vary greatly in their pathogenicity. For example, the bacteria responsible for cholera, plague, and typhoid fever cause serious human diseases in nearly

mals. In the early 1980s, Stanley Prusiner and colleagues made a very interesting discovery while studying scrapie. Scrapie is the TSE that causes infected sheep to scrape their bodies raw on fences or other surfaces. Prusiner and colleagues isolated an unusual protein from scrapie-infected tissue that they thought represented the infectious agent. Prusiner called the infectious particle a **prion**. Based on more research and study, Prusiner and colleagues proposed the protein-only hypothesis. This hypothesis says that prions are composed solely of protein and contain no nucleic acid.

Since the gene (called PrP) that codes for the prion protein is primarily expressed in the brain, it was proposed that there are two types of prion proteins: (1) normal prion proteins that are found in all uninfected brain cells where they have a specific but still unidentified role and (2) abnormal prions, which are similar to the normal prions in many ways except in their shape.

Abnormal prions are found in infected cells and are the suspected infectious agents of diseases like mad cow disease. With the wrong shape, the abnormal proteins cannot carry out their specific role and eventually brain cell function falters; cell death soon follows. Based on his work with prions, Prusiner won the Nobel Prize in Physiology or Medicine in 1997. Nevertheless, some scientists believe a virus is somehow involved because the prion hypothesis "challenges" the basic belief of DNA and inheritance.

If the protein-only hypothesis is correct, how can an infectious agent without a shred of nucleic acid spread and cause disease? Many researchers believe that prion diseases are spread by the infectious abnormal prion binding to normal prion proteins, causing the latter to change shape. In a domino-like scenario, they hypothesize that the newly converted prion proteins cause other normal prion proteins to become abnormal. According to the hypothesis, all this occurs without any need for nucleic acid to replicate.

The 1990s outbreak of mad cow disease supports this idea. Scientists originally thought BSE made its way into cattle through feed that contained bone meal made from the carcasses of scrapie-infected sheep. However, in October 2000, the British BSE inquiry concluded that BSE is a new disease, possibly arising from a mutation in one cow's genes. The inquiry concluded that the remains of this one cow were used in cattle feed, thus starting the infection of Britain's cattle herds.

Many people then ate beef or beef byproducts that were unknowingly contaminated either directly from nervous tissue or from instruments that had contacted nervous tissue that contained the abnormal prions. These abnormal proteins eventually crossed into the nervous system. Once the infectious agent enters the brain, it can lie dormant for several years (even as long as 10 to 15 years). Once activated, the disease usually runs its course in less than one year.

Scientists do not understand how abnormal prion proteins produce the clinical symptoms of the disease. It is known that infected nerve cells attempt to get rid of abnormal prions by clumping them together into plaques for cellular digestion. Because the nerve cells cannot digest the abnormal prions properly, however, they accumulate in the cells. When the cells die, the abnormal prions are released and then they infect other nerve cells. Over time, large, sponge-like holes develop in the brain where groups of nerve cells have died. Death of the animal occurs from the numerous nerve cell deaths that led to loss of brain function.

To protect the American beef industry, the federal government has banned the import of live cattle and animal parts from BSE-afflicted countries. However, with the December 2003 discovery of a cow with BSE in Washington state, the U.S. Department of Agriculture has implemented new protective measures. These include condemning animals having signs of neurological illness and holding from slaughter any cows suspected of having BSE until test results are known. In addition, "downer cattle"—those unable to walk on their own—cannot be used for human food or animal feed.

www.jbpub.com/humanbiology/5e

Visit Human Biology's Internet site for links to web sites offering more information on this topic.

all individuals infected by them. Other pathogens, however, only cause disease when a host's defenses are suppressed. These organisms are said to be **opportunistic**. They invade the body and cause disease only when the right circumstances or opportunity is present. Many of the pathogens that infect AIDS patients, as well as the bacterium *Listeria* that is responsible for listeriosis, are examples of opportunistic infections.

The degree of pathogenicity expressed by a pathogen is called **virulence**. The bacterium responsible for typhoid fever is considered highly virulent because it causes the disease in virtually everyone who comes in contact with it. *Listeria* is considered moderately virulent because it causes disease in some individuals. Finally, avirulent organisms are those that rarely, if ever, cause disease. This group includes all of the nonpathogenic microorganisms.

The Signs and Symptoms of Human Disease

A pathogen produces a set of signs and symptoms.

Pathogens usually follow a disease cycle from their invasion until the expression of symptoms. First, the pathogen must invade the body at a specific entry point. The entry point depends on the microbe (Table 15-2). For example, the bacterium responsible for tetanus, *Clostridium tetani*, is found in soil. The bacterial spores usually enter the body via puncture wounds, for example, when one steps on a rusty nail. (The spore cling to the rough edges of the nail—the rust itself is of no consequence.) The injected bacterial spores germinate in the wound where there is little, if any, oxygen. The bacterium then spreads to the nerves that control voluntary muscles and causes painful muscle stiffness.

TABLE 15-2	Common Routes of Transmission and Some Disease Examples	
Entry Site	**Transmission Route**	**Disease Example**
Respiratory	Aerosol droplets	Influenza, tuberculosis, common cold, strep throat
	Mouth to hand/object to nose	
Salivary	Kissing	Infectious mononucleosis
	Animal bite	Rabies
Gastrointestinal	Stool to hand to mouth	Hepatitis A
	Stool to water/food to mouth	Salmonellosis
Skin	Skin discharge to air to respiratory tract	Chickenpox
	Skin puncture	Tetanus
Blood	Transfusion or needle prick	Hepatitis B, AIDS
	Insect bite	Malaria
Urogenital	Urethral or cervical secretions	Gonorrhea, herpes
	Urine to hand to catheter	Urinary tract infections
From animal	Animal bite	Rabies
	Arthropod (flea, tick)	Plague, Lyme disease

Interestingly, though, tetanus will not develop if the bacterial spores are consumed with food because the spores cannot germinate in the human intestinal tract. This is why one can eat a freshly picked radish without fear of developing tetanus. Other pathogens enter the body as the result of a cut or abrasion of the skin, while many viruses, such as those that cause the flu, enter via the respiratory tract by the mouth, nose, or eyes.

Once a pathogen invades a healthy body, the immune system springs into action. As you learned in Chapter 14, the immune system attacks infectious agents through antibodies or directly via T cells. The body also combats infectious agents via nonspecific mechanisms. For example, coughing and sneezing help to eliminate many viruses from the respiratory system, while fever raises the body temperature and slows viral replication.

Every disease that infects humans changes the body from the healthy state to a diseased state by altering body structures and functions in specific ways. Such changes are referred to as signs and symptoms. **Signs** are changes in body function that a physician can detect and measure. A mild fever and swelling of lymph nodes are examples. Most infectious diseases, however, are also accompanied by one or more **symptoms**. These are changes in the body that a patient can experience but an observer (i.e., physician) cannot—for example, a tired feeling or a headache.

Some diseases produce a specific group of signs and symptoms and are therefore referred to as a **syndrome**. AIDS is an example. AIDS, as you learned in the supplement following Chapter 14, is a viral disease that produces a characteristic set of signs, such as diarrhea, loss of certain white blood cells, weight loss, and pneumonia, as well as symptoms such as discomfort and fatigue.

The Course of a Disease

A disease typically follows a series of five stages.

If the immune system's first and second lines of defense fail to eliminate a pathogen, the body typically goes through a series of disease stages (Figure 15-10). Consider the flu.

Each fall and winter season, an estimated 10 to 20 percent of the American population gets the flu (influenza). This disease begins with an **incubation period**, a period between a person's exposure to the flu virus and the appearance of the first symptoms. The incubation period for the flu virus is short, usually 2 to 3 days. Other diseases such as the measles, have longer incubation periods of one to two weeks. Still others, such as leprosy, have prolonged incubation periods of three to six years. The incubation period is determined by such factors as the number of organisms that enter the body, their rate of reproduction or replication, and the level of host resistance.

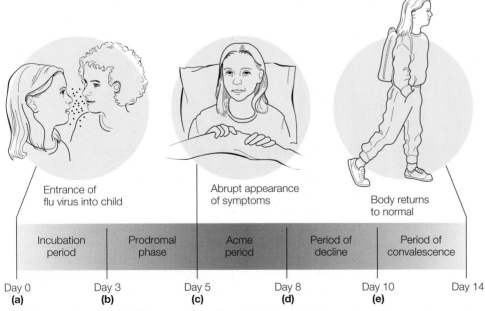

Incubation period	Prodromal phase	Acme period	Period of decline	Period of convalescence

Entrance of flu virus into child

Abrupt appearance of symptoms

Body returns to normal

Day 0   Day 3   Day 5   Day 8   Day 10   Day 14
(a)   (b)   (c)   (d)   (e)

FIGURE 15-10 **The Course of Disease, as Typified by the Flu** (a) A child is exposed to the flu viruses in respiratory droplets, and the incubation period begins. (b) At the end of this period, the child experiences fever, headache and tiredness as the prodromal phase ensues. (c) The acme period begins with the appearance of specific flu symptoms, such as cough, muscle pain, and shaking chills. (d) Sweating and normal skin color return as the period of decline takes place. (e) With the period of convalescence, the body returns to normal.

The second phase of a disease is a time of mild signs or symptoms and is called the **prodromal phase**. For many diseases, this period is characterized by fairly general symptoms such as fever, headache, and tiredness. They indicate that a competition between pathogen and host has begun.

During this phase, diseases often are characterized as subclinical or clinical. A **subclinical disease** is accompanied by few obvious symptoms. Many people, for example, have experienced subclinical cases of mumps or infectious mononucleosis. A **clinical disease** is one in which the symptoms are apparent (such as occurs with a common cold). Symptoms may be anywhere from mild to severe, depending on the pathogen.

The **acme period** or **climax** is the third stage in disease progression. This is the critical stage of the disease. Very specific symptoms appear during this phase. During a bad case of the flu, for instance, symptoms include a severe cough and muscle pain in the chest, back, and legs. Many patients experience high fever and shaking chills (chills are a result of the difference in temperature between the superficial and deep areas of the body). Individuals suffering from the flu are most contagious during this period.

Complications can arise during this phase, depending on the specific form of the disease and the state of the host. While many people contract the flu each year in the United States, approximately 115,000 of them require hospitalization because of complications—for example, because of a fever lasting more than five days (which is usually a sign of bacterial or viral pneumonia that has developed because the individual is in a weakened state). In many elderly individuals or people who are in poor health, complications such as this can lead to death, if treatment is not sought early on. In fact, each year, influenza kills more than 35,000 Americans. Most people who are in good health at the time of a flu infection survive.

As the symptoms of a disease subside, victims enter a **period of decline**. Sweating is common during this phase of the disease, as the body releases excessive amounts of heat. The recovery period varies in length. For individuals ill with the flu, it is often quite rapid. In other infectious diseases, however, recovery may take a long time.

The disease sequence concludes after the body passes through a **period of convalescence**, during which time the body's systems return to normal. Interestingly, during the period of decline—and even convalescence—individuals can transmit the disease to others. During this period, pathogens exit the body. Bacteria responsible for gastrointestinal illness may be pass through the feces, while flu viruses can be released by coughing and sneezing.

15-3 How Pathogens Cause Disease

Disease involves a complex series of interactions between a pathogen and a host. In this section, we will examine some of the factors that determine whether disease can occur, focusing most of our attention on the properties of microorganisms that allow them to overcome a host's immune defense.

Infection is dependent on the pathogen's ability to adhere to cells in specific tissues. The bacterium that causes the sexually transmitted disease known as *gonorrhea*, for example, only adheres to the lining of the urogenital tract. Many human viruses, including the flu virus, have protein structures called *spikes* that protrude from their capsids or envelopes. The spikes attach to specific host cells. Once a pathogen binds to a cell membrane, it is usually engulfed by the cell by phagocytosis (Figure 15-11).

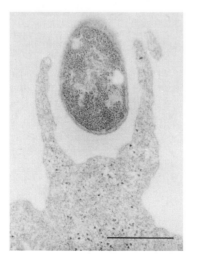

FIGURE 15-11 Tissue Invasion
A bacterium is being engulfed by an epithelial cell (bars = 1 μm.).

Enzymes and Toxins

Pathogenicity and virulence depend on key metabolic characteristics of a pathogen.

The ability of many pathogens to penetrate tissues and cause damage depends on certain factors, such as enzymes or toxins released by the microorganism. Certain enzymes, for instance, help pathogens resist body defenses and thus increase the virulence of a microbe. These enzymes may interfere with certain cellular functions or alter barriers that are meant to thwart invasion. Microbial enzymes, such as coagulase, for example, produce blood clots that encompass clusters of pathogens, protecting them from immune system attack. Other bacterial enzymes target and destroy cells of the immune system. The bacterium *Staphylococcus*, which is responsible for a variety of illnesses including pneumonia, food poisoning, and skin infections, produces enzymes that allow the bacteria to penetrate deep into body tissues. These enzymes digest the "cellular cement" that holds cells of tissues together.

Several human diseases involve the formation of **biofilms** (impenetrable colonies) in tissues that are very difficult for the immune system or even medical intervention to eliminate. Health Note 15-2 describes biofilms.

Some bacteria cause disease by producing toxins. **Toxins** are microbial poisons and fall into two categories: endotoxins and exotoxins.

Exotoxins are protein molecules, often enzymes, that are manufactured by bacteria in body tissues. When released from the bacteria, the toxins may be transported throughout the body.

15-2   Infectious Disease and Biofilms

In your reading of this chapter, you may have the impression that pathogens act as independent agents to cause disease. In a few cases, that might be true. However, for the most part, pathogens do not act as individuals; rather, they survive in complex communities called a biofilm. A **biofilm** consists of an immobilized population of bacteria (or other microorganisms) caught in a sticky web of tangled polysaccharide fibers that adheres to various surfaces.

Biofilms develop on virtually all surfaces in contact with a watery environment. This includes the surfaces of aquatic plants or animals, water pipes, and stones. Contact with a fluid environment ensures a plentiful supply of nutrients, and as the bacteria grow, they secrete sticky polysaccharides.

Researchers have found that biofilms are highly organized structures. They contain water channels that serve to deliver nutrients to the bacteria and remove wastes. The bacteria share these passageways, interact metabolically, and benefit from each other's metabolic byproducts. They also form communities that store nutrients and resist predators such as protozoa and viruses. Some scientists estimate that in nature, 99 percent of all microbial activities occur in biofilms.

Bacteria in biofilms act very differently than individual cells and are extremely difficult to treat. In biofilm, for example, the bacteria often are impervious to drugs such as antibiotics, disinfectants, and antiseptics, which are designed to attack individual cells (Figure 1). Pathogens in biofilms may resist the immune system's attempts to eliminate them and therefore can be extremely virulent. For example, white blood cells have difficulty reaching the pathogens in this slimy conglomeration of armor-like material.

Health officials at the Centers for Disease Control and Prevention (CDC) estimate that more than 65 percent of human infections involve biofilms. If you or someone you know has had a middle ear infection, the cause was a bacterial biofilm. Dental plaque is a type of biofilm as well. Brushing and flossing teeth and seeing a dentist regularly are essential to reduce and control this biofilm. Eye doctors are also concerned about the formation of biofilms on contact lenses, which can produce serious eye infections. Proper cleaning and storage of contact lenses is important to prevent biofilm formation.

Biofilms can also form on catheters and medical devices such as artificial hearts. Biofilms in urinary catheters often provide starting points for urinary tract infections, as bacteria creep up the catheters. Once in body tissue, the bacteria sequestered in the slimy conglomerates are shielded from attack by the body's immune system, and they are difficult to kill with antibiotics. In males, biofilms also have been implicated in prostate gland infections, which are accompanied by chronic pain and sexual dysfunc-

---

Usually, only minute amounts of an exotoxin are needed to cause disease. The exotoxin that causes botulism is among the most lethal toxins known. One pint of the pure toxin would be sufficient to destroy the entire human population of around 6.5 billion people.

Exotoxins destroy cellular structures or inhibit essential metabolic functions. Thus, the disease symptoms vary depending on the exotoxin involved. Table 15-3 highlights several examples.

Endotoxins are a lipid portion of the cell wall of many bacteria. They usually are released only upon disintegration (death) of the bacterial cell. When released, endotoxins cause chills,

fever, weakness and aches, and general malaise. Endotoxins may also damage the circulatory system, causing a massive increase in the permeability of the blood vessels. This, in turn, causes blood to leak into the intercellular spaces, where it is useless. Tissues swell, the blood pressure drops, and the patient may lapse into a coma.

Like exotoxins, endotoxins add to the virulence of pathogens and enhance their ability to cause disease. However, they are required in larger doses than are exotoxins. Typhoid fever and some urinary tract infections are the result of endotoxin production.

TABLE 15-3	Diseases Caused By Exotoxins	
Disease	Bacterium	Signs or Symptoms; Mechanism of Action
Botulism	*Clostridium botulinum*	Muscle paralysis; Inhibits the release of acetylcholine at the synaptic junction
Whooping cough (pertussis)	*Bordetella pertussis*	Spasmodic and recurrent coughing, ending with a loud inspiratory whoop with chocking on mucus; Paralyzes ciliated cells and impairs mucus movement
Cholera	*Vibrio cholerae*	Severe diarrhea; Causes massive loss of water from the intestines, which can produce a rapid drop in blood pressure
Diphtheria	*Corynebacterium diphtheriae*	Thick membrane coating the upper respiratory mucous membranes. Fever, sore throat, cough; interferes with protein synthesis in the cytoplasm of epithelial cells of the upper respiratory tract; respiratory blockage
Anthrax	*Bacillus anthracis*	Inflammation, hemorrhage, shock (inhalational anthrax); three exotoxins generate an accumulation of fluid and killing of host cells.

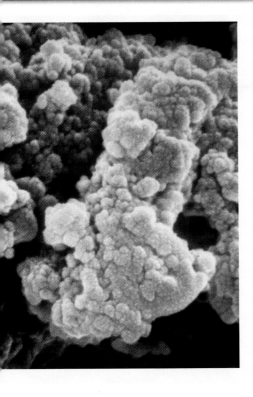

tion. Finding a way to penetrate this polysaccharide barrier is important in order to stem the tide of urinary tract infections.

Besides being involved in human disease, biofilms can cause problems in industry. Biofilms, for example, can corrode water pipes. When they contaminate computer chips, biofilms act as conductors and thereby interfere with electronic signals. Indeed, one researcher has called biofilms the "venereal disease of industry."

Biofilms can also be used to benefit humans. For example, they can be used in a process called *bioremediation*, the degradation of toxic wastes and other dangerous products of industry. Biofilms can be used to degrade organic compounds, thus retarding pollutant buildup.

FIGURE 1 Biofilms Biofilms are communities of microorganisms. Some biofilms can cause infectious diseases, such as these *Pseudomonas* bacteria that have formed a biofilm on the lungs of a cystic fibrosis patient.

www.jbpub.com/humanbiology/5e

Visit Human Biology's Internet site for links to web sites offering more information on this topic.

15-4 How Infectious Diseases Are Transmitted

For a disease to spread, pathogens must be transmitted to other hosts. In some cases, an infectious disease spreads only within a given region. If this occurs, and the level of infection is relatively low, microbiologists and health officials consider it an **endemic disease**. Plague in the American Southwest is an example of an endemic disease. If, on the other hand, the disease breaks out in explosive proportions within a population, it is considered an **epidemic**. Influenza often causes epidemics. A more contained occurrence is considered an **outbreak**. An abnormally high number of measles cases in one U.S. city would be classified as an outbreak, whereas it would be classified as an epidemic if it occurred in several states. A **pandemic** disease (or pandemic) occurs worldwide. The most recent example of a pandemic would be AIDS, although other pandemics have occurred (Table 15-4).

So, how are the microbial agents that can cause outbreaks, epidemics, and pandemics spread? Although diseases can be transmitted in many ways, all modes of transmission fall into two broad categories: direct and indirect transmission methods.

Direct Transmission Methods

Direct transmission may occur through direct physical contact.

Direct transmission occurs in a variety of ways. The most common involves person-to-person contact during which time bacteria, viruses, or other infectious agents are transferred from people who have the disease to uninfected individuals. Direct physical contact usually occurs during hand-shaking, kissing, or exchanging body fluids—for example, from sexual intercourse or blood transfusions (Figure 15-12). Infectious mononucleosis, a disease common among college students, typically is transmitted by kissing. Gonorrhea and genital herpes are transmitted directly during sexual contact. Hepatitis B is transmitted by blood transfusions.

TABLE 15-4	Some of the Major Historical Pandemics and Resulting Human Deaths	
Pandemic	Estimated Deaths
Bubonic plague/"Black death" (6th, 14th, 17th centuries)	137 million
Smallpox (1900–1977)	300–500 million
Influenza "Spanish flu" (1918–1919) "Asian flu (1957–1958) "Hong Kong flu" (1968–1969)	20–50 million 1–4 million 1–4 million
AIDS (1981–2004)	23 million

Infectious diseases can also spread from a mother to her unborn child through the placenta or via the vagina during childbirth. Disease can also be transmitted after birth through breast milk. The AIDS virus is an example of an infectious agent that can be passed from mother to her fetus.

Diseases can also spread from animals to people. A pet cat or dog, for example, can carry disease-causing microbes such as rabies. Rabies can be transmitted to a pet by a bite from another animal (e.g., raccoons, bats, cattle, rabbits, skunks, and foxes) that has rabies. A bite from the infected cat or dog can then spread the viral disease to a human.

Toxoplasmosis, which is sometimes called *litter box disease*, results from contact with a protozoan parasite in cat feces. Although the infected animal lacks signs and symptoms, initial symptoms in a human after contact are similar to those of the flu (swollen lymph glands, fatigue, fever, and headache). To be safe, women who are pregnant and have a cat should allow someone else to clean the litter box because the disease can cause miscarriage, premature births, and mental retardation in newborns.

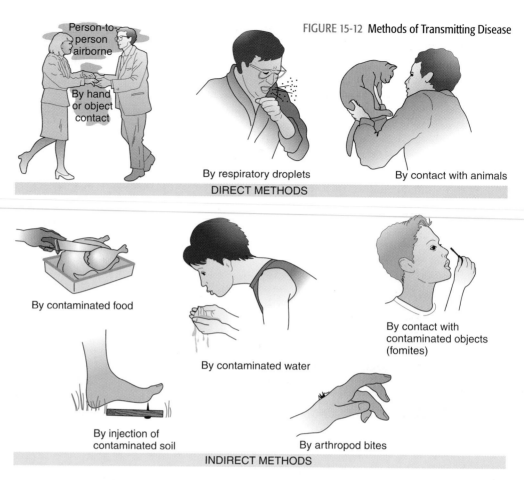

FIGURE 15-12 **Methods of Transmitting Disease**

By respiratory droplets

By contact with animals

DIRECT METHODS

By hand or object contact

Person-to-person airborne

By contaminated food

By contaminated water

By contact with contaminated objects (fomites)

By injection of contaminated soil

By arthropod bites

INDIRECT METHODS

Infectious diseases also are transmitted by less common pets such as reptiles and birds. All reptiles, especially turtles and iguanas, can carry the bacterium *Salmonella*. If transmitted to humans, the bacterium can cause gastroenteritis, an inflammation of the stomach and intestine, which results in diarrhea and vomiting.

Infectious disease also can be spread directly by respiratory droplets and airborne particles. For example, when you cough or sneeze, you expel droplets into the air around you. If you have a cold, the flu, or another contagious respiratory illness, these droplets contain the infectious agent. Respiratory droplets travel only about 3 feet because they are too large to stay suspended in the air for a long period. However, if a droplet contacts the eyes, nose, or mouth of someone else, that individual may become infected and may experience signs and symptoms of the disease. Crowded places, especially indoor environments—including classrooms, airports, and airplanes—can increase the chances of respiratory droplet contact, which may explain the increase in such respiratory infections in the winter months when more people congregate in enclosed places.

Indirect Transmission

Diseases also can be transmitted by indirect contact.

Disease-causing organisms may also be transmitted indirectly (see Figure 15-12). **Indirect transmission** occurs when a person comes in contact with pathogens on inanimate objects, such as handkerchiefs, doorknobs, or faucet handles. For example, if you touch a doorknob contacted by someone who had the flu, you may pick up the viruses he or she left behind. If you then touch your eyes, mouth, or nose before washing your hands, you may become infected. Skin punctures by contaminated objects also can spread a disease-causing organism, as mentioned earlier for tetanus.

Some pathogens travel through the air on much smaller airborne particles, known as **aerosols**. Aerosols consist of moisture droplets and fine dust particles. They can remain suspended in the air for extended periods and can travel in air currents. Aerosols containing pathogens such as viruses or bacteria may be inhaled, causing disease in the host. Tuberculosis and SARS are examples of diseases that usually spread through the air as respiratory droplets and aerosols.

Arthropods, such as mosquitoes, flies, fleas, lice, and ticks, are also responsible for the indirect transmission of disease. Such carriers are called vectors. Some mosquitoes, for example, may carry the malaria parasite or the West Nile virus. Being bit by a mosquito carrying West Nile virus can transfer the virus into the blood and lead to West Nile fever. Deer ticks can carry the bacterium responsible for Lyme disease. Fleas may carry the plague bacterium. Even common houseflies may carry diseases. Thus, when a housefly lands on your dinner plate, it may transfer pathogens to the food. If a sufficient dose is transmitted, eating the contaminated food may make you sick. Finally, some infectious agents can spread through the food we eat and the water we drink. Poor food processing or food preparation can introduce pathogens into meats and other foods. The food then serves as the vehicle by which the pathogens are spread. Toxin-producing strains of *Escherichia coli* often make the news because improper food-processing procedures have accidentally introduced the bacterium into a food product, often hamburger. In 1998, the U.S. Department of Agriculture recalled 25 million pounds of raw hamburger contaminated with a toxin-producing strain called *E. coli* O157:H7. If contaminated raw foods are not cooked to the proper temperature to kill the bacteria, chances are people will become ill and, in extreme cases, even die.

15-5 Emerging Infectious Diseases and Bioterrorism

In the latter half of the twentieth century, many infectious diseases that once ravaged human populations seemed to be under control. After millions upon millions of deaths, humans had finally triumphed over infectious disease. Our success stemmed from the use of antibiotics, the development of vaccines, a vigilant public health system, better sanitation, and water purification. In fact, in 1980, the virus that causes smallpox was eradicated from the face of the Earth through a massive global vaccination effort.

Over the last several decades, however, a number of new infectious diseases, such as AIDS and SARS, have emerged (Figure 15-13). We have also witnessed a resurgence of some infectious diseases, such as tuberculosis and dengue fever, that health officials thought were under control. Why are new infectious diseases emerging and other infectious diseases reemerging?

Emerging and Reemerging Infectious Diseases

Several infectious diseases once thought to be under control are reemerging, and new infectious diseases are emerging for a variety of reasons.

Emerging infectious diseases are those that have recently surfaced in a population. Among the more newsworthy have been AIDS, hantavirus pulmonary syndrome, Lyme disease, Ebola hemorrhagic fever, mad cow disease and, most recently, SARS and West Nile fever (in the Americas).

One of the major reasons for the appearance of new diseases is the expanding world population. As populations grow and expand into previously uninhabited areas, humans are exposed to insects and other animals that harbor infectious agents. Human contact with infected animals may result in a direct transmission of pathogens. Such a scenario probably explains how HIV "jumped" to humans from chimpanzees, which suffer a similar disease, and how the Ebola virus results in sporadic outbreaks of hemorrhagic fever in parts of Africa.

Another reason for the appearance of new diseases is the increased worldwide transport of animals—especially animals for the pet trade. The SARS virus probably was transmitted to humans as a result of animal transport in Asia. In fact, the trade in exotic animals into the United States caused the 2003 outbreak of monkeypox.

Increased international travel also can spread diseases to new geographical areas. It is believed that the West Nile virus that emerged in New York City in 1999 came from an individual or animal that was infected in the Middle East, where the virus is endemic. Finally, changes in food handling or processing can also be the cause for an emergent disease. As Health Note 15-2 explains, prions, which are responsible for mad cow disease, spread from "infected" beef carcasses that were used to make cattle feed.

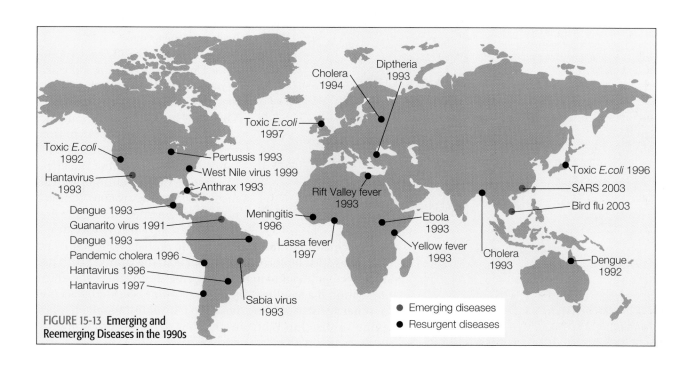

FIGURE 15-13 Emerging and Reemerging Diseases in the 1990s

Reemerging infectious diseases are ones that have existed in the past but are now showing a resurgence in frequency or geographic range. Some of the more prominent reemerging diseases are cholera, tuberculosis, and dengue fever. One reason diseases are reemerging is antibiotic resistance, described in Chapter 14. Antibiotic-resistant strains of the bacterium that causes tuberculosis, for example, have evolved over time. Because they are resistant to many of the antibiotics used to treat the disease, the disease is spreading. Yet another cause for the reemergence of pathogenic organisms is that large segments of the human population, notably AIDS victims, suffer from lowered immunity to infectious disease, allowing many formerly rare diseases to reappear.

Diseases are also reemerging because of lax public health programs. In the 1980s, the collapse of the Soviet Union and subsequent economic depression that hit many of the countries resulted in a decline in public health programs, leaving many members of the population unvaccinated and thus susceptible to infectious disease. In one case, diphtheria vaccinations disappeared in parts of the former Soviet Union. In a three-year period, diphtheria became epidemic in those areas.

The emergence of new diseases and the reemergence of others once thought to be under control may also be due to climate change (Chapter 23). Careful scientific studies have shown that warming in several regions has resulted in the spread of vector-borne infectious diseases such as dengue fever into neighboring areas, which were previously too cool to support the vectors. Although the magnitude of such infections is yet to be determined, global warming may become a prime factor in the emergence and reemergence of infectious disease in the decades ahead. Because additional emerging or reemerging infectious diseases are inevitable in the future, public health systems around the world need to be vigilant and join together in to combat emerging infectious diseases and prevent their spread.

FIGURE 15-14 **Bioterrorism** (a) Combating the threat of bioterrorism often requires special equipment and protection, because many organisms seen as possible bioweapons are spread through the air. (b) This photo, taken in 1967, shows a smallpox patient.

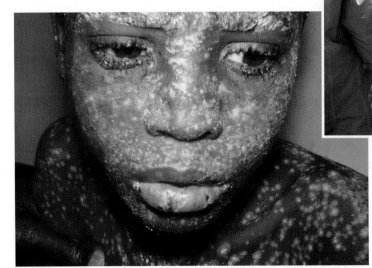

(b)

Bioterrorism

Bioterrorism is an attempt to use infectious disease agents to frighten and inflict pain and suffering in populations.

As you have seen, infectious disease in the human population results from the presence of infectious disease agents and many human factors, such as increasing population density, increased travel, and settlement of previously uninhabited areas. Infectious agents are also being used by terrorists to cause fear or inflect pain and suffering and even death on large populations. This threat is called **bioterrorism**. The anthrax attacks that occurred in the eastern United States in October 2001 demonstrate the potential for such agents to cause fear and anxiety.

Potential bioterrorists could rely on a large number of infectious agents, including pathogenic bacteria, fungi, and viruses and bacterial toxins. The potential impact of the biological agents depends on their virulence and the ease with which they can be disseminated. The pathogens of most concern, called the *Category A Select Agents*, are those that can be spread via aerosols (Figure 15-14). This group includes anthrax and smallpox. Category A also includes bacterial toxins, poisonous substances that can be added to food or water supplies, such as the toxin that causes botulism.

Biological weapons offer rogue nations and terrorist groups several advantages over conventional weapons of mass destruction and terror. Perhaps most important, biological weapons are much cheaper to produce than chemical and nuclear weapons. In addition, they provide a deterrent every bit as dangerous and deadly as the nuclear weapons of the more developed nations.

In May 2000, Ken Alibek, a scientist who once worked in the Soviet bioweapons program before defecting to the United States, testified before the U.S. House Armed Services Committee that the best defense against biological weapons is to develop appropriate medical defenses such as vaccines. These could minimize the impact of bioterrorism agents, making them useless as weapons.

The threat of biological weapons in the United States is being addressed, in part, by careful monitoring of sudden and unusual disease outbreaks. That way, health officials hopefully can react quickly to stop them from spreading. Extensive research is being carried out to determine the effectiveness of various antibiotic treatments against such weapons. Research is also being carried out on the best ways to develop effective vaccines.

(a)

Vaccination of the general public probably offers the best defense against bioterrorism. As of 2003, the United States had stockpiled sufficient smallpox vaccine to vaccinate the entire population if a smallpox bioterrorist event occurred. Vaccines for other agents are being developed. In addition, scientists are trying to develop instruments that detect microbiological weapons before they can cause disease.

15-6 Health and Homeostasis—Staying Healthy

In this chapter, we have described several human diseases that result from microbial pathogens in our environment. Just how susceptible are we to contracting an infectious disease? The answer depends on many factors, including one's age, one's health status, and where one lives. In the tropical regions of the world, for example, where malaria and cholera are endemic, a person is much more likely to contract these diseases. In the United States, these diseases pose little or no threat.

In general, a person's risk of contracting one of the emerging diseases, such as SARS and West Nile fever, is relatively low—especially if he or she is in good health. It is much more likely that illnesses will result from pathogens found in the home or the neighborhood. That is, you are much more likely to get sick from contaminated surfaces in the home kitchen than from the bite of an infected mosquito.

As an example, one of the most common food-borne illnesses is salmonellosis, caused by *Salmonella* bacteria found in contaminated foods. Salmonellosis is caused by eating meat that wasn't cooked long enough or that was cooked at a lower-than-optimal temperature. It is also caused by consuming food that has been improperly stored after cooking. Salmonellosis is characterized by diarrhea, fever, and abdominal cramps. Nearly 40,000 cases of salmonellosis are reported annually in the United States. However, the Centers for Disease Control and Prevention (CDC) estimate that the actual number of salmonella illnesses each year is greater than 1 million. That's because most mild cases are not reported to physicians. Careful attention to food preparation—such as making sure eggs and meats such as poultry are thoroughly cooked and food preparation surfaces are cleaned with soap and hot water first—can reduce the risk of food-borne pathogens.

Although it can be difficult to avoid surfaces that have been contaminated with flu or cold viruses, some simple steps can help minimize the chances of infection. Washing one's hands frequently, for example, is one of the easiest and most effective means of preventing the spread of infectious disease. As a precaution, you should wash your hands thoroughly before preparing or eating food, after coughing or sneezing, after changing a diaper, and after using the toilet. When soap and water are not readily available, use a hand-sanitizing gel.

Do not handle an animal that looks ill or appears to have an infection, and do not share a drinking glass or eating utensils with someone who is sick. It is important to also stay up to date on infectious diseases that might have the greatest impact in your area, such as West Nile fever.

Certain medicines can keep you from contracting an infectious disease. For example, if you are traveling in an area where malaria is common, you may want to ask your doctor for an antiparasitic medication. In areas where you will encounter mosquitoes, ticks, and other biting insects that carry disease, doctors recommend the use of insect sprays containing the repellent DEET.

Some infectious diseases, such as the common cold, usually do not require a visit to the doctor. However, if you think you have encountered a serious infectious disease, contact your doctor. He or she can perform the necessary tests to determine if you are infected, the seriousness of the infection, and how best to treat it. Even if you only have a virus, physicians may prescribe antibiotics to lower the risk of secondary bacterial infections. Remember, however, that although antibiotics may be a short-term answer for some bacterial diseases, long-term or indiscriminate use of antibiotics may result in the appearance of more resistant, harder-to-treat strains of bacteria.

Vaccination is the best line of defense for many infectious diseases. Over time, the list of vaccine-preventable diseases has continued to expand. Currently, there are more than a dozen vaccines available. These include vaccines for childhood diseases such as chickenpox, measles, mumps, and rubella. In addition, many vaccines or boosters are available for adults to prevent illnesses such as tetanus and diphtheria. Yearly flu shots usually provide a high degree of protection for you, your family, and the population in general. They are especially recommended for youngsters and adults over 50 who are more likely to die from the flu.

SUMMARY

The Infectious Agents of Human Disease

1. Microorganisms are found in nearly every environment on Earth. Microbes range from submicroscopic viruses to macroscopic parasitic worms. Although they are found everywhere, only a very small percentage of the microbes found in our environment is capable of causing human disease.

2. Pathogens are the microbes capable of causing disease. If they can be transmitted from one organism to another, they are considered contagious; if they invade and cause disease, they are considered infectious.

3. Our present day understanding of pathogens as the cause of human infectious disease arose largely from the work of Louis Pasteur and Robert Koch, who formulated the germ theory of disease in the late 1800s.

4. Viruses are composed of genetic information (DNA or RNA) and a protein capsid. The capsid may be surrounded by a viral envelope.

5. Viral replication requires the virus to (1) adhere to the appropriate host cells; (2) be brought into the cell; (3) replicate its genetic information and manufacture capsid proteins; (4) assemble genetic information and capsids into new virus particles; and (5) be released from the host cell.

6. Most bacteria, which can be seen with a light microscope, are composed of rods, spheres, or spirals. They reproduce by replicating the DNA and then splitting to form two separate cells. Such reproduction can occur at a very rapid rate.

7. Eukaryotic microorganisms also can cause infectious disease. Fungi can cause athlete's foot, oral thrush, and serious respiratory infections. Protozoa are responsible for wa-

terborne diseases. Malaria, one of the major infectious disease killers, takes the lives of more people worldwide than any other protozoan disease. The helminthes consist of flatworms that cause diseases such as schistosomiasis and roundworms that cause pinworm disease and others.

The Course of a Human Disease

8. The ability of a pathogen to enter the body and cause disease is called *pathogenicity*. It varies from one pathogen to another. Some pathogens are opportunistic, only causing disease when the host's immune system is suppressed or unable to mount a defense. Virulence describes the degree of pathogenicity.

9. Most pathogens produce a set of signs and symptoms that are characteristic of the disease. Signs are what the physician can detect (e.g., mild fever and tissue swelling) and symptoms are what the patient feels (e.g., headache and sore throat). A syndrome is a specific set of signs and symptoms characteristic of some diseases, such as AIDS.

10. Most diseases go through a five-stage course, including: (a) the incubation period, (b) the prodromal phase, (c) the acme period, (d) the period of decline, and (e) the period of convalescence. Some convalescent patients may still be carriers for the disease from which they are recovering.

How Pathogens Cause Disease

11. Most pathogens must adhere to specific cells or tissues for an infection to progress. Adhesive proteins may be found on many bacteria and virus cell surfaces.

12. Many bacterial pathogens depend on enzymes to increase their virulence. Enzymes may protect the pathogen from destruction by the immune system, destroy immune cells, or facilitate the penetration of deeper host tissues.

13. Several bacterial pathogens produce toxins that also increase virulence. Exotoxins are products of cell metabolism that can destroy host cell structures or interfere with their metabolism. Their effects vary depending on the specific toxin ingested.

14. Endotoxins are parts of the cell wall in some bacteria. Their effects require a higher dose of toxin than exotoxins. The signs and symptoms are similar for all endotoxins.

How Infectious Diseases Are Transmitted

15. Infectious diseases can be endemic, that is, they remain localized within a small number of individuals. Infectious agents can also create epidemics, explosive increases in the numbers of diseased individuals. An explosive but localized increase in numbers of infected individuals is called an outbreak. A disease that produces explosive numbers of ill individuals worldwide results in a pandemic.

16. Diseases can be transmitted by direct contact of pathogens between an infected individual and another person, for example, from a mother to her unborn child.

17. Animals can carry human diseases, such as rabies and toxoplasmosis, and transmit them directly to humans. Direct contact transmission also occurs through respiratory droplets that can carry an infectious disease a few feet in the air.

18. Indirect contact is also responsible for the transmission of disease. Inanimate objects, for example, can become contaminated with disease agents. Contact with such objects can result in the transmission of the disease to another individual.

19. Aerosols, consisting of moisture droplets and dust contaminated with pathogens can carry disease farther than respiratory droplets.

20. Arthropods, such as mosquitoes, fleas, and ticks, can carry infectious diseases. Infectious disease also can be spread through food or water.

Emerging Infectious Diseases and Bioterrorism

21. Emerging infectious diseases are those that have appeared in a population for the very first time, such as AIDS in the 1980s and SARS in 2003. Causes for such emergence include a growing world population that has become exposed to insects and animals harboring such infectious diseases; the worldwide transport of animals and exotic animals that harbor infectious disease; and the increased numbers of individuals who travel internationally.

22. Reemerging infectious diseases are those that were under control but now are increasing in incidence. The increase in antibiotic resistance, a lack of a strong public health system, reductions in vaccinations, and global warming are all responsible for the increasing numbers of such diseases.

23. Bioterrorism is a strategy used by terrorist organizations and rogue states to cause fear, inflict harm, or cause death in a population using pathogens or other harmful biological agents. Pathogens that can be dispersed relatively easily, such as the smallpox virus or anthrax bacteria or bacterial toxins that can be placed in water supplies are some of the most dangerous agents. Minimizing such threats requires strong medical defenses (vaccines) and improved methods of detection.

Health and Homeostasis—Staying Healthy

24. The chance of contracting an emerging infectious disease is not high for individuals who have good, healthy immune systems. It is much more likely that an individual will contract an infection or illness from pathogens in his or her kitchen. Therefore, to stay healthy means practicing those methods that eliminate and reduce the chances for disease transmission: keeping one's kitchen and bathroom clean, washing one's hands often, and making sure all vaccinations are up to date.

critical thinking

Thinking Critically—Analysis

This analysis corresponds to the Thinking Critically scenario presented at the beginning of this chapter.

There are several important points concerning you friend's decision to keep and eat the "potentially contaminated" luncheon meat. First, the disease is an example of an outbreak; it appears to be limited (at the time of the newscast) to a few individuals in the town. Second, because your friend is in apparently good health, he is not at great risk. Remember that poor health makes an individual more susceptible to many potentially harmful microbes.

Another point to remember is that if the packaged meat is contaminated with *Listeria*, this disease is not contagious. A person with the disease cannot transmit it directly to anyone else. However, the meat does represent a mechanism for indirect transmission. You should advise your friend not to eat it or offer it to anyone. If he objects, you should note that it is possible that the luncheon meat is contaminated with *Listeria,* but that the number of bacteria at the time he ate his first sandwich might have been low. Leaving the meat in the refrigerator longer provides time for the bacteria to grow and produce significant numbers (a dose) that could be dangerous if ingested at a later date.

Tell your friend no thanks for the lunch invitation! And advise him not to eat it or offer it to anyone else. It's best if he disposed of it immediately.

KEY TERMS AND CONCEPTS

acme period, p. 295
aerosol, p. 294
bacterium, p. 284
biofilm, p. 291
bioterrorism, p. 296
capsid, p. 280
clinical disease, p. 291
contagious, p. 279
direct transmission, p. 293
disease, p. 288
emerging infectious disease, p. 295
endemic, p. 293
endotoxin, p. 292
epidemic, p. 293
exotoxin, p. 291
flatworm, p. 286

fungus, p. 285
germ theory of disease, p. 279
helminths, p. 286
host, p. 280
incubation period, p. 290
indirect transmission, p. 294
infection, p. 288
infectious, p. 279
microorganism, p. 278
opportunistic, p. 289
outbreak, p. 293
pandemic, p. 293
parasite, p. 286
pathogen, p. 279
pathogenicity, p. 288
period of convalescence, p. 291

period of decline, p. 291
prion, p. 289
prodromal phase, p. 291
protozoan, p. 285
reemerging infectious diseases, p. 296
roundworm, p. 286
sign, p. 290
subclinical disease, p. 291
symptom, p. 290
syndrome, p. 290
Tapeworm, p. 286
toxin, p. 291
vector, p. 286
viral envelope, p. 280
virulence, p. 289
virus, p. 280

CONCEPT REVIEW

1. List and briefly describe the five steps that must take place for a virus to infect a cell, replicate its kind, and spread to other cells. p. 281
2. Identify several diseases caused by fungi, protozoa, and helminths. Have you had any of them or do you know anyone who has? pp. 282–283
3. Define the terms *pathogenicity* and *virulence*. pp. 288–289
4. List and describe the five stages in the course of an infectious disease. When is an individual likely to spread the disease? pp. 290–291

5. How are infectious diseases transmitted? Give examples of each major type. pp. 293–295
6. What is a vector? Describe the importance of vectors in disease transmission. pp. 294–295
7. Describe the difference between an epidemic and an outbreak. p. 293
8. How do enzymes help pathogens overcome host defenses? pp. 291–292
9. What are exotoxins and endotoxins? pp. 291–292
10. Identify and explain the reasons for the emergence of "new" infectious diseases and the

reemergence of infectious diseases once thought to be under control. pp. 295–296
11. Make a list of the five most likely areas in your home where illness and disease could arise. Why did you select these areas? p. 297
12. Define bioterrorism and identify a viral disease and two bacterial diseases that are considered potential bioterror agents. Why are these agents of particular concern? p. 296

SELF-QUIZ: TESTING YOUR KNOWLEDGE

1. Microbes that cause disease are called _____. They are _____ diseases if they can be transmitted easily between humans. p. 279
2. Two scientists, _____ and _____, were responsible for proposing and developing the germ theory of disease. p. 279
3. The viral genetic information is surrounded by a protein coat, called a/an _____, which in turn may be surrounded by a/an _____. p. 280
4. In order to replicate, a virus must infect a/an _____ cell. p. 280
5. The three variations in bacterial shape include _____, _____, and _____. p. 284
6. Among the eukaryotic pathogens, yeast-like infections are caused by _____, the _____ often are parasites of the intestinal tract, and the _____ are parasitic worms. pp. 285–286

7. A _____ is a change from the healthy state of the body, which depends on the _____ of the infecting organism. p. 288
8. Those pathogens that only cause disease when the host's immune system is suppressed are called _____. p. 289
9. The term _____ is used to describe those changes in body function that a patient experiences while _____ is used to identify those changes that can be detected and measured by a physician. p. 290
10. The period between a person's exposure to a pathogen and the appearance of the first symptoms is called the _____ period, while the _____ period is the stage characterized by the development of very specific symptoms characteristic of the disease. p. 291
11. The two different types of bacterial toxins can be recognized by the fact that the _____ are produced usually only after the death or disintegration of bacterial cells while the _____ are

produced by live bacteria in the host tissues. pp. 291–292
12. The occurrence of rabies in several dozen residents in a city would be considered a/an _____ because the disease had not spread throughout the state or neighboring states. p. 293
13. Being exposed to the flu virus from touching a contaminated door knob would be an example of _____ transmission while being exposed to the virus from respiratory droplets in the air would be an example of _____ transmission. p. 294
14. Examples of recently emerging infectious diseases would include _____ and _____; reemerging infectious diseases are represented by _____ and _____. pp. 295–296
15. The best defense against the potential use of a bioterrorism agent is through _____ of the population. p. 296

www.jbpub.com/humanbiology/5e

The site features eLearning, an online review area that provides quizzes, chapter outlines, and other tools to help you study for your class. You can also follow useful links for in-depth information, research the differing views in the Point/Counterpoints, or keep up on the latest health news.

Chromosomes, Cell Division, and the Cell Cycle

Electricity flowing through wires produces a magnetic field. In 1979, two researchers from the University of Colorado discovered that magnetic fields such as these could produce serious health effects. Their studies suggested that magnetic fields generated by high-voltage power lines may increase the incidence of leukemia (a cancer of the white blood cells) in children who live nearby. Cancer death rates in these children were twice what would have been expected in the general public.

In the 1970s, environmentalists began warning people of the potential dangers from the release a class of chemicals called chlorofluorocarbons (CFCs) into the atmosphere. Once used in refrigerators, freezers, and as blowing agents to make rigid foam like Styrofoam, CFCs escape from these products and drift into the upper atmosphere. Here, sunlight causes them to break apart, producing a chemical that destroys ozone molecules in the ozone layer. The ozone layer protects us from dangerous ultraviolet radiation. Exposure to ultraviolet radiation can cause skin cancer, cataracts, and other problems. As the ozone layer thinned, environmentalists and others warned that skin cancer rates and mortality from skin cancer could increase dramatically.

Against this gloom and doom, a loud dissenting voice was raised on the airways of America, led in large part by conservative radio and television celebrity Rush Limbaugh. He called the ozone threat a "scam." He said scientists were behind it, in part, because they viewed it as a way to enhance their funding. He and others alleged—and continue to allege—that claims by environmentalists and scientists were wrong.

Environmentalists countered that Limbaugh and others like him were spreading misinformation. How would you go about analyzing this debate?

In 1986, researchers from the University of North Carolina published reports showing a fivefold increase in childhood cancer (particularly leukemia) in families living 25–50 feet from such wires (Figure 16-1). Experimental work in the laboratory also sug-

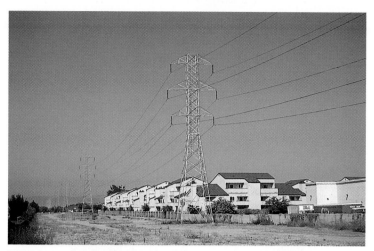

FIGURE 16-1 Too Close for Comfort? Some studies suggest that extremely low frequency radiation from power lines may increase the incidence of cancer, especially childhood leukemia, in nearby residents. A recent analysis of the research, however, indicates that this link between cancer and electrical lines is not real.

gests a link between magnetic fields and cancer. For example, one researcher discovered that ELF fields increased the growth rate of cancer cells in tissue culture. They also rendered the cells 60%–70% more resistant to the immune system's naturally occurring killer cells, which attack cancer cells.

Despite these and other studies, a panel of scientists convened by the prestigious National Research Council concluded in 1986 that electromagnetic fields associated with power lines, appliances, and other sources do not pose a health risk to humans. Their conclusion was based on an exhaustive analysis of over 500 studies on the subject performed over a 17-year period. They concluded that although some studies do show an effect, the bulk of the evidence shows no credible link. In those studies that demonstrated a link, scientists suggested that other unmeasured factors may have been responsible. Today, however, not everyone agrees. Some scientists, public health officials, and activists have raised new fears about the health effects of our electronic world, notably concern over the potential effect of cell phones. Some fear that prolonged use of cell phones may lead to brain cancer. To date, however, there is not enough long-term research to confirm or refute this hypothesis.

This chapter provides basic information about chromosomes and cell division that will help you understand this debate as well as the information on cancer presented in Chapter 20.

16-1 The Cell Cycle

The cell cycle is divided into two parts, interphase and cell division.

As noted in Chapter 1, all cells arise from preexisting cells via cell division. This process is the basis of growth and is therefore essential to all forms of life. In humans and many other organisms, cell division also plays an important role in the repair of damaged tissues and organs.

Cell division is one part of a cell's life cycle, called the **cell cycle**. As illustrated in Figure 16-2, the cell cycle consists of two parts: cell division and interphase. During **cell division**, the nucleus and the cytoplasm divide, splitting the cell more or less equally, thus forming two new cells. **Interphase** is the period between cell divisions.

Interphase

Interphase is a period of intense activity between cell divisions and is divided into three parts.

Once thought to be a resting period, interphase is now known to be a time of intense metabolic activity. During interphase, for example, cells replicate their DNA. They also replicate many cytoplasmic components (including organelles), a process that prepares cells for division. Thus, when cells divide, their offspring, or daughter cells, receive an adequate supply of the organelles and molecules needed to function normally.

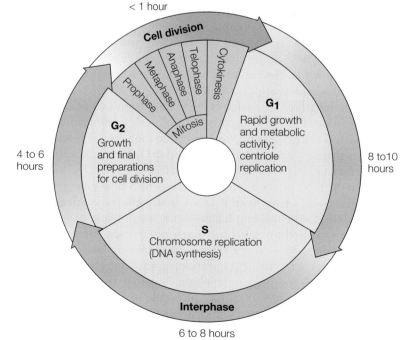

FIGURE 16-2 The Cell Cycle The cell cycle consists of two parts: interphase and cell division. Interphase is divided into three phases. The times given here represent the approximate time that mammalian cells spend in each phase when grown in the laboratory.

TABLE 16-1	Phases of the Cell Cycle
Phase	**Characteristics**
Interphase	
G_1 (gap 1)	Stage begins immediately after mitosis.
	RNA, protein, and other molecules are synthesized.
S (synthesis)	DNA is replicated.
	Chromosomes become double stranded.
G_2 (gap 2)	Mitochondria divide; precursors of spindle fibers are synthesized.
Mitosis	
Prophase	Chromosomes condense.
	Nuclear envelope disappears.
	Centrioles divide and migrate to opposite poles of the dividing cell.
	Spindle fibers form and attach to chromosomes.
Metaphase	Chromosomes line up on equatorial plate of the dividing cell.
Anaphase	Chromosomes begin to separate.
Telophase	Chromosomes migrate or are pulled to opposite poles.
	New nuclear envelope forms.
	Chromosomes uncoil.
Cytokinesis	Cleavage furrow forms and deepens.
	Cytoplasm divides.

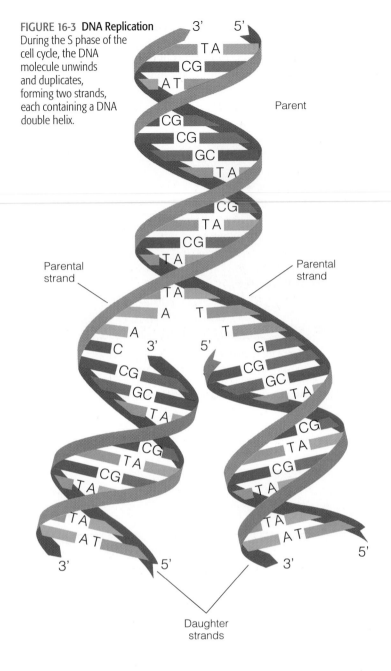

FIGURE 16-3 DNA Replication During the S phase of the cell cycle, the DNA molecule unwinds and duplicates, forming two strands, each containing a DNA double helix.

Parent

Parental strand

Parental strand

Daughter strands

Interphase is divided into three phases: G_1, S, and G_2 (Figure 16-2 and Table 16-1). G_1 (**gap 1**) begins immediately after a cell divides. During G_1, the cell makes RNA, proteins, and other molecules. These molecules, which are vital for cell function, are stockpiled by the cell to ensure that each daughter cell receives an adequate supply of nutrients.

The length of G_1 varies considerably from one cell to another. Some cells in an organism spend only a few minutes or hours in G_1; others spend weeks or even months. Cells such as nerves that never divide remain locked in G_1 for life. In general, mammalian cells in culture spend 8–10 hours in G_1 (Figure 16-2).

In the next phase of the cell cycle—the **S phase** or **synthesis phase**—DNA replicates. **DNA** is shorthand for **deoxyribonucleic acid**. This complex molecule is a polymer of smaller molecules called **nucleotides**. Each nucleotide consists of a sugar, a phosphate, and a nitrogen-containing base. They're joined together by covalent bonds forming a long chain that intertwines with another similar chain. Together, the intertwined chains form a single spiral-like molecule of DNA (Figure 16-3). It's called a double helix. (The structure of DNA is discussed at in length in Chapter 18.)

As a general rule, mammalian cells grown in the laboratory in containers with proper nutrients spend 6–8 hours in the S phase. Immediately following this phase is a period known as

G_2 (**gap 2**), which averages 4–6 hours in cultured mammalian cells. During this phase, mitochondria divide and the precursors of the spindle fibers (microtubular structures essential to nuclear division) form. The chromosomes also begin to condense during G_2.

Cell Division

Nuclear and cytoplasmic division occur separately.

Interphase is followed by cell division, a rapid event that lasts only about 1 hour. During cell division, the nucleus and cytoplasm both divide. However, nuclear division, commonly known as **mitosis** (my-TOE-siss), precedes cytoplasmic division.

Mitosis involves a series of dramatic structural changes in chromosomes. In order for the replicated chromosomes to divide, they must first condense. During this process, the chromosomes become compact structures clearly visible in ordinary light microscopic preparations (Figure 16-4). Condensation facilitates chromosomal segregation. Without it, cell division would be impossible. Dividing the 46 unraveled chromosomes in humans would result in a tangled mass of broken chromosomes. It would be like trying to separate a plate of spaghetti into two equal piles without breaking a single strand.

During mitosis the chromosomes line up in the center of the cell. From this position, the two chromatids making each chromosome are easily drawn apart with one-half going to each new daughter cell. This alignment in the center of the cell allows chromosomes to be equally divided.

Cytoplasmic division, or **cytokinesis** (literally, "cell movement," pronounced SITE-toe-ki-NEE-siss), occurs independently of mitosis and usually toward the end of it. With this overview of cell division, we now turn our attention to the chromosome.

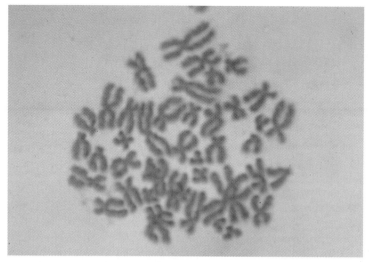

FIGURE 16-4 **Metaphase Chromosomes** Taken from a human cell during metaphase, these chromosomes are clearly visible under a light microscope.

16-2 The Chromosome

Human body cells contain 46 chromosomes that contain all of the genetic information required to control cellular activity.

Chromosomes contain the genetic information of cells, DNA, which directs most cellular events. (For a discussion of the scientific work that led to the discovery of DNA, see **Scientific Discoveries 16-1**; Table 16-2 reviews terminology.) All organisms have a set number of chromosomes. All of the cells in your body, for example, contain 46 chromosomes, except for the gametes (or germ cells—the sperm and egg), which contain half that number.

Each type of chromosome contains very specific genetic information and is different from every other type. Chromosomes also differ in several other key respects, such as size, location of the centromere, and banding patterns (Figure 16-5). By studying these features, geneticists (scientists who study chromosomes, genes, and heredity) have found that chromosomes in most of the cells of the body exist in pairs. In humans, for example, geneticists can distinguish 23 structurally similar pairs. As shown in Figure 16-5, each pair has the same length, banding

pattern, and centromere position. Such chromosomes are described as **homologous** (ha-MOLL-ah-GUS).

Homologous chromosomes contain genes that control the same inherited traits. For example, if a gene for hair color is located on one chromosome of a pair, another gene of similar structure will be found at the same location on the other member of the pair.

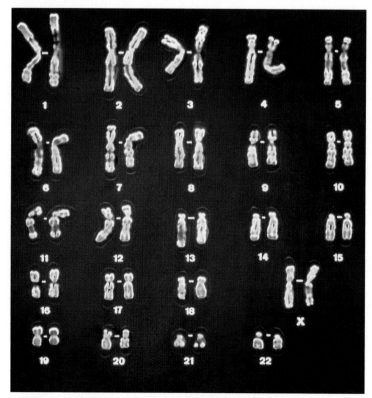

FIGURE 16-5 **Karyotype** Chromosomes from a normal human female arranged in order of decreasing size. Note that there are 46 chromosomes—or 23 pairs—in somatic cells. Each pair is similar in size, banding pattern, and location of the centromere.

TABLE 16-2	Review of Terminology
Chromatin	General term referring to a strand of DNA and associated histone protein
Chromatin fiber	Strand of chromatin
Chromosome	Structure consisting of one or two chromatin fibers
Chromatid	Generally used to refer to one of the chromatin fibers of a replicated chromosome
Centromere	Region of each chromatid to which a sister chromatid attaches

Scientific Discoveries that Changed the World

16-1	**Unraveling the Structure and Function of DNA** Featuring the Work of Watson, Crick, Wilkins, and Franklin

In 1953, two scientists, James Watson and Francis Crick published a brief paper that proposed a model (hypothesis) for the structure of DNA (Figure 1). They suggested that DNA was a double helix and explained how the nucleotides fit together.

Like many other scientific advances that changed our understanding of the world and, in some cases, our way of life, the discovery of the structure of DNA resulted from the efforts of many scientists over many years. This work began in the mid-1940s and culminated in the publication of Watson and Crick's paper in 1953. Watson and Crick used available information from other researchers. One vital piece came from Rosalind Franklin, who worked in the laboratory of British researcher Maurice Wilkins. Franklin produced X-ray photographs of crystals of highly purified DNA (Figure 2). Her remarkable photographs suggested that the DNA molecule was helical.

Another vital piece of evidence came from the laboratory of Erwin Chargaff and his colleagues, who had spent years studying the chemical nature of DNA. Chargaff's lab showed that the amount of purine in DNA always equals the amount of pyrimidine and that the ratios between adenine and thymine and between cytosine and guanine are always 1 : 1.

Armed with these and other vital details, Watson and Crick devised an ingenious model that fit the chemical and physical data. They published their results in 1953 in the journal *Nature*, and 2 months later they published a second paper proposing that the complementary strands of their model could explain replication. In addition, Watson and Crick hypothesized that genetic information could be stored in the sequence of bases and that mutations might result from a change in that sequence.

In the ensuing years, numerous scientists have provided hard evidence that supports the model proposed by Watson and Crick. And in 1962, Watson, Crick, and Wilkins were awarded the Nobel Prize for their discovery. Interestingly, although much of the X-ray data on which the Watson-Crick model was based had come from Franklin, she was not included in the celebrations. She couldn't be. Unfortunately, Rosalind Franklin died of cancer in 1958, and Nobel Prizes are awarded only to living scientists. Had she been alive, it is likely that she too would have shared in this coveted prize and been a household name as well.

FIGURE 1 The Double Helix Model for DNA Watson and Crick and their double helix model.

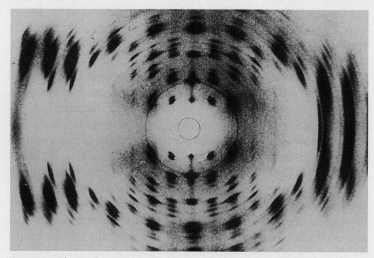

FIGURE 2 Discovering DNA's Helical Shape The famous X-ray diffraction photograph of DNA by Rosalind Franklin.

The presence of homologous chromosome pairs results from sexual reproduction—the uniting of a sperm and an ovum. During sexual reproduction, a new individual is created when the sperm and egg combine. Both the sperm and the egg contain 23 chromosomes.

Cells that contain the full number of chromosomes—that is, 46 chromosomes—are said to be **diploid** (diploid = twofold). In humans, all body cells, like those of skin and muscle, are diploid. Such cells are often called **somatic cells** to distinguish them from **germ cells**, the sperm and ovum, which carry paternal and maternal chromosomes, respectively. Germ cells are **haploid** because they contain half the chromosomes of somatic

cells. Germ cells are produced by a special kind of cell division known as meiosis (my-OH- siss), which occurs in the gonads (ovaries and testes). Meiosis is discussed in Chapter 17.

Chromosomal Condensation

Chromosomes condense after replication, which facilitates mitosis.

Soon after they duplicate, the chromosomes of the cell begin to coil, compacting in much the same way that a stretched phone cord shortens and compacts when the tension is removed. Unlike a phone cord, which functions just as well when

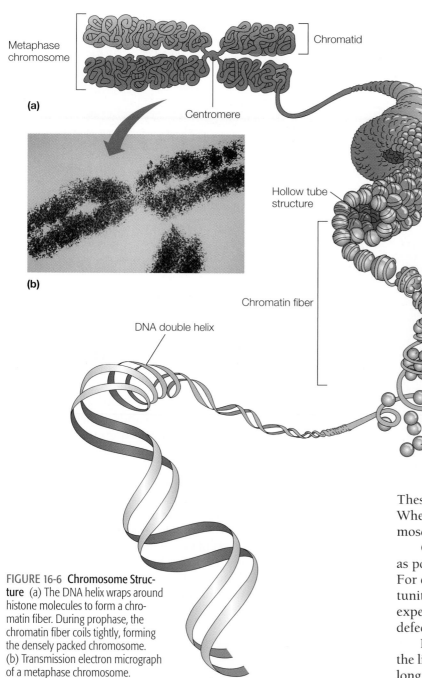

Metaphase chromosome

Chromatid

(a)

Centromere

(b)

Hollow tube structure

Chromatin fiber

DNA double helix

Histones

FIGURE 16-6 **Chromosome Structure** (a) The DNA helix wraps around histone molecules to form a chromatin fiber. During prophase, the chromatin fiber coils tightly, forming the densely packed chromosome. (b) Transmission electron micrograph of a metaphase chromosome.

it is stretched or compacted, chromosomes in the condensed state are metabolically inactive—that is, unable to produce either RNA or DNA.

The structure of a condensed chromosome is shown in Figure 16-6b. To understand this structure, begin at the bottom and work up. As illustrated, each chromatid consists of a DNA molecule in a double helix and associated proteins called **histones** (HISS-tones). The histones are globular proteins. They are thought to play a role in regulating the DNA's activities (Chapter 18). As illustrated in Figure 16-6, loops of the double helix of DNA encircle the histones, forming small clusters. The small clusters along the DNA molecule compact to form hollow tubules.

These tubules, in turn, form coils like those of a phone cord. When the coils are compacted, they form the body of the chromosomes.

Chromosomal condensation is of great importance to cells, as pointed out earlier. It also has many practical applications. For example, it provides physicians and geneticists an opportunity to search for genetic defects. This is especially useful to expectant parents who are concerned about possible genetic defects in their babies.

In order to acquire fetal cells, physicians extract fluid from the liquid-filled cavity surrounding the growing fetus through a long needle inserted through the mother's abdomen (Figure 16-7). This process is called *amniocentesis* (AM-knee-oh-cen-TEE-siss).

Fetal skin cells present in the fluid are separated from it, and then grown in culture dishes. There, they proliferate, dividing by mitosis. After the number of cells has increased, the tissue culture is treated with a chemical substance that stops cell division. The cells are then removed from the culture, placed on a glass slide, and stained. A small piece of glass or plastic is placed over them and compressed, causing the cells to flatten out. The chromosomes are then photographed through a camera mounted on the microscope. Using a computer program, chromosomes are then arranged in homologous pairs with the largest first, as shown in Figure 16-5. The resulting display is called a **karyotype** (CARE-ee-oh-TYPE).

Lining the chromosomes up like this helps geneticists locate obvious structural defects—for example, extra chromosomes or missing segments. More subtle changes cannot be

FIGURE 16-7 **Amniocentesis** (a) A patient undergoing amniocentesis. (b) A needle is inserted into the fluid-filled space surrounding the fetus. Fluid containing fetal cells is then withdrawn. Cells are subjected to chromosomal and biochemical analyses. Ultrasound monitors are used to direct the needle and prevent injury to the fetus.

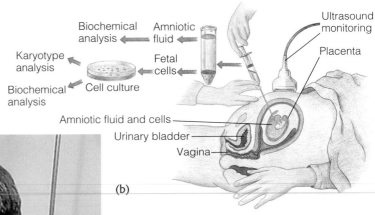

(b)

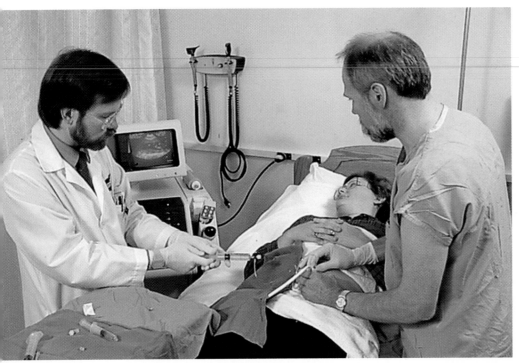

(a)

detected this way. As the next chapter explains, gross defects in chromosomes can have serious consequences. In many cases, chromosomal analysis assures parents that their child has no obvious genetic defects, alleviating worry during a period of heightened anxiety. It has the added benefit of helping parents learn the sex of their baby.

Chromatids and Chromosomes

The number of chromatids in chromosomes varies, depending on the stage of the cell cycle.

The study of genetics can be quite confusing at first, in part, because chromosomes may contain one or two chromatids, depending on the stage of the cell cycle. For example, as noted earlier, during G_1 of interphase, each chromosome consists of a single DNA molecule (a double helix) and associated protein. The 46 unreplicated chromosomes in the human cell are packed loosely in the nucleus. During the S phase, however, the DNA strands replicate (Figure 16-3). Each chromosome now consists of two identical chromatids held together at the centromere. Each chromatid is a molecule of DNA, arranged in a double helix, with its associated proteins. When a cell divides, these chromosomes are split in half, with one chromatid going to each daughter cell.

16-3 Cell Division

With this information in mind, let's explore cell division, beginning with mitosis, nuclear division.

Mitosis

Mitosis is divided into four stages.

The four stages of mitosis are prophase, metaphase, anaphase, and telophase (Figure 16-8).

Prophase. Prophase begins immediately after interphase. During **prophase**, the replicated chromosomes condense—that is, shorten and thicken—as shown in Figures 16-8a and 16-8b. In addition, the nucleoli disappear. The nucleoli are regions of active rRNA synthesis.

In the cytoplasm, the cell's centrioles separate. Shown in Figure 16-9, each **centriole** consists of two small, cylindrical structures identical to the basal bodies (Chapter 3). Like other organelles, centrioles replicate during interphase. During prophase, the centrioles separate and migrate to opposite ends of the nucleus.

During prophase, the mitotic spindle forms. The **mitotic spindle** is an elaborate array of microtubules (Figure 16-10). It is responsible for movement of the chromosomes. Each spindle fiber is a bundle of microtubules composed of tubulin molecules. These molecules are derived from the disassembly of the cell's cytoskeleton during mitosis. Also associated with the centrioles in animal cells is a star burst of microtubules called the **aster**. Its role in cell division remains unclear, although it may anchor the spindle to the plasma membrane.

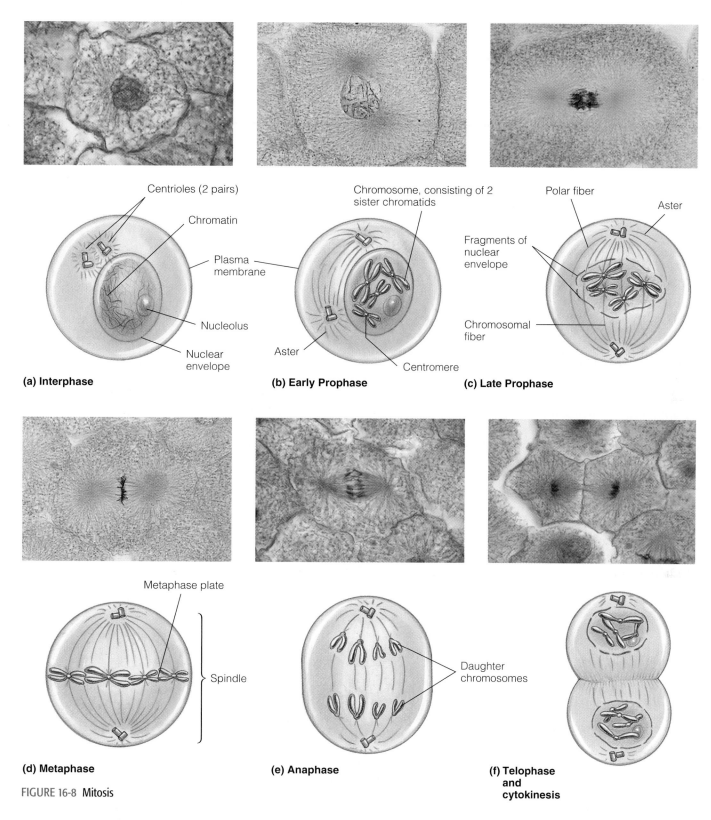

Centrioles (2 pairs)

Chromatin

Plasma membrane

Nucleolus

Nuclear envelope

(a) Interphase

Chromosome, consisting of 2 sister chromatids

Aster

Centromere

(b) Early Prophase

Polar fiber

Aster

Fragments of nuclear envelope

Chromosomal fiber

(c) Late Prophase

Metaphase plate

Spindle

(d) Metaphase

Daughter chromosomes

(e) Anaphase

(f) Telophase and cytokinesis

FIGURE 16-8 Mitosis

Late in prophase (Figure 16-8c), the nuclear envelope begins to break down. This permits the chromosomal fibers of the mitotic spindle to attach to the chromosomes, setting the stage for the next step, metaphase.

Metaphase. Before chromosomes can be divided equally, they are lined up in the center of the cell with their centromeres located along the equatorial plane (Figure 16-8d). This alignment

occurs during **metaphase** and greatly facilitates the separation of chromatids.

Anaphase. When the chromatids of each chromosome begin to separate, the cell enters **anaphase**. The shortest part of mitosis, anaphase lasts only a few minutes (Figure 16-8e). During anaphase, the chromatids of the homologous chromosomes are drawn toward opposite poles of the mitotic spindle. As shown

FIGURE 16-9 **Centriole** (a) The centriole, like the basal body, consists of nine sets of microtubules with three in each set. Centrioles are found in the cytoplasm and replicate during interphase. (b) They migrate to opposite poles of the nucleus during mitosis.

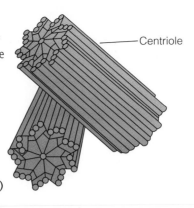

Centriole

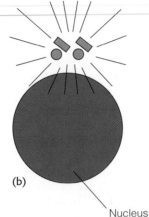  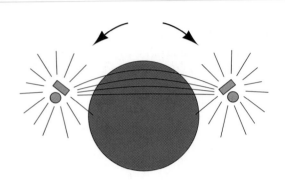

in Figure 16-8e, the chromosomes are dragged centromere-first, with their arms trailing behind. All of the chromosomes begin to separate simultaneously.

(b)

Nucleus

Telophase. The final stage of nuclear division is **telophase** (Figure 16-8f). Telophase begins when the chromosomes complete their migration to the poles. This stage involves a series of changes in the nuclei that is essentially the reverse of prophase.

During telophase, membranes form around the chromosomes of each new (daughter) cell. They soon fuse into one nuclear envelope. Also during telophase, the spindle fibers disappear. The chromosomes uncoil, regaining a threadlike appearance. Nucleoli reappear. Telophase ends when the nuclei of the daughter cells appear to be in interphase.

Cytokinesis

Cytokinesis, the division of the cytoplasm, is made possible by contraction of microfilaments beneath the cell membrane.

Cytokinesis usually begins late in anaphase or early in telophase. Cytokinesis is brought about by a dense network of contractile fibers, or microfilaments, that lies beneath the plasma membrane. These microfilaments, which are part of the cytoskeleton of the cell, are composed of the same proteins found in muscle cells.

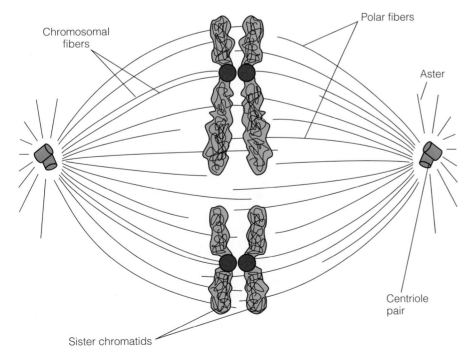

Chromosomal fibers

Polar fibers

Aster

Centriole pair

Sister chromatids

FIGURE 16-10 **Mitotic Spindle** The mitotic spindle, made of microtubules, forms during prophase in the cytoplasm of the cell. Chromosomal fibers connect to the chromosomes, and polar fibers extend from a pole to the equatorial region of the spindle. In ways not currently understood, the mitotic spindle separates each double-stranded chromosome during mitosis.

FIGURE 16-11 **Cytokinesis** (a) The cytoplasm of a cell is divided in two by the contraction of a ring of contractile microfilaments. Cytokinesis begins late in anaphase or early in telophase. (b) Cell undergoing cytokinesis.

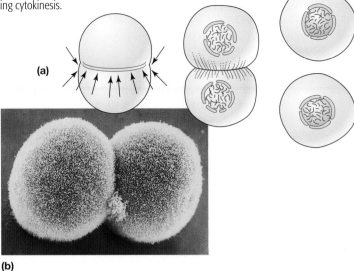

(a)

(b)

During cytokinesis, the microfilaments along the midline of the cell contract and pull the membrane inward (Figure 16-11). The membrane furrows as if it were being constricted by a thread tied around the cell, then tightened. As cytoplasmic division proceeds, the furrow deepens, eventually pinching the cell in two. This results in two daughter cells with more or less equal amounts of cytoplasm and cellular organelles.

16-4 Control of the Cell Cycle

Many factors appear to control the cell cycle.

Some experiments suggest that chemical messages from the cytoplasm control the cell cycle. Research shows that certain external signals may exert some control over cytoplasmic events. Studies of cell cultures—that is, cells grown in the laboratory under precise conditions—have shown that normal cells migrate and divide until they spread evenly across the bottom surface of culture dishes. At this point, the cells stop dividing, even though nutrients are plentiful. The contact of cells with one another apparently inhibits further growth; this process is called **contact inhibition**. Contact inhibition also occurs in the tissues of the body.

Interestingly, one of the reasons cancer cells grow uncontrollably is that they lose contact inhibition. Thus, in a culture dish, cancer cells proliferate wildly, growing on top of one another and quickly utilizing nutrients in the nutrient bath they're immersed in. In the body, cancer cells grow in a similar way, forming large growths, called tumors, that may eventually destroy vital organs and cause death.

Hormones affect the cell cycle as well. During each menstrual cycle, the hormone estrogen released from a woman's ovaries stimulates the growth of breast tissue and the uterine lining in preparation for pregnancy. Several other hormones also stimulate the division of body cells.

Growth-promoting and growth-inhibiting factors likewise play a role in controlling the cell cycle. Skin cells, for example, produce a growth-inhibiting factor that prevents cell division. However, damage to the skin—say, a cut or burn—reduces the number of skin cells and presumably reduces the amount of inhibiting factor at the site of injury. This, in turn, releases cells from inhibition. Cells proliferate and repair the damaged area. When the tissue is repaired, the rate of cell division returns to normal.

Obviously, many factors play a role in controlling the cell cycle. These factors may act directly on the cytoplasm but more likely stimulate changes in the genes in the nucleus that regulate the cell cycle.

SUMMARY

The Cell Cycle

1. The life cycle of a cell is called the cell cycle and consists of two parts: interphase and cell division.
2. Interphase is a period of active cellular synthesis and is divided into three phases: G_1, S, and G_2.
3. During G_1, the cell carries out its day-to-day activities. It also produces RNA, proteins, and other molecules and is a period of preparation for division.
4. During the S phase, the DNA replicates. After replication, each chromosome in the nucleus of the cell contains two chromatin fibers, or chromatids.
5. G_2 is a much shorter period and is relatively inactive.
6. During interphase, the cell replicates its organelles and molecules needed by the two daughter cells.
7. Cell division follows interphase. It requires two separate but related processes: mitosis, or nuclear division, and cytokinesis, or cytoplasmic division.

The Chromosome

8. Each organism has a set number of chromosomes. All body cells, except the germ cells, are called somatic cells; they contain a full complement of chromosomes and are described as diploid.
9. Germ cells or gametes contain half the number of chromosomes of somatic cells and are

referred to as haploid cells. Gametes are produced by a special type of cell division known as meiosis, which occurs in the gonads (ovaries and testes).

10. The chromosomes are loosely arranged in the nucleus during interphase, but condense during prophase. Condensation facilitates chromosome separation and also allows for the study of chromosomes. Chromosomes arranged according to size and other features form a karyotype. It is used by geneticists to count the chromosomes and to locate potential abnormalities.

Cell Division: Mitosis and Cytokinesis

11. During cellular division, the cell divides its chromosomes equally and distributes its cytoplasm and organelles more or less evenly between the two daughter cells.

12. Cell division is essential to reproduction and growth. It is the basis of tissue repair and provides replacements for cells lost through normal wear and tear.

13. Mitosis, nuclear division, is divided into four stages: prophase, metaphase, anaphase, and telophase, and is summarized in **Table 16-1**.

14. Successful mitosis depends on the presence of the mitotic spindle, an array of microtubules that forms during prophase. Some of the spin-

dle fibers attach to the chromosomes and help pull them apart during anaphase.

15. Cytokinesis, the division of the cytoplasm, begins in late anaphase or early telophase. Cytokinesis results from the contraction of microfilaments lying beneath the plasma membrane.

Control of the Cell Cycle

16. The cell cycle is controlled in part by substances produced in the cytoplasm. These chemicals cause changes in the nucleus of the cell. External controls, such as hormones, growth regulators, cell contact inhibition, and others, are also imposed on the cell.

critical thinking

Thinking Critically—Analysis

This Analysis corresponds to the Thinking Critically scenario that was presented at the beginning of this chapter.

When analyzing statements, one should first consider the source and look for bias. You will find that Rush Limbaugh is not a scientist and, say critics, has little knowledge of the atmospheric science necessary to understand ozone depletion. Furthermore, Limbaugh is a powerful advocate of free enterprise and takes a dim view of government regulation. In other words, he's not exactly an unbiased observer.

Furthermore, if you read his work, you'll find that he gets most of his facts from the late Dixie Lee Ray, also an avowed anti-environmentalist. According to author Robert Boyle, she, in turn, obtained much of her information from

Rogelio Maduro, an associate editor of *21st Century*. Maduro coauthored a book entitled *The Holes in the Ozone Scare*. It asserts that efforts to ban CFCs stem in large part from a corporate plot led by Du Pont, the world's leading manufacturer of CFCs. According to Maduro, they are eager to profit from substitutes for CFCs, which they developed.

Unfortunately, this characterization is in error. Du Pont and other chemical companies actually opposed bans on ozone-depleting CFCs for many years. Only when the evidence that CFCs were damaging the ozone layer became irrefutable did Du Pont back down and support a ban.

Articles in two reputable scientific journals, *Science* and *Chemical and Engineering News,* have systematically refuted the claims made by Limbaugh, Ray, and Maduro. Unfortunately, few people will ever read these journals to explore the scientific treatment of this subject.

KEY TERMS AND CONCEPTS

Anaphase, p. 307
Aster, p. 306
Cell cycle, p. 301
Cell division, p. 301
Centriole, p. 306
Contact inhibition, p. 309
Cytokinesis, p. 303
Deoxyribonucleic acid (DNA), p. 302

Diploid, p. 304
Germ cell, p. 304
Haploid, p. 304
Histone, p. 305
Homologous chromosome, p. 303
Interphase (G$_1$, S, and G$_2$), p. 301
Karyotype, p. 305

Metaphase, p. 307
Mitosis, p. 302
Mitotic spindle, p. 306
Nucleotides, p. 302
Prophase, p. 306
Somatic cells, p. 304
Telophase, p. 308

CONCEPT REVIEW

1. Describe the cell cycle, and explain what happens during each part. pp. 301–302

2. Draw a diagram of the cell cycle, and note how many chromosomes are found in the cell at each stage and whether they are single- or double-stranded. p. 301

3. List the stages of mitosis, and describe the major nuclear changes occurring in each stage. When does cytokinesis occur? What are the major cytoplasmic changes? pp. 306–309

4. Discuss the factors that control the cell cycle. Is it likely that many factors play a role in determining when and how often a cell divides? p. 309

SELF-QUIZ: TESTING YOUR KNOWLEDGE

1. The cell cycle is divided into two major parts: cell division and _____. p. 301

2. During the _____ phase of interphase, the cell produces DNA. p. 302

3. Cytoplasmic division is also known as _____. p. 303

4. Cytoplasmic division is brought about by _____ proteins lying beneath the plasma membrane of cells. p. 308

5. Two chromosomes that contain the same kinds of genes are known as _____ chromosomes. p. 303

6. A germ cell is called _____ because it contains half the number of chromosomes of somatic cells. p. 304

7. The proteins associated with the DNA is known as _____. p. 305

8. During _____ in mitosis, the condensed chromosomes line up in the center of the cell. p. 307

9. Microtubules that help to separate chromosomes during mitosis make up the _____. p. 306

10. During _____ a new nuclear envelope forms around the chromosomes. p. 308

 www.jbpub.com/humanbiology/5e

The site features eLearning, an online review area that provides quizzes, chapter outlines, and other tools to help you study for your class. You can also follow useful links for in-depth information, research the differing views in the Point/Counterpoints, or keep up on the latest health news.

CHAPTER

17

Principles of Human Heredity

thinking critically

Nearly a hundred and fifty years after his death, President Abraham Lincoln has become the subject of a scientific controversy. The debate among scientists is not over who killed him or why but, rather, over the possibility that he suffered from a genetic disorder called Marfan's syndrome.

Marfan's syndrome is an extremely rare disorder that affects the skeletal system, the eyes, and the cardiovascular system. It is difficult to diagnose and is usually identified only after its victims die.

Marfan's victims typically have exceedingly long arms and legs. Nearsightedness and lens defects are also common among Marfan's patients. The most serious problem, however, is enlargement and weakening of the aorta, the large artery that carries blood away from the heart. Weakening occurs in the aortic arch, the very first segment, and is caused by a degeneration of the connective tissue in the wall of this vessel. If untreated, the aorta can burst, causing instant death.

Evidence suggests that Lincoln had Marfan's syndrome. He was tall and lanky, for example, and he wore glasses. And, his great-great-grandfather had the disease, making it possible that Abe had received the Marfan's gene. What additional evidence might you look for to confirm this hypothesis? What critical thinking rule(s) do(es) this exercise illustrate?

For most of us, emotions tend to fluctuate with changing circumstances. Some events bring elation, while others evoke anger, fear, or other emotions. Such peaks and troughs are normal. But for some people, the emotional roller coaster ride is intense—their feelings swing erratically for little or no reason. One minute, they are in a state of elated overactivity (mania); the next, they fall into a state of depressed inactivity. This condition, once known as *manic depression*, but now called *bipolar disease*, is characterized by cyclic mood swings that are unrelated to external events.

Bipolar disease varies widely from one person to the next. In some, it persists for only a short time. In others, it continues for years. During the animated and highly energetic state, bipolar patients often act irrationally or obnoxiously. Such behavior can ruin relationships and can lead to financial disaster, especially if it interferes with decision making. When in a state of deep depression, victims sometimes threaten suicide, although their lack of energy often prevents them from following through. In some cases, however, the self-destructive yearning persists after the period of depression and results in suicide.

Bipolar disease occurs in about 3% of the U.S. population and tends to run in families, suggesting a genetic link. Researchers, in fact, recently located a gene near the tip of chromosome 11 that may be responsible.

Genes on chromosomes determine the structure and function of cells and organs. Growing evidence suggests that they also influence behavior, sometimes profoundly. This chapter examines how genes are passed from parent to offspring. We begin with an overview of meiosis, a special kind of cell division responsible for gamete formation in humans. We then focus our attention on some basic principles of heredity.

17-1 Meiosis and Gamete Formation

During gamete formation, germ cells undergo a special kind of division called meiosis, which reduces the number of chromosome by half.

Human body cells or somatic cells contain 46 chromosomes. The male and female gametes have half that number. Thus, when the male and female gametes unite, the offspring will have 46 chromosomes. The mechanism responsible for reducing the number of chromosomes in sex cells is called *meiosis* (my-OH-siss).

Meiosis is a type of nuclear division that occurs in germ cells and yields haploid gametes, that is, gametes with half the number of chromosomes of a normal (diploid) body cell. Germ cells are formed in the reproductive organs, the ovaries (OH-var-ees) of women and the testes (TESS-tees) of men. When germ cells unite, they form a zygote with 46 chromosomes, 23 from the mother and 23 from the father. The zygote continues to divide and eventually forms an embryo, then a fetus, and after birth, a baby who grows to become an adult. Each of the body or somatic cells contains 46 chromosomes.

As you may recall from Chapter 16, chromosomes occur in pairs, called *homologous pairs*, because they are the same length, have the same appearance, and contain the same genes. During meiosis, the homologous pairs in the cells that give rise to gametes are split so that each gamete ends up with one chromosome from each pair.

Meiosis is a fascinating process that resembles mitosis, discussed in Chapter 16. If you haven't read that section or have forgotten some of the details, you may want to read it now. Unlike mitosis, meiosis involves two nuclear divisions. The two nuclear divisions are referred to as *meiosis I and II*. Let's consider the main details of each one.

Meiosis I

Meiosis I is a reduction division.

Meiosis I is illustrated in Figure 17-1. Take a moment to look at the figure. One of the first things you will note is that meiosis contains the same four steps as mitosis: prophase, metaphase, anaphase, and telophase. Just as in mitosis, chromosomes duplicate in interphase, so that each chromosome contains two chromatids. During prophase I, the chromosomes condense, as they do in mitosis. However, something unusual occurs in prophase I of meiosis; that is, homologous chromosomes "pair up." In other words, chromosome 1 from the mother and chromosome 1 from the father pair up—that is, become attached. Chromosome 2 from the mother pairs with chromosome 2 of the father, and so on.

During metaphase I of meiosis I, the homologous pairs line up along the equatorial plate. During anaphase I, the pairs separate, with one member of each pair going to each daughter cell. As a result, each daughter cell ends up with half the number of chromosomes (23) of the parent cells (46). Each chromosome contains two chromatids. Because the number of chromosomes decreases by half during this process, meiosis I is called a *reduction division*. In mitosis, in contrast, duplicated chromosomes simply line up "single file" along the equatorial plate and split down the middle so that each daughter cell contains the same number of chromosomes as the parent cell.

Meiosis II

Meiosis II is like mitosis in many ways.

The second division, **meiosis II**, also resembles mitosis, except for one difference: The cells start out as haploid. As illustrated in Figure 17-1, the chromosomes of the haploid cells condense during prophase II, then line up in single file along the equatorial plate during metaphase II. During anaphase II, the chromatids of each chromosome dissociate, one going to each pole. The result is two cells, each containing a haploid number of single-stranded (unreplicated) chromosomes. It is these cells that give rise to the haploid gametes.

Although this may sound confusing, take a few minutes to review mitosis. Then go over the steps in meiosis, paying attention primarily to two things: the number of chromosomes and the number of chromatids in each one. Here's a recap of what you'll find: In meiosis I, the cells begin with 46 chromosomes—each with two chromatids. During meiosis I, each daughter cell ends up with 23 chromosomes, and each chromosome has two chromatids. During meiosis II, the 23 chromosomes split apart, and each daughter cell ends up with 23 chromosomes—each with one chromatid.

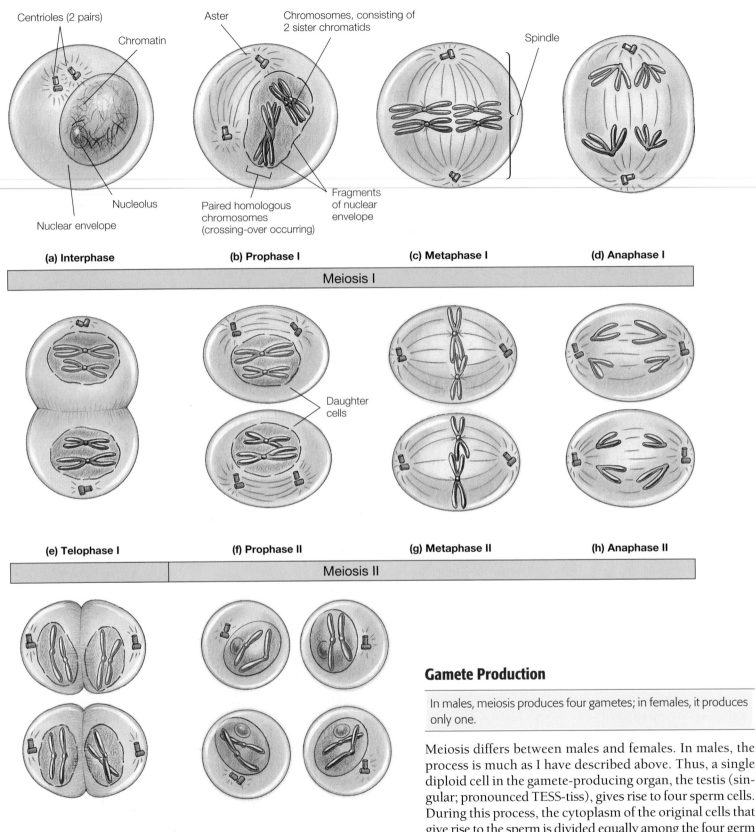

(a) Interphase

Centrioles (2 pairs)
Chromatin
Nucleolus
Nuclear envelope

(b) Prophase I

Aster
Chromosomes, consisting of 2 sister chromatids
Paired homologous chromosomes (crossing-over occurring)
Fragments of nuclear envelope

(c) Metaphase I

Spindle

(d) Anaphase I

Meiosis I

(e) Telophase I

Daughter cells

(f) Prophase II

(g) Metaphase II

(h) Anaphase II

Meiosis II

(i) Telophase II

(j) Haploid daughter cells

FIGURE 17-1 **Meiosis** Note that meiosis is divided into two phases: meiosis I and meiosis II.

Gamete Production

In males, meiosis produces four gametes; in females, it produces only one.

Meiosis differs between males and females. In males, the process is much as I have described above. Thus, a single diploid cell in the gamete-producing organ, the testis (singular; pronounced TESS-tiss), gives rise to four sperm cells. During this process, the cytoplasm of the original cells that give rise to the sperm is divided equally among the four germ cells. However, during the final stages of sperm development, much of the cytoplasm is discarded to produce a highly streamlined sperm capable of swimming up the female reproductive tract (Figure 17-2).

FIGURE 17-2 **Meiosis in Sperm Formation** Note that four haploid cells are produced during sperm formation. The cytoplasm is discarded in the final conversion to sperm.

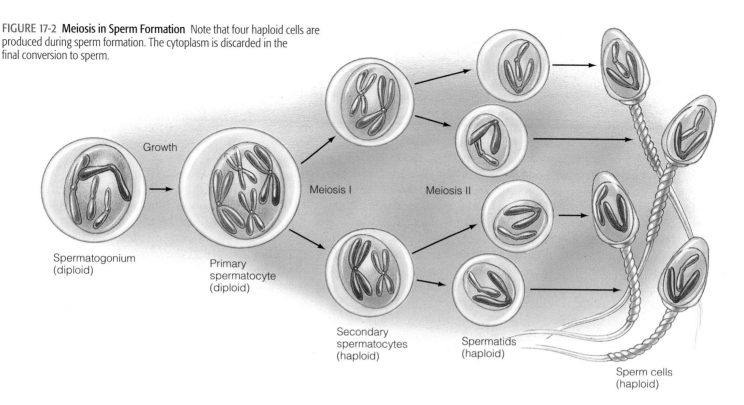

Spermatogonium
(diploid)

Growth

Primary
spermatocyte
(diploid)

Meiosis I

Secondary
spermatocytes
(haploid)

Meiosis II

Spermatids
(haploid)

Sperm cells
(haploid)

In females, the single diploid cell gives rise to only one gamete, the egg or ovum (Figure 17-3). Thus, all of the cytoplasm is retained in one cell. The extra chromosomal material is systematically "discarded" during the two meiotic divisions. This difference reflects the fact that the egg is fairly sedentary. It's the sperm that must do most of the work during fertilization! This early division of labor also ensures that adequate cellular organelles (supplied by the female) are available for the fertilized ovum.

FIGURE 17-3 **Meiosis in Ovum Formation** Note that only one haploid cell is formed, containing all of the cytoplasm. Nuclear material is discarded along the way.

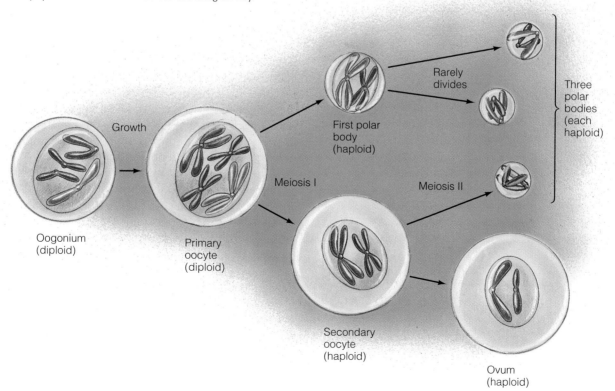

Oogonium
(diploid)

Growth

Primary
oocyte
(diploid)

Meiosis I

First polar
body
(haploid)

Rarely
divides

Meiosis II

Three
polar
bodies
(each
haploid)

Secondary
oocyte
(haploid)

Ovum
(haploid)

17-2 Principles of Heredity

Much of our early knowledge on heredity came from the work of Gregor Mendel.

With these basics in mind, we turn our attention to heredity—the transmission of genes from parents to offspring. Heredity is one aspect of the study of genetics. The formal study of genetics began with an examination of heredity at the organismic level that was carried out by a nineteenth-century monk, Gregor Mendel, working in his garden (Figure 17-4).

Born in Eastern Europe in 1821, Mendel entered an Augustinian monastery in Brno, in what is now the Czech Republic, at the age of 21. He later enrolled at the University of Vienna, where he studied botany, mathematics, and other sciences. Returning to the monastery, he began a series of experiments on garden peas—studies that would reveal several fundamental principles of genetics.

During Mendel's time, many scientists believed that the traits of a child's parents were blended in the offspring, producing a child with intermediate characteristics. In addition, because the ova of sexually reproducing organisms are much larger than the sperm, some scientists believed that the female had a greater influence on the characteristics of the offspring than the male. Mendel designed studies to examine these assumptions. Over a 10-year period, he studied inheritance in pea plants.

Blending of Traits

Mendel discovered that the traits he was studying did not blend.

When Mendel bred a pea with a white flower to a pea with a purple flower, he did not produce pea plants with intermediate pink flowers as many would have hypothesized, based on existing assumptions. Instead, these crosses produced plants with purple flowers, for reasons discussed shortly. Experiments that Mendel performed on the inheritance of other traits showed similar results—refuting the notion that parental traits blended in offspring.

Principle of Segregation

Parents contribute equally to the characteristics of their offspring.

Mendel's research led him to conclude that adult plants contain pairs of hereditary factors for each trait (for example, seed color) and that each pair governed the inheritance of a single trait. Today, we refer to Mendel's hereditary factors as the *genes*. Research has shown that the genes are located in the DNA of chromosomes.

Mendel reasoned that because a male gamete and a female gamete combine to form a new organism and an adult contains only two hereditary factors for any given trait, each gamete must contain only one hereditary factor (gene) for each trait. Mendel's studies therefore suggested that the paired hereditary factors of the mother and father must separate during gamete formation so that each gamete contributes one and only one hereditary factor to the zygote. The separation of hereditary factors during gamete formation is known as the **principle of segregation**.

Because the gametes of the parents combine to produce an offspring and because each gamete contains one hereditary factor for each trait, Mendel concluded that the contributions of the parents must be equal. That is, each parent contributes one hereditary factor, or gene, for each trait.

These conclusions may seem unremarkable to us today, especially with our knowledge of meiosis and gamete formation. However, in the 1850s and 1860s when Mendel ran his experiments, chromosomes had not yet been discovered!

Dominant and Recessive Genes

Mendel also discovered the principle of dominance.

Mendel postulated that hereditary factors (genes) are either dominant or recessive (Table 17-1). When a dominant factor and a recessive factor are present in a pea, Mendel concluded, the **dominant** factor is always expressed. A recessive factor is expressed only when the dominant factor is missing.

(b)

(a)

FIGURE 17-4 **Mendel and His Garden** (a) Gregor Mendel worked out some of the basic rules of inheritance in the mid-1800s in his research on garden peas. (b) Mendel's garden at the monastery in Czechoslovakia as it appears today.

TABLE 17-1	Genetic Traits Studied by Mendel	
Structure Studied	Dominant Trait	Recessive Trait
Seeds	Smooth	Wrinkled
	Yellow	Green
Pods	Full	Constricted
	Green	Yellow
Flowers	Axial (along stems)	Terminal (top of stems)
	Purple	White
Stems	Long	Short

An alternative form of the same gene is called an **allele** (ah-LEEL). Thus, dominant and recessive factors are known today as dominant and recessive alleles. Dominant alleles are designated by capital letters; recessive alleles are signified by lowercase letters. In Mendel's peas, for instance, the gene for flower color is the P gene. The dominant form is P (purple flowers), and the recessive form is p (white flowers). The P and p genes are both alleles of the flower-color gene.

For genes with two alleles, three combinations are possible during fertilization. The first consists of two dominant genes—for example, PP. This individual is said to be **homozygous dominant** (hoe-moe-ZYE-gus) for that particular trait. The second consists of two recessive alleles—in this example, pp—and the individual is said to be **homozygous recessive** for that trait. The third occurs when dominant and recessive alleles are present (Pp), and the individual is **heterozygous**. Take a moment to memorize these terms. They're essential to your understanding of much of the rest of the material in this chapter.

Genotype and Phenotype

The genotype of an organism is its genetic makeup; the phenotype is its appearance.

As noted earlier, Mendel determined that blending did not occur in the inheritance of traits, such as flower color. Thus, a pea plant with purple flowers (PP) bred with a pea plant with white flowers (pp) produced offspring with purple flowers and not pink flowers, which would occur if blending took place. Mendel proposed that the purple-flowered offspring of this cross were heterozygous—Pp. In such cases, he argued, the recessive hereditary factors (genes) are masked by the dominant hereditary

factors (genes). Because of dominance, Mendel noted, the outward appearance of a plant—its **phenotype** (FEEN-oh-type)—does not always reflect its genetic makeup, or **genotype** (JEAN-oh-type). As a general rule, a homozygous dominant individual and a heterozygous individual are indistinguishable on the basis of outward appearance or phenotype.

Determining Genotypes and Phenotypes

Genotypes and phenotypes can be determined by the Punnett square.

To track the genotypes and phenotypes of breeding experiments, such as those that Mendel performed, geneticists use a relatively simple tool known as the **Punnett square** (Figure 17-5). To illustrate the process, consider one of the traits that Mendel studied—seed-coat texture. The gene for seed-coat texture has two alleles: one that codes for smooth seeds (S) and one that codes for wrinkled seeds (s). As indicated by the capital S, the smooth-seed allele is dominant.

Suppose that you bred a homozygous-dominant plant (SS) with a homozygous-recessive plant (ss). Figure 17-5 lists the genotypes of each parent and the possible genotypes of the gametes. Determining the genotype is a rather simple affair. If a diploid parent is SS, its haploid gametes are all S. If a parent is ss, the gametes are all s.

To determine the outcome of crossing these two plants, all of the possible gametes produced from one plant are listed along the top of the Punnett square, and all of the possible gametes from the other are listed along the side. The gametes are then combined as shown, producing all of the possible genotypes present in the offspring.

As this example illustrates, a cross between a homozygous-dominant (SS) individual and a homozygous-recessive (ss) individual produces only heterozygous (Ss) offspring. Thus, the offspring are phenotypically uniform, and all of them resem-

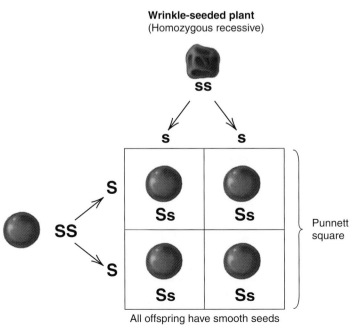

Wrinkle-seeded plant
(Homozygous recessive)

Smooth-seeded plant
(Homozygous dominant)

Punnett square

All offspring have smooth seeds

FIGURE 17-5 **Mendel's Early Experiments** When studying the inheritance of seed conformation, Mendel first crossed homozygous-recessive (ss) and homozygous-dominant (SS) plants. The offspring were all heterozygotes (Ss).

ble the homozygous-dominant parent—they have a smooth seed coat.

In the language of genetics, when one organism is bred to another to study the transmission of a single trait, the procedure is called a **monohybrid cross** (MON-oh-HYE-brid).

In a monohybrid cross, the organisms that are bred initially constitute the P_1 generation (P for parents). The first set of offspring constitutes the F_1 generation (first filial generation; *filialis* is Latin for "of a son or daughter"). In this example, the F_1 offspring could be bred to produce additional offspring, the F_2 generation (second filial generation).

To determine the genotypes and phenotypes of such a cross, the Punnett square can be used once again. As Figure 17-6 shows, the gametes of the F_1 generation are S and s because all of the parents are heterozygous (Ss). All of the possible gametes from the father are placed along the top of the Punnett square, and all of the possible gametes of the mother are placed along the side. The gametes are combined as before in the appropriate boxes.

As shown in Figure 17-6, this cross produces three different genotypes: SS, Ss, and ss. The ratio of genotypes of the F_2 generation is one SS to two Ss to one ss, or 1:2:1. Thus, if 100 offspring are produced, you would expect 25 of them to be SS (homozygous dominant), 50 to be Ss (heterozygous), and 25 to be ss (homozygous recessive). Because of dominance, though, this cross yields only two phenotypes: plants with smooth seeds (SS and Ss) and plants with wrinkled seeds (ss). The ratio of phenotypes, or the phenotypic ratio, is three smooth to one wrinkled, or 3:1

The Principle of Independent Assortment

Genes on different chromosomes segregate independently during gamete formation.

In his later experiments, Mendel tracked two traits at the same time, a procedure geneticists refer to as a **dihybrid cross** (DIE-HIGH-brid) (Figure 17-7). This work led to the idea that genes located on different chromosomes segregate independently during meiosis. To understand what this principle means and how Mendel arrived at it, consider another example that examines seed-coat texture and seed color.

As noted above, peas contain a gene for seed-coat texture, the S gene. The dominant form is S (smooth), and the recessive form is s (wrinkled). Peas also contain a gene for seed color; the dominant form (Y) yields yellow seeds, and the recessive form (y) yields green seeds.

To study the inheritance of these two genes, Mendel crossed homozygous-dominant plants with smooth, yellow seeds (SSYY)

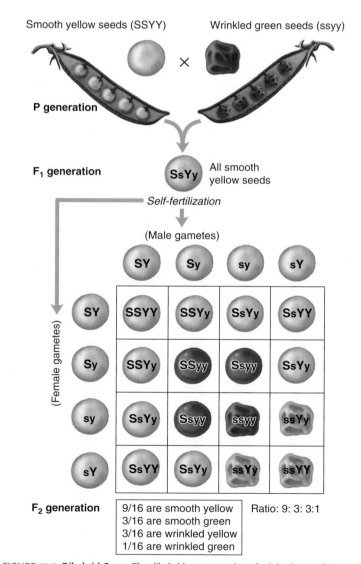

FIGURE 17-7 Dihybrid Cross The dihybrid cross examines the inheritance of two traits. Independent assortment of genes produces a large number of genotypes.

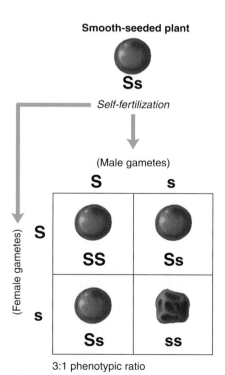

FIGURE 17-6 Monohybrid Cross In Mendel's early work, he crossed offspring from his F_1 generation (Ss) and found a variety of genotypes and phenotypes.

with homozygous-recessive plants (*ssyy*). As Figure 17-7 shows, the genotype of the F$_1$ generation was *SsYy*. Consequently, the offspring were phenotypically identical, and all displayed the dominant characteristics (smooth, yellow seeds).

When Mendel crossed two of the F$_1$ offspring (genotype *SsYy*), however, he got a mixture of phenotypes (Figure 17-7). To produce this combination of offspring, Mendel concluded, the hereditary factors *S* and *s* must have segregated independently of *Y* and *y* during gamete formation. Translated into modern terms, this means that the *Y* and *S* genes (and their alleles *y* and *s*) were on different chromosomes and separated

independently of each other during gamete formation (meiosis). Because of independent assortment, the gametes of the F$_1$ generation contain all combinations of the alleles in equal proportions: *SY*, *Sy*, *sY*, and *sy*. Fertilization also occurs at random, giving rise to 16 possible genetic combinations (see F$_2$ genotypes in Figure 17-7).

This research led Mendel to propose the **principle of independent assortment**, which states that the segregation of the alleles of one gene on one chromosome is independent of the segregation of the alleles of another gene on a second chromosome during gamete formation.

17-3 Mendelian Genetics in Humans

What do studies on garden peas have to do with humans? Interestingly, much of what Mendel discovered in garden peas pertains to human genes and human inheritance. Table 17-2 lists a number of human traits and diseases whose inheritance follows basic Mendelian principles. This section describes the inheritance of those traits, using several common diseases as examples.

TABLE 17-2	Traits and Diseases Carried on Human Chromosomes
Autosomal recessive	
Albinism	Lack of pigment in eyes, skin, and hair
Cystic fibrosis	Pancreatic failure, mucus buildup in lungs
Sickle-cell anemia	Abnormal hemoglobin leading to sickle-shaped red blood cells that obstruct vital capillaries
Tay-Sachs disease	Improper metabolism of a class of chemicals called gangliosides in nerve cells, resulting in early death
Phenylketonuria	Accumulation of phenylalanine in blood; results in mental retardation
Attached earlobe	Earlobe attached to skin
Hyperextendable thumb	Thumb bends past 45° angle
Autosomal dominant	
Achondroplasia	Dwarfism resulting from a defect in epiphyseal plates of forming long bones
Marfan's syndrome	Defect manifest in connective tissue, resulting in excessive growth, aortic rupture
Widow's peak	Hairline coming to a point on forehead
Huntington's disease	Progressive deterioration of the nervous system beginning in late twenties or early thirties; results in mental deterioration and early death
Brachydactyly	Disfiguration of hands, shortened fingers
Freckles	Permanent aggregations of melanin in the skin

Autosomal-Recessive Traits

Autosomal-recessive traits are expressed only when both alleles are recessive.

Human cells contain 23 pairs of chromosomes. They can be divided functionally into one pair of sex chromosomes and 22 pairs of autosomes (AU-toe-zomes).

The **sex chromosomes** are involved in determining whether we turn out male or female—as well as a few other functions. Two types of sex chromosomes exist, X and Y. Females have two X chromosomes; their genotype is therefore XX. Males have one X chromosome and one Y chromosome so their genotype is XY.

The remaining 22 pairs of chromosomes are called the **autosomes**. The autosomes contain numerous genes that control a variety of traits. This section examines **several autosomal-recessive traits**. These are traits expressed only when both recessive alleles are present.

Over 600 traits in humans have been identified as autosomal recessive, and another 800 are strongly suspected of being autosomal recessive. Some autosomal-recessive traits are the cause of abnormalities, among them albinism and cystic fibrosis.

Albinism. Albinism is one of the most common genetic defects known to science. It occurs in 1 of every 38,000 Caucasian births and in 1 of every 22,000 African-American births.

Albinism in humans occurs when two recessive genes are inherited from one's parents. The recessive genes result in a deficiency in the metabolic pathways leading to the production of melanin. Melanin is the brown pigment responsible for coloration of the eyes, skin, and hair. Individuals who are homozygous-recessive for albinism have either no melanin or reduced levels of melanin (Figure 17-8). Consequently, the skin of an albino is pale, and the hair is white. The eyes of albinos are pink, because there is no pigment in the iris or retina.

Melanin in the skin protects against the effects of ultraviolet radiation. Its absence in albinos makes them highly susceptible to sunburn and skin cancer. Moreover, the lack of pigment in the eyes may result in damage in the light-sensitive region, the retina, which is essential to vision. This, in turn, may result in blindness.

FIGURE 17-8 Albinism Albinism is an autosomal-recessive trait. An albino man with his wife in India.

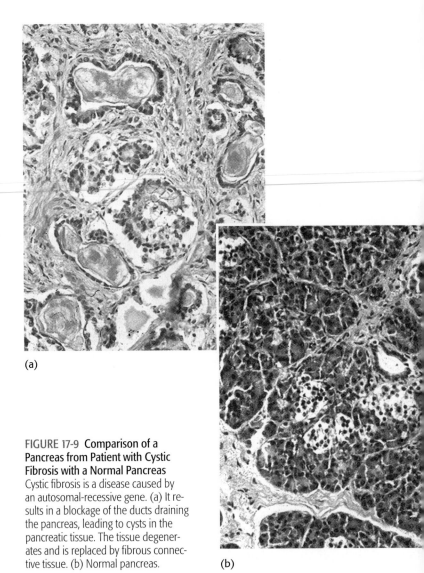

(a)

Cystic Fibrosis.   Cystic fibrosis (SISS-tick fie-BRO-siss) is an autosomal-recessive disease that leads to early death. The presence of two recessive alleles for the disease alters the function of sweat glands of the skin, mucous glands in the respiratory system, and ducts in the pancreas. Defective sweat glands, for example, release excess amounts of salt, a marker that helps physicians diagnose the disease. The most significant symptoms occur in the pancreas and lungs.

In individuals with cystic fibrosis, the ducts that drain digestive enzymes from the pancreas into the small intestine become clogged. This not only impairs digestion but also causes a buildup of enzymes in the organ that results in the formation of cysts. Over time, the pancreas begins to degenerate, and fibrous tissue replaces glandular tissue—hence, the name *cystic fibrosis* (Figure 17-9).

Despite adequate nutritional intake, individuals with this disease often show signs of malnutrition. To enhance digestion, patients must eat powdered or granular extracts of animal pancreases containing digestive enzymes. Massive doses of vitamins and nutrients must also be taken.

The respiratory systems of most victims of cystic fibrosis produce copious amounts of mucus. Mucus blocks the passages, making breathing difficult. Patients must be treated several times a day to remove the mucus and prevent infections from bacteria trapped in it (Figure 17-10). Despite treatment, most cystic fibrosis patients live only into their late teens or early twenties.

Cystic fibrosis is one of the most common genetic diseases. One of every 22 Caucasians carries a gene for this disease, and approximately 1 of every 2000 Caucasians born in the United States suffers from it. In the African-American population, only about 1 in 100,000–150,000 individuals is a carrier.

FIGURE 17-9 Comparison of a Pancreas from Patient with Cystic Fibrosis with a Normal Pancreas Cystic fibrosis is a disease caused by an autosomal-recessive gene. (a) It results in a blockage of the ducts draining the pancreas, leading to cysts in the pancreatic tissue. The tissue degenerates and is replaced by fibrous connective tissue. (b) Normal pancreas.   (b)

Autosomal-Dominant Traits

Autosomal-dominant traits are expressed in all individuals.

Many human traits are autosomal dominant. These traits are carried on the autosomes and are expressed in heterozygous (Aa) and homozygous-dominant (AA) individuals. To date,

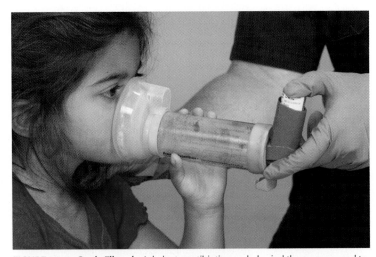

FIGURE 17-10 Cystic Fibrosis Inhalants, antibiotics, and physical therapy are used to treat patients with cystic fibrosis. To remove mucus from the lungs, parents or physical therapists must treat the patient two or three times a day. Pounding on the rib cage with a cupped hand (clopping) loosens the mucus.

nearly 1200 human traits have been identified as autosomal dominant, and 1000 others are suspected. Freckles, cleft chin, and drooping eyelids are all **autosomal-dominant traits**, as is Marfan's syndrome, discussed in the Thinking Critically section in this chapter. Let's consider an autosomal-dominant trait called *widow's peak*.

Take a moment to examine the hairlines of your friends and classmates. In some individuals, the hairline runs straight across the forehead. In others, it juts forward in the center, forming a "widow's peak" (Figure 17-11). Widow's peak is determined by an autosomal-dominant gene, indicated by W. Because the W allele is dominant, this phenotype is expressed in both homozygous dominant individuals (*WW*) and heterozygotes (*Ww*). Individuals with the genotype *ww* have a straight hairline.

(a)

(b)

FIGURE 17-11 **Widow's Peak** (a) The widow's peak is a dominant trait carried on one of the autosomes. (b) A straight hairline is a recessive trait. Simple Mendelian genetics can be used to determine the genotype of the offspring.

17-4 Variations in Mendelian Genetics

Since Mendel's time, researchers have uncovered additional modes of inheritance and gene expression, discussed below.

Incomplete Dominance

Incomplete dominance results in a kind of blending of traits.

One example of another form of inheritance is a phenomenon called **incomplete**, or **partial, dominance**. Incomplete dominance occurs when heterozygous offspring exhibit intermediate phenotypes. In other words, incomplete dominance produces F_1 offspring with phenotypes intermediate to the parental phenotypes. This is the kind of blending that Mendel said didn't exist!

Incomplete dominance occurs in the snapdragon flower. Figure 17-12 shows a cross between a plant with red flowers (*II*) and one with ivory flowers (*ii*). Even though the red color is dominant, this cross produces offspring with pink flowers, an intermediate phenotype (Figure 17-12). That's because the *I* gene does not exert complete dominance. Incomplete dominance also occurs in a number of human disorders, including sickle-cell disease.

Sickle-Cell Disease. Sickle-cell disease affects the blood, and chiefly afflicts African Americans and Caucasians of Mediterranean descent. Sickle-cell disease is caused by a genetic defect that leads to abnormal hemoglobin formation. Hemoglobin is a protein found in red blood cells, which carry oxygen to body cells.

Sickle-cell disease occurs in individuals who are homozygous recessive for this trait. The abnormal hemoglobin causes red blood cells to convert to sickle-shaped cells when they encounter low oxygen levels in the blood—for example, when blood cells flow through capillaries in metabolically active tissues.

Sickle cells are larger and less flexible than red blood cells. As a result, they clog capillaries, reducing oxygen flow to brain cells, heart cells, and other organs. In homozygous-recessive individuals, sickle-cell disease is usually lethal. In fact, most individuals with the disease die by their late twenties.

Individuals who are heterozygous for the trait are referred to as **carriers**, because they can pass the gene on to their children. Carriers generally lead relatively normal lives. Although their RBCs appear normal, they contain 50% normal and 50% abnormal hemoglobin. Moderate sickling occurs when these cells are exposed to low oxygen levels in the blood.

Approximately 1 in every 500 African Americans born in the United States is homozygous recessive, and about 1 of every 12 is heterozygous (a carrier) for the sickle-cell trait. Why is this allele so prevalent?

In tropical climates like much of Africa, the sickle-cell allele protects carriers and homozygous-recessive individuals from malaria, a deadly disease prevalent in humid, tropical regions. Malaria is caused by a microscopic parasite called *Plasmodium* (plaz-MOW-dee-um), which is transmitted from one person to the next by certain mosquitos. Inside the body, the parasites invade and colonize the liver, where they multiply rapidly. New parasites leave the liver and enter the bloodstream, where they attack and destroy red blood cells. The presence of altered hemoglobin changes the plasma membrane of RBCs and prevents the parasite from entering. Interestingly, although homozygous-recessive individuals die earlier, the selective advantage that carriers enjoy has caused the allele to increase in the population.

Multiple Alleles

Some genes have multiple alleles.

Mendel studied seven characteristics of peas. Each characteristic is determined by a gene with two alleles—two alternative forms. In humans and other species, however, researchers have shown that some genes can have more than two alleles; such a gene is said to have **multiple alleles.**

Consider blood types. One gene that controls blood type is called the **I gene** (for isoagglutinin; pronounced EYE-sa-AH-glue-TEH-nin). The I gene is located at a particular site (or locus) on one pair of chromosomes. But unlike the genes Mendel studied, which had only two alleles, the I gene exists in one of three distinct forms. The three alleles are I^A, I^B, and I^O. However, even though there are three possible alleles in human beings, an individual can have only two of the alleles in his or her genes, one on each homologous chromosome. It's like having three mittens. You can only wear two at a time.

Four possible blood types exist: A, B, AB, and O. Table 17-3 lists the four blood types and the six different genotypes that give rise to them. Take a moment to study them.

As shown, the A blood type occurs when an individual has two A alleles of the I gene or an A and an O. Type B occurs when an individual has two B alleles of the I gene or a B and an O. Type AB has both A and B alleles. Type O occurs when two O alleles of the I gene are present. The I gene is responsible for the production of glycoproteins (proteins with carbohydrate attached) that project from the surface of red blood cells, discussed in Chapter 7.

Both A and B are dominant genes, and the O allele is recessive. The I^A and I^B genes are said to be codominant. **Codominant genes** are expressed fully and equally. So not only are blood types an example of multiple genes, but also they are an example of codominance, another type of inheritance.

Polygenic Inheritance

Some traits are determined by more than one gene pair.

For many years, geneticists believed that each gene controls a single trait. In humans and other animals, however, many traits are controlled by a number of genes—from a few to perhaps hundreds. Skin color, for example, is controlled by as many as eight genes. This type of inheritance is called poly-genic inheritance (polly-JEAN-ick). Polygenic inheritance results in incredible phenotypic variation.

To see how polygenic inheritance leads to such wide phenotypic variation, consider an example using two genes for skin color, designated A and B. In this example, we will say that the genotype of an African American is AABB and the genotype of a Caucasian is aabb. Table 17-4 lists all of the possible genotypes in this example with possible phenotypes; Figure 17-13 shows what they look like. Take a moment to study the table and the figure.

Polygenic inheritance is also responsible for height, weight, intelligence, and a number of behavioral traits. But in each instance, the genes are not the only factors controlling these traits. As a rule, the genotype establishes the range in which a phenotype will fall, but environmental factors (for example, diet) determine exactly how much of the potential will be realized for

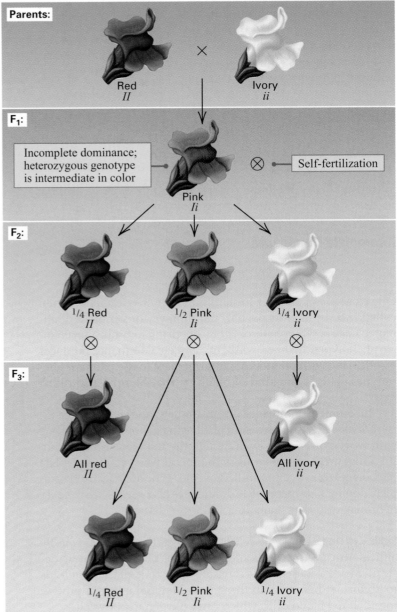

FIGURE 17-12 Incomplete Dominance Incomplete dominance involves two alleles, neither of which is dominant over the other. The result is an intermediate phenotype as shown here in the snapdragon flower.

TABLE 17-3	Phenotype and Genotype in ABO system
Phenotype (blood type)	Genotypes
Type A	I^AI^A, I^AI^O
Type B	I^BI^B, I^BI^O
Type AB	I^AI^B
Type O	I^OI^O

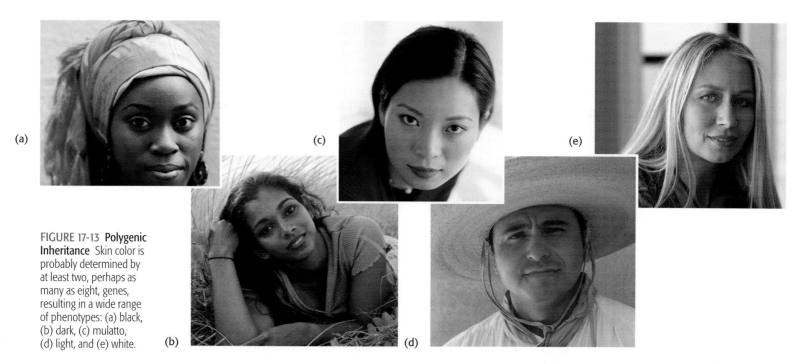

FIGURE 17-13 **Polygenic Inheritance** Skin color is probably determined by at least two, perhaps as many as eight, genes, resulting in a wide range of phenotypes: (a) black, (b) dark, (c) mulatto, (d) light, and (e) white.

many genes. For example, although genes determine height, this trait is also influenced by diet. The better a child eats from infancy to adolescence, the taller he or she will be.

Crossing Over

Crossing over is a genetic swapping that occurs in meiosis.

As pointed out earlier, Mendel found that during gamete formation, the genes under study in his pea plants were segregated independently. Also noted earlier, independent assortment generally occurs only when the genes under study are on different chromosomes. As a rule, if two genes are on the same chromosome, they do not segregate independently.

Humans have an estimated 100,000 genes on their 46 chromosomes. Those genes located on the same chromosome tend to be inherited together and are said to be **linked**. However, linkage can be disrupted by a phenomenon called crossing-over.

Crossing-over occurs during meiosis: when the homologous chromosomes come together, or "pair up," in prophase I. During this process, the chromatids of homologous chromosomes fuse at one or more points, as shown in Figure 17-14. Later, when the chromosomes separate, arms of one chromosome may end up with the other. In this process, then, chromosomes exchange segments of their chromatin with chromatin on homologous

TABLE 17-4	Possible Skin-Color Genotypes and Phenotypes with Two Skin-Color Genes	
Genotype	Phenotype	Number of Recessive Genes*
AABB	Black	0
AABb	Dark	1
AaBB	Dark	1
AaBb	Mulatto	2
AAbb	Mulatto	2
aaBB	Mulatto	2
Aabb	Light	3
aaBb	Light	3
aabb	White	4

*Skin color probably involves many more genes.

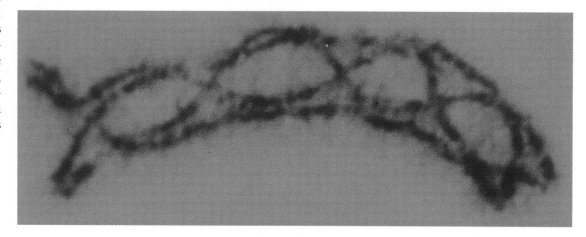

FIGURE 17-14 **Crossing-Over** The crossing-over shown here increases genetic variation in gametes and offspring and occurs during meiosis.

chromosomes. When this occurs, genes that were once said to be linked are not inherited together.

Crossing-over "unlinks" genes and increases genetic variation in gametes; this, in turn, leads to genetic variation in offspring. The more genetic variation, the more possible phenotypes in a population. As Chapter 23 points out, genetic variation may lead to characteristics that give one organism an advantage over another. This advantage can be passed to an organism's offspring, resulting in evolutionary changes.

Crossing-over can occur anywhere along the length of a chromosome. However, the greater the distance between two genes on the same chromosome, the more likely it is that a crossover will occur between them. This fact helped scientists map the human **genome**.

The Human Genome Project

The Human Genome Project seeks to determine the sequence of bases in human DNA and the location of all of the genes.

In the 1990s, scientists the world over embarked on what is conceivably the largest unified biological research project of its kind in history. It was designed to determine the location of all of the genes on all 46 human chromosomes as well as the sequence of bases of the human genes and intervening DNA in the chromosomes. With an estimated 100,000 genes and three billion bases in the human DNA, costs of this monumental task were estimated to be $3 billion dollars. Time to complete the task was estimated to be 15 years or longer.

The **Human Genome Project**, as it is called, began in earnest in 1991. Scientists found that the first part of the project—mapping the location of the 100,000 or so genes—occurred at a much faster pace than anticipated. By 2000, they had completed this task—six years ahead of schedule!

Knowing the location of genes on human chromosomes adds to our knowledge of human genetics. By locating genes and determining the sequence of bases in them, scientists can begin to synthesize specific genes—for example, the gene that controls the production of blood-clotting factors. Synthesized in the lab in massive quantities, these genes could be used to treat various illnesses, the topic of Chapter 19. Manufactured genes, for instance, could be spliced into the DNA of a person to replace defective genes. A parent, faced with the possibility of having a severely physically impaired child with a genetic disorder, for example, might opt to have gene replacement therapy on the unborn fetus in hopes of having a healthy child.

Researchers are currently experimenting with ways to replace mutated tumor suppressor genes. As noted in Chapter 20, tumor suppressor genes are normal genes that hold cells in check. When mutated, cells can undergo rapid division. By replacing the defective gene in tumor cells with intact tumor suppressor genes, researchers hope someday to tame malignancies, bringing the cells back under control.

Although the Human Genome Project could result in many improvements in health care, some critics argue that knowledge gained from the project could lead to a situation in which a physician can diagnose an illness through genetic means long before cures are available. Some individuals are concerned that insurance companies could access a person's records and, after examining them, might deny coverage. Other critics worry about the possibility that the knowledge gained from the project could result in manipulation of human genes.

One outcome of this project has been the mapping of chromosomes in nonhuman species, a step that could help bring important genetic improvements in the case of livestock; in the case of disease organisms, chromosome mapping could help scientists find better tools to combat them.

17-5 Sex-Linked Genes

The presence or absence of the Y chromosome determines one's sex.

Earlier, I noted that the sex chromosomes determine an individual's sex. In truth, sex is determined by the Y chromosome. If the Y chromosome is present, an individual becomes a male. If absent, the individual is a female.

The Y chromosome exerts its effect during embryonic development. Interestingly, early in the embryo's development, the gonads of males and females are identical. When the Y chromosome is present, however, the embryonic gonad becomes a testis—thanks, at least in part, to the presence of the *t* gene (testis-determining gene). The testis, in turn, produces testosterone and other androgens, which are responsible for the male secondary sex characteristics such as male hair growth, body shape and size, and so on (Chapter 21). The absence of the Y chromosome results in the development of an ovary with its hormones and the resultant female characteristics.

The X and Y chromosomes also carry genes that determine many other traits. Genes situated on the X and Y chromosomes are therefore known as **sex-linked genes**. So far, the majority of the sex-linked genes are located on the X chromosome. These are known as **X-linked genes**. The following sections describe the inheritance of some common sex-linked genes.

Recessive X-Linked Genes

In order for a recessive gene to be expressed, it must be located on both X chromosomes in females, but only one in males.

With hundreds of genes, the X chromosome plays an important role in determining our appearance. It also is responsible for certain diseases. Color blindness, discussed in Chapter 11, for instance, and certain forms of hemophilia are **recessive** traits carried on the X chromosome.

In order for a female (XX) to display a recessive sex-linked trait, each of her X chromosomes must carry the recessive gene. For males (XY), however, only one recessive allele is required. Why? Because the Y chromosome is not genetically equivalent to the X chromosome. Thus, in males, only one recessive gene is needed to exhibit a recessive X-linked trait.

Figure 17-15 shows four possible genetic combinations leading to color blindness. This illustration also introduces you to a genetic-tracking system used to follow traits in families, known as a **pedigree**. In a pedigree, squares represent men and circles represent women. The horizontal line linking a square to a circle indicates a mating. Offspring are shown below their parents. When a square is lightly shaded, the individual is a carrier of the gene. He or she does not have the disease but does carry the recessive gene and can pass it on to his or her offspring. A darkly shaded square or circle indicates the person has the trait.

In Figure 17-15a, for example, a man and a woman have four children, two boys and two girls. The woman (light blue circle) is a carrier of color blindness. Her cells contain one X chromosome with a recessive allele for color blindness (designated X^c) and another X chromosome with the normal allele. Consequently, half of her ova will contain the recessive allele. As shown, the woman's husband (white square) is not color-blind. He produces sperm containing either an X or a Y chromosome. When one of his X-bearing sperm unites with one of her ova bearing an X chromosome with the allele for color blindness, the result is a daughter who is a carrier, indicated by a light blue circle. When one of his X-bearing sperm unites with an ovum carrying a normal X chromosome, the result is a daughter who is neither a carrier nor color blind (white circle).

Males are produced when a Y-bearing sperm unites with an ovum carrying an X chromosome. If the X chromosome carries the recessive allele for color blindness, the boy is color-blind (green square). If the X chromosome is normal, the boy's color vision is unimpaired (white square). Take a moment to study the other possibilities in Figure 17-15.

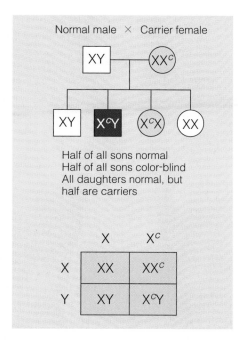

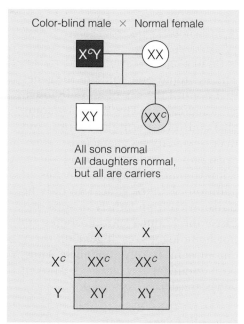

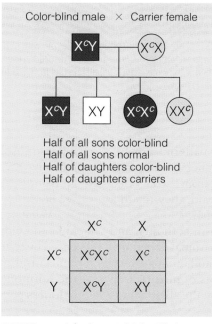

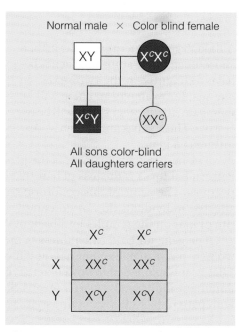

FIGURE 17-15 **Inheritance of Color Blindness** Four possible genetic ways a sex-linked recessive gene such as color blindness can be passed to offspring. Males are indicated by boxes and females by circles. In this scheme, green boxes and circles represent men and women with color blindness. Light blue boxes represent men and women who are carriers. White boxes and circles represent men and women without the color blindness gene. The notation X^c indicates an X chromosome carrying the gene for color blindness.

Dominant X-Linked Genes

When present, a dominant gene is expressed in males and females.

Recessive X-linked genes are the most common sex-linked trait in humans. However, there are a few noteworthy examples of dominant X-linked genes. One of the best understood is a disorder with a tongue-twisting name of **hypophosphatemia** (high-poe-FOS-fuh-TEEM-ee- uh). This disease is characterized by low phosphate levels in the blood and tissues of the body. This, in turn, results in a form of rickets (RICK-its), or bowleggedness (Figure 17-16). Rickets usually results from a dietary deficiency of vitamin D or in-

sufficient exposure to sunlight (Chapter 5). Alleviating the dietary deficiency usually solves the problem. In this genetic disease, however, vitamin D cannot reverse the symptoms.

Hypophosphatemia occurs when either the male (XY) or the female (XX) has one X chromosome containing the dominant gene, indicated as X' (X prime). Figure 17-17 illustrates the pattern of inheritance when a woman who is heterozygous ($X' X$) for the trait mates with a man who does not carry the trait (XY). As you can see, whenever the dominant gene is present, it is expressed—regardless of the sex of the individual.

Y-Linked Genes

> Both dominant and recessive Y-linked genes are always expressed.

The Y chromosome contains the genes that control the sex of an individual, as noted earlier. The Y chromosome also controls sperm production and male secondary sex characteristics. Because only males have Y chromosomes, Y-linked traits only appear in males. In addition, Y-linked genes can only be transmitted from fathers to sons. Because the X and Y chromosomes are not homologous, each gene on the Y chromosome has only one allele. Thus, both dominant and recessive Y-linked genes are always expressed.

Sex-Influenced Genes

> The action of some genes is influenced by the sex of an individual.

Although genes determine traits in men and women, some autosomal genes behave differently in the two sexes. In one sex, for example, an allele will be dominant; in the other sex, it will be recessive. These genes are known as **sex-influenced genes**.

The best-known example is the gene for pattern baldness. **Pattern baldness** is the loss of hair that often begins in a man's twenties. The name of this condition results from the fact that affected individuals do not go completely bald; they retain a rim of hair on the temples and back of the head.

The gene for pattern baldness is present in both men and women, but in men the gene acts as if it is autosomal dominant. That is, it is expressed in both heterozygous and homozygous-dominant individuals, which means that lots of men end up with pattern baldness. In contrast, in women, the allele acts as

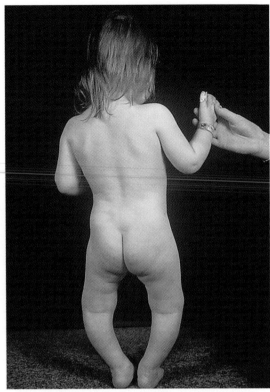

FIGURE 17-16 **Hypophosphatemia** People with hypophosphatemia, a dominant X-linked genetic disorder, resemble this child with rickets, which usually results from inadequate vitamin D intake.

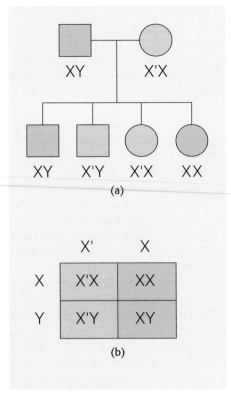

FIGURE 17-17 **Inheritance of a Sex-Linked Dominant Gene** (a) Pedigree. (b) Corresponding Punnett square.

if it is autosomal-recessive. Only women who are homozygous recessive for the trait exhibit baldness. Thus, very few women go bald. For a discussion of the causes and cures of baldness, see **Health Note 17-1**.

What accounts for the different behavior of these genes in men and women? Geneticists believe that the genes are influenced by testosterone, a sex steroid hormone found in far greater concentration in the blood of men than women. In this case, the hormone somehow affects the expression of the pattern-baldness gene.

17-6 Chromosomal Abnormalities and Genetic Counseling

> Abnormal chromosome numbers generally result from a failure of chromosomes to separate during gamete formation.

Besides defective genes, some defects occur as a result of an abnormal number of chromosomes.

During meiosis I, homologous chromosomes pair, then separate—with one member of each pair going to each daughter cell (Figure 17-18a). If a homologous pair fails to separate during meiosis, however, one of the new cells will end up with an extra chromosome (Figure 17-18b). The other cell will be short one chromosome. The failure of homologous chromosomes to separate is called **nondisjunction** (non-diss-JUNK-shun).

Nondisjunction can also occur in the second meiotic division (Figure 17-18c). In this division, you may recall, the 23 replicated chromosomes split apart, with one chromatid going to each daughter cell. If a chromosome fails to separate into its two chromatids, the result is the same as nondisjunction in meiosis I—a daughter cell with an extra chromosome and another daughter cell missing one chromosome.

When a gamete with one extra chromosome unites with a normal gamete, the zygote produced will contain 47 chromosomes. The zygote may be able to divide successfully by mitosis, producing an embryo whose cells have an additional chromosome. Thus, instead of the normal 23 chromosome pairs, each cell in

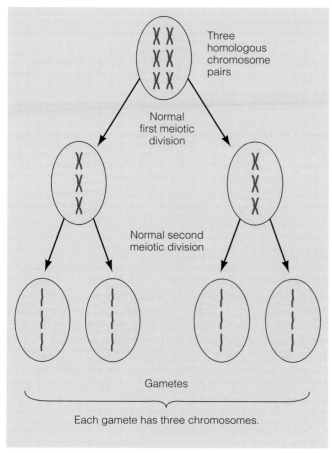

(a)

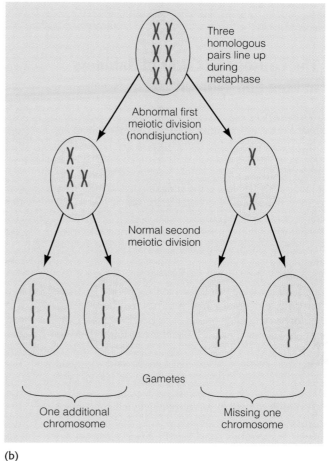

(b)

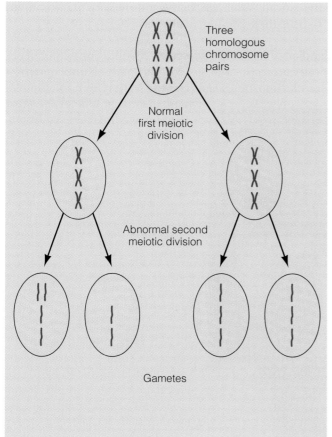

(c)

FIGURE 17-18 Meiosis and Abnormal Chromosome Numbers (a) A simplified version of meiosis. During the first meiotic division, the homologous pairs line up, then separate, producing daughter cells with one-half the number of chromosomes. In the second meiotic division, the chromosomes line up single file and separate, with one chromatid going to each daughter cell. (b) Nondisjunction in the first meiotic division. A chromosome pair may fail to separate during meiosis I, resulting in abnormal gametes. Half are missing a chromosome, and the other half have an extra chromosome. (c) Nondisjunction in second meiotic division, resulting in two normal gametes, one gamete with two chromosomes, and one gamete with four.

the embryo contains 22 pairs and one triplet. This condition is called **trisomy** (TRY-sew-mee; literally, "three bodies").

Gametes with a missing chromosome can also unite with normal gametes. This results in individuals with 45 chromosomes—22 chromosome pairs and a chromosome singlet. This condition is called **monosomy** (MON-oh-SO-me). Surprisingly, one of every two conceptions contains an abnormal chromosome number. Most of these embryos and fetuses die inside the mother.

Down Syndrome Is Trisomy 21. One of the most common trisomies is **Down syndrome**, or **trisomy 21**. Approximately 1 of every 700 babies born in the United States has Down syndrome. Down syndrome children typically are short with round, moonlike faces (Figure 17-19). Their tongues protrude forward, forcing their mouths open, and their eyes slant upward at the corners. Their IQs are rarely over 70. A significant number of Down syndrome babies die from heart defects and respiratory infections in the first

::::healthnote

17-1 The Causes and Cures of Baldness in Men and Women

As we age, we all lose hair. However, some people lose a lot more than others. Doctors recognize three types of hair loss. Partial hair loss involves the loss of small, isolated patches of hair. The second type, pattern baldness, involves the loss of most or all of the hair on the head. Total body hair loss is the complete loss of hair from every part of the body: head, eyelids, eyebrows, arms, legs, and so on.

Hair loss strikes both sexes, all ethnic groups, and all age groups. In fact, two of every three American men will develop some form of balding. An even larger percentage of men and women will lose some hair.

Hair loss, while not life-threatening, can have severe emotional impacts on individuals. Men whose hair begins to thin in their twenties often feel self-conscious and anxious in public. They may feel isolated from their cohorts, for hair loss often makes them look five to ten years older than their peers. Friends may poke fun at them, further adding to their dismay. Even in older men, whose friends are balding, hair loss can result in depression, feelings of low self-esteem, and a general sense of inadequacy. In women, hair loss can have even more devastating effects.

A surprising number of factors can cause hair to fall out. One of the most widely recognized is chemotherapy—chemical treatments used to fight cancer. These drugs attack rapidly dividing cells in the body such as cancer cells and the cells in hair follicles. High fever, severe infections, and severe cases of the flu also cause hair loss, starting one to twelve months after the illness. Many prescription drugs also cause hair loss. Birth control pills, iron deficiencies, and protein-deficient diets can have a similar effect in some patients. Many women report excessive hair loss after pregnancy. But the most common cause of all is genetic.

Hereditary balding (pattern baldness) or thinning, described in the chapter, results from genes that arise from either the mother's or the father's side of the family. In men, this condition results in near total or complete loss of hair. Women, on the other hand, tend to experience excessive thinning, but rarely go completely bald. What can be done to combat balding?

Two lines of attack are possible: medicinal and surgical. Consider the medicinal approach first. Several medicines, some of which are sold over the counter, are available. One of the most widely publicized is Rogaine. In a study of 2300 patients with male pattern baldness, Rogaine treatment resulted in moderate to marked hair growth in 39% of the patients, compared to 11% of those in the control group. For Rogaine to be effective, however, this drug must be applied twice a day every day of one's life. Stopping treatment causes bald spots to reappear. Costing $600 to $1000 per year, Rogaine is recommended for younger men from ages 20 to 30 who have begun to lose hair within the last five years. Men who are completely bald are poor candidates for treatment.

Antiandrogens can be given to women. These drugs inhibit the binding of androgens (male hormones found in women's blood) to the hair follicles. For reasons not well understood, this treatment can result in a com-

---

year of infancy. Modern medical care, especially antibiotics, has reduced early death, and many individuals with Down syndrome live to age 20 or beyond.

The incidence of Down syndrome (and many other monosomies and trisomies) increases with maternal age. As Figure 17-20 shows, a woman's chances of having a Down syndrome baby in-

crease dramatically after age 35. For this reason, many couples choose to have their children at an earlier age.

Nondisjunction of the Sex Chromosomes. Nondisjunction can occur in both autosomes and sex chromosomes. Nondisjunction of the sex chromosomes can lead to a variety of nonlethal genetic disorders. One of the most common is Klinefelter syndrome.

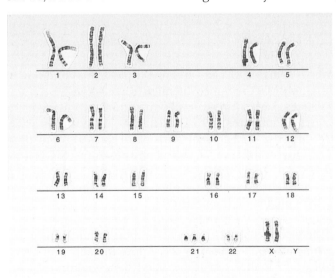

Female with Down Syndrome (47,XX,+21)

(a)

(b)

FIGURE 17-19 Down Syndrome (a) Karyotype of Down syndrome girl. Note trisomy of chromosome 21. (b) Notice the distinguishing characteristics described in the text.

plete reversal of hair loss, but only if it is begun within two years of the onset of hair loss. Men can also be treated with antiandrogens, but not without significant problem, for these drugs cause a loss of sex drive and an undesirable elevation of the voice.

For those who do not respond to drugs, surgery is an option. But surgery can be quite expensive, costing as much as $15,000. Despite its potentially high price tag, an estimated 250,000 American males elect to have one of several different types of surgical procedures each year.

One of the most common is hair transplantation (Figure 1). In this operation, small plugs of hair are taken from the sides and back of the scalp where hair grows thickly. These plugs are placed in the bald spot.

Another common procedure is scalp reduction. In this procedure, surgeons remove well-defined bald spots (up to 2 by 7 inches) in the top of the scalp. The edges of the incision are then drawn together, thus reducing the area of baldness. Hair transplants may also be performed in conjunction with this procedure to fill in the remaining area.

Another option is a wig or toupee. Although wigs or toupees have improved dramatically and well-crafted hair pieces made from real or synthetic hair are very difficult to distinguish from the real thing, there are enough bad ones around that this option is often looked upon with disfavor by many men. Hair pieces can even be sutured in place.

People who are experiencing baldness should seek medical attention quickly to determine the causes. With their doctor's help, they can plot a strategy, if they so desire, to ward off balding. Or, they can simply accept their fate and learn to live with—or even appreciate—it.

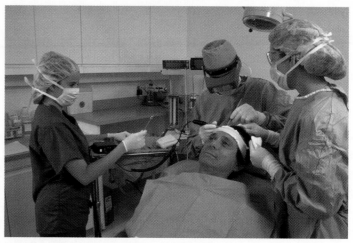

FIGURE 1 Patient undergoing hair transplant surgery

www.jbpub.com/humanbiology/5e

Visit Human Biology's Internet site for links to web sites offering more information on this topic.

Klinefelter syndrome occurs when an ovum with an extra X chromosome is fertilized by a Y-bearing sperm, resulting in an XXY genotype (Figure 17-21). This condition occurs in about 1 of every 700 to 1000 newborn males. Although Klinefelter syndrome patients are males, masculinization is incomplete. The males' external genitalia and testes are unusually small, and about 50% of them develop breasts. Spermatogenesis is abnormal, and Klinefelter patients are generally sterile.

Another common disorder resulting from nondisjunction of the sex chromosomes is **Turner syndrome**, a monosomy. Turner syndrome results when an ovum lacking the X chromosome is fertilized by an X-bearing sperm. It may also result when a genetically normal ovum is fertilized by a sperm lacking an X or Y chromosome (Figure 17-21). The result is an offspring with 22 pairs of autosomes and a single, unmatched X chromosome (XO) (Figure 17-22).

Turner syndrome patients look like females and are characteristically short with wide chests and a prominent fold of skin on their necks. Because their ovaries fail to develop at puberty, Turner syndrome patients are sterile, have low levels of estrogen, and have small breasts. For the most part, they lead fairly normal lives. Mental retardation is not associated with the disorder. Turner syndrome occurs in 1 of every 10,000 female births. The rarity of this condition, compared with Klinefelter's syndrome, is due to the fact that the XO embryo is more likely to be spontaneously aborted.

Defects in Chromosome Structure

Genetic disorders may also result from variations in chromosome structure.

Another defect in chromosomes involves alterations in chromosome structure, the most common of which are (1) **deletions**, the loss of a piece of chromosome, and (2) **translocations**, breakage followed by reattachment elsewhere.

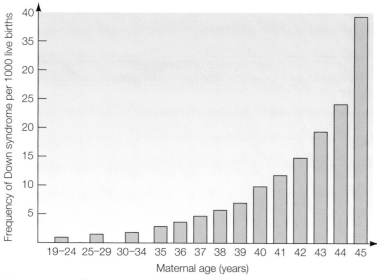

FIGURE 17-20 **Incidence of Down Syndrome Babies at Various Maternal Ages** The incidence of Down syndrome rises quickly after the maternal age of 35.

Deletions.   Most deletions are deleterious, and embryos with them are usually eliminated early in pregnancy. Nevertheless, some embryos whose cells contain deletions do survive. Table 17-5 lists a few disorders caused by deletions. One of the more striking is called Praeder-Willi syndrome (PRAY-der Will-ee).

Praeder-Willi syndrome occurs in 1 in 10,000–25,000 births. It is caused by the loss of one of the arms of chromosome 15 during gamete formation and is characterized by slow infant growth, compulsive eating, and obesity. Babies born with the syndrome have a poor suckling reflex and do not feed well. By age 5 or 6, however, these children become compulsive eaters. Parents must lock their cupboards and refrigerators. Neighbors must be warned to discourage begging and must keep their garbage cans under lock and key. The urge to eat results in obesity, which often leads to diabetes. If food intake is not restricted, victims can literally eat themselves to death. Researchers believe that the eating disorder may result from an endocrine imbalance caused by the deletion.

Translocations.   Translocations occur when a segment of a chromosome breaks off and reattaches to another site on the same chromosome or to another chromosome. Although this might at first seem innocuous, movement of a segment of a chromosome to another site can upset the delicate balance of gene expression and alter homeostasis. Translocations, for example, may be the cause of certain forms of leukemia.

Genetic Screening

> Genetic screening allows parents to determine whether they will have a genetically normal baby.

Chapter 16 noted that parents can now find out the sex of their child well before birth via amniocentesis. This procedure also permits geneticists to search for abnormal chromosome numbers, as well as deletions and translocations. Furthermore, biochemical tests can also be run to pinpoint metabolic diseases, although only a dozen or so are routinely screened.

Amniocentesis is usually recommended only if one or more of the following conditions are met: (1) a woman is over 35, (2) she has already delivered a baby with a genetic defect, (3) she is a carrier of an X-linked genetic disorder, or (4) she or the father has a known chromosomal or genetic abnormality.

As a rule, amniocentesis is usually not performed until the sixteenth week of pregnancy. Before this time, there is not enough fluid surrounding the fetus, and the needle could damage the fetus. Analysis of the fetal cells withdrawn from the amnion requires an additional 10–15 days.

FIGURE 17-22 **Turner Syndrome** In a person with Turner syndrome only one X chromosome is present.

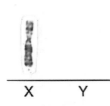

X   Y

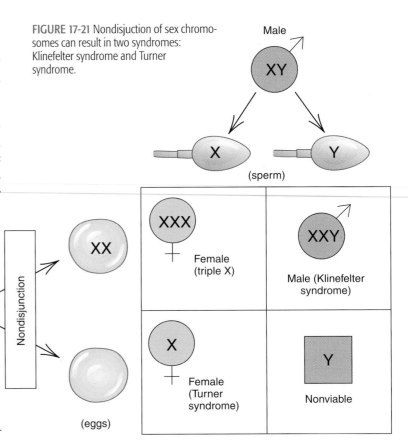

FIGURE 17-21 Nondisjuction of sex chromosomes can result in two syndromes: Klinefelter syndrome and Turner syndrome.

To permit earlier detection of genetic defects, a new procedure, known as chorionic villus biopsy (CORE-ee-on-ick VILL-us BYE-op-see), has been developed. Chorionic villi are composed of embryonic tissue that form the fetal portion of the placenta (plah-SEN- tah), a structure that nourishes the growing fetus. To perform this procedure, physicians remove a small sample of a villus from the uterus through the vagina. The cells of the villus are then examined, much the same way as those removed during amniocentesis. Although chorionic villus biopsy allows for earlier detection, it poses a slightly higher risk to the mother and her fetus.

TABLE 17-5	Chromosome Deletions
Syndrome	Phenotype
Wolf-Hirschhorn syndrome	Growth retardation, heart malformation, cleft palate; 30% die within 24 months.
Cri-du-chat syndrome	Infants have catlike cry, some facial anomalies, severe mental retardation.
Wilm's tumor	Kidney tumors, genital and urinary tract abnormalities
Retinoblastoma	Cancer of eye, increased risk of other cancers
Praeder-Willi syndrome	Infants are weak and grow slowly; children and adults are obese and are compulsive eaters.

Mitochondrial DNA Abnormalities

DNA abnormalities may also occur in mitochondria, resulting in diseases.

Mitochondria contain a tiny fraction (about 0.3%) of a cell's DNA. Until recently, though, the significance of mitochondrial DNA was not well understood.

Researchers now know that a rare form of blindness in humans is linked to a defect in the mitochondrial DNA. The defective gene in the mitochondria that leads to blindness codes for a protein required in the first step of ATP production. The absence of this protein in the neurons of the optic nerve results in their death. This, in turn, leads to blindness usually by age 20.

Because mitochondria are passed on by the mother, all of the children of a woman with the defective gene will inherit it. Only a small fraction of the children who inherit the defective gene actually go blind, so it is thought that other factors also contribute to blindness.

Several rare genetic diseases may also result from mitochondrial DNA defects. Researchers suggest that even some of the more common diseases may be caused by genetic defects in mitochondrial DNA. Some cases of heart, kidney, and central nervous system failure, whose causes are now unknown, may one day be linked to defective mitochondrial DNA.

17-7 Health and Homeostasis

Throughout this book, the idea that homeostasis and human health are profoundly influenced by our environment has emerged repeatedly. In this section, we will look at genes and behavior and the role of our environment in determining our personality and behavior and, perhaps, our health.

At one time, psychologists viewed a baby as a blank slate. A child's personality, they said, develops through interaction with its environment—its parents, friends, teachers, and so on. However, extensive research now suggests that our personalities are also influenced by our genes. Thus, our genes and our environment probably operate together in determining personality.

Michael Lewis, a researcher at the Robert Wood Johnson Medical School, studies infant response to stress. His research shows that newborn babies differ markedly in how they respond to the stress of a blood test performed well before their environment could have affected their personality. Lewis found that some children wail when poked with a needle during routine blood tests in the first few days of life; others hardly seem to notice. Of the newborns who cry, some quickly dampen their response. Others seem to go on forever.

Lewis believes that the difference in response to stress, also seen later in life, is genetically based and that the inherent differences will persist. Lewis has also found that babies differ in how they react to frustration. He performed a series of experiments to test infant response to a frustrating situation. Most children responded with anger. Some, however, showed no response at all, and others displayed sadness. These innate differences in behavior, occurring too early to stem from differences in upbringing, probably result from genetic differences. They may help account for the profound differences in responses to stress seen in adults.

Psychologist Nathan Fox has performed some intriguing studies on shy and extroverted children. His studies show that shy children cling to their mothers when a clown suddenly appears; extroverted children eagerly engage the clown in play. Fox has found that shy children show greater electrical activity in the right part of the brain; extroverted children show a higher level of activity on the left side. Genetics, he suspects, is the reason for this result.

Allison Rosenberg, a researcher at the National Institutes of Health, has also studied shy and outgoing children. Her work shows marked differences between these two groups in heart rate and the release of the hormone cortisol, further supporting the notion that there are inherent physiological differences in children from the outset. These are related to early differences in personality and, very possibly, to differences in genetic makeup. Interestingly, geneticists have recently discovered a gene dubbed the Prozac gene. It is not responsible for the production of Prozac, an antidepressant, but rather is believed to be responsible for mild-mannered behavior in those who have it. Similar findings have been reported in infant monkeys. Important as these genes are, environmental effects, especially events very early in life, can modify genetically programmed behavior.

Another example of the genetic basis of behavior and the role of one's environment in modifying genetically predisposed behavioral patterns comes from research on thrill seeking. Psychologists believe that some individuals are naturally born thrill seekers, or type T people (Figure 17-23). Some researchers believe that thrill seeking may be genetically based. One theory is that risk takers, the type T or "big T" individuals, may be hard to excite and, there-

FIGURE 17-23 **Type T Behavior** Thrill seeking, or type T behavior, is thought to have a genetic basis.

fore, may attempt extraordinary feats for excitement. At the opposite end of the spectrum are "small t" people, risk avoiders, who are easily aroused. They seek to avoid stimulation. Presumably, there's a whole spectrum of folks in the middle.

Some researchers believe that type T individuals have an imbalance of a neurotransmitter known as monoamine oxidase (MAO) in the brain. Thrill seeking supposedly increases the MAO levels in the brain, creating a feeling of exhilaration.

Type T behavior can be modified by an individual's upbringing and turned in a negative or positive direction, say some psychologists. A positive direction might lead an individual to play for the Green Bay Packers football team. A negative direction might lead that same individual, under different environmental conditions, into gang fighting and crime in the streets.

Peers, teachers, relatives, ministers, parents, and others make up the environment. Their influence may turn the type T child to healthy, constructive opportunities or unhealthy, destructive ends. Environmental influences, then, may affect health in a roundabout way. Children who respond abnormally to stress, for instance, may be steered toward mechanisms that reduce stress, resulting in a more healthy lifestyle. Homeostasis is thus broadly affected by the psychological environment.

SUMMARY

Meiosis and Gamete Formation

1. During meiosis, the number of chromosomes is reduced by half. Meiosis involves two nuclear divisions. In males, meiosis produces four gametes, but in females it produces only one.

Principles of Heredity

2. Gregor Mendel, a nineteenth-century monk, derived several important principles of inheritance from his work on garden peas. He determined that in peas, traits do not blend and suggested that each adult has two hereditary factors for a given trait. These factors are called genes today. Mendel suggested that these factors (genes) separate during gamete formation, which is the principle of segregation.

3. Mendel also postulated that a hereditary factor was either dominant or recessive. The dominant and recessive genes are alternative forms of the gene, or alleles. A dominant factor masks a recessive factor. A recessive factor is expressed only when the dominant factor is missing.

4. Three genetic combinations are possible for a given trait: heterozygous, homozygous dominant, and homozygous recessive.

5. The genetic makeup of an organism is called its genotype. The physical appearance, which is determined by the genotype and the environment, is the phenotype.

6. From his studies, Mendel concluded that the hereditary factors were separated independently of one another during gamete formation. This is the principle of independent assortment and holds true only for nonlinked genes.

Mendelian Genetics in Humans

7. Human cells contain 23 pairs of chromosomes: 22 pairs of autosomes and 1 pair of sex chromosomes. Chromosomes carry dominant and recessive traits, and inheritance of these traits is consistent with Mendel's principles of inheritance, although additional mechanisms are at work in humans and other organisms.

8. Sickle-cell disease, cystic fibrosis, and albinism are autosomal-recessive traits and are expressed only in homozygous-recessive individuals.

9. Widow's peak and Marfan's syndrome are autosomal-dominant traits and are expressed in heterozygous and homozygous-dominant genotypes.

Variations in Mendelian Genetics

10. Genetic research since Mendel's time has turned up several additional modes of inheritance. One mode is incomplete dominance. It occurs when an allele exerts only partial dominance, producing intermediate phenotypes.

11. Some genes have more than two possible alleles. Multiple alleles result in more genotypes and phenotypes in a population. Because chromosomes exist in pairs, individuals can have only two of the possible alleles.

12. Codominance occurs in multiple-allele genes. Codominant genes are expressed fully and equally.

13. Some traits are controlled by many genes. This phenomenon is referred to as polygenic inheritance.

14. Genes that are found on the same chromosome are said to be linked. If crossing-over does not occur, these genes are inherited together.

Sex-Linked Genes

15. The X and Y chromosomes are commonly referred to as the sex chromosomes. However, studies suggest that the real determinant of sex is the Y chromosome. If it is absent, the individual (XX) is a female. If it is present, the individual (XY) is a male.

16. The sex chromosomes also carry genes that determine physical traits. A trait determined by a gene on a sex chromosome is a sex-linked trait. Most sex-linked traits occur on the X chromosome.

Chromosomal Abnormalities and Genetic Counseling

17. Abnormalities in the human genome arise from mutations (changes in DNA structure), abnormalities in chromosome number, and alterations in chromosome structure.

18. Alterations in the number of chromosomes result chiefly from errors in gamete formation when chromosomes fail to separate during meiosis, a process called nondisjunction.

19. Variations in chromosome structure result from two occurrences: deletions, or the loss of a piece of chromosome, and translocations, or breakage followed by reattachment elsewhere.

20. Embryos with abnormal chromosome numbers or abnormal chromosome structure are likely to die and be aborted spontaneously.

Health and Homeostasis

21. At one time, psychologists thought that a child's personality developed principally through interaction with the environment—parents, friends, and teachers. New research suggests, however, that our personalities are also influenced by our genes.

critical thinking

Thinking Critically—Analysis

This Analysis corresponds to the Thinking Critically scenario that was presented at the beginning of this chapter.

To begin, reread the previous material in this exercise and make a list of the symptoms characteristic of Marfan's syndrome. They are aortic weakening, abnormally long legs and arms, and nearsightedness. Although you won't be able to find any information about aortic weakening, you could possibly find informa-tion on Lincoln's health—perhaps old medical files. In fact, Lincoln's medical records are available, and they show that the former president displayed no signs of cardiovascular disease. That doesn't really prove anything, by itself. However, Lincoln's lanky limbs were within the normal dimensions of tall people. As for the president's eyesight, it turns out that Lincoln was farsighted, not near-sighted. This evidence strongly suggests that Abe Lincoln did not have this ge-netic disorder. This exercise illustrates the importance of a close examination of the facts.

KEY TERMS AND CONCEPTS

Allele, p. 317
Autosomal-dominant traits, p. 321
Autosomal-recessive traits, p. 319
Autosome, p. 319
Carrier, p. 321
Codominant gene, p. 322
Crossing-over, p. 323
Deletion, p. 329
Dihybrid cross, p. 318
Dominant gene, p. 316
Down syndrome, p. 327
Genome, p. 324
Genotype, p. 317
Heterozygous, p. 317
Homozygous dominant, p. 317

Homozygous recessive, p. 317
Human Genome Project, p. 324
Hypophosphatemia, p. 325
I gene, p. 322
Incomplete dominance, p. 321
Klinefelter syndrome, p. 329
Linked, p. 323
Meiosis, p. 313
Meiosis I, p. 313
Meiosis II, p. 313
Monohybrid cross, p. 318
Monosomy, p. 327
Multiple alleles, p. 322
Nondisjunction, p. 326
Pattern baldness, p. 326

Pedigree, p. 325
Phenotype, p. 317
Polygenic inheritance, p. 322
Principle of independent assortment, p. 319
Principle of segregation, p. 316
Punnett square, p. 317
Recessive, p. 324
Reduction division, p. 313
Sex chromosome, p. 319
Sex-influenced gene, p. 326
Sex-linked gene, p. 324
Translocation, p. 329
Trisomy, p. 327
Turner syndrome, p. 329
X-linked gene, p. 324

CONCEPT REVIEW

1. Explain the process of meiosis in general terms. Where does it occur? What does it accomplish? p. 313
2. Draw a diagram showing the various stages of meiosis I and meiosis II. Make a note of the number of chromosomes at each stage and their condition—that is, whether they have one chromatid or two. Which division is the reduction division? p. 314
3. How is mitosis different from meiosis? How is it similar? pp. 313–315
4. Mendel's research was designed to answer two basic questions. What were the questions and what were his findings? p. 316
5. Define the following terms: principle of segregation, principle of independent assortment, allele, phenotype, genotype, heterozygous, homozygous, monohybrid cross, and dihybrid cross. pp. 316–319

6. Freckles are an autosomal-dominant trait. A woman with freckles (*Ff*) marries and has a baby by a man without freckles (*ff*). What are the chances that their children will have freckles? pp. 317–318
7. Attached earlobes (*A*) are an autosomal-dominant trait. The *A* allele is dominant over the *a* allele, which produces unattached earlobes in homozygous-recessive individuals. A woman with freckles and attached earlobes (*FfAa*) marries a man who has freckles and attached earlobes (*FfAa*). Draw a Punnett square showing the various gametes as well as the genotypes of the offspring. List all possible phenotypes and the genotypes that correspond to them. p. 318
8. What is sickle-cell disease? What causes it? Why can a person be a carrier of the disease but not display outward symptoms? p. 321

9. How do incomplete dominance and codominance differ? Give examples of each. pp. 321–322
10. Assuming that two genes (A and B) control height, list all of the possible genotypes, and indicate the phenotype associated with each. p. 322
11. Describe how crossing-over works. What effect does it have on the genotype of a person's gametes? pp. 323–324
12. Color blindness is a recessive, X-linked gene. A color-blind man and his wife have four children, two boys and two girls. One boy and one girl are color-blind, and the other two are normal. What is the genotype of the woman? p. 325
13. What is a sex-influenced gene? Give an example. p. 326

SELF-QUIZ: TESTING YOUR KNOWLEDGE

1. Meiosis I is also known as a _____ division. During this process, the cell goes from _____ to _____ chromosomes (give a number). p. 313
2. During meiosis II, the _____ chromosomes each contain ____ chromatids. p. 313
3. In males, meiosis results in the formation of _____ gametes, each with _____ (number) single-stranded chromosomes. p. 314
4. Mendel postulated that hereditary factors are either _____ or recessive. p. 316
5. The physical appearance of an organism is known as its _____ while its genetic make up is known as its _____. p. 317

6. The principle of independent assortment only applies to genes on _____ chromosomes. p. 319
7. The human somatic cell contains _____ autosomes and two sex chromosomes. p. 319
8. The sex chromosomes of women are _____. p. 319
9. In an autosomal recessive disease such as cystic fibrosis, both alleles must be _____ for the disease to appear. p. 320
10. Blending of traits does occur as a result of _____ dominance. p. 321

11. The ABO blood types are the result of three different alleles. The A and B alleles are expressed fully and equally. This phenomenon is called _____. p. 322
12. The mapping of the human genome was the result of the _____ project. p. 324
13. A recessive gene on the Y chromosome of a man is _____ expressed. p. 326
14. The failure of two chromosomes to separate during meiosis is known as _____. p. 326
15. A _____ occurs when a segment of a chromosome breaks off and attaches to another part of the chromosome. p. 330

www.jbpub.com/humanbiology/5e

The site features eLearning, an online review area that provides quizzes, chapter outlines, and other tools to help you study for your class. You can also follow useful links for in-depth information, research the differing views in the Point/Counterpoints, or keep up on the latest health news.

18-1 DNA and RNA: Macromolecules with a Mission

18-2 How Genes Work: Protein Synthesis

18-3 Controlling Gene Expression

18-4 Health and Homeostasis

How Genes Work and How Genes are Controlled

thinking critically

One of America's leading drug companies spent $300 million to construct a facility to produce synthetic growth hormone to be sold to dairy farmers. The hormone will be generated by genetically engineered bacteria that contain growth-hormone genes isolated and transplanted into the bacteria from dairy cattle. Administered to cows, the hormone dramatically increases milk production. Business economists believe that an increase in milk production will reduce the cost of milk to the consumer. Based on this information, does introducing synthetic growth hormone seem like a good idea?

In 1988, a U.S. military court sentenced a serviceman in Korea to 45 years in prison for rape and attempted murder. Ten days later, a Florida court convicted a man on two counts of first-degree murder. What makes these two convictions important is that prosecutors relied heavily on the results of a technique called "DNA fingerprinting."

In this procedure, criminologists analyze the composition of the genetic material DNA in samples of hair, semen, or blood recovered from crime scenes. Then they compare the DNA in these samples with the DNA of the accused. If it matches, prosecutors can make a strong case for guilt. Without this procedure, prosecutors believe, the two convictions cited above could not have been won.

Many people believe that DNA fingerprinting is revolutionizing criminal investigations. Many police departments are now taking DNA samples from all convicted sex offenders. Like fingerprints, the results of DNA analysis will be kept on file, readily available for future cases. The FBI is also using this technique in its work.

DNA fingerprinting holds great promise, say supporters, because there is only 1 chance of a mistaken identity in 4 or 5 trillion. In contrast, the best conventional methods, such as blood typing, run the risk of error in about 1 in every 1000 cases.

Although DNA fingerprinting sounds promising, some geneticists believe that the probabilities (noted above) are based on improper assumptions (violating one of the critical thinking rules) and comparatively little scientific data on the genetic composition of populations.

DNA fingerprinting is one of many offshoots of research in genetics. This chapter covers some of the basic research that led to the discovery of this technique. You will learn about the structure of DNA and RNA, how genes work, and how genes are controlled in body cells.

18-1 DNA and RNA: Macromolecules with a Mission

DNA consists of a double helix held together by hydrogen bonds.

In 1953, James Watson, an American biologist, and Francis Crick, a British biologist, proposed a model for the structure of the DNA molecule. This model was based on research by Rosalind Franklin, Maurice Wilkins, and many other scientists, as outlined in Scientific Discoveries That Changed the World 16-1. Their work opened the doors to a new field of research known as *molecular genetics*, the study of the structure and function of RNA and DNA at the molecular level. We begin this chapter by looking first at the structure of DNA.

DNA is a molecule that consists of two separate strands that intertwine to form a double helix, a structure that resembles a spiral staircase (Figure 18-1). Each strand of the double helix contains millions of small molecules known as *nucleotides*. Like several other important biological molecules, then, DNA is a polymer, a molecule made up of many smaller ones.

Each nucleotide in turn consists of three even smaller molecules: a nitrogenous base, a phosphate group, and a monosaccharide, deoxyribose (Figure 18-2). The nucleotides join by covalent bonds to form polynucleotide chains.

In the DNA molecule, the two polynucleotide chains are held in place by hydrogen bonds. As shown in Figure 18-1, the hydrogen bonds that form between the bases of the nucleotides of each chain project inward and therefore lie inside the helix. Hydrogen bonds between the bases are indicated by the dotted lines in Figure 18-1. These bonds are much weaker than covalent bonds and can be easily broken. This, in turn, allows the molecule to unzip for replication.

DNA contains two types of nitrogen-containing bases. They are the purines (PURE-eens) and pyrimidines (pa-RIM-a-DEENS) (Figure 18-2). The **purines** consist of two rings. The **pyrimidines** contain only one ring. In DNA, you will find two purine bases. They are adenine (A) (AD-ah-neen) and guanine (G) (GUAN-neen). The pyrimidines in DNA are cytosine (C) (sigh-toe-seen) and thymine (T) (thigh-mean). My genetics prof came from the agricultural college of my university, and, evidently proud of his roots, gave us the mnemonic "pure AG" to remember these.

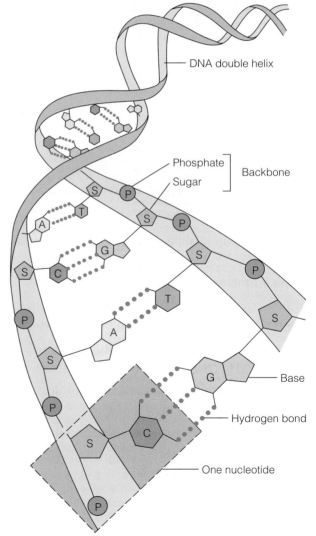

FIGURE 18-1 **DNA** DNA consists of two intertwined polynucleotide chains that form a double helix. Sugars and phosphates form the backbone of each chain, with the bases projecting inward. The bases on opposite strands are connected by hydrogen bonds.

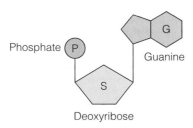

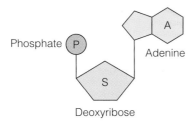

(a) DNA nucleotides containing purine bases

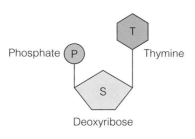

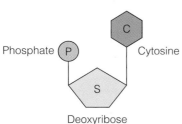

(b) DNA nucleotides containing pyrimidine bases

FIGURE 18-2 **Nucleotides Containing Purine and Pyrimidine Bases** All nucleotides consist of three subunits: a phosphate group; an organic, nitrogen-containing base; and a five-carbon sugar. DNA nucleotides contain the sugar deoxyribose. Two types of bases are found: purines and pyrimidines. (a) The purines are adenine and guanine. (b) The pyrimidines are cytosine and thymine.

Purines on one strand bind (via hydrogen bonds) to pyrimidines on the opposite strand. But the relationship between the two is even more specific. If you look at Figure 18-3b, you will see that the adenine binds only to the pyrimidine thymine. Guanine binds only to the pyrimidine cytosine. Adenine and thymine are therefore said to be complementary bases, as are guanine and cytosine. As you will soon see, this specific coupling, called **complementary base pairing**, ensures the accurate replication of DNA and the reliable transmission of the genetic information from a parent cell to its daughter cells during cell division.

DNA Replication

DNA unwinds and then serves as a template for the production of new DNA strands.

As noted in Chapter 3, before a cell can divide, it must first make an exact copy of all of its DNA. DNA replication ensures that a cell about to divide has two identical sets of genetic information, one for each daughter cell.

During the S phase of interphase, cells replicate their DNA by unzipping each DNA double helix along the hydrogen bonds that unite the complementary bases. Each polynucleotide strand then serves as a template (described below) on which a new strand is produced (Figure 18-3a). The new strand is called a *complementary strand*. The template is called the *original strand*.

DNA replication is referred to as a semiconservative process because each polynucleotide chain of the DNA double helix serves as a template for the production of a new strand of DNA.

DNA replication during the S phase begins when special enzymes start to pull apart, or unzip, the DNA double helix. As the two strands are separated, the bases of the polynucleotide chains are exposed. They are then free to form hydrogen bonds with complementary nucleotides in the nucleus (Figure 18-3). Because adenine binds only to thymine and guanine binds only to cytosine, the original strands (the templates) are said to direct the synthesis of new strands. Synthesis on the DNA templates occurs one base at a time, and the accuracy of replication is ensured by complementary base pairing.

During replication, incoming nucleotides must first be aligned properly so that the hydrogen bonds can form between complementary bases and so that the phosphate group of the incoming nucleotide can bond to the sugar of the nucleotide already in place, as illustrated in Figure 18-4. Alignment and attachment are aided by an enzyme known as **DNA polymerase** (PUL-yi-merr-ace) (Figure 18-4). DNA polymerase slides along the template, aligning nucleotides one at a time, then linking each new nucleotide to the one already in place. As noted above, the enzyme joins the phosphate group of one nucleotide to the deoxyribose molecule (sugar) of its neighbor.

When DNA synthesis is finished, two new DNA molecules exist, each of which is a double helix. Each double helix, in turn, contains one strand from the original molecule and one new polynucleotide chain. The chromosome, once a single molecule of DNA and protein, now consists of two DNA molecules with associated proteins. As noted in Chapter 16, each strand of DNA and associated protein is called a *chromatid*. The two chromatids of each chromosome are joined at their centromeres after synthesis.

RNA

Three types of RNA exist, each of which is involved in protein synthesis.

DNA contains the genetic information of the cell that determines the structure and controls most of its functions. Like a commander in the army, DNA does not exert its influence directly. Its work is carried out by other molecules, notably RNA. Three types of RNA are involved in this process: ribosomal RNA (rRNA), messenger RNA (mRNA), and transfer RNA (tRNA).

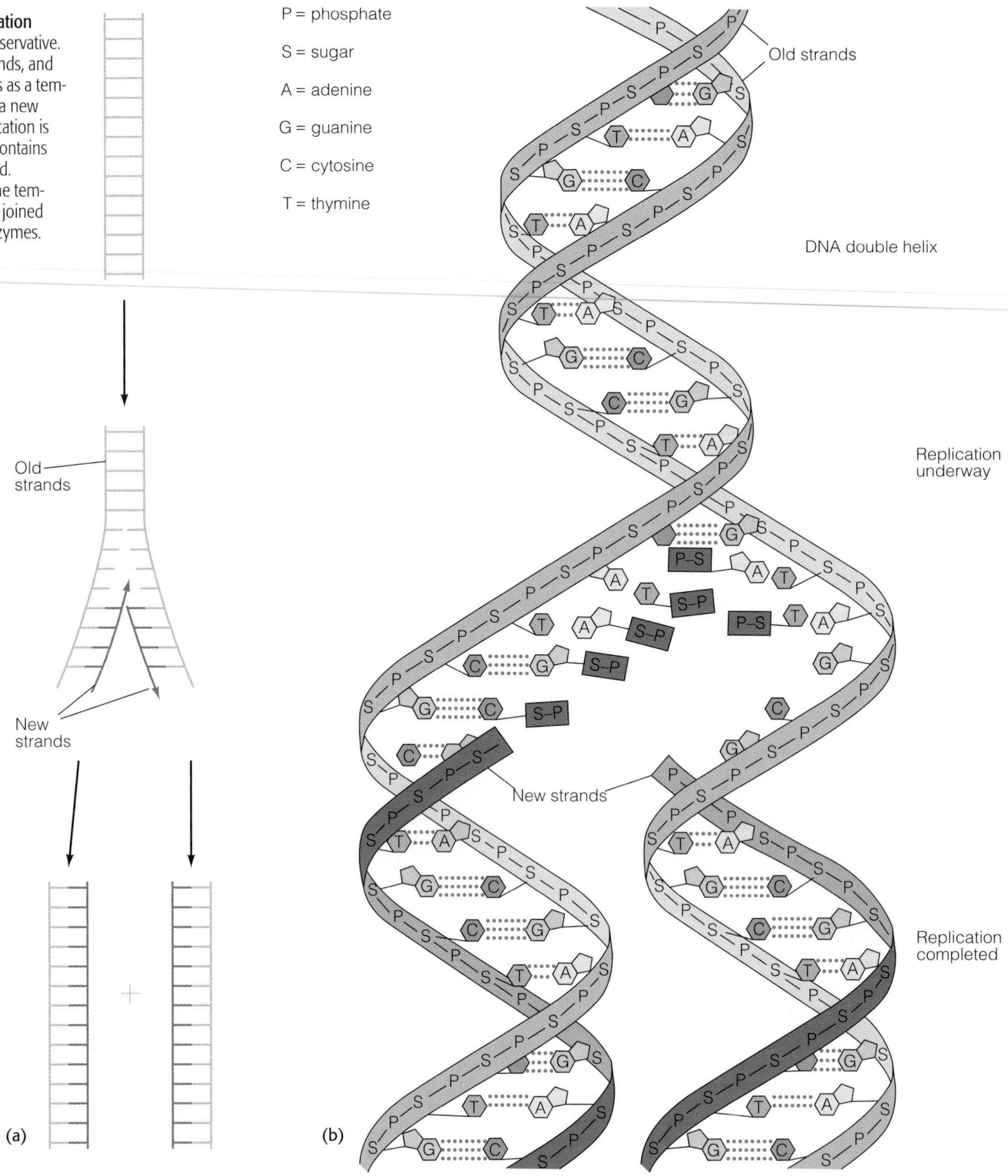

FIGURE 18-3 DNA Replication
DNA replication is semiconservative.
(a) Each double helix unwinds, and each half of the helix serves as a template for the production of a new strand of DNA. When replication is complete, each new helix contains one old and one new strand.
(b) Nucleotides attach to the template one at a time and are joined together with the aid of enzymes.

P = phosphate

S = sugar

A = adenine

G = guanine

C = cytosine

T = thymine

Old strands

DNA double helix

Old strands

Replication underway

New strands

New strands

Replication completed

(a)

(b)

Each has a unique function in protein synthesis (Table 18-1). Despite major functional differences in these molecules, all three RNA molecules are biochemically similar. In humans, for example, all RNA molecules are single-stranded, and all are polynucleotides (Figure 18-5). RNA nucleotides consist of three molecules: a sugar, a nitrogenous base, and a phosphate group. Unlike the DNA nucleotides, however, the RNA nucleotides contain the sugar ribose instead of deoxyribose. They also contain the pyrimidine uracil instead of thymine. Table 18-2 summarizes the key differences between RNA and DNA.

RNA Synthesis

RNA synthesis takes place on a DNA template in the nuclei of cells.

Unlike DNA, which is self-replicating, all three types of RNA must be synthesized on DNA templates. The synthesis of RNA on a DNA template permits the genetic information coded in the DNA to be transferred to RNA. As you will soon see, the genetic code in DNA is transferred to only one of the three types of RNA, messenger RNA. Messenger RNA is free to leave the nucleus. It therefore serves as a kind of shuttle service (or mes-

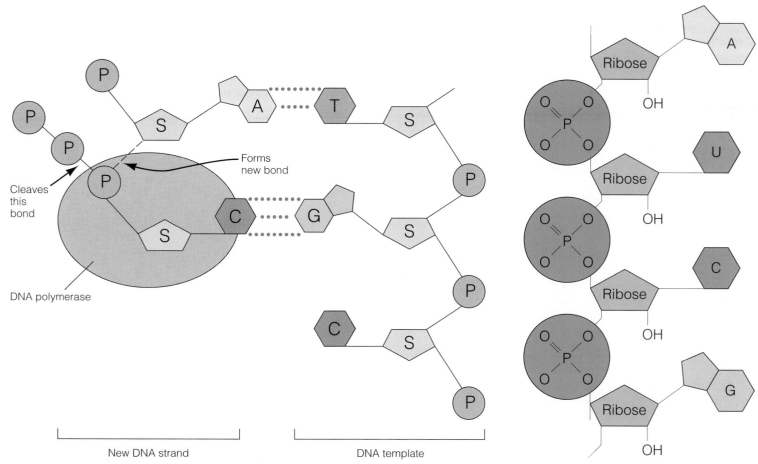

FIGURE 18-4 **Role of DNA Polymerase** DNA polymerase binds loosely to the DNA template and to the nucleotide, aligning it for insertion. The enzyme catalyzes the formation of the bond between the phosphate of the new nucleotide and the sugar of the one previously inserted. The enzyme also cleaves off two phosphates of each nucleotide.

FIGURE 18-5 **RNA** RNA is a single-stranded molecule consisting of many RNA nucleotides. A small section of an RNA molecule is shown here.

senger) that transfers the genetic information necessary for protein synthesis into the cytoplasm, where proteins are made.

RNA synthesis on a DNA template is called **transcription**. Geneticists chose this name because this process is much like the transcribing process students perform during a lecture where they transfer information from one medium (speech) to another (written form). In this case, genetic information housed in the DNA is simply written in another form, RNA.

Transcription occurs during interphase of the cell cycle. During transcription, small sections of the DNA helix unzip temporarily with the aid of special enzymes. This creates a DNA template on which RNA can be made (Figure 18-6). But RNA is produced on only one of the DNA strands, and only a small portion (less than 1%) of a cell's DNA is used to make RNA. (This helps explain why most mutations in the DNA have no effect on cell function.)

During RNA synthesis, an enzyme called **RNA polymerase** helps align the RNA nucleotides on the DNA template in much the same way that DNA polymerase functions during DNA synthesis. RNA polymerase also catalyzes the formation of covalent

TABLE 18-1	**Role of RNA Molecules**
Molecule	Role
Messenger RNA (mRNA)	Carries the genetic information that is needed to make proteins in the cytoplasm from the nucleus
Transfer RNA (tRNA)	Binds to specific amino acids, transports them to the mRNA, and inserts them in the correct location on the mRNA
Ribosomal RNA (rRNA)	Component of the ribosome

TABLE 18-2	**Differences Between RNA and DNA in Prokaryotic and Eukaryotic Cells**
DNA	RNA
Double-stranded	Single-stranded
Contains the sugar deoxyribose	Contains the sugar ribose
Contains adenine, guanine, cytosine, and thymine	Contains adenine, guanine, cytosine, and uracil
Functions primarily in the nucleus	Functions primarily in the cytoplasm

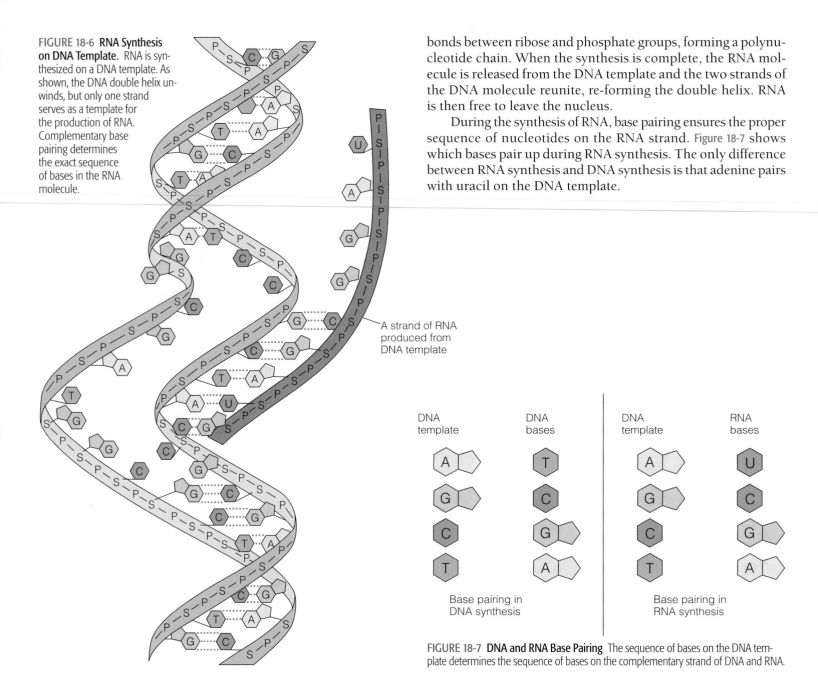

FIGURE 18-6 **RNA Synthesis on DNA Template.** RNA is synthesized on a DNA template. As shown, the DNA double helix unwinds, but only one strand serves as a template for the production of RNA. Complementary base pairing determines the exact sequence of bases in the RNA molecule.

A strand of RNA produced from DNA template

bonds between ribose and phosphate groups, forming a polynucleotide chain. When the synthesis is complete, the RNA molecule is released from the DNA template and the two strands of the DNA molecule reunite, re-forming the double helix. RNA is then free to leave the nucleus.

During the synthesis of RNA, base pairing ensures the proper sequence of nucleotides on the RNA strand. Figure 18-7 shows which bases pair up during RNA synthesis. The only difference between RNA synthesis and DNA synthesis is that adenine pairs with uracil on the DNA template.

DNA template	DNA bases		DNA template	RNA bases
A	T		A	U
G	C		G	C
C	G		C	G
T	A		T	A

Base pairing in DNA synthesis

Base pairing in RNA synthesis

FIGURE 18-7 **DNA and RNA Base Pairing** The sequence of bases on the DNA template determines the sequence of bases on the complementary strand of DNA and RNA.

18-2 How Genes Work: Protein Synthesis

Protein is synthesized on an mRNA template.

The information required to synthesize protein is transported out of the nucleus as messenger RNA. In the cytoplasm, mRNA serves as a template for protein synthesis, which occurs either on the endoplasmic reticulum or free within the cytoplasm, as noted in Chapter 3. The synthesis of protein on an mRNA template is called **translation** (Figure 18-8). The RNA message (the genetic code) is translated into a new "molecular language"— that of the protein.

Protein synthesis requires two additional "players," transfer RNA (tRNA) and ribosomes. **Transfer RNA** molecules are relatively small strands of RNA that bind to specific amino acids

in the cytoplasm and transport them to the mRNA. Here they are incorporated into the protein that's being constructed on the mRNA molecule (Figure 18-9). A tRNA molecule bound to an amino acid is generally written like this: tRNA–AA.

Ribosomes, mentioned briefly in Chapter 3, are small granular structures found in the cytoplasm of cells. These structures play an important role in the production of protein. They are composed of **ribosomal RNA (rRNA)** and protein. Ribosomes consist of two subunits, a small one and a large one. The subunits are often detached in the cytoplasm, by unit when protein is synthesized on the mRNA template.

Protein synthesis consists of three stages: (1) chain initiation, (2) chain elongation, (2) and chain termination. During

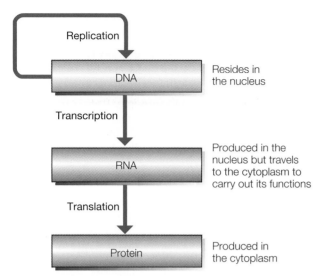

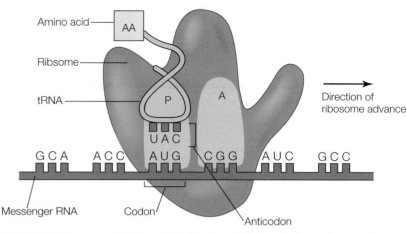

FIGURE 18-8 The Central Tenet of Molecular Genetics The DNA controls the cell through protein synthesis, that is, the production of enzymes and structural proteins. RNA serves as an intermediary, carrying genetic information to the cytoplasm, where protein is synthesized.

FIGURE 18-9 Messenger RNA The mRNA, either free within the cytoplasm or bound to the endoplasmic reticulum, serves as a template for protein synthesis. The codons, each consisting of three bases on the mRNA, determine the sequence of amino acids by binding to complementary anticodons on the tRNA molecules. Each tRNA, in turn, delivers a specific amino acid to the mRNA template. The ribosome provides binding sites for tRNA–AA molecules, catalyzes the formation of the peptide bonds, and slides down the mRNA template to permit chain elongation.

chain initiation, the small subunit of the ribosome attaches to the mRNA at a specific site called the **initiator codon** (COE-dawn). The initiator codon is a sequence of three bases on the mRNA molecule that marks where protein synthesis should begin. After the small subunit attaches, the large subunit of the ribosome links up. As shown in Figure 18-10a, the ribosome contains two binding sites for tRNA–AA.

Soon after the ribosome attaches to the mRNA, a tRNA bearing a specific amino acid enters the first binding site. But how does the cell know which amino acid to insert? At the base of all tRNA molecules is a sequence of three bases. Called an **anticodon**, it contains complementary bases that pair with the bases of the initiator codon of the mRNA template.

After the first tRNA–amino acid is in place, a second tRNA–AA enters the scene, inserting itself into the second binding site on the ribosome. The second binding site is located above the next codon, the next three bases on the mRNA template (Table 18-3). In the example shown in Figure 18-10, the codon

TABLE 18-3	Codons on mRNA and Their Corresponding Amino Acids						
Codon	**Amino Acid**	**Codon**	**Amino Acid**	**Codon**	**Amino Acid**	**Codon**	**Amino Acid**
AAU	Asparagine	CAU	Histidine	GAU	Aspartic acid	UAU	Tyrosine
AAC		CAC		GAC		UAC	
AAA	Lysine	CAA	Glutamine	GAA	Glutamic acid	UAA	Terminator codon*
AAG		CAG		GAG		UAG	
ACU	Threonine	CCU	Proline	GCU	Alanine	UCU	Serine
ACC		CCC		GCC		UCC	
ACA		CCA		GCA		UCA	
ACG		CCG		GCG		UCG	
AGU	Serine	CGU	Arginine	GGU	Glycine	UGU	Cysteine
AGC		CGC		GGC		UGC	
AGA	Argenine	CGA		GGA		UGA	Terminator codon*
AGG		CGG		GGG		UGG	Tryptophan
AUU	Isoleucine	CUU	Leucine	GUU	Valine	UUU	Phenylalanine
AUC		CUC		GUC		UUC	
AUA		CUA		GUA		UUA	Leucine
AUG	Methionine	CUG		GUG		UUG	

*Terminator codons signal the end of the formation of a polypeptide chain.

FIGURE 18-10 **Protein Synthesis**

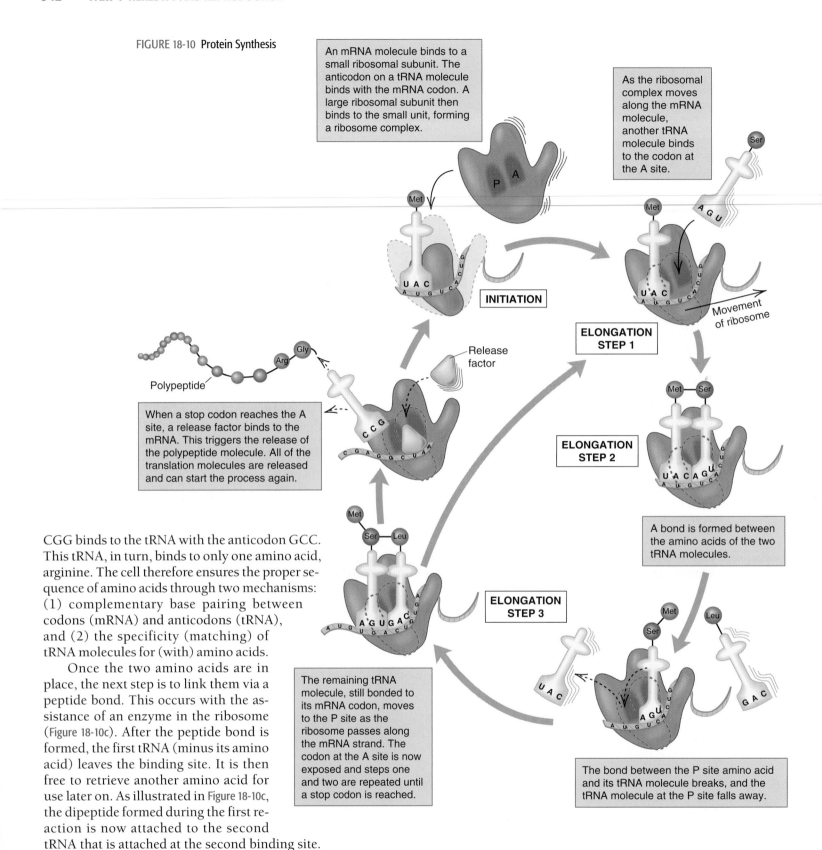

An mRNA molecule binds to a small ribosomal subunit. The anticodon on a tRNA molecule binds with the mRNA codon. A large ribosomal subunit then binds to the small unit, forming a ribosome complex.

As the ribosomal complex moves along the mRNA molecule, another tRNA molecule binds to the codon at the A site.

INITIATION

Movement of ribosome

ELONGATION STEP 1

Release factor

ELONGATION STEP 2

When a stop codon reaches the A site, a release factor binds to the mRNA. This triggers the release of the polypeptide molecule. All of the translation molecules are released and can start the process again.

Polypeptide

A bond is formed between the amino acids of the two tRNA molecules.

ELONGATION STEP 3

The remaining tRNA molecule, still bonded to its mRNA codon, moves to the P site as the ribosome passes along the mRNA strand. The codon at the A site is now exposed and steps one and two are repeated until a stop codon is reached.

The bond between the P site amino acid and its tRNA molecule breaks, and the tRNA molecule at the P site falls away.

CGG binds to the tRNA with the anticodon GCC. This tRNA, in turn, binds to only one amino acid, arginine. The cell therefore ensures the proper sequence of amino acids through two mechanisms: (1) complementary base pairing between codons (mRNA) and anticodons (tRNA), and (2) the specificity (matching) of tRNA molecules for (with) amino acids.

Once the two amino acids are in place, the next step is to link them via a peptide bond. This occurs with the assistance of an enzyme in the ribosome (Figure 18-10c). After the peptide bond is formed, the first tRNA (minus its amino acid) leaves the binding site. It is then free to retrieve another amino acid for use later on. As illustrated in Figure 18-10c, the dipeptide formed during the first reaction is now attached to the second tRNA that is attached at the second binding site.

In order for the chain to grow, the ribosome must move down the mRNA strand. This is accomplished with the aid of a contractile protein in the ribosome. This protein permits the ribosome to slide along mRNA one codon at a time.

As illustrated in Figure 18-10d, after the first tRNA is released, the ribosome moves down the mRNA, and the dipeptide is shifted to the first binding site. This frees up the second site, permitting it to accept another tRNA–AA. A new tRNA–AA then enters the vacant site and is joined to the dipeptide, thus forming a tripeptide.

Chain elongation takes place by the addition of one amino acid at a time, but this does not mean that protein synthesis is a slow process. Quite the contrary, protein synthesis occurs with remarkable speed. In bacteria, many proteins containing numerous amino acids are synthesized in 15–30 seconds. In hu-

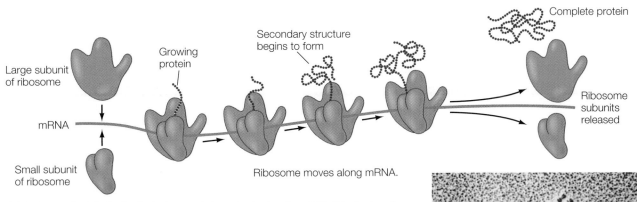

FIGURE 18-11 **Protein Synthesis** As the protein is synthesized on the mRNA, it begins to coil and bend, forming its secondary structure. Several ribosomes may "work" a single strand of mRNA simultaneously.

mans, the large subunits of the very large hemoglobin molecule are synthesized in 3 minutes.

As the peptide chain is formed, hydrogen bonds between amino acids on different parts of the chain cause it to bend and twist, forming the secondary structure of the protein or peptide, described in Chapter 3 (Figure 18-11). When the ribosome reaches the terminator codon, the sequence of bases that signals the end of the protein chain, the protein is released. If the protein is produced on mRNA on the surface of the rough endo-

plasmic reticulum, it is transferred into its interior. The protein may then be chemically modified and packaged into transport vesicles, which pinch off and transport their cargo to the Golgi complex. If protein synthesis occurs on mRNA molecules in the cytoplasm, the protein is released into the cytoplasm.

18-3 Controlling Gene Expression

In humans, genetic expression is controlled at four levels.

In the previous sections, you've seen how RNA and protein are produced. In this section, we look at how genes are controlled. This information is useful to geneticists because it helps them discover new ways to treat, or even cure, diseases such as cancer that originate in the genes of an organism.

Genetic control occurs at four levels: (1) control at the chromosome level, (2) control at transcription, (3) control after transcription before translation, and (4) control at translation (Figure 18-12).

Control at the Chromosome Level. Chapter 16, you may recall, noted that chromatin fibers in the nucleus condense and become inactive during prophase of mitosis in preparation for cell division. In the condensed state, they are inactive and cannot produce RNA or new DNA.

During interphase, cells also inactivate some of their chromatin. Why? Most cells have far more DNA than they need. Muscle cells, for instance, don't need all of the genes that a liver cell needs. Since the muscle cells do not need this DNA, they inactivate it.

Chromosomal condensation or coiling provides a crude way of regulating genetic expression. More precise controls occur at other levels.

Control of Transcription. In humans, transcription (RNA production) can be controlled by certain factors, for example, the end

products of metabolic pathways or substrates of metabolic pathways. The presence of these substances may turn on (activate) some genes. In other cases, they turn off (repress) genes. For more on this process, see Scientific Discoveries that Changed the World 18-1.

Human cells also control gene activity via a third mechanism, known as *enhancement*. **Enhancement** is a process in which already active genes increase their production of mRNA.

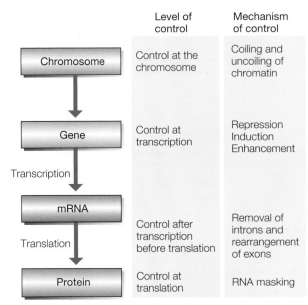

FIGURE 18-12 **Gene Expression** In humans, gene expression is regulated at four levels.

Scientific Discoveries that Changed the World

18-1 Unraveling the Mechanism of Gene Control
Featuring the Work of Jacob and Monod

FIGURE 1 **Jacob and Monod**
These two French scientists proposed the operon hypothesis.

One of the questions that intrigued geneticists for many years—and still does—is how the genetic information contained in a cell's DNA is controlled. Many experiments have been performed over the years in an attempt to answer this question, but none so important as those of two French molecular geneticists, Francois Jacob and Jacques Monod, on the common intestinal bacterium *E. coli* (Figure 1). Today, much of what is known about gene regulation in bacteria and humans comes from their studies.

This occurs as a result of the action of nearby segments of the DNA called *enhancers*. These are protein molecules that bind to enhancer regions of the chromosome. These proteins increase RNA production by DNA. Thus, enhancers do not turn nearby genes on and off; they amplify their activity.

Control before Translation. Genetic studies show that DNA in humans contains two types of genetic material: sections that produce mRNA and sections that do not. As Figure 18-13 shows, the noncoding segments of DNA are called *introns* and are interspersed within the functional segments of DNA (functional genes), the *exons*. The name *intron* signifies that these are *in*tervening segments of noncoding DNA. Exons are *ex*pressed segments—sections of the DNA that can be used to produce the mRNA that is used to make protein.

Both introns and exons are transcribed during the cell cycle, producing an RNA transcript that is a complete "readout" of the DNA. Cells then process the transcript, cutting out the introns, and then joining the exons together to produce a functional mRNA molecule. Put another way, in order for a functional mRNA molecule to be synthesized, it must be "cut and pasted" together. The exons can be linked differently, producing slightly different products.

Control at Translation. Eukaryotic cells can also control gene expression at the level of translation (protein synthesis). Messenger RNA, for example, which is produced in the nucleus, may be transferred to the cytoplasm in an inactive state. The mRNA is said to be *masked*. Masking allows the cell to build up large supplies of mRNA in preparation for a sudden burst of protein synthesis. When the time is right, the mRNA is activated and begins producing large quantities of protein. This phenomenon occurs in the human ovum, the female sex cell produced in the ovaries. The ovum produces an enormous amount of masked mRNA as it awaits fertilization. When the sperm arrives and fertilizes the ovum, the mRNA that has accumulated in the cytoplasm is unmasked, and there is an explosion of protein synthesis necessary for subsequent cell divisions.

In closing, gene expression in humans and other eukaryotes is a complex phenomenon. Like the cell cycle, it is no doubt under the influence of many controlling factors.

Exon   Intron   Exon   Intron   Exon

DNA

Posttranscriptional modifications:

Transcription

mRNA

Introns are cut out.

Section removed   Section removed

mRNA

Exons are joined together.

The mRNA transcript is transported to cytoplasm for translation.

FIGURE 18-13 **Posttranscriptional Control of Human Genes** One key mechanism of control occurs after transcription. DNA contains noncoded segments (introns) and useful segments (exons). An mRNA transcript is produced from the DNA, and the segments produced by introns are removed. The final product is a strand of mRNA containing only RNA copies of the exons. The exon copies can be linked differently, producing slightly different products.

Working at the Pasteur Institute in the late 1950s, Jacob and Monod focused much of their attention on the bacterial uptake and breakdown of lactose, a disaccharide also known as milk sugar. *E. coli* absorb lactose and break it down with the aid of an enzyme known as galactosidase (ga-lack-TOSE-uh-DACE). The breakdown of lactose results in the release of energy.

E. coli living in a lactose-free medium, however, contain very little of the enzyme. It is only when lactose is added that the enzyme is synthesized.

The introduction of lactose also increases the production of a carrier molecule (galactoside permease) found in the plasma membrane. It transports lactose into the bacterium. In addition, the presence of lactose increases the intracellular concentration of another enzyme, which may play a role in lactose breakdown, but which is not essential to the process. The carrier protein and two enzymes are part of what scientists call an inducible enzyme system. The production of these chemicals is induced by the presence of lactose.

Mapping studies have shown that three adjacent genes on the bacterial DNA are responsible for the production of these proteins. These studies led Jacob and Monod to hypothesize that the three genes belong to a single unit, which they called an operon. They defined an operon as a cluster of genes with related functions that is regulated in such a way that all the genes in the group are activated and inactivated simultaneously.

In a paper published in 1961, the researchers proposed a mechanism by which these genes might be controlled. This intriguing model is "one of the truly important conceptual advances in biology," according to University of Wisconsin cell biologist Wayne Becker.

Extensive studies of bacteria helped Jacob and Monod piece together a picture of gene regulation in *E. coli*. As a testimony to the thoroughness of their work, the original model has undergone very little change in over 30 years.

18-4 Health and Homeostasis

One of the leading causes of death in humans is a disease called *cancer*. Not one disease, cancer actually consists of over 100 different diseases such as liver cancer, lung cancer, breast cancer, and prostate cancer. Cancer is caused by numerous biological, chemical, and physical agents in a two-step process. One cause of cancer is mutations, that is, alterations of parts of the genetic material.

Researchers have made a startling discovery about such mutations that may explain the cause of most cancers. They have found that humans and other species contain a group of genes called **proto-oncogenes** (pro-toe-ON-co-JEANS). Proto-oncogenes control functions related to cellular replication, among them cell adhesion and the production of the plasma membrane receptors that bind to growth factors or hormones. When mutations in these genes are not repaired, the result can be uncontrolled cellular proliferation, cancer. (For a discussion of methods used to assess the potency of potential cancer-causing agents, see the Point/Counterpoint.)

Certain viruses also cause cancer. Cancer-causing viruses fall into two groups. Some viruses stimulate cancer by activating proto-oncogenes in body cells. Others contain cancer-causing genes themselves, which they insert into the DNA of the nuclei of infected cells. These genes are known as **oncogenes**. After being inserted into a cell's DNA, oncogenes turn on the cell's genes that control cell division.

The discovery of proto-oncogenes and viral oncogenes has resulted in a quantum leap in our understanding of cancer. Some scientists believe that the key to treating cancer may ultimately lie in finding ways to turn off activated proto-oncogenes. Such changes could halt replication of cancer cells. Who knows, maybe someday you or someone you know will benefit directly from this idea. (For more on cancer treatment, see Health Note 20-2.)

Thousands of mutations occur in each cell of your body every day, many of which could lead to serious problems, including cancer. As noted earlier, the cells of your body repair much of the damage. This repair is brought about by special enzymes in the nucleus that "snip off" damaged sections of the DNA, then rebuild the molecule, thus helping to ensure normal cell function.

Remaining healthy, though, requires that we not stretch that resiliency to the breaking point. Some researchers fear that modern industrial society may be doing just that. How?

Today, 80,000 chemical substances are in commercial use. The National Academy of Sciences notes that few of these substances have been adequately tested for their ability to cause mutations, cancer, and birth defects. Exposure to some of these chemicals or combinations of chemicals, some fear, may increase our chances of developing cancer.

Others believe that the extensive use of chemicals and the widespread publicity over the ill effects of some have created a chemical paranoia—sometimes referred to as *chemophobia*—among some members of our society. New research suggests that while the threat of chemical carcinogens is real, many people are overreacting to it. In fact, numerous studies suggest that naturally occurring chemicals in our food, like one type of mold that grows on peanut butter, probably cause more cancer than pesticide residues. (For a debate about chemical contamination and the relative importance of pesticides and other industrial chemicals on human health, see the Point/Counterpoint in Chapter 19.)

Some critics argue, however, that the debate about human health is misleading, for it ignores the potentially widespread effects of toxic chemicals on plants, animals, and microorganisms that share this planet with us.

Are Current Procedures for Determining Carcinogens Valid?

Animal Testing for Cancer Is Flawed by Philip H. Abelson

The principal method of determining potential carcinogenicity of substances is based on studies where huge doses of chemicals are administered daily to inbred rodents for their lifetime. Then by questionable models, which include large safety factors, the results are used to extrapolate the effects of minuscule doses in humans. Resultant stringent regulations and attendant frightening publicity have led to public anxiety and chemophobia. If current ill-based regulatory levels continue to be imposed, the cost of cleaning up phantom hazards will be in the hundreds of billions of dollars with minimal benefit to human health. In the meantime, real hazards are not receiving adequate attention.

The current procedures for gauging carcinogenicity are coming under increasing scrutiny and criticism. A leader in the examination is Bruce Ames, who with others has amassed an impressive body of evidence and arguments. Ames and Gold summarized some of their recent data and conclusions in *Science* (31 August 1990, p. 970). Three articles in the *Proceedings of the National Academy of Sciences* provide an elaboration of the information with extensive bibliographies. The articles also provide data about other pathologic effects of natural chemicals.

Philip Abelson was the Deputy Editor of *Science*. This essay is excerpted from "Testing for Carcinogens with Rodents," *Science 249*: September 21, 1990. Copyright © 1990 by the American Association for the Advancement of Science.

A limited number of chemicals tested, both natural and synthetic, react with DNA to cause mutations. Most chemicals are not mutagens, but when the maximum tolerated dose (MTD) is administered daily to rodents over their lifetime, about half of the chemicals give rise to excess cancer, usually late in the normal life span of the animals. Experiments in which synthetic industrial chemicals were administered in the MTD to both rats and mice resulted in 212 of 350 chemicals being labeled as carcinogens. Similar experiments with chemicals naturally present in food resulted in 27 of 52 tested being designated as carcinogens. These 27 rodent carcinogens have been found in 57 different foods, including apples, bananas, carrots, celery, coffee, lettuce, orange juice, peas, potatoes, and tomatoes. They are commonly present in quantities thousands of times as great as the synthetic pesticides.

The plant chemicals that have been tested represent only a tiny fraction of the natural pesticides. As a defense against predators and parasites, plants have evolved a large number of chemicals that have pathologic effects on their attackers and consumers. Ames and Gold estimate that plant foods contain 5,000–10,000 natural pesticides and breakdown products. In cabbage alone, some 49 natural pesticides have been found. The typical plant contains one percent or more of such substances. Compared with the amount of synthetic pesticides we consume, we eat about 10,000 times more of the plant pesticides.

It has long been known that virtually all chemicals are toxic if ingested in sufficiently high doses. Common table salt can cause stomach cancer. Ames and others have pointed out that high levels of chemicals cause large-scale cell death and replacement by division. Dividing cells are much more subject to mutations than quiescent cells. Much of the activity of cells involves oxidation, including formation of highly reactive free radicals that can react with and damage DNA. Repair mechanisms exist, but they are not perfect. Ames has stated that oxidative DNA damage is a major contributor to aging and to cancer. He points out that any agent causing chronic cell division can be indirectly mutagenic because it increases the probability of DNA damage being converted to mutations. If chemicals are administered at doses substantially lower than MTD, they are not likely to cause elevated rates of cell death and cell division and hence would not increase mutations. Thus, a chemical that produces cell death and cancer at the MTD could be harmless at lower dose levels.

Diets rich in fruits and vegetables tend to reduce human cancer. The rodent MTD test that labels plant chemicals as cancer-causing in humans is misleading. The test is likewise of limited value for synthetic chemicals. The standard carcinogen tests that use rodents are an obsolescent relic of the ignorance of past decades. At that time, extreme caution made sense. But now tremendous improvements of analytical and other procedures make possible a new toxicology and far more realistic evaluation of the dose levels at which pathological effects occur.

Current Methods of Testing Cancer Are Valid by Devra Davis

The vast majority of the scientific community endorses the conduct of experimental studies in order to try to identify those materials that could cause disease and prevent harmful exposures. In the typical toxicologic study of 50 rodents, each animal is a stand-in for 50,000 people. Because rodents live about 2 years on average, they are usually exposed to amounts of the suspect agent that approximate what a human would encounter in an average lifetime of 70 years.

Some have argued that the use of the maximum tolerated dose (MTD) in these studies produces tissue damage and cell proliferation, which, in turn, lead to cancer. This is biological nonsense that ignores a fundamental characteristic of cancer biology that has been known for more than 50 years: cancer is a multi-stage disease, with multiple causes, which arises by a stepwise evolution that involves progressive genetic changes, cell proliferation, and clonal expansion. Thus, swamping of tissues with high doses alone may well kill an animal or damage its tissues, but high doses alone are not sufficient to cause cancer. A 2-year study at the National Institute of Environmental Health Sciences (NIEHS) by David Hoel and colleagues provides good evidence on this point. They looked for signs of damage in tissues taken from rats and mice used in cancer studies. Cancers occurred in organs that did not show apparent damage, and some damaged organs were completely free of tumors.

In addition, most compounds tested do not cause cancer only in the highest dose group tested but typically produce a dose-response relationship, where the amount of cancer developed is proportional to the dose administered. Some chemicals cause toxicity only, others cause only cancer. Not all of those that cause cancer do so through organ toxicity. In fact, almost 90% of the substances shown to cause cancer in the National Toxicology Program do so without producing any increased cellular toxicity, and they also cause cancer at both lower and higher doses.

Animal studies are evolving and being further refined, as is our understanding of differences between species that need to be taken into account in conducting these studies. Every compound known to cause cancer in humans also produces cancer in animals, when adequately tested. For 8 of the 54 known human cancer-causing agents, evidence of carcinogenicity was first obtained in laboratory animals; in many cases, the same target organs and doses have been involved in producing cancer in both animals and humans.

The NIEHS has carried out nearly 400 long-term rodent studies, which have been published following peer review by specialists in the field. Chemicals nominated for testing usually represent a sample of potentially "problematic" materials. About 40% of the "suspect" pesticides evaluated to date have been found to cause cancer. As to the role of so-called natural pesticides, rodent diets are also loaded with many of these materials. Nevertheless, a number of test compounds added to these diets markedly increase the amount of tumors produced. Thus, animals are more sensitive to certain synthetic compounds than to the background level of natural materials. Humans are also likely to have acquired some resistance to these natural materials through evolution.

Devra Davis is a senior adviser in the office of the Assistant Secretary for Health, Department of Health and Human Services.

To date, only about 20% of all synthetic organic chemicals have been adequately tested for their potential human toxicity. Those who must set public policy on the use of chemicals need a rational basis on which to stake their actions. Continued advances in animal testing provide an important contribution to environmental health sciences and to public health efforts to predict, and then prevent, the development of environmentally caused disease.

In summary, the current system should be used until it can be replaced by a demonstrably better one. There is no scientific basis for rejecting the MTD as capable of inducing cancer.

Sharpening Your Critical Thinking Skills

1. Summarize each author's main points and supporting data. Do you see any inconsistencies or examples of faulty logic in either essay? If so, where?

2. Why is it that two scientists can disagree on an issue such as this?

3. Given the disagreement, what course of action would you recommend?

www.jbpub.com/humanbiology/5e

Visit Human Biology's Internet site for links to web sites offering more information on this topic.

SUMMARY

DNA and RNA: Macromolecules with a Mission

1. DNA houses the genetic information of the cell. This molecule consists of two polynucleotide strands joined by hydrogen bonds between purine and pyrimidine bases on the opposite strands.

2. Each nucleotide in the DNA molecule consists of a purine or pyrimidine base, the sugar deoxyribose, and a phosphate group. The nucleotides are joined by covalent bonds.

3. Complementary base pairing is an unalterable coupling in which adenine on one strand of the DNA molecule always binds to thymine on the other and guanine always binds to cytosine. Complementary base pairing ensures accurate replication of the DNA and accurate transmission of genetic information from one cell to another and from one generation to another.

4. To replicate, DNA must first unwind with the aid of a special enzyme. After unwinding, the strands provide templates for the production of complementary DNA strands, a process aided by the enzyme DNA polymerase.

5. The cell contains three types of RNA: transfer RNA, ribosomal RNA, and messenger RNA. All play a role in the synthesis of protein in the cytoplasm.

6. All three RNA molecules are single-stranded polynucleotide chains. RNA nucleotides contain the sugar ribose instead of deoxyribose, which is found in DNA. RNA nucleotides contain four bases: adenine, guanine, cytosine, and uracil (instead of thymine).

7. RNA is synthesized in the nucleus on a template of DNA. The synthesis of RNA is called transcription.

How Genes Work: Protein Synthesis

8. The genetic information contained in the DNA molecule is transferred to messenger RNA. Messenger RNA molecules carry this information to the cytoplasm, where proteins are synthesized. Messenger RNA serves as a template for protein synthesis. Ribo-somes are required to produce proteins on the mRNA template.

9. Transfer RNA molecules deliver amino acid molecules to the mRNA and insert them in the growing chain. Each tRNA binds to a specific amino acid and delivers it to a specific codon, a sequence of three bases on the mRNA. Thus, the sequence of codons determines the sequence of amino acids in the protein. Proteins are synthesized by adding one amino acid at a time.

10. During protein synthesis, the ribosome first attaches to the mRNA at the initiator codon. The large subunit then attaches. A specific tRNA bound to an amino acid binds to the initiator codon and the first binding site of the ribosome. A second tRNA–amino acid then enters the second site.

11. An enzyme in the ribosome catalyzes the formation of a peptide bond between the two amino acids. After the bond is formed, the first tRNA (minus its amino acid) leaves the first binding site.

12. The ribosome moves down the mRNA, shifting the tRNA bound to its two amino acids to the first binding site and opening the second site for another tRNA–amino acid. This process repeats itself many times in rapid succession.

13. As the peptide chain is formed, hydrogen bonds begin to form between the amino acids, and the chain begins to bend and twist, forming the secondary structure of the protein or peptide. When the ribosome reaches the terminator codon, the peptide chain is released.

Controlling Gene Expression

14. Gene control in humans occurs at four levels: at the chromosome level, at transcription, after transcription but before translation, and at translation.

15. At the chromosome level: Condensation, or coiling, of the chromatin fibers inactivates genes. How the cell condenses a segment of the chromatin is not clearly understood.

16. At transcription: Certain chemicals either turn on or turn off genes. These chemicals are either the end products of metabolic reactions or substrates in these pathways. Some segments of the DNA, called enhancers, can greatly increase the activity of nearby genes. Geneticists think that protein molecules bind to enhancer regions of the chromosome and increase gene activity.

17. After the production of RNA before translation: Human DNA contains far more genetic material than it needs. Those segments of the DNA used to produce the RNA that will be used to make protein are called exons. The intervening segments of noncoding DNA are called introns. Both introns and exons are transcribed during the cell cycle. The cell, however, removes the RNA from the introns and joins the RNA fragments produced by exons to create a functional mRNA molecule. The exon-produced RNA can be spliced together in different ways; the resulting mRNAs produce different proteins.

18. At translation: Messenger RNA may be transferred to the cytoplasm in an inactive, or masked, state. Masking permits the cell to build up large supplies of mRNA in preparation for a sudden burst of protein synthesis.

Health and Homeostasis

19. Humans and other organisms contain specific genes, called proto-oncogenes, that lead to cancer when mutated. Proto-oncogenes are normal genes that code for cellular structures and functions, such as cell adhesion, mitotic proteins, and the production of plasma-membrane receptors for growth factors or hormones.

20. Chemical, physical, and biological agents can mutate these genes, resulting in uncontrolled cellular proliferation (cancer).

21. Viruses can also cause cancer. Some viruses, for example, possess oncogenes, which enter human body cells and are incorporated in the cell's DNA, stimulating uncontrolled cellular division. Other viruses carry genes that stimulate cancer by activating human proto-oncogenes.

www.jbpub.com/humanbiology/5e

The site features eLearning, an online review area that provides quizzes, chapter outlines, and other tools to help you study for your class. You can also follow useful links for in-depth information, research the differing views in the Point/Counterpoints, or keep up on the latest health news.

critical thinking

Thinking Critically—Analysis

This Analysis corresponds to the Thinking Critically scenario that was presented at the beginning of this chapter.

After you have thought about this issue for a while, you may want to consider some additional facts. The U.S. dairy industry already produces an excess of milk. The federal government buys the surplus, dehydrates some of it, and makes cheese out of the rest. The government stockpiles dehydrated milk and cheese, which are given to needy families. Will increasing the surplus produce more food for the poor?

Before you draw any conclusions, consider another issue—the effects on small dairy operations of increasing the surplus of milk. Many owners of small dairy herds believe that the large producers will be the primary beneficiaries of the hormone. It will allow them to increase their milk production and probably drive down the cost of milk and milk products. In the process, it could put many small dairy farmers out of business. This loss could adversely affect the economies of many rural regions.

Some consumer groups are concerned about the potential health effects of using synthetic growth hormone. A trace of growth hormone is present in milk produced normally. Will the use of synthetic growth hormone increase this concentration in milk? What effect will that have on your health or the health of children?

Make a list of the pros and cons of using synthetic growth hormone. Then make a list of questions—those that you can answer and those that you cannot answer. Do you need more information to decide?

If you are able to make up your mind, what factors swayed your opinion? Did the concerns of one group outweigh the concerns of another? What critical thinking rules did you use in this exercise?

KEY TERMS AND CONCEPTS

Anticodon, p. 341
Complementary base pairing, p. 337
DNA polymerase, p. 337
Enhancement, p. 343
Enhancer, p. 343
Initiator codon, p. 341

Oncogene, p. 345
Purine, p. 336
Proto-oncogene, p. 345
Pyrimidine, p. 336
Ribosomal RNA, p. 340

Ribosome, p. 340
RNA polymerase, p. 339
Transcription, p. 339
Transfer RNA, p. 340
Translation, p. 340

CONCEPT REVIEW

1. What is the genetic code? How is the genetic code, housed in the DNA, translated into instructions the cell can understand? p. 341
2. Describe how DNA is synthesized. How does the cell ensure the accurate replication of DNA? p. 337
3. List the three types of RNA, and briefly describe their function in protein synthesis. pp. 337–339
4. In what ways are RNA and DNA similar? In what ways are they different? p. 339
5. Describe the production of protein on mRNA. Be sure to note the enzymes involved and the role of the ribosome. pp. 340–343
6. Discuss the various levels at which human genes are controlled. Give an example of each. pp. 343–344
7. What is a proto-oncogene? How is it affected by a mutagen, a chemical or physical agent that causes mutation? p. 345

SELF-QUIZ: TESTING YOUR KNOWLEDGE

1. The DNA molecule is a double helix; its strands are held together by _____ bonds between bases. p. 337
2. In DNA, the base adenine is joined to the base _____ on the complementary strand. p. 337
3. In DNA, the sugar molecules in the nucleotides are known as _____. p. 336
4. The exact replication of DNA is made possible by _____ base pairing. p. 337
5. The formation of RNA on a DNA template is called _____. p. 339
6. The formation of protein on an mRNA template is known as _____. p. 340
7. RNA is produced on a DNA template inside the cell's _____. p. 340
8. _____ attaches to amino acids in the cytoplasm and inserts them in newly forming protein chains. p. 340
9. In cells during interphase, chromosomal _____ results in a tight coiling of the DNA, inactivating genes in that section. p. 343
10. _____ is a process in which already active genes increase their production of mRNA. p. 343

CHAPTER

19

Genetic Engineering and Biotechnology

In the movie *Jurassic Park*, scientists removed DNA from dinosaur blood found in the gut of mosquitoes preserved in amber (hardened sap of pine trees) during the age of the dinosaurs. They took this DNA and inserted it into the nuclei of frog cells and produced living, breathing, lawyer-devouring dinosaurs.

Although this scenario is pure science fiction, it illustrates the kinds of genetic manipulations that scientists are engaged in. One procedure, in which a segment of the DNA of one species is cut out and transferred to another species, is popularly referred to as **genetic engineering**. For reasons explained shortly, scientists typically refer to it as **recombinant DNA technology**.

thinking critically

A chef at a famous restaurant in New York City organized fellow chefs in the city to protest the sale of genetically engineered produce, such as tomatoes. His main concern is that the government has refused to require farmers who are growing genetically engineered tomatoes to label their products. Consumers, therefore, won't know if they're getting genetically engineered vegetables or ones produced naturally. How would you analyze this issue? What critical thinking rules might apply?

19-1 Genetic Manipulation: An Overview

Genetic engineering is the newest form of genetic tinkering.

Although we like to think of genetic engineering as a modern invention, it is not new at all. In fact, plants, animals, and microorganisms have been "performing" their own genetic experiments for at least 3.5 billion years. They've been undergoing mutations, exchanging segments of genes (via crossing-over), and combining genes in new ways through sexual reproduction. Evidence suggests that some species of microbes actually pick up genes from others. These manipulations are natural; most are key elements of evolution itself that result in new genetic combinations. Over the course of evolution, nature's genetic tinkering has resulted in a diverse array of life-forms.

Humans took genetic experimentation a step further by **selective breeding**—intentionally crossing certain plants or animals to produce offspring with desirable traits. Domestic animals, such as dogs, came about by selective breeding, as have cattle, cats, fancy pigeons, and other animals (Figure 19-1). Crops, such as corn and tomatoes, are here today thanks to the efforts of early geneticists who selected plants with desirable traits and then pollinated them from other plants with similar traits. The goal was to produce offspring with good-tasting fruits and seeds.

Some consider genetic engineering, the newest method of genetic manipulation, to be nothing more than an advanced form of the genetic tinkering that occurs in nature and in human societies through selective breeding. Others see it as an intrusion into the natural process of evolution. Still others view it as a dangerous gamble that has potential health effects. Some opponents warn that genetically engineered organisms could spread throughout the world, causing disease and environmental problems.

FIGURE 19-1 **Fancy Pigeon** This beautiful bird barely resembles its relative the rock dove, or common pigeon. Selective breeding produced this bird, as it has many other new genetic combinations.

19-2 Cloning

Cloning offers us an opportunity to produce identical replicas of existing or extinct organisms.

Yet another growing controversy has to do with a process called **cloning**. In this technique, cells from an organism are used to develop another, genetically identical organism—a clone. Nuclei from the cells of prize-winning cattle, for example, can be transferred to enucleated eggs. The eggs are then transferred to other cows, where they develop. The result is a genetically superior offspring.

In 1997, much furor arose when a Scottish researcher reported the birth of a cloned sheep. The clone, named Dolly, was derived from a cell taken from the mammary gland of another sheep. The announcement stirred considerable controversy, as people began to wonder whether this technique would be used for humans. In 1998, scientists announced yet another clone—this one in mice.

In 2000, the first cloned cow was sold at auction. Produced by a company in Wisconsin, called Infigen, this cow was an exact copy of a prize-winning Holstein. It was produced when an ear cell from the mother was fused with an enucleated egg. The egg was then activated using a chemical that stimulates the release of calcium in the egg, causing it to divide. The newly "fertilized" egg was transplanted into another cow, where it developed fully.

Researchers at Texas A & M University have used cloning to produce a disease-resistant cow. They used genetic material from a bull that was naturally resistant to several common diseases. These diseases not only affect cattle but also can be passed to humans in uncooked beef or unpasteurized milk. One of the things that was remarkable about this feat is that the donor cells from the disease-resistant bull had been frozen 15 years earlier!

Cloning has also been used to produce an Asian gaur, an endangered oxlike animal native to the bamboo forests of India and

FIGURE 19-2 **Asian Gaur** Scientists are using cloning to help increase the numbers of this and other endangered species.

Burma (Figure 19-2). A skin cell from the gaur was fused with an enucleated egg cell. Researchers hope that this technique may help increase populations of endangered wildlife and may even bring back recently extinct species. They are beginning cloning work to help the Giant Panda, another species in danger of extinction. They're hoping that eggs of black bears will prove to be up to the task.

In 2003, a group of researchers reported cloning of a human, although evidence of this feat has not been forthcoming. Even Hollywood got into the act in 2004 with a film called *Godsend*, a movie about human cloning.

This chapter examines recombinant DNA technology and its real and potential applications. It also outlines the basic controversy over the use of this technology.

19-2 Recombinant DNA Technology

In recombinant DNA technology, genes are removed from one organism, replicated, then spliced into the cells of another of the same or a different species.

Genetic engineering got its start over 40 years ago as the result of research by Stanley Cohen at Stanford University and Herbert Boyer at the University of California at San Francisco. In 1973, these scientists cut segments of DNA from one bacterium using a special enzyme and spliced them into the genetic material of another. The recipient bacterium then multiplied and, in the process, produced many copies of the altered genetic material. The process Cohen and Boyer pioneered still provides the basis for much of the genetic engineering research and development being done today. To understand how it works, let's look at the individual steps, beginning with the slicing of a piece of DNA from a chromosome.

Removing Genes

Enzymes are used to cut out segments of genes and insert them elsewhere.

Genetic engineering was made possible by the discovery of a group of enzymes that snip off segments of the DNA molecule. Known as **restriction endonucleases**, these enzymes cut through both strands of the DNA, as shown in Figure 19-3.

Restriction endonucleases are found in many species of bacteria. In bacteria, these enzymes serve a protective function, destroying the DNA of viruses that enter them. In so doing, they prevent these cellular pirates from taking over. To date, scientists have discovered hundreds of different restriction endonucleases, each one specific for a particular sequence of bases on the DNA.

Scientists cut DNA from bacteria or any other organisms, including humans, using restriction endonucleases. They then use the same enzyme to slice open the recipient cell's DNA. The gene segments are then inserted into the recipient.

After a slice of DNA is added to the bacterial DNA, another enzyme is used to seal it in place. This enzyme is called **DNA ligase**. At this point, the DNA splicing is complete. Geneticists have formed a **recombinant DNA** molecule—a molecule that contains DNA from two different organisms (a combination of DNAs).

Mass Producing Genes

Plasmids from bacteria can be used to clone DNA fragments for mass production.

As noted earlier, recombinant DNA technology began with experiments in bacteria. As you may recall from Chapter 3, bacteria contain a single, circular "strand" of DNA (Figure 19-4). They also contain smaller circular strands, called **plasmids** (PLAZ-mids). Plasmids carry a few genes—such as those that confer

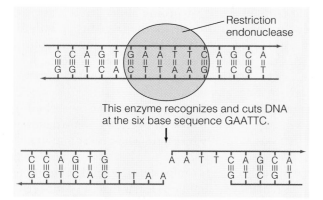

FIGURE 19-3 **DNA Scissors** Restriction endonucleases extracted from bacteria selectively slice through different segments of the DNA double helix, producing a staggered cut as shown here.

FIGURE 19-4 **Plasmids** Rupturing a bacterium releases circular chromosome and numerous plasmids, indicated by arrows.

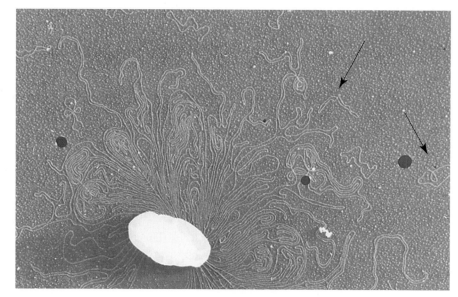

antibiotic resistance to bacteria. They also replicate independent of the chromosome. Plasmids are much smaller than bacterial DNA and are also relatively easy to extract and are therefore often used in genetic engineering as a recipient of foreign genes. That is foreign genes can be transplanted into them.

As Figure 19-5 illustrates, plasmids can be sliced with restriction endonucleases to open the circular DNA. Foreign genes can be inserted into the gap in the plasmid and sealed in place with DNA ligase.

Plasmids carrying foreign genes can then be added to culture dishes containing bacteria or yeast cells. In culture, the plasmids are taken up by the new hosts. Then, as the host cells divide, the plas-

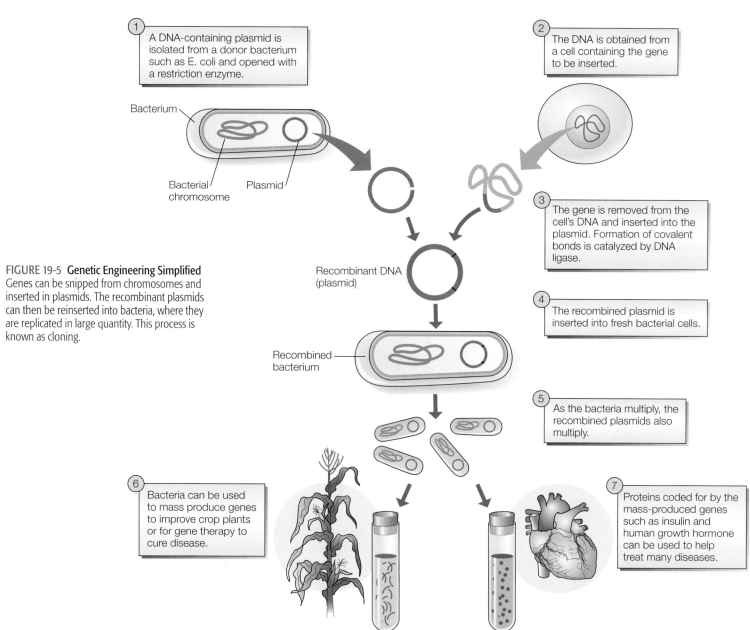

1 A DNA-containing plasmid is isolated from a donor bacterium such as E. coli and opened with a restriction enzyme.

2 The DNA is obtained from a cell containing the gene to be inserted.

3 The gene is removed from the cell's DNA and inserted into the plasmid. Formation of covalent bonds is catalyzed by DNA ligase.

4 The recombined plasmid is inserted into fresh bacterial cells.

5 As the bacteria multiply, the recombined plasmids also multiply.

6 Bacteria can be used to mass produce genes to improve crop plants or for gene therapy to cure disease.

7 Proteins coded for by the mass-produced genes such as insulin and human growth hormone can be used to help treat many diseases.

Bacterium

Bacterial chromosome

Plasmid

Recombinant DNA (plasmid)

Recombined bacterium

FIGURE 19-5 **Genetic Engineering Simplified** Genes can be snipped from chromosomes and inserted in plasmids. The recombinant plasmids can then be reinserted into bacteria, where they are replicated in large quantity. This process is known as cloning.

mids replicate. This produces multiple copies of the gene, which is then said to be cloned. Cloning produces numerous identical copies of the foreign gene. These can be used to produce useful products, such as hormones needed to treat patients with hormonal deficiencies.

Producing Genes on mRNA

Genes can be mass produced on strands of mRNA.

Geneticists can produce multiple copies of DNA in other ways, too. One common way of doing this is by extracting messenger RNA from cells and making copies of DNA from them. How?

Interestingly, scientists have discovered an enzyme in certain viruses that makes DNA from messenger RNA. This process is called **reverse transcription** because it is the reverse of transcription. It relies on an enzyme called **reverse transcriptase**.

To make DNA, geneticists remove mRNA from cells and then add reverse transcriptase (Figure 19-6). The enzyme then creates a single strand of DNA on the mRNA template. The hybrid DNA/RNA molecule is then treated with a chemical that destroys the RNA, leaving a single strand of DNA. Additional enzymes convert the single-stranded DNA to a double-stranded DNA molecule. DNA molecules mass produced from mRNA are referred to as cDNA.

The Polymerase Chain Reaction

Genes can also be mass produced using the polymerase chain reaction.

Perhaps the most widely used method of gene amplification is called the **polymerase chain reaction**, the brainchild of biochemist Kary Mullis. In this technique, geneticists extract a segment of the DNA double helix from a cell's nucleus. They then split the DNA molecules into two strands by heating them. Next, they cool the solution and add DNA polymerase enzymes. These enzymes catalyze the formation of complementary DNA molecules.

The DNA molecules are then split again with heat and more DNA polymerase is added. The new polymerase triggers an-

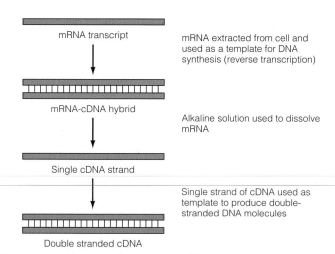

FIGURE 19-6 **Making Genes from mRNA** In this process, mRNA is extracted and then used as a template for the production of a complementary strand of DNA (cDNA), a process called reverse transcription. The mRNA molecule is then destroyed, leaving behind a single-stranded molecule of cDNA, which is then used to produce complementary cDNA.

other round of DNA replication. This procedure, which is carried out by machine, permits scientists to produce a million or more copies of the original DNA, that is, many copies of genes.

Mass production permits scientists to produce large quantities of DNA, which are needed to chemically analyze the DNA of various species. In so doing, geneticists can determine the exact sequence of bases that comprise the various genes. This procedure has helped scientists map the entire 100,000 or more genes of the human **genome**, discussed in Chapter 17. Interestingly, this study has shown that the human genome contains many genes found in other organisms, such as bacteria, mice, and snakes.

Knowing the sequence of bases in a gene permits scientists to synthesize genes. Manufacturing genes permits geneticists to create normal copies of human genes, which can be inserted into human cells to correct genetic defects. Mass-producing genes also gives scientists the needed material to mass-produce gene products—notably, proteins.

19-4 Applications of Recombinant DNA Technology

Recombinant DNA technology has resulted in a number of practical, though sometimes controversial, applications.

Treating Diseases

Hormones and other proteins created from mass-produced genes can be used to treat many diseases.

By removing the human genes that code for hormones and then transferring those genes into bacterial cells that are grown in large tanks, drug companies can generate huge quantities of hormones (Figure 19-7). They can also produce other important proteins such as blood clotting factors that are missing in some individuals. This technique has resulted in a dramatic increase in the supply of protein hormones such as insulin and growth hormone for medical use. Table 19-1 lists some of the products and their uses.

Here's an example of the benefit of this mass production of hormones and proteins. Human growth hormone that is used to treat people with glandular defects was once extracted from the pituitary glands of human cadavers. In order to extract enough growth hormone for one dose, technicians had to remove glands from 50 human cadavers. This process was costly and time-consuming.

Pituitaries from pigs have also been used to obtain growth hormone. Although pig growth hormone acts like human growth hormone, it is different enough chemically to cause an immune reaction in some patients.

FIGURE 19-7 **Mass Producing Genetically Engineered Bacterial Products** Large vats like this one contain hundreds sometimes thousands of gallons of culture medium and genetically modified bacteria that produce valuable proteins such as human growth hormone for treating human diseases.

Recombinant DNA technology offers an opportunity to produce a growth hormone that is much safer than pig growth hormone, because it avoids potential immune reactions.

Protein hormones, like growth hormone, can also be produced for other uses, notably for livestock production. Cows, for instance, can be injected with bovine growth hormone produced by genetically engineered bacteria. This treatment greatly increases milk production. (For a discussion of the pros and cons of this, see the Thinking Critically section in Chapter 18.)

Vaccines

Recombinant DNA techniques can be used to produce vaccines.

Drug companies can also use genetic engineering to mass produce vaccines at a lower cost than is possible by conventional means. As noted in Chapter 14, vaccines are inactivated bacteria or viruses. When injected in an individual, a vaccine stimulates an immune response, providing protection from the infectious forms of bacteria or viruses.

Two types of genetically engineered vaccines are being produced. The first consists of proteins or polypeptides identical to those in the capsids of viruses. These vaccines, produced from viral genes created by genetic engineering, stimulate an immune reaction.

The other type of genetically engineered vaccine consists of genetically modified but inactive viruses. These, too, stimulate an immune reaction. However, some researchers are concerned that the genetic changes could have unpredictable effects when the vaccines are injected into humans. The altered viruses of the vaccine could also affect other species not currently affected by the viruses. There is also concern that the genetically altered viruses could exchange genetic material with live viruses with potentially adverse effects in humans.

At present, numerous vaccines are under development. One vaccine, for an infectious disease that strikes the liver (hepatitis B), has been produced via this method. But geneticists hope to develop others, perhaps even one for the AIDS virus.

Transgenic Organisms

Gene splicing can be used to transfer genes from one organism to another.

In what some people view as a major intrusion into the evolution of life, scientists are now able to transfer single genes from one organism into another, creating a **transgenic organism.** As shown in Figure 19-8, foreign genes can be injected directly into

TABLE 19-1	Some Useful Products of Genetic Engineering
Products	Used to
Growth hormone	Treat patients with dwarfism
Insulin	Treat patients with diabetes
Tumor necrosis factor	Treat patients with cancer
tPA (tissue plasminogen activator)	Dissolves clots formed during heart attacks
Erythropoietin	Stimulate blood cell formation for patients with some forms of anemia
Clotting factors	Treat people with hemophilia
Surfactant	Treat babies with blue baby syndrome, caused by collapse of air sacs in lungs, which is caused by a lack of surfactant
Interferons	Treat some forms of cancer and some viral infections
Vaccines	Prevent hepatitis A, B, and C, as well as AIDS*, malaria*, and herpes*
DNA probes	Map human chromosomes, perform DNA fingerprinting, detect infectious diseases, and detect genetic disorders

*Under development, but not currently available.

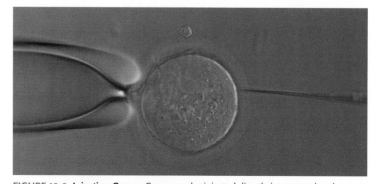

FIGURE 19-8 **Injecting Genes** Genes can be injected directly into eggs via microneedles (on right). The large structure on the left is a glass rod that holds the egg so the genes can be injected. The egg is then placed in the oviduct of a surrogate mother where it can develop.

the egg of another animal. The egg is then placed in the oviduct of a female, where it can develop into an embryo. Another technique is shown in Figure 19-9. In this technique, fibroblast cells are removed from a cow fetus. Foreign genes are then added to a culture dish containing the cells. The genes are taken up by some of the fibroblasts. The fibroblasts are then fused with enucleated ova from cows and treated with an electrical current. This causes the incorporation of the fibroblast DNA into the egg and stimulates division, creating many identical embryos that are then transplanted in surrogate mothers.

Transgenic organisms don't differ much from their original form; they may, in fact, display only one new trait, such as a larger body size. One of the most successful attempts to date involves the transplantation of the human gene that codes for growth hormone into pigs and mice (Figure 19-10). When successful, this results in offspring that are much larger than their siblings who didn't receive the gene.

Researchers are looking for ways to re-engineer cows so that the milk they produce will be chemically more similar to human milk. Scientists, for example, have transplanted a human gene that codes for the production of a chemical found in mother's milk, called *lactoferin*, into cows. This substance helps babies fight off infections.

One of the most exciting developments is the production of mice containing genes identical to those that cause human diseases. This could advance medical research on diseases and speed the development of treatments and cures.

FIGURE 19-9 Creating Transgenic Cows This diagram shows another means of introducing foreign genes into ova to create transgenic organisms.

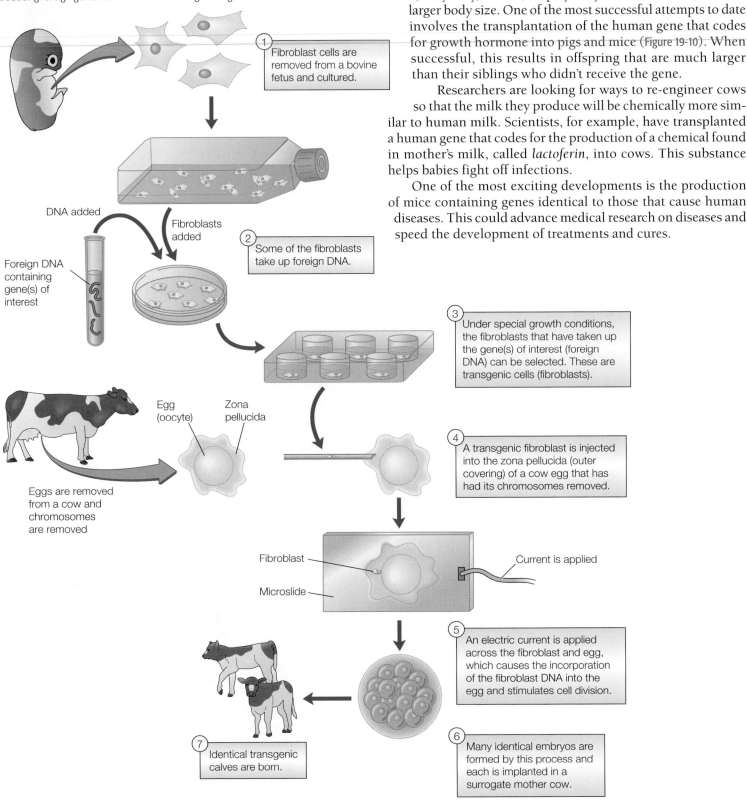

1 Fibroblast cells are removed from a bovine fetus and cultured.

DNA added

Fibroblasts added

Foreign DNA containing gene(s) of interest

2 Some of the fibroblasts take up foreign DNA.

3 Under special growth conditions, the fibroblasts that have taken up the gene(s) of interest (foreign DNA) can be selected. These are transgenic cells (fibroblasts).

Egg (oocyte)

Zona pellucida

Eggs are removed from a cow and chromosomes are removed

4 A transgenic fibroblast is injected into the zona pellucida (outer covering) of a cow egg that has had its chromosomes removed.

Fibroblast

Current is applied

Microslide

5 An electric current is applied across the fibroblast and egg, which causes the incorporation of the fibroblast DNA into the egg and stimulates cell division.

7 Identical transgenic calves are born.

6 Many identical embryos are formed by this process and each is implanted in a surrogate mother cow.

FIGURE 19-10 **Supermice** The mouse on the right, pictured with its littermate, has received a gene for human growth hormone.

In 1997, scientists from the Medical College of Georgia announced the successful transplantation of genes that code for a green fluorescing protein into zebra fish. It turned the red blood cells green under blue light, allowing scientists to study blood cell formation in the zebra fish embryos. The scientists had performed the same process with nerve cells and hope that this will allow them to study how nerve cells form connections in the developing brain.

Although such techniques sound simple, they're actually quite difficult. Scientists can't just inject a foreign gene into a baby pig or cow. The gene generally has to be introduced at the embryonic stage when the organism is in the one-cell stage. This is accomplished by injecting genes into the nuclei of ova via tiny needles. Some nuclei incorporate the foreign gene, which is then transmitted from one cell to the next during embryonic development. Even so, there's no guarantee that the gene will end up in the proper location.

Scientists are also experimenting with ways to transplant normal genes into cancer cells or the cells lining the respiratory tract of people suffering from cystic fibrosis. In the latter case, they're inserting genes into viruses that infect the cells lining the respiratory passageways, hoping that it will become part of these cells' genome and correct the problems.

Transgenic Plants

Transgenic plants are easier to produce than transgenic animals.

Plants are easier to perform gene transplants on than animals. One of the first success stories involved the transplantation of a gene that codes for an enzyme called luciferase (lew-SIFF-er-ace) into a tobacco plant. This enzyme is normally found in

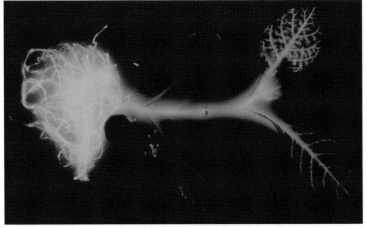

FIGURE 19-11 **Glow Little Tobacco Plant, Glimmer Glimmer** This tobacco plant contains firefly genes for the enzyme luciferase. When the plant is sprayed with luciferin, the enzyme breaks it down, releasing light.

fireflies and is responsible for catalyzing a light-generating reaction. In the tobacco plant, the enzyme didn't do much unless the plant was sprayed with a chemical called *luciferin*. At this point, the plant glowed (Figure 19-11).

This experiment was not performed to create tobacco plants that glow in the dark so they could be harvested at night, but only to show that foreign genes could be transplanted from one species to another. It opened up the door to more practical experiments—for instance, transplanting genes that make plants produce more nutritious seeds and fruit or that permit the roots of some plants to synthesize their own nitrogen fertilizer.

One of the more promising developments involves the introduction of genes that allow oats to grow in salty soil. Transplanting these genes into commercial crop species could allow farmers to use vast acreages in the United States and other countries idled by the buildup of salts. This is caused by the irrigation of poorly drained soils with water high in salts.

Researchers are also developing ways to make crop plants resistant to herbicides, chemicals applied to crops to control weeds. Although herbicides generally act only on weed species, they sometimes impair the growth of crop species. Proponents hope that herbicide-resistant crops will help farmers increase yields. Critics argue that if herbicide-resistant crops are used, farmers may apply larger amounts of herbicide to their fields. These chemicals, in turn, could be washed into nearby waterways, poisoning aquatic life. They could also poison soils, killing microorganisms essential for the long-term health of the soil. (For a discussion, see the Point/Counterpoint in this chapter.)

Geneticists are also working on ways to alter plants themselves to increase their resistance to pests. By transferring the genes that protect wild species from insects into crop species, scientists may be able to produce many varieties that require little, if any, chemical pesticides.

Efforts to produce genetically engineered vegetables such as tomatoes have also proven successful. Critics, however, fear that such experiments may result in unforeseen reactions such as allergies in people who consume the **genetically modified organisms** (GMOs). There is evidence to suggest that some people do indeed develop allergies to genetically engineered crops.

Transgenic Microorganisms

Transgenic microbes are some of the most controversial of all creations.

Genetic engineering began with microorganisms, and work in this area continues today. For example, scientists have successfully transplanted a bacterial gene that produces an insect-killing toxin into a species of bacteria that lives on the roots of crop plants. Root-eating insects ingest the lethal bacteria when feeding on the roots of treated plants. This kills the insects and protects crops. This technique benefits the farmer, who has to use fewer chemicals to control root-eating insects, and also benefits the environment, because it reduces the use of potentially dangerous chemicals that could poison soil microbes, leach into groundwater, or be washed into nearby lakes and streams.

Controversy over Herbicide Resistance Through Genetic Engineering

The Benefits of Genetically Engineered Herbicide Resistance by Charles J. Arntzen

POINT ■ COUNTERPOINT

One of the principal outcomes of plant genetic engineering has been a rapid expansion in our understanding of plant cell biology, genetics, and molecular controls over the structure and function of plants. This understanding is now helping to solve practical problems such as developing insect- and disease-resistant crops, thereby decreasing needs for chemical pesticides. Other approaches include creating herbicide-resistant plants, which will allow farmers more choices in weed control, and to switch to newer, rapidly biodegradable herbicides.

Herbicides are chemicals sprayed on crops to control weeds. Their effect is selective—that is, they kill weeds without significantly harming crops. All herbicides currently available to farmers began with an evaluation of the sensitivity of weeds and crops to experimental chemicals. This evaluation is accomplished by applying potentially useful chemicals to test samples of weed and crop seedlings. Compounds that kill weeds but not crops are then subjected to animal toxicology studies. Those deemed acceptable are developed for farmers' use.

The success of herbicides is based on the fact that crop species contain enzymes that convert chemical compounds, including herbicides, to inactive forms. These enzymes tend to be crop-specific. For instance, the enzymes that render soybeans immune to 30 or more commercial herbicides are different from the enzymes that make corn resistant to other herbicides.

Dr. Charles J. Arntzen is a professor in the Alkek Institute of Biosciences and Technology in the Texas Medical Center in Houston, Texas. Dr. Arntzen was elected to the U.S. National Academy of Sciences in 1983 as a result of his pioneering research in photosynthesis and plant molecular biology. He was the chairman of the National Biotechnology Policy Board of the National Institutes of Health from 1991–1993 and is a member of the Editorial Board of *Science*. This article is adapted by the author from a 1991 publication entitled *The Genetic Revolution, Scientific Prospects and Public Perceptions* with the permission of the publisher, The Johns Hopkins University Press.

This can be a benefit when corn or soybeans are grown in rotation. In such instances, both the unwanted "volunteer" corn and weeds that germinate in a soybean field can be controlled by the soybean herbicide.

Interestingly, certain weeds have the same enzymes as the crop species they invade. This is often true when the weeds and crops are somewhat related, such as with wild oats in a field of wheat or barley. Thus, the oats are resistant to the same herbicides as the wheat or barley. Consequently, the farmer currently has few or no choices in chemical weed control since herbicides that would kill the weeds would also kill the crop.

Through genetic engineering, researchers have devised ways to make crops resistant to new classes of herbicides. That is, they have found ways to give crops such as soybeans additional enzymes to augment their natural herbicide resistance. By transferring genes that code for enzymes conferring herbicide resistance, scientists can give farmers a broader set of options when selecting herbicides. If successful, these options will solve problems where there is currently no weed control, or where the herbicides available are prohibitively expensive or environmentally damaging.

This line of research may also give the farmer a simpler means to control weeds through the use of a single, more effective herbicide selected on the basis of several traits, including herbicide resistance, reduced cost, and, ideally, greater safety in the environment.

This explanation of the benefits of genetic research aimed at making crops more resistant to herbicides may not satisfy the reader who is asking, "Why don't they simply stop using all herbicides?" The main answer is economics. Weed control is essential for all crops. Left unchecked, weed populations can reduce crop yields to near zero. Prior to the 1950s, weeds were controlled strictly by mechanical means—that is, by cultivation or hand pulling. This was a labor-intensive and costly process. By the 1980s, more than 95% of all major row crops in the United States were produced using herbicides. Farmers based this decision (to apply herbicides rather than using a cultivator or a hoe) on costs. An application of herbicides at the time of planting, costing as little as $10 per acre, is many times cheaper than mechanical cultivation. Weed control through herbicides is a factor in the low food prices we enjoy in the United States. It also saves on energy use and, when properly used, does not harm the environment.

With the continuing development of herbicide-resistant crops as part of an integrated weed-control strategy, crop production costs can be held at low levels while new methods for weed control are being developed.

Genetic researchers have also developed a bacterium that retards the formation of frost on plants. The genetic variant is found naturally in the environment but not in sufficient quantities to protect crops. Thus, researchers have cloned the bacterium and have begun testing its efficiency in the field. Should it prove successful and safe, the bacterium could be sprayed on crops late in the growing season when frosts can be devastating, potentially saving farmers millions of dollars a year. Some critics worry that this bacterium could escape into the environment and displace normally occurring bacteria. Who knows what detrimental effects this might have on the function of natural systems?

The Perils of Genetically Engineered Herbicide Resistance by Margaret Mellon

Environmentalists have been active in the debate about genetic engineering since the beginning. The current focus of environmental interest in genetic engineering is based on the prospect of commercial production of a broad range of genetically engineered organisms—for example, bacteria and viruses. These organisms could have direct and adverse impacts on our environment and our health.

Public concern over genetic engineering is also focused on indirect environmental consequences, the best example of which is herbicide resistant plants. The development of major crops that are tolerant of chemical herbicides seems likely to lead to increased or prolonged use of these dangerous agricultural chemicals.

Herbicide-resistant crops, like most genetically engineered agricultural products, are being developed by large, transnational chemical companies. Worldwide, at least 28 enterprises have launched more than 65 research programs to develop herbicide-resistant crops. All of the major crop plants are involved, including cotton, corn, soybeans, wheat, and potatoes. Engineered crops being made resistant to a company's own herbicides will surely increase that company's market share in chemicals. Thus, industry analysts expect herbicide-resistant plants to be big business, mainly for pesticide and herbicide producers.

Herbicide-resistant products run directly counter to the promise of a reduced dependence on agricultural chemicals put forward by biotechnology advocates. Rather than weaning agriculture from chemicals, these crops will be shackled to herbicide use for the foreseeable future. Herbicides represent an estimated 65% of the chemical pesticides used in agriculture. By continuing to promote the use of herbicides, the biotechnology industry greatly restricts its potential for reducing overall chemical use.

While herbicide-resistant crops offer no hope of environmental benefits in agriculture, there are alternatives to these products that do promise substantial reduction in overall pesticide use. These alternatives fall under the rubric of sustainable agriculture.

Sustainable agriculture employs a variety of agricultural practices such as crop rotation, intercropping, and ridge tillage (that produce long ridges of soil on which plants are grown). These practices control pests without the application of synthetic pesticides and fertilizers. Used by knowledgeable farm managers, these techniques work to make farms more profitable and environmentally sound. Growing different crops in successive growing seasons, for example, dramatically reduces pests by sequentially removing the hosts on which the pests depend. With fewer weeds or insects to contend with, the farmers reduce the need for costly chemical pesticides. Once thought impractical, sustainable agriculture is rapidly gaining support in national policy forums, including the National Academy of Sciences.

The environmental advantage of such an approach is clear. Crop rotations that reduce pests reduce the need for pesticides now and into the future. Thus, developing sustainable practices should be the highest priority for agriculture.

Although a minor part of the sustainable agriculture picture up to now, biotechnology could help by developing new crop varieties for use in sustainable systems—for example, faster germinating, cold-tolerant crop varieties could enhance low-input sustainable systems by enabling more effective weed control. Engineered products such as these would fulfill the promise of biotechnology, and would benefit farmers, the environment, and the rural economy alike.

Margaret Mellon is the director of the Agriculture and Biotechnology Program at the Union of Concerned Scientists (UCS). UCS works at the interface of technology and society to promote sustainable agriculture and evaluate applications of biotechnology. Dr. Mellon lectures widely on biotechnology issues and has appeared frequently on television and radio talk shows.

Sharpening Your Critical Thinking Skills

1. State the main thesis of each author in your own words.
2. List the data or arguments used to support each hypothesis and the data used by the authors to refute the other's point of view, if any.
3. Can you determine whether there are any flaws in the reasoning in either essay?
4. Which viewpoint do you agree with? Why?

www.jbpub.com/humanbiology/5e

Visit Human Biology's Internet site for links to web sites offering more information on this topic.

Gene Therapy

Recombinant DNA technology can be used to cure genetic diseases.

As noted earlier, researchers are also tinkering with ways in which genetic engineering can be used to insert normal human genes into genetically defective body cells, curing serious genetic diseases. This technique is called *gene therapy*.

Approximately 1 of every 100 children born in the United States suffers from a serious genetic defect, such as sickle-cell disease or hemophilia. Thanks to advances in genetic engineering, scientists may someday be able to replace their defective genes with normal genes. Instead of developing a treatment to reduce suffering, this technique could, if successful, permit physicians to cure diseases previously considered incurable. Many of these diseases result in enormous human suffering and cost our society millions of dollars a year. The largest obstacle, however, is getting the genes into the DNA of body cells where they are missing. For instance, to cure a genetic defect in brain cells, the genes must somehow be delivered to the cells of the brain.

One promising means of delivering genes to sites where they're required is the bone marrow transplant. Consider an example. Krabbe's disease is a rare genetic disorder resulting from a deficiency in one enzyme in human brain cells. The absence of this enzyme allows fat to accumulate in the nervous system, causing nerve cells to degenerate. Seizures and visual problems occur early in life, and most victims die within the first 2 years of life.

To test a means of correcting this disease, scientists used a strain of mice that suffers from a similar condition. They injected them with bone marrow cells containing the gene that codes for the missing enzyme. Scientists found that the transplanted cells became established in the lungs and liver where they restored enzymatic activity. Some even found their way into the brain!

Another potentially promising technique involves the use of microspheres—tiny lipid spheres. Called *liposomes*, these tiny lipid spheres can be packed with genes and coated with antibodies for specific target cells. If successful, this will allow the liposomes to deliver their contents to the correct cells.

Researchers have also developed a novel transplantation technique that could enable surgeons to introduce genetically engineered cells into specific organs in the human body. In laboratory studies, scientists have injected genetically altered liver cells into a foamlike material. The foam was impregnated with a hormone that stimulates the growth of blood vessels from nearby larger vessels. The foam was then transplanted into rats. Within a week after implantation, the artificial tissue was rid-

dled with a network of blood vessels, allowing the cells to live on. This technique could someday be used to introduce genetically normal cells into numerous tissues and organs of the body, providing a cure for disorders such as diabetes mellitus and Parkinson's disease.

DNA Probes

DNA probes are used for DNA analysis.

DNA from infectious diseases or blood left at a crime scene can be analyzed using **DNA probes**, small fragments of single-stranded DNA produced by machines called *DNA synthesizers*. Probes bind to complementary base pairs of sample DNA. If scientists know the composition of the DNA probes, they can determine the sequence of bases in the DNA under study. This allows technicians to identify DNA in samples of blood, hair, or semen to identify perpetrators of crime. DNA probes can also be used to identify infectious organisms, even anthrax and other potentially lethal biological agents.

Mapping the Human Genome

DNA probes were used to help map human chromosomes.

Geneticists have long been interested in mapping human chromosomes—that is, in determining which genes are on which chromosomes. For years, the mapping process has been rather crude, relying on studies of patterns of linked traits in families.

Another, more refined technique has also been used. In this procedure, scientists fuse a human cell with a mouse cell (Figure 19-12). The new hybrid cell containing 96 chromosomes divides, but when the cell divides, it begins to lose human chro-

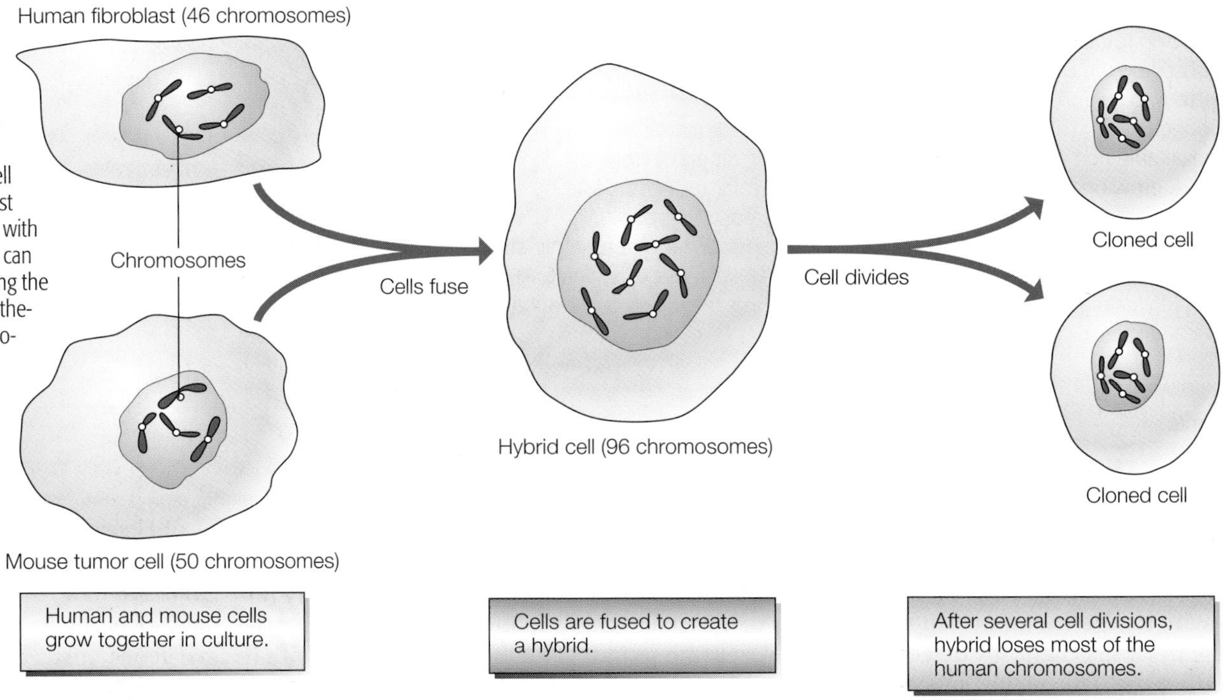

FIGURE 19-12 **Mapping by Fusion** This technique permits scientists to determine the chromosomal address of certain genes. Human and mouse cells are fused, creating a new cell with excess chromosomes. This cell divides and, in so doing, loses most of the excess chromosomes. Cells with one or two human chromosomes can be grown in culture. By determining the unique human proteins being synthesized in these cells and the chromosomes present, geneticists can determine the chromosomal "address" of certain genes. More precise techniques are needed to determine the exact location on the chromosome.

Human fibroblast (46 chromosomes)

Chromosomes

Mouse tumor cell (50 chromosomes)

Cells fuse

Hybrid cell (96 chromosomes)

Cell divides

Cloned cell

Cloned cell

Human and mouse cells grow together in culture.

Cells are fused to create a hybrid.

After several cell divisions, hybrid loses most of the human chromosomes.

mosomes. The daughter cells divide again and lose more. Eventually, cells are produced with all mouse chromosomes and one or two human chromosomes. By studying the resulting protein products, geneticists can associate certain genes with certain chromosomes.

Recombinant DNA technology has permitted a dramatic refinement in chromosome mapping. The DNA probe helped geneticists determine what genes are on what chromosomes, but also their exact location on human chromosomes.

19-5 Controversies over Genetic Engineering—Ethics and Safety Concerns

Despite its promises, genetic engineering is fraught with controversy over safety and ethics.

In October, 1997, several scientists issued a worldwide alert asking all governments to ban imports of a genetically engineered soybean from the United States. The soybeans were genetically altered to be resistant to the herbicide Roundup. The scientists were concerned that the use of this herbicide would increase the level of plant estrogens in soybeans, thus increasing the levels of these potentially harmful chemicals in humans.

In the year 2000, taco shells, tortillas, and numerous snack foods produced from an unapproved genetically engineered corn were recalled from U.S. markets as a result of health concerns. Approximately 300 types of taco shells, tortillas, and snack chips made by the Texas-based Mission Foods were affected by the recall, with an estimated value of about $10 million. This genetically engineered corn (and the resulting products made from it) had not been approved because of questions about its ability to cause allergic reactions. Interestingly, "normal" corn crops were also affected because pollen from the genetically engineered crops was dispersed by the wind into their fields, thus altering the neighboring "normal" corn crops.

At this time, many safety questions remain unanswered. As noted earlier, perhaps the greatest safety concern is the possibility of unleashing genetically altered bacteria or viruses. At least two studies now indicate that genetically engineered bacteria that are applied to seeds take up residence on the roots but migrate very little from the site of application in the short term. Critics, however, are concerned with long-term consequences. Some individuals fear that a genetically altered strain could spread through the environment, wreaking havoc on ecosystems and, possibly, human populations. Once unleashed, it would be impossible to retrieve. The genetically altered bacterium that retards frost formation, for example, could enter the atmosphere on dust particles, reducing cloud formation and altering global climate. No one knows for sure how serious this threat really is. The Point/Counterpoint in Chapter 2 discusses the safety of genetically engineered food.

Other critics object to genetic tinkering, especially the transfer of genes from one species to another, on ethical grounds. Do humans have the right, they ask, to interfere with the course of evolution?

Proponents of genetic tinkering argue that livestock and plant breeders have been selectively breeding hardy animals to produce genetically superior livestock and crops for hundreds, if not thousands, of years. In so doing, people have been performing a kind of genetic engineering—albeit a slow one—for millennia. Genetic engineering, say proponents, merely offers a quicker way of achieving the same goals.

Genetic engineering and selective breeding do indeed achieve the same end points in some instances. Selective breeding simply alters gene frequencies—that is, the percentage of organisms in a population carrying a specific allele. Genetic engineering does the same, as in the case of the frost-retarding bacteria. In other words, geneticists are simply producing large quantities of a mutant already found in nature. But that in and of itself may be reason for concern. In nature, for reasons not well understood, the mutant is found in small quantity. Could shifting the allele frequency cause some ecological catastrophe?

In other instances, the two techniques produce quite different results. That is, in some cases, genetic engineering introduces new genes into species, producing genetic combinations never before encountered in life. Introducing the human growth hormone gene into cattle embryos is an example of this form of tinkering. This manipulation violates a fundamental law of nature: Different species do not interbreed. Is it ethical to blur nature's naturally maintained boundaries? Who can say? And can we put a lid on our ambition? Will we know when to stop?

Unfortunately, experience with genetic engineering is too limited to answer the safety concerns of critics. Preliminary work suggests that the dangers have been exaggerated and that genetically engineered bacteria are not a threat to ecosystem stability. Still, further research is needed to be certain. Ethical questions, like political questions that have been plaguing some countries for centuries, may never be answered to the satisfaction of everyone.

Another point that is rarely mentioned in the debate is the potential for wrong-doing—for creating superinfectious organisms that could wipe out huge numbers of people. Imagine for instance, the catastrophe that might arise if some mad scientist introduced the deadly AIDS virus genes into the genes of the flu virus, which is rapidly transmitted around the world.

Scientists who have been exploring designer pathogens, highly infectious agents, have found that most potential superbugs turn out to be less potent than their natural forms. That is, virulence was typically lost.

In 1998 and 1999, however, Australian researchers made a startling discovery while working to find a way to render mice infertile, in an effort to protect world grain supplies. Mice and rats ravage grain supplies in many countries, reducing food available for people. The researchers had spliced a mouse immune system gene into the mousepox virus, which is similar to the human smallpox virus. Much to their surprise, the mice did not become infertile—they died—even mice that had been immunized against mousepox virus. The researchers published their work with a warning that this technique could potentially be used to produce deadlier forms of human pathogens for biological warfare.

Whatever the outcome of the debate, genetic engineering is here to stay. Future disasters, should they occur, may compel us to prohibit certain forms of genetic tinkering, but in cases where genetic engineering reduces suffering, advances agriculture, and facilitates environmental cleanup, public support may be hard to contain.

SUMMARY

Genetic Manipulation: An Overview

1. Genetic engineering, or recombinant DNA technology, is a procedure by which geneticists remove segments of DNA from one organism and insert them into the DNA of another.

Cloning

2. Cloning is the production of organisms to produce exact replicas using DNA from one (existing or extinct) organism and the cell of another. Cloning also refers to the replication of genes via various methods.

Recombinant DNA Technology

3. Geneticists remove and transplant segments of DNA from an organism using a special enzyme called *restriction endonuclease*. After a segment of DNA is transplanted, a second enzyme, DNA ligase, is used to seal it in place, forming a recombinant DNA molecule—a molecule that contains DNA from two different organisms.

4. Bacteria contain small, circular strands of DNA, called plasmids, which are separate from the chromosome. Foreign genes can be spliced into plasmids, and the plasmid carrying a foreign gene can be reinserted into bacteria or yeast cells in culture. As these cells divide, the plasmids replicate and produce multiple copies of the gene, which is then said to be cloned. Cloning produces numerous identical copies of the foreign gene, which can be used for genetic studies or to produce useful products, such as hormones.

5. Geneticists can produce multiple copies of DNA by extracting messenger RNA from cells and using it to produce DNA using the enzyme reverse transcriptase.

6. The most widely used method of gene amplification is the polymerase chain reaction. In this technique, DNA is alternately heated to split the double helix, then cooled. DNA polymerase enzymes are added to catalyze the formation of complementary DNA molecules.

Applications of Recombinant DNA Technology

7. Recombinant DNA technology and other related techniques have resulted in a number of practical and sometimes controversial applications.

8. Hormones and other proteins can be mass-produced in genetically engineered microorganisms and used to treat a variety of disorders. Recombinant DNA techniques can be used to produce vaccines.

9. Gene splicing can be used to transfer genes from one organism to another, creating transgenic organisms. These organisms rarely differ much from their original form; they may, in fact, only display one new trait, such as a larger body size.

10. Transgenic plants are easier to produce than transgenic animals, and today, efforts are underway to create plants that produce more nutritious seeds and fruits or are more resistant to herbicides and pests.

11. Recombinant DNA technology may also be used to cure genetic disease, a treatment called gene therapy. In this procedure, scientists hope to be able to insert normal human genes into genetically defective body cells.

12. Another practical outcome of research in recombinant DNA technology is the DNA probe. DNA probes are tiny segments of genes that bind to complementary base pairs of sample DNA—for example, hair, blood, or semen taken from crime scenes. DNA probes can also be used to detect infectious organisms and genetic disorders. DNA probes are also being used to determine the location of genes on human chromosomes.

Controversies Over Genetic Engineering—Ethics and Safety Concerns

13. Although genetic engineering has spawned a great deal of enthusiasm, it does have its critics who are concerned with safety and ethical issues. Perhaps the greatest safety concern is the possibility of intentionally or unintentionally releasing potentially dangerous genetically altered bacteria or viruses into the environment.

14. Some critics object to genetic tinkering, especially the transfer of genes from one species to another, on ethical grounds, wondering if we have the right to interfere with the course of evolution.

15. Unfortunately, experience with genetic engineering is too limited to answer the safety concerns of critics. Preliminary work suggests that the dangers have been exaggerated and that genetically engineered bacteria are not a threat to ecosystem stability. Further research is needed to be certain.

critical thinking

Thinking Critically—Analysis

This Analysis corresponds to the Thinking Critically scenario that was presented at the beginning of this chapter.

Before you accept or reject an argument, such as the one made by the chef, it's important to gather all the facts. Reading this chapter will give you some perspective on genetic engineering. I'd also suggest reading some specific articles on genetically engineered tomatoes. An excellent one was published in *Science News*, November 28, 1992. This article explains the ways that geneticists are trying to alter tomatoes, using genetic engineering. Upon careful analysis, it appears that several of the improvements they're seeking could have been acquired by selective breeding—that is, cross-breeding tomatoes that have desired traits with those that have other traits, producing a "super" tomato with good taste,

long shelf life, and other desirable features. In this instance, there's nothing magical about genetic engineering. It's just a faster way of getting desirable genes into a tomato.

But you'll also learn that geneticists are transplanting bacterial genes into tomato plants. One gene in particular retards the production of ethylene gas, which normally stimulates ripening. When the bacterial gene is present, it slows ripening, allowing farmers to leave the tomato on the vine several extra days. This increases the tomato's flavor with obvious benefits to anyone who has had the opportunity to compare a store-bought tomato with one grown in a garden.

The questions here might be: Will this harm consumers? Can you think of any other concerns and ways to address them? Are the chefs merely reacting to the phrase "genetic engineering"? Could this be one of those thought stoppers described in the Critical Thinking section in Chapter 1?

KEY TERMS AND CONCEPTS

Cloning, p. 351
DNA ligase, p. 352
DNA probes, p. 360
Genetically modified organism, p. 357
Gene therapy, p. 359

Genetic engineering, p. 350
Genome, p. 354
Plasmid, p. 352
Polymerase chain reaction, p. 354
Recombinant DNA technology, p. 350

Restriction endonuclease, p. 352
Reverse transcriptase, p. 354
Reverse transcription, p. 354
Selective breeding, p. 351
Transgenic organism, p. 355

CONCEPT REVIEW

1. Describe the methods of recombinant DNA technology. Your answer should explain how genes are removed, spliced, and cloned. pp. 352–354

2. Describe three methods of gene amplification. pp. 352–354

3. What barriers lie in the way of creating transgenic organisms? pp. 355–358

4. List and describe three practical applications of recombinant DNA technology. pp. 354–358

5. List the major concerns for genetic engineering and describe each one. pp. 361–362

6. Debate the statement: Genetic engineering is morally wrong and should be discontinued. It violates the laws of nature and gives humans a power beyond their control. pp. 361–362

SELF-QUIZ: TESTING YOUR KNOWLEDGE

1. Genetic engineering is also known as _____ DNA technology. p. 350

2. _____ is the process of producing an exact replica of an organism. p. 351

3. The enzyme used to slice off a piece of DNA belongs to a group of enzymes known as restriction _____. p. 352

4. The enzyme DNA _____ is used to insert segments of DNA into an opening in the structure. p. 353

5. _____ are small circular strands of DNA in bacteria that can be used to insert genes into bacteria for mass production. p. 352

6. DNA can be produced on mRNA thanks to an enzyme known as reverse _____. p. 354

7. The _____ chain reaction is yet another way to mass produce DNA for genetic studies and other purposes. p. 354

8. An organism containing genes transplanted from another is known as a _____ organism. p. 357

9. Inserting genes into the cells of diseased individuals to cure diseases is known as _____. p. 359

10. A DNA _____ is used to identify DNA from evidence taken from the scene of various crimes to identify potential perpetrators. p. 360

www.jbpub.com/humanbiology/5e

The site features eLearning, an online review area that provides quizzes, chapter outlines, and other tools to help you study for your class. You can also follow useful links for in-depth information, research the differing views in the Point/Counterpoints, or keep up on the latest health news.

thinking critically

Cancer kills nearly 600,000 Americans each year and is a disease that most of us will have close experience with. In fact, according to national statistics, one of every three Americans will contract cancer during his or her lifetime, and one of every four will die from it. (For a debate on whether America is in a cancer epidemic see the Point/Counterpoint discussion in this chapter). Current estimates suggest that 20%–40% of all cancers arise from environmental and workplace pollutants. The remaining cases are caused by smoking, diet, and natural causes.

Researchers at a local medical school have just completed a two-year study on mice and rats that indicates that a common chemical to which we are exposed daily causes cancer in these laboratory animals. They suggest that the chemical should be carefully scrutinized and perhaps banned from public use to protect human health. What research would you suggest needs to be done before this chemical is retired from use? Why?

FIGURE 20-1 Cancer in Plants? Cancer occurs in many different plant and animal species. This tree suffers from multiple tumors.

Although we often think of cancer as a human disease, in actuality, cancer occurs in a wide range of animals and plants (Figure 20-1). In fact, cancer is turning up in many fish, which scientists believe is the result of chemical pollutants in fresh and salt water ecosystems. This chapter looks at cancer in humans. It will familiarize you with terms and concepts that you will encounter throughout your lifetime.

20-1   Benign and Malignant Tumors

Abnormal cell division may lead to benign tumors, which stop growing, or malignant tumors, which can invade other parts of the body.

Cancer is a disease in which cells divide uncontrollably, often invading other parts of the body. But not all abnormal cellular proliferation is cancerous. Some cells, in fact, form small masses that reach a certain size, then stop growing. These are called **benign tumors**. As a rule, they pose no significant medical problems unless they put pressure on nerves or block blood vessels.

Of great concern, however, are the **malignant tumors**, which continue to grow and often spread to other parts of the body, where they may destroy vital organs. The term *cancer* is reserved for malignant tumors, of which there are over 100 different types, such as brain cancer, lung cancer, and liver cancer. (Table 20-1 lists other types of cancer.) Even within these general categories are subtypes; for example, there are several different types of skin cancer.

Almost every type of cell in the body can become malignant. Notable exceptions include neurons and muscle cells, two types of cells that cannot divide. What is more, each type of cancer has its own unique properties.

Metastasis

Cancer cells often migrate to other sites.

Malignant tumors spread when individual cells or clusters of cells break loose from the original tumor, known as the *primary tumor* (Figure 20-2). They then can travel through the body in the circulatory system, or blood vessels, and in the lymphatic sys-

tem, a network of vessels that drains excess fluid from body tissues (Chapter 6). These cells may settle in other sites. In their new residence the cancer cells can divide uncontrollably to form secondary tumors. The spread of cells from one region of the body to another is called **metastasis** (ma-TASS-tah-SISS).

TABLE 20-1	Types of Cancer
Cancer Type	Estimated Deaths
Lung	170,000
Colon/Rectum	55,000
Breast	48,000
Prostate	36,000
Lymphoma	28,000
Leukemia	22,000
Liver	12,000
Skin	10,000
Cervix	5,000
Total	**600,000**

From Lauren Sompayrak, *How Cancer Works,* Sudbury, MA; Jones and Bartlett, 2004, p. 3. Used with permission.

Are We Facing an Epidemic of Cancer?

America's Epidemic of Chemicals and Cancer by Lewis G. Regenstein

America is in the throes of an unprecedented cancer epidemic, caused in part by the pervasive presence in our environment and food chain of deadly, cancer-causing pesticides and industrial chemicals.

Today, significant levels of hundreds of toxic chemicals known to cause cancer, miscarriages, birth defects, immune and central nervous system damage, and other health effects are found regularly in our food, our air, our water—and our own bodies. Accompanying this widespread pollution has been a dramatic and alarming rise in the cancer rate in recent decades (which has only now begun to abate for some cancers, with the restricting of some of the most dangerous chemicals).

Over 4.6 billion pounds of active pesticide ingredients are used in the United States each year—mainly on farms—which amounts to some 18 pounds for every man, woman, and child in the country. Some 75 million pounds of pesticides are used on just lawns and turf. Of the 36 most widely used lawn chemicals, 13 can cause cancer; 14 can cause birth defects, 11 can cause reproductive effects, 15 can damage the liver, and 21 can harm the nervous system.

Lewis Regenstein, an Atlanta writer, is author of *Cleaning Up America the Poisoned* and is president of the Interfaith Council for the Protection of Animals and Nature.

Indeed, pesticides are used almost everywhere in our society: in homes, schools, restaurants, hospitals, parks, offices, hotels—they are virtually inescapable. One EPA study found 23 pesticides in indoor dust and air, many of which had not been used on the premises; another found that most households had at least five pesticides in indoor air, often at levels ten times higher than outdoors.

So it should not be surprising that by the time restrictions were placed on some of the deadliest compounds, such as DDT, dieldrin, and BHC, these carcinogens were being found in the flesh tissues of literally 99 percent of all Americans tested, including mother's milk.

Numerous studies link exposure to pesticides to incidences of cancer, especially in children. Higher rates of childhood brain cancers, leukemia, and soft tissue sarcomas are found in homes where pesticides are used. Indeed, a child in a household where home and garden pesticides are used has a 6.5-fold increased risk of contracting leukemia.

The U.S. National Research Council and the EPA have estimated that pesticide residues in food may be responsible for 20,000 to 60,000 excess cancer cases in America each year. And the American Academy of Pediatrics has warned that infants and children are at particular risk, since "the government is permitting 100 to 500 times as much chemicals in the food as a health basis number would dictate."

Cigarette smoking is by far the biggest cause of cancer, mainly of the lung; and diet can also be an important risk factor, especially the frequent consumption of meat and other high fat foods. But the constant, unavoidable, lifelong chemical onslaught to which we are subjected, from conception on, also plays a role in this disease, once considered rare, that now strikes 1.2 million Americans a year. (This does not count the rapidly growing number of usually non-fatal skin cancers, now a million cases a year. These are mainly caused by exposure to sunlight and increased ultraviolet radiation resulting from the ongoing depletion of the earth's protective ozone layer, which is also caused by pesticides and other synthetic chemicals.) One American in three can now expect to contract a potentially fatal cancer, which kills some 570,000 of us every year—more Americans than were killed in combat in World War II, Korea, and Vietnam combined.

And cancer has now become a common disease of the young as well as the old, with the actual incidence (and not just detection) of childhood cancers, especially leukemia and brain tumors, mounting sharply in recent years.

The response of the U.S. government, through various administrations, has been largely weak or non-existent enforcement of the nation's health and environmental protection laws. EPA, under pressure from Congress and chemical and agricultural interests, has, with few exceptions, refused to carry out its legal duty to ban or restrict cancer causing pesticides.

Thus, we are even now acting to ensure that the current cancer epidemic will continue long into the future. Only time will tell what will be the effect on this generation of Americans, and future ones—the chemical industry's ultimate guinea pigs.

America's Epidemic of Chemicals and Cancer—Myth or Fact? by David L. Eaton, Ph.D.

There is no debate that cancer is a devastating and deadly disease. One in three people living in the United States today will contract some form of cancer in his or her lifetime, and one in four will die from it, if current rates continue. Are cancer rates increasing in epidemic proportions? The total number of people and the fraction of all deaths attributable to cancer have increased dramatically in the past 75 years. However, cancer is largely a disease of old age, and thus it is necessary to adjust such statistics for changes in the age distribution of our population. Annual mortality statistics from both the American Cancer Society and the National Cancer Institute indicate that age-adjusted deaths from all causes of cancer have not increased significantly in the past several decades if lung cancer mortality is excluded. There is no question that we experienced an epidemic of lung cancer in the 20th century. The age-adjusted mortality for lung cancer in both males and females has increased over 50-fold since the early 1900s, peaking in the late 1980s in men and late 1990s in women. About 85-90% of all lung cancer is directly attributable to smoking. Per capita consumption of cigarettes increased 5-fold in men from 1900–1960, and with it concomitant increase in lung cancer. This pattern was repeated, 20 years later, in women. Tobacco products also increase the risk of several other types of common cancers (e.g., cancers of the bladder, kidney, esophagus and mouth). In contrast, there was an equally dramatic decline in the incidence of stomach cancer in the US, during the first half of the 20th century. This decline was mostly likely due to changes in dietary habits that came with the widespread use of refrigeration and possibly antioxidant food preservatives (versus the older methods of smoking and salting methods used to preserve foods). Although mortality statistics have not risen sharply, the incidence (number of cases per 100,000 population) of certain cancers has been increasing in the past several decades. For example, the incidence of childhood leukemia and brain cancers, and adult cases of non-Hodgkin's lymphoma, have increased significantly. While at least part of these increases are likely to be due to better disease surveillance and reporting, the cause(s) of these increases are not known and may be related to as yet unidentified environmental factors. Infectious agents also cause many cancers. For example, nearly all cases of cervical cancer are thought to result from papilloma virus infection that is spread by sexual activity. Likewise, many cases of liver cancer are from hepatitis virus infections, and stomach cancer risk is increased by certain bacterial infections.

Finally, recent advances in the understanding of the biology of cancer suggest that spontaneous or background alterations in DNA may explain much of the cause of cancer. The use of modern techniques in molecular biology has revealed that DNA is inherently unstable and can be altered by normal errors in DNA replication. Within our life span, our cells undergo about 10 trillion cell divisions. Spontaneous errors in this process, which lead to mutations and cancer, accumulate with age. DNA is also subject to extensive oxidative damage from processes associated with normal cellular metabolism that occurs in each cell every day of our lives. Although most of this damage is repaired in the cells, over many years, small amounts of unrepaired damage to DNA accumulates. It is not surprising then that cancer is a frequent outcome of old age. One reason that diet is an important risk factor for many types of cancer is because of the presence of naturally occurring "antioxidants" found in fruits and vegetables. These chemicals help protect against DNA damage from normal cellular metabolism, and may also offer protection against other DNA dam-

aging chemicals found in cigarette smoke, our diet, and other environmental sources of exposure.

David L. Eaton is Professor of environmental health and Associate Dean for Research in the School of Public Health and Community Medicine at the University of Washington. He has an active research program on the molecular mechanisms by which chemicals cause cancer.

The vast majority of cancer researchers throughout the world do not support the view that we are in an overall cancer epidemic caused largely by industrial chemicals. Unfortunately, it will take some time for the political arena, influenced greatly by public fears, to come to grips with the fact that further reduction in exposure of the general public to trace amounts of synthetic chemicals will not have a measurable impact on overall cancer incidence. Hundreds of billions of dollars are spent each year on pollution reduction in the US alone, often under the rubric of reducing cancer risk to the public. I believe that much of these expenditures are justified to enhance the quality of our environment, protect valuable natural resources, and ensure the habitability of our planet for our children and our children's children. But these expenditures for pollution prevention are not going to have a significant impact on cancer incidence and mortality in the US. Certainly we must continue our efforts to identify potentially cancer-causing industrial chemicals and take appropriate actions to minimize occupational and environmental exposures to such chemicals. However, if our society is truly concerned about reducing the human tragedy from cancer, more efforts should be focused on eliminating smoking and alcohol abuse and increased public education about dietary risk factors. More research into the biochemical and molecular events that lead to cancer will ultimately lead to more effective prevention and treatment of cancer.

Sharpening Your Critical Thinking Skills

1. Regenstein asserts that America is in the midst of a cancer epidemic caused in large part by pesticides and industrial chemicals. Eaton argues the opposite. He says that the death rate for all types of cancer except lung cancer is actually declining. Summarize the supporting data of each author.

2. Using the critical thinking skills described in Chapter 1, analyze each argument.

www.jbpub.com/humanbiology/5e

Visit Human Biology's Internet site for links to web sites offering more information on this topic.

FIGURE 20-2 Cancer Growth and Metastasis Cancers grow by cell division. Cells can break free from the tumor and spread in the blood and lymphatic systems to other parts of the body where they establish secondary tumors. Secondary tumors often develop in the liver, lungs, and lymph nodes.

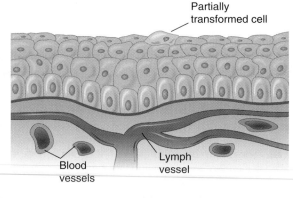

1 An epithelial cell becomes partially transformed.

Partially transformed cell

Blood vessels

Lymph vessel

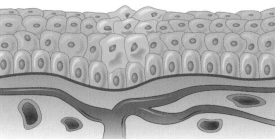

2 This cell multiplies, forming a mass of dysplastic cells.

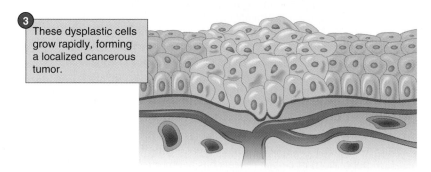

3 These dysplastic cells grow rapidly, forming a localized cancerous tumor.

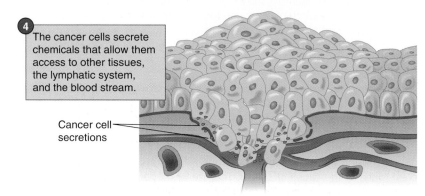

4 The cancer cells secrete chemicals that allow them access to other tissues, the lymphatic system, and the blood stream.

Cancer cell secretions

The ability of cancer cells to metastasize is one of the main reasons cancers are often so difficult to treat and so lethal. Acquiring the ability to metastasize is, therefore, a key event in the development of most cancers. What allows cancer cells to break away from the primary tumor?

Most cells in the body stay in place. They're joined in tissues and organs by molecules on their surfaces, known as *cell-cell adhesion molecules*. In cancer cells, these molecules are either lacking or altered. Cancer cells therefore lose their anchorage. (Interestingly, researchers found that experimentally restoring these molecules in cancer cells grown in culture dishes can render cancer cells ineffective [unable to spread] when they are transplanted into mice.)

Not all cells that break away from the primary tumor survive to form a secondary tumor, however. In fact, researchers estimate that only one in 10,000 cancers cells that enter the circulatory system is successful in establishing a secondary tumor.

The most common site of metastasis is the lungs. Why? After leaving the organ in which the primary tumor is located, the first vascular bed most cells encounter is in the lungs. The lungs have extensive networks of capillaries that are involved in the exchange of oxygen and carbon dioxide. Cancer cells quickly attach to the lining of the small blood vessels, where they set up residence. The liver is the second most common site for secondary tumors. However, some types of cancer cells show a preference for certain organs other than the lung and liver. Prostate cancer cells, for instance, tend to settle in the bones. It is believed that molecules on the prostate cancer cells bind specifically to molecules on the cells lining the blood vessels in bone.

20-2 How Cancers Form

The conversion of a normal cell to a cancerous one involves two steps.

A growing body of research suggests that the production of a malignant tumor (cancer) involves two steps that may take place over a long period of time: (1) conversion and (2) development and progression (Figure 20-3).

The conversion of a normal cell to a cancerous one typically begins with a mutation, a change in the DNA of cells. Mutations are caused by chemical, physical, or biological agents (viruses, for example). Those agents that cause the initial transformation are known as *co-carcinogens*.

FIGURE 20-3
Stages Leading
to Cancer

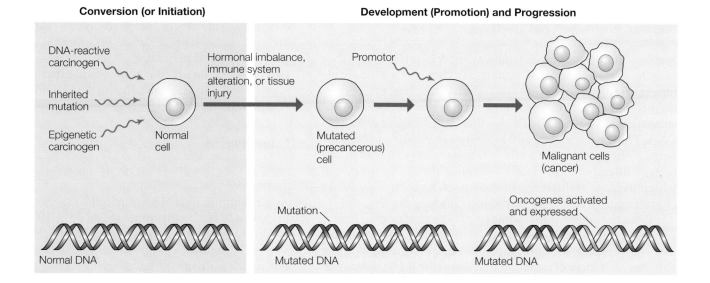

Mutations and Cancer

> Mutations in two groups of genes are primarily responsible for cancer.

Mutating just any gene won't lead to cancer, however. The mutation must occur in one of the 50 genes that control the replication of cells. These genes are responsible for regulating the cell cycle and fall into one of two groups.

One group is the **proto-oncogenes**, a group of genes that control functions related to cell replication. For example, they control the production of plasma membrane receptors that bind to growth factors or hormones. When mutations in these genes are not repaired, the result can be uncontrolled cellular proliferation—cancer. The second group, the **tumor suppressor genes**, naturally inhibit growth in the cell. described more fully later in the chapter.

When mutations occur in either of these genes, cells may be partially or completely released from the normal growth controls. Partially released cells, called *precancerous cells*, often remain dormant for long periods. They are very likely repressed by chemicals released from nearby cells, called *tissue factors*. This homeostatic control keeps the cancer cell from dividing.

The second stage leading to cancer is called *development and progression*. It occurs when a chemical substance, even a naturally occurring hormone, stimulates the growth of a precancerous cell. Those chemical substances that influence the second stage of carcinogenesis are called *promoters*. Promoters stimulate cells to divide uncontrollably. Once released, cells begin a frenzy of growth that, if unchecked, usually kills the host.

Mutations, like those that result in cancer, occur with great frequency. Fortunately, though, cells have evolved an intricate mechanism to repair genetic damage. If a mutation that transforms a cell into a cancerous one can be repaired before the cell divides, the cancer is eliminated. These mechanisms must also be considered part of the body's homeostatic weaponry. The immune system can also locate and destroy tumor cells. Unfortunately, not all cancerous mutations are repaired in time or eliminated by the immune system.

Other Causes of Cancer

> Several physical and biological agents are also responsible for producing cancer.

Chemical agents are not the only cause of cancer. X-rays that physicians use to diagnose disease, ultraviolet radiation from the sun or a tanning salon, and radioactive radon gas given off by soils in some parts of the country can lead to mutations that result in cancer (Figure 20-4).

Biological agents also cause cancer. Researchers, in fact, have discovered many viruses that cause cancer in a wide variety of animals, including humans. Unlike physical and chemical carcinogens, which alter DNA structure, carcinogenic viruses contain genes that, when inserted into the genetic material of the cells, may cause uncontrollable cell division.

FIGURE 20-4 **Fun in the Sun** As much fun as beach life is, excess exposure to harmful ultraviolet radiation in sunlight can lead to skin cancer later in life.

Epigenetic Carcinogens

Some carcinogens exert their effect outside of the DNA.

Many chemical carcinogens induce cancer by altering the DNA of cells; that is, they cause mutations. These agents are therefore generally referred to as *DNA-reactive carcinogens*. A growing body of evidence, however, indicates that many chemical substances act outside the DNA—that is, they do not react with DNA. These carcinogens and known as *epigenetic carcinogens*. Some of them may cause hormonal imbalances that lead to cellular proliferation. Others may suppress the immune system, which permits cellular proliferation. Still others may cause chronic (persistent) tissue injury that leads to cancer.

Evidence suggests that some epigenetic carcinogens stimulate changes that lead to modifications in the DNA that cause tumor development from precancerous cells. For example, alpha particles are a form of radiation that can cause direct damage to DNA. However, when alpha particles miss the nucleus, they can cause the formation of free radicals, highly reactive chemicals that possess an extra electron. These highly reactive substances can diffuse into the nucleus, where they affect the DNA.

Interestingly, seemingly innocuous chemicals can turn into dangerous substances inside the body. Nitrites, for example, are used as a food preservative, for example, in sandwich meats like bologna and in hot dogs. Although normally harmless, nitrites are converted to carcinogenic nitrosamines inside the cells of the body. (See the Health and Homeostasis section in Chapter 2.)

Latent Period

Cancers develop many years after the initial exposure to a carcinogen.

As a rule, mutations do not manifest themselves immediately in cancer. In fact, many tumors form 20 or 30 years after the mutation. A few cancers such as leukemia occur within 5 years. The period between exposure and the emergence of a cancerous tumor is called the *latent period*.

Because the latent period is long, it has been extremely difficult for medical researchers to determine the exact causes of many cancers, a fact that has hindered public health research considerably.

20-3 Cancer Treatments

Cancer varies in its lethality. Some forms, such as lung cancer or the skin cancer called *melanoma*, are highly lethal; other types of skin cancer are not. Fortunately, new drugs have helped boost the survival rate of individuals suffering from many types of cancer, including childhood leukemia. Surgery also works effectively for some cancers, for example, the common types of skin cancer. Radiation treatment is successful in others. Many times physicians use a combination of measures, such as surgery followed by chemotherapy or radiation.

Surgery, Chemotherapy, and Radiation

Tumors can be removed surgically or treated with chemotherapy and radiation.

Primary tumors can often be removed surgically (Figure 20-5) or destroyed by radiation. The removal of the original tumor, however, is fruitless unless secondary tumors are destroyed. Unfortunately, finding secondary tumors is extremely difficult, for there are literally thousands of places where the cells can become established. Cancer cells from the breast, for example, may lodge in the brain or lungs or in lymph nodes, organs interspersed along the lymphatic system to filter the fluid coursing through the lymphatic vessels. Physicians can take tissue samples of lymph nodes to assess the spread of the cancer and can remove cancerous nodes. Even so, it is often difficult to locate secondary tumors.

When a secondary tumor is difficult to remove, physicians generally use radiation or chemotherapeutic drugs, drugs that attack rapidly dividing cells of tumors. This shotgun approach has many side effects, including nausea, hair loss, and general sickness, but it may be a patient's only hope. New developments in cancer treatment are discussed below.

FIGURE 20-5 **Surgery** Removing a tumor by surgery can prove effective in eliminating cancers unless a part of the tumor is missed or malignant cells from the tumor have spread to other parts of the body. Because of the high likelihood that cancer cells have metastasized, surgery is typically followed up by a course of radiation or chemotherapy.

New Treatments for Cancer

Hardly a year goes by without headlines announcing promising new treatments for some form of cancer. Let's look at some of the more promising ones.

Anticancer Vaccines

Stimulating the body to produce antibodies and T cells that attack tumors can help eliminate existing tumors.

Efforts to treat cancers with a vaccine date back to the beginning of the field of immunology. Early researchers used malignant cells, either from the patient's own tumors or from other patients. The tumor cells were irradiated so they would not continue to divide. They were then injected into the bloodstream. The goal behind this was to stimulate the immune system to make antibodies and T cells that would attack not just the cells being injected, but the tumor itself. Such efforts were never very successful.

In 1998, however, researchers reported preliminary results of a clinical study using a "vaccine" against lung cancer, a disease that annually kills 170,000 Americans (mostly cigarette smokers). The new vaccine contained antigens taken from the plasma membranes of tumor cells.

The lung cancer vaccine is not really a vaccine in the traditional sense, because it does not prevent the disease. This treatment is actually an immune system boost that helps attack an existing tumor. In the study, 34 patients with early-stage lung cancer were treated with a plasma membrane protein extracted from lung cancer cells. When given to patients in early stages of lung cancer, the protein appears to stimulate the patient's immune system. The immune system, in turn, mounts an assault on the foreign protein in the blood. However, because the protein is also attached to the plasma membrane of lung cancer cells, the immune system attacks and kills the cancer cells, especially those that have spread to other sites.

In clinical studies, this experimental therapy greatly increased the subjects' 5-year survival rates. Normally, only 1 patient in 10 who is diagnosed with lung cancer is alive 5 years after diagnosis and conventional treatment (surgery followed by chemotherapy). Five years after the immune system boost, 5 of 10 patients were still alive. Spurred by this research and others, scientists now have a rapidly growing list of tumor antigens, many of which are being tested in clinical trials.

Microspheres

Liposomes containing cancer-killing chemicals or monoclonal antibodies attached to chemotherapeutic agents can deliver a lethal dose of these chemicals directly to cancers.

In recent years, some scientists have been working with lipid microspheres (liposomes) containing cancer-killing chemicals. Liposomes can be designed to target cancer cells, in the process delivering small doses of the lethal drugs to them, thus sparing the rest of the body. Targeting cancer cells is the job of antibodies, special proteins incorporated in the membrane of liposomes. These antibodies bind to proteins in the plasma membrane of cancer cells, allowing the "toxic bullet" to find its target.

Cancer-killing agents (oncotoxins) may also be attached directly to monoclonal antibodies. Monoclonal antibodies are antibodies synthesized by the immune system of animals that have been injected with human cancer cells. The antibodies are isolated, then chemically bonded to oncotoxins, either a chemotherapeutic drug or a radioactive compound. They are then administered to patients. Inside the body of the patient, the antibody-oncotoxin complex binds to the cell surface of cancer cells, killing them. Laboratory and clinical trials have shown some promising results.

Attacking a Cancer's Blood Supply

Reducing blood vessel development in tumors successfully stops cancers from growing.

Another promising development in the treatment of cancer involves techniques that attack its blood supply. Early on, cancerous tumors remain fairly small. For example, it may take a tumor a decade to reach the size of a pea. Blood vessels on the periphery of the tumor supply blood, but the tumor cannot grow larger because of a lack of blood to its interior.

But then something happens. New blood vessels begin to form. They invade the tumor, previously held in check by a lack of internal capillaries. The formation of new blood vessels is called *angiogenesis*.

New blood vessels supply the tumor cells with nutrients and growth factors that stimulate growth of the tumor. In fact, researchers believe that the development of new blood vessels is one of the crucial steps in a tumor's transition from a small, harmless mass of mutated cells to a large, malignant growth that can spread to other organs of the body. Growth-promoting factors may come from a variety of sources: the tumor cells themselves, nearby tissue cells, or perhaps even macrophages.

Researchers are also testing a variety of antiangiogenic drugs in hopes that they can shrink tumors. Numerous compounds are under study in laboratories around the world. One of the most promising is a protein known as *angiostatin*. It stops angiogenesis in large primary tumors and secondary tumors in laboratory animals. Used in conjunction with chemotherapeutic agents or radiation, angiostatin or other similar drugs could help treat cancer. They may eventually be used for long-term treatment of cancer patients after they've undergone more traditional treatments such as surgery, chemotherapy, or radiation.

Attacking Cancer at the Level of the Gene

Blocking genes involved in cell replication or altering their products could help doctors kill cancer cells.

Research is also underway to treat tumors by attacking oncogenes and tumor suppressor genes, and others. **Oncogenes** are a mutated version of genes that regulate cell growth, called *proto-oncogenes*. Tumor suppressor genes are naturally occurring genes that normally keep cells from becoming malignant. By blocking mutated forms of these genes or altering their products, researchers hope to find ways to target cancer cells specifically. Their hope is that such treatments at the genetic level will help them stop cancer dead in its tracks, while reducing toxic side effects seen with other forms of treatment such as chemotherapy.

Some researchers are even studying ways to correct the genetic defects in cancer cells that lead to uncontrolled replication. Take the tumor suppressor gene, for example. Tumor suppressor genes code for the production of a suppressor protein that, in normal cells, serves as a brake to DNA replication. Mutations of the gene can render the protein inactive. Cells that are normally held in check are unleashed to proliferate madly. By inserting normal tumor suppressor genes in tumor cells, researchers hope to return them to normalcy. Encouraging results are being produced in laboratory studies.

Treating Cancer with Light

Drugs combined with light therapy may prove to be an effective treatment in some forms of cancer.

Scientists have found that certain drugs, when activated by laser light, can kill cancer cells. In one therapy, patients with tumors in the lining of the esophagus and the bronchi of the lung are given the drug Photofrin. After the drug circulates in the blood and bathes the cancer cells, the researchers insert a small fiberoptic fiber in the affected area. A laser beam is trained on the cancer and, when turned on, activates the chemical, which selectively destroys the cancer cells. This treatment has been approved in the United States for certain forms of advanced esophageal cancer. One of the chief advantages is this therapy is that it poses little risk to adjacent tissues and can be repeated several times and combined with other therapies.

Early Detection

The earlier a cancer is detected and treated, the better one's chances of survival.

One of the most effective "treatments" of cancer is not a drug or radiation; it is early detection. The earlier a cancer is detected, the greater one's chances of survival. When a cancer is detected early, there's less chance that it will have spread to other parts of the body.

New technologies make it possible to detect cancer cells in the blood of patients after the cells have broken away from the primary tumor, long before any clinical signs of a secondary tumor have appeared. Additional tests permit researchers to identify the origin of cancer cells, too.

Cancer Prevention

Preventing cancer may be the best approach.

Despite the promising new developments in cancer treatment, prevention is always the best approach. By quitting smoking or not taking it up in the first place, by limiting exposure to X-rays and UV radiation, and by choosing a safe occupation and dwelling, you can greatly reduce your likelihood of contracting cancer.

As noted in Chapter 5, proper diet may also help prevent cancer. Dietary fiber in grains and vegetables, for example, may help reduce your chances of contracting colon cancer, a leading killer of men (Figure 20-6).

Other substances in food known as *phytochemicals* may also help reduce the risk of cancer. Many phytochemicals are antioxidants, as explained in Chapter 5. They remove free radicals that can trigger early changes in the DNA that may lead to cancer.

Recent research suggests that a chemical substance called *ellagic acid*, which is found in raspberries, blackberries, strawberries, and other fruit and nuts, may destroy carcinogenic molecules in the body, reducing the likelihood of developing cancer. Scientists studying carcinogenic agents found in tobacco smoke, auto exhaust, and foods noted that in laboratory animals, ellagic acid reduced DNA damage caused by one prevalent carcinogen by 45%–70%.

Despite gains from years of intensive research and billions of dollars spent studying cancer, modern science may actually be losing the war against many types of cancer. For example, long-term survival rates in cancer patients over the past decade or so have increased only 4.2% compared with an overall increase of 5.1% in survival rates for all other diseases. Because of this statistic, many agree that the cheapest and most effective cure is prevention.

FIGURE 20-6 Reducing Cancer by Eating Right Fruits, vegetables, and grains provide fiber and phytochemicals that help reduce the incidence of colon cancer.

SUMMARY

1. Cancer is a disease characterized by uncontrolled cell division. Current estimates suggest that 20%–40% of all cancers arise from environmental and workplace pollutants. The remaining cases are caused by smoking, diet, and natural causes.
2. No matter what the cause, cancer is a disease that most of us will have close experience with. One out of three Americans will contract the disease, and one out of four will die from it.

Benign and Malignant Tumors

3. Cells that divide uncontrollably form tumors. Tumors that fail to grow and spread are benign and rarely cause medical problems.
4. Malignant tumors are those that continue to grow, often spreading to other parts of the body and invading vital organs. The term cancer is reserved for malignant tumors.
5. Malignant tumor cells spread (metastasize) through the body via the circulatory and lymphatic systems, often becoming established in distant sites, where they form secondary tumors. The lungs are one of the main sites of metastasis.

How Cancers Form

6. The transformation of a normal cell into a cancerous one often results from a mutation, a change in the DNA resulting from a chemical, biological, or physical agent. These agents are called co-carcinogens.
7. The initial mutation caused by a co-carcinogen may or may not result in a cancerous growth. A second factor, a promoter, however, may stimulate the cell to proliferate uncontrollably.

Cancer Treatments

8. Tumors can be removed surgically or destroyed with radiation, but successful treatment depends on the location and destruction of secondary tumors.

9. Patients can also be treated with chemotherapeutic agents, chemical substances that attack rapidly dividing cells.
10. Many new treatments are also underway. Medical researchers are, for example, studying ways to target cancer cells specifically, for example, by activating the immune system so that it produces antibodies and T cells that attack cancer cells. They are also researching ways to deliver lethal cancer-fighting agents by microspheres.
11. Some researchers have found that preventing the growth of blood vessels stops cancer growth. Others are working on ways to remedy the underlying genetic defects in cancer cells by inserting healthy genes to replace the mutated ones.
12. Despite all of the advances, early detection and prevention remain two of the most important measures in the battle against cancer.

critical thinking

Thinking Critically—Analysis

This Analysis corresponds to the Thinking Critically scenario that was presented at the beginning of this chapter.

Just because a chemical causes cancer in rats and mice doesn't mean that it will have the same effect on humans. To study this chemical, it would be wise to examine the scientific literature to see if there are any studies of workers exposed to this particular chemical through their workplace. If so, it would be prudent to determine the level of exposure—that is, how high concentrations were at their workplaces. If they match those of the experiment and are close to those levels of exposure in our daily lives, perhaps it might be wise to consider regulating the chemical or finding ways of dramatically reducing our exposure. It might also be wise to study the effects of the chemical in our daily lives, by looking for potential health effects in people already exposed to it. Studies such as these can be costly and may take several years to complete, but, combined with other research, they could give us an indication of what steps, if any, need to be taken to protect the public from unnecessary exposure.

KEY TERMS AND CONCEPTS

Benign tumor, p. 365
Cancer, p. 365
Malignant tumor, p. 365
Metastasis, p. 365

Oncogene, p. 372
Proto-oncogene, p. 369
Tumor suppressor gene, p. 369

CONCEPT REVIEW

1. In what ways are cancer cells different from normal cells? p. 365
2. What is the difference between a benign and a malignant tumor? p. 365
3. Describe the two stages of cancer development. pp. 368–370

4. Why don't more mutated cells become cancerous? p. 368
5. Cancer is often considered a disease of aging. Given the fact that cancer is caused by DNA mutation, why is this assertion correct? p. 370

6. Can you think of some reasons why the incidence of some cancers has risen in the past 20 years? pp. 372–373

SELF-QUIZ: TESTING YOUR KNOWLEDGE

1. A _____ tumor is one that grows uncontrollably. p. 365
2. The conversion of a normal cell to a precancerous cell is known as _____. p. 369

3. A chemical substance that stimulates a precancerous cell to divide is known as a _____. p. 369
4. Mutations of proto-oncogenes and tumor _____ genes can lead to cancer. p. 369

5. Chemicals that alter the DNA, causing mutations that lead to cancer, are known DNA-_____ carcinogens. p. 370

 www.jbpub.com/humanbiology/5e

The site features eLearning, an online review area that provides quizzes, chapter outlines, and other tools to help you study for your class. You can also follow useful links for in-depth information, research the differing views in the Point/Counterpoints, or keep up on the latest health news.

thinking critically

Imagine that you are a journalist for a newspaper. One day, you receive a press release from a medical school announcing that one of its researchers has discovered that a chemical found in a common household cleaning agent reduces fertility in mice. Large doses were given to males and females before conception. The results showed a statistically significant decline in the litter size as well as several physical abnormalities in the offspring. The press release quotes the researcher, who says that the chemical should be banned from use in homes. Using your critical thinking skills, what questions would you ask before writing your article? What other information would you seek?

Sandra Collins woke one day with a pain in her abdomen that persisted throughout the morning. Instead of calling her doctor, though, she shrugged off the pain and went Christmas shopping with her husband. A few hours later, while she was browsing through a bookstore, the pain grew worse and she blacked out. Her husband rushed her to the emergency room, where doctors discovered that Sandra was suffering from internal bleeding caused by an ectopic (eck-TOP-ick) pregnancy—that is, a fertilized ovum that had developed in the upper part of her reproductive tract, outside the uterus. Surgeons whisked her into the operating room, where they surgically removed the fetus, placenta, and surrounding tissue and repaired torn blood vessels.

Reproduction is one of the most basic body functions. However, as this account shows, it doesn't always operate smoothly. This chapter examines human reproduction and related topics. Sexually transmitted diseases are covered in A Closer Look at the end of this chapter.

21-1 The Male Reproductive System

In humans, the male reproductive system (shown in Figure 21-1a; see also Table 21-1), consists of two gonads, the **testes** (TESS-teas). The testes produce sperm and male sex hormones. The testes are suspended in the **scrotum**, a sack of skin attached to the body below the attachment of the penis. The scrotum provides a slightly cooler environment that is necessary for sperm to survive.

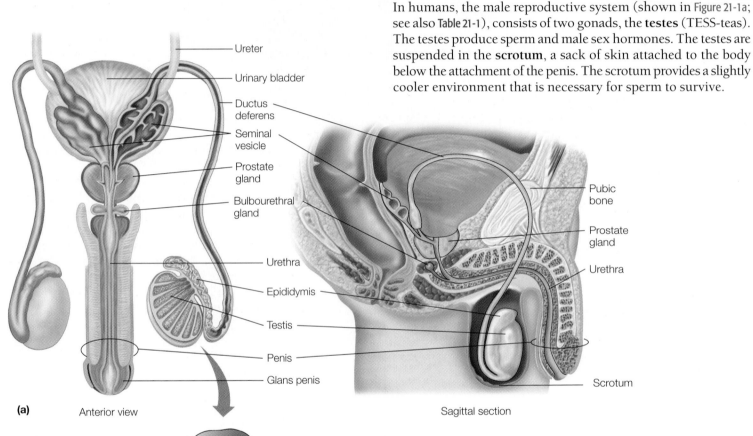

(a) Anterior view

Sagittal section

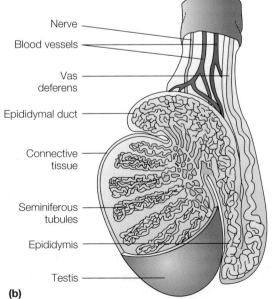

(b)

FIGURE 21-1 **The Male Reproductive System** (a) Organs of the male reproduction system. (b) Interior view of the testis.

TABLE 21-1	The Male Reproductive System
Component	Function
Testes	Produce sperm and male sex steroids
Epididymes	Store sperm
Vasa deferentia	Conduct sperm to urethra
Sex accessory glands	Produce seminal fluid that nourishes sperm
Urethra	Conducts sperm to outside
Penis	Organ of copulation
Scrotum	Provides proper temperature for testes

The Testes

Sperm are produced in the seminiferous tubules and stored in the epididymis until ejaculation.

As Figure 21-1b shows, each testis contains numerous highly convoluted **seminiferous tubules** (SEM-in-IF-er-uss), in which sperm are formed.

Sperm produced in the seminiferous tubules empty into a network of connecting tubules in the "back" of the testes. These tubules, in turn, empty into the **epididymal duct** (ep-eh-DID-eh-mal), where sperm are stored until they are released during **ejaculation**, the ejection of sperm. While being stored, sperm fully mature, a process that makes them capable of swimming. As Figure 21-1b shows, the epididymal duct forms the **epididymis**. During ejaculation, the epididymal duct empties into the **vas deferens** (plural, **vasa deferentia**, pronounced VAH-sah DEAF-er-en-she-ah). These ducts pass from the scrotum into the body cavity through the **inguinal canals**, small openings in the wall of the abdomen (Figure 21-2a). The inguinal canals are potential weak spots in the abdominal wall. In some men, loops of intestine can bulge through weakened inguinal canals (Figure 21-2b). This condition, known as an **inguinal hernia**, can be corrected surgically.

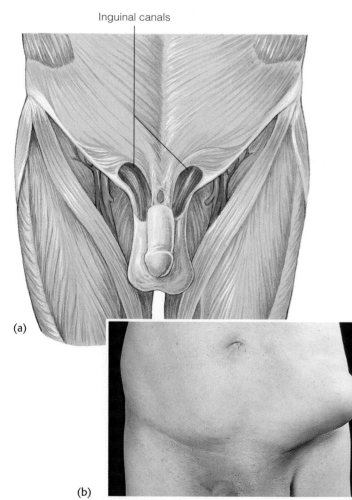

(a)

(b)

FIGURE 21-2 The Inguinal Canal and Hernia (a) During development, the testis descends through the inguinal canal, an opening in the musculature in the lower abdominal wall. In adults, the inguinal canal provides a route for the vas deferens, blood vessels, and nerves that supply each testis. (b) Loops of intestine may push through the weakened musculature surrounding the inguinal canal.

The vasa deferentia then empty into the urethra. From here, sperm course through the penis to the outside of the body during ejaculation, the release of sperm (discussed shortly).

The Sex Accessory Glands

Semen consists mostly of fluids produced by the sex accessory glands.

During ejaculation, sperm are combined with fluids produced by small glands located near the neck of the urinary bladder, the **sex accessory glands**. These include the seminal vesicles, the prostate gland, and the Cowper's glands (Figure 21-1).

The paired seminal vesicles empty into the vasa deferentia and produce the largest portion of the ejaculate. The Cowper's glands are a pair of pea-sized glands located below the prostate on either side of the urethra. The prostate gland surrounds the neck of the bladder and empties its contents directly into the urethra.

The sex accessory glands produce fluid that, when combined with sperm produced by the testis, form **semen**. Sperm, the male sex cells, constitute only 1% of the volume of the semen.

Fluid from the sex accessory glands has many chemical components. For example, it contains the simple sugar fructose, which is used by sperm to generate energy needed to help propel themselves through the female reproductive tract. Semen also contains a chemical buffer, a substance that neutralizes the lethal (to sperm) acidic secretions of the female reproductive tract. Yet another component of the semen is prostaglandin. This is a chemical that causes the muscle of the womb (uterus) to contract. Muscle contractions are believed to be primarily responsible for the movement of sperm up the female tract.

Prostate Diseases

The prostate enlarges with age and is the site of a common form of cancer in men.

Routine medical examinations of men over the age of 45 show that nearly all of them have enlarged prostates. This condition results from the formation of small nodules inside the gland. In some cases the nodules grow so large that they block the flow of urine and make urination painful. This condition requires surgery.

The prostate is also a common site for cancer in men. Prostate cancers are detected by digital rectal exams that allow doctors to feel the prostate. A blood test is also now commonly used to determine levels of prostate-specific antigen, circulating levels of antigen produced by cancer cells.

Sperm Formation

Sperm are formed in the seminiferous tubules.

The formation of sperm in the seminiferous tubules is illustrated in Figure 21-3. Sperm are formed from special cells in the periphery of the tubules, known as **spermatogonia** (sper-MAT-oh-GO-nee-ah). The spermatogonia divide by mitosis, producing a constant supply of new cells to make sperm.

Some spermatogonia, however, undergo meiosis, a process that involves two cellular divisions that result in the formation of four cells, called *spermatids*. Each spermatid contains one-half the number of chromosomes of the spermatogonia. The spermatids then differentiate, becoming **sperm**. During this process, the nucleus condenses and the cytoplasm is eliminated. A long-whiplike tail (a flagellum) forms from the centriole. The mitochondria of the spermatid congregate around the first part of the tail, where they can provide energy for propulsion. These changes streamline the cell so it is able to swim and to fertilize an ovum. The Golgi apparatus enlarges and forms an enzyme-filled cap that fits over the condensed nucleus like a stocking cap. This structure, the **acrosome** (ACK-row-sohm), will help the sperm digest its way through the coatings surrounding the ovum during fertilization.

On average, men produce 200–300 million sperm every day. The average 3-milliliter ejaculate contains 240 million or more. Such large numbers no doubt evolved because many

sperm are required to ensure fertilization. In fact, nearly all sperm are eliminated as they travel through the female reproductive tract.

In humans, each sperm formed during meiosis contains 23 single-stranded (unreplicated) chromosomes—half the number in a normal somatic cell. Thus, when the sperm unites with an ovum (also containing 23 unreplicated chromosomes), they produce a **zygote** containing 46 single-stranded chromosomes. One-half of its chromosomes come from each parent.

Interstitial Cells

The interstitial cells produce male sex steroid, primarily testosterone.

The testes also produce male sex hormones. These hormones are produced in cells found in spaces lying between the seminiferous tubules. Here you will find clumps of large cells known as **interstitial cells** (in-ter-STISH-al). These cells produce a group of sex steroid hormones known as **androgens**, so named because they exert a masculinizing effect. The most important androgen is **testosterone** (discussed shortly).

Testosterone

Testosterone stimulates sperm formation and many other functions.

Testosterone stimulates the formation of sperm. Testosterone is also transported in the bloodstream throughout the body and stimulates cellular growth in bone and muscle. This hormone

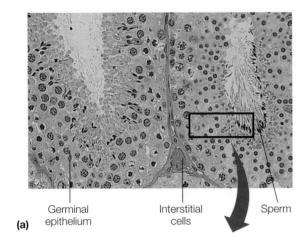

(a) Germinal epithelium   Interstitial cells   Sperm

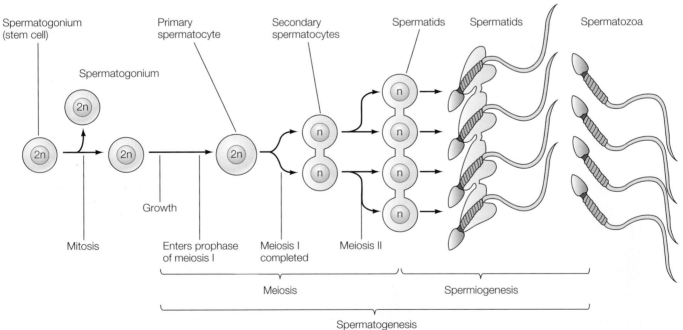

Each spermatogonium contains 46 single-stranded (unreplicated) chromosomes.

Each primary spermatogonium contains 46 double-stranded (replicated) chromosomes.

Each secondary spermatogonium contains 23 double-stranded chromosomes.

Each spermatid contains 23 single-stranded chromosomes.

FIGURE 21-3 Formation of the Sperm (a) Cross section through two seminiferous tubules showing the germinal epithelium where sperm are formed and the interstitial cells where testosterone is produced. (b) Details of sperm formation.

(b)

accounts in part for the fact that men are generally taller and more massive than women. In addition, testosterone promotes facial hair growth and thickening of the vocal cords, typically giving men deeper voices than women. Testosterone also adversely affects the hair follicles on the heads of many men as they grow older, leading to pattern baldness.

Acne and Testosterone

Testosterone secretion may lead to acne.

Testosterone also stimulates the oil glands of the skin of in both sexes. (Women secrete lesser amounts of testosterone.) Oil-producing glands, known as **sebaceous glands** (seh-BAY-schuss),

secrete oil onto the skin, moisturizing it (Figure 21-4a). During puberty (sexual maturation) in boys, rising levels of testosterone dramatically increase sebaceous gland activity. This itself is not a problem. However, if dead skin cells block the pores that normally carry the oil to the skin's surface, oil known as sebum may collect inside the gland (Figure 21-4b). Bacteria on the skin may invade the gland where they proliferate. This results in inflammation, pus formation, and swelling. The skin protrudes, forming an acne pimple.

Mild **acne** can be treated by gently washing the skin twice a day with warm water and a mild, unscented soap. The skin should also be treated with an acne cream containing benzoyl peroxide. Moderate and severe acne can be successfully treated with special ointments and antibiotics prescribed by a doctor.

The Penis

The penis contains erectile tissue that fills with blood during sexual arousal.

Sperm are deposited in the female reproductive tract with the aid of the **penis**. The penis consists of a shaft of varying length and an enlarged tip, the glans penis (Figure 21-5). The glans is covered by a sheath of skin at birth, the foreskin. The foreskin

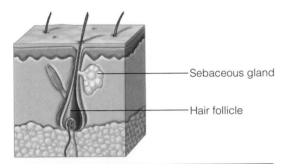

(a) Sebaceous glands associated with hair follicles secrete sebum, an oily substance that lubricates the skin.

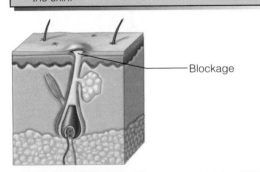

(b) A follicle may become blocked by excess sebum and dead skin cells. Unable to escape, the sebum builds up in the hair follicle.

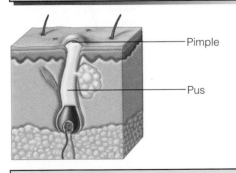

(c) Bacteria present on the skin may infect the sebum, causing inflammation; pus and swelling form an acne pimple.

FIGURE 21-4 **Formation of an Acne Pimple** (a) Testosterone stimulates oil production in the sebaceous glands. (b) If the outlet is blocked, sebum builds up in the gland and (c) the gland may become infected.

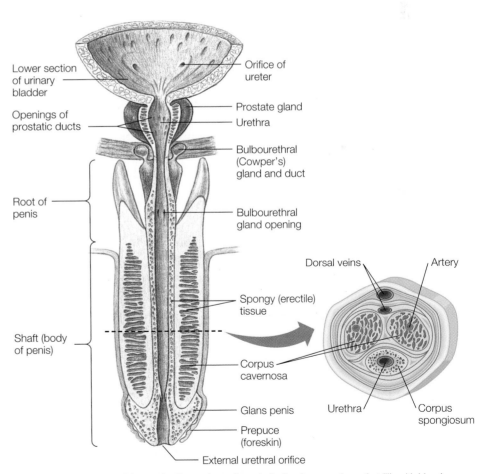

FIGURE 21-5 **Anatomy of the Penis** The penis consists principally of spongy tissue that fills with blood during sexual arousal. The urethra passes through the penis, carrying urine or semen.

gradually becomes separated from the glans in the first 2 years of life. At puberty, the inner lining of the foreskin begins to produce an oily secretion. Bacteria can grow in the protected, nutrient-rich environment created by the foreskin, so special precautions must be taken to keep the area clean.

Because of potential health problems or religious reasons, many parents opt to have the foreskin removed in the first few days of their son's life. The operation, called *circumcision*, may help reduce penile cancer in men and may also reduce cervical cancer in the wives or sexual partners of circumcised men.

During sexual arousal, nerve impulses traveling to the penis from the spinal cord cause arterioles in the organ to open. Blood flows into a spongy **erectile tissue** (eh-REK-tile) in the shaft of the penis, making it harden. The growing turgidity (swelling) compresses a large vein on the dorsal surface of the penis, blocking the outflow of blood and further stiffening the organ.

Some men lose their ability to achieve or to sustain an erection. This condition is known as **erectile dysfunction** (formerly called impotence). Most men experience impotence at some time in their life, but it is usually temporary. Persistent impotence, however, is a more serious condition, and is more common in middle-age and elderly men, especially smokers.

Persistent impotence may be caused by marital conflict, stress, fatigue, and anxiety. Naturally low testosterone levels and nerve damage may also cause impotence. Alcohol ingestion and some medications may contribute to the condition as well. Blockages (due to atherosclerosis) in the arteries leading to the penis may be the cause in some cases, as is smoking.

Numerous treatments are possible, depending on the cause of the problem. One popular treatment is the drug Viagra (and, more recently, other similar chemicals). Viagra works by causing the muscle in the walls of the arteries supplying the penis to relax. This permits blood to flow into the organ more readily when stimulated.

Ejaculation

Ejaculation is a reflex mechanism.

Ejaculation occurs when the penis is erect and is a reflex triggered by sexual stimulation. Sexual stimulation causes motor neurons in the spinal cord to send impulses along nerves to the smooth muscle in the walls of the epididymis, vasa deferentia, and sex accessory glands. These nerve impulses cause the smooth muscle to contract, which propels sperm and fluid into the urethra.

Hormonal Control of Male Reproduction

The male reproductive system is controlled by three hormones.

As noted earlier, the testes produce sex steroid hormones—mainly, testosterone. Testosterone secretion, however, is controlled by a hormone from the pituitary gland. This hormone is known as **luteinizing hormone** (**LH**). In males, LH is also known as **interstitial cell stimulating hormone** (**ICSH**) be-

cause it stimulates the interstitial cells to produce androgen hormones.

ICSH secretion is controlled by a releasing hormone produced by the hypothalamus, known as **gonadotropin releasing hormone** (**GnRH**). As Figure 21-6 shows, the secretion of GnRH and ICSH is controlled by testosterone levels in the blood in a negative feedback loop. When testosterone levels in the blood decline, receptors in the hypothalamus detect the change and

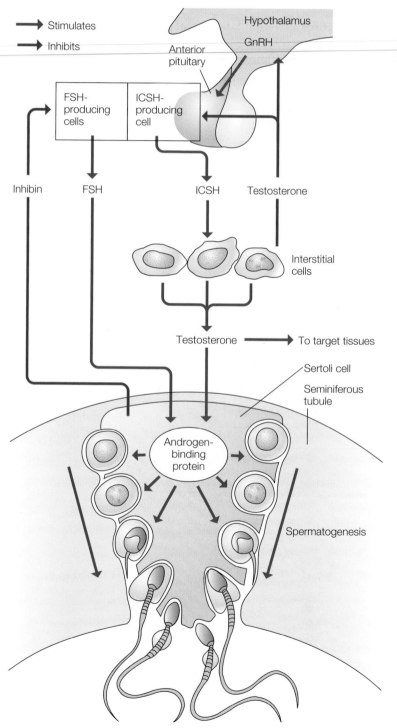

FIGURE 21-6 Hormonal Control of Testicular Function Testosterone, FSH, and ICSH participate in a negative feedback loop. The testes also produce a substance called inhibin, which controls GnRH secretion.

signal an increase in GnRH secretion. This hormone flows in the blood to the anterior pituitary, where ICSH is produced. GnRH stimulates the production and release of ICSH. It travels in the bloodstream to the testes, where it stimulates testosterone production. But when testosterone levels return to normal, GnRH release subsides, as does ICSH secretion.

The pituitary also produces a gonadotropin known as **follicle-stimulating hormone** or **FSH**. Like testosterone, FSH stimulates sperm formation. FSH secretion is controlled by GnRH and a peptide hormone called *inhibin*, produced by the testes. Inhibin gets its name from the fact that it inhibits the activity of the cells in the anterior pituitary that produce FSH.

21-2 The Female Reproductive System

Figure 21-7 illustrates the female reproductive system, which consists of two parts: the reproductive tract (internal organs) and the external genitalia. Let's start with the structures of the female reproductive tract. Table 21-2 summarizes the role of each.

Anatomy of the Female Reproductive System

> The female reproductive tract consists of the ovaries, uterus, and oviducts.

The **ovaries** are paired, almond-shaped organs located in the pelvic cavities of women. They produce the female gametes, the ova or eggs. The ovaries also produce several important reproductive hormones, discussed shortly.

An egg released from an ovary is taken up by one of two hollow, muscular tubes, known as the **uterine tubes** (YOU-ter-in), or **oviducts** (OH-va-ducts). As Figure 21-7a shows, the ends of the uterine tubes are widened and fit loosely over the ovaries. Currents created by cilia in the lining of the uterine tubes draw the egg inside and down the tubes to the uterus.

Fertilization occurs in the upper third of the uterine tubes. The fertilized ovum is then transported down the uterine tubes to the uterus. Inside the uterus, the fertilized ovum attaches to the lin-

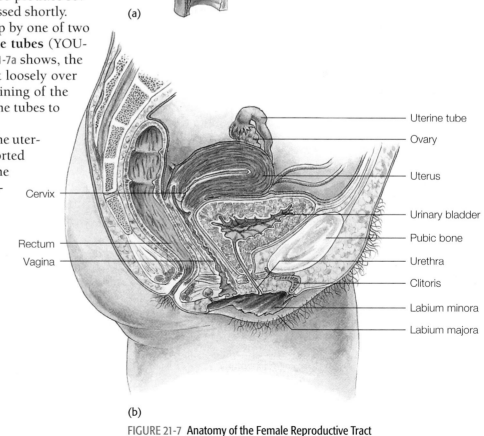

(a)

(b)

FIGURE 21-7 **Anatomy of the Female Reproductive Tract** (a) Frontal view. (b) Midsagittal view.

TABLE 21-2	The Female Reproductive System
Component	Function
Ovaries	Produce ova and female sex steroids
Uterine tubes	Transport sperm to ova; transport fertilized ova to uterus
Uterus	Nourishes and protects embryo and fetus
Vagina	Site of sperm deposition, birth canal

ing, or the **endometrium** (EN-doh-MEE-tree-um), and embeds itself there, remaining for the duration of pregnancy.

The womb or **uterus** is a pear-shaped organ about 7 centimeters (3 inches) long and about 2 centimeters (less than 1 inch) wide at its broadest point in nonpregnant women. The wall of the uterus contains a thick layer of smooth muscle cells. The uterus houses and nourishes the developing fetus.

At birth, the baby is expelled from the uterus through the **cervix** (SIR-vix), the lowermost portion of the uterus. As Figure 21-7b shows, the cervix protrudes into the vagina (vah-GINE-ah).

The **vagina** is a distensible, 3-inch, tubular organ that leads to the outside of the body. Although it is quite narrow, the cervix stretches considerably at birth to allow the passage of the baby into the vagina. The vagina also serves as the receptacle for sperm during sexual intercourse. To reach the ovum, sperm must travel through a tiny opening and narrow canal of the cervix that leads into the uterus. From here, sperm move up both uterine tubes.

The External Genitalia

The external genitalia are the externally visible parts of the female reproductive system.

The external genitalia are known as the **vulva**. It consists of two flaps of skin on either side of the vaginal and urethral openings (Figure 21-8). The outer folds are known as the **labia majora** (LAY-bee-ah ma-JOR-ah). These large folds of skin extend from the mons pubis, a mound of fatty tissue lying over the pubic bone. The labia are covered with hair on the outer surface and contain numerous sebaceous glands on the inside. The inner flaps are the **labia minora** (meh-NOR-ah). Anteriorly, they meet to form a hood over a small knot of tissue called the *clitoris* (CLIT-er-iss). The **clitoris** is a highly sensitive organ involved in female sexual arousal. It consists of erectile tissue and becomes filled with blood during sexual arousal. It is formed from the same embryonic tissue as the penis. In fact, some women are born with greatly elongated clitorides (plural).

Sexual Arousal and Orgasm

Many changes occur in women during sexual arousal and orgasm.

The sexual response of women involves four stages. During the first phase, the excitement phase, cervical glands produce a secretion that lubricates the vagina. Erectile tissue in the labia fills with blood, causing them to expand. The nipples become erect. Heart rate, muscle tension, and blood pressure increase.

During the next phase, breathing, heart rate, blood pressure, and muscle tension increase more. Blood vessels in the wall of the vagina and clitoris fill with blood, becoming engorged. Glands near the opening of the vagina are activated, producing a lubricating fluid.

Continued sexual arousal can lead to orgasm, rhythmic contractions of the uterus and vagina. These contractions are often accompanied by intense physical pleasure. Women often report feelings of warmth throughout their bodies after orgasm.

In the fourth stage, blood drains from the clitoris and labia. Blood pressure, heart rate, and respiration return to normal.

The Ovaries

The ovaries produce ova and release them during ovulation.

During each menstrual cycle, one ovary releases an **ovum**, the female gamete. This process is called **ovulation** (OV-you-LAY-shun). The release of an ovum occurs approximately once a month in women during their reproductive years—from puberty (age 11–15) to menopause (age 45–55). Ovulation is temporarily halted when a woman is pregnant and may be suppressed by emotional and physical stress.

The structure of an ovary is shown in Figure 21-9. As illustrated, several ovarian landmarks are visible. One is the female

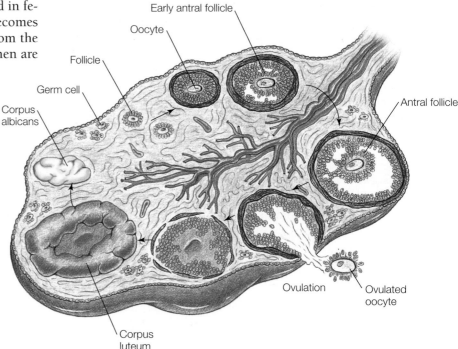

FIGURE 21-9 **Structure of the Ovary** This drawing illustrates the phases of follicular development and also shows the formation and destruction of the corpus luteum (CL). Antral follicles give rise to the CL. A fully formed CL and antral follicle would not be found in the ovary at the same time.

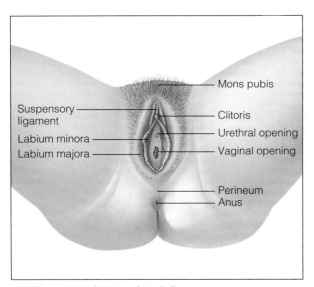

FIGURE 21-8 **Female External Genitalia**

germ cells—immature ova. They are surrounded by varying numbers of cells. Together, the immature ova (also known as *oocytes*) and the cells surrounding them are called **follicles**. As illustrated, follicles vary in size. The earliest ones contain only a few follicle cells. The more developed follicles contain many

layers. Figure 21-10 (right side) illustrates the growth and development of follicles. Each month, a dozen or so follicles begin to develop.

In the largest follicles, a clear liquid begins to accumulate between the follicle cells. Eventually, so much liquid accu-

FIGURE 21-10
Oogenesis and Follicle Development

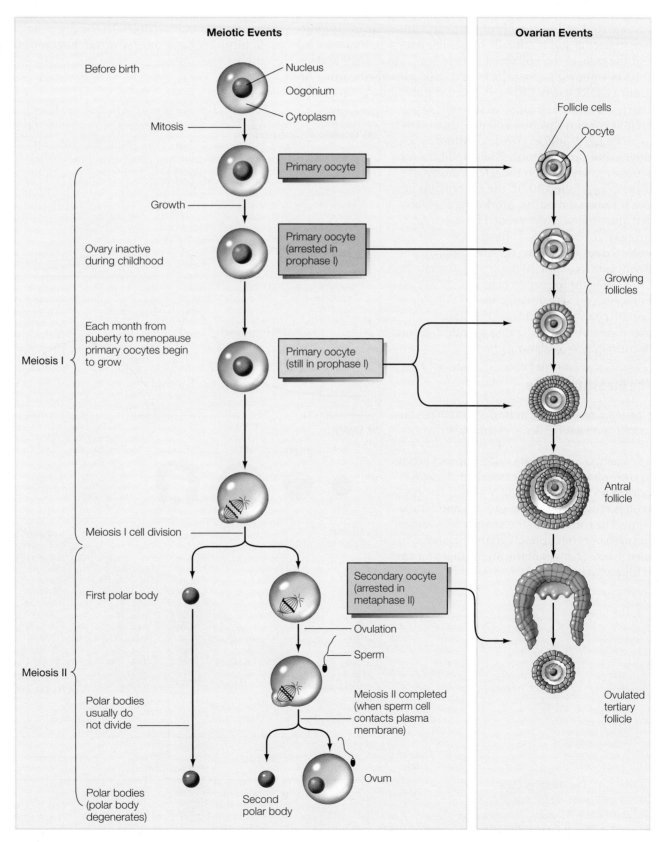

mulates that one central cavity is formed. At this point, the follicle is called an **antral follicle** (AN-tril). Although a dozen or so follicles begin developing during each cycle, as a rule, only one makes it to ovulation. The rest stop growing and degenerate.

The follicle (or follicles) that survives continues to enlarge by accumulating more fluid. As the fluid builds up, the antral follicle begins to bulge from the surface of the ovary. The pressure exerted on the outside of the ovary causes the ovary's surface to stretch and eventually burst. The oocyte is then released and the cells of the collapsed follicle begins to enlarge and secrete hormones. These cells form a structure called **corpus luteum** (CORE-puss LEU-tee-um; "yellow body") or CL for short—so named because of the yellow pigment it contains in cows and pigs (Figure 21-9).

The CL produces two sex hormones, estrogen and progesterone. The fate of the CL ultimately depends on the fate of the oocyte. If the oocyte is fertilized, the CL remains active for several months, producing the estrogen and progesterone needed for a successful pregnancy. If fertilization does not occur, the CL disappears in about 10 days.

During the formation of follicles, oocytes also undergo important changes. It is during this time that meiosis begins. The first meiotic division occurs right at ovulation.

The Menstrual Cycle

The menstrual cycle consists of cyclic changes in hormone levels, the uterus, and the ovaries.

The changes in the ovary each month, known as the *ovarian cycle*, are caused by changes in hormones from the pituitary gland. The hormones stimulate follicle growth and ovulation. They also cause changes in ovarian hormone production. Ovarian hormones, in turn, cause changes in the uterus that prepare it for pregnancy.

Together, the hormonal, ovarian, and uterine changes occurring in women each month constitute the **menstrual cycle** (MEN-strell). On average, the menstrual cycle repeats itself every 28 days. Ovulation usually occurs approximately at the midpoint of the 28-day cycle. Although the average length of the menstrual cycle is 28 days, in some women it lasts 25 days; in others it may last 35 days. The length of the cycle may also vary from month to month in the same woman.

Figure 21-11, starting with the top panel, shows the changes in pituitary hormones, ovarian hormones, the ovary, and the uterus.

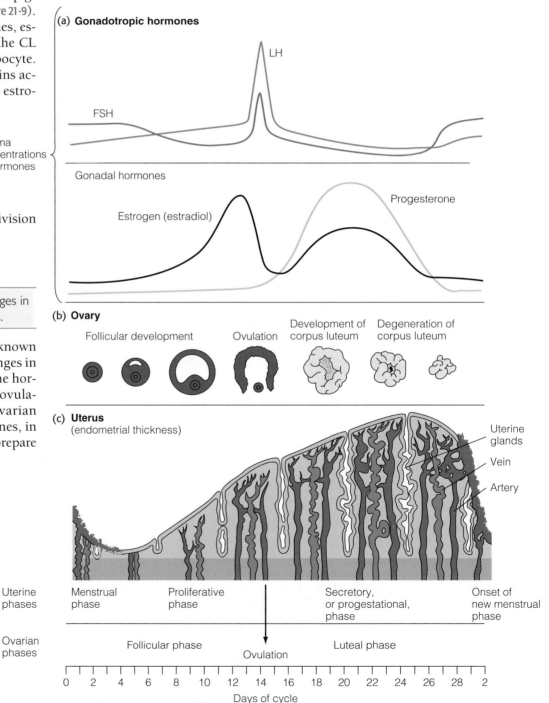

FIGURE 21-11 **The Menstrual Cycle**
(a) Hormonal cycles. (b) The ovarian cycle. (c) The uterine cycle.

Hormonal Control of Female Reproduction

Pituitary hormones stimulate follicle development, ovulation, and ovarian hormone production.

As illustrated in the top panel, you see that two pituitary gonadotropic hormones, FSH and LH, are released during the first half of the menstrual cycle. These hormones peak in the middle of the menstrual cycle, just before ovulation.

As its name implies, FSH stimulates follicular development by stimulating the mitotic division of the follicle cells. LH stimulates estrogen production by follicles. Estrogen also stimulates mitotic division of follicle cells, promoting growth. The first half of the menstrual cycle is appropriately known as the *follicular phase*, because it is during this time that the follicles grow toward ovulation.

LH and FSH secretion are controlled in a negative feedback mechanism involving estrogen and GnRH produced by the hypothalamus. Just before ovulation, LH and FSH secretion surge. This peak in production, however, is the result of one of the body's rarest events, a positive feedback loop, caused when estrogen levels reach a certain critical level. At this point, the hypothalamus responds with a sudden outpouring of GnRH. The pituitary responds with a sudden massive release of LH and FSH that stimulate ovulation.

After ovulation, FSH and LH release decline sharply. The corpus luteum, shown in the third panel, forms from the collapsed follicle and releases both estrogen and progesterone. The second half is of the menstrual cycle is called the **luteal phase** (LU-tee-al), so named because the corpus luteum forms during this time.

Uterine Changes

The uterine lining, or endometrium, also undergoes cyclic changes during the menstrual cycle.

Changes in ovarian hormones that occur during the menstrual cycle also have profound effects on the uterine lining, as shown in the bottom panel in Figure 21-11. As illustrated, the lining or endometrium thickens throughout much of the cycle in preparation for pregnancy. In the absence of fertilization, most of the thickened endometrium is shed, a process called **menstruation** (MEN-strew-A-shun).

To understand how the endometrium responds to hormonal changes, we begin on day 1 of the average 28-day menstrual cycle. Day 1 of the cycle marks the first day of menstruation. During the next 4 or 5 days, the uterine lining is shed—that is, the tissue that formed in the previous menstrual cycle sloughs (sluffs) off from the lining. It then passes out of the uterus into the vagina along with a considerable amount of blood—on average, about 50–150 milliliters. (This loss of blood is the reason that women are more prone to develop anemia than men and why women should take iron supplements or eat iron-rich foods.)

As soon as the endometrium has been shed, the lining of the uterus begins to rebuild—to prepare for the possibility of a pregnancy in the new cycle. Initial regrowth is stimulated by estrogen released by the ovaries. Estrogen stimulates the growth of glands in the endometrium and causes them to fill with a nutritive secretion that will nourish an embryo should fertilization occur. It also promotes cell division in the deepest layer of the endometrium, which causes it to grow thicker.

After ovulation, the endometrium continues to thicken under the influence of estrogen and progesterone—again in preparation for pregnancy. The uterine glands become swollen with a glycogen-rich secretion.

If fertilization does not occur, however, the uterine lining starts to shrink approximately 4 days before the end of the cycle. It then begins to slough off, starting menstruation. The shedding of the uterine lining is triggered by a decline in estrogen and progesterone concentrations in the blood.

When progesterone levels fall, the uterus begins to undergo periodic contractions. That's because progesterone inhibits smooth muscle contraction in the myometrium. These contractions propel the detached tissue out of the uterus and are responsible for the cramps many women experience during menstruation.

Effects of Fertilization

Fertilization prevents the uterine lining from being sloughed off.

If fertilization occurs, the fertilized egg passes into the uterus and embeds in the thickened lining of the uterus, from which it derives its nutrients early in pregnancy. If pregnancy is to continue, ovarian hormones must continue to be secreted. Ovarian hormone production, however, is now stimulated by a hormone produced not from the pituitary but from the newly formed embryo. This hormone is known as **human chorionic gonadotropin** (KO-ree-ON-ick), or **HCG**.

HCG has the same effect as LH—that is, it stimulates estrogen production by the ovary (in particular by the CL). Continued estrogen production keeps the uterine lining from deteriorating. HCG shows up in detectable levels in a woman's blood and urine about 10 days after fertilization. If the ovum is fertilized and the zygote successfully embeds in the uterine lining, HCG will maintain the CL for approximately 6 months.

Pregnancy tests available through a doctor's office or drugstore detect HCG in a woman's urine. The home pregnancy tests are relatively inexpensive, fairly reliable, and fast.

Estrogen and Progesterone

The female sex steroids, estrogen and progesterone, have numerous effects.

Like testosterone in boys, estrogen secretion in girls increases dramatically at puberty. As the level of estrogen in the blood increases, the hormone begins to stimulate follicle development in the ovaries. Menstrual cycles begin. Estrogen also stimulates the growth of the external genitalia and the breasts. It also stimulates growth of internal structures including the uterus, uterine tubes, and vagina.

Estrogen's influence extends far beyond the reproductive system. For example, estrogen promotes rapid bone growth in the early teens. Because estrogen secretion in girls usually occurs earlier than testosterone secretion in boys, girls typically experience a growth spurt before similarly aged boys. However, estrogen also stimulates the closure of the growth zones of the bones, putting an end to the female growth much earlier than in boys. Thus, most girls reach their full adult height by the age of 15–17. Boys continue growing until the age of 20–21. Besides promoting bone growth, estrogen stimulates the deposition of fat in women's hips, buttocks, and breasts, giving the female body its characteristic shape.

Premenstrual Syndrome

Premenstrual syndrome is a condition afflicting many women.

For reasons not yet fully understood, many women suffer from irritability, depression, fatigue, and headaches just before menstruation. Many also complain of bloating, tension, joint pain, and swelling and tenderness of the breasts. These complaints are symptomatic of a condition known as **PMS**, or **premenstrual syndrome**.

Premenstrual syndrome is a clinically recognizable condition characterized by one or more of the symptoms noted above. Four of every 10 women of reproductive age experience PMS in varying degrees. Although scientists are still trying to determine its causes, dozens of "cures," ranging from massive doses of progesterone to vitamin B6, have been prescribed by clinics specializing in PMS. Buyers beware, however, for very little good scientific evidence is available to indicate which, if any, of the "cures" really work.

Fortunately, work is now underway to test various treatments to see if any consistently bring relief. In the meantime, physicians recommend that women suffering from PMS see their family doctor to be certain that the symptoms are not caused by some other medical problem. Doctors recommend relaxation and avoidance of stress to help avoid or relieve PMS. Light exercise may help, as may warm baths. More frequent light meals with plenty of carbohydrates and fiber may be beneficial, too. Reductions in salt intake and avoiding excess chocolate consumption are also recommended. Cutting out caffeine drinks and taking vitamin B6 supplements are also generally advised.

Menopause

Menopause is the cessation of menstruation.

The menstrual cycle continues throughout the reproductive years. As a woman ages, however, estrogen levels decline. This decline is brought about by a decrease in the number of ovarian follicles. Ovulation and menstruation become increasingly erratic. Between the ages of 45 and 55, most of the follicles that were in the ovary at puberty have been stimulated to grow and have either degenerated or ovulated. This results in complete cessation of the menstrual cycle and is called **menopause**.

The decline in estrogen production results in several important body changes. For example, it causes the breasts and internal reproductive organs to begin to shrink. Vaginal secretions often decline, and, in some women, sexual intercourse becomes painful.

The decline in estrogen levels may also result in behavioral disturbances. Many women, for instance, become more irritable and suffer bouts of depression. Headaches, insomnia, and sleepiness may occur in some women. Very noticeable physical changes may also occur. Three quarters of all women suffer "hot flashes" and "night sweats" induced by massive vasodilation of vessels in the skin. For some women, however, the symptoms are quite mild. Fortunately for all people concerned, these symptoms usually pass.

As noted in Chapter 12, declining estrogen levels also accelerate osteoporosis, a softening of the bone due to a reduction in calcium. To counter osteoporosis and relieve other unwelcome changes taking place during menopause, physicians can prescribe small amounts of estrogen or a combination of estrogen and progesterone. This is known as **hormone replacement therapy**. Most women are treated for 5 to 10 years after the onset of menopause, but not longer, for long-term treatment may increase the risk of breast cancer. Osteoporosis can also be prevented, even reverse, with new drugs and exercise.

21-3 Birth Control

Birth control is any method or device that prevents births. Birth control measures fit into two broad categories: (1) contraception, ways of preventing pregnancy, and (2) induced abortion, the deliberate expulsion of a fetus.

Contraception

Contraceptive measures help prevent pregnancy.

Figure 21-12 summarizes the effectiveness of the most common means of contraception. Effectiveness is expressed as a percentage. A 95% effectiveness rating means that 95 women out of 100 using a certain method in a year will not become pregnant.

The most effective means of contraception is abstinence, refraining from sexual intercourse. This form of birth control not only reduces unwanted pregnancy, it prevents the transmission of sexually transmitted diseases (see the Chapter 21 supplement).

The next most effective means of preventing unwanted pregnancies is surgical sterilization. It is also the leading method of contraception practiced by married couples in the United States.

In women, sterilization is performed by cutting or cauterizing the uterine tubes. This technique is called *tubal ligation* and is shown in Figure 21-13a. Tubal ligations, performed in a hospital, are performed by making tiny incisions in the abdomen to access the uterine tubes.

FIGURE 21-12 **Effectiveness of Contraceptive Measures** Percent effectiveness is a measure of the number of women in a group of 100 who will not become pregnant in a year.

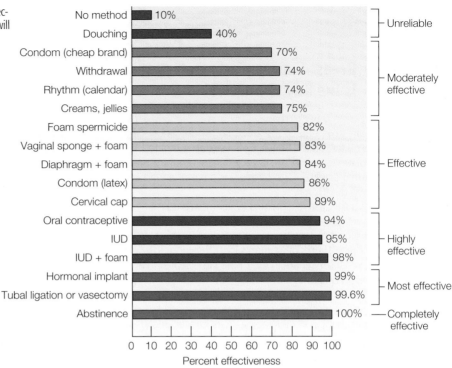

Method	Percent effectiveness	Category
No method	10%	Unreliable
Douching	40%	
Condom (cheap brand)	70%	Moderately effective
Withdrawal	74%	
Rhythm (calendar)	74%	
Creams, jellies	75%	
Foam spermicide	82%	Effective
Vaginal sponge + foam	83%	
Diaphragm + foam	84%	
Condom (latex)	86%	
Cervical cap	89%	
Oral contraceptive	94%	Highly effective
IUD	95%	
IUD + foam	98%	
Hormonal implant	99%	Most effective
Tubal ligation or vasectomy	99.6%	
Abstinence	100%	Completely effective

Male sterilization is known as a *vasectomy* (vah-SECK-toe-me), and can be carried out in a physician's office under local anesthesia (Figure 21-13b). To perform a vasectomy, a physician makes a small incision in the scrotum. Each vas deferens is exposed, then cut, and the free ends are tied off or cauterized.

Vasectomies only prevent the sperm from passing into the urethra during ejaculation. They do not impair sex drive, and because they do not block the flow of the sex accessory glands, which produce 99% of the volume of the ejaculate, they have virtually no effect on ejaculation. Vasectomy and tubal ligation can be reversed, although the operation is not always successful.

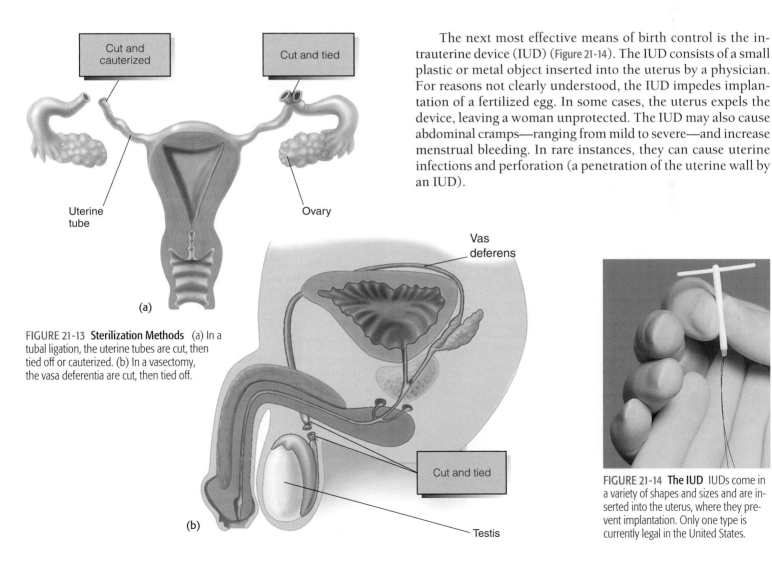

FIGURE 21-13 **Sterilization Methods** (a) In a tubal ligation, the uterine tubes are cut, then tied off or cauterized. (b) In a vasectomy, the vasa deferentia are cut, then tied off.

The next most effective means of birth control is the intrauterine device (IUD) (Figure 21-14). The IUD consists of a small plastic or metal object inserted into the uterus by a physician. For reasons not clearly understood, the IUD impedes implantation of a fertilized egg. In some cases, the uterus expels the device, leaving a woman unprotected. The IUD may also cause abdominal cramps—ranging from mild to severe—and increase menstrual bleeding. In rare instances, they can cause uterine infections and perforation (a penetration of the uterine wall by an IUD).

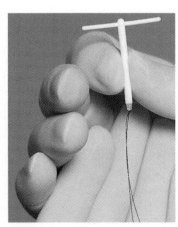

FIGURE 21-14 **The IUD** IUDs come in a variety of shapes and sizes and are inserted into the uterus, where they prevent implantation. Only one type is currently legal in the United States.

21-1   Breast Cancer: Early Detection Is the Key to Survival

Breast cancer is one of the most common cancers affecting women in the United States. Each year, an estimated 180,000 cases of breast cancer are detected in this country. Breast cancer is a fairly lethal disease. One of every five women who contracts breast cancer will die from it.

Doctors and medical researchers have uncovered many risk factors associated with this disease. Numerous studies, for instance, show that the risk of breast cancer increases as a woman ages. In fact, the risk doubles every ten years of woman's life. Most women who are diagnosed with breast cancer are in their 50s; the disease is very rare in women under 30 years of age.

Studies also indicate that estrogen is a risk factor for breast cancer. Generally speaking, the longer a woman is exposed to estrogen, the higher are her chances of developing breast cancer. Because of this, the early onset of menstruation is another risk factor, as is the late onset of menopause. Both of these risk factors are thought to be related to overall estrogen exposure.

Estrogen in birth control pills may also slightly increase a woman's chances of developing a cancerous breast tumor. Estrogen given to postmenopausal women for ten years or more can elevate a woman's risk, too.

Obesity also increases a woman's risk. Excess body fat causes an increase in estrogen. For reasons not completely understood, the age of a woman at the time of her first child's birth also affects her risk of developing breast cancer. If her first child is born after a woman's thirtieth birthday, her chances of developing breast cancer are two times higher than for a woman whose first child was born before her twentieth birthday. If she never has children, her chances of developing breast cancer are also twice as high as a woman who gives birth to a child before age 20.

Some women inherit a higher risk for breast cancer. There are at least two abnormal genes that can be linked to this disease. Women who inherit these genes often elect to have their breasts removed as a kind of pre-emptive strike.

Researchers have also discovered factors that help reduce the risk of breast cancer. A low animal-fat diet is one of them. Breast feeding for two years or more also seems to lower risk.

Early detection is a key to beating this disease. The earlier it is detected, the more likely a woman is to survive. Because of this, many doctors recommend that women over 40 have a routine mammogram, an X-ray of the

---

Another highly effective approach involves the insertion of matchstick-sized capsules containing a potent form of progesterone (Norplant) inserted under the skin of a woman's arm (Figure 21-15). These prevent pregnancy for up to 5 years.

The birth control pill is another effective means of birth control. Birth control pills come in several varieties, but the most common contains a mixture of synthetic estrogen and progesterone. These hormones inhibit the release of LH and FSH by acting on the pituitary and hypothalamus. The lack of LH and FSH, in turn, inhibits follicle development and ovulation. A minipill containing progesterone alone is also available. Even though it is less effective than the combined pill, the minipill is more suitable for some women because it has fewer side effects.

Birth control pills must be taken throughout the menstrual cycle. Skipping even a few days may release the pituitary and hypothalamus from the inhibitory influences of estrogen and progesterone, resulting in ovulation and possible pregnancy.

Although effective, birth control pills may have adverse health effects in some women. Even though the incidence of these adverse side effects is small, a woman considering different birth control options should study them carefully before making a decision.

One very rare side effect is death. Deaths from the use of birth control pills may result from heart attacks, strokes, or blood clots. To reduce this risk, pharmaceutical companies have dramatically lowered the estrogen content of the combined pill, because estrogen is responsible for most of the adverse effects.

Early studies showed a positive correlation between the use of birth control pills and cancers of the breast and cervix (see Health Note 21-1). More recent studies suggest that the new generation of low-estrogen pills is less likely to cause cancer of the breast or cervix.

Even with reduced estrogen levels, however, women who take birth control pills are more likely to develop cervical cancer than women who do not. Physicians therefore recommend annual Pap smears for women on the pill.

During a Pap smear, the cervical lining is swabbed. The swab picks up cells sloughed off by the lining of the cervix. The cells are then examined under a microscope for signs of cervical cancer. Early diagnosis increases a woman's chances of survival.

The next most effective means of birth control are the barrier methods—the diaphragm, condom, and vaginal sponge—all of which prevent the sperm from entering the uterus. The

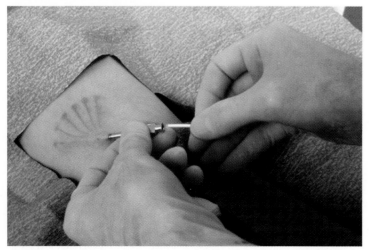

FIGURE 21-15 Subcutaneous Progesterone Implant Inserted under the skin, this tiny device releases a steady stream of progesterone, blocking ovulation for months.

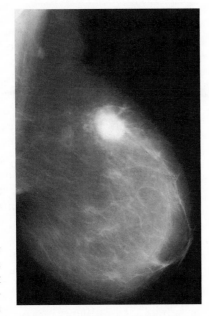

FIGURE 1 **Mammogram** showing a small breast tumor.

breast, every 1 to 2 years. Women over 50 should have a mammogram every year (Figure 1).

Women should also perform regular self-exams. The best time to perform a self-exam is immediately after a menstrual period. Doctors will provide instructions on how to do this.

Most lumps women detect are harmless, but they should be checked out immediately by a physician.

Breast cancer is typically treated by surgery. If the tumor is small, the surgeon will remove it along with some of the surrounding tissue. This procedure is known as a *lumpectomy*. If the tumor is

large, or if there are multiple tumors, all of the tissue in the breast will be removed. This procedure is known as a *mastectomy* (mass-teck-toe-me). The surgeon usually removes a few lymph nodes from the armpits as well, because breast cancer frequently spreads to the lymph nodes.

Surgery is typically followed up by radiation therapy, especially if the tumor was large or if the cancer had spread to the lymph nodes. Women with estrogen-sensitive tumors are typically treated with the drug tamoxifen. This drug blocks estrogen's effect and causes the tumor to shrink. In other instances, doctors may recommend chemotherapy.

Studies show that a combination of drug treatment and surgery enhance a woman's chances of survival. Nine of every ten women whose tumors are detected early and treated immediately survive ten years or more. If a tumor has spread to other organs, however, chances of survival are much lower.

www.jbpub.com/humanbiology/5e

Visit Human Biology's Internet site for links to web sites offering more information on this topic.

diaphragm (DIE-ah-FRAM) is a rubber cup that fits over the end of the cervix (Figure 21-16). To increase its effectiveness, a spermicidal (sperm-killing) jelly, foam, or cream should be applied to the rim and inside surface of the cup.

Smaller versions of the diaphragm, called *cervical caps*, are also available. Fitting over the end of the cervix, the cervical cap most often used is held in place by suction. When used with spermicidal jelly or cream, the caps are as effective as full-sized diaphragms.

Condoms are thin, latex rubber sheaths that fit onto the erect penis (Figure 21-17). Sperm released during ejaculation are trapped inside (in a small reservoir at the tip of the condom) and are therefore prevented from entering the vagina. Besides preventing fertilization, condoms also protect against sexually

transmitted diseases, a benefit not offered by any other birth control measure except abstinence.

Yet another barrier method is the vaginal sponge (Figure 21-18). This small absorbent piece of foam is impregnated with spermicidal jelly. Inserted into the vagina, the sponge is positioned over the end of the cervix. The sponge is effective immediately after placement and remains effective for 24 hours. Cervical sponges can be purchased without a doctor's prescription. Like the condom, one size fits all.

As mentioned earlier, spermicidal jellies, creams, foams, and films contain chemical agents that kill sperm but are apparently harmless to the woman. Spermicidal preparations are most often used in conjunction with diaphragms, condoms, and cervical caps. They can also be used alone, but are only 75% effective.

FIGURE 21-16 **The Diaphragm** Worn over the cervix, the diaphragm is coated with spermicidal jelly or cream and is an effective barrier to sperm.

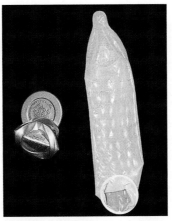

FIGURE 21-17 **The Condom** Worn over the penis during sexual intercourse, it prevents sperm from entering the vagina.

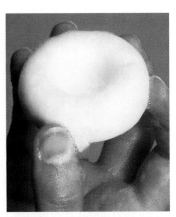

FIGURE 21-18 **The Vaginal Sponge** Impregnated with a spermicidal chemical, the vaginal sponge is inserted into the vagina and is effective for up to 24 hours.

One of the oldest, but least successful, means of birth control is withdrawal, disengaging the penis before ejaculation. This method requires tremendous willpower and frequently fails, for three reasons: caution is often tossed to the wind in the heat of passion, because the penis is withdrawn too late, or because of preejaculatory leakage—the release of a few drops of sperm-filled semen before ejaculation.

Abstaining from sexual intercourse around the time of ovulation—the rhythm, or natural, method—can help couples reduce the likelihood of pregnancy. If a couple knows the exact time of ovulation, they can time sexual intercourse to prevent pregnancy more precisely.

To determine when ovulation occurs, women can keep records on the length of their menstrual cycles, sampling cervical secretions or taking body temperature just after awakening each morning. In most women, body temperature rises one half to one degree right after ovulation. By keeping a temperature record over several menstrual cycles, a woman can determine the length of her cycle and the time of ovulation. Once the length of the cycle and the time of ovulation have been determined, the days a couple should practice abstinence can be determined.

Another method used to time ovulation involves taking samples of the cervical mucus. Cervical mucus varies in consistency during the menstrual cycle. By testing its thickness on a daily basis, a woman can tell fairly accurately when she has ovulated. Because ova remain viable 12–24 hours after ovulation and sperm may remain alive in the female reproductive tract for up to 3 days, abstinence 4 days before and 4 days after the probable ovulation date should provide a margin of safety. This minimizes the chances of a viable sperm reaching a viable ovum. Unfortunately, some women experience the greatest sexual interest around the time of ovulation. Sexual intercourse after a period of abstinence may also advance the time of ovulation.

Abortion

Abortion is the surgical termination of pregnancy.

Some couples may elect to terminate pregnancy through abortion.

Abortion is not suitable or morally acceptable to all people. Pro-life advocates argue that abortion should be outlawed or severely restricted—that is, allowed only in cases of rape, incest, and threat to the life of the mother. These individuals advise unmarried women to abstain from sexual intercourse or, if they become pregnant, to give birth and either keep the baby or put it up for adoption.

Pro-choice advocates, on the other hand, support abortion. They argue that women should have the freedom to choose whether to terminate a pregnancy or have a child. Abortion, they say, reduces unwanted pregnancies and untold suffering among unwanted infants, especially in poor families, and gives women more options than motherhood. Nonetheless, pro-choice advocates often point out that abortion should not be practiced as a primary means of birth control. Abstinence and contraception is less costly, less traumatic, and more morally acceptable.

In the first 12 weeks of pregnancy, abortions can be performed surgically in a doctor's office via vacuum aspiration. In this procedure, the cervix is first dilated by a special instrument. Next, the contents of the uterus are drawn out through an aspirator tube. Vacuum aspiration is a fairly simple and relatively painless procedure. Usually no anesthesia is given. Although women bleed for a week or so after the procedure, they generally experience few complications.

From weeks 13 to 16, aspiration is supplemented by physical scraping of the lining of the uterus to ensure complete removal of the fetal tissue. After 16 weeks, abortions are more difficult and more risky. Solutions of salt, urea, or prostaglandins, which stimulate uterine contractions, are injected into the sac of fluid surrounding the fetus to induce premature labor. The hormone oxytocin may be administered to the woman with the same effect.

Implantation can also be impaired by use of a pill called RU486, incorrectly dubbed "the abortion pill," because it simply prevents implantation. In the United States, doctors also recommend another birth control pill. Although it is called the "morning after pill," it actually consists of two to four pills that can be taken one to several days after unprotected sex. One product (Preven) consists of four pills, two taken within three days of sex, and the other two taken 12 hours later. Containing synthetic progesterone, this pill prevents implantation and is about 85% effective.

21-4 Infertility

Infertility is fairly common among reproductive age couples.

A surprisingly large percentage (about one in six) of American couples cannot get pregnant. The inability to conceive is called **infertility**. According to some statistics, in about 50% of the couples, infertility results from problems occurring in the woman. Approximately 30% of the cases are due to problems in the man alone, and about 20% are the result of problems in both partners.

If after a year of actively trying to conceive, a couple is unsuccessful, they can consult a fertility specialist who will first check obvious problems such as infrequent or poorly timed sex. If timing is not the problem, and it rarely is, the physician tests

the man's sperm count. A low sperm count is one of the most common causes of male infertility and is easy to test.

A low sperm count may result from overwork, emotional stress, and fatigue. Excess tobacco and alcohol consumption can also contribute to the problem. Tight-fitting clothes and excess exercise, both of which raise the scrotal temperature, also tend to reduce the sperm count. The testes are also sensitive to a wide range of chemicals and drugs that reduce sperm production.

If infertility appears to be caused by a low sperm count, a couple may choose to undergo artificial insemination, using sperm from a sperm bank. These sperm are generally acquired from anonymous donors and are stored frozen. When thawed, the sperm are reactivated, then deposited in the woman's vagina or cervix around the time of ovulation.

If sperm production and ejaculation appear normal, a physician checks the woman's reproductive tract. Physicians first assess ovulation. If ovulation is not occurring, fertility drugs may be administered. Several kinds of drugs are available. One of the more common is HCG, which, as noted earlier, is an LH-like hormone that induces ovulation. Unfortunately, fertility drugs often result in superovulation (the ovulation of many fertilizable ova), leaving couples with a "litter" of 4–6 babies, instead of the one child they had hoped for. Most of the multiple births you hear about on the news are the result of fertility drugs.

If tests show that ovulation is occurring normally, the physician examines the uterine tubes to determine whether they are obstructed. In some instances, a previous sexually transmitted diseases (for example, gonorrheal or chlamydial infections) that spread into the tubes may have caused scarring that obstructed the passageway. In such instances, couples may be advised to adopt a child or to try *in vitro* (in VEE-trow) fertilization.

During *in vitro* fertilization, ova are surgically removed from the woman and fertilized by the partner's sperm outside her body. The fertilized ovum is then implanted in the uterus of the woman, where it can grow and develop successfully. Besides being expensive and time-consuming, this procedure has a low success rate.

20-5 Health and Homeostasis

The human reproductive system is not essential to homeostasis or the survival of an individual. It is, however, essential to the survival of our species.

Like other systems, the reproductive system is susceptible to environmental conditions—to pollution and social and psychological conditions that create stress. The presence of many synthetic chemicals in the environment has been receiving a great deal of attention in recent years. Numerous studies show that sperm count has been declining in young men in the United States since the 1950s.

Prior to 1950, the average sperm count was about 110 million per ml. By 1980 and 1981, sperm counts had dropped to about 60 million per ml. Further studies show continuing decline to under 50 million in the late 1980s. Statistical studies suggest that the decline may be related to growing pesticide use, air pollution, and other factors. Especially harmful may be chemicals in the environment that mimic the female sex hormone estrogen, such as dioxins, generated in paper production and waste

incineration. Phthalates, which are added to plastics and are present in many foods and beverages, may also be responsible.

Research suggests that exposing males to such chemicals early *in utero*, when they are extremely chemically sensitive, could have dramatic effects on their testes. At least 60 such chemicals have been identified to date.

Ten commonly prescribed antibiotics can also reduce sperm count. Tagamet, a drug that is used to relieve stress and treat excess stomach acidity, reduces sperm count by over 40%. By one estimate, at least 40 commonly used drugs depress sperm production. Thousands of other drugs and environmental pollutants have not been tested.

These facts do not necessarily mean that the United States is in a sperm crisis, but they do suggest the need for further research to determine the potential impacts, if any, of the many thousands of chemicals now commonly used or released into the environment. Research may prove that we need to clean up our act, or it may show that these fears are unwarranted.

SUMMARY

The Male Reproductive System

1. The male reproductive tract consists of seven basic components: (a) testes, (b) epididymis, (c) vasa deferentia, (d) sex accessory glands, (e) urethra, (f) penis, and (g) scrotum.
2. The testes reside in the scrotum, which provides a suitable temperature for sperm development.
3. Sperm produced in the seminiferous tubules are stored in the epididymis. During ejaculation, sperm pass from the epididymis to the vas deferens, then to the urethra. Secretions from the sex accessory glands are added to the sperm at this time, forming semen.
4. Between the seminiferous tubules are the interstitial cells, which produce the hormone testosterone. Testosterone secretion is controlled by luteinizing hormone (LH) from the anterior pituitary.

5. Testosterone stimulates the formation of sperm as well as facial hair growth, thickening of the vocal cords, sebaceous gland secretion, and bone and muscle development.
6. The penis is the organ of copulation. It contains erectile tissue, which fills with blood during sexual arousal, making it rigid.
7. Ejaculation is under reflex control. Motor neurons in the spinal cord send impulses to the smooth muscle in the walls of the epididymis, the vasa deferentia, the sex accessory glands, and the urethra, causing ejaculation.

The Female Reproductive System

8. The female reproductive system consists of two basic components: the reproductive tract and the external genitalia.

9. The reproductive tract consists of (a) the uterus, (b) the two uterine tubes, (c) the two ovaries, and (d) the vagina.
10. The external genitalia consist of two flaps of skin on both sides of the vaginal opening, the labia majora and the labia minora.
11. Female germ cells are housed in follicles in the ovary.
12. A dozen or so follicles enlarge during each menstrual cycle, but most follicles degenerate. In humans, usually only one follicle makes it to ovulation during each cycle.
13. The oocyte is released during ovulation and drawn into the uterine tubes.
14. The menstrual cycle consists of a series of changes occurring in the ovaries, uterus, and endocrine system of women.
15. The first half of the menstrual cycle is called the follicular phase. It is during this period

that FSH from the pituitary stimulates follicle growth and development. LH stimulates estrogen production.

16. Estrogen levels rise slowly during the follicular phase, then trigger a positive feedback mechanism that results in a surge of FSH and LH, which triggers ovulation.

17. The oocyte is expelled from the antral follicle at ovulation. The follicle then collapses and is converted into a corpus luteum (CL), which releases estrogen and progesterone.

18. In the absence of fertilization, the CL degenerates. If fertilization occurs, however, HCG from the embryo maintains the CL for approximately 6 months.

19. During the menstrual cycle, ovarian hormones stimulate growth of the uterine lining, which is necessary for successful implantation. If fertilization does not occur, the uterine lining is sloughed off during menstruation, which is triggered by a decline in ovarian estrogen and progesterone.

20. Like testosterone levels in boys, estrogen levels in girls increase at puberty. Estrogen promotes growth of the external genitalia, the reproductive tract, and bone. It also stimulates the deposition of fat in women's hips, buttocks, and breasts.

21. Progesterone works with estrogen to stimulate breast development. It also promotes growth of the uterine lining and inhibits uterine contractions.

22. Many women suffer from premenstrual syndrome or PMS, which is characterized by irritability, depression, tension, fatigue, headaches, bloating, swelling and tenderness of the breasts, and joint pain.

23. The menstrual cycle continues throughout the reproductive years, but as a woman ages estrogen levels decline. Ovulation and menstruation become erratic as a woman approaches 45. Between the ages of 45 and 55, ovulation and menstruation cease. The end of reproductive function in women is known as menopause.

24. The decline in estrogen levels results in the atrophy of the reproductive organs and other symptoms such as irritability, depression, and hot flashes and night sweats induced by intense vasodilation of vessels in the skin.

Birth Control

25. Birth control refers broadly to any method or device that prevents births and includes two general strategies: contraception and induced abortion.

26. The most effective form of birth control is abstinence. Another highly effective measure is sterilization—vasectomy in men and tubal ligation in women.

27. The intrauterine device or IUD is a plastic or metal coil that is placed inside the uterus, where it prevents the fertilized ovum from implanting.

28. Progesterone implants (under the skin) also prevent pregnancy and offer long-lasting protection.

29. The pill is a highly effective means of birth control. The most common pill in use today contains a mixture of estrogen and progesterone that inhibits ovulation. In some women, however, estrogen causes adverse health effects.

30. The diaphragm, condom, and vaginal sponge are less effective measures of birth control.

31. The condom is a thin, latex rubber sheath worn over the penis that prevents sperm from entering the vagina during sexual intercourse.

32. The vaginal sponge is a tiny, round sponge worn by the woman. It is impregnated with a spermicidal chemical.

33. One of the oldest, but least effective, methods of birth control is withdrawal, removing the penis before ejaculation. Spermicidal chemicals used alone are about as effective as withdrawal.

34. Abstaining from sexual intercourse around the time of ovulation, known as the rhythm method, is another way to prevent pregnancy. Statistics on effectiveness show that the natural method is one of the least successful of all birth control measures.

Infertility

35. Infertility, the inability to conceive, can result from a variety of problems in men and women, such as poorly timed or infrequent sex, low sperm count in men, and obstruction of the uterine tubes in women.

Health and Homeostasis

36. A variety of drugs and chemical pollutants affect sperm development and may be causing a decline in the sperm count of American men.

---

critical thinking

Thinking Critically—Analysis

This Analysis corresponds to the Thinking Critically scenario that was presented at the beginning of this chapter.

Before writing your article, you might want to check the experimental design. Were adequate sample sizes used in the experiments? Did the researcher have proper control groups? Were the animals particularly prone to these abnormalities? What dosages were given? Are these at all similar to dosages people might be exposed to?

You might also want to see if these results have been reported elsewhere. Moreover, you would probably want to see whether there is any research data linking this chemical to human abnormalities. Having done this, you would have a strong base from which to piece together a good article.

KEY TERMS AND CONCEPTS

CONCEPT REVIEW

1. You are a fertility specialist. A young woman arrives in your office complaining that she has been trying to get pregnant for 2 years, but to no avail. Describe how you would go about determining whether the problem was with her, her husband, or both of them. pp. 390–391

2. Describe the anatomy of the male reproductive system. Where are sperm produced? Where are they stored? What structures produce the semen? pp. 376–378

3. List the hormones that control testicular function. Where are they produced, and what effects do they have on the testes? pp. 380–381

4. Trace the pathway for a sperm from the seminiferous tubule to the site of fertilization. p. 376

5. Describe the process of ovulation and its hormonal control. pp. 382–384

6. What is the corpus luteum? How does it form? What does it produce? Why does it degenerate at the end of the menstrual cycle if fertilization does not occur? p. 384

7. What is menstruation? What triggers its onset? p. 385

8. Describe the effects of estrogen and progesterone on the reproductive tract and the body. pp. 385–386

9. A 50-year-old female friend of yours complains of irritability and depression. She says that she wakes up in the middle of the night in a sweat. Would you give her the name of a psychiatrist? Why or why not? If not, what would you do? p. 386

10. Describe each of the following birth control measures, explaining what they are and how they work: the pill, IUD, diaphragm, cervical cap, condom, spermicidal jelly, hormonal implants, and natural method. pp. 386–390

SELF-QUIZ: TESTING YOUR KNOWLEDGE

1. Sperm are produced in the testes inside the _____ tubules. p. 377

2. The bulk of the ejaculate consists of fluids produced by the sex _____ glands. p. 377

3. Sperm are stored in the _____, where they mature before ejaculation. p. 377

4. The hormone _____ produced by the pituitary stimulates testosterone production in the testes. p. 380

5. Testosterone is produced by the _____ cells in the testes. p. 378

6. In women, ova are produced by the _____ located in the pelvic cavity. p. 381

7. Fertilization takes place in the upper one third of the _____. p. 381

8. The fertilized egg implants in the lining of the _____, where it develops. p. 381

9. Female germ cells are found in structures inside the ovaries called _____. p. 383

10. Ovulation is stimulated by the release of two hormones from the pituitary, LH and _____. p. 385

11. A collapsed follicle is converted into a structure called a _____, which produces estrogen and progesterone. p. 384

12. Human _____ gonadotropin is an LH-like hormone released by the newly developed embryo. p. 385

13. Hormone replacement therapy in women helps treat symptoms of _____, including osteoporosis. p. 386

14. The most effective contraceptive method is _____. p. 386

15. Surgical sterilization in women is called _____ ligation; in men it is known as _____. p. 386–387

www.jbpub.com/humanbiology/5e

The site features eLearning, an online review area that provides quizzes, chapter outlines, and other tools to help you study for your class. You can also follow useful links for in-depth information, research the differing views in the Point/Counterpoints, or keep up on the latest health news.

Sexually Transmitted Diseases

Certain bacteria and viruses are transmitted from one person to another by sexual contact. These organisms penetrate the lining of the reproductive tracts of men and women and thrive in the moist, warm environment of the body, creating local infections. These organisms can cause numerous health problems, including pain during urination and puslike discharges.

Diseases caused by bacteria and viruses that are transmitted by sexual contact are called **sexually transmitted diseases** (**STDs**). Most of the infectious agents that cause STDs are spread by vaginal intercourse, but other forms of sexual contact are also responsible for the transmission of these microbes. Syphilis, for example, is an STD caused by a bacterium that is spread by oral, anal, and vaginal sex. (For more on infectious diseases, see Chapter 15.)

Although STDs pass from one person to another during sexual contact, the symptoms are not confined to the reproductive tract. In fact, several STDs, including syphilis and AIDS, are primarily systemic diseases—that is, they affect entire body systems. Syphilis affects the nervous system and cardiovascular system. while AIDS affects the immune system.

One complicating factor in controlling STDs is that occasionally some diseases such as gonorrhea produce no obvious symptoms in many men and women. As a result, the disease can be transmitted without a person knowing he or she is infected. In others such as AIDS, symptoms may not appear for weeks or even years after the initial infection. Thus, sexually active individuals who are not monogamous can transmit the AIDS virus to many people before they are aware that they are infected. As noted in A Closer Look at the end of Chapter 14, new blood tests are being used to detect HIV earlier and may help to reduce the spread of this deadly disease. In this section we will examine the most common STDs, except AIDS, which was discussed in A Closer Look at the end of Chapter 14.

Gonorrhea

Gonorrhea is a bacterial infection that can spread to many organs.

Gonorrhea (GON-or-REE-ah; referred to colloquially as the "clap") is caused by a bacterium that commonly infects the urethras of men and the cervical canals of women. Gonorrhea often causes no symptoms. When they do appear, painful urination and a puslike discharge from the urethra are common complaints in men. Women may experience a cloudy vaginal discharge and lower-abdominal pain. If a woman's urethra is infected, urination may be painful. Symptoms of gonorrhea usually appear about 1–14 days after sexual contact.

Gonorrhea is treated with antibiotics and clears up quickly, usually within 3 to 4 days, if treatment begins early. If left untreated, however, gonorrhea in men can spread to the prostate gland and the epididymis. They can be more difficult to treat. Moreover, infections in the urethra lead to the formation of scar tissue. This may narrow the urethra and make urination even more difficult. In some women, bacterial infection spreads to the uterus and uterine tubes, causing the buildup of scar tissue. Scar tissue in the uterine tubes may block the passage of sperm and ova, resulting in infertility.

Gonorrheal infections can also spread into the abdominal cavity through the opening of the uterine tubes. If the infection enters the bloodstream in men or women, it can travel throughout the body. Gonorrhea can also be transferred from mother to her offspring during birth—that is, as the baby passes through the vagina during birth.

Syphilis

Syphilis is caused by a bacterium and can be extremely debilitating if untreated.

Syphilis is an STD caused by a bacterium that penetrates the linings of the oral cavity, vagina, and penile urethra. It may also enter through breaks in the skin. It can also be transmitted from a mother through the placenta. If untreated, syphilis proceeds through three stages. In stage 1, between 1 and 8 weeks after exposure, a small, painless red sore develops, usually in the genital area. Easily visible when on the penis, these sores often go unnoticed when they occur in the vagina or cervix (Figure 1). The sore heals in 1–5 weeks, leaving a tiny scar.

Approximately 6 weeks after the sore heals, individuals complain of fever, headache, and loss of appetite. Patients often display prominent nonitchy rashes that appear throughout the body, even the palms of the hands and soles of the feet. Lymph nodes in the neck, groin, and armpit swell as the bacteria spread throughout the body. This is stage 2, and it lasts for about 4 to 12 weeks.

As a rule, the symptoms of stage 2 syphilis disappear for several years. In some individuals, the disease progresses no

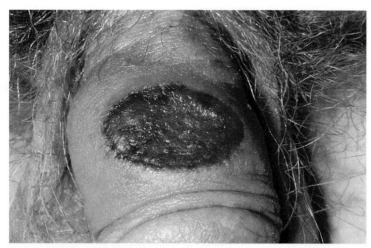

FIGURE 1 **Stage 1 Syphilis** Sores like this are common during stage 1 of syphilis. They disappear, sometimes leaving little evidence.

further. That is, it spontaneously resolves. In others, the disease flares up again, creating additional symptoms that persist until the individual dies. This is stage 3.

During stage 3, an autoimmune reaction occurs. Patients experience a loss of their sense of balance and a loss of sensation in their legs. As the disease progresses, individuals experience paralysis, senility, and even insanity. Some patients go blind. Huge ulcers may develop on the skin or on internal organs (Figure 2). In some cases, the bacterium weakens the walls of the aorta, causing aneurysms (Chapter 6).

Children infected in utero may be still born or may be born blind with numerous structural birth defects.

Syphilis is diagnosed by the symptoms and by examining the pus under a microscope. Blood tests are also useful in stages 1 and 2. Syphilis can be successfully treated with antibiotics, but only if the treatment begins early. Suspicious sores in the mouth and genitals should be brought to the attention of a physician. In stage 3, antibiotics are useless. Tissue or organ damage is permanent.

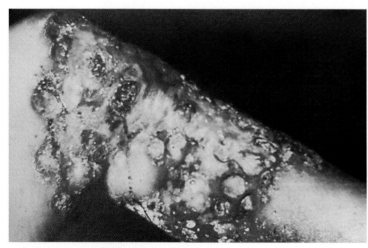

FIGURE 2 **Stage 3 Syphilis.** This large open sore appeared on the skin of the patient during Stage 3 syphilis. These ulcerated sores may also appear on internal organs.

Chlamydia

Chlamydial infections are extremely common among college students.

One of the most common sexually transmitted diseases, affecting 3 to 10 million people each year—many of them college students—is known as **chlamydia** (clam-ID-ee-ah). Caused by a bacterium, this disease is characterized, in men, by a burning sensation during urination and a discharge from the penis. Women also experience a burning sensation during urination and a vaginal discharge. If the bacterium spreads, it can cause more severe infection and infertility. Like other STDs, many people experience no symptoms at all and therefore risk spreading the disease to others. Children born to mothers with chlamydia can develop eye infections and pneumonia.

Chlamydia bacteria often migrate to the lymph nodes, where they cause considerable enlargement and tenderness in the affected area. Blockage of the lymph nodes may result in tissue swelling in the surrounding tissue. Doxycycline and other antibiotics are effective in treating this disease.

Genital Herpes

Genital herpes is caused by a virus, is extremely common, and is essentially incurable.

Genital herpes (HER-peas) is another common sexually transmitted disease. Contracted by 200,000–300,000 people each year, genital herpes is caused by a virus that enters the body and remains there for life. The first sign of viral infection is pain, tenderness, or an itchy sensation on the penis or female external genitalia. These symptoms usually occur 6 days or so after contact with someone infected by the virus. Soon afterward, painful blisters appear on the external genitalia, thighs, buttocks, and cervix, or in the vagina (Figure 3).

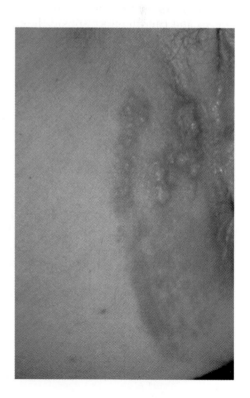

FIGURE 3 **Genital Herpes** Blisters on the external genitalia and inner thigh.

The blisters break open and become painful ulcers that last for 1–3 weeks, then disappear. Unfortunately, the herpes virus is a lifelong resident of the body. New outbreaks can occur from time to time, especially when an individual is under stress. Recurrent outbreaks are generally not as severe as the initial one, and, in time, the outbreaks generally cease.

Herpes can be transmitted to other individuals during sexual contact only when the blisters are present or (as recent research suggests) just beginning to emerge. When the virus is inactive, sexual intercourse can occur without a partner becoming infected.

Although herpes cannot be cured, physicians can suppress outbreaks with antiviral drugs such as acyclovir (A-sigh-CLOE-ver). These drugs not only reduce the incidence of outbreaks but also accelerate healing of the blisters.

Herpes is not a particularly dangerous STD, except in pregnant women, who run the risk of transferring the virus to their infants at birth. Because the virus can be fatal to newborns, these women are often advised to deliver by cesarean section (an incision made just above the pubic bone) if the virus is active at the time of birth.

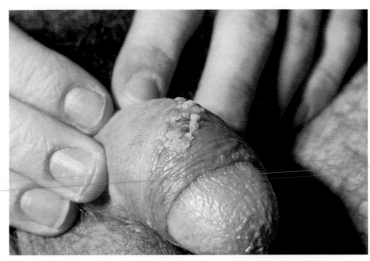

FIGURE 4 **Genital Warts** Genital warts on the penis are caused by infection of the skin by papilloma viruses. Genital warts can be removed by a variety of treatments but sometimes recur.

Nongonococcal Urethritis

NGU is an extremely common disease caused by several types of bacteria.

Nongonococcal urethritis (YUR-ee-THRIGHT-iss), or NGU for short, is the most common sexually transmitted disease. Moreover, NGU is one of several STDs whose incidence is steadily rising in the United States. Caused by any of several different bacteria, this infection is generally less threatening than gonorrhea, syphilis, and chlamydia, although some infections can result in sterility.

Many men and women often exhibit no symptoms whatsoever and can therefore spread the disease without knowing it. In men, when symptoms occur, they resemble those of gonorrhea—painful urination and a cloudy mucous discharge from the penis. In women, urination becomes painful and more frequent. NGU can be treated by antibiotics, but individuals should seek treatment quickly to avoid the spread of the disease and more serious complications.

Genital Warts

Genital warts are caused by human papillomavirus (HPV).

The vast majority of Americans carry a virus known as *human papillomavirus* (pap-ILL-oh-mah) or HPV. Transmitted by sexual contact, this virus can cause genital warts. Genital warts are benign growths that appear on the external genitalia and around the anuses of men and women (Figure 4). Warts also grow inside the vagina of women. Warts generally occur in individuals whose

immune systems are suppressed, for example, after long periods of stress.

These warts can remain small or can grow to cover large areas, creating cosmetically unsightly growths. They may cause mild irritation, and certain strains of HPV are associated with cervical cancer in women.

Genital warts can be treated with chemicals or removed surgically—although rates of recurrence are quite high. In 20%–30% of the cases, genital warts disappear spontaneously. Getting rid of the virus, however, is impossible, for it resides in the body forever.

Preventing STDs

Sexually transmitted diseases caused by bacteria can, for the most part, be treated with antibiotics, eliminating the disease organism. Even syphilis can be eliminated if it is caught early. STDs caused by viruses, however, are another story. Although drugs are available to suppress the outbreak of viruses like the one that causes genital herpes, the virus remains in the body forever.

The best medicine for sexually transmitted diseases is prevention, through abstinence. Those who chose to partake in sex are advised to practice contraception. Contraceptive measures prevent pregnancy, but only one type of contraception prevents STDs—condoms. Condoms, discussed in Chapter 21, are thin latex sheaths worn over the penis during sexual intercourse. They prevent the exchange of body fluids and direct contact of the penis with the vaginal or anal canal lining, in the case of anal intercourse, preventing disease organisms from passing from one individual to another.

SUMMARY

1. Certain viruses and bacteria can be transmitted from one individual to another during sexual contact. Infections spread in this way are called sexually transmitted diseases (STDs).

2. Gonorrhea is caused by a bacterium that commonly infects the urethra in men and the cervical canal in women. Overt symptoms of the infection are frequently not present, so people can spread the disease without knowing it. If left untreated, gonorrhea can spread to other organs, causing considerable damage.

3. Syphilis is a serious STD caused by a bacterium that penetrates the linings of the oral cavity, vagina, and penile urethra. If untreated, syphilis proceeds through three stages. It can be treated with antibiotics during the first two stages, but in stage 3, when damage to the brain and blood vessels is evident, treatment is ineffective.

4. Chlamydial infections, an extremely common bacterial STD, resemble gonorrhea.

5. Genital herpes is also a very common STD. It is caused by a virus. Once the virus enters the body, it remains for life. Blisters form on the genitals and sometimes on the thighs and buttocks. The blisters break open and become painful ulcers. At this stage, an individual is highly infectious. New outbreaks of the virus may occur from time to time, especially when an individual is under stress.

6. Nonspecific urethritis (NSU), the most common sexually transmitted disease, is caused by several different bacteria, but most commonly by chlamydia. It is less threatening than gonorrhea or syphilis. Many men and women show no symptoms of NSU and can therefore spread the disease without knowing it. Symptoms, when they occur, resemble those of gonorrhea.

7. Genital warts occur in men and women whose immune systems are suppressed and are caused by HPV.

CHAPTER

22

Human Development and Aging

thinking critically

In a hospital room in a major city, two parents bring a newborn child into the world with the assistance of a nurse midwife. The wife holds their new daughter in the hospital bed while three floors up in the geriatric ward, an elderly man quietly passes away.

Birth and death are two of life's most dramatic acts. They're the subject of this chapter. We will begin our journey with a discussion of fertilization, and then the turn to the development of the fertilized ovum into an embryo and, then into a fetus. We will conclude with a look at aging and death.

"Thanks to improvements in medicine, people are living longer." You may have heard this statement dozens of times in one form or another. It is repeated so often that most of us believe it implicitly. But is it true? Are people really living longer than they used to?

22-1 Fertilization

Fertilization occurs in the upper third of the oviduct.

Fertilization, the uniting of sperm and egg, is the first step in human development. This process occurs in the upper third of the oviducts. To get there, sperm must travel from the vagina, the site of ejaculation, through the cervical canal. They must then travel through the uterus into the oviducts. Thirty minutes after ejaculation, those sperm that have successfully found their way into the cervical canal arrive at the junction of the uterus and oviducts or uterine tubes. Sperm then travel up the oviducts, where they may encounter a secondary oocyte released from the ovary during ovulation.

Passage through the Female Reproductive Tract

Very few sperm make it to the site of fertilization.

Although many millions of sperm are deposited in the vagina, only a tiny fraction actually make it into the oviducts (Figure 22-1). The rest are killed by the acidic secretions of the vagina or get lost along the way. Some sperm, for instance, get lost in the vagina and fail to find their way into the cervix. Many sperm that enter the cervix get lost in the folds of the inner lining. The same fate faces sperm that enter the uterus.

Studies based on laboratory animals suggest that in humans only 1 in 3 sperm makes it through the cervical canal and 1 in 1000 makes it to the uterine tubes (Figure 22-1). Those that are killed by acidic secretions of the vagina or get lost in the vagina, cervix, uterus, or oviducts are eventually engulfed by cells in the lining of these organs.

Sperm Mobility

Sperm travel by swimming and muscular contraction of the female reproductive tract.

Sperm travel through the reproductive tract of women partly on their own, by swimming. This action is permitted by the whip-like tail of the sperm. Most transport, however, is the result of muscular contraction occurring in the walls of the uterus and uterine tubes. These contractions are stimulated by prostaglandins, hormone-like substances in the semen.

Sperm Penetration

Sperm contain enzymes that help them penetrate the protective layers surrounding the egg.

After swimming around the secondary oocyte for a while, a process that dissolves away the protective coating on the sperm cells, sperm begin to wiggle their way through the layer of cells surrounding the female gamete. The cells surrounding the oocyte form the corona radiata (radiating crown). Getting through the closely attached cells is made possible by enzymes released from the acrosome, a stocking-caplike structure pulled down over the head of the sperm (Figure 22-2).

After passing through the corona radiata, sperm must digest their way through another barrier, the zona pellucida (the clear zone), a clear gel-like layer immediately surrounding the oocyte. Sperm penetrate it with the aid of additional acrosomal enzymes.

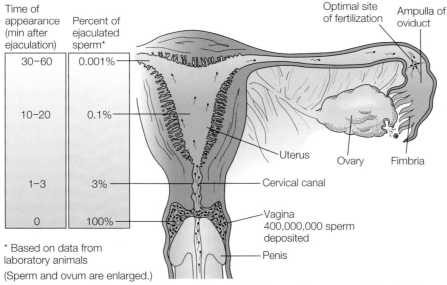

FIGURE 22-1 **Sperm Transport in the Female Reproductive System** Sperm move rapidly up the female reproductive tract of humans and other vertebrates principally as a result of contractions in the muscular walls of the uterus and uterine tubes. Notice the rapid decline in sperm number along the way.

FIGURE 22-2 **Fertilization** (a) The plasma membrane of the sperm and the outer membrane of the acrosome fuse and the membranes break down, releasing enzymes that allow the sperm to penetrate the corona radiata. Sperm digest their way through the zona pellucida via enzymes associated with the inner acrosomal membrane. Sperm are engulfed by the oocyte plasma membrane. Cortical granules are released when the sperm cell contacts the membrane. These granules cause other sperm in contact with the membrane to detach. (b) Although many sperm gather around the egg, only one will enter. The fertilizing sperm are phagocytized by the egg.

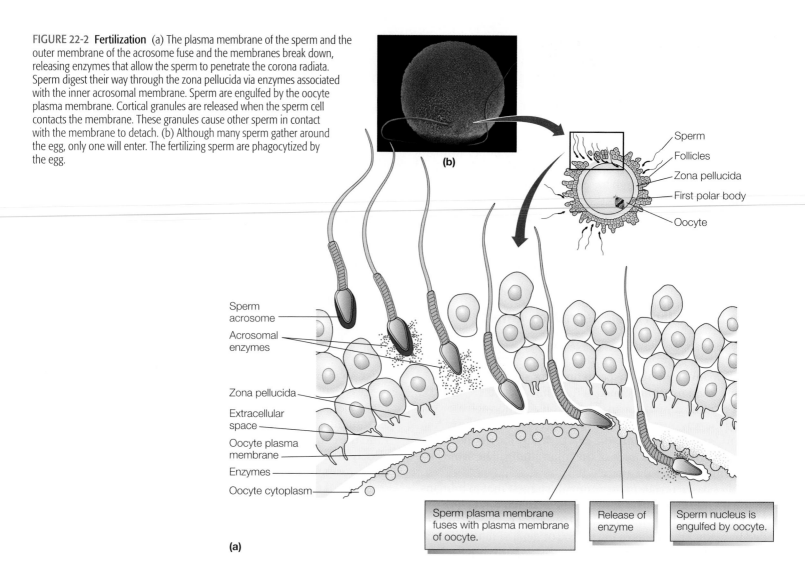

Sperm cells that penetrate the zona pellucida enter the space between the zona pellucida and the plasma membrane of the oocyte. As a rule, the first sperm cell to come in contact with the plasma membrane of the oocyte will fertilize it; all other sperm are excluded.

The Block to Polyspermy

Only one sperm can fertilize an egg.

Preventing multiple sperm entry (called polyspermy) into an oocyte results from a rapid membrane depolarization of the oocyte's plasma member, that is, a change in the resting potential of the plasma membrane. Sperm are also prevented from entering the oocyte as a result of the release of enzymes from membrane-bound vesicles lying beneath the plasma membrane of the oocyte (Figure 22-2a). These enzymes cause the zona pellucida to harden, blocking other sperm from reaching the oocyte. These secretions may also cause the "extra" sperm that have at-

tached to the plasma membrane of the oocyte to detach. Without these mechanisms, eggs might quickly overload with extra nuclear material, which would impair cell division and result in embryonic death.

Zygote Formation

Sperm and egg nuclei join to create a zygote with 46 chromosomes.

Sperm contact with the plasma membrane of the oocyte triggers the second meiotic division, thus converting the secondary oocyte into an **ovum** or egg. But the life of an ovum is short-lived, for once the sperm nucleus enters, the ovum is called a **zygote**.

Once inside the ovum, the sperm and egg nuclei combine (Figure 22-3). The result is a zygote with 46 chromosomes, half from each parent. The zygote is now ready for the first mitotic division, one of many billions of cell divisions that will occur in the long journey of human growth and development.

FIGURE 22-3 Fertilization and the First Mitotic Division

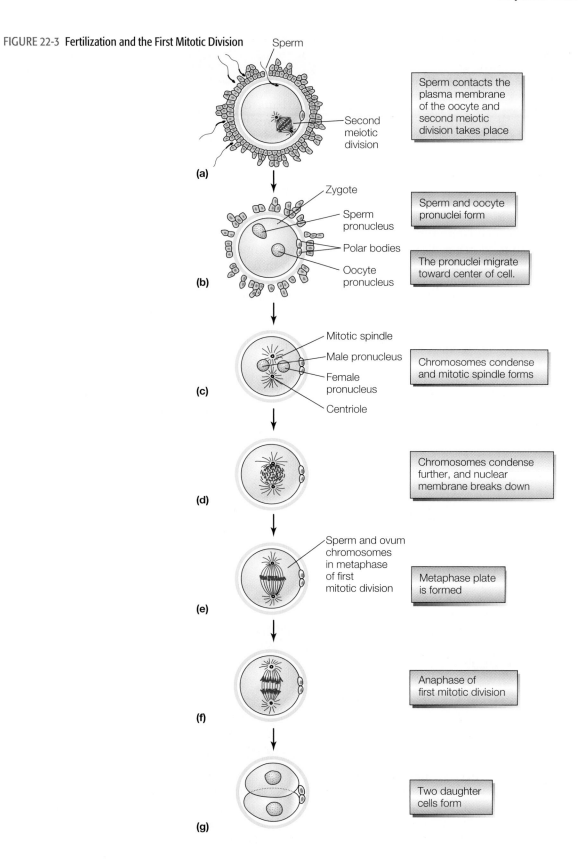

(a) Sperm contacts the plasma membrane of the oocyte and second meiotic division takes place

Second meiotic division

(b) Sperm and oocyte pronuclei form

The pronuclei migrate toward center of cell.

Zygote
Sperm pronucleus
Polar bodies
Oocyte pronucleus

(c) Chromosomes condense and mitotic spindle forms

Mitotic spindle
Male pronucleus
Female pronucleus
Centriole

(d) Chromosomes condense further, and nuclear membrane breaks down

(e) Metaphase plate is formed

Sperm and ovum chromosomes in metaphase of first mitotic division

(f) Anaphase of first mitotic division

(g) Two daughter cells form

22-2 Pre-Embryonic Development

Pre-embryonic development begins at fertilization and ends just after implantation.

Human development consists of three stages: pre-embryonic, embryonic, and fetal.

Pre-embryonic development includes all the changes that occur from fertilization to the time just after an embryo implants in the uterine wall. During this phase, the zygote undergoes rapid cellular division and is soon converted into a solid ball of cells not much bigger than the fertilized ovum (Figure 22-4). This structure is called a **morula** from the Latin *morus* meaning mulberry. The morula is nourished by secretions produced by the epithelium of the uterine tubes. Approximately 3–4 days after ovulation, the morula enters the uterus.

Fluid soon begins to accumulate in the morula, converting it into a **blastocyst**, a hollow sphere of cells (Figure 22-4). The blastocyst consists of a clump of cells, the inner cell mass, which will become the **embryo**, and a ring of flattened cells, the trophoblast ("to nourish the blastocyst"). The trophoblast gives rise to the embryonic portion of the **placenta**, an organ that supplies nutrients to and removes wastes from the embryo. The blastocyst remains unattached inside the uterus for 2–3 days. During this period, it is nourished by secretions of uterine glands.

Implantation

The embryo embeds in the wall of the uterus.

The blastocyst eventually attaches to the uterine lining and then digests its way into this thickened layer using enzymes released by its cells. This process is called *implantation*. It occurs 6–7 days after fertilization (Figure 22-5).

Most embryos implant high on the back wall of the uterus. The cells of the trophoblast first contact the endometrium, then adhere to it, but only if the uterine lining is healthy and properly primed by estrogen and progesterone. If the endometrium is not ready or is "unhealthy"—for example, because of the presence of an IUD, use of a "morning after pill," or an endometrial infection—the blastocyst cannot implant. Blastocysts may also fail to implant if their cells contain certain genetic mutations.

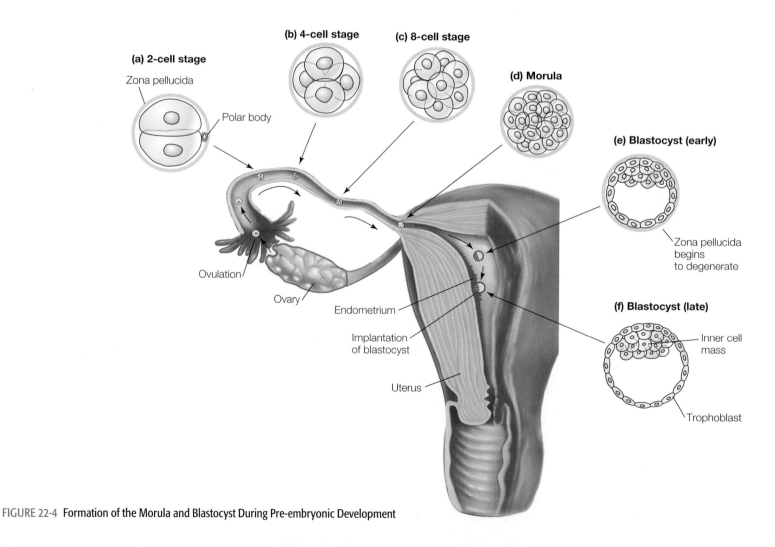

FIGURE 22-4 Formation of the Morula and Blastocyst During Pre-embryonic Development

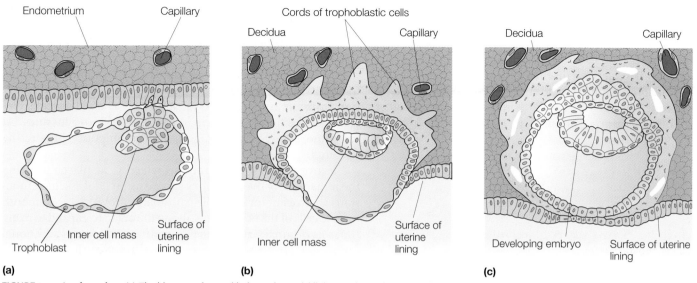

FIGURE 22-5 Implantation (a) The blastocyst fuses with the endometrial lining. Endometrial cells proliferate, forming the decidua. (b) The blastocyst digests its way into the endometrium. Cords of trophoblastic cells invade, digesting maternal tissue and providing nutrients for the developing blastocyst. (c) The blastocyst soon becomes completely embedded in the endometrium.

Unimplanted blastocysts are absorbed (phagocytized) by the cells of the endometrium or expelled during menstruation.

When implantation does occur, the cells of the endometrium at and around the point of contact enlarge. Enzymes released by the cells of the trophoblast digest a small hole in these cells, and the blastocyst "bores" its way into the deeper tissues of the uterine lining (Figure 22-5b). During this process, the blastocyst feeds on nutrients released from the cells it digests.

By day 14, the uterine endometrium grows over the blastocyst, enclosing it completely and walling it off from the uterine cavity. Endometrial cells respond to the invasion of the blastocyst by producing prostaglandins. These substances stimulate an increase in the development of uterine blood vessels, which ensures an ample supply of blood and nutrients for the blastocyst.

Soon thereafter, endometrial and embryonic tissue combine to form a structure known as the *placenta* (Figure 22-6). The placenta performs many important functions. It nourishes the embryo and fetus. It also provides oxygen and removes the wastes they produce throughout their residence inside the womb.

Ectopic Pregnancy

Implantation occasionally occurs outside of the uterus.

Fertilized ova do not always make their way to the uterus for implantation. They may end up implanting in the uterine tubes, for example, because scar tissue blocks their passage into the uterus. They may also implant outside the reproductive tract altogether in the body cavity. Such an occurrence is called an *ectopic pregnancy*. About 1 in every 200 pregnancies is ectopic.

In a tubal pregnancy (where the embryo implants in the uterine tubes), placental development typically damages the uterine tubes. This causes internal bleeding and severe abdominal pain. Because the uterine tube cannot sustain the embryo, a tubal pregnancy cannot generally proceed to term. Surgery is required to remove the embryo.

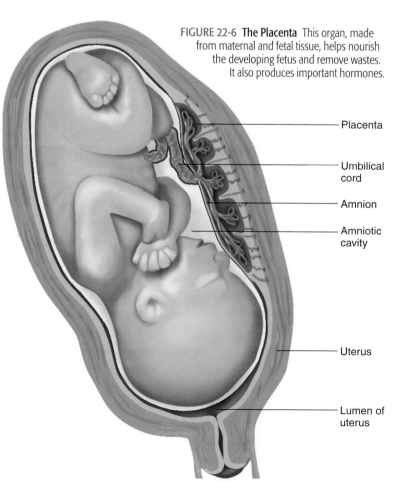

FIGURE 22-6 The Placenta This organ, made from maternal and fetal tissue, helps nourish the developing fetus and remove wastes. It also produces important hormones.

Placental Hormones

Placental hormones are essential to reproduction.

The placenta also produces several hormones vital to a successful pregnancy. This section discusses three of the placenta's hormones: human chorionic gonadotropin (HCG), estrogen, and progesterone (Table 22-1). Figure 22-7 shows blood levels of these three hormones during pregnancy.

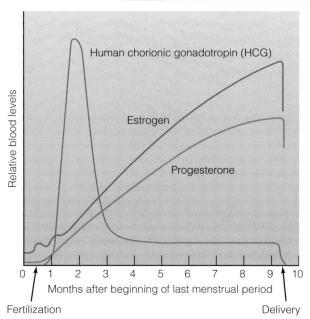

FIGURE 22-7 **Blood Levels of Placental Hormones** Human chorionic gonadotropin levels peak in the second month of pregnancy, then drop off by the end of the third month. Levels of estrogen and progesterone, produced chiefly by the placenta, continue to rise.

HCG is a hormone produced by the embryo early in pregnancy. Like LH from the pituitary of women, HCG stimulates estrogen and progesterone production by the corpus luteum or CL, a hormone-producing "gland" that forms in the ovary from the collapsed follicle after ovulation. These hormones are essential to maintaining pregnancy. However, as illustrated, HCG secretion from the embryonic tissue lasts only about 10 weeks. Because of the natural decline in HCG, the CL degenerates. At this time, estrogen and progesterone production shifts to the placenta.

Estrogen and progesterone secreted by the placenta during the remainder of the pregnancy serve a number of essential functions. Estrogen, for example, stimulates growth of the smooth muscle cells of the uterus. This allows the organ to expand to many times its original size during pregnancy and provides additional propulsive force needed to expel the child at birth.

Progesterone calms the uterine musculature during pregnancy, preventing the embryo and fetus from being expelled. It also stimulates the production of cervical mucus, which plugs the cervical canal, preventing bacteria from entering the uterus and infecting the growing embryo.

The Amnion

The amnion forms from the inner cell mass.

As the placenta begins to form, the inner cell mass (ICM) of the blastocyst undergoes some remarkable changes. Early in development, a layer of cells separates from the ICM to form the **amnion** (AM-knee-on). A small cavity, the amniotic cavity, forms between the ICM and the amnion. The amniotic cavity fills with a liquid called *amniotic fluid*. It will eventually form a cushion around the baby during development, helping to protect it from injury (Figure 22-6).

TABLE 22-1	Hormones Produced by the Placenta
Hormone	**Function**
Human chorionic gonadotropin (HCG)	Maintains corpus luteum of pregnancy
	Stimulates secretion of testosterone by developing testes in XY embryos
Estrogen (also secreted by corpus luteum of pregnancy)	Stimulates growth of myometrium, increasing uterine strength for parturition (childbirth)
	Helps prepare mammary glands for lactation
Progesterone (also secreted by corpus luteum of pregnancy)	Suppresses uterine contractions to provide quiet environment for fetus
	Promotes formation of cervical mucous plug to prevent uterine contamination
	Helps prepare mammary glands for lactation
Human chorionic somatomammotropin	Helps prepare mammary glands for lactation
	Believed to reduce maternal utilization of glucose so that greater quantities of glucose can be shunted to the fetus
Relaxin (also secreted by corpus luteum of pregnancy)	Softens cervix in preparation of cervical dilation at parturition
	Loosens connective tissue between pelvic bones in preparation for parturition

22-3 Embryonic Development

FIGURE 22-8 **Formation of the Spinal Cord** A cross section of a 17-day human embryo showing the relationship between the three embryonic tissues and the amnion and yolk sac.

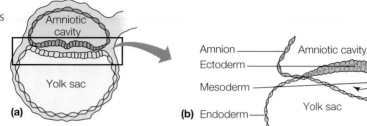

TABLE 22-2	End Products of Embryonic Germ Layers	
Ectoderm	**Mesoderm**	**Endoderm**
Epidermis	Dermis	Lining of the digestive system
Hair, nails, sweat glands	All muscles of the body	Lining of the respiratory system
Brain and spinal cord	Cartilage	Urethra and urinary bladder
Cranial and spinal nerves	Bone	Gallbladder
Retina, lens, and cornea of eye	Blood	Liver and pancreas
Inner ear	All other connective tissue	Thyroid gland
Epithelium of nose, mouth, and anus	Blood vessels	Parathyroid gland
Enamel of teeth	Reproductive organs Kidneys	Thymus

Embryonic development ends when the organs are more or less formed.

After the amnion forms, the cells of the inner cell mass differentiate, forming three distinct germ cell layers, the **ectoderm**, **mesoderm**, and **endoderm**. These are known as the *primary germ layers*. The formation of the primary germ layers marks the beginning of **embryonic development**.

The primary germ layers of the embryo give rise to the organs in a process called **organogenesis**. Organogenesis therefore is the main event of embryonic development. Table 22-2 shows the organs that form from each layer.

One of the first events of organogenesis is the formation of the central nervous system (the spinal cord and brain). It arises from ectoderm from the ICM. Early in embryonic development, the ectoderm along the back of the embryo folds inward. This creates a long trench, the neural groove, that runs the length of the back surface of the embryo (Figure 22-8b). Over the next few weeks, the

The neural groove begins to form from ectoderm

The neural groove deepens

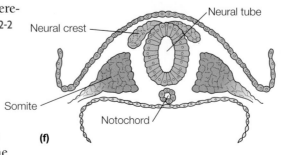

The neural tube forms. Note the presence of the neural crest, ectodermal cells that give rise to nerves

neural groove deepens and eventually closes off, creating the neural tube (Figure 22-8d). The walls of the neural tube thicken and form the spinal cord. In the head region, the neural tube expands to form the brain.

As you may recall from your study of the nervous system, numerous nerves attach to the spinal cord and brain. They're known as the *spinal nerves* and *cranial nerves*, respectively. These nerves develop from small aggregations of ectodermal cells, the neural crest, lying on either side of the neural tube, as shown in Figure 22-8d. These cells give rise to axons that grow into the body and attach to organs, muscle, bone, and skin. The ectoderm also gives rise to the outer layer of skin (the epidermis).

The middle germ layer, the mesoderm, gives rise to deeper structures—the muscle, cartilage, bone, and others. Much of the

mesoderm first aggregates in blocks, called the *somites* (SO-mights) (Figure 22-8d). The somites form the backbone and the muscles of the neck and trunk. Mesoderm lateral to the somites becomes the dermis of the skin, connective tissue, and the bones and muscles of the limbs.

The endoderm, the "lowermost" germ layer of the ICM, forms a large pouch under the embryo called the **yolk sac** (Figures 22-8a and 22-9). The uppermost part of the yolk sac becomes the lining of the intestinal tract. The yolks sac also gives rise to blood cells and primitive germ cells. During organogenesis, the germ cells migrate from the wall of the yolk sac to the developing testes and ovaries, which are located near the kidneys. These cells become spermatogonia or oogonia.

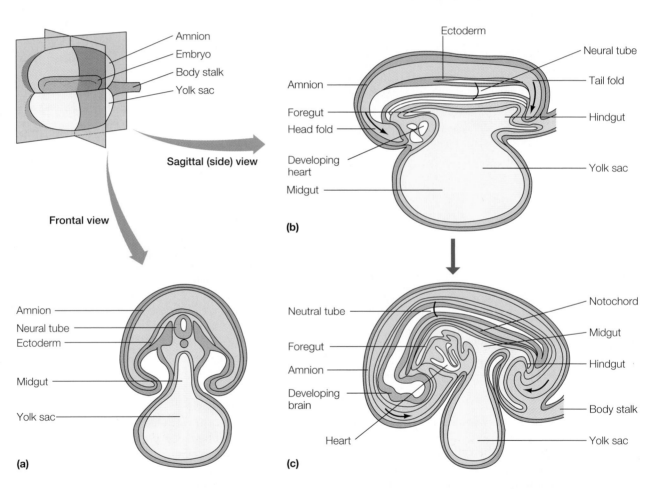

FIGURE 22-9
The Yolk Sac (a) A cross section of the embryo. The yolk sac forms from embryonic endoderm. (b) A longitudinal section showing how the upper end of the yolk sac forms the embryonic gut, (c) which will become the lining of the intestinal tract.

22-4 Fetal Development

During fetal development, organs continue to develop and grow.

Fetal development begins in the eighth week of pregnancy and ends at birth. It involves (1) continued organ development and growth and (2) changes in body proportions—for example, elongation of the limbs.

The fetus grows rapidly during the fetal period. It increases in length from about 2.5 centimeters (1 inch) to 35–50 cen-

timeters (14–21 inches) and increases in weight from 1 gram to 3000–4000 grams. The fetus also undergoes considerable change in physical appearance, becoming more humanlike as each month passes (Figure 22-10). Organ development that commenced during the embryonic stage reaches completion during the fetal stage.

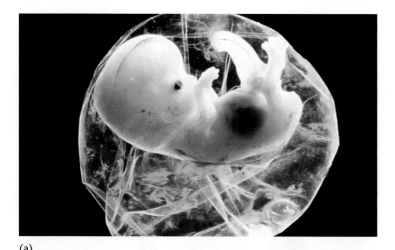

(a)

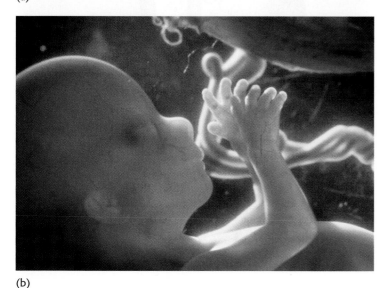

(b)

(c)

FIGURE 22-10 **Fetal Development** (a) Fetus at 5 to 6 weeks. (b) Fetus at 4 months. (c) Fetus at 5 months.

The Fetal Circulatory System

The fetal circulatory system contains three bypasses around the lungs and liver.

One organ system that does not quite reach completion during fetal development is the circulatory system. As shown in Figure 22-11a, the fetal circulatory system is much like the circulatory system of newborns (and adults) except for three bypasses—pathways that divert blood around two organs that are still growing and not yet functional: the lungs and liver.

Fetal blood circulates to and from the placenta in the **umbilical cord**. The cord contains two umbilical arteries and a single, large umbilical vein. The umbilical vein, however, carries oxygen- and nutrient-rich blood from the placenta to the fetus. Some of the blood flows into and through the fetal liver, as shown in Figure 22-11a. From the liver, the blood passes into the inferior vena cava and on to the heart. Because the liver is not yet functional, most of the blood bypasses the organ. It does so via a small shunt, a blood vessel known as the *ductus venosus*. It connects the umbilical vein directly to the inferior vena cava.

Blood from the inferior vena cava flows into the right atrium of the heart. In an adult, all of the blood flows from the right atrium to the right ventricle, then is pumped to the lungs where it is oxygenated. In a fetus, however, only a small portion of the blood goes into the right atrium, and then to the lungs where it supplies nutrients required for growth and development. The majority of the blood bypasses the lungs. It flows directly from the right ventricle to the left ventricle through a small hole in the wall between these two chambers. The hole is called the *foramen ovale*—literally, oval hole. Blood entering the left atrium is then pumped to the left ventricle. From there, it is pumped to the head and the rest of the body through the aorta and its many branches.

A third shunt also exists. Known as the *ductus arteriosus*, it lies between the pulmonary artery and the aorta. Like the foramen ovale, it helps to divert blood away from the lungs.

Blood pumped throughout the body of the fetus via the aorta returns to the heart by the inferior and superior vena cavae. However, a large percentage of the blood that has been stripped of oxygen and charged with wastes of cellular activities returns to the placenta via the umbilical arteries. In the placenta, the blood is rid of its waste and recharged with oxygen and vital nutrients picked up from the mother's bloodstream.

At birth, the fetal circulatory system quickly changes to the adult pattern. The foramen ovale normally closes at birth, establishing the adult circulation pattern. The two other shunts shrivel and close up, then eventually disappear. In some children, however, the foramen ovale remains open. Babies born with this defect usually appear blue because blood is unable to get to the lungs to be oxygenated. Blood deprived of oxygen appears slightly blue. Surgery is required to repair the defect.

FIGURE 22-11 Fetal Circulation
(a) before and (b) after birth.

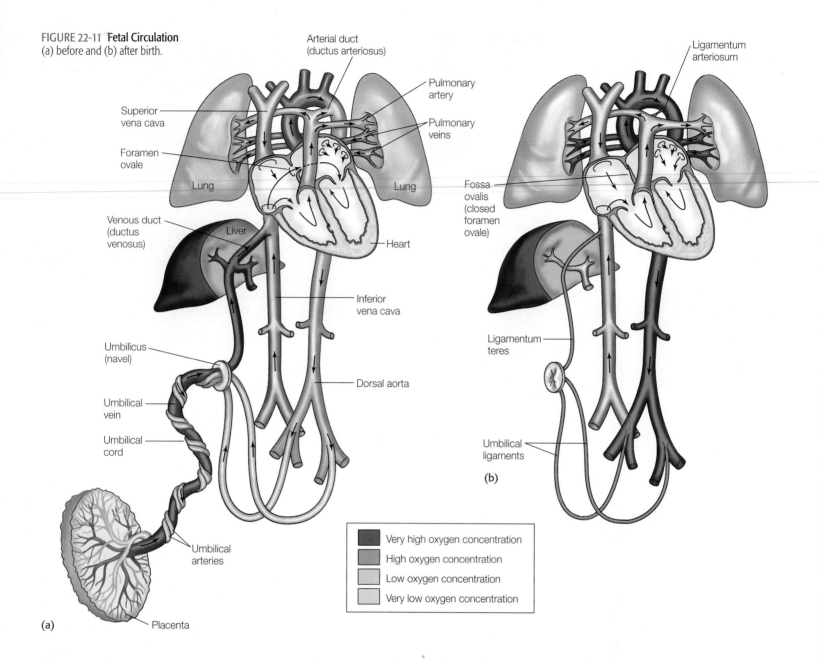

Arterial duct (ductus arteriosus)

Pulmonary artery

Superior vena cava

Pulmonary veins

Foramen ovale

Lung

Lung

Venous duct (ductus venosus)

Liver

Heart

Inferior vena cava

Umbilicus (navel)

Dorsal aorta

Umbilical vein

Umbilical cord

Umbilical arteries

Ligamentum arteriosum

Fossa ovalis (closed foramen ovale)

Ligamentum teres

Umbilical ligaments

(b)

Very high oxygen concentration

High oxygen concentration

Low oxygen concentration

Very low oxygen concentration

(a)   Placenta

22-5 Birth Defects

Birth defects include structural and functional defects that appear in the newborn.

By several estimates, 31% of all successful fertilizations end in a miscarriage—that is, a spontaneous abortion. Two of every three of these miscarriages occur before a woman is even aware that she is pregnant. Why such a high rate? Biologists hypothesize that early miscarriage is nature's way of "discarding" defective embryos.

Despite this culling that helps our species remain genetically fit, many children are born each year with **birth defects**, physical or physiological abnormalities. By various estimates, 10%–12% of all newborns have some kind of birth defect, ranging from minor biochemical or physiological problems, which are not

even noticed at birth, to gross physical defects (Figure 22-12). Scientists believe that most birth defects arise from chemical, biological, and physical agents. Table 22-3 lists known and suspected agents.

The effect of a birth-defect causing agent on the developing embryo is related to (1) the time of exposure, (2) the nature of the agent, and (3) the dose. Consider time first.

Because the organ systems develop at different times, the timing of exposure determines which systems are affected by a given agent. Organ systems are usually most sensitive to potentially harmful agents early in their development, as indicated by the pink bars in Figure 22-13. The central nervous system, for example, begins to develop during the third week of pregnancy. Because most women do not know they are pregnant for 3–4

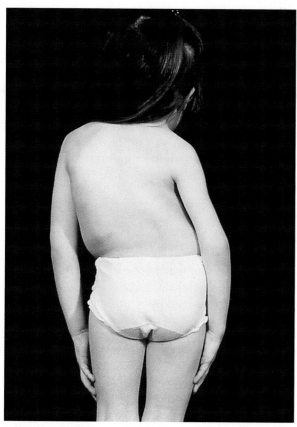

FIGURE 22-12 **A Common Birth Defect** Scoliosis is a lateral curvature of the spine.

TABLE 22-3	Known and Suspected Human Teratogens
Known Agents	**Possible or Suspected Agents**
Progesterone	Aspirin
Thalidomide	Certain antibiotics
Rubella (German measles)	Insulin
Alcohol	Antitubercular drugs

Irradiation	Antihistamines
	  Barbiturates
	  Iron
	  Tobacco
	  Antacids
	  Excess vitamins A and D
	  Certain antitumor drugs
	  Certain insecticides
	  Certain fungicides
	  Certain herbicides
	  Dioxin
	  Cortisone
	  Lead

weeks after fertilization, the central nervous system is at risk. In contrast, the teeth, palate, and genitalia do not begin to form until about the sixth or seventh week of pregnancy. Exposure during the seventh week might therefore affect them, but have little effect on the CNS, which has entered a less sensitive phase.

The nature of a birth-defect causing agent also influences the outcome of exposure. Some chemicals, for instance, affect a variety of developing systems. Others affect only one system. Ethanol in alcoholic beverages, for example, is a broad-spectrum chemical. Thus, children who are born to alcoholic mothers–or even women who have consumed one or two drinks early in pregnancy–may have a variety of physical defects. Behavioral problems and learning disabilities are also common in children of alcoholic mothers. These symptoms are part of a condition called *fetal alcohol syndrome.*

In contrast, other agents are more selective, targeting only one system. Methyl mercury found in fish and shellfish, for instance, damages the CNS, creating abnormalities in the brain and spinal cord, but has little effect on other systems.

Finally, like most toxic substances, chemicals generally follow a dose-response relationship: The greater the dose, the greater the effect.

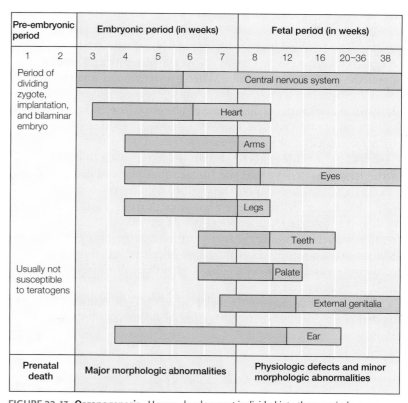

FIGURE 22-13 **Organogenesis** Human development is divided into three periods, or stages: pre-embryonic, embryonic, and fetal. Organogenesis occurs during the embryonic stage. Each bar indicates when an organ system develops. The pink-shaded area indicates the periods most sensitive to teratogenic agents (agents that can cause birth defects).

One of the best-known birth-defect causing agents is not a chemical substance, but the virus that causes German measles or rubella (rue-BELL-ah). If a pregnant woman contracts the disease during the first 3 months of pregnancy, she has a one in three chance of giving birth to a baby with a serious birth defect. Deafness, cataracts, heart defects, and mental retardation are common. Fifteen percent of all these babies die before the age of one.

Many birth defects can be avoided. German measles vaccine given to young girls usually protects them throughout their childbearing years. Vaccinating boys also reduces the risk to so-ciety. Other precautions also help prevent birth defects. Pregnant women or women who are trying to get pregnant should carefully control what they eat or drink.

Once the period of high sensitivity has passed, good nutrition and a healthy environment remain essential. This is because fetal development is also influenced by a number of physical and chemical agents in our homes and places of work. Toxic chemicals can stunt fetal growth and, in higher quantities, can even kill fetuses, resulting in stillbirths. Proper nutrition is essential because nutrient, vitamin, and mineral deficiencies can retard growth.

22-6 Childbirth and Lactation

Uterine muscle contractions are stimulated by a change in hormonal levels.

Childbirth begins with mild uterine contractions. During labor, contractions increase in strength and frequency until the baby is born.

Several factors stimulate uterine contractions. A rise in estrogen levels late in pregnancy is thought to stimulate the production of oxytocin receptors in the smooth muscle cells of the uterus. As you may recall from your study of the endocrine system, oxytocin is a hormone produced by the posterior pituitary of women. This hormone causes contraction of the smooth muscle of the uterus. The increase in the number of receptors in the smooth muscle cells renders them increasingly responsive to the small amounts of oxytocin.

Oxytocin first comes from the fetal pituitary gland, not the mother. The fetal pituitary releases small quantities right before birth. Fetal oxytocin travels across the placenta and circulates in the mother's bloodstream. When it arrives at the sensitized uterine musculature, it stimulates muscle contraction. Fetal oxytocin also stimulates the release of placental prostaglandins that act on smooth muscle. Together, these hormones stimulate more frequent and powerful uterine contractions.

Maternal oxytocin also plays an important role in childbirth. Scientists believe that uterine contractions and discomfort create stress that triggers the release of maternal oxytocin. It, in turn, augments muscle contractions caused by fetal oxytocin and prostaglandins (Figure 22-14). As uterine muscle contraction increases, maternal oxytocin release increases. This stimulates even stronger contractions, resulting in additional oxytocin release, a positive feedback loop that continues until the baby is delivered.

Research also suggests that the high levels of estrogen at the end of pregnancy block the "calming" influence of placental progesterone. Consequently, the uterus begins to contract at irregular intervals.

Physicians (and expectant mothers) recognize two types of labor contractions. The first are false labor contractions. These irregular uterine contractions usually begin a month or two before childbirth and are also known as **Braxton-Hicks contractions**. As the due date approaches, false labor contractions occur with greater frequency, causing many anxious couples to race to the hospital, only to be told to go home and wait for a few weeks for the real thing. In contrast, true labor contractions are more intense than those of false labor and occur at regular intervals.

Contractions are not all that's needed for childbirth. Some changes must also take place to enable the large baby to emerge from the uterus and pass through the uterus and vagina without causing severe damage. Much of the success of birth depends on the hormone **relaxin**. Produced by the ovary and the placenta, relaxin is released near the end of pregnancy. As its name implies, relaxin softens the fibrocartilage uniting the pubic bones. This, in turn, allows the pelvic cavity to widen and thus greatly facilitates childbirth. Relaxin also softens the cervix, allowing it to expand so the baby can pass through it without causing severe damage.

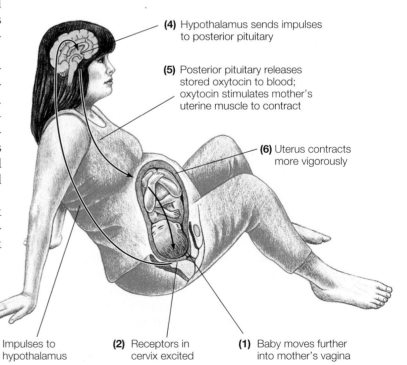

(4) Hypothalamus sends impulses to posterior pituitary

(5) Posterior pituitary releases stored oxytocin to blood; oxytocin stimulates mother's uterine muscle to contract

(6) Uterus contracts more vigorously

(3) Impulses to hypothalamus

(2) Receptors in cervix excited

(1) Baby moves further into mother's vagina

FIGURE 22-14 Oxytocin Positive Feedback Mechanism in Birth
The mechanism continues to cycle until interrupted by the birth of the baby.

Stages of Childbirth

Childbirth occurs in three stages.

Figure 22-15 illustrates the three stages of childbirth, dilation, expulsion, and placental.

Stage 1, the dilation stage, gets its name from the dilation of the cervix. This phase begins when uterine contractions start and generally lasts 6–12 hours, but it can last much longer.

At the beginning of stage 1, uterine contractions typically last only 30 seconds and may come every half hour or so. As time passes, however, contractions become more frequent and powerful. Uterine contractions generally rupture the amnion early in the dilation phase, causing the release of the amniotic fluid, an event commonly referred to as "breaking the water."

Uterine contractions push the infant's head against the relaxin-softened cervix, causing it to stretch and become thin (Figure 22-15b). By the end of stage 1, the cervix has dilated to about 10 centimeters (4 inches), approximately the diameter of a baby's head. During this phase, the baby is pushed downward by the uterine contractions and descends into the pelvic cavity. When its head is "locked" in the pelvis, the baby is said to be engaged.

Stage 2, the expulsion stage, begins after the cervix is dilated to 10 centimeters and the baby is engaged (Figure 22-15c). At this time, uterine contractions usually occur every 2 or 3 minutes and last 1–1.5 minutes each. For most first-time mothers, stage 2 lasts 50–60 minutes. If a woman is delivering her sec-

ond child, it lasts about 20–30 minutes on average. The expulsion stage ends when the child is pushed through the vagina into the waiting hands of a doctor, midwife, proud father, or bewildered cab driver.

To facilitate the delivery, an incision is often made in the skin to widen the vaginal opening. The incision prevents unnecessary tearing and allows the infant to pass quickly. The incision is sutured immediately after the baby is born.

Once the baby's head emerges from the vagina, the rest of the body slips out quickly. However, the baby is still attached to the placenta inside the mother via the umbilical cord. The cord pulsates for a minute or more, continuing to deliver blood to the newborn. When the pulsating stops, the cord is tied off and cut.

Most babies (95%) are delivered head first with their noses pointed toward the mother's tailbone (Figure 22-15c). Occasionally, however, babies may be oriented in other positions—making delivery more difficult, time-consuming, and hazardous to both mother and baby. The most common alternative delivery is the breech birth, in which the baby is expelled rear-end first. Because breech births require more time, they can cause extreme fatigue in the mother and brain damage in the baby. In breech babies, the umbilical cord sometimes wraps around the infant's neck during birth. This can cut off the fetus' blood supply, causing brain damage or death.

To avoid such complications, physicians often "turn" breech babies before birth by applying pressure to the woman's abdomen. If a fetus cannot be turned, the baby is usually deliv-

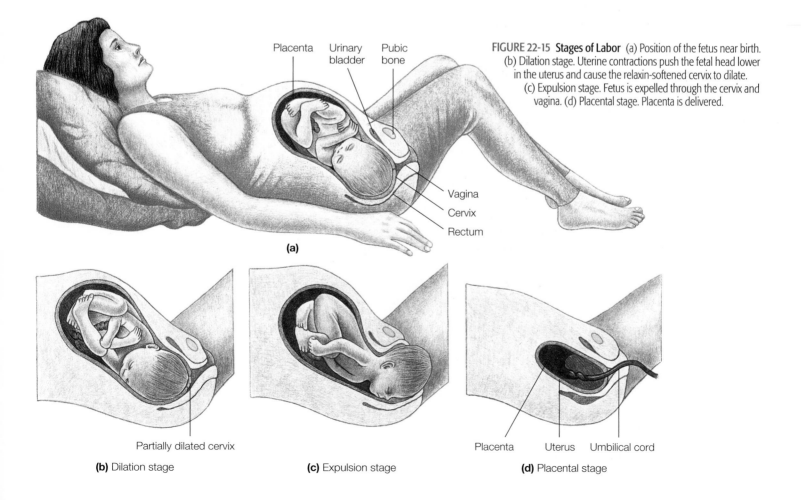

FIGURE 22-15 **Stages of Labor** (a) Position of the fetus near birth. (b) Dilation stage. Uterine contractions push the fetal head lower in the uterus and cause the relaxin-softened cervix to dilate. (c) Expulsion stage. Fetus is expelled through the cervix and vagina. (d) Placental stage. Placenta is delivered.

Placenta   Urinary bladder   Pubic bone

Vagina
Cervix
Rectum

(a)

Partially dilated cervix

Placenta   Uterus   Umbilical cord

(b) Dilation stage   **(c)** Expulsion stage   **(d)** Placental stage

ered by a **cesarean section**, a horizontal incision through the abdomen just at the pubic hair line and then through the uterus. Cesarean sections are also performed if labor is prolonged or if the mother has an active infection caused by herpes or some other sexually transmitted disease that might be transferred to her child as it passes through the vagina.

The final stage of delivery is the placental stage (Figure 22-15d). The placenta is expelled by uterine contractions, usually within 15 minutes of childbirth. The uterine blood vessels then clamp shut, preventing hemorrhage, although the mother continues to lose some blood for 3–6 weeks after delivery. After delivery, the uterus gradually returns to its normal size.

Lactation

Milk production, or lactation, is controlled by the hormone prolactin.

A baby emerges into a novel environment at birth. No longer connected to the placenta, which has provided nutrients from the maternal bloodstream, the newborn must acquire food from another source, its mothers' breasts.

The **breasts** contain milk-producing glandular tissue. The glands are drained by ducts that lead to the nipple, a surface structure where the milk ducts converge (Figure 22-16). During pregnancy, the ducts and glands proliferate under the influence of estrogen and progesterone from the placenta and ovary. Milk production is itself stimulated by the hormone prolactin and a placental hormone.

Although the breasts do not produce milk for 2 or 3 days after childbirth, they immediately begin producing small quantities of colostrum. **Colostrum** (co-LOSS-trum) is a clear fluid rich in protein and lactose (a sugar), but lacking fat. Colostrum also contains antibodies that protect the infant from bacteria. (Health Note 14-2 describes the benefits of colostrum and breast-feeding.) Within 2–3 days, the breasts start to produce large quantities of milk.

Milk production continues as long as the baby suckles. If nursing is interrupted for 3 or 4 days, however, the breasts stop producing milk. In most women, milk production begins to decline by the seventh to ninth month of lactation, a time at which babies begin to feed on semisolid food and breast-feed less frequently. The breasts decrease in size and typically become smaller than they were before pregnancy. (For a discussion of breast implants sometimes used to correct this condition, see Health Note 22-1.)

The production of milk during suckling is brought about by prolactin, released from the mother's pituitary. Its release is stimulated by suckling. Milk does not flow out of the breasts; however, it is actively propelled from the glandular units through the ducts. This is brought about by the hormone oxytocin, whose release is controlled by a neuroendocrine reflex stimulated by suckling. Oxytocin, which is released from the posterior pituitary, causes muscles in the milk glands to contract and eject their milk.

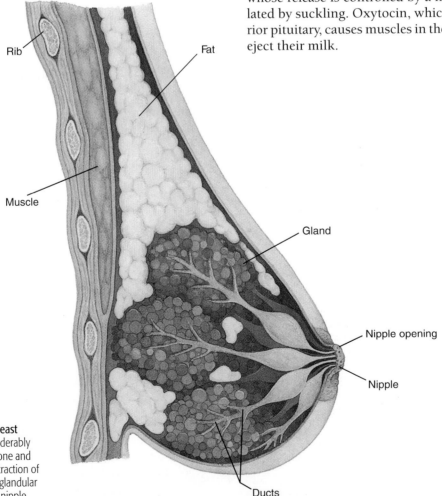

FIGURE 22-16 The Lactating Breast
The glandular units enlarge considerably under the influence of progesterone and prolactin. Milk is expelled by contraction of musclelike cells surrounding the glandular units. Ducts drain the milk to the nipple.

:::: healthnote

22-1 Breast Augmentation: Balancing the Benefits and Risk

Marilyn Allbright is a 37-year-old mother of two children, both of whom she breast fed. After her second pregnancy, Marilyn noticed a dramatic decrease in the size of her breasts and decided to undergo breast augmentation surgery.

Marilyn is not alone. Each year, over 125,000 women in the United States undergo this procedure. Some women have implants to increase the size of their breasts, which like Marilyn's, have shrunk after breast-feeding or pregnancy. Others have implants because they feel their breasts are too small or because their breasts are markedly different sizes. Still others undergo the surgery because they've had breasts removed, either to halt breast cancer or as a preventative measure. Some women who have a high genetic risk for breast cancer opt to have their breasts removed pre-emptively.

Breast augmentation is most often performed in clinics or surgical suites, not hospitals, under general anesthesia. The breast implant is a soft, pliant roughly breast-shaped plastic bag made from an elastic-type silicone material (Figure 1). Most are filled with a salt solution. The implant is placed under the existing breast tissue, either directly under it or beneath the chest muscle. Most plastic surgeons believe that implants under the muscle turn out much better. The breast is considerably softer and more natural looking.

In some instances, however, the implant is better placed over the chest muscle. For example, if a woman is suffering from droopy breasts, the surgeon may remove some skin and then may place the implant directly over the muscle, creating a rounder, firmer breast.

Surgeons insert breast implants via an incision in the skin. Four main incision points are currently used: in the armpit, under the breast, at the darker skin surrounding the nipple, or, more recently, just above the belly button, known as a *transumbilical insertion*. Transumbilical insertion, the latest technique, is designed to reduce scarring. During this procedure, a tiny half-moon-shaped incision is made in the navel. Using a special device, the surgeon tunnels under the skin to reach the breast. Here, the surgeon creates a pocket for the breast implant, using a tissue separator. The empty implant is rolled up, inserted into the tunnel, and then moved into position. It is then filled with saline.

The transumbilical approach results not only in less scarring, but also less pain and a faster recovery. It may also reduce the chances of damaging nerves that supply the breast, which can lead to a loss of sensation. The major drawback is that this approach can only be used to place an implant on top of the chest muscle.

Despite the benefits of breast augmentation, including greater self-esteem, the operation does have its downsides. For one, most breast implants do not appear normal. They tend to be rounder and firmer than natural breasts. To address this complaint, some companies are now manufacturing tear-shaped breast implants that have a more natural appearance.

In some instances, for example, in the case of heavy, droopy breasts, the appearance of the breast after implantation may actually be much worse! Breast augmentation, like any surgery, is also traumatic. After the operation, the breast and surrounding tissue swells. Some women report intense pain and bruising that may last for a couple weeks. It may take several months before the breasts look normal again. Infections can also develop.

FIGURE 1 Implants

One of the most common problems is that the body forms a connective tissue layer (internal scar) around the implant. When the scar tissue shrinks, it causes hardening of the breasts that may cause the implants to deform. It can also be quite painful. As a result, the implant may need to be removed and replaced. To reduce this problem, some companies have begun producing implants that have a textured silicone shell.

Breast implants may also rupture. Although saline-containing implants are not considered to pose any threat, the silicone-containing implants that were once commonly used are of greater concern. Over the years, many women with silicone-filled implants have complained of unusual health symptoms, usually following the rupture of the implants. They have reported muscle spasms and pain, swollen and painful joints, rashes, and hair loss, among other symptoms. Although no one knows what causes these problems, some researchers think that they may be the result of low-grade infections that develop around the implants. Many of the women suffering from these problems joined a massive class action lawsuit against the maker of the silicone implants, Dow Corning, which settled the claim for $3.2 billion, causing the company to go into bankruptcy. Interestingly, extensive clinical research studies show that the incidence of symptoms in breast implant patients is the same as in the general population of women.

Scarring is not generally a problem. If the incision is made around the areola, the scar is usually barely noticeable; if it is made under the breast, the breast itself hides any scar. Incisions in the armpit or axilla are generally only noticeable when the arm is raised. However, if a woman forms keloids, thick scars, breast scars can be quite noticeable.

Some procedures result in a loss of sensation around the nipple and on the breast—for example, the insertion of the implant through an incision in the areola. Breast implants may also crinkle or crease around the edge, especially in women who had very little natural breast tissue at the outset. In addition, implants may need to be replaced at a later date. No one knows how long breast implants will last or how often they may rupture.

The decision should be made after considering many factors. Women considering a breast augmentation procedure should seek a competent surgeon and learn as much as possible about the procedures before making their decision.

www.jbpub.com/humanbiology/5e

Visit Human Biology's Internet site for links to web sites offering more information on this topic.

22-7 Aging and Death

Aging is a progressive deterioration of the body.

Aging is a part of life. Technically, **aging** is defined as a progressive deterioration of the body's structure and function. The most notable changes occurring with age are physical shifts—wrinkling, loss of hair, and stooping. Less obvious is the gradual deterioration of the function of body organs. These often occur so slowly as to seem imperceptible (Figure 22-17). For example, as people age, their vision and hearing both decline, usually by small increments. Muscular strength also ebbs as a result of a decrease in the number of myofibrils (bundles of contractile filaments) in the skeletal muscles. Bones tend to thin, and joints often show signs of wear.

Aging is also accompanied by a gradual reduction in heart and lung function. The decrease in lung function results from a reduction in the amount of air that can be inhaled and exhaled. As a result, oxygen absorption decreases. As we age, the number of functional nephrons in the kidney decreases, as does renal function.

The immune system also becomes less able to respond to antigens as we age. The nervous system does not escape the aging process, either. Memory deteriorates and reaction time decreases.

Although these changes are part of the aging process, not all of them begin at the same time. Table 22-4 lists some important age-related changes in various organ systems along with their causes.

Another function that deteriorates as we age is homeostasis. As homeostatic mechanisms falter, the body becomes less resilient. Healing takes longer. Individuals become more susceptible to disease.

Decline in Cell Number and Cell Function

Aging occurs as a result of a decline in cell numbers and their function.

As we age, the number of cells in body organs like the brain declines. Laboratory experiments show that body cells grown in culture divide a certain number of times, then stop. The number of divisions a cell can undergo is directly related to the age of the organism from which they were taken: The older the organism, the fewer divisions are possible from that point in time on.

Laboratory studies like these have also uncovered a correlation between the number of divisions in culture and the life span of the species from which the cells were taken. In other words, cells from species with long life spans undergo more divisions than cells from species with short life spans. These and other data suggest that the end of cell division and, therefore, aging, may be genetically programmed.

If cells are programmed to divide a given number of times, why don't all humans live to be the same age? One answer is that people differ genetically, which may contribute to the differences in life span. In addition, people live markedly different lifestyles and eat very different diets. Research shows that the better one lives, the longer one lives. Diet, exercise, and stress management are keys to a healthy life.

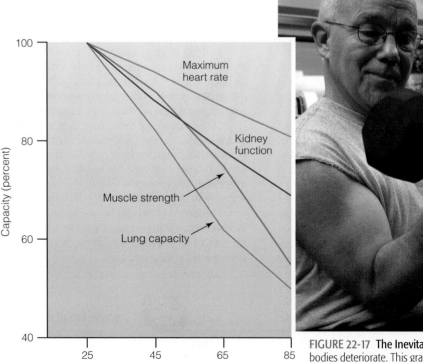

FIGURE 22-17 **The Inevitable Slide** As we age, our bodies deteriorate. This graph shows the dramatic decline in four essential body functions.

TABLE 22-4	Age-Related Changes in the Body	
Organ System	Change	Cause
Skin	Thinning	Reduced cell division in epidermis
	Loss of elasticity	Elastic fibers in dermis replaced by collagen fibers
	Wrinkling	Same as above plus loss of fat in hypodermis
	Drying	Loss of sweat glands and sebaceous glands
	Loss of body hair	Loss of hair follicles
Bones	Become weaker, more porous, and more brittle	Loss of calcium is the primary reason for this, but also caused by loss of collagen fibers
Joints	Often become difficult to move and sometimes arthritic	Deterioration of cartilage on bone surfaces
Muscles	Loss of mass and muscle strength	Loss of motor neurons that stimulate muscles and a loss of muscle protein; fat and collagen fibers tend to invade muscles
Respiratory	Decreased oxygen absorption	Breakdown of alveoli
Cardiovascular	Decreased ability to pump blood so less blood is supplied to other tissues and organs	Reduction in the amount of heart muscle
	Hardening of the arteries and high blood pressure	Loss of elastic fibers in arteries and deposition of calcium, fat, and cholesterol in arterial walls
Nervous	Brain shrinkage	Loss of nerve cells throughout life as a natural process and as a result of alcohol use and trauma
	Deterioration of short-term memory and longer time to process various forms of information	Same as above
Reproductive	Loss of menstrual cycle in women	Dramatic decline in estrogen secretion by ovaries
	Reduced fertility in men	Decline in testosterone production
Urinary	Urinary incontinence	Weakening of the muscles of the urinary sphincter in men and women and weakening of the muscles of the pelvic floor in women who have given birth in earlier years

Scientists are beginning to discover why proper diet and exercise can increase life span. Laboratory studies, for instance, show that by manipulating the chemical environment of cells, one can produce more cell divisions than normal. For example, vitamin E in large quantities increases the number of cell divisions in tissue cultures and increases a cell's life span. Whether vitamin E extends a person's life is not known.

Aging also results from a decrease or deterioration of cellular function. Studies show that the decline in cell function results from problems arising in the DNA, RNA, and proteins. Cells are exposed to many potentially harmful factors in the course of a lifetime. Natural and human-produced radiation and other potentially harmful agents (such as chemicals in the home and at work) may damage cells and impair their function. The cells of the body also produce oxygen free radicals, which are oxygen atoms that carry unpaired electrons. These chemical species are natural byproducts of cellular metabolism and can react with (oxidize) proteins, DNA, lipids, and other important molecules, causing damage. As we age, our cells accumulate oxidized lipids and proteins, including important enzymes, which cause a deterioration in cell function.

Additional evidence shows that cells may lose the ability to repair DNA as they age, making matters worse. As a result, cells would be unable to fix mutations in genes that produce important structural and functional proteins such as enzymes. Cancer, the incidence of which increases with age, could result. As the damage accumulates, body function gradually deteriorates.

Some researchers think that aging is largely the result of a gradual decline in immune system function caused by a reduction in the number of cells and a loss of cell function. Although the immune system does falter, this decline surely cannot explain the other signs of aging.

∷∷ healthnote

22-2 Can We Reverse the Process of Aging?

Suppose a drug were invented that would help you live to be 100. Would you take it?

If you are like most people, you probably would, as long as your additional years would be healthy and fairly trouble free. Because most of us would like to beat aging, we secretly hope for a medical breakthrough that would prolong our lives.

Unfortunately, the search for a "cure" for old age has been fraught with frustration. Over the years, numerous treatments have been tried and have failed. There is some encouraging news, however. Scientists recently discovered a protein called *stomatin* (stow-MA-tin), which they isolated from fibroblasts that have stopped dividing in tissue culture. Scientists hope that stomatin will stop other cells from dividing as well. If this turns out to be true, researchers may have discovered an important clue in the aging puzzle: the signal that ends cell division. Moreover, they may have found a way to prolong human life.

If the stomatin gene can be located, researchers may be able to find a way to inactivate it. A drug, for example, could be injected in people to block the gene's effect. The tissues of these people might continue to regenerate beyond the genetically programmed life span.

Efforts to reverse the aging process have met with some success in another arena as well, such as slowing down the aging of skin. Because age-related changes in the skin mirror those occurring in other aging tissues, scientists have long studied the skin to expand their understanding of the aging process. Their studies have shown that skin aging results from two processes: intrinsic or chronological aging—which may be genetically programmed—and extrinsic aging, which results from accumulated environmental damage from sunlight and other factors.

Many skin scientists doubt whether anything can be done about intrinsic aging. Their studies suggest that after a skin cell has lived out its lifetime, its plasma membrane receptors become insensitive to growth factors that stimulate DNA replication and cell division.

New research suggests, however, that extrinsic aging, especially that induced by sunlight, is preventable and even reversible. Unknown to many, sunlight damages epithelial cells of the epidermis and the fibroblasts in the underlying dermis. The plasma membranes of skin cells, in fact, can be damaged within a few minutes of exposure to sunlight (**Figure 1**). Sunlight energy can also be absorbed by oxygen in the tissue, resulting in the formation of oxygen free radicals, highly destructive chemicals that are primarily responsible for extrinsic aging.

Preventing Diseases of Old Age

The likelihood of contracting "diseases of old age" depends largely on lifestyle.

Many diseases are associated with old age. Osteoporosis, arthritis, atherosclerosis, Alzheimer's disease, and cancer are examples. Although these diseases are more prevalent in older people, they are not the inevitable consequence of aging. That is, they are not unavoidable signs of aging like gray hair. As previous chapters have shown, the likelihood of developing most of these diseases can be greatly reduced by exercise, diet, and other lifestyle adjustments. In other words, people can age in a healthy manner by living healthy lives. Although our bodies will eventually fail, we do not have to suffer from the chronic illnesses that plague so many people. Health Note 22-2 offers some additional information on efforts now underway to slow the process of aging.

Death

Death may be defined by the loss of brain or heart and pulmonary function.

Aging results in a deterioration of function that eventually leads to death. But death can also result from traumatic injury to the body—for example, severe damage to the brain or a sudden loss of blood—and many other causes. What is death?

Although we can all agree that death is the cessation of life, trouble begins when people try to define it in instances requiring life-support systems. In 25 states, a person is considered dead when his or her heart stops beating and breathing ceases. In the remaining 25 states, however, death is more liberally defined as an irreversible loss of brain function. A woman in a coma, for example, who shows little or no brain activity other than that needed to keep the heart beating and the lungs working is alive in some states and legally dead in others.

Maintaining a person on life support is extremely costly to a family and to society if, for example, the patient is receiving federal benefits through Medicare. The money spent on maintaining a life could arguably be used to save dozens of other lives through preventive measures. These issues inevitably lead to another biomedical dilemma, the controversy over euthanasia, a word derived from the Greek meaning "easy death."

Euthanasia (you-thin-A-zah) refers to any act or method of causing death painlessly and may be either passive or active. Passive euthanasia consists of decisions and actions to withhold treatment that might prolong life. Active euthanasia consists of decisions and actions that actively shorten a person's life—for example, injecting a lethal substance to terminate a patient's life. The Michigan physician, Dr. Jack Kevorkian, has been engaged in active euthanasia and is currently serving time in jail for his actions.

FIGURE 1

A stimulates the growth of new blood vessels in the dermis. This, in turn, may nurture the regeneration of damaged skin cells. Retin-A also detoxifies oxygen free radicals in tissues and can inhibit the destruction of collagen fibers in the dermis.

Despite the discovery of stomatin and Retin-A, there is as yet no evidence that medical scientists can prevent or even retard the overall rate of aging. The best route to a long and healthy life is to live it well: Learn ways to reduce stress, eat well, exercise regularly, get enough sleep, take alcohol in moderation or not at all, and avoid harmful practices such as smoking.

www.jbpub.com/humanbiology/5e

Visit Human Biology's Internet site for links to web sites offering more information on this topic.

In 1988, medical researchers found that Retin-A, a derivative of vitamin A used to treat severe cases of acne, also reduces wrinkles, age spots, and roughness as long as it is applied regularly. Studies show that Retin-

Opinions on euthanasia vary widely, and the controversy will, no doubt, be debated for many years. (For a debate on physician-administered euthanasia, see the Point/Counterpoint in this chapter.) As a final note, individuals who want to save their relatives emotional and legal turmoil can sign a living will. This is a legal document stipulating conditions under which doctors should allow a person to die.

21-8 Health and Homeostasis

Reproduction has nothing to do with maintaining homeostasis, but homeostasis is absolutely essential to normal reproduction, as this case study on the reproductive effects of computer monitors suggests.

Computer monitors, or video display terminals (VDTs), produce numerous types of radiation with frequencies ranging from X-rays to radio waves. Fortunately, manufacturers have installed protective shields that prevent most of this radiation from escaping and bombarding the user. It is the lower-frequency radiation (radio frequencies) that has people concerned. Why?

Laboratory experiments show that electromagnetic fields generated by extremely low-frequency radiation can alter fetal development in chickens, rabbits, and swine. In addition, researchers in Oakland, California, recently published results of

a medical study on the incidence of miscarriage in nearly 1600 women. The study showed that women who sit in front of a computer monitor for more than 20 hours a week during the first 3 months of pregnancy are nearly twice as likely to miscarry as women in similar jobs who do not use computers.

Researchers suspect that miscarriages result from radiation emitted from the VDTs, but believe that other factors may also be involved—for example, the stress of using a computer. Preferring to err on the conservative side, some scientists advise pregnant women to minimize their exposure to radiation from computer monitors. Should the link between radiation from VDTs and miscarriage be substantiated by further research, it would provide yet another link in the growing case for the importance of a healthy environment to human health.

Physician-Assisted Euthanasia

The Right to a Physician Who Will Not Kill by Rita Marker

Should physician-administered euthanasia—the direct and intentional killing of a patient by a physician—be legalized? Common sense has always said NO. And laws have wisely banned euthanasia.

Now, however, an effort is underway to change these laws. Those who seek to legalize euthanasia have framed their arguments in terms of personal rights and freedom. They've clearly recognized the importance of words in molding public opinion: carefully crafted verbal engineering is being employed to transform the appalling crime of mercy killing into an appealing matter of patient choice.

Proposals appearing in state after state have such benign titles as "Death With Dignity Act," and the deceptively soothing term "aid-in-dying" is often substituted for "euthanasia."

Rita L. Marker is director of the International Anti-Euthanasia Task Force and the author of *Deadly Compassion*.

As efforts of euthanasia activists increase, careful examination of what is at stake has become increasingly important. This examination should include answers to key questions.

Who will be the recipients of new "rights" if euthanasia is legalized? The law would benefit doctors, not patients. Competent adults already have the legal right to refuse medical treatment as well as the right to make their wishes known about such interventions for the future. Additionally, neither attempting nor carrying out the very personal and tragic decision to commit suicide is a criminal offense. Bluntly put, people can refuse medical care, or can kill themselves without a doctor's help.

Doctors, however, are prevented by law from killing their patients. Legalization of euthanasia would give doctors the right to directly and intentionally kill patients. An action that is now considered homicide would become a "medical service."

What about safeguards? Efforts to legalize euthanasia have failed in Washington and California. Opponents pointed out the real dangers inherent in the proposals. Advocates, however, insisted that their measures contained sufficient safeguards. Yet, after each version failed, proponents acknowledged the lack of necessary safeguards. It's as though each and every attempt to legalize euthanasia is treated as a dress rehearsal. Yet we're not dealing with a play or musical event. A "learn as you go" attitude in law and public policy could lead to the deaths of millions of vulnerable people.

What changes in law are now being promoted? Recognizing that the dangers of their initial proposals were too apparent, euthanasia proponents have repackaged their agenda, calling for changes that would allow physician "assisted" death. Often referred to as "assisted suicide" measures, these new proposals would permit a physician to prescribe medication with the express purpose that it be used to end a patient's life. Such a change would lead to classification of such an action as a medical intervention. The reality would be simple: The physician would have prescribed medication to kill.

Logic and common sense make it readily apparent that "prescribing" encompasses far more than writing a prescription that is taken to the corner drug store. Medication prescribed for a patient in a hospital or care facility is often provided by health professionals. The current "change" allowing "only" for the death prescription is yet another deception intended to mask reality in meaningless safeguards.

What has happened where euthanasia has been practiced? One need only look to the experience in Holland for a glimpse of where the euthanasia road leads. Although it is widely practiced in Holland, euthanasia remains technically illegal. Yet, even with "safeguards," which are far tighter than those which have been proposed in the U.S., a 1991 Dutch government study released horrifying information. The study found that in tiny Holland—a country with only one-half the population of the State of California—Dutch physicians deliberately and intentionally end the lives of more than 11,000 people each year by lethal overdoses or injections. And more than half of those killed had not requested euthanasia.

What will happen if euthanasia is legalized? With increasing emphasis being placed on cost containment, individuals who are disabled are particularly vulnerable. Leaders in the disability rights community are becoming increasingly alarmed by suggestions such as that of Jack Kevorkian that their "choice" of death would "enhance public health and welfare."

Euthanasia proponents have been clear in describing their goals. For example, Derek Humphry has called for expansion of euthanasia to include those who are physically and mentally disabled. He has also written that there would be a means for handling the dilemma of "terminal old age" once euthanasia is legalized. Jack has outlined plans for designated death zones with special clinics where "planned death" could be carried out.

Those who think that euthanasia, once unleashed, can be controlled are making a deadly mistake. Euthanasia is not a means of dying with dignity, nor is it a matter of liberty. It does not legislate compassion. It only legalizes killing.

The Right to Choose to Die by Derek Humphry

If we are truly free people and if our bodies belong to ourselves and not to others, then we have the right to choose when and how to die. Death comes to us all in the end, although modern medicine can often help us improve and extend our life spans. Thus, given today's high technology and ruinous cost of medicine, it is wise if we all give advance thought to the manner of our dying. Such decisions should be transferred to paper (a living will and durable power of attorney for health care), because 46 states now legally recognize one's wishes in respect to the withdrawal of life-support equipment.

Advance-declaration documents, of course, deal only with the legal and ethical problems of medical equipment and treatments. Less than half of dying people are connected to such equipment. For many more patients, technology does nothing for their terminal illness, so—in effect—there is no "plug to pull." Therefore, the cutting edge of the right-to-die debate in the twenty-first century centers on assisted death and voluntary euthanasia.

Hemlock Society supporters feel not only that hopelessly sick people should not only have an unfettered right to accelerate their end to save them pain, distress, and indignity, but also that willing doctors should be able to help them die. [To clarify some terms: self-deliverance (suicide) is ending your own life to be free of suffering; assisted suicide is helping another die; euthanasia is the direct ending of another's life by request. Currently, suicide in any form, for any reason, is legal, but assisted suicide and euthanasia are technically crimes.]

Most people at the close of life do not wish to suffer, desiring a quick and painless demise. The sophisticated pain-management drugs now available, when properly administered, control some 90% of terminal pain. But they do nothing to alleviate the indignities, the psychic pain, and the loss of quality of life associated with some debilitating terminal illnesses.

A complaint that any hastening of the end interferes with God's authority can be answered for many through a belief that their God is tolerant and charitable and would not wish to see them suffer. Other people have no faith in God. Pious people differ with this view, of course, and I respect that.

Numerous opinion polls in the United States and other Western countries indicate that two-thirds of people want the right to have a doctor lawfully assist them to die. People in the states of Washington, Oregon, Maine, Michigan, and California have engaged in political action to achieve this law reform. The proposed law in these states was called the Death With Dignity Act, and the broad outline of its purpose is as follows:

1. The patient wanting physician-assisted dying would have to be a mature adult suffering from a terminal illness likely to cause death within about 6 months.

2. People with emotional or mental illness (especially depression) could not get help to die under this law.

3. The request would have to be in writing, and the signature would have to be witnessed by two independent persons.

4. The physician could decline to help the patient die on grounds of conscience but would then cease to be the treating physician. The patient could then seek a physician who was willing.

5. The family would have to be informed and its views, if any, taken into account. But the family could neither promote nor veto the patient's request to die.

6. The physician would have to be satisfied that the patient was fully aware of his or her condition, had been informed of all possible alternatives, and was competent to make this request.

7. If the physician was unsure of the patient's mental state, a mental-health professional could be called to make an evaluation.

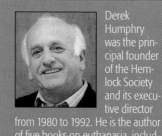

Derek Humphry was the principal founder of the Hemlock Society and its executive director from 1980 to 1992. He is the author of five books on euthanasia, including the best-seller *Final Exit* (1991). Web site: @www.finalexit.org:

8. The time and manner of the assisted dying would have to be negotiated between patient and physician, with the patient's wishes paramount.

9. At any time the patient could orally or in writing revoke the request for assisted dying.

10. Any person who pressured another person to get assistance in dying, who forged such a request, or who ignored a revocation would be subject to prosecution.

11. After helping the patient die, the physician would have to report the action in confidence to a state health agency.

The right-to-die movement feels that these conditions, plus others in the Death With Dignity Act too numerous to describe here, are an intelligent and humane approach to euthanasia with the necessary safeguards against abuse. Only Oregon has passed such a law.

Dying on one's own terms is not only an idea whose time has come, but also the ultimate civil and personal liberty.

Sharpening Your Critical Thinking Skills

1. Summarize the key points of each author and their supporting arguments.

2. Which viewpoint corresponds to yours? Why?

www.jbpub.com/humanbiology/5e

Visit Human Biology's Internet site for links to web sites offering more information on this topic.

SUMMARY

Fertilization

1. The sperm and egg of humans unite during fertilization, which usually takes place in the upper third of the uterine tube.
2. Sperm deposited in the vagina reach the site of fertilization partly on their own (by swimming) but mostly with the aid of muscular contractions in the walls of the uterus and uterine tube.
3. Sperm dissolve away the cells surrounding the oocyte and then bore through the zona pellucida and contact the plasma membrane. The first one to contact the membrane fertilizes the oocyte. Further sperm penetration is blocked. The chromosomes of the sperm and oocyte duplicate and merge in the center of the cell, where mitosis begins.
4. Human development is divided into three stages: pre-embryonic, embryonic, and fetal.

Pre-embryonic Development

5. During pre-embryonic development (from fertilization to implantation) the zygote undergoes rapid cellular division, forming a morula.
6. The morula is then converted into a blastula, a structure with a hollow cavity, called a blastocyst in humans. The blastocyst consists of a clump of cells, the inner cell mass (ICM)—which becomes the embryo, and the trophoblast, which gives rise to the embryonic portion of the placenta.
7. While the placenta is forming, a layer of cells from the ICM of the blastocyst separates from it and forms the amnion. The amnion fills with fluid and enlarges during embryonic and fetal development, eventually surrounding the entire embryo and fetus, protecting them during development.

Embryonic Development

8. The ICM differentiates into the three germ cell layers: ectoderm, mesoderm, and endoderm. The formation of the three primary germ layers marks the beginning of embryonic development. The organs develop from these three primary tissues during organogenesis. Table 22-2 lists the organs and tissues formed from each of the layers.

Fetal Development

9. Fetal development begins 8 weeks after fertilization and is primarily a period of fetal growth because most of the organ systems have developed or are under development.
10. The placenta produces several hormones that play an important part in reproduction. HCG maintains the corpus luteum during pregnancy. Progesterone and estrogen stimulate uterine growth and the development of the glands and ducts of the breast.
11. The placenta delivers oxygen and nutrients to fetal blood and removes waste products from it.
12. The fetal circulatory system is very much like the adult circulatory system, except that it contains three pathways that allow blood to bypass the lungs and liver, which are developing but not functioning. At birth these bypasses close up and the adult pattern of circulation develops.

Birth Defects

13. Birth defects arise from chemical, biological, and physical agents. The effect of these agents is related to the time of exposure, the nature of the agent, and the dose. A defect is most likely to arise if a woman is exposed during the embryonic period when the organs are forming.

Childbirth and Lactation

14. Labor consists of intense and frequent uterine contractions believed to be caused by the release of small amounts of fetal oxytocin prior to birth. Fetal oxytocin stimulates the release of prostaglandins by the placenta. Oxytocin and prostaglandins stimulate contractions in the sensitized uterine musculature.
15. Maternal oxytocin is also released, augmenting muscle contractions. As uterine contractions increase, they cause more maternal oxytocin to be released, which stimulates stronger contractions and more oxytocin release, a positive feedback that continues until the baby is born.
16. Labor consists of the dilation, expulsion, and placental stages.
17. The breasts consist primarily of fat and connective tissue interspersed with milk-producing glandular tissue and ducts.
18. During pregnancy, the glands and ducts proliferate under the influence of placental and ovarian estrogen and progesterone.
19. Milk production is induced by maternal prolactin. Before milk production begins, the breasts produce small quantities of a protein-rich fluid, colostrum. A newborn can subsist on colostrum for the first few days; the baby derives antibodies from colostrum that help protect it from bacteria.
20. Suckling causes a surge in prolactin secretion. Each surge stimulates milk production needed for the next feeding.

Aging and Death

21. Aging is the progressive deterioration of the body's homeostatic abilities and the gradual deterioration of the function of body organs. These changes result from at least two factors: a decrease in the number of cells in the organs and a decline in the function of existing cells.
22. Death results from aging, traumatic injury, or infectious disease.

critical thinking

Thinking Critically—Analysis

This Analysis corresponds to the Thinking Critically scenario that was presented at the beginning of this chapter.

Medical scientists measure longevity by a statistic called life expectancy at birth. Technically, it is the number of years, on average, a person lives after he or she is born. Life expectancy at birth has increased dramatically in the last century. In 1900, for example, on average, white American females lived only 50 years. Today, life expectancy is 81 years. For males, a similar trend is observed. In 1900, for example, the life expectancy of a white American male was 47 years; today, it is 74 years.

Most people take these statistics to mean that men and women are actually living to a much older age. In reality, something very different is happening: The increase in the life expectancy at birth is largely the result of declining infant mortality. That is, thanks to improvements in medicine, more children are living through the first year of life and this has dramatically increased life expectancy. In the early 1900s, 100 of every 1000 babies died during the first year of life. Today, that number has fallen to 12. The impact on average life expectancy is incredible.

This is not to say that all of the gain in life expectancy results from a decline in infant mortality. Medical advances have certainly increased life expectancy after infancy, but these changes are small in comparison to those brought about by lowering infant mortality. In fact, about 85% of the increase in life span in the last century is the result of decreased infant mortality.

KEY TERMS AND CONCEPTS

Aging, p. 414
Amnion, p. 404
Birth defect, p. 408
Blastocyst, p. 402
Braxton-Hicks contractions, p. 410
Breast, p. 412
Cesarean section, p. 412
Colostrum, p. 412
Ectoderm, p. 405

Embryo, p. 402
Embryonic development, p. 405
Endoderm, p. 405
Euthanasia, p. 416
Fertilization, p. 399
Fetal development, p. 406
Mesoderm, p. 405
Morula, p. 402

Organogenesis, p. 405
Ovum, p. 400
Placenta, p. 402
Pre-embryonic development, p. 402
Relaxin, p. 410
Umbilical cord, p. 407
Yolk sac, p. 406
Zygote, p. 400

SELF-QUIZ: TESTING YOUR KNOWLEDGE

1. Fertilization occurs in the upper one-third of the _____. p. 399
2. Enzymes in the head of the sperm in a structure called the _____ help it dissolve its way through the layers of cells and the zona pellucida surrounding the egg. p. 399
3. The fertilized egg or _____ divides and forms a solid ball of cells known as a _____. p. 402
4. The blastocyst is a hollow sphere of cells that gives rise to the embryo and the trophoblast, which forms the embryonic portion of the _____. p. 402
5. The blastocyst attaches to the lining of the uterus and digests its way into it; this process is called _____. p. 402
6. The placenta produces three hormones, HCG, estrogen, and _____. p. 404
7. The _____ is filled with fluid that protects the embryo and fetus from damage. p. 404
8. Organogenesis, the formation of organs, occurs primarily during _____ development. p. 405
9. An _____ pregnancy occurs when a fertilized ovum implants in the oviduct or inside the body cavity, not the uterus. p. 403
10. Milk production in the breast is stimulated by the hormone _____ from the pituitary gland. p. 412

CONCEPT REVIEW

1. Give an overview of the process of development in humans. Describe what happens during each stage. pp. 406–407
2. Why is the formation of endoderm, mesoderm, and ectoderm so important during embryonic development? pp. 405–406
3. Describe the process of fertilization in humans. Use drawings to elaborate your points, and label all drawings. pp. 399–401
4. Define the terms inner cell mass and trophoblast. p. 402
5. Where does the human embryo acquire nutrients before the placenta forms? p. 402
6. What are the major functions of the human placenta? p. 403
7. Describe the flow of blood from the placenta to the fetus and back. What is the role of the various bypasses? pp. 407–408
8. What are birth defects? What causes them? How can they be prevented? p. 408–410
9. Describe the factors that trigger labor. p. 410
10. What factors are responsible for cervical dilation during labor? p. 411
11. Define the term aging, and describe some of the hypotheses that attempt to explain it. p. 414
12. Thanks to modern medicine, men and women are living longer, says a friend. Using your understanding of average life expectancy data and the reason for an increase in life expectancy in the last century, explain why this statement is not quite true. p. 420

 www.jbpub.com/humanbiology/5e

The site features eLearning, an online review area that provides quizzes, chapter outlines, and other tools to help you study for your class. You can also follow useful links for in-depth information, research the differing views in the Point/Counterpoints, or keep up on the latest health news.

Human Evolution

thinking critically

The timing and origin of the emergence of modern humans is a matter of considerable debate among scientists. Two basic hypotheses exist. The first, the Multiregion Evolution model, says that modern humans derive from small populations living in many regions of Europe, Africa, China, and Indonesia. These populations gave rise to present-day populations. Thus, modern Chinese came from an ancient Chinese population, and so on.

The second hypothesis holds that although archaic regional populations existed, they were displaced by a single population of humans that arose in Africa approximately 200,000 years ago. This new species supposedly spread into new territories, replacing regional human populations.

The second hypothesis, the African Origins model, is based on genetic studies of DNA in mitochondria. In these analyses, scientists have found that only the African population exists in an unbroken line. Genetic lineages of archaic peoples (regional populations) seem to have faded into oblivion.

Using your critical thinking skills, make a list of questions that might be useful in analyzing the African Origins model.

Scientists believe that the atoms that make up both the living world and the nonliving world (rocks, for example) once existed in outer space as a giant cloud of cosmic dust and gas. About 4.5 billion years ago, this huge cloud began to condense. No one knows what triggered the condensation, but some think that a nearby exploding star (supernova) may have been the catalyst. As the huge cloud condensed, scientists speculate, dust particles in the center compacted more rapidly than outlying particles. This region would eventually give rise to the sun. Scientists hypothesize that when the mass of the forming sun reached a critical density, heat and pressure began to cause hydrogen and helium in it to chemically combine to form larger atoms. This process, known as *fusion*, not only forms larger atoms, it also releases large amounts of energy.

This energy is found in many forms, including heat and light. The process continues today in our sun.

Cosmic dust lying outside the forming sun also condensed, creating the planets (Greek for "wanderers"). The third planet from the sun was the Earth, the only piece of real estate in our solar system that supports life.

When it originated, the Earth was a solid mass of rock and ice. Scientists believe that over time, radioactive decay, intense solar heat, and other sources of heat caused the Earth to melt, turning it into a mass of molten material. As the Earth grew hotter, water and the lighter elements such as hydrogen escaped into the atmosphere.

In the millennia that followed, the Earth gradually cooled. A thick, rocky crust formed around a molten core. As the Earth cooled, water in the atmosphere began to rain down from the skies, creating lakes and oceans. Today, the oceans cover nearly 70% of the planet. It is in these oceans, or on their shores, that many scientists believe life began.

In this chapter we will trace the emergence of life on Earth and study evolution. We'll conclude by looking at human evolution.

23-1 The Evolution of Life

In the most general terms, **evolution** is a process in which existing life-forms change in response to changing environmental conditions. In some cases, they change so dramatically that they produce entirely new species.

The evolution of life on Earth can be divided into three phases. During the first phase, organic molecules formed. During the second phase, cells evolved. And during the third phase, multicellular organisms arose. This section outlines the key events in each phase.

Chemical Evolution

The formation of organic molecules from inorganic molecules is chemical evolution.

When they first formed, the seas were lifeless bodies of water. The land masses, so richly carpeted today with plants, were barren rock. How could life have formed from such an unpromising start?

Research suggests that **chemical evolution** probably began about 4 billion years ago—or 500–600 million years after the Earth formed. At that time, the Earth's atmosphere contained a mixture of simple chemicals such as water vapor, methane, ammonia, and hydrogen. Water vapor condensed to form rain and washed many of these molecules from the sky.

In the shallow waters of the newly formed seas, sunlight, heat from volcanoes, or lightning energized these molecules. This energy caused them to react with one another, producing a variety of simple organic molecules, including simple sugars (monosaccharides) and amino acids.

According to the theory of chemical evolution, these organic molecules, in turn, began to react with one another. In the process, they formed small polymers—large molecular weight molecules consisting of many smaller molecules. As a result, primitive proteins and simple RNA or DNA molecules formed. Slowly but surely, over tens of thousands of years, all of the organic molecules necessary for life began to emerge.

But that's not the end of the process. Scientists believe that polymers, in turn, combined to form the very first, albeit primitive, cells.

Evidence for Chemical Evolution Theory

The theory of chemical evolution is supported by several convincing experiments.

Although the theory of chemical evolution may seem wild, many studies have been performed over the years demonstrating that it is possible. The first direct evidence in support of the theory of chemical evolution came from an American graduate student, Stanley Miller. While studying for his Ph.D. in chemistry at the University of Chicago in the early 1950s, Miller devised an apparatus to test part of the theory (Figure 23-1). To his closed, sterilized glass apparatus, Miller added three gases thought to have existed in the Earth's primitive atmosphere: methane (CH_4), ammonia (NH_3), and hydrogen (H_2). A sparking device simulated lightning and provided energy. Boiling water created steam that circulated through the device, carrying with it the reactant gases and the products of the reactions occurring inside.

Several days later, Miller found that the water in the apparatus had turned brown. A chemical analysis of the liquid revealed the presence of several biologically important organic compounds, including amino acids, urea, and lactic acid (Table 23-1). Other researchers soon found that a variety of organic molecules, including the building blocks of DNA and RNA, could be created in conditions similar to those believed to be present early in the Earth's history. Further experiments showed that these smaller molecules spontaneously assembled into long-chain molecules, giving birth to proteins, polysaccharides, DNA, and RNA.

TABLE 23-1	Chemical Products Produced Abiotically in Miller's Experiment	
Glycine	Glutamic acid	Urea
Alanine	Acetic acid	Succinic acid
Aspartic acid	Formic acid	Aldehyde
Butyric acid	Lactic acid	Hydrogen cyanide

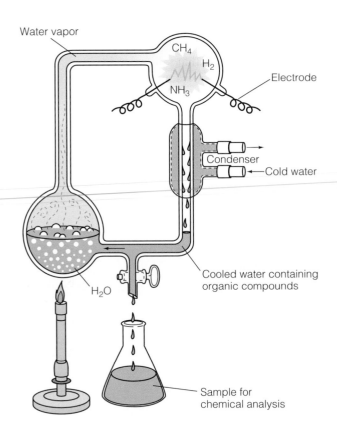

FIGURE 23-1 **Miller's Apparatus** This device showed that organic molecules could be produced from the chemical components of the Earth's early atmosphere.

The First Cells

The earliest cells were probably aggregations of polymers that acquired nutrients directly from the environment.

Scientists believe that the earliest cells were simple aggregations of the polymers and lipids formed in water. (Lipids and proteins may have formed a primitive cell membrane.) Scientists also hypothesize that some of the proteins and perhaps even some of the RNA molecules in these primitive cells, called **protocells**, may have served as primitive enzymes. These enzymes may have allowed these primitive cells to synthesize some of their own molecules and break down others to generate energy. In other words, they may have made metabolism possible. Researchers also hypothesize that small molecules of DNA that formed during chemical evolution may have been incorporated into protocells, providing a primitive mechanism of heredity.

Many biologists believe that the protocells probably received nourishment from organic molecules such as glucose that they absorbed from their environment. As such, these organisms were **heterotrophs** (literally, nourished by others). They may have evolved simple mechanisms to acquire energy through fermentation (the breakdown of carbohydrates in the absence of oxygen) since oxygen was not yet present.

Over time, these bacterialike heterotrophs gave rise to a group of single-celled organisms capable of making their own food. Appropriately called **autotrophs** (self-nourish), they very likely produced organic molecules using energy from inorganic molecules in the environment.

Later in evolution, another group of autotrophs arose. They synthesized their own food from carbon dioxide, water, and so-

lar energy through a primitive form of photosynthesis. Although early forms of photosynthesis did not release oxygen, in time, as the process became more complex, oxygen began to be created.

As populations of these oxygen-releasing organisms grew, oxygen levels in the water and atmosphere increased. Many of the anaerobic organisms either perished or retreated to oxygen-free environments such as the mud or sediment beneath lakes and oceans. Many of their descendants (the anaerobic bacteria) remain in such locations today. Still others probably evolved mechanisms that allowed them to live in an oxygen-rich environment. These organisms formed the aerobic (oxygen requiring) bacteria that are so common today.

The Emergence of Eukaryotes

Eukaryotes probably arose by endosymbiotic evolution.

All of the organisms up to this point in evolution belonged to a group known as *prokaryotes*. **Prokaryotes** are cells that lack nuclei.

The prokaryotes gave rise to an entirely new group of organisms made up of one or more nucleated cells. These cells are known as **eukaryotes**. The very first eukaryotes, which evolved about 1.2 billion years ago, were single-celled organisms like the ameoba. Not only did they have nuclei, they also developed numerous organelles, distinct structures that carry out functions of the cell. (For a timetable of these events, see Figure 23-2.)

Evolution of Plants and Animals

Eukaryotes gave rise to multicellular organisms.

Over many millions of years, eukaryotic organisms gave rise to a variety of multicellular plants and animals. They evolved first in the oceans, protected from ultraviolet radiation from the sun. Over time, oxygen produced by photosynthesis created the ozone layer, which blocked harmful ultraviolet radiation. This, in turn, allowed life to evolve on land.

FIGURE 23-2 Summary of the Evolution of the Earth and Life

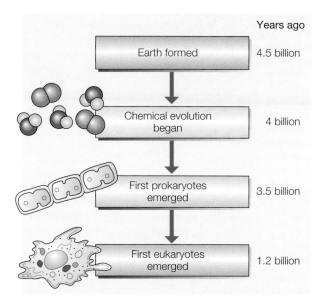

	Years ago
Earth formed	4.5 billion
Chemical evolution began	4 billion
First prokaryotes emerged	3.5 billion
First eukaryotes emerged	1.2 billion

The first species to invade the land from the oceans were very likely plants. They thrived in moist environments. Soon after the land plants invaded, animals came ashore. (The animals now had a food source, so terrestrial life was possible.) The very first of these animals were probably scorpionlike creatures. In time, millions of new plant and animal species evolved.

23-2 How Evolution Works

Many new life-forms have emerged over several billion years of the Earth's history. And many have changed, or evolved. The theory of evolution explains the emergence of new life-forms and changes in existing life-forms and is quite different from a once competing theory, discussed in Scientific Discoveries that Changed the World 23-1.

Genetic Variation

Genetic variation is the raw material of evolution.

During evolution, three types of changes have occurred in species: structural (anatomical), functional (physiological), and behavioral. How do such changes occur?

These changes occur because of changes in the genetic material of populations. More specifically, they occur because of changes in the frequency of genes in populations of organisms. Evolutionary biologists have identified a number of different processes that are responsible for changes in gene frequency.

Basically, though, here's how evolution works. In all populations, organisms differ from one another. These differences in structure, function, and behavior are caused by genetic variation—slight differences in the genes of organisms. These variations may result from mutations that occur naturally and randomly. Mutations occur in the body cells (somatic cells) and germ cells. In the evolution of multicellular organisms, somatic mutations are meaningless. They cannot be passed to future generations. It is germ cell mutations that are of importance.

Beneficial germ cell mutations, however, produce characteristics in an organism's offspring that may give it an advantage over other members of the population. Genetically based characteristics that increase an organism's ability to survive and reproduce and thus pass on its genes—are called **adaptations**. For example, a random mutation in the germ cell of a fish may make its offspring more efficient in catching prey. This, in turn, increases the likeli-

hood that their offspring will survive and reproduce. Members of the same population that do not share this trait are less likely to survive and reproduce and will very likely produce fewer offspring. Over time, the better adapted fish will leave a larger number of offspring. These offspring are also more likely than others to survive and reproduce. Over time, their success results in a shift in the gene pool so that future populations contain proportionately more offspring of the fish with the beneficial mutation. On a genetic level, then, the gene that gave some fish an advantage over others increased in frequency in the population.

Beneficial mutations, which produce genetic variation in populations, are often called the raw material of evolution. Genetic variation comes from other processes as well. For example, when a sperm and egg unite, the genes from the parents combine to produce a new offspring that may enjoy an advantage over other members of the population. If the new combination of genes produces advantageous adaptations, the genes may increase in frequency in the population.

Natural Selection

Natural selection is a process by which organisms become better adapted to their environment.

Sociologist Andrew Schmookler once wrote that "evolution employs no author, only an extremely patient editor." What he meant is that evolution is not directed. There is no author. Instead, genetic variation arises through mutations and sexual reproduction. Mutations and sexual reproduction may produce adaptations that may persist over time, creating organisms better suited to a particular environment. In some cases, new species may evolve.

The shifts in gene pools are brought about by evolution's patient "editor," in a process called **natural selection**. **Charles Darwin**, a nineteenth-century British naturalist, and **Alfred**

Scientific Discoveries that Changed the World

23-1 Debunking the Notion of the Inheritance of Acquired Characteristics

The eighteenth-century French naturalist Jean-Baptiste Lamarck argued that life had been created in a relatively simple state. Over time, he said, life-forms gradually improved as a result of an innate drive for perfection. He further hypothesized that this inherent drive was centered in nerve cells and that these cells released a "fluida" (chemical substance) that traveled to different body parts needing improvement.

Lamarck's explanation of the evolution of the giraffe illustrates his hypothesis. According to Lamarck, the giraffe of years past was a short-necked animal. Over time, the giraffe's neck lengthened as a result of the simple act of stretching of its neck—a feat required to feed on leaves that were out of the reach of other animals. The act of stretching, said Lamarck, directed flu-

ida to the neck. This stimulus made the neck grow longer. What is more, the slightly stretched neck that an adult giraffe acquired was then transmitted to its offspring. The offspring, in turn, stretched their necks to reach food, causing further elongation. According to Lamarck then, generation after generation of giraffes, each stretching to find food, led to the modern giraffe.

Lamarck proposed an evolutionary theory based on the "use and disuse" of organs. It stated that an individual acquired traits during its lifetime and that such traits were in some way incorporated into the hereditary material and passed to the next generation. This explained how a species could change over time. Conversely, not using an organ could lead to its disappearance.

The inheritance of acquired characteristics was doubted by many of Lamarck's contemporaries, in large part because he failed to support his hypothesis with observations or experiments. Lamarck's concept is based on the mistaken belief that all of the organs of the parents produce hereditary factors that form corresponding parts in the offspring. Thus, any change that occurred in an organ (an acquired characteristic) would result in the production of a hereditary factor that would reflect the change. This hered-

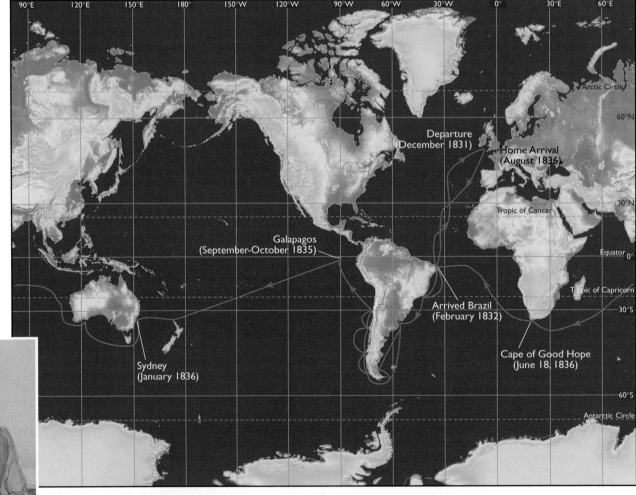

(b)

FIGURE 23-3 "Evolutionary Voyage" (a) Darwin as a young man proposed the theory of evolution by natural selection, helping solve one of the key puzzles of biology: how species evolve. (b) From 1831 to 1836, Darwin made a famous voyage on the *Beagle*, collecting and cataloging thousands of diverse species.

(a)

itary factor would somehow then be transmitted to the sperm of the male or egg of a female. Thus when a new offspring was formed, it would have the acquired characteristics.

After the discovery of fertilization, it was assumed that the organs transmitted parcels of hereditary information to the gametes. In this way, acquired traits were passed on to the offspring.

August Weismann, a German scientist in the late 1800s, attempted to bring some sense to the debate in a series of extraordinary essays. Although Weismann did not perform experiments to disprove Lamarck's notions, his essays were based on the research of others. Based on these studies, he argued that the germ cells contained the hereditary information responsible for the transmission of traits from one generation to the next. He also argued that the cells of an organism did not dispatch small hereditary particles to germ cells. His ideas helped debunk Larmarck's erroneous views of genetics and evolution.

Twenty years later, two scientists, W. E. Castle and John C. Phillips, published the results of studies that helped to support Weismann's work. They transplanted ovaries from black guinea pigs (homozygous dominant) into albino guinea pigs (homozygous recessive) whose ovaries had been removed. Later, the albino females were bred to albino males.

Previous studies had shown that matings between albino males and females resulted in 100% albino offspring. Therefore, if acquired characteristics were inherited, Castle and Phillips argued, the offspring of these matings should be white, having acquired their coloration from the host. When bred to an albino male, however, the females produced only black offspring. The young, they said, "are such as might have been produced by the black guinea pig herself, had she been allowed to grow to maturity and been mated with the albino male used in the experiment."

Lamarck did have many good ideas—for instance, that species changed over time and that the environment was a factor in this change. He also contributed to the science of biology in other ways. For example, he was the first scientist to distinguish between animals with backbones (vertebrates) and those without backbones (invertebrates). Unfortunately, he's best known for his erroneous concepts of inheritance and evolution.

Wallace independently proposed the idea of natural selection. Darwin described it as a process in which slight variations, if useful, are preserved (Figure 23-3). Thus, natural selection is a process by which organisms often become better adapted to their environment.

Two principal factors contribute to natural selection: **biotic factors**, or other living organisms, and **abiotic factors**, the physical and chemical environment (temperature, rainfall, and so on). Abiotic and biotic factors influence survival and reproduction by "selecting" the fittest—those best able to reproduce and pass on their genes. If these conditions change, those organisms best adapted to the new conditions tend to remain and pass their genes on to subsequent generations. This process causes a shift in the frequency of certain genes in a population.

In Darwin's time, evolution was widely discussed among naturalists and other scientists, but the mechanism by which it occurred remained a mystery. Darwin dedicated many years to the search for an answer, even traveling around the world by ship and cataloging species (Figure 23-3b). It was on this lengthy voyage that Darwin reportedly came up with the idea of natural selection.

A careful scientist, though, Darwin spent the next 20 years looking for flaws in his own reasoning. In 1858, much to his surprise, Darwin received a paper from Wallace, a respected naturalist who had proposed the same concept. Darwin sent Wallace's paper to some of his colleagues and suggested that they pub-

lish it. Fortunately, Darwin's colleagues, who were aware of his own theory, prevailed on him to write a paper of his own. In 1858, Wallace's and Darwin's papers were both presented to the Linnaean Society. In 1859, Darwin published his now-famous book, *On the Origin of Species by Means of Natural Selection*.

As in the case of Mendel, it took many years for Darwin's and Wallace's ideas to be understood and accepted. Not until the 1940s, about 80 years after the men presented their ideas to the world, did natural selection become widely accepted. Today, it is one of the central tenets of biology.

Survival of the Fittest

Natural selection ensures that the fittest organisms in a population survive and reproduce.

Darwin used the phrase "survival of the fittest" to describe how natural selection worked. Survival of the fittest is commonly interpreted as "survival of the strongest." To a biologist, however, **fitness** is a measure of reproductive success, which can be attributed to many different features, not just strength. Fitness does not necessarily result from speed, either. The ability to hide, digest food more effectively, or use water more efficiently can be just as important, if not more important, than standard measures of strength.

23-3 The Evidence Supporting Evolution

Darwin's theory of evolution by natural selection can be summarized as follows. Natural variations exist in all species. Those inherited (genetic) variations that provide an advantage to certain members of a species survive and reproduce. Natural selection determines which of these organisms survive and reproduce. While evolution is commonly referred to as a theory, like cell theory and many other key tenets of modern biology, there is a great deal of evidence that supports it.

The Fossil Record

The fossil record yields some of the best supporting evidence for evolution.

One of the most important sources of information on evolution is the fossil record. **Fossils** consist of the remains of organisms that lived on the Earth, such as the bones of dinosaurs, shells, or teeth. Impressions are also part of the fossil record. For example, some creatures such as dinosaurs left footprints in the mud that hardened into stone (Figure 23-4a). Imprints of leaves make up part of the fossil record (Figure 23-4b). Some ancient organisms (frogs, insects, and flowers) were preserved in amber, resins released by ancient trees (Figure 23-4c).

The fossil remains of species no longer in existence provide evidence that supports the theory of evolution. Because fossils and the rocks they come from can be dated using radioactive techniques, scientists can determine when different species lived.

Studying when a species lived and tracking the types of life-forms present at different times during the Earth's history have permitted scientists to piece together an evolutionary history of the planet. These studies show that the oldest rock contains relatively simple single-celled organisms and that successively younger rock houses fossils representing increasingly more complex life-forms.

To date, scientists have discovered fossils belonging to about 250,000 species, most of them from rock formed within the last 600 million years. In this still-growing record, some lineages are nearly complete.

The fossil record not only shows the changes that occur to a particular animal—for example, how horses evolved—they also suggest the origin of new species. Studies of the fossil record show that, during evolution, a primitive fish probably gave rise to the amphibians. Amphibians gave rise to the reptiles. Birds and mammals evolved from reptiles.

(a)

(b)

(c)

FIGURE 23-4 **Fossils** (a) Dinosaur tracks made in a Texas streambed about 120 million years ago. (b) Imprint of a leaf. (c) Insect embedded in amber about 40 million types ago.

Homologous Structures

Common anatomical features in different species support the theory of evolution.

Today's organisms are adapted to a wide range of conditions and exhibit a dazzling diversity of appearance. Despite the diversity of life, many organisms exhibit similar anatomical features. Structures thought to have arisen from common ancestors are known as **homologous structures**. The presence of homologous structures among life's diverse forms represents yet another piece of evidence in support of evolution. Figure 23-5 illustrates some homologous structures: wings, flippers, arms, and legs. As shown, the bones in the wing of a bird and bat, the flipper of a whale, the arm of a human being, and the leg of a horse and cat are quite similar, even though these appendages perform quite different functions. Similarities such as these correspond to what you would expect if birds and mammals evolved from a common ancestor.

Biochemical Similarities

The common biochemical makeup of organisms supports the theory of evolution.

Organisms as distantly related as roses and rhinos are made of the same basic biochemicals—ATP, DNA, RNA, and protein. All living organisms store genetic information in DNA, and all use ATP to collect, store, and transport energy. In addition, many biochemical pathways in organisms (such as those involved in producing energy) bear a remarkable similarity.

To many biologists, the common biochemistry of the Earth's diverse organisms lends further support to the notion of a common ancestry and thus to the theory of evolution. (To creationists, it supports the notion of a designer using similar blueprints.)

Although many species have similar enzymes and other proteins, differences do exist. Hemoglobin, for example, is found in vertebrate red blood cells, that is, the red blood cells of animals with backbones. However, differences in the composition of hemoglobin are common among vertebrates. These differences have proven quite useful to researchers. By comparing the amino acid sequence of certain proteins (such as hemoglobin) from different species, evolutionary biologists can determine how closely related they are. Chimpanzees and humans, for instance, share about 98% of the same genes, suggesting that they are indeed very closely related. The more differences there are, the more distant the evolutionary relationship.

Analysis of the amino acid sequence of proteins and the composition of the DNA has enabled evolutionary biologists to create their own evolutionary trees. Not surprisingly, the evolutionary trees they have developed correspond well to those created by comparative anatomists.

Biochemical analyses of DNA and protein have recently been expanded to extinct organisms (museum specimens) and even fossils. DNA extracted from the cells of extinct species can be cloned via genetic engineering. This process provides sufficient amounts of DNA to determine its sequence and make comparisons with other organisms, both living and dead, thus allowing evolutionary biologists to study the relationships among many more species.

Embryologic Development

Studies of common developmental patterns support the theory of evolution.

Studies show that the embryos of many different groups of organisms develop similarly. Thus, the embryos of chickens, turtles, mice, fishes, and humans—all vertebrates—bear a re-

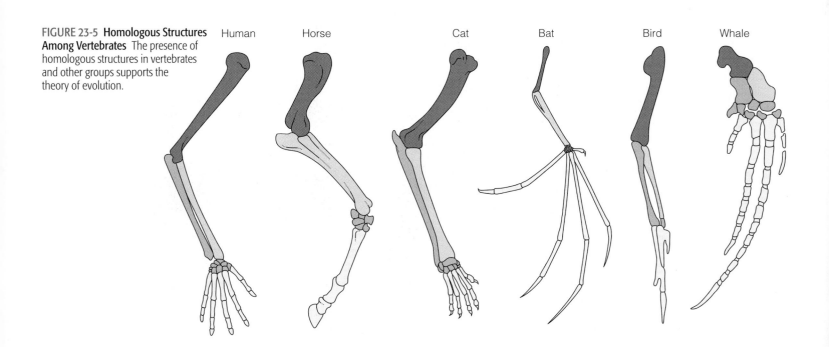

FIGURE 23-5 **Homologous Structures Among Vertebrates** The presence of homologous structures in vertebrates and other groups supports the theory of evolution.

Human   Horse   Cat   Bat   Bird   Whale

markable resemblance during their early stages of development. For example, each of them has a tail and gill slits, even though only adult fishes and amphibian larvae have gills.

One plausible explanation for the similarity of vertebrate embryos is that all of these species contain the genes that control the development of tails and gills and that these genes were passed on from a common ancestor. In humans, however, these genes are active only during embryonic development. The structures they produce either become inconspicuous, as in the case of the tail (all that is visible is the tailbone), or become other structures, as in the case of the tissue lying between the gill slits. It forms various structures in the neck.

Experimental Evidence

Experimental evidence supports the theory of evolution by natural selection.

A long list of experiments supports the theory of evolution by natural selection. An exhaustive series of field and laboratory studies, for example, shows how genetic differences in wild populations of fruit flies can be attributed to natural selection.

Biogeography

The study of biogeography helps explain evolution.

Biogeography is the study of the distribution of plants and animals throughout the world. It seeks to understand why certain species occur in specific places and not in others. Why, for example, are the egg-laying mammals found in Australia, Tasmania, and New Guinea and nowhere else in the world? It also seeks to understand why are very similar species found in distant lands.

The answer lies in understanding that 195 to 240 million years ago, all the continents were part of a very large land mass called *Pangea* (Figure 23-6a). These land masses split, however, and separated, migrating to new positions, creating distinct continents (Figure 23-6 b and c). That's because continents are part of large plates in the Earth's crust, called *tectonic plates*, that are capable of moving around as a result of activity in the Earth's molten core. Tectonic plate movement continues today and is responsible for earthquakes and volcanic eruptions.

Tracing the origin of continents helps scientists understand why two very similar species exist in different parts of the world—for example, South America and Africa. You can see from Figure 23-6 that these two continents fit together like two pieces of a puzzle, having been joined in Pangea. This explains why two similar species exist today in these very distant continents.

240 MILLION YEARS AGO
PANGEA

70 MILLION YEARS AGO

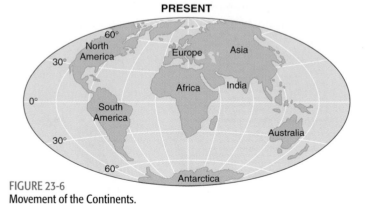

PRESENT

FIGURE 23-6
Movement of the Continents.

23-4 Early Primate Evolution

Primates evolved from an insect-eating mammal.

Humorist Will Cuppy once quipped that "all modern men are descended from a wormlike creature, but it shows more in some people." In reality, humans belong to a group or, more correctly, an order known as the *primates*. The primates are believed to have evolved not from worms, but from an insect-eating mammal that probably resembled the modern-day tree shrew of Southeast Asia (Figure 23-7).

To understand human evolution, we need to briefly examine the evolution of primates, beginning with the mammalian insectivores.

As shown in Figure 23-8, primates evolved from an insect-eating animal that resembled the modern day tree shrew. This an-

FIGURE 23-7 **Look Familiar?** An organism resembling this tree shrew is believed to have been the early ancestor of the primates.

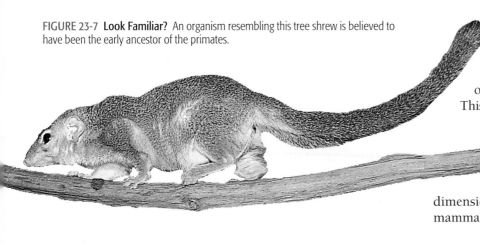

imal gave rise to a group called the **prosimians**, the very first primates to inhabit the Earth. Today, the prosimians consist of tree shrews, lemurs, and tarsiers. Figure 23-9 shows a tarsier and a lemur. During the course of evolution, the prosimians gave rise to the **anthropoids**. This group includes monkeys, apes, and humans. As shown, humans evolved from an apelike animal.

Primates are characterized by several features, including grasping hands, which permit them to pick up objects, and forward-directed eyes, which enhance three-dimensional vision. Primates also have the largest brains of all mammals in proportion to their body size.

FIGURE 23-8 **The Evolution of Primates** (adapt from Mader)

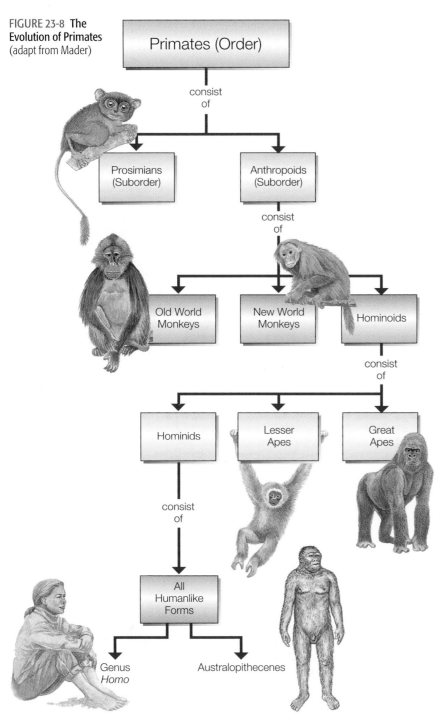

FIGURE 23-9 **Prosimians** The prosimians (premonkeys) were probably the first primates. The earliest ones probably resembled modern-day (a) tarsiers and (b) lemurs.

Fossil Evidence

The fossil evidence of primate evolution is limited, making it difficult to determine the exact progression of early human ancestors.

Unlike the dinosaurs, early primates did not live in habitats conducive to fossil preservation. Whereas the mud in swamps or along rivers preserved the bones of many a giant dinosaur, the forests and grasslands in which primates lived were not environments where bones of a dead animal would be readily covered with sediment and preserved. Thus, the primate fossil record is less than complete. Another reason for the scarcity of early primate fossils is that dead or dying animals were probably eaten by scavengers, and their bones no doubt were scattered across the landscape and even eaten.

Because the primate fossil record is incomplete, the origin of primates remains speculative. But this has not deterred many scientists from making tentative conclusions about primate evolution. The scheme presented here represents a synthesis of current thinking on the evolution of primates.

23-5 Evolution of the Australopithecines

The earliest hominids were members of the genus *Australopithecus*.

Archaeological evidence suggests that the first **hominids**, the first humanlike primates, belonged to the genus *Australopithecus* (awe-STRAY-loh-PITH-eh-cuss; meaning "southern apeman") (Figure 23-10). They evolved, as just noted, from an apelike ancestor.

The oldest known australopithecine skeleton was unearthed in Africa and is about 3.2 million years old. The specimen named Lucy is one of the most complete fossils of early hominids yet found (Figure 23-11). Known as *Australopithecus afarensis* (*afarensis* means from the Afar region of Ethiopia), these hominids stood only about 3 feet high and had brains only slightly larger than those of apes. The skulls of *A. afarensis* have many apelike features, including massive brow ridges, low foreheads, and forward-jutting jaws. The shape of its pelvis suggests that *A. afarensis* walked erect.

About 3 million years ago, *A. afarensis* vanished and was replaced by another species, *A. africanus*, which lived for about 1.5 million years (Figure 23-12). *A. africanus* was slightly taller than its predecessor and had a slightly larger brain.

Studies of the fossil record suggest that about 2.3 million years ago, another species of *Australopithecus* emerged, *A. robustus*—so named because it was heavier and taller and had a larger brain than *A. africanus* (Figure 23-10). About 2.2 million years

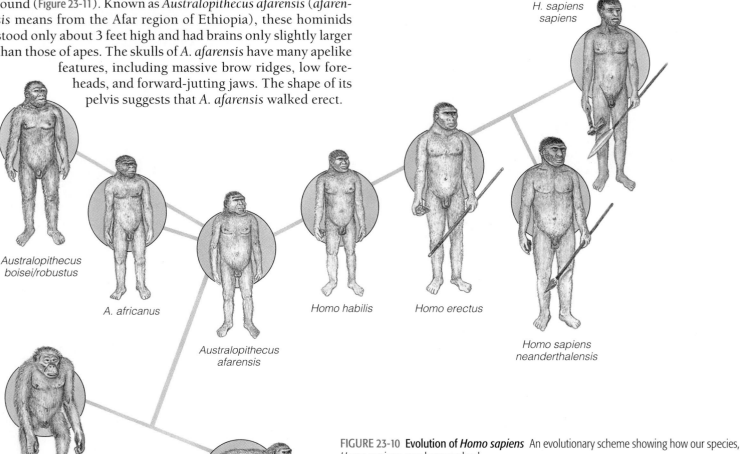

H. sapiens sapiens

Australopithecus boisei/robustus

A. africanus

Australopithecus afarensis

Homo habilis

Homo erectus

Homo sapiens neanderthalensis

Modern apes

Dryopithecus

FIGURE 23-10 Evolution of *Homo sapiens* An evolutionary scheme showing how our species, *Homo sapiens*, may have evolved.

FIGURE 23-11 **Skeleton of *Australopithecus afarensis*** This skeleton, estimated to be about 3.5 million years old, is popularly known as Lucy.

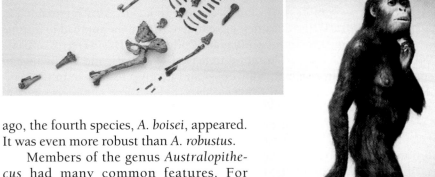

ago, the fourth species, *A. boisei*, appeared. It was even more robust than *A. robustus*.

Members of the genus *Australopithecus* had many common features. For example, they were all bipedal (buy-PED-al)—that is, they walked on two legs—and they all ranged in height from 1 meter to 1.7 meters (3.3 to 5.6 feet). Their brains were larger than those of chimpanzees but considerably smaller than those of modern humans. The differences among the four species are relatively minor, mostly a

matter of degree. As the genus evolved, their brains got larger, and their height increased. The enlarged brain is believed to be the result of natural selection.

Figure 23-12 indicates when the various species of *Australopithecus* lived. As illustrated, anthropologists believe that all three "younger" species coexisted, at least for a while, during the time span ranging from 1 million to 3 million years ago. Over 1 million years ago, however, australopithecines disappeared. Why they vanished, no one knows.

FIGURE 23-12 **Australopithecine Evolution** Conventional thinking holds that the genus *Homo* evolved from *Australopithecus afarensis*.

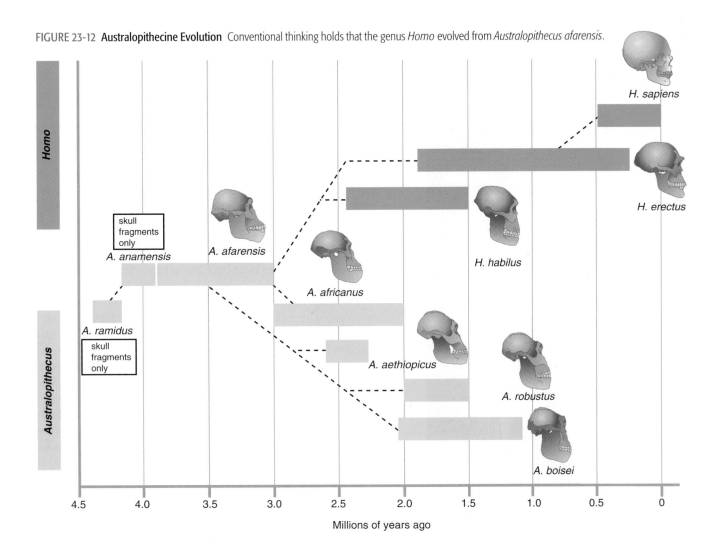

23-6 Evolution of the Genus *Homo*

The first truly humanlike organisms were called *Homo habilis.*

The earliest evidence of the genus *Homo* is a 1.8 million-year-old skull and partial skeleton found in Tanzania in 1960 by Mary Leakey. She and her husband, Louis, called this species *Homo habilis* ("skillful man"). Archaeological evidence indicates that *Homo habilis* apparently made primitive tools from fractured rocks and butchered large animals.

What was *Homo habilis* like physically? The skull of *Homo habilis* is like that of *Australopithecus*, suggesting an evolutionary relationship. However, the brain of *Homo habilis* is 50% larger than its presumed australopithecine relatives. The skeleton of *H. habilis* displays many apelike features, especially the relatively long arms and small body. Because of this, some paleontologists contend that skeletons classified as belonging to *Homo habilis* are really members of the genus *Australopithecus*.

Homo Erectus

Homo habilis gave rise to *Homo erectus.*

Two hundred thousand years after the emergence of *Homo habilis*, the first unmistakable member of the genus *Homo*, known as **Homo erectus** ("upright man"), arose (see Figure 23-10). Skeletons of *Homo erectus* appear in geological formations 300,000 to 1.6 million years old in the Old World. The most complete fossilized human skeleton that has been found is that of a 12-year-old boy (Figure 23-13).

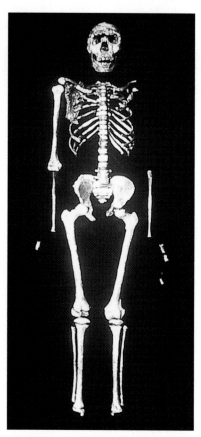

FIGURE 23-13 **Skeleton of *Homo erectus*** This, the most complete fossilized skeleton of *H. erectus* ever found, belonged to a boy who lived about 1.6 million years ago.

Unlike *Australopithecus* and *H. habilis*, which remained in Africa, *H. erectus* first appeared in Africa, then spread to Europe and Asia (India, China, and Indonesia). *Homo erectus* appears to be the first hominid to leave the warmth and abundance of the tropics and subtropics and take up residence in the temperate zone, a region characterized by warm summers but cold winters.

Homo erectus stood about 5 feet tall, used fire, and made more sophisticated tools and weapons than its predecessor, *Homo habilis*. With a brain slightly smaller than ours, *Homo erectus* is believed to be the direct ancestor of *Homo sapiens*, the self-proclaimed "thinking man."

Homo Sapiens

Modern humans belong to *Homo sapiens*

Homo sapiens emerged about 300,000 years ago. *Homo sapiens* consists of two subspecies in order of their appearance: **Homo sapiens neanderthalensis**, the Neanderthals, and *Homo sapiens sapiens*, modern humans (Figure 23-10).

Widely distributed in Europe and Asia, the Neanderthals lived in caves and camps. Their name comes from the Neander Valley in Germany, where the first specimens of this group were discovered. Archaeological evidence shows that the Neanderthals gathered fruits, berries, grains, and roots and hunted animals with weapons. They cooked some of their food on fires. Neanderthals stood erect and walked upright. Evidence also indicates that they lived in small clans and buried their dead in elaborate rituals.

Although the skeletons of the Neanderthals resemble those of humans, they differ in several ways. For example, the skulls of Neanderthals were, on average, larger than those of modern humans. Their skeletons were also more massive and heavily muscled than those of modern humans, and they had rather short lower limbs, much like Inuits (Eskimos) and other cold-adapted people.

Many people think of the Neanderthals as dimwitted and brutish, shuffling along bent over like an ape. This view is based on an interpretation of a Neanderthal skeleton found in 1908 in France. At the time of the discovery, however, the researchers failed to recognize that the skeleton under study was bent over because the individual suffered from arthritis of the hip and had diseased vertebrae. It is interesting, though, how this hasty conclusion from one observation spawned such a long-standing myth.

Some anthropologists wonder whether Neanderthals, despite many humanlike behaviors, should really be considered members of *Homo sapiens*. The differences in their physical appearance are quite striking. With broad faces, large projecting brow ridges, and heavily built bodies, they should perhaps be placed in a species all their own (Figure 23-14). Nonetheless, if one compares skeletons of australopithecines, early species of the genus *Homo*, and modern *Homo sapiens*, the Neanderthal does seem to fit well in the progression from *Homo erectus* to *Homo sapiens sapiens* (Figure 23-15).

FIGURE 23-14 **Neanderthal** Notice the projecting brow ridges of this reconstruction of a Neanderthal man.

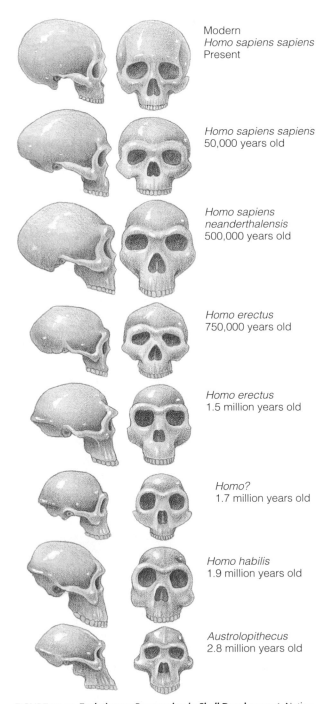

Modern
Homo sapiens sapiens
Present

Homo sapiens sapiens
50,000 years old

Homo sapiens
neanderthalensis
500,000 years old

Homo erectus
750,000 years old

Homo erectus
1.5 million years old

Homo?
1.7 million years old

Homo habilis
1.9 million years old

Austrolopithecus
2.8 million years old

FIGURE 23-15 **Evolutionary Progression in Skull Development** Notice the changes that occurred over time in the shape of the human skeleton, starting with the apelike *Australopithecus* at the bottom. The dates indicate age of skulls in the drawing, not the evolutionary emergence of each species.

Neanderthals disappeared approximately 40,000 years ago for reasons still not understood. Some archaeologists believe that they were replaced by the earliest known members of *Homo sapiens sapiens*, the **Cro-Magnons** (CROW-MAN-yens). Cro-Magnons first appeared in Africa and then spread out across Europe and northern Asia, perhaps wiping out the Neanderthals or possibly interbreeding with them.

Archaeological evidence shows that Cro-Magnons used sophisticated tools and weapons, including the bow and arrow, and were highly skilled nomadic hunters, following great herds of animals during their seasonal migrations. They may have had a well-developed language. Cro-Magnons lived in caves and rock shelters, in groups of 50–75 people, and are best known for the elaborate artwork that adorned the walls of their caves (Figure 23-16).

Approximately 10,000 years ago, truly modern humans emerged, but the changes were not great. In fact, over the past 40,000 years, human evolution has produced little noticeable

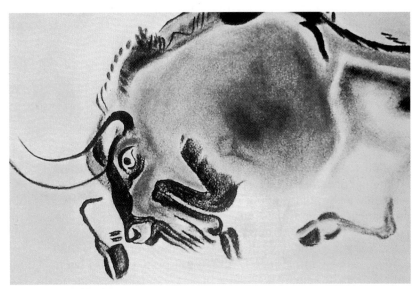

FIGURE 23-16 **Cro-Magnon Art**

(a)

(b)

(c)

(d)

(e)

FIGURE 23-17 **Human Races** Humans, no matter what they look like, belong to one species, *Homo sapiens*. The various races are phenotypic variants of that species. Many scientists recognize five distinct races: (a) Mongoloid, (b) Negroid, (c) Caucasian, (d) Native American, and (e) Australian Aborigine.

change in the physical appearance of humans. A Cro-Magnon on the streets of Los Angeles, in fact, would probably go unnoticed. Those 40,000 years have not been without change, however, for during this period, *Homo sapiens* has developed a rich and varied culture, complex language, and extraordinarily sophisticated tools.

Human Races

Human races result from variations caused by geographic separation.

Studies in evolution show that species that become subdivided into isolated populations undergo changes in response to their environment. If the changes are profound, the subpopulations often lose the ability to interbreed and produce new species.

Over the course of time, the human population has wandered far and wide on the planet and has fragmented into distinct subpopulations, living in very different environments. This fragmentation has resulted in the formation of a number of phenotypically distinct subpopulations, or **races** (Figure 23-17). It is important to remember, however, that races are not distinct species, just regional variations of the one species *Homo sapiens*.

Most of us are familiar with regional characteristics. We know, for instance, that Mexicans tend to have brown eyes, dark skin, and black hair and that Scandinavians tend to have blue eyes and blond hair. Many differences found in the races are adaptations to differing environmental conditions. The darker skin that evolved in tropical and subtropical populations, for instance, may be an adaptation that protects people from harmful ultraviolet radiation.

How many races are there? Some scientists recognize three main ones: Mongoloids, Negroids, and Caucasians. Others recognize two others: Native Americans and Australian Aborigines (Figures 23-17d and 23-17e). Still others recognize 30 different races. What all this tells us is that the races are arbitrary categories. As people move about the planet and interbreed, the races are mixing, blurring the lines among the world's peoples. As time goes on, then, the distinction among various races will probably decrease.

23-7 Health and Homeostasis

Evolution has resulted in the presence of many homeostatic mechanisms—in the bodies of organisms like us and in the environment on which we depend. As pointed out previously, humans have adopted practices that unknowingly upset those mechanisms. One of these ways—the use of pesticides—has potentially far-reaching impacts.

Chemical pesticides kill insects and other organisms that damage crops. Unfortunately, the use of pesticides has been shown to create **genetic resistance** in pest populations. Genetic resistance results from naturally occurring genetic variation—

the presence of a mutation that makes a small percentage (about 5%) of any insect population resistant to chemical pesticides.

When farmers spray their fields with chemical pesticides, they kill 95% of the pests. The survivors are primarily those that are naturally resistant to the pesticide. They, in turn, breed and, over time, produce new populations that are even more resistant to pesticides. To kill these pests, farmers must apply higher doses of pesticides or switch to another type. However, a small segment of the population is genetically resistant to the higher dose or the new chemical preparation. That group sur-

healthnote

23-1 Well Done, Please: The Controversy over Antibiotics in Meat

Approximately 50 years ago, livestock growers began adding antibiotics to cattle and pig feed to protect animals confined to pens from disease as they were being fattened for market (Figure 1). Antibiotics were first given to control disease, which can run rampant under crowded conditions. However, farmers soon found that the drugs had an unanticipated beneficial effect: For reasons still not fully understood, antibiotics accelerated the rate of body growth.

Not surprisingly, today 70% of all cattle and 90% of all veal calves and pigs are reared on feed treated with penicillin or tetracycline. Nearly half of the antibiotics sold in the United States, in fact, are used for livestock feed. This practice is also popular in Europe.

The addition of antibiotics to feed has been criticized by microbiologists and health officials. They fear that it could promote the evolution of superstrains of bacteria immune to antibiotics. In an editorial in the *New England Journal of Medicine*, a Tufts University microbiologist, Stuart Levy, noted that "every animal . . . taking an antibiotic . . . becomes a factory producing resistant strains" of bacteria. Resistant bacteria, in turn, could transfer their resistance to other bacteria, creating highly lethal strains that could infect humans.

Scientists are also concerned that some resistant bacteria in cattle (such as *Salmonella*) could be transmitted directly to people in meat or milk. The effects could be grave.

Medical researchers are also concerned that antibiotics in livestock grains may also end up in the meat of the animal. It may then be passed on to the meat-eating public. In humans, the antibiotics could result in the production of antibiotic resistant bacteria, adding to the already growing problem caused by antibiotics used to treat infections (Chapter 14).

Despite these concerns, efforts to reduce the use of antibiotics in feed dating back to the late 1970s have been defeated. Livestock producers, feed producers, and the multimillion-dollar drug industry, however, argued back that microbiologists' concerns had not been proven.

Since then, scientific evidence has begun to accumulate, confirming the suspicions of the medical profession. A study by researchers at the Centers for Disease Control, published in *Science* in 1984, showed that antibiotic-resistant bacteria artificially (created by adding antibiotics to cattle feed) could be transferred directly from meat to humans. The study also showed that 20%–30% of the *Salmonella* outbreaks in the United States involved antibiotic-resistant strains. About 4% of the people contracting the antibiotic-resistant bacteria died, compared with only 0.2% of the victims of the nonresistant *Salmonella*.

In 1999, a team of Minnesota researchers published results of a study that provided additional evidence. The researchers found that only 1.3 % of bacteria isolated from humans in 1992 were resistant

FIGURE 1 **Feed Lot Cattle** Antibiotics reduce infections in cattle in tight quarters and also accelerate growth. Microbiologists, however, worry that they may be stimulating the evolution of super strains of bacteria.

to a certain type of antibiotic. In 1998, three years after the drug was approved for use in animal feed, 10.2 % of the bacteria were resistant to the antibiotic. The researchers concluded that the practice of giving the drug to poultry "created a reservoir" of resistant bacteria. In 2000, Danish researchers found that antibiotic resistant bacteria could be transmitted from meat to humans.

Proponents of antibiotic use in feed believe that the link between antibiotics and human disease is still weak and that further research is needed. Even if these findings are substantiated by further research, proponents believe, the benefits of using antibiotics outweigh the potential health effects. Banning antibiotics or cutting back on their use, they say, could have enormous economic impacts that must be weighed against sickness and loss of life. But health experts hope that the United States, like Europe, which strictly limited the use of antibiotics in animal feed in the early 1970s, will find the political will to end this activity.

Individuals can take direct action, too. You can buy "organically grown" beef—that is, beef fed a diet that contains no hormones or antibiotics. You can also change your eating habits. Instead of asking for rare meat or a half-cooked burger, you may want to consider asking for a well-done piece of meat. If ground beef has been heated in the center to over 160 °F, you are pretty safe; chicken and poultry need to be heated to more than 170 °F or until the fluids run clear.

www.jbpub.com/humanbiology/5e

Visit Human Biology's Internet site for links to web sites offering more information on this topic.

vives the spraying, and is selected for and produces an even more resistant population, thus continuing an ever-escalating cycle often referred to as the "pesticide treadmill."

The pesticide treadmill is an excellent example of artificial selection at work. Some scientists warn farmers that they can never win in the battle against pests. Thanks to genetic variation in a population, no matter what pesticide they use, or how

much they use, there will always be a resistant strain. Ever more powerful and ever more frequent applications at higher doses will be necessary just to stay even. Today, over 550 species of insects are resistant to one or more chemical pesticides. Twenty of the worst pests are now resistant to all types of insecticide.

Don't despair. There are ways to control pests without the use of pesticides and without creating genetically resistant strains. Organic farmers are practicing these methods quite successfully. One technique is crop rotation. Crop rotation requires farmers to alternate the crops they plant in a given field. One year, corn might be grown in the field, the next year, beans, and the third year, alfalfa. How does it work?

Many insect pests are specialists, preferring one crop to another. A field of corn, for example, provides an abundant supply of food for corn borers. At the end of the season, corn borers lay their eggs in the soil or in organic material on the surface. If the same crop is planted the next year, the newly hatched insects will have an abundant food supply. The insect population will increase quickly, causing considerable damage. If a different crop is planted the second year, however, the corn borer population will decline. Continued rotation helps to hold down pests year after year without pesticides.

Chemical resistance is a rather common occurrence in the modern world. Even weeds can become resistant to herbicides. Microorganisms also develop resistance to antibiotics, which is one reason many physicians prescribe antibiotics sparingly. Health Note 23-1 discusses this problem.

SUMMARY

The Evolution of Life: An Overview

1. Scientists believe that the Earth and the sun—and the rest of the system—came from an enormous cloud of cosmic dust and gas that gave rise to the Earth and the sun.

2. The Earth formed about 4.5 billion years ago.

3. The evolution of life probably began in the sea and is divided into three phases: chemical evolution, cellular evolution, and the evolution of multicellular organisms.

4. Chemical evolution, scientists hypothesize, began about 4 billion years ago. Simple inorganic chemicals that were dissolved in the seas combined to form small organic molecules. Over time, the organic molecules combined to form polymers—small proteins and nucleic acids. These polymers combined to form aggregates that may have been the precursors of cells.

5. The first true cells arose from the primitive aggregates. They contained primitive enzymes and simple genes, and may have derived nourishment from organic molecules they absorbed from the environment.

6. Autotrophs, organisms capable of synthesizing their own food, arose next. The earliest autotrophs probably acquired energy from chemicals in the environment. With the evolution of chlorophyll, photosynthetic autotrophs arose.

7. Over time, scientists believe, photosynthesis evolved further, and the new photosynthetic organisms began to produce oxygen.

8. Prokaryotes (bacterialike organisms) emerged about 3.5 billion years ago, and eukaryotes (cells with nuclei) evolved about 1.2 billion years ago. The evolution of eukaryotes opened the door for the evolution of multicellular organisms.

9. A variety of multicellular plants and animals evolved from single-celled eukaryotes in the oceans. As the ozone layer developed, life on land became possible.

How Evolution Works

10. Evolution has produced a great diversity of organisms.

11. Evolution takes place because of genetic variation and natural selection.

12. Genetic variation in a species arises from many factors such as mutations and from new genetic combinations resulting from sexual reproduction.

13. Genetic variation results in variations in traits that may offer some organisms an advantage over others, giving them a better chance of surviving and reproducing and passing the genes on to future generations.

14. Beneficial traits are preserved in a population by natural selection.

The Evidence Supporting Evolution

15. The scientific knowledge in support of evolution is rich and varied. The fossil record, anatomical similarities in groups of organisms, the common biochemical makeup of organisms, similar embryological development among many groups of organisms, and experimental evidence all support the theory of evolution by natural selection.

Early Primate Evolution

16. Humans belong to the order called primates. Today, most primate species live in tropic and subtropic forests. The main exception is humans, who inhabit a wide range of habitats.

17. Primates are characterized by grasping hands, forward-looking eyes, and large brains (in proportion to body size).

18. Based on fossil evidence, it appears that the primates evolved from a mammalian insectivore that resembled the modern-day tree shrew and lived about 80 million years ago.

19. Humans evolved from an early apelike creature.

Evolution of the Australopithecines

20. The first truly human primate belonged to the genus *Australopithecus*.

21. The oldest known australopithecine skeleton was unearthed in Africa and is believed to be about 3.5 million years old. It belongs to a group called *Australopithecus afarensis*. It stood about 3 feet high and had a brain only slightly larger than an ape's, but it probably walked erect.

22. About 3 million years ago, *A. afarensis* was replaced by another species, *A. africanus*, which was slightly taller than its predecessor and had a slightly larger brain.

23. About 2.3 million years ago *A. robustus* emerged. It was taller and heavier and had a larger brain than its predecessors.

24. About 2.2 million years ago, the fourth species, *A. boisei*, appeared.

Evolution of the Genus *Homo*

25. Many paleontologists believe that *A. afarensis* also gave rise to the genus *Homo*, the ancestors of modern humans, *Homo sapiens*. Others believe that the genus *Homo* may have evolved from an as-yet-unidentified ancestor.

26. The earliest discovered evidence of the genus *Homo* is a 1.8 million-year-old skull and partial skeleton found in Tanzania. It belongs to *Homo habilis*.

27. Two hundred thousand years after the emergence of *Homo habilis*, *Homo erectus* arose. Skeletons of *Homo erectus* appear in geological formations 300,000 to 1.6 million years old in the Old World.

28. Unlike *Australopithecus* and *H. habilis*, which remained in Africa, *H. erectus* moved from Africa to Europe and Asia.

29. *Homo erectus* stood about 5 feet tall, used fire, and made more sophisticated tools and weapons than its predecessors. With a brain slightly smaller than ours, *Homo erectus* is believed to be the direct ancestor of *Homo sapiens*.

30. *Homo sapiens* emerged about 300,000 years ago and consists of two subspecies: *Homo sapiens neanderthalensis* (the Neanderthals) and *Homo sapiens sapiens*.

31. The Neanderthals lived in caves and camps in Europe and Asia until approximately 40,000 years ago, when they disappeared. Some archaeologists believe that they were replaced by modern humans, the Cro-Magnons, the earliest known members of *H. sapiens sapiens*.

32. The Cro-Magnons first appeared in Africa and then spread across Europe and northern

Asia, perhaps wiping out the Neanderthals or possibly interbreeding with them.

33. Over the course of time, the human population has wandered far and wide on the planet and has come to inhabit a wide range of climatic zones. This has resulted in the formation of a number of distinct subpopulations, or races. Many differences found in the races are thought to be adaptations to differing environmental conditions.

Health and Homeostasis

34. Homeostasis at the organismic and environmental levels can be upset by pesticides. Pesticides can create a form of artificial selection because they select resistant species.

35. Farmers who spray their fields to kill insects leave behind a genetically resistant subpopulation that breeds and repopulates farm fields. A second application at a higher dose or an application of another pesticide kills off more susceptible insects, but this leaves behind another subset that often becomes a further pest, forcing farmers to use higher doses or switch to another pesticide. This escalation in the war against pests is called the pesticide treadmill.

36. Chemical resistance is a rather common occurrence in the modern world. Weeds can become resistant to herbicides, and even microorganisms develop resistance to antibiotics.

critical thinking

Thinking Critically—Analysis

This Analysis corresponds to the Thinking Critically scenario that was presented at the beginning of this chapter.

One of the first questions you might ask is this: Does the archaeological data support the genetic data? In other words, does the fossil record support the African Origins model? Or does it support the Multiregion Evolution model?

Some archaeological evidence suggests that the Multiregion Evolution model might be the most valid hypothesis. For example, studies of facial features of the Chinese skulls illustrate commonalities in modern and archaic Chinese. More importantly, these features are not present in the skulls of early modern Africans. One would expect the skulls of early modern Africans to resemble the Chinese skulls if the modern Chinese populations did indeed come from an ancestral African population. A similar study of skulls of modern and archaic European populations shows no similarities with either the earliest modern Africans or archaic African populations.

Another question that might be useful is this: Is there any archaeological evidence of the spread of African populations into other areas at the proper time? Again, the answer is no. Nor is there evidence of the spread of African culture (as witnessed by stone tools) as one might expect.

Given these facts, can you think of any reason to question the Multiregion Evolution model?

One common criticism of the Multiregion Evolution model is that archaeologists make many subjective judgments and often base their conclusions on small numbers of skulls. Geneticists wonder if such small samples are truly representative of an entire population. In contrast, the genetic data offers quantitative indicators of evolutionary change that are relatively free of biases.

Given the controversy, it is clear that neither hypothesis can be ruled out. More research is needed, and many years will no doubt pass before scientists can agree on which is correct.

KEY TERMS AND CONCEPTS

Abiotic factor, p. 427
Alfred Wallace, p. 425
Anthropoid, p. 431
Adaptation, p. 425
Australopithecines, p. 432
Autotroph, p. 424
Biogeography, p. 430
Biotic factor, p. 427
Charles Darwin, p. 425

Chemical evolution, p. 423
Cro-Magnons, p. 435
Eukaryote, p. 424
Evolution, p. 423
Fitness, p. 427
Fossil, p. 428
Genetic resistance, p. 436
Heterotroph, p. 424
Hominids, p. 432

Homo erectus, p. 434
Homo habilis, p. 434
Homo sapiens, p. 434
Homologous structure, p. 429
Natural selection, p. 425
Prokaryote, p. 424
Prosimian, p. 431
Protocells, p. 424
Race, p. 436

CONCEPT REVIEW

1. Describe the theory of chemical evolution. p. 423

2. Describe how the first cells may have arisen during evolution. What critical requirements must have been met for life to begin? p. 424

3. The development of photosynthesis and the emergence of eukaryotes were pivotal events in evolution. Why? p. 424

4. Numerous plants and animals had evolved in the sea before life emerged on land. Why? p. 424

5. A critic of evolution says, "Life-forms are too diverse to have come from a common ancestor." How would you respond? pp. 425–427

6. How do random mutations in germ cells contribute to the evolutionary process? p. 425

7. Define the following terms: adaptation, variation, natural selection, biotic factors, abiotic factors, selective advantage, and fitness. pp. 425–427

8. Discuss the following statement: Natural selection is nature's editor. pp. 425–427

9. Discuss the evidence supporting the theory of evolution. pp. 427–430

10. Why is the primate fossil record incomplete? What problems does this gap create in tracing the evolutionary history of primates? p. 432

11. Briefly describe the evolution of modern humans, *Homo sapiens*. pp. 434–436

12. Describe the term pesticide treadmill. How can farmers avoid this problem? p. 437

1. Inorganic molecules in the oceans may have reacted chemically to produce the very first _____ molecules. p. 423

2. The very first cells were known as _____ and probably consisted of aggregations of polymers and other molecules formed during chemical evolution. p. 424

3. An organism capable of producing its own food is called a(n) _____. p. 424

4. Cells with nuclei belong to a group known as _____. p. 424

5. Life first evolved in the _____. p. 424

6. A genetically based characteristic that increases an organism's chance of surviving and reproducing is called a(n) _____. p. 425

7. Abiotic and biotic factors influence survival and reproduction by "selecting" the fittest—those best able to reproduce and pass on their genes. This process is called natural _____. p. 425

8. _____ is a measure of reproductive success that can be attributed to a variety of factors such as greater strength or cunning. p. 427

9. Structures thought to have arisen from a common ancestor like the arm of a human and the fin of a whale are known as _____ structures. p. 429

10. Primates are believed to have evolved from a(n) _____ mammal. p. 430

11. The first primate species to evolve was the _____. p. 430

12. Humans are believed to have evolved directly from a(n) _____ mammal. p. 431

13. The first truly humanlike animals were members known as *Homo* _____. p. 434

14. Modern humans belong to a species known as *Homo* _____, which emerged about 10,000 years ago. p. 435

15. _____ resistance refers to naturally occurring resistance among pest species, bacteria, and other organisms to pesticides, herbicides, and antibiotics. p. 436

www.jbpub.com/humanbiology/5e

The site features eLearning, an online review area that provides quizzes, chapter outlines, and other tools to help you study for your class. You can also follow useful links for in-depth information, research the differing views in the Point/Counterpoints, or keep up on the latest health news.

CHAPTER

24

Ecology and the Environment

thinking critically

A newspaper article notes that "on the issue of global warming the scientific community is divided." In support of this assertion, the author quotes two scientists. One says that he's "convinced the Earth's temperature is increasing and this increase is being caused by pollution." The other argues that "there's not enough evidence to support this conclusion." The article goes on to say that because of the uncertainty among the scientific community, it makes no sense to launch a global effort to reduce carbon dioxide emissions. Can you detect any problem in this reporting? What critical thinking rules were helpful to you in this examination?

Most of us live our lives seemingly apart from nature. We make our homes in cities and towns surrounded by concrete and steel, and drown out the sound of birds with our noise. The closest many of us get to nature is a romp with the family dog on the grass in our backyard.

Raymond Dasmann, a world-renowned ecologist, wrote that despite what many of us may think, a human apart from nature is an abstraction. No such being exists. Human life depends on the environment. The clothes we wear, the breakfast cereal we eat, the energy we consume, even the oxygen we breathe are all products of nature.

Nature provides other free services as well. For example, plants protect the land, preventing flooding and erosion. Swamps help purify the water we drink. Birds control insect populations. Clearly, nature "serves" us well. Thus, although we may have isolated ourselves from nature, we are extremely dependent on it. This chapter will help you understand why.

24-1 An Introduction to Ecosystems

Ecology is the study of the interaction between organisms, including humans, and their environment.

This chapter discusses ecology and environmental issues. **Ecology** is a branch of science. Its main focus is on how organisms interact with one another. It also focuses on how organisms are affected by the physical and chemical environment. In addition, ecology examines the ways organisms affect their environment.

Ecology, like all disciplines in science, is a body of knowledge and a process of inquiry. However, *ecology* probably ranks as one of the most misused words in the English language. Banners proclaim, "Save Our Ecology." Speakers argue that "our ecology is in danger," and others talk about the "ecological movement." These common uses of the word ecology are incorrect. Why?

Ecology is a discipline of science. It is not synonymous with the word *environment*. It does not mean the web of interactions in the environment. We can save our ecology department and ecology textbooks, but we cannot save our ecology. Our ecology is not in danger; our environment is. You cannot join the ecology movement, but you can join the environmental movement.

The Biosphere

The biosphere is the zone in which all life exists on Earth.

The science of ecology, unlike many other branches of scientific endeavor, often focuses on systems. The largest biological system on Earth is the **biosphere** (BUY-oh-sfear). The biosphere can be thought of as the thin skin of life on the planet. As shown in Figure 24-1, the biosphere forms at the intersection of air, water, and land. Organisms consist of components derived from all three. The carbon atoms in the proteins in your body, for instance, come from carbon dioxide in the atmosphere captured by plants. The minerals in your bones come from the soil in which plants grow. Water comes from streams and lakes.

The biosphere extends from the bottom of the ocean to the tops of the highest mountains. Although that may seem like a long way, it's not—at least, in comparison with the size of the Earth. In fact, if the Earth were the size of an apple, the biosphere would be about the thickness of its skin. Although life exists throughout the biosphere, it is rare at the extremes, where conditions for survival are marginal.

The biosphere is a closed system, much like a sealed terrarium. By definition, a closed system receives no materials from the outside. The only outside contribution to the biosphere is sunlight.

As noted previously in this book, sunlight is the source of energy for virtually all life. Even the energy released when coal, oil, and natural gas are burned owes its origin to the sunlight that fell on the Earth several hundred million years ago. Coal, for instance, is made from ancient plants that lived in and along swamps. Leaves and other plant matter fell into the water and were covered by sediment. Over time, pressure and heat converted the plant matter to coal.

Because the Earth is a closed system, all materials necessary for life must be recycled over and over, as you shall soon see.

Biomes and Aquatic Life Zones

The biosphere is divided into biomes and aquatic life zones.

Viewed from outer space, the Earth resembles a giant jigsaw puzzle, consisting of large landmasses and vast expanses of ocean. The landmasses, or continents, can be divided into large biological subregions or biomes. A **biome** is a region characterized by a distinct climate and specific plants and animals adapted to it. Some of the most familiar biomes include the desert, grassland, and tundra (Figure 24-2).

The oceans also can be divided into subregions, known as **aquatic life zones**. Aquatic life zones are the aquatic equivalent of biomes. Like their land-based counterparts, each of these regions has a distinct environment and characteristic plant and animal life adapted to conditions of the zone. Four major aquatic life zones exist: coral reefs, estuaries (the mouths of rivers where fresh and salt water mix), the deep ocean, and the continental shelf.

Humans inhabit all biomes on Earth. Within these regions, humans benefit in numerous ways from the microbes, plants, animals, soil, water, and air that compose the biome. We often tend to think of biomes in terms of their natural resources (coal, timber, oil). Although vital to human society's economic health, these resources are but a fraction of the services we receive from the biome. Trees, for instance, not only provide lumber to build homes and make paper, they also provide oxygen and protect watersheds. In addition, they may remove certain pollutants from the air. Many species besides humans also depend on these services.

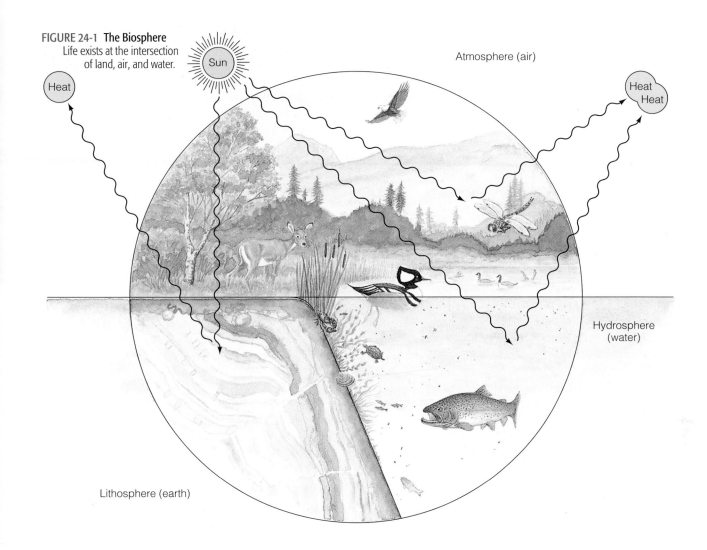

FIGURE 24-1 The Biosphere
Life exists at the intersection of land, air, and water.

Heat

Sun

Atmosphere (air)

Heat
Heat

Hydrosphere
(water)

Lithosphere (earth)

Ecosystems

Ecosystems consist of organisms and their environment.

The biosphere is a global **ecological system**, or **ecosystem** for short. The term *ecosystem* is used to describe a community of organisms and their physical and chemical environment. Numerous interactions are possible within an ecosystem.

Ecosystems consist of two basic components: abiotic and biotic. The **abiotic components** of an ecosystem are the physical and chemical factors necessary for life. They include sunlight, precipitation, temperature, and nutrients.

Humans often alter the physical and chemical components of the environment. These changes can cause serious biological impacts. If conditions change drastically, for instance, some species perish.

The **biotic components** of an ecosystem are the organisms that live there. Within biomes and aquatic life zones, organisms of a given species often occupy very specific regions to which they are well adapted. A group of organisms of the same species occupying a specific region constitutes a **population**. In any given ecosystem, several populations exist together and form a **biological community**, an interdependent network of plants, animals, and microorganisms.

Habitat

Habitat is where an organism lives.

If asked to give a brief description of yourself, you would probably begin by describing the place where you live. You would then likely discuss the work you do, the friends you have, and other important relationships that describe your place in society. A biologist would do much the same when describing an organism. He or she would start with a description of the place an organism lives—that is, its **habitat**.

Niche

An organism's niche consists of all of its relationships in an ecosystem.

Next, the biologist would describe how the organism "fits" into the ecosystem, its **ecological niche**, or simply **niche** (pronounced nitch). An organism's niche includes all of its relationships with its environment. For example, the niche includes what an organism eats, what eats it, and other important facts.

The concept of the niche is very important. For example, successful control of an insect pest is best achieved through an understanding of the pest's niche. Such an analysis might show that a particular species of bird or insect feeds on the pest. As another example, to protect a plant or animal you have to protect the ecosystem it lives in. Such efforts help protect the species' niche, which, in turn, helps protect the species itself.

(a)

(b)

(c)

Figure 24-2 **North American Biomes** (a) Tundra, (b) northern coniferous forests, (c) temperate deciduous forest, (d) prairie, and (E) desert.

(d)

(e)

24-2 Ecosystem Function

Producers

> Producers generate nutrients consumed by all other organisms.

Life on land and in water is possible principally because of the existence of a group of organisms known as **producers** (Figure 24-3). Producers primarily include the algae and plants. These organisms absorb sunlight. They then use its energy to make organic molecules from atmospheric carbon dioxide and water. This process is called **photosynthesis**. Organic molecules generated by photosynthesis nourish the producers, and also feed all the other organisms in ecosystems.

Another large group of organisms is the **consumers**. They reap the benefit of photosynthesis in producers. Ecologists place consumers into four general categories, depending on the type of food they eat. Some consumers, such as deer and cattle, feed directly on plants. They are called **herbivores**. Others, such as wolves, feed on herbivores and other animals and are known as **carnivores**. Humans and a many other animal species consume plants *and* animals and are known as **omnivores**. The final group feeds on animal waste or the remains of plants and animals. They are **detrivores** (DEE-treh-vores), or **decomposers**. This important group includes many bacteria, fungi, and insects.

Food Chains and Food Webs

> Organisms are part of food webs.

In ecosystems, each organism is a food source for some other organism. As a result, all organisms belong to one or more food chains. Technically, a **food chain** is a series of organisms, each one feeding on the organism preceding it (Figure 24-3). Biological communities consist of numerous food chains.

Biologists recognize two general types of food chains: grazer and decomposer. **Grazer food chains** begin with plants and algae, the producers. These organisms are consumed by herbivores, or grazers; hence, the name grazer food chain. Herbivores, in turn, may be eaten by carnivores. **Decomposer food chains** begin with dead material—either animal wastes (feces) or the remains of plants and animals (Figure 24-4).

As in most things in nature, decomposer and grazer food chains typically function together (Figure 24-4). For example, waste products from the grazer food chain enter the decomposer food chain. Nutrients liberated by the decomposer food chain enter the soil and water and are reincorporated into plants at the base of the grazer food chain.

FIGURE 24-3 **Simplified Food Chains** Two grazer food chains are shown, a terrestrial one and an aquatic one.

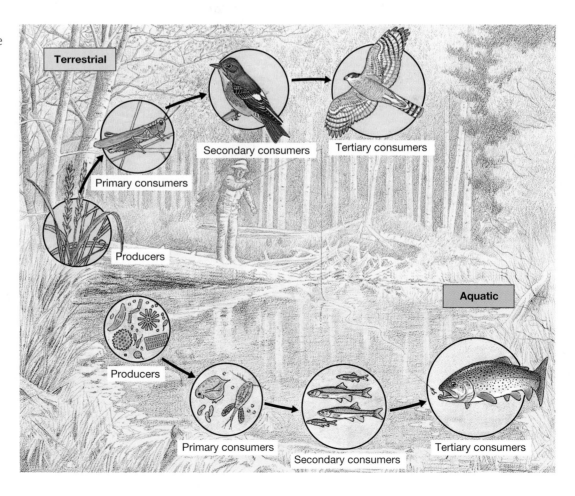

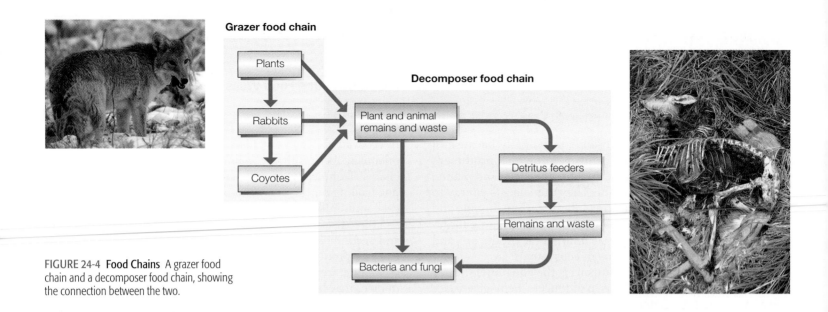

FIGURE 24-4 Food Chains A grazer food chain and a decomposer food chain, showing the connection between the two.

In a biological community, food chains are woven together into more complex networks, called **food webs** (Figure 24-5). Food webs present a complete picture of the feeding relationships in any given ecosystem.

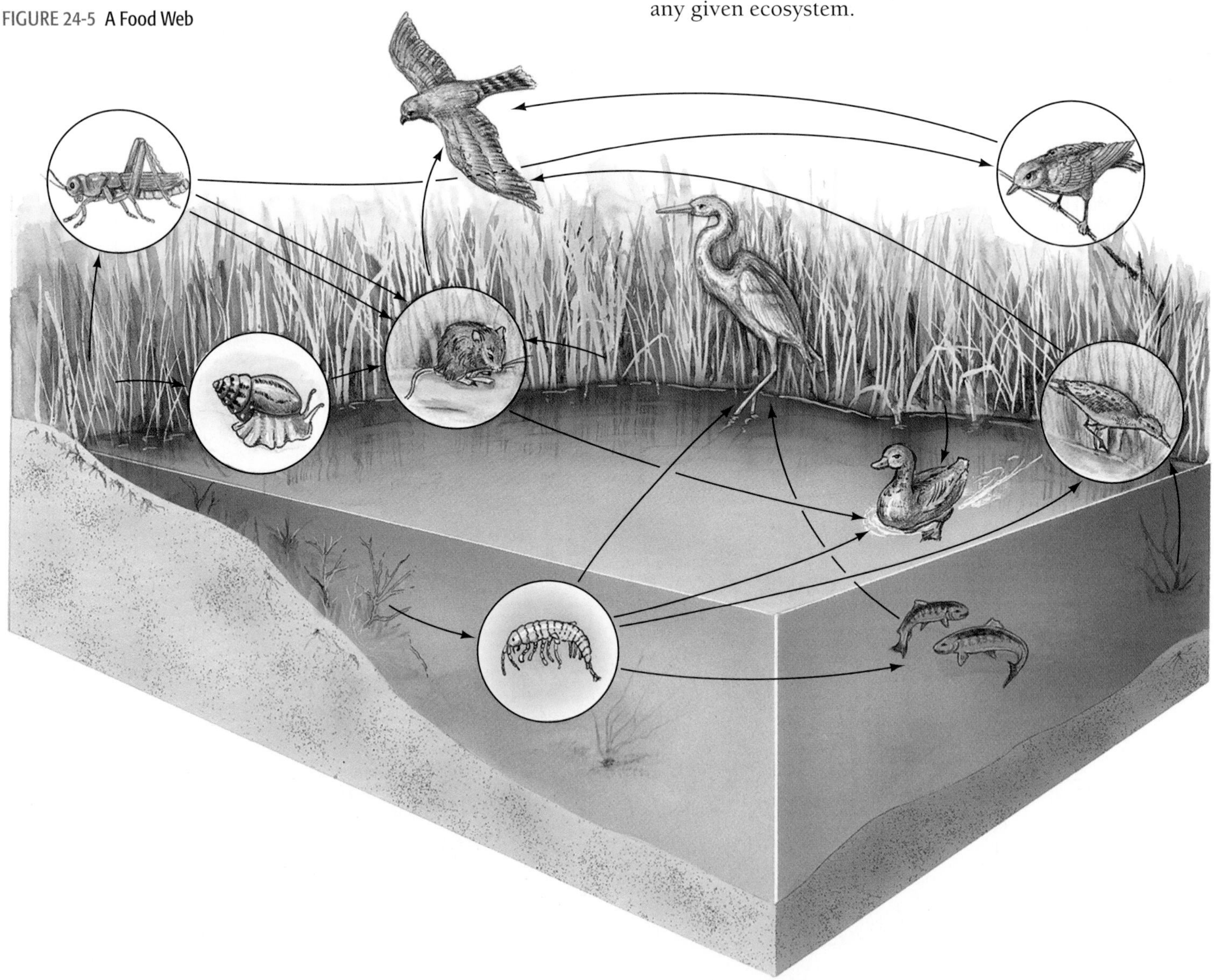

FIGURE 24-5 A Food Web

Energy and Nutrient Flows

Energy and nutrients flow through food webs, but in very different ways.

Although we tend to think of food chains simply as a map of feeding relationships, they are much more. Food chains are avenues for the flow of energy and the cycling of nutrients through the environment. Let's consider energy first.

In the biosphere, almost all of the energy needed by organisms comes from the sun. Solar energy is captured by plants and algae and is used to produce organic food molecules. This energy is stored in the molecules of organisms. In the food chain, organic molecules pass from plants to animals, where they are broken down in their mitochondria. They release stored solar energy, which is then used to power numerous cellular activities.

During cellular respiration, much of the energy stored in organic food molecules is lost as heat. Heat escaping from plants and animals is radiated into the atmosphere and then into outer space. It cannot be recaptured and reused by organisms. Because all solar energy is eventually converted to heat, energy is said to flow one way through food webs. Energy cannot be recycled.

In sharp contrast, nutrients are recycled over and over. Nutrients in the soil, air, and water are first incorporated into plants and algae, and then passed from plants to animals. Nutrients in the food chain eventually reenter the environment by one of three paths: (1) the excretion of wastes, (2) the decomposition of waste, and (3) the decomposition of dead organisms.

One way or another, all nutrients eventually make their way back to the environment for reuse. Each new generation of organisms therefore relies on the recycling of matter. The atoms in your body, for instance, have been recycled many times since the beginning of life on Earth. Who knows? Some of those atoms may have been part of the very first cells.

Trophic Levels

The organisms of a food chain exist on different trophic levels.

Ecologists classify the organisms in a food chain according to their position, or **trophic level** (TROE-fic; literally, "feeding" level). The producers comprise the base of the grazer food chain and are therefore members of the first trophic level. The grazers are members of the second trophic level. Carnivores that feed on grazers are members of the third trophic level, and so on.

Most terrestrial food chains are limited to three or four trophic levels. In fact, longer terrestrial food chains are quite rare. Food chains generally do not have a large enough producer base to support many levels of consumers.

When plotted, the biomass at the various trophic levels forms a pyramid, the **biomass pyramid** (Figure 24-6). Because biomass contains energy (stored in the chemical bonds), the biomass pyramid can be converted into a graph of the chemical energy in the various trophic levels. This graph is called an en-

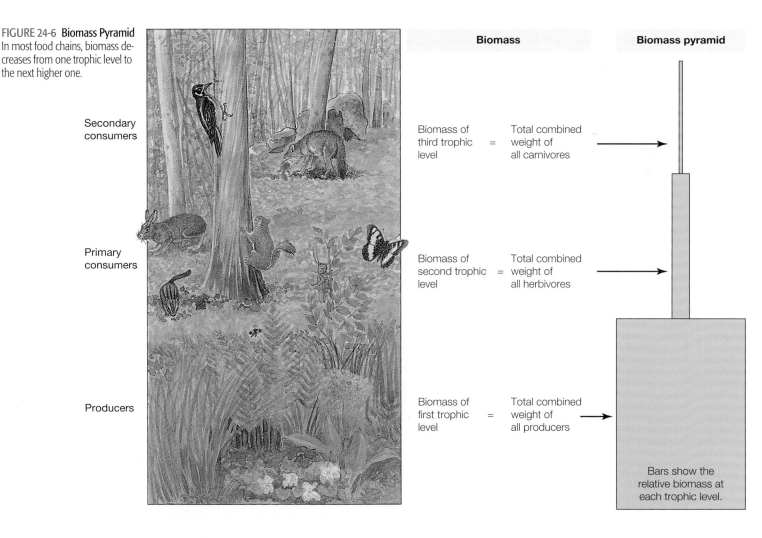

FIGURE 24-6 **Biomass Pyramid** In most food chains, biomass decreases from one trophic level to the next higher one.

Secondary consumers

Primary consumers

Producers

Biomass

Biomass of third trophic level = Total combined weight of all carnivores

Biomass of second trophic level = Total combined weight of all herbivores

Biomass of first trophic level = Total combined weight of all producers

Biomass pyramid

Bars show the relative biomass at each trophic level.

ergy pyramid. In most food chains, the number of organisms also decreases with each trophic level, forming a **pyramid of numbers**.

Nutrient Cycles

Nutrient cycles consist of two phases, organismic and environmental.

As noted earlier, nutrients flow from the environment through food webs, then are released back into the environment. This circular flow constitutes a **nutrient cycle**, or a **biogeochemical cycle**.

Nutrient cycles can be divided into two phases: environmental and organismic. In the environmental phase, a nutrient exists in the air, water, or soil or in two or more of them simultaneously. In the organismic phase, nutrients are found in living organisms.

Dozens of global nutrient cycles operate continuously to ensure the availability of chemicals vital to organisms. Unfortunately, many human activities disrupt nutrient cycles. These activities can profoundly influence the survival of species. This section looks at three important nutrient cycles.

The Water Cycle.   Water is part of a global recycling network known as the *hydrological cycle*, or **water cycle**. The hydrological cycle runs day and night, collecting, purifying, and distributing water throughout the planet. Along its way, it serves humans and other living creatures in a multitude of ways.

The water cycle is driven by two basic processes, evaporation and precipitation. Evaporation occurs when water molecules escape from surface waters, soils, and plants, becoming suspended in air (Figure 24-7). When water molecules depart, they leave behind impurities. Thus, evaporated water is free of contamination until it mixes with atmospheric pollutants from human and natural sources.

In the atmosphere, water is suspended as fine droplets known as *water vapor*. Although we're not usually aware of it, you can detect its presence on a cold day when moisture beads up on cold windows, or on a warm day, when moisture collects on the outside of a glass of ice-cold lemonade.

The amount of moisture (water vapor) air can hold depends on the temperature of the air. The warmer the air, the more moisture it can hold.

Water that has evaporated eventually returns to Earth through precipitation. Here's how it happens. On a warm summer day, sunlight warms the Earth. This causes water to evaporate from surface waters, land, and plants. Warm, moisture-laden air then rises. As it rises, the air expands because atmospheric pressure decreases. As it expands, it cools. As it cools, its ability to hold water decreases, and clouds begin to form. (This same phenomenon occurs when moisture-laden air is pushed upward by mountain ranges.)

Clouds form as water molecules in air begin to attach onto various small particles suspended in the air such as salts from the sea, dusts, or particulates from factories, power plants, and vehicles.

In tropical and semitropical regions, fine cloud droplets collide and combine, forming raindrops. Over a million fine water droplets must come together to produce a single drop of rain. In temperate climates and at the poles, the temperature of the air in clouds is often well below the freezing point, even in the warmer months of the year. As a result, minute ice crystals tend to form in the clouds. When the crystals reach a certain critical mass, they fall from the sky as snowflakes. In the spring, summer, and fall, however, the snowflakes generally melt as they fall, producing rain.

Clouds move about on the winds, and deposit their moisture throughout the globe as rain, drizzle, snow, hail, or sleet. This process, called *precipitation*, returns water to lakes, rivers, oceans, and land.

Water that falls on the land may evaporate again, or it may flow into lakes, rivers, streams, or groundwater. Eventually it returns to the ocean, from which it may once again evaporate.

At any single moment 97% of the Earth's water is in the oceans. The remaining 3% is freshwater. Of this, most (99%) is locked up

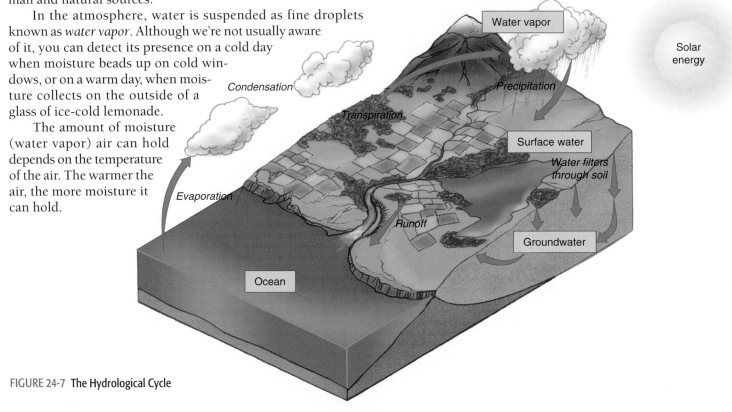

FIGURE 24-7 **The Hydrological Cycle**

in polar ice, in glaciers, and in deep, inaccessible aquifers—underground reservoirs usually in some permeable material such as sandstone. This leaves a paltry 0.003% available for use by humans, but most of this water is hard to reach and much too costly to be of any practical value. At present, human civilization and all land-based forms of life are maintained by only a tiny fraction of the Earth's water supply, a fact that underscores the importance of treating our freshwater supplies with care.

The Carbon Cycle. The **carbon cycle** is illustrated in Figure 24-8. Let's begin with free carbon dioxide. In the environmental phase of the cycle, carbon dioxide resides primarily in the atmosphere and surface waters (oceans, lakes, and rivers). As illustrated, atmospheric carbon dioxide is absorbed by plants and other photosynthetic organisms. These organisms convert carbon dioxide into organic food materials, which travel along the food chain from one trophic level to the next. Carbon dioxide reenters the environmental phase via cellular energy production (cellular respiration) of the organisms in the grazer and decomposer food chains and from the decomposition of waste and dead organisms.

For tens of thousands of years, the carbon cycle was in balance. With the advent of the Industrial Revolution, however, production of carbon dioxide began to exceed the planet's ability to absorb it. This was due in large part to widespread combustion of fossil fuels, which releases carbon dioxide, as well as deforestation. Because trees absorb enormous amounts of carbon dioxide for photosynthesis, deforestation reduced the amount of atmospheric carbon dioxide they could incorporate.

Today, over 7 billion tons of carbon—in the form of carbon dioxide—is added to the atmosphere each year. Three quarters of the carbon dioxide comes from the combustion of fossil fuels such as the gasoline in our cars. The remaining quarter stems from deforestation. As noted later, carbon dioxide traps heat escaping from Earth and reradiates it to the Earth's surface. As carbon dioxide levels increase, global temperature rises.

The Nitrogen Cycle. Nitrogen is an element essential to many important biological molecules, including amino acids, DNA, and RNA. The Earth's atmosphere contains enormous amounts of it. However, atmospheric nitrogen exists as nitrogen gas (N_2), which is unusable to all but a few organisms. To be incorporated into living organisms, atmospheric nitrogen must first be converted to a usable form–either nitrate or ammonia.

The conversion of nitrogen to ammonia is known as **nitrogen fixation**. As Figure 24-9 shows, nitrogen fixation partly occurs in the roots of certain plants called legumes. These include peas, beans, clover, alfalfa, and others. Inside small nodules in their roots live symbiotic bacteria that convert atmospheric nitrogen to ammonia. (Ammonia is also produced by certain bacteria that live in the soil.)

Once ammonia is produced, other soil bacteria convert it to nitrite and then to nitrate. Nitrates are incorporated by plants and are used to make amino acids and nucleic acids. All consumers therefore ultimately receive the nitrogen they require from plants.

Nitrate in soil is also produced indirectly from the decay of animal waste and the remains of plants and animals.

Humans alter the nitrogen cycle in at least four ways: (1) by applying excess nitrogen-containing fertilizer on farmland, much of which ends up in waterways; (2) by disposing of nitrogen-rich municipal sewage in waterways; (3) by raising cattle in feedlots adjacent to waterways; and (4) by burning fossil

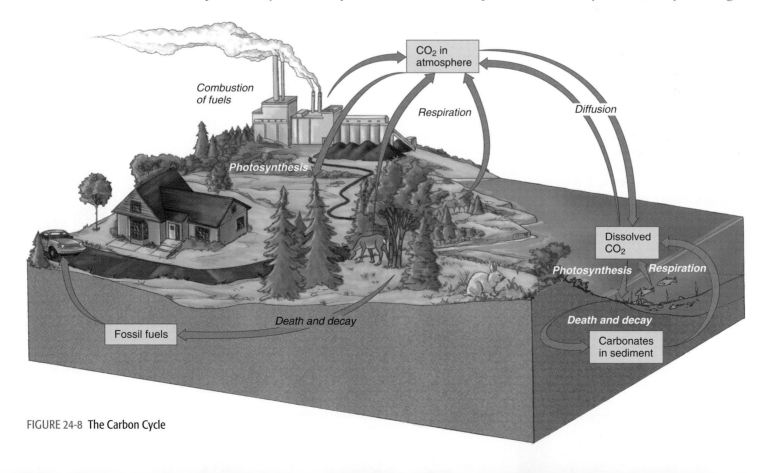

FIGURE 24-8 The Carbon Cycle

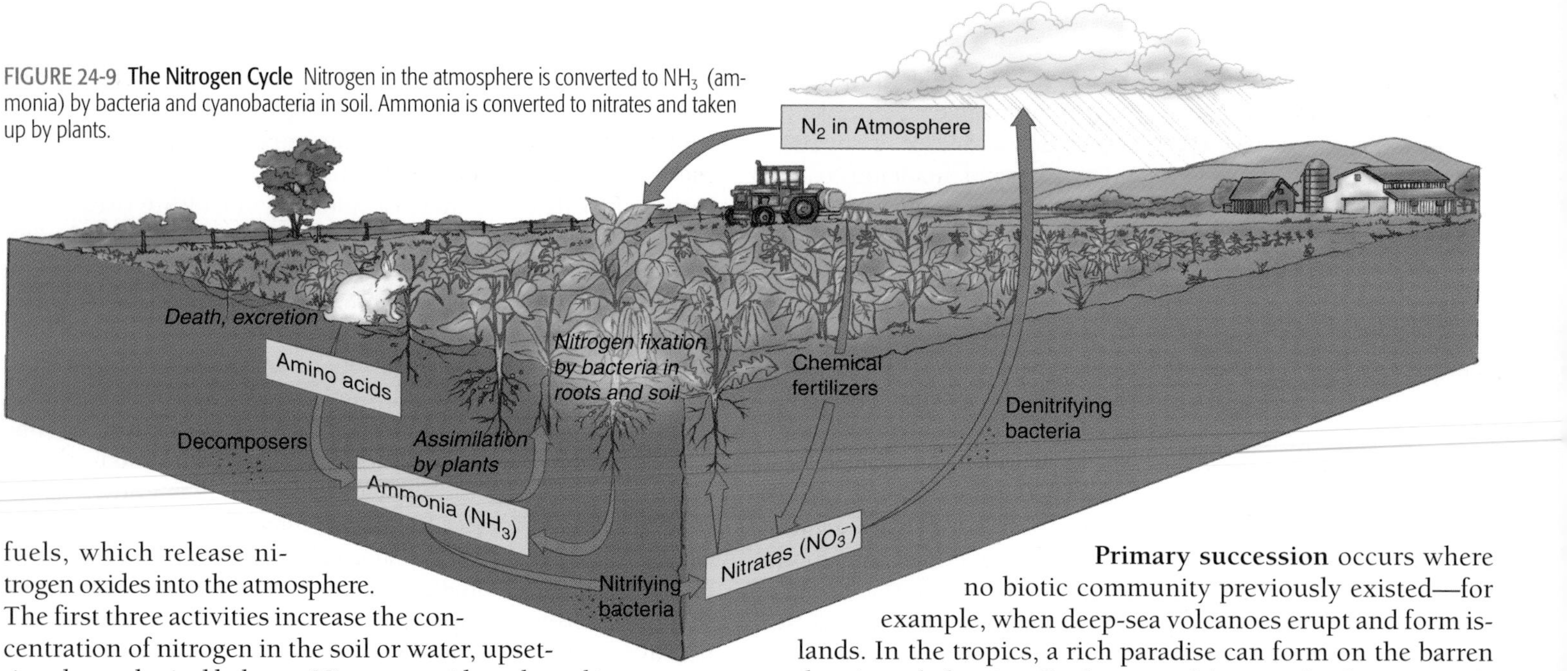

FIGURE 24-9 **The Nitrogen Cycle** Nitrogen in the atmosphere is converted to NH_3 (ammonia) by bacteria and cyanobacteria in soil. Ammonia is converted to nitrates and taken up by plants.

N₂ in Atmosphere

Death, excretion

Amino acids

Nitrogen fixation by bacteria in roots and soil

Chemical fertilizers

Denitrifying bacteria

Decomposers

Assimilation by plants

Ammonia (NH_3)

Nitrates (NO_3^-)

Nitrifying bacteria

fuels, which release nitrogen oxides into the atmosphere. The first three activities increase the concentration of nitrogen in the soil or water, upsetting the ecological balance. Nitrogen oxides released into the atmosphere by power plants, automobiles, and other sources are converted to nitric acid, which falls with rain or snow. Acid rain and snow can have devastating effects on terrestrial and aquatic ecosystems, as described shortly.

Nitrogen is a plant nutrient. It stimulates the growth of aquatic plants. As a result, nitrogen-polluted rivers and lakes may become congested with dense mats of vegetation, making them unnavigable. Sunlight penetration to deeper water levels is also impaired by the growth of plants. This causes oxygen levels in deeper waters to decline. In the autumn, when aquatic plants die and decay, oxygen levels can fall further, killing aquatic life.

Biological Succession

Succession is the progressive development of biological communities.

Biological communities, with their complex food webs and nutrient cycles, may develop in areas devoid of life or in areas that have been severely disturbed. Because the development of new communities takes place over time in a series of changes, biologists refer to this process as succession. **Succession** is a process of change in which one community is replaced by another until a mature ecosystem is formed. A mature ecosystem is one that has reached a state of long-term dynamic balance. Two types of succession exist: primary and secondary.

Primary succession occurs where no biotic community previously existed—for example, when deep-sea volcanoes erupt and form islands. In the tropics, a rich paradise can form on the barren volcanic rock, but it will take tens of thousands of years. Seeds for plants may be carried to the islands by waves or may be dropped by birds. Over time, the plants take root, then spread to cover the entire island. New species may arise as a result of genetic changes in these species. Birds may settle on the island as well. In the ensuing years, new species may evolve. The process of primary succession on rock exposed by the retreat of glaciers is shown in Figure 24-10.

FIGURE 24-10 **Primary Succession** As rock is exposed by a retreating glacier, biological communities develop in a process that can take hundreds or thousands of years.

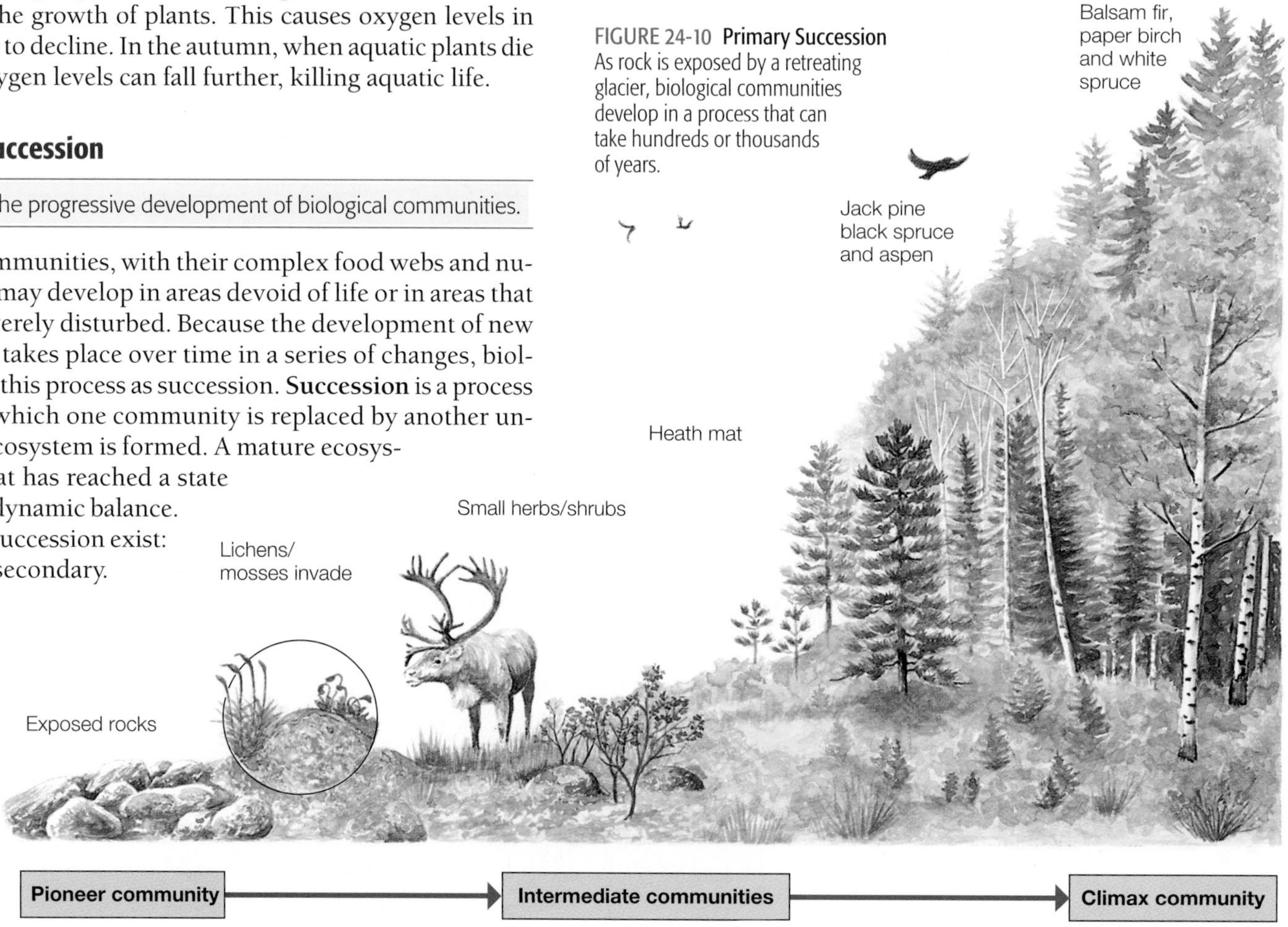

Balsam fir, paper birch and white spruce

Jack pine black spruce and aspen

Heath mat

Small herbs/shrubs

Lichens/ mosses invade

Exposed rocks

| Pioneer community | Intermediate communities | Climax community |

Secondary succession occurs when a disturbed area, such as a farm field, is no longer used. As in primary succession, it also goes through a series of changes that can take decades to complete. Eventually the original ecosystem is restored.

Secondary succession occurs much more rapidly than primary succession because soil is already present. In primary succession, soil must be formed from rock, gravel, or sand, and this can take 100–1000 years.

24-3 Overshooting the Earth's Carrying Capacity

The carrying capacity of the environment is the number of organisms it can sustain indefinitely.

Environmental problems of significance occur in both the rich, industrialized nations and the poor, less-developed nations. Although the problems vary from one nation to another, they all are signs of a common root cause: human society is exceeding the Earth's carrying capacity.

Carrying capacity is the number of organisms an ecosystem can support indefinitely—that is, the number it can sustain. Carrying capacity for all organisms, including humans, is determined by three factors: (1) food production, (2) resource supply, and (3) the environment's ability to assimilate pollution.

Exceeding the Carrying Capacity

In many countries, the human population is exceeding food production.

As the global human population grows, many nations are finding it more and more difficult to meet rising demands for food. Starvation abounds in the less-developed nations. There, an estimated 12 million people, many of them children, perish each year from malnutrition and starvation or from diseases worsened by hunger, a sure sign of populations living beyond the local carrying capacity (Figure 24-11).

Resources

Many resources are in short supply and will be depleted in the near future.

Human populations require many resources such as fuel, fiber, and building materials. Those resources that are finite such as oil, natural gas, and minerals are known as **nonrenewable resources**. Resources that replenish themselves via natural biological and geological processes are called **renewable resources**. Wind, hydropower, trees, and fishes are examples.

Today, some important nonrenewable resources such as oil and natural gas are on the decline. In fact, global oil production is expected

to reach a peak from 2004 to 2010. Global natural gas production is expected to peak by around 2015. Once production of a finite natural resource such as oil and natural gas peaks, production is no longer able to keep up with demand. Serious economic problems could ensue.

While some nonrenewable resources are beginning to decline, others are just about used up.

Renewable resources are also in danger. Some of them, such as tropical forests, are being depleted faster than they are being replanted. The rapid destruction of renewable and nonrenewable resources is another sign that we're living beyond the Earth's carrying capacity.

Pollution

Pollution from human activities exceeds the environment's assimilative capacity.

Carrying capacity is also determined by the environment's ability to assimilate and degrade pollutants. In natural ecosystems, wastes are usually diluted to harmless levels or broken down and recycled in nutrient cycles. Unfortunately, human populations often produce amounts of wastes that overwhelm these cycles, resulting in pollution of air and water that can be detrimental to life.

FIGURE 24-11 **Food Shortages** Overpopulation, drought, political turmoil, and mismanagement of farmland are the chief causes of hunger and starvation. Children die by the thousands every day.

24-4 Overpopulation: Problems and Solutions

> Overpopulation is a condition in which populations exceed the ability of the environment to supply resources and/or assimilate wastes.

One reason people are living beyond the carrying capacity, say many scientists, is overpopulation. **Overpopulation** can be explained in six words: too many people, reproducing too rapidly. Currently, the world population is about 6.4 billion and is increasing at a rate of 1.4% per year. Although the growth rate may seem small, it translates into 84 million new people every year, or about 240,000 people being added to the world population every day. If the current rate continues, world population could reach 9 billion people by the year 2050.

baby born in the United States will use 20–40 times as many resources as a newborn in India. This means the 290 million Americans alive in 2005 caused as much environmental damage as 5.8 to 11.6 billion people in the Third World!

The human population has not always been so large, nor has it always grown so rapidly. As Figure 24-12a shows, it was not until the last 200 years that global human population began to skyrocket. This growth was stimulated by better sanitation, improvements in medicine, and advances in technology. Pollution, resource depletion, hunger, starvation, and many other environment problems have also grown in concert with the skyrocketing human population.

Where Is Overpopulation Occurring?

> Overpopulation is a problem in virtually all countries, rich and poor.

Today, the most rapid growth is occurring in three areas: Africa, Asia, and Latin America. In Europe, the population is shrinking. In the United States, population growth is among the fastest of all more-developed nations.

Most people view overpopulation as a problem of the less-developed nations. Actually, overpopulation is a problem in all countries. It is as serious in the United States as it is in Bangladesh. The reason for this apparent irony is the high standard of living in the rich, industrialized countries. The higher the standard of living, the greater the energy and resource consumption and the greater the impact on the environment. In fact, each

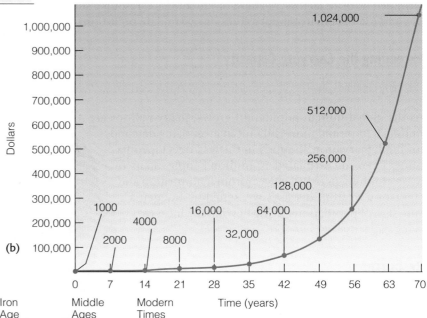

(b)

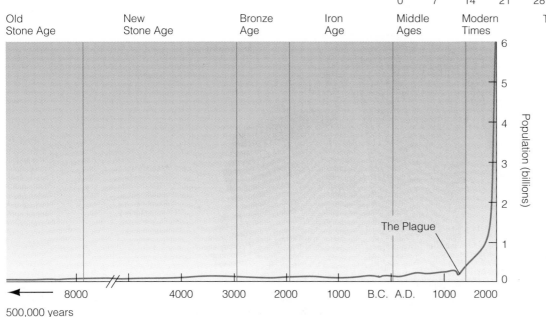

(a) 500,000 years

FIGURE 24-12 **Exponential Growth of the World Population** (a) World population. (b) Exponential growth of a bank account starting with $1000 at 10% interest.

Benefits of Reducing Population Growth

Reducing population growth will help reduce environmental problems.

Most world leaders agree that reducing the rate of population increase will help both less-developed and more-developed nations solve many pressing social and environmental problems such as hunger, environmental deterioration, habitat loss, species extinction, and resource shortages.

Slowing the growth of the human population may not be sufficient to create an enduring human presence, however. Over the next 50 years, some experts believe that it will very likely be necessary to reduce the size of the human population. The principal means of reducing population size is through attrition—reducing the birth rate so that it falls below the death rate. Under such conditions, populations will begin to shrink.

24-5 Resource Depletion: Eroding the Prospects of All Organisms

The human population depends on a variety of resources for its survival and well-being—among them, forests, soil, water, minerals, and oil. Today, however, many of these resources are in danger.

Forests

Humanity is destroying the world's forests, but massive replanting efforts could reduce, or even halt, the problem.

At one time, tropical rain forests covered a region about the size of the United States (Figure 24-13). Today, about half of those forests are gone. Because of the rapid increase in population size and the resulting timber harvests, most of the remaining tropical rain forests could be destroyed within your lifetime. Along with them, perhaps as many as a million species could vanish. Making matters worse, in the tropics, only 1 tree is replanted for every 10 trees that are cut down. In tropical Africa, this ratio is 1 to 29.

The decline of the world's forests is not limited to the tropics. In the United States, for example, much of the forested land has been cut down and converted to other uses.

Because global deforestation exceeds reforestation and because population and demand for resources continue to climb, many environmentalists are urging sharp cutbacks in demand for timber and wood products and other strategies that will ease the pressure on forests. Smaller homes, paper recycling, widespread tree planting, and better forest management could all help avert a shortage of timber in the coming years and protect remaining forests.

Saving forests is not just a matter of ensuring a steady supply of wood and wood products. Forest conservation also helps purify air and reduces carbon dioxide buildup in the atmosphere. It protects watersheds, reduces soil erosion, and maintains recreational opportunities as well. Tropical forests are a potential source of new medicines and food plants that could help feed the world's people. Forest conservation also protects the habitat of wild species, many of which are endangered. (For a discussion of extinction, see Point/Counterpoint.)

Soil

Soil erosion, like deforestation, is also a worldwide phenomenon.

Although soil erosion occurs naturally, it is often greatly accelerated by human activities such as farming, construction, and mining (Figure 24-14). In the past 100 years, about one-third of the fertile topsoil on American farmland has been eroded away by

FIGURE 24-13 **Tropical Rain Forests** Deforestation: a desperate effort to stave off poverty results in an environmental disaster. These farmers live near the Andasibe reserve in Madagascar, an island where the per capita income is less than $250 a year. Clearing the rain forest allows these impoverished farmers to plant rice and graze cattle. But the environmental price may be too steep. In Madagascar alone, 80% of the rain forest has been destroyed.

FIGURE 24-14 **Soil Erosion** Millions of acres of farmland are destroyed each year because of poor land management practices that lead to severe soil erosion.

Why Worry About Extinction?

Humans Are Accelerating Extinction by David M. Armstrong

Evolution is the process of change in gene pools. When one gene pool becomes reproductively independent of another, a new species has formed. Such speciation generates species; extinction takes them away. Simply put, extinction is a failure to adapt to change, the termination of a gene pool, and the end of an evolutionary line.

Extinction is a natural process. Most of the species that have lived on this planet are now extinct. The 3–30 million species on Earth today are no more than 1%–10% of the species that have evolved since life began about 3.5 billion years ago. Given these facts, why are thoughtful people concerned about endangered species? After all, history makes it clear that—given enough time—all species will become extinct.

The basis for concern is that today the natural process of extinction is proceeding at an unnatural rate. Let us estimate by how much human activity has accelerated rates of extinction. The lifespan of species seems to average from 1 million to 10 million years. Assume (to be conservative) that the average longevity of a species of higher vertebrates is 1 million years. In round numbers (to make calculations easy), there are 10,000 species of birds and mammals. So, on average, one species ought to go extinct each century. However, between 1600 and 1980, at least 36 species of mammals and 94 species of birds became extinct. That is about 0.29 species per year, 29 times the natural rate.

David M. Armstrong teaches science for non-scientists at the University of Colorado at Boulder and has written several books on the mammals and ecology of the Rocky Mountain region.

What does it mean to increase a rate by 29 times? The speed limit is 55 miles per hour. Exceed the speed limit by 29-fold, and you are moving 1,595 miles per hour, over twice the speed of sound. The difference between natural rates of extinction and present, human-influenced rates is analogous to the difference between a casual drive and Mach 2! Is that a problem? You decide: concern is a moral construct, not a scientific one.

Several human activities have contributed—mostly inadvertently—to accelerating rates of extinction. The dodo and the passenger pigeon were extinguished by overhunting. Wolves and grizzly bears were exterminated over much of their ranges as threats to livestock. The black-footed ferret was driven to the verge of extinction because prairie dogs, its staple food, were poisoned as agricultural pests. The smallpox virus was exterminated in the "wild" (but survives in a half-dozen laboratories).

Habitat change is the most important cause of endangerment and extinction. Clearing forests for agriculture has decimated the lemurs of Madagascar. Chemical pesticides led to the decline of the peregrine falcon. Introducing exotic species (like goats on the Galapagos and mongooses in Hawaii) displaces native animals and plants. Developing the Amazon Basin is a habitat alteration, and a cause of extinction, on an unprecedented scale.

Many urge saving species for their aesthetic value. Whooping cranes are beautiful, and part of the beauty is that they are products of a marvelous evolutionary process. Most concern about accelerated extinction, however, stresses economic value. A tiny fraction of Earth's seed plants are used commercially. Perhaps an obscure plant like jojoba will become a source of oil more reliable than that beneath the sands of Saudi Arabia. Wild grasses have furnished genes that improved disease-resistance in wheat. Numerous wild animals (like musk ox, kudu, and whales) could contribute protein to the human diet. Wild species may have medical value; penicillin, after all, was once merely an obscure mold on citrus fruit. Some sensitive species are useful monitors of environmental quality, and the presence of healthy populations of many species may promote greater stability or resilience of ecosystems. Naturalist Aldo Leopold noted that we humans have a way of "tinkering" with the ecosphere to see how it works. Given that, we ought to remember the first rule of tinkering: never throw away any of the parts.

"Extinction is forever," and extinction impoverishes both Earth's ecosystems and the potential richness of human life. Borrowing again from Aldo Leopold, I believe we should be concerned about unnaturally rapid extinction because such concern is part of a "right relationship" between people and the landscapes that nurture and inspire them.

Biologist Sir Julian Huxley noted that "we humans find ourselves, for better or worse, business agents for the cosmic process of evolution." We hold power over the future of the biosphere, the power to destroy or to preserve. German philosopher George Hegel noted that freedom (including the power to destroy species) implies responsibility (to preserve them). I agree. The question of human-accelerated extinction boils down to a simple ethical question, "Does posterity matter?" Some of us have ethics that are human-centered. We ask simply, "Do my children deserve a life as rich, with as much opportunity, as mine?"

If they do, then we have the responsibility to choose restraint. (Perhaps my life has been rich enough without the dodo, but I am reluctant to make that judgment for future generations.)

Extinction Is the Course of Nature by Norman D. Levine

Evolution is the formation of new species from preexisting ones by a process of adaptation to the environment. Evolution began long ago and is still going on. During evolution, those species better adapted to the environment replaced the less well adapted. It is this process, repeated year after year for millennia, that has produced the present mixture of wild species. Perhaps 95% of the species that once existed no longer exist.

Human activities have eliminated many wild species. The dodo is gone, and so is the passenger pigeon. The whooping crane, the California condor, and many other species are on the way out. The bison is still with us because it is protected, and small herds are raised in semicaptivity. The Pacific salmon remains because we provide fish ladders around our dams so it can reach its breeding places. The mountain goat survives because it lives in inaccessible places. But some thousands of other animal species, to say nothing of plants, are extinct, or soon will be.

Some nature lovers weep at this passing and collect money to save species. They make lists of animals and plants that are in danger of extinction and sponsor legislation to save them.

I don't. What the species preservers are trying to do is to stop the clock. It cannot and should not be done.

Extinction is an inevitable fact of evolution, and it is needed for progress. New species continually arise, and they are better adapted to their environment than those that have died out.

Extinction comes from failure to adapt to a changing environment. The passenger pigeon did not disappear because of hunting alone, but because its food trees were destroyed by land clearing and farming. The prairie chicken cannot find enough of the proper food and nesting places in the cultivated fields that once were prairie.

And you cannot necessarily introduce a new species, even by breeding it in tremendous numbers and putting it out into the wild. Thousands of pheasants were bred and set out year after year in southern Illinois, but in the spring of each year there were none left. Another bird, the capercaillie, is a fine, large game bird in Scandinavia, but every attempt to introduce it into the United States has failed. An introduced species cannot survive unless it is preadapted to its new environment.

A few introduced species are preadapted and some make spectacular gains. The United States has received the English sparrow, the starling, and the house mouse from Europe, and also the gypsy moth, the European corn borer, the Mediterranean fruit fly, and the Japanese beetle. The United States gave Europe the gray squirrel and the muskrat, among others. The rabbit took over in Australia, at least for a time.

The rabbit and the squirrel were successful on new continents because their requirements are not as narrow as those of species that failed. Today, adjustment to human-made environments may be just as difficult as adjustment to new continents. The rabbit and the squirrel have succeeded in adjusting to the backyard habitat, but most wild animals have disappeared.

Human-made environments are artificial. People replace mixed grasses, shrubs, and trees with rows of clean-cultivated corn, soybeans, wheat, oats, or alfalfa. Variety has turned into uniform monotony, and the number of species of small vertebrates and invertebrates that can find the proper food to survive has become markedly reduced. But some species have multiplied in these environments and have assumed economic importance; the European corn borer in this country is an example.

Would it improve Earth if even half of the species that have died out were to return? A few starving, shipwrecked sailors might be better off if the dodo were to return, but I would not be. The smallpox virus has been eliminated, except for a few strains in medical laboratories. Should it be brought back? Should we bring panthers back into the eastern states? Think of all the horses that the automobile and tractors have replaced, and of all the streets and roads that have been paved and the wild animals and plants killed as a consequence. Before people arrived in America about 10,000 years ago, the animal-plant situation was quite different. What should we do? Should we all commit suicide?

Evolution exists, and it goes on continually. People are here because of it, but people may be replaced someday. It is neither possible nor desirable to stop it, and that is what we are trying to do when we try to preserve species on their way out. It can be done, I think, but should we do it to them all? Or to just a few, as we are doing now?

Norman D. Levine was a professor emeritus at the College of Veterinary Medicine and Agricultural Experiment Station, University of Illinois at Urbana. His research interests covered parasitology, protozoology, and human ecology. This essay is from *"Evolution and Extinction,"* BioScience 39:38, and is copyright © 1989 by the American Institute of Biological Sciences.

Sharpening Your Critical Thinking Skills

1. Summarize the key points of both authors.
2. Do you see any flaws in the reasoning of either author?
3. Which viewpoint do you adhere to? Why?

www.jbpub.com/humanbiology/5e

Visit Human Biology's Internet site for links to web sites offering more information on this topic.

FIGURE 24-15 **Farmland Conversion** Cities are often surrounded by excellent farmland soils that drain well and are flat and highly productive. Many of the features that make soils suitable for farmland also make them suitable for building. This scene, unfortunately, is all too common in expanding urban areas.

water and wind, largely because of poor land management. Although efforts have been made to reduce soil erosion in the United States and other countries, high levels of erosion still occur.

Globally, annual erosion rates are 18–100 times greater than rates at which soil is regenerated. (The average renewal time is 500 years, but it ranges from 200 to 1000 years.) Making matters worse, millions of acres of farmland are lost to urban and suburban sprawl, highway construction, and other human activities worldwide (Figure 24-15). In the United States, approximately 3500 acres of rural land are lost every day. That's equivalent to 1.25 million acres a year, or a strip 0.3 mile wide extending from New York City to San Francisco.

These figures paint a rather grim picture for the long-term future of food production here and abroad, especially when viewed in light of increasing populations. The loss of topsoil can be stopped, and soils can be replenished. However, worldwide conservation efforts are needed, as are measures to reduce population growth, which help to reduce the conversion of farmland to other uses such as housing. Additional solutions shown in Table 24-1 could alleviate world hunger and could also reduce the loss of farmland and rangeland.

Water

Many areas of the world are facing water shortages or will soon face them as the human population and demand for water increase.

Water shortages are becoming acute worldwide. Many of the most populous nations, including China, India, and the United States are facing severe shortages in agricultural areas. Water

shortages generally result because too many people are drawing on limited water supplies. Global warming may also be causing shortages.

Many steps can be taken to conserve water and stretch available supplies. In the agricultural sector, for instance, lining irrigation ditches with concrete or using pipes rather than open ditches to transport water to fields can dramatically reduce losses due to evaporation. More efficient sprinklers, computerized

TABLE 24-1	Some Solutions to Alleviate World Hunger
Reduce population growth.
Reduce soil erosion.
Reduce desertification.
Reduce farmland conversion.
Improve yield through better crop strains.
Improve yield through better soil management.
Improve yield through fertilization.
Improve yield through better pest control.
Reduce spoilage and pest damage after harvest.
Use native animals for meat production.
Tap farmland reserves available in some countries.

systems that monitor soil moisture so that farmers know exactly how much irrigation water is required, and other measures can also help.

Oil

Most students alive today will see the end of oil in their lifetimes.

Oil is the lifeblood of modern society. In the United States, for example, oil supplies approximately 40% of our annual energy demand. But the supply of oil is finite. No one knows how much oil is left in Earth's crust, but as noted earlier, most experts agree that oil production worldwide will peak between 2004 and 2010. At this point, demand for oil products including gasoline and jet fuel will outstrip production. This, in turn, could cause major economic problems including runaway inflation and economic stagnation.

The first step in meeting future demand is energy efficiency. The efficient use of oil and its many important byproducts, such as gasoline, diesel fuel, and home heating oil, can help us stretch current oil supplies considerably. More efficient vehicles and greater reliance on mass transit in urban areas can also extend oil supplies.

Alternative fuels can help meet future demand. Ethanol produced from corn, wheat, and other crops, can power automobiles and trucks. Another potentially important fuel is biodiesel, a fuel produced from vegetable oils and hydrogen. Hydrogen is made from water. It burns cleanly and is available from an abundant resource.

24-6 Pollution

As noted earlier, waste and pollution from human society is overwhelming many nutrient cycles, poisoning other species (and ourselves). This section recaps four of the most serious waste problems.

Global Warming

Global warming results from the release of carbon dioxide and other air pollutants

Carbon dioxide is produced during the combustion of all organic materials—most importantly, fossil fuels. In normal concentrations in the atmosphere, carbon dioxide has a warming effect on the planet. Acting much like the glass in a greenhouse, it traps heat escaping from the Earth and radiates it back to the surface. Carbon dioxide is, therefore, known as a **greenhouse gas**. A little bit of carbon dioxide is essential to life on Earth. In fact, without carbon dioxide in the atmosphere, the planet would be about 55°F (30°C) cooler than it is. Too much, however, may lead to overheating, a phenomenon called **global warming**.

Over the past 100 years, global carbon dioxide levels have increased about 32.5%, principally as a result of industrialization powered by the combustion of fossil fuels and deforestation (Figure 24-16a). Several other pollutants also contribute to global warming, however. These include methane from livestock and ozone-destroying chlorofluorocarbons (CFCs) and their replacements, the hydrochlorofluorocarbons (HCFCs), both of which are used in refrigerators, air conditioners, and freezers.

Because of the dramatic increase in the release of greenhouse gases, many atmospheric scientists believe that global temperature is on the rise and may rise dramatically in the coming decades, causing an extraordinary shift in climate. Figure 24-16b shows average global temperature over the past fifty years.

Impacts of Global Warming

The environmental impacts of global warming could be severe.

According to latest projections based on the present rate of increase in greenhouse gases, global temperatures could be 5° to 10°F hotter by 2100. The models suggest that global rainfall patterns will shift dramatically as a result of warming. Computer simulations of U.S. climate suggest that the Midwest and much of the western United States will be drier and hotter than they are today. If that happens, many Midwestern farmers will be driven out of business. As rainfall declines in this agriculturally productive region, farming may intensify in the northern states. Overall agricultural productivity in the United States may fall, however, because northern soils are not as rich as those in the Midwest. Food shortages caused by the decline in U.S. agricul-

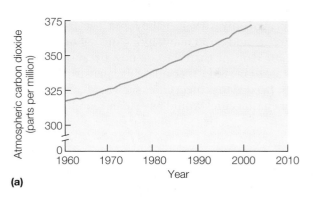

(a)

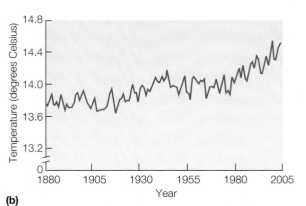

(b)

FIGURE 24-16 Global Carbon Dioxide and Temperature Trends (a) Carbon dioxide levels in the atmosphere have risen dramatically since 1958. (b) Graph of average global temperature since 1950.

FIGURE 24-17   Mount Rainier Glaciers   This enormous glacier and others like it could melt as the Earth gets warmer, raising the sea level.

ture in the Midwest and rising food prices could affect the economy in profound ways. Computer climate models predict that the southern United States and Pacific Coast may be wetter but hotter.

A rise in global temperature is expected to melt glaciers and the polar ice caps (Figure 24-17). Warmer temperatures would also expand the volume of the seas. Together, melting ice and expanding oceans would cause the sea level to rise. In the past 50 years, sea level has risen 10–12 centimeters (4–6 inches). By 2100, the computer models suggest that the sea level will rise 50 centimeters (2 feet) because of global warming.

In the United States, approximately half of the population lives within 50 miles of the ocean, and many large cities such as Miami are located only a few feet above sea level (Figure 24-18). A rising sea level would flood many low-lying regions. Storms could cause more damage than they do now because high water would be able to move farther inland. Expensive dikes and levees would be needed to protect cities such as Miami and New Orleans. Other coastal cities would have to be rebuilt on higher

FIGURE 24-18   Miami Underwater?   A rising sea level due to global warming could flood many coastal cities the world over.

ground, at a staggering cost. A rising sea level would be particularly hard on Bangladesh and other Asian countries with extensive lowland rice paddies.

In 2000, a report by the National Research Council predicted that rising temperatures will cause the tropical climate of equatorial regions to shift northward into the lower-tier states. Their climate, in turn, will shift to the mid-tier states and the climate of the top-tier states would move into Canada. The effect of such a rapid shift on vegetation and wildlife could be devastating.

Rising global temperatures could also result in the spread of tropical diseases into other areas. Warmer climates are already causing the spread of insects that carry malaria to higher altitudes and higher latitudes in Africa and Central America.

Global warming could cause considerable human suffering, with many deaths attributed to heat waves. Global warming may also trigger more violent weather, especially hurricanes and tornados that cause billions of dollars worth of damage and loss of human life. A recent study of severe storms in the United States showed that violent downpours are on the rise.

These predictions sound dire—and indeed they are. But a small number of scientists believe that we don't know enough about the atmosphere and the impact of greenhouse gases to know for certain whether the predictions will come true. Global climate is extremely complex and several factors could dampen the warming trend. Other factors could worsen it.

Solutions to Global Warming

Solving global warming requires massive action, and soon.

While scientists and politicians debate global warming, the Earth appears to be getting hotter. In fact, the 1990s was the hottest decade on record in the past 100 years.

Reducing the many impacts of global warming, say experts, will require sharp reductions in fossil-fuel consumption by energy efficiency and conservation, and the use of alternative fuels will be required. Global reforestation is essential. Additional strategies are shown in Table 24-2. We as individuals can also help (Table 24-3).

Acid Deposition

Large portions of the world are threatened by acid rain and snow.

Throughout the world, thousands of lakes are turning acidic, killing fishes and other aquatic organisms (Figure 24-19). Acidification is most prevalent in the northeastern United States, southeastern Canada, and in Sweden and Norway, where dying lakes number in the thousands.

The acidification of lakes is a result of acids falling from the skies in rain or snow. They are produced from two atmospheric pollutants: sulfur dioxide and nitrogen dioxide. These pollutants arise chiefly from the combustion of fossil fuels: coal, oil, gasoline, natural gas, and jet fuel. In the atmosphere, these gases combine with water and oxygen to form sulfuric and nitric acids, which are responsible for **acid rain** and **acid snow**.

In the United States and Europe, acid deposition has been worsening for five decades, largely as a result of increased fossil-fuel combustion. Today, acid rain and snow are commonly

TABLE 24-2	Measures to Reduce Global Climate Change

Reduce the rate of population growth.

Switch from coal- and oil-fired power plants to natural gas, which produces much less carbon dioxide per unit of electricity.

Implement the technologies that burn coal more efficiently.

Boost automobile efficiency.

Expand mass transit.

Develop alternative liquid fuels for transportation.

Improve the efficiency of industry.

Make new and existing homes more energy-efficient.

Build many new homes that use solar energy for space heating.

Reduce global deforestation.

Begin a massive global reforestation effort.

Reduce consumption of unnecessary items.

Expand recycling efforts.

FIGURE 24-19 **Victims of Modern Society** These fish were killed by acids from acid deposition.

encountered downwind from virtually all major population centers. The acids come from pollutants produced by power plants, motorized vehicles, factories, and homes.

Acid deposition changes the pH of lakes and streams, killing fishes and other aquatic organisms. Acids falling on land also dissolve toxic minerals such as aluminum from the soil and wash them into surface waters, killing fish.

Figure 24-20 shows areas of North America most susceptible to acid deposition. These regions are generally mountainous and contain soils with little capacity to neutralize acids. Consequently, acids that fall on the land quickly wash into nearby lakes and streams.

Acids also damage crops and trees, directly and indirectly. Some scientists believe that massive forest diebacks occurring throughout the world may be the result of acidic rainfall and acidic fog that often blankets forests (Figure 24-21). Finally, acid deposition also damages buildings, statues, and other structures. The estimated cost in the United States is about $5 billion per year.

TABLE 24-3	Individual Actions That Can Reduce Global Climate Change

Automobile energy savings
  Buy energy-efficient vehicles.
  Reduce unnecessary driving.
  Car-pool, take mass transit, walk, or bike to work.
  Combine trips.
  Keep your car tuned and your tires inflated to the proper level.
  Drive at or below the speed limit.

Home energy savings
  Increase your attic insulation to R30 or R38.
  Caulk and weather-strip your house.
  Add storm windows and insulated curtains.
  Install an automatic thermostat.
  Turn the thermostat down a few degrees in winter, and wear
    warmer clothing.
  Replace furnace filters when needed.
  Lower water heater setting to 120°–130°F.
  Insulate water heater and pipes, install a water heater insulation
    blanket, and repair or replace all leaky faucets.

Home energy savings—*continued*
  Take shorter showers.
  Use cold water as much as possible.
  Avoid unnecessary appliances.
  Buy energy-efficient appliances.
  Use low-energy light bulbs.

Reducing waste and resource consumption
  Recycle at home and at work.
  Avoid products with excessive packaging.
  Reuse shopping bags.
  Refuse bags for single items.
  Use a diaper service instead of disposable diapers.
  Reduce consumption of throwaways.
  Donate used items to Goodwill, Disabled American
  Veterans, the Salvation Army, or other charities.
  Buy durable items.
  Give environmentally sensitive gifts.

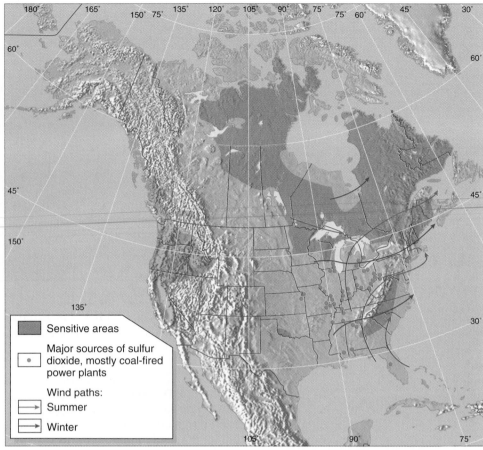

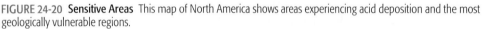

FIGURE 24-20 **Sensitive Areas** This map of North America shows areas experiencing acid deposition and the most geologically vulnerable regions.

FIGURE 24-21 **Forest Die-off** Ghostly remains of trees killed by acid deposition in the Pisgah National Forest, North Carolina.

Reducing the deposition of acids will require a dramatic reduction in the release of sulfur dioxide and nitrogen dioxide from power plants, factories, and automobiles. One way of reducing sulfur dioxide is the smokestack scrubber, a device that traps sulfur dioxide gas escaping from power plants, removing up to 95% of this pollutant. Installing and operating a scrubber is rather expensive, and some utilities have objected to this strategy. So far, most utilities have chosen the cheaper low-sulfur coal strategy.

Another even cheaper strategy is energy efficiency. By using fossil-fuel energy more efficiently, we reduce the overall rate of fossil fuel combustion. This reduces sulfur dioxide and nitrogen dioxide release. Renewable energy is another more important strategy for reducing acid deposition. Solar and wind energy, for example, meet our needs, but release no acid precursors.

The Ozone Layer

The ozone layer is endangered by human pollutants.

Encircling the Earth, 20–30 miles above its surface, is a region of the atmosphere, the **ozone layer**, containing a slightly elevated level of ozone gas. The ozone layer shields the Earth from harmful ultraviolet radiation from the Sun. Early in the evolution of life, in fact, the formation of the ozone layer probably allowed the colonization of land by plants and animals.

Today, however, the ozone layer is being destroyed by chemicals released into the atmosphere by modern society. The most potent destroyer of ozone is a class of chemicals known as the chlorofluorocarbons (CFCs). Once used as spray can propellants, refrigerants, blowing agents for plastic foam, and cleansing agents, CFCs are highly stable molecules.

In the early 1970s, scientists discovered that CFCs released from human activities gradually drift into the upper atmosphere. Here they are broken down by sunlight, where their byproducts react with and destroy ozone molecules.

In 1988, a panel of atmospheric scientists reviewing 20 years of satellite data on ozone levels concluded that the ozone layer is on the decline. (Figure 24-22). Ozone depletion was particularly evident at the poles. Scientists found that each year a giant hole in the ozone layer about the size of the United States formed over Antarctica. Ozone levels in the hole declined by as much as 50%.

Declining ozone levels increase ultraviolet radiation striking the Earth. Like many natural components of our environment, ultraviolet radiation is beneficial. In small amounts, ultraviolet radiation tans light skin and stimulates vitamin D

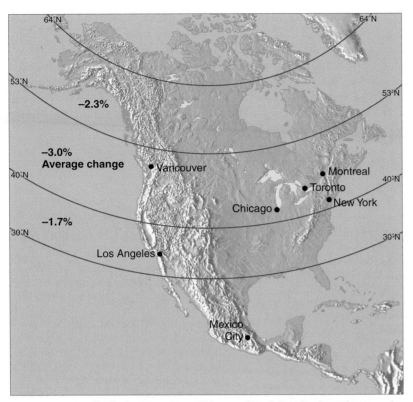

FIGURE 24-22 Decline in the Ozone Layer This map of North America shows the ozone depletion at different latitudes between 1969 and 1988.

production in the skin. However, excess ultraviolet exposure can cause problems. In humans, it can cause serious skin burns, cataracts (clouding of the eye's lens), skin cancer, and premature aging. Over the next five decades, the EPA estimates that ozone depletion will result in approximately 200,000 cases of skin cancer in the United States.

Studies of skin cancer show that light-skinned people are much more sensitive to ultraviolet radiation than more heavily pigmented individuals. Land- and water-dwelling plants could also suffer. Intense ultraviolet radiation is usually lethal to plants. Smaller, nonfatal doses damage leaves and inhibit photosynthesis. They can also cause mutations and stunt growth. Declining ozone and increasing ultraviolet radiation could also cause dramatic declines in commercial crops, costing billions of dollars a year. It may also damage certain commercially valuable tree species.

Finally, ultraviolet light is harmful to paints, plastics, and other materials. Losses from further decreases in the ozone layer could cost society enormous amounts of money.

Fortunately, ozone depletion has been addressed head on by the nations of the world. They have signed three treaties to ban ozone-depleting chemicals. Despite the bans, it could take the ozone layer 50 to 100 years to recover.

24-7 Health and Homeostasis

The trends and problems discussed in this chapter could, say many experts, spawn serious problems. What many people forget when they deal with such issues is that these problems could combine to produce impacts far greater than anticipated. That is, they produce a synergistic effect. For example, soil erosion, global warming, farmland conversion, acid deposition, ozone depletion, and population growth together could produce massive food shortages.

Solving these problems will require tougher laws and regulations, reductions in population growth, more efficient technologies, and changes in human behavior. Many people believe that broader societal changes are required. One suggestion is that we build a sustainable society, one that lives within the carrying capacity of the environment. Living within the limits of nature means achieving a population size and a way of life that do not exceed the planet's ability to supply food and other resources and to handle wastes. It means promoting human societies that manage their affairs in ways that protect the Earth's homeostatic mechanisms.

Building a sustainable society will require a shift to the patterns seen in nature. It will require that we become more efficient in our use of resources; recycle to the maximum extent possible; shift to renewable resources, especially energy; restore ecosystem damage; and stabilize, if not reduce, world population.

Building a sustainable society will require a lifetime of commitment on the part of businesses, governments, and individuals. Some changes required to create an enduring human presence may come as a result of legislative actions–new laws and regulations.

Technological innovations will also help us become a sustainable society. For example, a new line of compact fluorescent light bulbs is now available. These bulbs fit into a regular light socket and use one-fourth as much energy as a regular incandescent bulb (see Figure 24-23). Photovoltaics, thin silicon wafers that produce electricity from sunlight, could also help us make the transition to solar energy. That's how I power my home in Colorado.

**FIGURE 24-23
Energy Efficient
Lighting** Many new
light bulbs use only 24%
of the energy of standard light bulb
and last 10 times longer, saving $20 to $40 over their lifetime.

But new technologies are not the only, or even the most important, answer. Many existing technologies simply need to be installed. Insulation, weather stripping, flow restrictors for shower heads and faucets, and efficient appliances already on the market can make tremendous inroads into waste.

Many proponents also believe that we will eventually need to reduce consumption and alter our lifestyles. Taking shorter showers, shaving without the water running, turning off lights, and hundreds of other small actions on the part of individuals, when combined with similar actions by millions of other people, can result in significant cuts in resource demand.

Ending the waste and pollution—and soon—is crucial if we are to protect the environment and ourselves. The balance of nature is, after all, the balance that sustains us. We risk upsetting it at our own peril. Protecting the planet is the ultimate form of health-care.

SUMMARY

An Introduction to Ecosystems

1. Humans depend on the environment for many free services and resources such as food, oxygen, clothing, materials, and much else.
2. Ecology is the study of ecosystems. It examines the relationship of organisms to their environment and the many interactions between the abiotic and biotic components of ecosystems.
3. The living "skin" of the planet is called the biosphere. It extends from the bottom of the oceans to the top of the highest mountains.
4. The biosphere is a closed system in which materials are recycled over and over. The only outside contribution is sunlight, which powers virtually all biological processes.
5. The Earth's surface is divided into large biological regions, or biomes, each with a characteristic climate and characteristic plant and animal life. The oceans can also be divided into biological regions, known as aquatic life zones.
6. An ecosystem consists of a community of organisms, its environment, and all of the interactions between them.
7. A group of organisms of the same species living in a specific region constitutes a population. Two or more populations occupying that region form a community.
8. The physical space a species occupies is called its habitat. A species' niche includes its habitat, its position in the food chain, and so on.

Ecosystem Function

9. Virtually all life on Earth depends on plants and other producers, organisms that synthesize organic materials from sunlight, carbon dioxide, and water via photosynthesis.
10. Organisms dependent on producers and other organisms for food are called consumers. Four types of consumers are present: herbivores, carnivores, omnivores, and detrivores.
11. All organisms are part of one or more food chains. Food chains that begin with plants consumed by grazers (herbivores) are known as grazer food chains. Those that begin with

animal waste or the remains of dead organisms are decomposer food chains. Food chains are parts of larger food webs.
12. The position of an organism in a food chain is called its trophic level.
13. Nutrients necessary for life flow through nutrient cycles. Humans can interrupt nutrient cycles, locally and globally.
14. The term *succession* refers to a series of changes in an ecosystem in which one community replaces another until a mature ecosystem is produced. Primary succession occurs where no community previously existed. Secondary succession occurs where a community was destroyed by natural or human events.

Overshooting the Earth's Carrying Capacity

15. Although environmental problems vary from one nation to another, they are all the result of human populations exceeding the carrying capacity of the environment.
16. Carrying capacity is the number of organisms an ecosystem can support indefinitely and is determined by food and resource supplies and by the capacity of the environment to assimilate or destroy waste products of organisms.

Overpopulation: Problems and Solutions

17. Overpopulation occurs anytime a population exceeds the carrying capacity. It is manifest in shortages of food, lack of other resources, or excessive pollution—sometimes all three.
18. The most rapid growth in human population is occurring in the developing world in parts of Africa, Asia, and Latin America. Resource depletion and food shortages are the most common problems in these regions. Many experts believe that the industrialized countries are also overpopulated. Judging from the quality of our air, water, and soils, we are clearly exceeding the Earth's carrying capacity.
19. The human population problem can be summed up in six words: too many people, reproducing too rapidly, which results in resource shortages, excessive pollution, and poverty.

Resource Depletion: Eroding the Prospects of All Organisms

20. The human population requires a variety of renewable and nonrenewable resources for survival, and many of these resources are being depleted.
21. In many parts of the world, forests are being cut down faster than they can be replaced. To prevent the destruction of the world's forests, tree planting, paper recycling, and other strategies are needed.
22. Worldwide, agricultural soils are being eroded from rangeland and farmland at an unsustainable rate, and millions of acres of farmland are being destroyed by human encroachment. The destruction of productive soils threatens the long-term prospects for food production.
23. Soil conservation and population control measures can help ensure an adequate supply of soil, but serious efforts must begin soon.
24. Many areas of the world suffer water shortages. The rise in population and the rise in demand are likely to cause further shortages in the future. The growth of the human population must be reduced and strict water conservation is needed.
25. Oil supplies, like mineral supplies, are also finite. Global oil production is expected to peak between 2004 and 2010, at which time production may not be able to meet demand.
26. By using oil much more efficiently and seeking clean, renewable alternative fuels, modern society can make a smooth transition to a sustainable energy supply.

Pollution

27. Pollution from human activities is overwhelming nutrient cycles, poisoning other species (and ourselves), and destroying the planet's homeostatic mechanisms.
28. One of the most serious threats from pollution comes from carbon dioxide. Carbon dioxide is a greenhouse gas, trapping heat in the Earth's atmosphere. This, in turn, could increase the planet's surface temperature, altering its climate, shifting rainfall patterns and agricultural zones, flooding low-lying regions, and

destroying many species that cannot adapt to the sudden change in temperature.

29. To slow global warming, many scientists recommend massive reforestation projects and dramatic improvements in the efficiency of fossil-fuel combustion and development of alternative fuels.

30. Sulfur dioxide and nitrogen dioxide are two gaseous pollutants released from power plants, factories, automobiles, and other sources. In the atmosphere, they are converted to sulfuric and nitric acid, respectively.

31. Acids fall from the sky in rain and snow. Acids alter the pH of lakes and streams, killing fish and other organisms. They also leach (dissolve and remove) toxic metals from soils that kill fishes. They destroy trees and crops and deface buildings, costing society billions of dollars a year.

32. Acid rain and snow can be reduced by installing pollution-control devices on factories and power plants and by employing preventive measures such as energy conservation and alternative fuels.

33. The ozone layer encircles the Earth, trapping ultraviolet light. It is being destroyed by chlorofluorocarbons, or CFCs, and other pollutants.

Health and Homeostasis

34. Solving our problems will not only require tougher laws and tighter regulations but also a profound change in the way we live and conduct business.

35. A sustainable society is built on five operating principles: conservation, recycling, renewable resources, restoration, and population control.

critical thinking

Thinking Critically—Analysis

This Analysis corresponds to the Thinking Critically scenario that was presented at the beginning of this chapter.

Critical thinking rules encourage us to question sources of information and their conclusions. In this example, we find that the author of this article is grossly in error when he claims that the scientific community is divided on the issue. About 99% of the nation's 700 atmospheric scientists believe that global warming is a reality; only a handful embrace the opposite view. Critics say they're on the payroll of oil companies, too. The evidence the author introduces to support his assertion—a quote from each side of the issue—is terribly misleading.

This type of reporting is quite common in newspapers, television, and magazines. Although it is intended to give a balanced view of issues, in reality, it provides an extremely unbalanced view.

What would have been a more accurate conclusion? Perhaps that not all scientists agree that global warming is occurring, but that the vast majority do. It's important to note that even though the majority of the world's atmospheric scientists agree that global warming is happening, this doesn't mean they're right.

Digging a little deeper, finding out more, often throws conclusions into question.

KEY TERMS AND CONCEPTS

Abiotic components, p. 443
Acid rain and snow, p. 458
Aquatic life zone, p. 442
Biogeochemical cycle, p. 448
Biological community, p. 443
Biomass pyramid, p. 447
Biome, p. 442
Biosphere, p. 442
Biotic components, p. 443
Carbon cycle, p. 449
Carnivores, p. 445
Carrying capacity, p. 451
Community, p. 443
Consumers, p. 445
Decomposer food chains, p. 445

Detrivores, p. 445
Ecological system, p. 443
Ecology, p. 442
Ecosystem, p. 443
Energy pyramid, p. 447
Food chain, p. 445
Food web, p. 446
Global warming, p. 457
Grazer food chains, p. 445
Greenhouse gas, p. 457
Habitat, p. 443
Herbivores, p. 445
Niche, p. 443
Nitrogen cycle, p. 449
Nitrogen fixation, p. 449

Nonrenewable resources, p. 451
Nutrient cycle, p. 448
Omnivores, p. 445
Overpopulation, p. 452
Ozone layer, p. 460
Photosynthesis, p. 445
Population, p. 443
Primary succession, p. 450
Producers, p. 445
Pyramid of numbers, p. 448
Renewable resource, p. 451
Secondary succession, p. 451
Succession, p. 450
Trophic level, p. 447
Water cycle, p. 448

CONCEPT REVIEW

1. Define the term ecology, and give examples of its proper and improper use. p. 442
2. The Earth is a closed system. What are the implications of this statement? p. 442
3. Define the following terms: biosphere, biome, aquatic life zone, and ecosystem. pp. 442–443
4. Define the following terms: habitat, niche, producer, consumer, trophic level, food chain, and food web. pp. 443–446
5. Outline the flow of carbon dioxide through the carbon cycle, and describe ways in which humans adversely influence the carbon cycle. p. 449
6. With the knowledge you have gained, explain why it is beneficial to set aside habitat to protect an endangered species. p. 454

7. Human populations in both the industrialized and nonindustrialized countries are overshooting the Earth's carrying capacity. Do you agree or disagree with this statement? Be sure to describe what is meant by overshooting the carrying capacity. p. 451
8. Why is overpopulation as much a problem in the United States as it is in Bangladesh? p. 452
9. Describe key trends in the use of forests, soils, water, minerals, and oil suggesting that humanity is on an unsustainable course. List and discuss solutions to each of the problems. pp. 453–457

10. Given trends in natural resource use, many experts believe that global population growth must stop. Do you agree or disagree? Why? Is your position supported by scientific fact or based more on a general feeling? p. 451
11. Describe the cause and impacts of each of the following: global warming, acid deposition, stratospheric ozone depletion, and hazardous wastes. pp. 457–461
12. Make a list of solutions for each of the problems you identified in Question 11. How could conservation (efficiency), recycling, renewable resources, and population control factor into the solutions? pp. 458–462

SELF-QUIZ: TESTING YOUR KNOWLEDGE

1. A biologically distinct region such as a grassland is called a _____. p. 442
2. The term _____ is used to describe a community of organisms and their physical and chemical environment. p. 443
3. A group of organisms belonging to the same species and occupying a specific area is called a _____. p. 443
4. All of the organisms in an ecosystem form a biological _____. p. 443
5. The area where an organism lives is known as its _____. p. 443
6. An animal that feeds entirely on other organisms is called a _____. p. 445
7. A _____ food chain begins with detrivores feeding on dead material. p. 445

8. The position of an organism in a food chain is known as its _____ level. p. 447
9. Carbon dioxide is returned to the atmosphere as a result of the decay of waste and also _____. p. 449
10. The conversion of atmospheric nitrogen into ammonia in the roots of plants and elsewhere is known as nitrogen _____. p. 449
11. The development of a biological community in a previously barren environment is known as _____ succession. p. 450
12. The _____ capacity is the number of organisms an ecosystem can support indefinitely. p. 451

13. The accumulation of _____ in the atmosphere, primarily from the combustion of fossil fuels and deforestation, is resulting in a warming trend known as _____ warming. p. 457
13. Sulfuric and nitric acids falling with the rain is known as _____ rain. p. 459
14. Ozone is present in the upper layer of the atmosphere. It is being depleted by chemicals once used in spray cans and cooling systems known as _____. p. 460
15. Ozone depletion could result in an increase in the amount of _____ radiation striking the Earth, which could increase skin cancer among humans. p. 460

 www.jbpub.com/humanbiology/5e

The site features eLearning, an online review area that provides quizzes, chapter outlines, and other tools to help you study for your class. You can also follow useful links for in-depth information, research the differing views in the Point/Counterpoints, or keep up on the latest health news.

Periodic Table of Elements

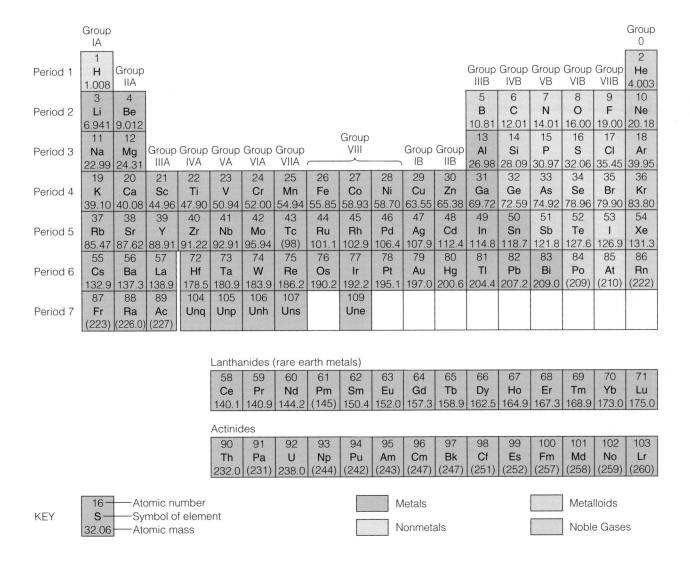

Lanthanides (rare earth metals)

58	59	60	61	62	63	64	65	66	67	68	69	70	71
Ce	Pr	Nd	Pm	Sm	Eu	Gd	Tb	Dy	Ho	Er	Tm	Yb	Lu
140.1	140.9	144.2	(145)	150.4	152.0	157.3	158.9	162.5	164.9	167.3	168.9	173.0	175.0

Actinides

90	91	92	93	94	95	96	97	98	99	100	101	102	103
Th	Pa	U	Np	Pu	Am	Cm	Bk	Cf	Es	Fm	Md	No	Lr
232.0	(231)	238.0	(244)	(242)	(243)	(247)	(247)	(251)	(252)	(257)	(258)	(259)	(260)

KEY

16 —— Atomic number
S —— Symbol of element
32.06 —— Atomic mass

Metals

Nonmetals

Metalloids

Noble Gases

The Metric System

Most Common English Units and the Corresponding Metric Units		
Measure	**English Unit**	**Metric Unit**
Weight	tons pounds ounces	metric tons kilograms grams
Length	miles yards inches	kilometers meters centimeters
Square Measure	acres square miles	hectares square kilometers
Volume	quarts and gallons fluid ounces	liters milliliters

Converting English Units to Metric Units		
Measure	**English Unit**	**Metric Unit**
Weight	1 ton (2000 pounds) 1 pound 1 ounce	0.9 metric tons 0.454 kilograms 28.35 grams
Length	*1 mile 1 yard *1 inch	1.6 kilometers 0.9 meters 2.54 centimeters
Square Measure	1 acre 1 square mile	0.4 hectares 2.59 square kilometers
Volume	1 quart *1 gallon 1 fluid ounce	0.95 liters 3.78 liters 29.58 milliliters

*Most useful conversions to know.

Converting Metric Units to English Units		
Measure	**Metric Unit**	**English Unit**
Weight	*1 metric ton 1 metric ton *1 kilogram 1 gram	2204 pounds 1.1 tons 2.2 pounds 28.35 ounces
Length	*1 kilometer 1 meter 1 centimeter	0.6 miles 1.1 yards 0.39 inches
Square Measure	*1 hectare 1 square kilometer	2.47 acres 0.386 square miles
Volume	1 liter 1 liter 1 milliliter	1.057 quarts 0.26 gallons 0.0338 fluid ounces

*Most useful conversions to know.

Glossary

Acetylcholine Neurotransmitter substance in the central and peripheral nervous systems of humans.

Acetylcholinesterase Enzyme that destroys the neurotransmitter acetylcholine in the synaptic cleft.

Acid Any chemical substance that adds hydrogen ions to a solution, such as sulfuric acid.

Acrosome Enzyme-filled cap over the head of a sperm. Helps the sperm dissolve its way through the corona radiata and zona pellucida of the ovum.

Actin microfilaments Contractile filaments made of protein and found in cells as part of the cytoskeleton. Especially abundant in the microfilamentous network beneath the plasma membrane and in muscle cells.

Action potential Recording of electrical change in membrane potential when a neuron is stimulated.

Active immunity Immune resistance gained when an antigen is introduced into the body either naturally or through vaccination.

Active site The region of an enzyme molecule that binds substrates and performs the catalytic function of the enzyme.

Active transport Movement of molecules across membranes using protein molecules and energy supplied by ATP. Moves molecules and ions from regions of low to high concentration.

Adaptation Genetically based characteristic that increases an organism's chances of passing on its genes to its offspring.

Adenine One of the purine bases found in nucleotides such as ATP.

Adenosine diphosphate (ADP) Precursor to ATP; consists of adenine, ribose, and two phosphates; *see also* Adenosine triphosphate.

Adenosine triphosphate (ATP) A molecule composed of ribose sugar, adenine, and three phosphate groups. The last two phosphate groups are attached by "high-energy bonds" that require considerable energy to form but release that energy again when broken. ATP serves as the major energy carrier in cells.

Adipose tissue Type of loose connective tissue containing numerous fat cells. Important storage area for lipids.

Adrenal cortex Outer portion of the adrenal gland. Produces a variety of steroid hormones, including cortisol and aldosterone.

Adrenal gland Endocrine gland located on top of the kidney. Consists of two parts: adrenal cortex and medulla, each with separate functions.

Adrenal medulla Inner portion of the adrenal glands. Produces epinephrine (adrenalin) and norepinephrine (noradrenalin).

Adrenalin (epinephrine) Hormone secreted under stress. Contributes to the fight-or-flight response by increasing heart rate, shunting blood to muscles, increasing blood glucose levels, and other functions.

Adrenocorticotropic hormone (ACTH) Polypeptide hormone produced by the anterior pituitary. Stimulates the cells of the adrenal cortex, causing them to synthesize and release their hormones, especially glucocorticoids.

Aerobic exercise Exercise, such as swimming, that does not deplete muscle oxygen. Excellent for strengthening the heart and for losing weight.

Agglutination Clumping of antigens that occurs when antibodies bind to several antigens.

Aging Inevitable and progressive deterioration of the body's function, especially its homeostatic mechanisms.

AIDS Acquired immunodeficiency syndrome. Fatal disease caused by the HIV virus, which attacks T helper cells, greatly reducing the body's ability to fight infection.

Albinism Genetic disease resulting in a lack of pigment in the eyes or the eyes, skin, and hair. An autosomal recessive trait.

Aldosterone Steroid released by the adrenal cortex in response to a decrease in blood pressure, blood volume, and osmotic concentration. Acts principally on the kidney.

Alga (algae, plural) Heterogeneous group of aquatic plants consisting of three major groups: green, red, and brown. Important producer essential to aquatic food chains.

Alleles Alternative form of a gene.

Allergen Antigen that stimulates an allergic response.

Allergy Extreme overreaction to some antigens, such as pollen or foods. Characterized by sneezing, mucus production, and itchy eyes.

Allosteric inhibition Enzyme regulation in which an inhibitor molecule binds to an enzyme at a site away from the active site, changing the shape or charge of the active site so that it can no longer bind substrate molecules.

Allosteric site Region of an enzyme where products of metabolic pathways bind, changing the shape of the active site. In some enzymes this prevents substrates from binding to the active site; in others, it allows them to bind. Thus, allosteric sites can either turn on or turn off enzymes.

Alveoli Tiny, thin-walled sacs in the lung where oxygen and carbon dioxide are exchanged between the blood and the air.

Ameboid motion Cellular locomotion common in single-celled organisms and some cells in the human body. The cells send out slender cytoplasmic projections that attach to the substrate "ahead" of the cell. The cytoplasm flows into the projections, or pseudopodia, advancing the organism.

Amino acid A small organic molecule that serves as a building block for proteins. Approximately 20 different amino acids are found in the proteins in the human body.

Amniocentesis Procedure whereby physicians extract cells and fluid from the amnion surrounding the fetus. The cells are examined for genetic defects, and the fluid is studied biochemically.

Amnion Layer of cells that separates from the inner cell mass of the embryo and eventually forms a complete sac around the fetus.

Amniotic fluid Liquid in the amniotic cavity surrounding the embryo and fetus during development. Helps protect the fetus.

Amylase Enzyme in saliva that helps break down starch molecules.

Anabolic steroids Synthetic androgen hormones that promote muscle development.

Analogous structures Anatomical structures that function similarly but differ in structure; for example, the wing of a bird and the wing of an insect.

Anaphase Phase of mitosis during which the chromatids of each chromosome begin to uncouple and are pulled in opposite directions with the aid of the mitotic apparatus.

Androgens Male sex steroids such as testosterone produced principally by the testes.

Anemia Condition characterized by an insufficient number of red blood cells in the blood or insufficient hemoglobin. Often caused by insufficient iron intake.

Aneurysm Ballooning of the arterial wall caused by a degeneration of the tunica media.

Anoxia Lack of oxygen.

Antagonistic Refers to hormones or muscles that exert opposite effects.

Anterior pituitary Major portion of the pituitary gland, which is controlled by hypothalamic hormones. Produces seven protein and polypeptide hormones.

Anthropoids Monkeys, the great apes, and humans.

Antibodies Proteins produced by immune system cells that destroy or inactivate antigens, including pollen, bacteria, yeast, and viruses.

Anticodon Sequence of three bases found on the transfer RNA molecule. Aligns with the codon on messenger RNA and helps control the sequence of amino acids inserted into the growing protein.

Anticodon loop Part of the transfer RNA molecule that bears three bases that bind to the three bases of the codon on messenger RNA.

Antidiuretic hormone (ADH) Hormone released by the posterior pituitary. Increases the permeability of the distal convoluted tubule and collecting tubules, increasing water reabsorption.

Antigens Any substance that is detected as foreign by an organism and elicits an immune response. Most antigens are proteins and large molecular weight carbohydrates.

Aorta Largest artery in the body; carries the oxygenated blood away from the heart and delivers it to the rest of the body through many branches.

Aquatic life zones Ecologically distinct regions in fresh water and salt water.

Aqueous humor Liquid in the anterior and posterior chambers of the eye.

Arteries Vessels that transport blood away from the heart.

Arteriole Smallest of all arteries; usually drains into capillaries.

Arthroscope Device used to examine internal injuries, such as in joints.

Association cortex Area of the brain where integration occurs.

Association neurons Nerve cells that receive input from many sensory neurons and help process them, ultimately carrying impulses to nearby multipolar neurons.

Aster Array of microtubules found in the cell in association with the spindle fibers during cell division.

Asthma Respiratory disease resulting from an allergic response. Allergens cause histamine to be released in the lungs. Histamine causes the air-carrying ducts (bronchioles) to constrict, cutting down airflow and making breathing difficult.

Astigmatism Unequal curvature of the cornea (sometimes the lens) that distorts vision.

Atom Smallest particles of matter that can be achieved by ordinary chemical means, consisting of protons, neutrons, and electrons.

Atomic mass unit Unit used to measure atomic weight. One atomic mass unit is 1/12 the weight of a carbon atom.

Atomic weight Average mass of the atoms of a given element, measured in atomic mass units.

Atrioventricular bundle Tract of modified cardiac muscle fibers that conduct the pacemaker's impulse into the ventricular muscle tissue.

Atrioventricular node (AV node) Knot of tissue located in the right ventricle. Picks up the electrical signal arriving from the atria and transmits it down the atrioventricular bundle.

Atrioventricular valves Valves between the atria and ventricles.

Auditory (eustachian) tube Collapsible tube that joins the nasopharynx and middle ear cavities and helps equalize pressure in the middle ear.

Auricle (or pinna) Skin-covered cartilage portion of the outer ear.

Autoimmune reaction Immune response directed at one's own cells.

Autonomic nervous system That part of the nervous system not under voluntary control.

Autosomal dominant trait Trait that is carried on the autosomes and is expressed in heterozygotes and homozygote dominants.

Autosomal recessive trait Trait that is carried on the autosomes and is expressed only when both recessive genes are present.

Autosomes All human chromosomes except the sex chromosomes.

Autotrophs Organisms such as plants that, unlike animals, are able to synthesize their own food.

Axial skeleton The skull, vertebral column, and rib cage. Contrast with appendicular skeleton.

Axon Long, unbranched process attached to the nerve cell body of a neuron. Transports nerve impulses away from the cell body.

Basal body Organelle located at the base of the cilium and flagellum. Consists of nine sets of microtubules arranged in a circle. Each "set" contains three microtubules.

Basilar membrane Membrane that supports the organ of Corti in the cochlea.

Benign tumor Abnormal cellular proliferation. Unlike a malignant tumor, the cells in a benign tumor stop growing after a while, and the tumor remains localized.

Bile Fluid produced by the liver and stored and concentrated in the gallbladder.

Bile salts Steroids produced by the liver, stored in the gallbladder, and released into the small intestine where they emulsify fats, a step necessary for enzyme digestion.

Biome Terrestrial region characterized by a distinct climate and a characteristic plant and animal community.

Biorhythms (biological cycles) Naturally fluctuating physiological process.

Biosphere Region on Earth that supports life. Exists at the junction of the atmosphere, lithosphere, and hydrosphere.

Bipedal Refers to the ability to walk on two legs.

Birth control Any method or device that prevents conception and birth.

Birth defects Physical or physiological defects in newborns. Caused by a variety of biological, chemical, and physical factors.

Blastocyst Hollow sphere of cells formed from the morula. Consists of the inner cell mass and the trophoblast.

Blood clot Mass of fibrin containing platelets, red blood cells, and other cells. Forms in walls of damaged blood vessels, halting the efflux of blood.

B-lymphocytes Type of lymphocyte that transforms into a plasma cell when exposed to antigens.

Bowman's capsule Cup-shaped end of the nephron that participates in glomerular filtration.

Brain stem Part of the brain that consists of the medulla and pons. Houses structures, such as the breathing control center and reticular activating systems, that control many basic body functions.

Braxton-Hicks contractions Contractions that begin a month or two before childbirth. Also known as false labor.

Breathing center Aggregation of nerve cells in the brain stem that controls breathing.

Breech birth Delivery of a baby feet first.

Bronchi Ducts that convey air from the trachea to the bronchioles and alveoli.

Bronchioles Smallest ducts in the lungs. Their walls are largely made of smooth muscle that contracts and relaxes, regulating the flow of air into the lungs.

Bronchitis Infection or persistent irritation of the bronchi characterized by cough.

Buffer A chemical substance that helps resist changes in acidity.

Bulimia Eating disorder characterized by recurrent binge eating followed by vomiting.

Calcitonin (thyrocalcitonin) Polypeptide hormone produced by the thyroid gland that inhibits osteoclasts and stimulates osteoblasts to produce bone, thus lowering blood calcium levels.

Cancer Any of dozens of diseases, all characterized by the uncontrollable replication of various body cells.

Capillaries Tiny vessels in body tissues whose walls are composed of a flattened layer of cells that allow water and other molecules to flow freely into and out of the tissue fluid.

Capillary bed Branching network of capillaries supplied by arterioles and drained by venules.

Capsid Protein coat of a virus.

Carbohydrate Organic compound consisting of carbon, hydrogen, and oxygen. A structural component of plant cells, it is used principally as a source of energy in animal cells.

Carcinogens Cancer-causing agents.

Cardiac muscle Type of muscle found in the walls of the heart; it is striated and involuntary.

Carnivores Meat eaters; organisms that feed on other animals.

Carrier proteins Class of proteins that transport smaller molecules and ions across the plasma membrane of the cell. Are involved in facilitated diffusion.

Carriers Individuals who carry a gene for a particular trait that can be passed on to their children, but who do not express the trait.

Carrying capacity The number of organisms an ecosystem can support on a sustainable basis.

Cartilage Type of specialized connective tissue. Found in joints on the articular surfaces of bones and other locations such as the ears and nose.

Catabolic reaction A reaction in which molecules are broken down; for example, the breakdown of glucose is a catabolic reaction.

Catalyst Class of compounds that speed up chemical reactions. Although they take an active role in the process, they are left unchanged by the reaction. Thus, they can be used over and over again. Catalysts lower the activation energy of a reaction. *See also* Enzyme.

Cataracts Disease of the eye resulting in cloudy spots on the lens (and sometimes the cornea) that cause cloudy vision.

Cell body Part of the nerve cell that contains the nucleus and other cellular organelles; the center of chemical synthesis.

Cell cycle Repeating series of events in the lives of many cells. Consists of two principal parts: interphase and cellular division.

Cell division Process by which the nucleus and the cytoplasm of a cell are split, creating two daughter cells. Consists of mitosis and cytokinesis.

Cellular respiration The complete breakdown of glucose in the cell, producing carbon dioxide and water. Composed of four separate but interconnected parts: glycolysis, the intermediate reaction, the citric acid cycle, and the electron transport system.

Central nervous system The brain and spinal cord.

Centriole Organelle consisting of a ring of microtubules, arranged in nine sets of three. Structurally identical to basal bodies but associated with the spindle apparatus. Gives rise to the basal body in ciliated cells.

Centromere Region on each chromatid that joins with the centromere of its sister chromatid.

Cerebellum Structure of the brain that lies above the pons and medulla. It has many important functions including synergy.

Cerebral cortex Outer layer of each cerebral hemisphere, consisting of many multipolar neurons and nerve cell fibers.

Cerebral hemisphere Convoluted mass of nervous tissue located above the deeper structures, such as the hypothalamus and limbic system. Home of consciousness, memory, and sensory perception; originates much conscious motor activity.

Cerebrum The largest part of the brain; consists of two cerebral hemispheres.

Cervix Lowermost portion of the uterus; it protrudes into the vagina.

Cesarean section Delivery of a baby via an incision through the abdominal and uterine walls.

Chemical equilibrium The point at which the "forward" reaction from reactants to products proceeds at the same rate as the "backward" reaction from products to reactants, so that there is no net change in chemical composition.

Chemical evolution Formation of organic molecules from inorganic molecules early in the history of the Earth.

Chlorofluorocarbons (CFCs) Chemical substances once used as spray can propellants, refrigerants, and blowing agents for foam products. These substances drift to the upper stratosphere and dissociate. Chlorine released by CFCs reacts with ozone, thus eroding the ozone layer.

Chlorophyll A green pigment that acts as the primary light-trapping molecule for photosynthesis.

Cholecystokinin (CCK) Hormone produced by cells of the duodenum when chyme is present; causes the gallbladder to contract, releasing bile.

Chromatid Strand of the chromosome consisting of DNA and protein.

Chromatin Long, threadlike fibers containing DNA and protein in the nucleus.

GLOSSARY

Chromosome Structure found in the nucleus of cells that houses the genetic information. Consists of DNA and protein.

Chromosome-to-pole fibers Microtubules of the spindle that extend from the centriole to the chromosome where they attach. They play a crucial role in separating the double-stranded chromosomes during mitosis.

Chronic bronchitis Persistent irritation of the bronchi, which causes mucous buildup, coughing, and difficulty breathing.

Chyme Liquified food in the stomach.

Cilia A cellular organelle that protrudes from the surface of many cells. Cilia beat in unison to cause fluids to move over their surface.

Circadian rhythm Biorhythm that occurs on a daily cycle.

Circulatory system Organ system consisting of the heart, blood vessels, and blood.

Circumcision Operation to remove the foreskin of the penis. Generally performed on newborns.

Citric acid cycle A cyclic series of reactions in which the pyruvates produced by glycolysis are broken down to CO_2, accompanied by the formation of ATP and electron carriers. Occurs in the matrix of mitochondria.

Clitoris Small knot of tissue located where the labia minora meet. Consists of erectile tissue.

Cloning Technique of genetic engineering whereby many copies of a gene are produced.

Cochlea Sensory organ of the inner ear that houses the receptor for hearing.

Codominant Refers to two equally expressed alleles.

Codon Three adjacent bases in the messenger RNA that code for a single amino acid.

Color blindness Condition that occurs in individuals who have a deficiency of certain cones. The most common form involves difficulty in distinguishing between red and green.

Colostrum Protein-rich product of the breast, produced for two to three days immediately after delivery and prior to milk production.

Community All of the plants, animals, and microorganisms in an ecosystem.

Compact bone Dense bony tissue in the outer portion of all bones.

Complement Group of blood proteins that circulate in the blood in an inactive state until the body is invaded by bacteria; then they help destroy the bacteria.

Complementary base pairing Unalterable coupling of the purine adenine to the pyrimidine thymine, and the purine guanine to the pyrimidine cytosine in DNA. Responsible for the accurate transmission of genetic information from parent to offspring. Also functions in RNA production.

Cone Type of photoreceptor that operates in bright light; is responsible for color vision.

Connective tissue One of the primary tissues. It contains cells and varying amounts of extracellular material. Holds cells together, forming tissues and organs.

Connective tissue proper Name referring to loose and dense connective tissue; supports and joins various body structures.

Consumers Organisms that eat plants and algae (producers) and other consumers.

Contact inhibition Cessation of growth that results when two or more cells contact each other. A feature of normal cells but absent in cancer cells.

Contraceptive Any measure that helps prevent fertilization and pregnancy.

Control group A group of subjects (plants, animals, etc.) that is used in an experiment; this group is identical to the experimental group in all regards except that it does not receive the experimental treatment.

Cornea Clear part of the wall of the eye continuous with the sclera; allows light into the interior of the eye.

Coronary bypass surgery Surgical technique used to reestablish blood flow to the heart muscle by grafting a vein to shunt blood around a clogged coronary artery.

Corpus luteum (CL) Structure formed from the ovulated follicle in the ovary; produces estrogen and progesterone.

Cortisol Glucocorticoid hormone that increases blood glucose by stimulating gluconeogenesis. Also stimulates protein breakdown in muscle and bone.

Cranial nerves Nerves arising from the brain and brain stem.

Critical thinking Process in which one seeks to determine the validity of conclusions based on sound reasoning. This often requires one to examine hidden bias, the appropriateness of experimental design, and other factors.

Crossing-over Exchange of chromatin by homologous chromosomes during prophase I of meiosis. Results in considerably more genetic variation in gametes and offspring.

Cushing's syndrome Disease that results from pharmacologic doses of cortisone usually administered for rheumatoid arthritis or allergies.

Cystic fibrosis Autosomal recessive disease that leads to problems in sweat glands, mucous glands, and the pancreas. Pancreas may become blocked, thus reducing the flow of digestive enzymes to the small intestine. Mucus buildup in the lungs makes breathing difficult.

Cytokinesis Cytoplasmic division brought about by the contraction of a microfilamentous network lying beneath the plasma membrane at the midline. Usually begins when the cell is in late anaphase or early telophase.

Cytoplasm Material occupying the cytoplasmic compartment of a cell. Consists of a semifluid substance, the cytosol, containing many dissolved substances, and formed elements, the organelles.

Cytoskeleton A network of protein tubules in the cytoplasmic compartment of a cell. Attaches to many organelles and enzyme molecules and thus helps organize cellular activities, increasing efficiency.

Cytotoxic T cells Type of T cell (T-lymphocyte) that attacks and kills virus-infected cells, parasites, fungi, and tumor cells.

Daughter cells Cells produced during cell division.

Decomposer food chain Series of organisms that feed on organic wastes and the dead remains of other organisms.

Defibrillation Procedure to stop fibrillation (erratic electrical activity) of the heart.

Dendrite Short, highly branched fiber that carries impulses to the nerve cell body.

Dense connective tissue Type of connective tissue that consists primarily of densely packed fibers, such as those found in ligaments and tendons.

Dermis Layer of dense irregular connective tissue that binds the epidermis to underlying structures.

Detrivores Organisms that feed on animal waste or the remains of plants and animals.

Diabetes insipidus Condition caused by lack of the pituitary hormone ADH. Main symptoms are excessive drinking and excessive urination.

Diabetes mellitus Insulin disorder either resulting from insufficient insulin production or decreased sensitivity of target cells to insulin. Results in elevated blood glucose levels unless treated.

Dialysis Procedure used to treat patients whose kidneys have failed. Blood is removed from the body and pumped through an artificial filter that removes impurities.

Diaphragm (birth control) Birth control device consisting of a rubber cup that fits over the end of the cervix. Used in conjunction with spermicidal jelly or cream.

Diaphragm (muscle) Dome-shaped muscle that separates the abdominal and thoracic cavities.

Diaphysis Shaft of the long bones. Consists of an outer layer of compact bone and an inner marrow cavity.

Diastolic pressure The pressure at the moment the heart relaxes. The lower of the two blood pressure readings.

Differentiation Structural and functional divergence from the common cell line. Occurs during embryonic development.

Diffusion Movement of molecules from a region of high concentration to a region of low concentration through a semipermeable membrane.

Dihybrid cross Procedure where one plant is bred with another to study two traits.

DNA polymerase Enzyme that helps align the nucleotides and join the phosphates and sugar molecules in a newly forming DNA strand.

Dominant Adjective used in genetics to refer to an allele that is always expressed in heterozygotes. Designated by a capital letter.

Double helix Describes the helical structure formed by two polynucleotide chains making up the DNA molecule.

Down syndrome Genetic disorder caused by an additional chromosome 21 that results in distinctive facial characteristics and mental retardation. Also known as Trisomy 21.

Dryopithecus Genus of apelike creatures that is thought to have given rise to the gibbons, gorillas, orangutans, and chimpanzees.

Duodenum First portion of the small intestine; site where most food digestion and absorption takes place.

E. coli Common bacterium that lives in the large intestine of humans and other mammals. Digests leftover glucose and other materials from food. Used in much genetic research.

Ecological niche An organism's habitat and all of the relationships that exist between that organism and its environment.

Ecological system (ecosystem) System consisting of organisms and their environment and all of the interactions that exist between these components.

Ecology Study of living organisms and the web of relationships that binds them together in the economy of nature. The study of ecosystems.

Ecosystem *See* ecological system.

Ecosystem balance Dynamic equilibrium in ecosystems. Maintained by the interplay of growth and reduction factors.

Ectoderm One of the three types of cells that emerges in human embryonic development. Gives rise to the skin and associated structures, including the eyes.

Edema Swelling resulting from the buildup of fluid in the tissues.

Effector General term for any organ or gland that is controlled by the nervous system.

Ejaculation Ejection of semen from the male reproductive tract.

Elastic arteries Arteries that contain numerous elastic fibers interspersed among the smooth muscle cells of the tunica media.

Elastic cartilage Type of cartilage containing many elastic fibers found in regions where support and flexibility are required.

Electron Highly energetic particle carrying a negative charge that orbits the nucleus of an atom.

Electron carrier A molecule that can reversibly gain and lose electrons. Electron carriers generally accept high-energy electrons produced during an exergonic reaction and donate the electrons to acceptor molecules that use the energy to drive endergonic reactions.

Electron transport system Series of protein molecules in the inner membrane of the mitochondrion that pass electrons from the citric acid cycle from one to another, eventually donating them to oxygen. The electrons come from the citric acid cycle. During their journey along this chain of proteins, the electrons lose energy, which is used to make ATP.

Elements Purest form of matter; substances that cannot be separated into different substances by chemical means.

Emphysema Progressive, debilitating disease that destroys the tiny air sacs in the lung (alveoli), caused by smoking and air pollution.

Endocrine glands Glands of internal secretion that produce hormones secreted into the bloodstream.

Endocrine system Numerous, small, hormone-producing glands scattered throughout the body.

Endocytosis Process by which cells engulf solid particles, bacteria, viruses, and even other cells.

Endoderm One of the three types of cells that emerges during embryonic development. Gives rise to the intestinal tract and associated glands.

Endolymph Fluid inside the semicircular canals that deflects the cupula, signaling rotational movement of the head and body.

Endometrium Uterine endothelium or lining.

Endoplasmic reticulum Branched network of channels found throughout the cytoplasm of many cells. Formed from flattened sheets of membrane derived from the nuclear membrane. *See* rough and smooth endoplasmic reticulum for functions.

Endothelium Single-celled lining of blood vessels.

End-product inhibition The inhibition of an enzyme by a product of the chemical reaction it catalyzes or by the product of a series of chemical reactions of which the enzyme is a part; may result from binding of the end product to the allosteric site or active site of the enzyme.

Energy carrier A molecule that stores energy in "high-energy" chemical bonds and releases the energy again to drive coupled endergonic reactions. ATP is the most common energy carrier in cells; NAD and FAD are others.

Energy pyramid Diagram of the amount of energy at various trophic levels in a food chain or ecosystem.

Enzyme A protein catalyst that speeds up the rate of specific biochemical reactions.

Epidermis Outermost layer of the skin that protects underlying tissues from drying out and from bacteria and viruses.

Epididymis Storage site of sperm. Located on the testis, it consists of a long, tortuous duct, the epididymal duct.

Epiglottis Flap of tissue that closes off the trachea during swallowing.

Epiphyseal plate Band of cartilage cells between the shaft of the bone and the epiphysis. Allows for bone growth.

Episiotomy Surgical incision that runs from the vaginal opening toward the rectum. Enlarges the vaginal opening, easing childbirth.

Epithelium One of the primary tissues. Forms linings and external coatings of organs.

Erectile tissue Spongy tissue of the penis that fills with blood during sexual excitement, making the penis turgid.

Erythropoietin Hormone produced by the kidney when oxygen levels decline. Stimulates red blood cell production in the bone marrow.

Esophagus Muscular tube that transports food to the stomach.

Essential amino acid One of nine amino acids that must be provided in the human diet.

Euchromatin Metabolically active chromatin.

Eukaryote Any cell containing a distinct nucleus and organelles. They are found in single-celled organisms of the kingdom Protista and all multicellular organisms of the kingdoms Plantae, Animalia, and Fungi.

Evolution Process that leads to structural, functional, and behavioral changes in species, making them better able to survive in their environment; also leads to the formation of new species. Results from natural genetic variation and environmental conditions that select for organisms best suited to their environment.

Exhalation Expulsion of air from the lungs.

Exocrine gland Gland of external secretion; empties its contents into ducts.

Exocytosis Process by which cells release materials stored in secretory vesicles. The reverse of endocytosis.

Exon Expressed segment of DNA.

Experiment Test performed to support or refute a hypothesis.

Experimental group—One of two groups in many scientific experiements (the other is the control group). Treated like the control except for one variable, the experimental variable.

Extension Movement of a body part (limbs, fingers, and toes) that opens a joint.

External auditory canal Channel that directs sound waves to the eardrum.

External genitalia External portion of the female reproductive system consisting of the clitoris, labia minora, and labia majora.

Extrinsic eye muscles Six muscles located outside the eye that are responsible for eye movement.

Facilitated diffusion Process in which carrier proteins shuttle molecules across plasma membranes. The molecules move in response to concentration gradients.

Feces Semisolid material containing undigested food, bacteria, ions, and water; produced in the large intestine.

Feedback mechanism A mechanism in which the product of one process regulates the rate of another process, either turning it on or shutting it off.

Fermentation Process occurring in eukaryotic cells in the absence of oxygen, during which pyruvic acid is converted to lactic acid. Also occurs in those prokaryotes that live in oxygen-free environments.

Fertilization Union of sperm and ovum.

Fiber Any of the indigestible polysaccharides in fruits, vegetables, and grains.

Fibrillation Cardiac muscle spasms occurring during heart attacks due to a loss of synchronized electrical signals.

Fibrin Fibrous protein produced from fibrinogen, a soluble plasma protein. Helps form blood clots.

Fibrinogen Protein in plasma that forms fibrin.

Fibroblast Connective tissue cell, found in loose and dense connective tissues that produces collagen, elastic fibers, and a gelatinous extracellular material; responsible for repairing damage created by cuts or tears to connective tissue.

Fibrocartilage Type of cartilage whose extracellular matrix consists of numerous bundles of collagen fibers. Principally found in the intervertebral disks.

Fight-or-flight response An automatic response of an organism to danger that enables it to flee or stand and fight; it is activated by the autonomic nervous system and results in an increase in heart rate and breathing and an increase in blood flow to the muscles.

Fitness Measure of reproductive success of an organism and, therefore, the genetic influence an individual has on future generations.

Flagellum Long, whiplike extension of the plasma membrane of certain protozoans and sperm cells in humans. Used for motility.

Flexion Movement of a limb, finger, or toe that involves closing a joint.

Follicle (ovary) Structure found in the ovary. Each follicle contains an oocyte and one or more layers of follicle cells that are derived from the loose connective tissue of the ovary surrounding the follicle.

Follicle (thyroid) Structure found in the thyroid gland. Consists of an outer layer of cuboidal cells surrounding thyroglobulin, a proteinaceous material from which thyroxine is formed.

Follicle-stimulating hormone (FSH) Gonadotropic hormone from the anterior pituitary that promotes gamete formation in both men and women.

Food chain Series of organisms in an ecosystem in which each organism feeds on the organism preceding it.

Food vacuole Membrane-bound vacuole in a cell containing material engulfed by the cell.

Food web All of the connected food chains in an ecosystem.

Foreskin Sheath of skin that covers the glans penis.

Fossil Remains or imprints of organisms that lived on Earth many years ago, usually embedded in rocks or sediment.

Gallbladder Sac on the underside of the liver that stores and concentrates bile.

Gastrin Stomach hormone that stimulates HCl production and release by the gastric glands.

Gene Segment of the DNA that controls cell structure and function.

Gene pool All the genes of all of the members of a population or species.

Gene therapy The use of artificially produced genes to treat, even cure, diseases.

Genetic engineering The artificial manipulation of genes in which certain genes from one organism are removed and transferred to another organism of the same or a different species. This permits scientists to transfer important genes to improve species—for example, to increase resistance to disease.

Genome All of the genes of an organism.

Genotype Genetic makeup of an organism.

Germ cell Refers to the sperm or ovum (egg) and the cells from which they are derived; germ cells contain half the chromosomes of somatic cells.

Germinal epithelium Germ cells in the wall of the seminiferous tubule that give rise to sperm.

Glans penis Slightly enlarged tip of the penis.

Glaucoma Disease of the eye caused by pressure resulting from a buildup of aqueous humor in the anterior chamber.

Glomerulus Tuft of capillaries that make up part of the nephron; site of glomerular filtration.

Glucagon Hormone released by the pancreas that stimulates the breakdown of glycogen in the liver and the release of glucose molecules, thus increasing blood levels of glucose.

Glucocorticoids Group of steroid hormones produced by the adrenal cortex that stimulate gluconeogenesis.

Glycolysis Metabolic pathway in the cytoplasm of the cell, during which glucose is split in half, forming two molecules of pyruvic acid. The energy released during the reaction is used to generate two molecules of ATP.

Glycoproteins Proteins that have carbohydrate attached to them.

Goiter Condition in which the thyroid gland enlarges due to lack of dietary iodide.

Golgi complex Organelle consisting of a series of flattened membranes that form channels. It sorts and chemically modifies molecules and repackages its proteins into secretory vesicles.

Golgi tendon organs Special receptors found in tendons that respond to stretch. Also known as neurotendinous organs.

Gonadotropin General term for FSH and LH, which are produced by the anterior pituitary and target male and female gonads.

Gonadotropin-releasing hormone or **GnRH** Hormone produced by the hypothalamus that controls the release of FSH (ICSH in males) and LH.

Gray matter Gray, outermost region of the cerebral cortex.

Grazer Herbivorous organism.

Grazer food chain Food chain beginning with plants and grazers (herbivores).

Greenhouse gas Gas, such as carbon dioxide and chlorofluorocarbons, that traps heat escaping from the Earth and radiates it back to the surface.

Growth hormone A protein hormone produced by the anterior pituitary that stimulates cellular growth in the body, causing cellular hypertrophy and hyperplasia. Its major targets are bone and muscle.

Habitat Place in which an organism lives.

Helper T cell Type of T-lymphocyte that stimulates the proliferation of T and B cells when antigen is present.

Hemoglobin Protein molecules inside red blood cells; binds to oxygen.

Hemophilia Disease caused by a gene defect occurring on the Y chromosome. Results in absence of certain blood-clotting factors.

Herpes One of the most common sexually transmitted diseases; caused by a virus.

Heterochromatin Inactive chromatin that is slightly coiled or compacted in the interphase nucleus.

Heterotrophs Organisms such as animals that, unlike plants, are unable to synthesize their own food. Consume plants and other organisms.

Heterozygous Adjective describing a genetic condition in which an individual contains one dominant and one recessive gene in a gene pair.

High-density lipoproteins (HDLs) Complexes of lipid and protein that transport cholesterol to the liver for destruction.

Histamine Potent vasodilator released by certain cells in the body during allergic reactions.

Histone Globular protein thought to play a role in regulating the genes.

Homeostasis A condition of dynamic equilibrium within any biological or social system. Achieved through a variety of automatic mechanisms that compensate for internal and external changes.

Hominid First humanlike creatures.

Hominoids Subgroup of anthropoids.

Homologous structures Structures thought to have arisen from a common origin.

Homozygous Adjective describing a genetic condition marked by the presence of two identical alleles for a given gene.

Hormone Chemical substance produced in one part of the body that travels to another, typically through the bloodstream, where it elicits a response.

Human chorionic gonadotropin (HCG) Hormone produced by the embryo that stimulates the corpus luteum in the mother's body to produce estrogen.

Humoral immunity Immune reaction that protects the body primarily against viruses and bacteria in the body fluids via antibodies produced by plasma cells.

Hyperglycemia High blood glucose levels.

Hypertension High blood pressure.

Hypothalamus Structure in the brain located beneath the thalamus. It consists of many aggregations of nerve cells and controls a variety of functions aimed at maintaining homeostasis.

Hypothesis Tentative and testable explanation for a phenomenon or observation.

Immune system Diffuse system consisting of trillions of cells that circulate in the blood and lymph and take up residence in the lymphoid organs, such as the spleen, thymus, lymph nodes, and tonsils, as well as other body tissues. Helps protect the body against invasion by foreign cells, such as bacteria and viruses, and protects against cancer cells.

Immunity Term referring to the resistance of the body to infectious disease.

Immunocompetence Process in which lymphocytes mature and become capable of responding to specific antigens.

Immunoglobulins Antibodies.

Implantation Process in which the blastocyst embeds in the uterine lining.

Impotence Inability of a male to achieve an erection.

Incomplete dominance Partial dominance. Occurs when an allele exerts only partial dominance over another allele, resulting in an intermediate trait.

Incontinence Inability to control urination.

Induced abortion Deliberate expulsion of a fetus or embryo.

Infectious mononucleosis White blood cell disorder caused by a virus. Characterized by a rapid increase in the number of monocytes and lymphocytes in the blood.

Inferior vena cava Large vein that empties deoxygenated blood from the body below the heart into the right atrium of the heart.

Infertility Inability to conceive; can be due to problems in either the male or the female or both partners.

Inflammatory response Response to tissue damage including an increase in blood flow; the release of chemical attractants, which draw monocytes to the scene; and an increase in the flow of plasma into a wound.

Inhalation Process of air being drawn into the lungs.

Inhibiting hormone Hormone from the hypothalamus that inhibits the release of hormones from the anterior pituitary.

Initiator codon Codon found on a messenger RNA strand that marks where protein synthesis begins.

Inner cell mass Cells of the blastocyst that become the embryo and amnion.

Insulin Hormone that stimulates the uptake of glucose by body cells, especially muscle and liver cells. Stimulates the synthesis of glycogen in liver and muscle cells.

Insulin-dependent diabetes Type of diabetes that can only be treated with injections of insulin. May be caused by an autoimmune reaction. Also known as early-onset diabetes.

Insulin-independent diabetes Type of diabetes that often occurs in obese people. In most patients, it can be controlled by diet. Also known as late-onset diabetes.

Integral protein Large protein molecules in the lipid bilayer of the plasma membrane.

Integration Process of making sense of various nervous inputs so that a meaningful response can be achieved.

Interferon Protein released from cells infected by viruses that stops the replication of viruses in other cells.

Interphase Period of cellular activity occurring between cell divisions. Synthesis and growth occur in preparation for cell division.

Interstitial cells Testosterone-producing cells located in the loose connective tissue between the seminiferous tubules of the testes.

Interstitial cell stimulating hormone (ICSH) Luteinizing hormone in males. Stimulates testosterone secretion.

Interstitial fluid Fluid surrounding cells in body tissues. Provides a path through which nutrients, gases, and wastes can travel between the capillary and the cells.

Intervertebral disks Shock-absorbing material between the bones of the spine.

Intron Segment of DNA that is not expressed. Lies between exons (expressed segments).

Ion Atom that has gained or lost one or more electrons. May be either positively or negatively charged.

Ionic bond Weak bond that forms between oppositely charged ions.

Iris Colored segment of the middle layer of the eye visible through the cornea.

Isotope Alternative form of an atom; differs from other atoms in the number of neutrons found in the nucleus.

Joint capsule Connective tissue that connects to the opposing bones of a joint and forms the synovial cavity. The inner layer of the joint capsule produces synovial fluid.

Kidney Organ that rids the body of wastes and plays a key role in regulating the chemical constancy of blood.

Kingdom A large grouping of organisms; scientists typically recognize five major kingdoms: Monera, Protista, Animalia, Plantae, and Fungi.

Klinefelter syndrome Genetic disorder that results from an XXY genotype.

Krebs cycle *See* citric acid cycle.

Labor The process or period of childbirth.

Lactation Milk production in the breasts.

Laparoscope Instrument used to examine internal organs through small openings made in the skin and underlying muscle.

Laryngitis Inflammation of the lining of the larynx, resulting in hoarseness. Caused by bacterial and viral infection and also excessive use of the voice.

Larynx Rigid but hollow cartilaginous structure that houses the vocal cords and participates in swallowing.

Lens Transparent structure that lies behind the iris and in front of the vitreous humor. Focuses light on the retina.

Leukemia Cancer of white blood cells.

Leukocytosis An increase in the concentration of white blood cells, which often occurs during a bacterial or viral infection.

Ligament Connective tissue structure that runs from bone to bone, located alongside and sometimes inside the joint. Offers support for joints.

Limbic system Array of structures in the brain that work in concert with centers of the hypothalamus. Site of instincts and emotions.

Lipid Commonly known as fats. Water-insoluble organic molecules that provide energy to body cells, help insulate the body from heat loss, and serve as precursors in the synthesis of certain hormones. A principal component of the plasma membrane.

Liver Organ located in the abdominal cavity that performs many functions essential to homeostasis. It stores glucose and fats, synthesizes some key blood proteins, stores iron and certain vitamins, detoxifies certain chemicals, and plays an important role in fat digestion by producing bile.

Loose connective tissue Type of connective tissue that serves primarily as a packing material. Contains many cells among a loose network of collagen and elastic fibers. Often contains cells that help protect the body from foreign organisms.

Low-density lipoproteins (LDLs) Complexes of protein and lipid that transport cholesterol, depositing it in blood vessels.

Lungs Two large saclike organs in the thoracic cavity where the blood and air exchange carbon dioxide and oxygen.

Luteinizing hormone (LH) Hormone produced by the anterior pituitary that stimulates gonadal hormone production. In men, LH stimulates the production of testosterone, the male sex steroid. In women, LH stimulates estrogen secretion.

Lymph Fluid contained in the lymphatic vessels. Similar to tissue fluid, but also contains white blood cells and may contain large amounts of fat.

Lymph node Small nodular organ interspersed along the course of the lymphatic vessels. Serves as a filter for lymph.

Lymphatic system Network of vessels that drains extracellular fluid from body tissues and returns it to the circulatory system.

Lymphocyte Type of white blood cell. *See also* B-lymphocyte and T-lymphocyte.

Lymphoid organs Organs, such as the spleen and thymus, that belong to the lymphatic system.

Lysosome Membrane-bound organelle that contains enzymes. Responsible for the breakdown of material that enters the cell by endocytosis. Also destroys aged or malfunctioning cellular organelles.

Macronutrients Nutrients required in relatively large amounts by organisms. Includes water, proteins, carbohydrates, and lipids.

Macrophage Phagocytic cell derived from monocytes that resides in loose connective tissues and helps guard tissues against bacterial and viral invasion.

Malignant tumor Structure resulting from uncontrollable cellular growth. Cells often spread to other parts of the body.

Marrow cavity Cavity inside a bone containing either red or yellow marrow.

Mast cell Cell found in many tissues, especially in the connective tissue surrounding blood vessels. Contains large granules containing histamine.

Matrix Extracellular material found in cartilage. Also the material in the inner compartment of the mitochondrion.

Matter Anything that has mass and occupies space.

Medulla Term referring to the central portion of some organs; for example, the adrenal medulla.

Megakaryocyte Large cell found in bone marrow that produces platelets.

Meiosis Type of cell division that occurs in the gonads during the formation of gametes. Requires two cellular divisions (meiosis I and meiosis II). In humans, it reduces the chromosome number from 46 to 23.

Meiosis I First meiotic division.

Meiosis II Second meiotic division.

Memory cells T or B cells produced after antigen exposure. They form a reserve force that responds rapidly to antigen during subsequent exposure.

Menopause End of the reproductive function (ovulation) in women. Usually occurs between the ages of 45 and 55.

Menstrual cycle Recurring series of events in the reproductive functions of women. Characterized by dramatic changes in ovarian and pituitary hormone levels and changes in the uterine lining that prepare the uterus for implantation. Ovulation occurs at the midpoint of the menstrual cycle.

Menstruation Process in which the endometrium is sloughed off, resulting in bleeding. Occurs approximately once every month.

Mesoderm One of the three types of cells that emerge in human embryonic development. Lies in the middle of the forming embryo. Gives rise to muscle, bone, and cartilage.

Messenger RNA (mRNA) Type of RNA that carries genetic information needed to synthesize proteins to the cytoplasm of a cell.

Metabolism The chemical reactions of the body, including all catabolic and anabolic reactions.

Metaphase Stage of cellular division (mitosis) in which chromosomes line up in the center of the cell.

Metastasis Spread of cancerous cells throughout the body, through the lymph vessels and circulatory system or directly through tissue fluid.

Microfilament Solid fiber consisting of contractile proteins that is found in cells in a dense network under the plasma membrane. Forms part of the cytoskeleton.

Micronutrients Nutrients required in small quantities. They include two broad groups, vitamins and minerals.

Microtubules Hollow protein tubules in the cytoplasm of cells that form part of the cytoskeleton. Also form spindles.

Microvilli Tiny projections of the plasma membranes of certain epithelial cells that increase the surface area for absorption.

Middle ear Portion of the ear located within a bony cavity in the temporal bone of the skull. Houses the ossicles.

Mineralocorticoids Group of steroid hormones produced by the adrenal cortex. Involved in electrolyte or mineral salt balance.

Mitochondrion Membrane-bound organelle where the bulk of cellular energy production occurs in eukaryotic cells. Houses the citric acid cycle and electron transport system.

Mitosis Term referring specifically to the division of a cell's nucleus. Consists of four stages: prophase, metaphase, anaphase, and telophase.

Mitotic spindle Array of microtubules constructed in the cytoplasm during prophase. Microtubules of the mitotic spindle connect to the chromosomes and help draw them apart during mitosis.

Molecule A structure formed by two or more atoms.

Monocyte White blood cell that phagocytizes bacteria and viruses in body tissues.

Monohybrid cross Procedure in which one plant is bred with another to study the inheritance of a single trait.

Morning sickness Nausea that often occurs in the first two to three months of pregnancy.

Morula Solid ball of cells produced from the zygote by numerous cellular divisions.

Motor unit Muscle fibers supplied by a single axon and its branches.

Mucus Thick, slimy material produced by the lining of the respiratory tract and parts of the digestive tract. Moistens and protects them.

Multipolar neuron Motor neuron found in the central nervous system. Contains a prominent, multiangular cell body and several dendrites.

Muscle fiber Long, unbranched, multinucleated cell found in skeletal muscle.

Muscle spindles Stretch receptors found in skeletal muscle. Also known as neuromuscular spindles.

Muscle tissue A contractile tissue found in varying amounts in all organs of the body. Consists of three types: skeletal, cardiac, and smooth.

Mutation Technically, a change in the DNA caused by chemical, physical, and biological agents. Also refers to a wide range of chromosomal defects.

Myelin sheath Layer of fatty material coating the axons of many neurons in the central and peripheral nervous systems.

Myofibril Bundle of contractile myofilaments in skeletal muscle cells.

Myometrium Uterine smooth muscle.

Myosin Protein filament found in many cells in the microfilamentous network. Also found in muscle cells.

Naked nerve ending Unmodified dendritic ending of the sensory neurons. Responsible for at least three sensations: pain, temperature, and light touch.

Natural selection Evolutionary process in which environmental abiotic and biotic factors "weed" out the less fit—those organisms not as well adapted to the environment as their counterparts.

Nephron Filtering unit in the kidney. Consists of a glomerulus and renal tubule.

Nerve Bundle of nerve fibers. May consist of axons, dendrites, or both. Carries information to and from the central nervous system.

Nervous tissue One of the primary tissues. Found in the nervous system and consists of two types of cells: conducting cells (neurons) and supportive cells.

Neuroendocrine reflex A reflex involving the endocrine and nervous systems.

Neuron Highly specialized cell that generates and transmits nerve impulses from one part of the body to another.

Neurosecretory neurons Specialized nerve cells of the hypothalamus and posterior pituitary that produce and secrete hormones.

Neurotransmitter Chemical substance released from the terminal ends (terminal boutons) of axons when a nerve impulse arrives. May stimulate or inhibit the next neuron.

Neutron Uncharged particle in the nucleus of the atom.

Neutrophil Type of white blood cell that phagocytizes bacteria and cellular debris.

Nitrogen fixation Process in which bacteria and a few other organisms convert atmospheric nitrogen to nitrate or ammonia, forms usable by plants.

Node of Ranvier Small gap in the myelin sheath of an axon; located between segments formed by Schwann cells. Responsible for saltatory conduction.

Nondisjunction Failure of a chromosome pair or chromatids of a double-stranded chromosome to separate during mitosis or meiosis.

Noradrenalin (norepinephrine) Hormone produced by adrenal medulla and secreted under stress. Contributes to the fight-or-flight response.

Nuclear envelope Double membrane delimiting the nucleus.

Nuclear pores Minute openings in the nuclear envelope that allow materials to pass to and from the nucleus.

Nucleic acids Refers to DNA and RNA.

Nucleoli Temporary structures in the nuclei of cells during interphase. Regions of the DNA that are active in the production of RNA.

Nucleotides The building blocks of DNA and RNA. Consist of a nitrogenous base, a sugar, and a phosphate group.

Nucleus (atom) Dense, center region of an atom that contains neutrons and protons.

Nucleus (cell) Cellular organelle that contains the genetic information that controls the structure and function of the cell.

Nutrient cycle Circular flow of nutrients from the environment through the various food chains back into the environment.

Olfactory membrane Receptor for smell; found in the roof of the nasal cavity.

Olfactory nerve Nerve that transmits impulses from the olfactory membrane to the brain.

Optic nerve Nerve that carries impulses from the retina to the brain.

Organ Discrete structure that carries out specialized functions.

Organ of Corti Receptor for sound; located in the inner ear within the cochlea.

Organ system Group of organs that participate in a common function.

Organogenesis Organ formation during embryonic development.

Osmosis Diffusion of water across a selectively permeable membrane.

Osmotic pressure Force that drives water across a selectively permeable membrane. Created by differences in solute concentrations.

Ossicles Three small bones inside the middle ear that transmit vibrations created by sound waves to the organ of Corti.

Osteoarthritis Degenerative joint disease caused by wear and tear that impairs movement of joints.

Osteoblast Bone-forming cell; secretes collagen.

Osteoclast Cell that digests the extracellular material of bone. Stimulated by the parathyroid hormone.

Osteocyte Bone cell derived from osteoblasts that has been surrounded by calcified extracellular material.

Osteoporosis Degenerative disease resulting in the deterioration of bone. Due to inactivity in men and women and loss of the ovarian hormone estrogen in postmenopausal women.

Outer ear External portion of the ear.

Oval window Membrane-covered opening in the cochlea where vibrations are transmitted from the stirrup to the fluid within the cochlea.

Ovary Female gonad. Produces ova (eggs) and steroid hormones, estrogen and progesterone.

Overpopulation Condition in which a species has exceeded the carrying capacity of the environment.

Ovulation Release of the oocyte from the ovary. Stimulated by hormones from the anterior pituitary.

Ovum (ova, plural) Germ cell containing 23 single-stranded chromosomes. Produced during the second meiotic division.

Oxaloacetate Four-carbon compound of the citric acid cycle. It is involved in the very first reaction of the cycle and is regenerated during the cycle.

Oxidation The loss of hydrogens or electrons from a substance.

Oxytocin Hormone from the posterior pituitary hormone. Stimulates contraction of the smooth muscle of the uterus and smooth-muscle-like cells surrounding the glandular units of the breast.

Ozone (O_3) Molecule produced from molecular oxygen. Accumulates in the stratosphere (upper layer of the atmosphere). Helps screen out incoming ultraviolet radiation. *See also* Ozone layer.

Ozone layer Region of the atmosphere located approximately 12 to 16 miles above the Earth's surface where ozone molecules are produced. Helps protect the Earth from ultraviolet light.

Pancreas Organ found in the abdominal cavity under the stomach, nestled in a loop formed by the first portion of the small intestine. Produces enzymes needed to digest foodstuffs in the small intestine and hormones that regulate blood glucose levels.

Pap smear Procedure in which cells are retrieved from the cervical canal to be examined for the presence of cancer.

Parasympathetic division (of the autonomic nervous system) Portion of the autonomic nervous system responsible for a variety of involuntary functions.

Parathyroid glands Endocrine glands located on the posterior surface of the thyroid gland in the neck. Produce parathyroid hormone.

Parathyroid hormone (PTH) Hormone that helps regulate blood calcium levels. Stimulates osteoclasts to digest bone, thus raising blood calcium levels. Also known as parathormone.

Passive immunity Temporary protection from antigen (bacteria and others) produced by the injection of immunoglobulins.

Penis Male organ of copulation.

Pepsin Enzyme released by the gastric glands of the stomach. Breaks down proteins into large peptide fragments.

Pepsinogen Inactive form of pepsin.

Perichondrium Connective tissue layer surrounding most types of cartilage. Contains blood vessels that supply nutrients to cartilage cells.

Periodic table of elements Table that lists elements by ascending atomic number. Also lists other vital statistics of each element.

Peripheral nervous system Portion of the nervous system consisting of the cranial and spinal nerves and receptors.

Peristalsis Involuntary contractions of the smooth muscles in the wall of the esophagus, stomach, and intestines, which propel food along the digestive tract.

Peritubular capillaries Capillaries that surround nephrons. They pick up water, nutrients, and ions from the renal tubule, thus helping maintain the osmotic concentration of the blood.

Pharynx Chamber that connects the oral cavity with the esophagus.

Phenotype Outward appearance of an organism.

Photoreceptors Modified nerve cells that respond to light. Located in the retina of humans and other animals.

Photosynthesis The series of chemical reactions in which the energy of light is used to synthesize high-energy organic molecules, usually carbohydrates, from low-energy inorganic molecules, usually carbon dioxide and water.

Pituitary gland Small pea-sized gland located beneath the brain. It produces numerous hormones and consists of two main subdivisions: anterior and posterior pituitary.

Placenta Organ produced from maternal and embryonic tissue. Supplies nutrients to the growing embryo and fetus and removes fetal wastes. Also produces hormones that help maintain pregnancy.

Plasma Extracellular fluid of blood. Comprises about 55% of the blood.

Plasma cell Cell produced from B-lymphocytes (B cells); synthesizes and releases antibodies.

Plasma membrane Outer layer of the cell. Consists of lipid and protein and controls the movement of materials into and out of the cell.

Plasmids Small circular strands of DNA found in bacterial cytoplasm separate from the main DNA.

Platelet Cell fragment produced from megakaryocytes in the red bone marrow. Plays a key role in blood clotting.

Polygenic inheritance Transmission of traits that are controlled by more than one gene.

Polyploidy Term referring to a genetic disorder caused by an abnormal number of chromosomes. Includes tetraploidy and triploidy.

Polyribosome Also known as polysome. Organelle formed by several ribosomes attached to a single messenger RNA. Synthesizes proteins used inside the cell.

Polysaccharide A carbohydrate molecule such as glycogen and starch that is made of many smaller molecules (monosaccharides).

Population Group of like organisms occupying a specific region.

Portal system Arrangement of blood vessels in which a capillary bed drains to a vein, which drains to another capillary bed.

Posterior chamber Posterior portion of the anterior cavity of the eye.

Posterior pituitary Neuroendocrine gland that consists of neural tissue and releases two hormones, oxytocin and antidiuretic hormone.

Predation Process in which an individual kills and feeds on another smaller individual.

Premenstrual syndrome (PMS) Condition that occurs in some women in the days before menstruation normally begins. Characterized by a variety of symptoms such as irritability, depression, fatigue, headaches, bloating, swelling and tenderness of breasts, joint pain, and tension.

Premotor area Region of the brain in front of the primary motor area. Controls muscle contraction and other less voluntary actions (playing a musical instrument).

Primary motor cortex Ridge of tissue in front of the central sulcus. Controls voluntary motor activity.

Primary oocyte Germ cell produced from oogonium in the ovary. Undergoes the first meiotic division.

Primary response Immune response elicited when an antigen first enters the body.

Primary sensory cortex Region of the brain located just behind the central sulcus. The point of destination for many sensory impulses traveling from the body into the spinal cord and up to the brain.

Primary succession Process of sequential change in which one community is replaced by another. Occurs where no biotic community has existed before.

Primary tissue One of the major tissue types, including epithelial, connective, muscle, and nervous tissue.

Primary tumor Cancerous growth that gives rise to cells that spread to other regions of the body.

Primates An order of the kingdom Animalia. Includes prosimians (premonkeys), monkeys, apes, and humans.

Primordial germ cells Cells that originate in the wall of the yolk sac and eventually become either spermatogonia or oogonia.

Principle of independent assortment Mendel's second law. Hereditary factors are segregated independently during gamete formation. Occurs only when genes are on different chromosomes.

Principle of segregation Mendel's first law, which states that hereditary factors separate during gamete formation.

Producers Generally refers to organisms that can synthesize their own foodstuffs. Major producers are the algae and plants that absorb sunlight and use its energy to synthesize organic foodstuffs from water and carbon dioxide.

Prokaryote A cell that has no nucleus and no organelles such as a bacterium. All prokaryotes are members of the Kingdom Monera.

Prolactin Protein hormone secreted by the posterior pituitary. In humans, it is responsible for milk production by the glandular units of the breast.

Prophase First phase of mitosis during which chromosomes condense, the nuclear membrane disappears, and the spindle forms.

Proprioception Sense of body and limb position.

Prosimians Premonkeys; tarsiers and lemurs.

Prostaglandins Group of chemical substances that have a variety of functions. Act on nearby cells.

Prostate gland Sex accessory gland that is located near the neck of the bladder. Produces fluid that is added to the sperm during ejaculation. Empties into the urethra.

Protein A polymer consisting of many amino acids.

Proton Subatomic particle found in the nucleus of the atom. Each proton carries a positive charge.

Proto-oncogenes Genes in cells that, when mutated, lead to cancerous growth.

Puberty Period of sexual maturation in humans.

Pulmonary circuit (or circulation) Short circulatory loop that supplies blood to the lungs and transports it back to the heart.

Pulmonary veins Veins that carry oxygenated blood from the lungs to the left atrium.

Pupil Central opening in the iris that allows light to penetrate deeper into the eye.

Purine Type of nitrogenous base found in DNA nucleotides. Consists of two fused rings.

Purkinje fiber Modified cardiac muscle fiber that conducts nerve impulses to individual heart muscle cells.

Pus Liquid emanating from a wound. Contains plasma, many dead neutrophils, dead cells, and bacteria.

Pyramid of numbers Diagram of the number of organisms at various trophic levels in a food chain or ecosystem.

Pyrimidine One of two types of nitrogen base found in DNA nucleotides. Consists of one ring.

Pyrogen Chemical released primarily from macrophages that have been exposed to bacteria and other foreign substances. Responsible for fever.

Radioactivity Tiny bursts of energy or particles emitted from the nucleus of some unstable atoms. Results from excess neutrons in the nuclei of some atoms.

Receptor Any structure that responds to internal or external changes. Three types of receptors are found in the body: encapsulated, nonencapsulated (naked nerve endings), and specialized (e.g., the retina and semicircular canals).

Recessive Term describing an allele of a gene that is expressed when the dominant factor is missing.

Recombinant DNA technology Procedure in which scientists take segments of DNA from an organism and insert them into DNA from other organisms.

Red blood cells (RBCs) Enucleated cells in blood that transport oxygen in the bloodstream.

Red marrow Tissue found in the marrow cavity of bones. Site of blood cell and platelet production.

Reflex Automatic response to a stimulus. Mediated by the nervous system.

Relaxin Hormone produced by the corpus luteum and the placenta. It is released near the end of pregnancy and softens the cervix and the fibrocartilage uniting the pubic bones, thus facilitating birth.

Releasing hormone Any of a group of hypothalamic hormones that stimulates the release of other hormones by the anterior pituitary.

Resting potential Minute voltage differential across the membrane of neurons. Also known as the membrane potential.

Restriction endonuclease Enzyme used in recombinant DNA technology. Cuts off segments of the DNA molecule for cloning and splicing.

Reticular activating system (RAS) Region of the medulla that receives nerve impulses from neurons transmitting information to and from the brain. Impulses are transmitted to the cortex, alerting it.

Retina Innermost, light-sensitive layer of the eye. Consists of an outer pigmented layer and an inner layer of nerve cells and photoreceptors (rods and cones).

Retrovirus Special type of RNA virus that carries an enzyme enabling it to produce complementary strands of DNA on the RNA template.

Reverse transcriptase Enzyme that allows the production of DNA from strands of viral RNA.

Rheumatoid arthritis Type of arthritis in which the synovial membrane of the joint becomes inflamed and thickens. Results in pain and stiffness in joints. Thought to be an autoimmune disease.

Ribosomal RNA (rRNA) RNA produced at the nucleolus. Combines with protein to form the ribosome.

Ribosome Cellular organelle consisting of two subunits, each made of protein and ribosomal RNA. Plays an important part in protein synthesis.

RNA polymerase Enzyme that helps align and join the nucleotides in a replicating RNA molecule.

Rod Type of photoreceptor in the eye. Provides for vision in dim light.

Root nodule Swelling in the roots of certain plants (legumes) containing nitrogen-fixing bacteria.

Rough endoplasmic reticulum (RER) Ribosome-coated endoplasmic reticulum. Produces lysosomal enzymes and proteins for use outside the cell.

Saccule Membranous sac located inside the vestibule. Contains a receptor for movement and body position.

Salivary gland Any of several exocrine glands situated around the oral cavity. Produces saliva.

Saltatory conduction Conduction of a nerve impulse down a myelinated neuron from node to node.

Sarcomere Functional unit of the muscle cell. Consists of the myofilaments, actin, and myosin.

Sarcoplasmic reticulum Term given to the smooth endoplasmic reticulum of a skeletal muscle fiber. Stores and releases calcium ions essential for muscle contractions.

Schwann cell Type of neuroglial cell or supportive cell in the nervous system. Responsible for the formation of the myelin sheath.

Science Body of knowledge on the workings of the world and a method of accumulating knowledge. *See also* Scientific method.

Scientific method Deliberate, systematic process of discovery. Begins with observation and measurement. From observations, hypotheses are generated and tested. This leads to more observation and measurement that supports or refutes the original hypothesis.

Sclera Outermost layer of the eye.

Scrotum Skin-covered sac containing the testes.

Sebum Oil excreted by sebaceous glands onto the surface of the skin.

Secondary sex characteristics Distinguishing features of men and women resulting from the sex steroids. In men, includes facial hair growth and deeper voices. In women, includes breast development and fatty deposits in the hips and other regions.

Secondary succession Process of sequential change in which one community is replaced by another. It occurs where a biotic community previously existed, but was destroyed by natural forces or human actions.

Secondary tumor Cancerous growth formed by cells arising from a primary tumor.

Secretin Hormone produced by the cells of the duodenum. Stimulates the pancreas to release sodium bicarbonate.

Secretory vesicles Membrane-bound vesicles containing protein (hormones or enzymes) produced by the endoplasmic reticulum and packaged by the Golgi complex of some cells. They fuse with the membrane, releasing their contents by exocytosis.

Selective permeability Control of what moves across the plasma membrane of a cell.

Semen Fluid containing sperm and secretions of the secondary sex glands.

Semicircular canal Sensory organ of the inner ear. Houses the receptors that detect body position and movement.

Seminal vesicles Sex accessory glands that empty into the vas deferens. Produce the largest portion of ejaculate.

Seminiferous tubule Sperm-producing tubule in the testis.

Sex accessory gland One of several glands that produce secretions that are added to sperm during ejaculation.

Sex chromosomes X and Y chromosomes that help determine the sex of an individual. They also carry a few other traits.

Sex-linked gene A gene that is on a sex chromosome.

Sex steroids Steroid hormones produced principally by the ovaries (in women) and testes (in men). Help regulate secretion of gonadotropins and determine secondary sex characteristics.

Sexually transmitted diseases (venereal diseases) Bacterial and viral infections that are transmitted by sexual contact.

Sickle-cell disease Genetic disease common in African Americans that results in abnormal hemoglobin in red blood cells, causing cells to become sickle shaped when exposed to low oxygen levels. Sickling causes cells to block capillaries, restricting blood flow to tissues.

Sinoatrial node The heart's pacemaker. Located in the wall of the right atrium, it sends timed impulses to the heart muscle, thus synchronizing muscle contractions.

Skeletal muscle Muscle that is generally attached to the skeleton and causes body parts to move.

Skeleton Internal support of humans and other animals. Consists of bones joined together at joints.

Sliding filament mechanism Sliding of actin filaments toward the center of a sarcomere, causing muscle contraction.

Smooth endoplasmic reticulum (SER) Endoplasmic reticulum without ribosomes. Produces phosphoglycerides used to make the plasma membrane. Performs a variety of different functions in different cells.

Smooth muscle Involuntary muscle that lacks striations. Found around circulatory system vessels and in the walls of such organs as the stomach, uterus, and intestines.

Somatic cell A body cell such as those of bone, muscle, liver, brain, and blood.

Species Group of organisms that is structurally and functionally similar. When members of the group breed, they produce viable, reproductively competent offspring. Also a subgroup of a genus.

Specificity The property of an enzyme allowing it to catalyze only one or a few chemical reactions.

Spermatozoan Sperm. Male germ cell.

Spinal nerve Nerve that arises from the spinal cord.

Spongy bone Type of bony tissue inside most bones. Consists of an irregular network of bone spicules.

Starch A polysaccharide found in plants and made of many glucose units.

Subatomic particles Electrons, protons, and neutrons. Particles that can be separated from an atom by physical means.

Substrate Molecule that fits into the active site of an enzyme.

Succession Process of sequential change in which one community is replaced by another until a mature or climax ecosystem is formed. *See also* Primary succession and Secondary succession.

Sulcus Indented region or groove in the cerebral cortex between ridges.

Surfactant Detergent-like substance produced by the lungs. Dissolves in the thin watery lining of the alveoli; helps reduce surface tension, keeping the alveoli from collapsing.

Suspensory ligament Zonular fibers that connect the lens to the ciliary body.

Sympathetic division Division of the autonomic nervous system that is responsible for many functions, especially those involved in the fight-or-flight response.

Synapse Juncture of two neurons that allows an impulse to travel from one neuron to the next.

Synaptic cleft Gap between an axon and the dendrite or effector (e.g., gland or muscle) it supplies.

Synergy Coordination of the workings of antagonistic muscle groups.

Systemic circulation System of blood vessels that transports blood to and from the body and heart, excluding the lungs.

Systolic pressure Peak pressure at the moment the ventricles contract. The higher of the two numbers in a blood pressure reading.

Taste bud Receptor for taste principally found in the surface epithelium and certain papillae of the tongue.

T cell *See* T-lymphocyte.

Telophase Final stage of mitosis in which the nuclear envelope reforms from vesicles and the chromosomes uncoil.

Tendons Connective tissue structures that generally attach muscles to bones.

Terminal boutons Small swellings on the terminal fibers of axons. They lie close to the membranes of the dendrites of other axons or the membranes of the effectors, and transfer nerve impulses from one cell to another.

Terminator codon Codon found on each strand of messenger RNA that marks where protein synthesis should end.

Testes Male gonads. They produce sex steroids and sperm.

Testosterone Male sex hormone that stimulates sperm formation and is responsible for secondary sex characteristics, such as facial hair growth and muscle growth.

Theories Principles of science—the broader generalizations about the world and its components. Theories are supported by considerable scientific research.

Theory of Natural Selection A theory proposed by Charles Darwin to explain how evolution works; it states that the fittest organisms of a population survive and reproduce, thus passing their genes on to future generations. Over time, this results in a shift in the genetic makeup of the population.

Thyroid gland U- or H-shaped gland located in the neck on either side of the trachea just below the larynx. Produces three hormones: thyroxin, triiodothyronine, and calcitonin.

Thyroid-stimulating hormone (TSH) Hormone produced by the pituitary gland. Stimulates production and release of thyroxine and triiodothyronine by the thyroid gland.

Thyroxine Hormone produced by the thyroid gland that accelerates the rate of mitochondrial glucose catabolism in most body cells and also stimulates cellular growth and development.

Tissue Component of the body from which organs are made. Consists of cells and extracellular material (fluid, fibers, and so on). *See* primary tissue.

T-lymphocyte Type of lymphocyte responsible for cell-mediated immunity. Attacks foreign cells, virus-infected cells, and cancer cells directly. Also known as T cell.

Trachea Duct that leads from the pharynx to the lungs.

Transcription RNA production on a DNA template.

Transfer RNA (tRNA) Small RNA molecules that bind to amino acids in the cytoplasm and deliver them to specific sites on the messenger RNA.

Transition reaction Part of cellular respiration in which one carbon is cleaved from pyruvic acid, forming a two-carbon compound, which reacts with Coenzyme A. The resulting chemical enters the citric acid cycle.

Translation Synthesis of protein on a messenger RNA template.

Translocation Process in which a segment of a chromosome breaks off but reattaches to another site on the same chromosome or another one.

Triiodothyronine Hormone produced by the thyroid gland. Nearly identical in function to thyroxin.

Trophic hormones Hormones that stimulate the production and secretion of other hormones. Also known as tropic hormones.

Trophic level Feeding level in a food chain.

Trophoblast Outer ring of cells of the blastocyst that form the embryonic portion of the placenta.

Tubal ligation Sterilization procedure in women. Uterine tubes are cut, preventing sperm and ova from uniting.

Tumor Mass of cells derived from a single cell that has begun to divide. In malignant tumors, the cells divide uncontrollably and often release clusters of cells or single cells that spread in the blood and lymphatic systems to other parts of the body. Benign tumors grow to a certain size, then stop.

Turner syndrome Genetic disorder in which an offspring contains 22 pairs of autosomes and a single, unmatched X chromosome. Phenotypically female.

Tympanic membrane (eardrum) Membrane between the external auditory canal and middle ear that oscillates when struck by sound waves.

Type I diabetes Form of diabetes that occurs mainly in young people and results from an insufficient amount of insulin production and release. Brought on by damage to insulin-producing cells of the pancreas. Also known as early-onset diabetes.

Type II diabetes Form of diabetes that occurs chiefly in older individuals (around the age of 40) and results from a loss of tissue responsiveness to insulin. Also known as late-onset diabetes caused by obesity.

Umbilical cord A structure that connects the fetus to the placenta and contains two blood vessels that carry blood to and from the embryo.

Umbilical vein Vein in the umbilical cord that carries blood from the placenta to the fetus.

Ureter Hollow, muscular tube that transports urine by peristaltic contractions from the kidney to the urinary bladder.

Urethra Narrow tube that transports urine from the urinary bladder to the outside of the body. In males, it also conducts sperm and semen to the outside.

Urinary bladder Hollow, distensible organ with muscular walls that stores urine. Drained by the urethra.

Urine Fluid containing various wastes that is produced in the kidney and excreted out of the urinary bladder.

Uterus Organ that houses and nourishes the developing embryo and fetus.

Utricle Membranous sac containing a receptor for body position and movement. Located inside the vestibule of the inner ear.

Vaccine Preparation containing dead or weakened bacteria and viruses that, when injected in the body, elicits an immune response.

Vagina Tubular organ that serves as a receptacle for sperm and provides a route for delivery of the baby at birth.

Variation Genetically based differences in physical or functional characteristics within a population.

Vas deferens Duct that carries sperm from the testis to the urethra. Contracts during ejaculation.

Vasectomy Contraceptive procedure in men in which the vas deferens is cut and the free ends sealed to prevent sperm from entering the urethra during ejaculation.

Vein Type of blood vessel that carries blood to the heart.

Vena cava One of two large veins that empty into the right atrium of the heart.

Venule Smallest of all veins. Empties into capillary networks.

Villi Fingerlike projections of the lining of the small intestine that increase the surface area for absorption.

Virus Nonliving entity consisting of a nucleic acid—either DNA or RNA—core surrounded by a protein coat, the capsid. Viruses are cellular parasites, invading cells and taking over their metabolic machinery to reproduce.

Vitamin Any of a diverse group of organic compounds. Essential to many metabolic reactions.

Vocal cords Elastic ligaments inside the larynx that vibrate as air is expelled from the lungs, generating sound.

White blood cells (WBCs) Cells of the blood formed in the bone marrow. Principally involved in fighting infection.

White matter The portion of the brain and spinal cord that appears white to the naked eye. Consists primarily of white, myelinated nerve fibers.

Yellow marrow Inactive marrow of bones in adults containing fat. Formed from red marrow.

Yolk sac Embryonic pouch formed from endoderm. Site of early formation of red blood cells and germ cells.

Zygote Cell produced by a sperm and ovum during fertilization. Contains 46 chromosomes.

Index

Note: Page numbers with f indicate figures; with t indicate tables.

A

Photo Credits

Chapter 6

Chapter opener. © Photos.com.
Figure 6-4b. © Phillipe Plailly/Photo Researchers, Inc.
Figure 6-4d. © Science Photo Library/Photo Researchers, Inc.
Figure 6-6a. © Ken Sherman/Medichrome.
Figure 6-7a. © D.W. Fawcett-Kuwabara/Visuals Unlimited.
Figure 6-8. © Cabisco/Visuals Unlimited.
Figure 6-10a. © Hattie Young/Photo Researchers. Inc.
Figure 6-11. © Robin L. Sachs/PhotoEdit.
Figure 6-15. © Fred Marsik/Visuals Unlimited.
Figure 6-16. © VAS Comm/Custom Medical Stock Photo.
Figure 6-17a. © William Ober/Visuals Unlimited.
Figure 6-18. © CNRI/Photo Researchers, Inc.
Scientific Discoveries 6-1. © Bettmann Archive.

Chapter 7

Chapter opener. Courtesy of Photographer's Mate 2nd Class Tiffini
M. Jones/U.S. Navy.
Figure 7-2a. © David M. Phillips/Visuals Unlimited.
Figure 7-2b. © Stanley Fleger/Visuals Unlimited.
Figure 7-3. © Stanley Flegler/Visuals Unlimited.
Figure 7-6. © Gladden Willis, M.D./Visuals Unlimited.
Figure 7-7. © G. Prance/Visuals Unlimited.
Figure 7-8. © John D. Cunningham/Visuals Unlimited.
Figure 7-9b. © David M. Phillips/Visuals Unlimited.
Figure 7-10 inset. © Yoav Levy/Phototake.
Table 7-2. © John D. Cunningham/Visuals Unlimited.

Chapter 8

Chapter opener. © 2003, Berta A. Daniels.
Figure 8-3b. © Chet Childs/Custom Medical Stock.
Figure 8-4b. © David M. Phillips/Visuals Unlimited.
Figures 8-6 a and b. © David M. Phillips/Visuals Unlimited.
Figure 8-6c. © Don Fawcett/Visuals Unlimited.
Figure 8-8. © R. Calentine/Visuals Unlimited.
Figures 8-9b left and right. © SIU/Visuals Unlimited.
Figure 8-10a. © SIU/Visuals Unlimited.
Health Note 8-2a. © American Cancer Society.
Health Note 8-2b. © Medical-on-Line/Alamy Images.

Chapter 9

Chapter opener. © LiquidLibrary
Figure 9-1. © SIU/Visuals Unlimited.
Figure 9-6 inset. © Photos.com.
Figure 9-8a. © NMSB/Custom Medical Stock.
Figure 9-8b. © Runk/Schoenberger/Medical Images, Inc.

Chapter 10

Chapter opener. © Imagine/Alamy Images.
Figure 10-3a. © David M. Phillips/Visuals Unlimited.
Figure 10-4c. © C. Raines/Visuals Unlimited.
Figure 10-6a. © Science VU/E.R. Lewis/Visuals Unlimited.
Figure 10-6c. © T. Reese-D.W. Fawcett/Visuals Unlimited.
Figure 10-13b. © Marcus Raichle, M.D.
Figure 10-15 inset. © AbleStock.
Figure 10-17b. © SIU/Visuals Unlimited.
Health Note 10-1. © TongueDisco/The Image Bank.
Health Note 10-2. © AbleStock.

Chapter 11

Chapter opener. © Don Hammond/Design Pics Inc./Alamy Images.
Figure 11-3 inset (bottom). © Cabisco/Visuals Unlimited.
Figure 11-3 inset (top). © Astrid & Hans-Frieder Michler/Photo
Researchers, Inc.
Figure 11-4d. © John D. Cunningham/Visuals Unlimited.
Health Note 11-1. © Photodisc.

Chapter 12

Chapter opener. © Photodisc.
Figure 12-1. © Donald Fawcett/Visuals Unlimited.
Figure 12-7a. © Dr. P. Marazzi/Photo Researchers, Inc.
Figure 12-7b. © CNRI/Photo Researchers, Inc.
Figure 12-8. © SIU/Visuals Unlimited.
Figures 12-9 a and b. © SIU/Visuals Unlimited.
Figure 12-11 inset. © Dennis MacDonald/Alamy Images.
Figures 12-12 a, b, c, and d. Reprinted with permission from *Calcified
Tissue Research*, 1967.
Figure 12-14. © Ed Reschke.
Figure 12-17a. © John D. Cunningham/Visuals Unlimited.
Figure 12-19. © Tony Freeman/PhotoEdit.
Health Note 12-1. © Yoav Levy/Phototake.

Chapter 13

Chapter opener. © Photodisc.
Figure 13-5 inset a. © LiquidLibrary.
Figure 13-5 inset b. © Photos.com.
Figures 13-6 a and b. © AP/Wide World Photos.
Figure 13-7. Reprinted with permission from *American Journal of
Medicine,* 20 (1956).
Figure 13-13b. © R. Calentine/Visuals Unlimited.
Figure 13-13c. © David M. Phillips/Visuals Unlimited.
Figure 13-14. © Ken Greer/Visuals Unlimited.
Figure 13-15b. © John D. Cunningham/Visuals Unlimited.
Figure 13-18b. © Photos.com.

Chapter 14

Chapter opener. © Frank Siteman/age footstock.
Figure 14-1c. © G. Musil/Visuals Unlimited.
Figure 14-2b. © David M. Phillips/Visuals Unlimited.
Health Note 14-2. © Photodisc/Getty Images.

A Closer Look

Figure 14s-1b. © Science VU/Visuals Unlimited.
Figure 14s-2a. © Gary Conner/PhotoEdit.
Figure 14s-2b. © Yoav Levy/Phototake/Alamy Images.
Figure 14s-4. Courtesy of National Cancer Institute.

Chapter 15

Chapter opener. © Photodisc.
Figure 15-1. © St. Bartholomew's Hospital/Photo Researchers, Inc.
Figure 15-3. © Dr. Steve Patterson/Photo Researchers, Inc.
Figure 15-6b inset. Courtesy of Tilney, from Trends in Microbiology
1(1), 1993.
Figure 15-7a. © 1987 American Society of Clinical Pathologists.
Reprinted with permission. Image courtesy of the CDC.
Figure 15-7b. Courtesy of CDC.
Figure 15-9a. Courtesy of the Armed Forces Institute of Pathology,
Neg. No. 69-77765-2.
Figure 15-9b. © R. Umesh Chandran, WHO/SPL/Photo Researchers,
Inc.
Figure 15-11. Courtesy of B. Finlay, from *ASM News* 58:486, 1992.
Figure 15-14a. © PhotoDisc.
Figure 15-14b. Courtesy of Dr. J. Noble, Jr./CDC.
Health Note 15-1. Courtesy of Stanley Prusiner/Institute for Neuro-
degenerative Diseases/University of California at San Fransisco.
Health Note 15-2. Bacterial Biofilm in a Chronic Respiratory Tract
Infection. © Hiroyuki Kobayashi. Liscensed for use, ASM
MicrobeLibrary.org.

Chapter 16

Chapter opener. © BananaStock/Alamy Images.
Figure 16-1. © Stan W. Elms/Visuals Unlimited.
Figure 16-4. © Cabisco/Visuals Unlimited.

Figure 16-5. © Custom Medical Stock Photography.
Figure 16-6b. © Science VU/Visuals Unlimited.
Figure 16-7a. © Yoav Levy/Phototake/Alamy Images.
Figures 16-8 a and b. © Michael Abbey/Visuals Unlimited.
Figure 16-8c. © Phototake/PNI.
Figure 16-8d. © John D. Cunningham/Visuals Unlimited.
Figures 16-8 e and f. © M. Abbey/Photo Researchers, Inc.
Figure 16-11b. © David M. Phillips/Visuals Unlimited.
Figure 16-12a. © Jack Bostrack.
Scientific Discoveries 16-1. © Cold Spring Harbour Laboratory.
Scientific Discoveries 16-2. © Science VU/NIHLBL/Visuals Unlimited.

■ Chapter 17
Chapter opener. © AbleStock.
Figure 17-4a. © The Bettmann Archive.
Figure 17-4b. © Malcolm Gutter/Visuals Unlimited.
Figure 17-8. © Joe McDonald/Visuals Unlimited.
Figures 17-9 a and b. © Science VU/Daniel V. Schidlow/Visuals Unlimited.
Figure 17-10. Courtesy of MIEMSS.
Figures 17-11 a and b. © Bruce Berg/Visuals Unlimited.
Figure 17-13a. © Photos.com.
Figure 17-13b. © LiquidLibrary.
Figure 17-13c. © Ryan McVay/Photodisc/Getty Images.
Figure 17-13d. © Photos.com.
Figure 17-13e. © LiquidLibrary.
Figure 17-14. © B. John/Cabisco/Visuals Unlimited.
Figure 17-16. © Biophoto Associates/Photo Researchers.
Figure 17-19a. © Science VU/Valerie Lindgren/Visuals Unlimited.
Figure 17-19b. © Bernd Wittich/Visuals Unlimited.
Figure 17-22a. © Science VU/Valerie Lindgren/Visuals Unlimited.
Figure 17-22b. © Dr. Ira Rosenthal/Department of Pediatrics, University of Illinois at Chicago.
Figure 17-23. © Photodisc.
Health Note 17-1. © Mauro Fermariello/Photo Researchers, Inc.

■ Chapter 18
Chapter opener. Courtesy of Bill Branson/National Cancer Institute.
Figure 18-11b. © E. Kifelva and D. Fawcett/Visuals Unlimited.
Scientific Discoveries 18-1. © AP/Wide World Photos.

■ Chapter 19
Chapter opener. © Photos.com.
Figure 19-1. © Tom McHugh/Photo Researchers.
Figure 19-2. © Ken Lucas/Visuals Unlimited.
Figure 19-4. © H. Potter and D. Pressler/Visuals Unlimited.
Figure 19-7. © Visuals Unlimited.
Figure 19-8. © M. Baret/Science Source/Photo Researchers, Inc.
Figure 19-10. © Science VU/© Jackson Lab/Visuals Unlimited.
Figure 19-11. © Keith V. Wood/Science VU/Visuals Unlimited.

■ Chapter 20
Chapter opener. © BananaStock/age footstock.
Figure 20-1. © John D. Cunningham/Visuals Unlimited.
Figure 20-4. © Photos.com.
Figure 20-5. © L. Gonzalez/Custom Medical Stock Photo.
Figure 20-6. Courtesy of National Cancer Institute.

■ Chapter 21
Chapter opener. © Photodisc.
Figure 21-2b. © Biophoto Associates/Photo Researchers.
Figure 21-14. © SIU/Visuals Unlimited.

Figure 21-15. © Hattie Young/Photo Researchers, Inc.
Figure 21-16. © SIU/Visuals Unlimited.
Figure 21-17. © M. Long/Visuals Unlimited.
Figure 21-18. © SIU/Visuals Unlimited.
Health Note 21-1. © SIU/Visuals Unlimited.

■ A Closer Look
Figure 21s-1. © Biophoto Associates/Photo Researchers, Inc.
Figure 21s-2. © Courtesy of Susan Lindsley/CDC.
Figure 21s-3. © Phototake/PNI.
Figure 21s-4. © Christopher Brown/Stock Boston.

■ Chapter 22
Chapter opener. © Photodisc.
Figure 22-2b. © Dr. David M. Phillips/Visuals Unlimited.
Figure 22-10a. © Biophoto Associates/Photo Researchers, Inc.
Figure 22-10b. © Nestle/Petit Format/Photo Researchers.Inc.
Figure 22-10c. © Neil Bromhall/Photo Researchers, Inc.
Figure 22-12. © SIU/Visuals Unlimited.
Figure 22-17b. © LiquidLibrary.
Health Note 22-1. © SIU/Visuals Unlimited.
Health Note 22-2. © Photos.com.

■ Chapter 23
Chapter opener. Courtesy of NASA, ESA, and J. Hester (ASU).
Figure 23-1b. © Dr. Stanley Miller.
Figure 23-3a. © The Granger Collection.
Figure 23-4a. Courtesy of Jerry Sintz/Bureau of Land Management.
Figure 23-4b. © Alex Kerstitch/Visuals Unlimited.
Figure 23-4c. © John D. Cunningham/Visuals Unlimited.
Figure 23-7. © Milton H. Tierney, Jr./Visuals Unlimited.
Figure 23-9a. © Michael Dick/Animals Animals.
Figure 23-9b. © Walt Anderson/Visuals Unlimited.
Figure 23-11 left. © Science VU/Visuals Unlimited.
Figure 23-11 right. © The Living World, St. Louis Zoo, St. Louis, MO.
Figure 23-13. © National Museum of Kenya/Visuals Unlimited.
Figure 23-14. © Cabisco/Visuals Unlimited.
Figure 23-16. © Science VU/Visuals Unlimited
Figures 23-17 a, b, c, and d. © Photos.com.
Figure 23-17e. © Suzy Bennett/Alamy Images.
Health Note 23-1. © USDA.

■ Chapter 24
Chapter opener. © Photodisc.
Figure 24-2a. © Thomas Kitchin/Tom Stack and Associates.
Figure 24-2b. © Charles Ott/Photo Researchers, Inc.
Figure 24-2c. © Corbis.
Figure 24-2d. © Tim McCabe/USDA Natural Resources Conservatiion Service.
Figure 24-2e. © John D. Cunningham/Visuals Unlimited.
Figure 24-4 left. © Ron Niebrugge/Alamy Images.
Figure 24-4 right. © Perihelion/Alamy Images.
Figure 24-11. © Reuters/Bettmann.
Figure 24-13. © Frans Lanting/Minten Pictures.
Figure 24-14. © Courtesy of Lynn Betts/NRCS USDA.
Figure 24-15. © Science VU/Visuals Unlimited.
Figure 24-17. © Courtesy of John Bortniak/NOAA Corps.
Figure 24-18. © Max Hunn/Visuals Unlimited.
Figure 24-19. © David S. Addison/Visuals Unlimited.
Figure 24-21. © Andre Jenny/Stock South/PNI.
Figure 24-23. © Courtesy General Electric.

Biological Roots

ONE OF THE MAJOR CHALLENGES of a biology course is mastering new terminology. One of the best ways of learning and remembering technical terms is to first learn their component parts—that is, their roots. But studying common Latin and Greek roots for biological terms before you begin your study can also help. Spend a few minutes early in your course studying these common roots. They will not only help you understand new terms but also make them easier to remember.

a-, an- [Gk. *an-*, not, without, lacking]: anaerobic, abiotic, anemia

ad- [L. *ad-*, toward, to]: adrenalin

amphi- [Gk. *amphi-*, two, both, both sides of]: amphibian

ana- [Gk. *ana-*, up, up against]: anaphase, anabolic, anatomy

andro- [Gk. *andros*, an old man]: androgen

anti- [Gk. *anti-*, against, opposite, opposed to]: antibiotic, antibody, antigen, antidiuretic hormone

arthro- [Gk. *arthron*, a joint]: arthropod, arthritis

auto- [Gk. *auto-*, self, same]: autoimmune, autotroph

bi-, bin- [L. *bis*, twice; *bini*, two-by-two]: binary fission, binocular vision, bicarbonate

bio- [Gk. *bios*, life]: biology, biomass, biome, biosphere, biotic

blasto-, -blast [Gk. *blastos*, sprout; pertains to embryo]: blastula, trophoblast, osteoblast

broncho- [Gk. *bronchos*, windpipe]: bronchus, bronchi, bronchiole, bronchitis

carb- [L. *carbo*, coal]: carbon, carbohydrate

carcino- [Gk. *karkin*, a crab, cancer]: carcinogen

cardio- [Gk. *kardia*, heart]: cardiac, myocardium, electrocardiogram

cat- [Gk. *kata*, down, downward]: catabolic

chloro- [Gk. *chloros*, green]: chlorophyll, chloroplast, chlorine

chromo- [Gk. *chroma*, color]: chromosome, chromatin

coelo, -coel [Gk. *koilos*, hollow, cavity]: coelom

com-, con-, col-, co- [L. *cum*, with, together]: coenzyme, covalent

cranio- [Gk. *kranios*, L. cranium, skull]: cranial, cranium

cuti- [L. *cutis*, skin]: cutaneous, cuticle

cyto-, cyte [Gk. *kytos*, vessel or container; now, "cell"]: cytoplasm, cytokinesis, erythrocyte, leukocyte

de- [L. *de-*, away, off, removal, separation]: deciduous, decomposer, dehydration

derm-, dermato- [Gk. *derma*, skin]: dermis, epidermis, ectoderm, endoderm, mesoderm

di- [Gk. *dis*, twice, two, double]: disaccharide, dioxide

dia- [Gk. through, passing through, thorough, thoroughly]: diabetes, dialysis, diaphragm

diplo- [Gk. *diploos*, two-fold]: diploid

eco- [Gk. *oikos*, house, home]: ecology, ecosystem, economy

ecto- [Gk. *ektos*, outside]: ectoderm

endo- [Gk. *endon*, within]: endoderm, endometrium

epi- [Gk. *epi*, on, upon, over]: epidermis, epididymis, epiglottis, epithelium

equi- [L. *aequus*, equal]: equilibrium

eu- [Gk. *eus*, good; eu, well, true]: eukaryote

ex, exo-, ec-, e- [Gk., L. out, out of, from, beyond]: emission, ejaculation, excretion, exergonic, exhale, exocytosis, exoskeleton

extra- [L. outside of, beyond]: extracellular, extraembryonic

-fer [L. *ferre*, to bear]: fertile, fertilization, conifer

gam-, gameto- [Gk. *gamos*, marriage; now usually in reference to gametes (sex cells)]: gamete

gastro- [Gk. *gaster*, stomach]: gastric, gastrin, gastrovascular cavity

gen- [Gk. *gen*, born, produced by; Gk. genos, race, kind; L. genus, generare, to beget]: polygenic genotype, geneology, glycogen, pyrogen, heterogenous

gluco-, glyco- [Gk. *glykys*, sweet; now pertaining to sugar]: glucose, glycogen, glycolysis, glycoprotein

hemo-, hemato-, -hemia, -emia [Gk. *haima*, blood]: hematology, hemoglobin, hemophilia

hepato- [Gk. *hepar, hepat-*, liver]: hepatitis, hepatic portal system

hetero- [Gk. *heteros*, other, different]: heterogeneous, heterozygote

histo- [Gk. *histos*, web of a loom, tissue; pertains to biological tissues]: histology, histamine, antihistamine

homo-, homeo- [Gk. *homos*, same; Gk. homios, similar]: homeostasis, homogeneous, homologous, homozygote

hydro- [Gk. *hydor*, water; pertains either to water or to hydrogen]: dehydration, carbohydrate

hyper- [Gk. *hyper*, over, above, more than]: hyperthyroid, hypertonic